T0248585

HANDBOOK OF

Media for Clinical Microbiology

SECOND EDITION

By
RONALD M. ATLAS
JAMES W. SNYDER

CRC Press
Taylor & Francis Group
Boca Raton London New York

CRC Press is an imprint of the
Taylor & Francis Group, an **informa** business
A TAYLOR & FRANCIS BOOK

First published 2006 by Taylor & Francis

Published 2019 by CRC Press
Taylor & Francis Group
6000 Broken Sound Parkway NW, Suite 300
Boca Raton, FL 33487-2742

First issued in paperback 2019

ISBN 13: 978-0-367-45360-2 (pbk)
ISBN 13: 978-0-8493-3795-6 (hbk)

Visit the Taylor & Francis Web site at
http://www.taylorandfrancis.com

and the CRC Press Web site at
http://www.crcpress.com

Preface

Almost 1,650 media are described in the second edition of the *Handbook of Media for Clinical Microbiology*, including newly described media for the cultivation of emerging pathogens. Diseases caused by emerging pathogens that are responsible for increased rates of morbidity and mortality rates, such as *Escherichia coli* O157:H7, methicillin resistant *Staphylococcus aureus*, and vancomycin resistant enterococci, have raised special concerns and various media included in the *Handbook* have been designed for the specific cultivation and identification of these pathogens.

Many of the new media included in the second edition of the *Handbook of Media for Clinical Microbiology* permit the cultivation of bacteria, fungi, and viruses that are currently causing major medical problems around the world. These media are very important for the rapid detection of pathogenic microorganisms and the diagnosis of individuals with specific infectious diseases. Several of the new media described in the second edition of the *Handbook of Media for Clinical Microbiology* include chromogenic or fluorogenic substrates that permit the rapid detection of specific pathogens.

An important function of the second edition of the *Handbook of Media for Clinical Microbiology* is to provide descriptions of the media that are used to cultivate and identify microorganisms from clinical specimens and to maintain reference cultures of human pathogens in various clinical settings. The *Handbook* provides a compilation of the formulations, methods of preparation, and applications for media used in the clinical microbiology laboratory. Each listing is alphabetical and includes medium composition, instruction for preparation, commercial sources, and intended uses.

The format of the *Handbook* allows easy reference to information needed to prepare media for the cultivation of microorganisms relevant to clinical diagnostics. The second edition of the *Handbook of Media for Clinical Microbiology* includes descriptions of expected results as they apply to microorganisms of importance for the examination of clinical specimens.

Importantly, the second edition of the *Handbook of Media for Clinical Microbiology* is user friendly and should save time and effort for anyone cultivating pathogenic microorganisms. It should be a valuable resource for anyone working in the area of clinical microbiology.

About the Authors

Ronald M. Atlas is Graduate Dean, Professor of Biology, and Professor of Public Health at the University of Louisville. He received his B.S. from the State University of New York at Stony Brook in 1968 and his M.S. and Ph.D. from Rutgers in 1970 and 1972, respectively. Dr. Atlas has received a number of honors including: The University of Louisville Excellence in Research Award, the Johnson and Johnson Fellowship for Biology, and the ASM Award in Applied and Environmental Microbiology. He has authored several microbiology textbooks and handbooks of microbiological media. His research has included development of diagnostic systems for pathogenic microorganisms. He has served on the editorial boards of *Applied and Environmental Microbiology, BioScience, Biotechniques, Environmental Microbiology, Biosecurity, and Bioterrorism,* and *Journal of Industrial Microbiology.* He is editor of *Critical Reviews in Microbiology.* He also has served as President of the American Society for Microbiology.

James W. Snyder is Professor of Pathology in the Department of Pathology and Laboratory Medicine at the University of Louisville School of Medicine and serves as the Director of Microbiology for the University of Louisville Hospital. He received his B.S. in 1969 and M.S. in 1970 from Eastern Kentucky University and Ph.D. in 1974 from the University of Dayton. Dr. Snyder is a diplomate of the American Board of Medical Microbiology (ABMM) and a Fellow in the American Academy of Microbiology. He has received a number of honors including: The South Central Association for Clinical Microbiology Outstanding Contribution to Microbiology Award, the Spirit of Louisville Foundation Award, Outstanding Service Award as Chairman of the Committee on Continuing Education of the American Society for Microbiology, and election to the Alpha Omega Alpha Honor Medical Society. He is the former Editor of *The Yearbook of Clinical Microbiology* and the *Certified Clinical Microbiology Data Base.* He is currently a member of the *Critical Reviews in Microbiology* editorial board and serves on the Professional Affairs Committee of the American Society for Microbiology. He maintains an active research program in both applied and basic clinical microbiology, including molecular techniques for the detection of pathogens and the assessment and comparison of automated and nonautomated diagnostic systems and tracking of microbial incidence and *in vitro* activity of antibiotics in support of local and national surveillance programs.

Table of Contents

Introduction

Diagnostic Microbiology: Isolation and Identification of Pathogens

The definitive diagnosis of an infectious disease is in part dependant on the detection and/or the isolation and identification of the pathogenic microorganism or the detection of antigens or antibodies specifically associated with the pathogen. Specific culture methods are used for the isolation and identification of a wide variety of pathogenic microorganisms. Traditional methods for the identification of pathogens depend on microscopic observations, phenotypic characteristics observed in culture, and metabolic changes following growth in a variety of substrates. Many differential media and selective media utilize color changes to highlight specific cultural features or substrate utilization. These growth-dependent methods provide a reliable and relatively accurate means for identifying the pathogen and diagnosing many infectious diseases.

A wide range of biochemical, serological, and nucleic acid-based procedures are available for the definitive identification of microbial isolates of clinical significance. Accuracy, reliability, and speed are important factors that govern the selection of clinical identification protocols. The selection of the specific procedures to be employed for the identification of pathogenic isolates is guided by the need for presumptive or definitive identification of the organism at the family or genus and species level, based on the observation of colonial morphology, phenotypic characteristics, and other key growth characteristics on primary isolation medium.

Identification of pathogenic filamentous fungi and protozoa is generally based on the morphological characteristics of the organism. Additionally, the identification of the former is based primarily on cultural and microscopic characteristics and a limited number of biochemical tests. Conventional identification schemes for bacteria and yeasts rely on the determination of a variety of biochemical features. In general, fewer than 20 biochemical tests are required to identify most clinical bacterial isolates to the species level. The primary objective is to differentiate the isolates present in the specimen using a minimal number of tests to accurately define distinct taxa.

Isolation and Culture Procedures

A variety of procedures are employed for the collection, isolation, and identification of pathogenic microorganisms from different anatomical sites including tissue, body fluids, pulmonary secretions, and blood. Various procedures are used for the isolation of different types of microorganisms. Laboratory procedures are designed to screen and facilitate the recovery of etiologic agents of disease that predominate in particular clinical syndromes. When the manifestation of a disease suggests that the disease may be caused by a rare pathogen and/or routine screening fails to detect a probable causative microorganism, additional specialized isolation procedures may be required.

As the clinical microbiology laboratory slowly enters into the new era of "molecular diagnostics," the dependency on traditional culture methods involving the production and preparation of growth supporting media will continue to play an integral role in our efforts to detect and characterize both known and "new" microbial pathogens. Furthermore, the major goal of the clinical microbiology laboratory, as stated by Dr. Raymond Bartlett over twenty years ago, has not and will not change, "to provide information of maximal and epidemiological usefulness as rapidly as is consistent with acceptable accuracy and minimal cost." The latter two criteria serve as the driving forces that account for the slow introduction and acceptance by the clinical microbiology community of methods that have been developed for either the noncultural detection or cultural confirmation of pathogenic microorganisms. Many procedures, including the more recent nucleic acid probes and gene amplification methods, remain cost prohibitive for many laboratories, lack acceptable sensitivity and/or specificity, and are not practical or adaptable for use in the service-oriented, clinical laboratory. Until such problems are resolved, microbiologists will continue to rely on the availability and utilization of a variety of media for the isolation, characterization, and identification of primary and opportunistic microbial pathogens. The selection of a specific procedure(s) to be employed for the definitive identification of a pathogen at the family or genus and species level is influenced by the information gleaned from observing colonial morphology, pigment production, and other characteristics that are observed following growth of the microorganism on either selective, differential, or general purpose media.

These concepts also apply to fungi and viruses. The limited availability of practical and cost-effective detection methods requires that these microorganisms be recovered in culture before confirmation of identity can be accomplished using key morphological and biochemical characteristics, the type of cytopathic effect (CPE), or the application of culture confirmation methods such as serological, direct fluorescent antibiody, nucleic acid probes, or amplification technologies. Culture media will continue to be developed and used for the cultivation of microorganisms despite the inevitable impact that noncultural methods will have on the rapid detection of microbial pathogens and ultimately affect, in a beneficial manner, patient outcome.

Media for the Isolation and Identification of Microorganisms from Clinical Specimens

The second edition of the *Handbook of Media for Clinical Microbiology* includes both classic and modern media used for the identification, cultivation, and maintenance of diverse bacteria described in the *Manual* for medically important microorganisms. Below are some of the primary media used in clinical microbiology laboratories for the isolation of pathogens.

Representative Microbiological Media that are used for Isolation and Identification of Microorganisms from Clinical Specimens

Medium	Use
Anaerobic Blood Agar	Cultivation of a wide variety of anaerobic pathogens
Anaerobic Kanamycin Vancomycin Laked Blood Agar	Cultivation of *Bacteroides* species
Azide Blood Agar	Cultivation of *Streptococcus pyogenes* which forms small colonies and demonstrates beta hemolysis
Bacteroides Bile Esculin Agar	Cultivation of *Bacteroides* species
Bile Esculin Agar	Cultivation of group D streptococci including *Enterococcus faecalis*
Bismuth Sulfite Agar	Cultivation of *Salmonella* including *S. typhi* from faecal specimens
Blood Agar	Cultivation of many pathogens
Blood Agar with Colistin and Nalixic Acid	Selective cultivation of Gram-positive bacteria
Brain Heart Infusion Medium	Cultivation of a wide variety of human pathogens
Brilliant Green Agar	Selective cultivation of *Salmonella* species
Campylobacter Agar	Selective cultivation of *Campylobacter* species
Chocolate Agar	Cultivation of a wide variety of human pathogens including *Haemophilus* and *Neisseria* species
Chopped Meat Glucose Medium	Cultivation of anaerobes
Deoxycholate Agar	Selective cultivation of Gram-negative enteric bacteria
Deoxycholate Citrate Agar	Selective cultivation of *Salmonella* and some *Shigella* species
Enterococcus Agar	Selective cultivation of enterococci
Eosin Methylene Blue Agar	Selective cultivation of Gram-negative enteric bacteria
GN Broth	Enrichment for *Salmonella* and *Shigella* species
Hektoen Enteric Agar	Selective cultivation of enteric bacteria including *Shigella* species
MacConkey Agar	Selective cultivation of Gram-negative enteric bacteria
Mannitol Salt Agar	Selective cultivation of *Staphylococcus* species
Martin–Lewis Agar	Selective cultivation of *Neisseria* species
New York City Agar	Selective cultivation of *Neisseria* and *Mycoplasma* species
Phenylethyl Alcohol Medium	Selective cultivation of *Staphylococcus* and *Streptococcus* species
Selenite F Broth	Enrichment for *Salmonella* species
Tetrathionate Broth	Enrichment for *Salmonella* and some *Shigella* species
Thayer Martin Agar	Selective cultivation of *Neisseria* species
Thioglycollate Broth	Cultivation of aerobes and anaerobes
Xylose Lysine Deoxycholate Agar	Selective cultivation of *Salmonella* and some *Shigella* species

Some Media and Procedures Used for the Diagnosis of Various Diseases

Anatomical Site	Specimen	Culture Media	Probable Organism	Syndrome
Upper respiratory tract: throat and nasopharyngeal cultures	Sterile dacron swabs	Blood agar	*Streptococcus pyogenes*	Pharyngitis, rheumatic fever
		Chocolate agar	*Haemophilus influenzae*	Epiglottitis
			Neisseria gonorrhoeae	Phayingitis
		Bordet-Gengou	*Bordetella pertussis*	Whooping cough
		Tellurite serum agar	*Corynebacterium diphtheriae*	Diphtheria
Lower respiratory tract	Sputum, transtracheal aspirate, bronchoalveolar lavage, biopsy	Blood agar	*Streptococcus pneumoniae, Staphylococcus aureus*	Pneumonia
		Chocolate agar	*Haemophilus influenzae*	
		MacConkey's agar	*Klebsiella pneumoniae, Haemophilus influenzae*	
		Sabouraud's agar	*Coccidioides immitis, Candida albicans*	
		Lowenstein-Jensen	*Mycobacterium tuberculosis*	Tuberculosis
Central nervous system	Lumbar puncture for cerebro–spinal fluid	Liquid enrichment media	*Streptococcus pneumoniae*	Meningitis
		Blood agar	*Neisseria meningitidis*	
		Chocolate agar	*Haemophilus influenzae*	
Circulatory system blood	Venus puncture	Tryptic Soy BHI	Various	Septicemia
Urinary tract	Midstream catch of voided urine	Blood agar	*Escherichia coli*	Urinary tract infections
		Cysteine lactose electrolyte-deficient agar	*Klebsiella, Proteus, Pseudomonas, Salmonella*	
		MacConkey's agar, EMB agar	*Serratia, E. coli, and other Gram-negative rods*	

Some Media and Procedures Used for the Diagnosis of Various Diseases

Anatomical Site	Specimen	Culture Media	Probable Organism	Syndrome
Genital tract	Urethral exudate (males) Swabs from cervix, vagina, and anal canal (females	Thayer-Martin medium and chocolate agar	*Neisseria gonorrhoeae*	Gonorrhea
Intestinal tract	Stool samples	Hektoen enteric media, xylose-lysine-desoxycholate media, brilliant green, EMB, Endo, and MacConkey's agar	*Salmonella-Shigella*	Gastroenteritis
Ears	Fluids	Blood, chocolate, and MacConkey's agar, Gram stain	*Haemonpilus influenzae*	Otitis media
Skin	Swabs, aspirates, or washings from lesions	Aerobic and anaerobic culture techniques	*Clostridium tetani,* *C. perfringens*	Tetanus Gas gangrene

References

Atlas of the Clinical Microbiology of Infectious Diseases, Volume 1: Bacterial Agents. 2003. Bottone, E.J. Parthenon Pub. Group, New York.

Atlas of the Clinical Microbiology of Infectious Diseases, Volume 2: Viral, Fungal, and Parasitic Agents, 2006. Bottone, E.J. Parthenon Pub. Group, New York.

Biohazardous Bloodborne and Airborne Pathogens: A Guide for Clinical and Laboratory Safety. 2004. Langerman, N. Lewis Pub., Boca Raton, FL.

Clinical Bacteriology. 2003. Struthers, J.K. and R.P. Westran. ASM Press, Washington DC.

Clinical Microbiology Procedures Handbook. 2004. Isenberg, H.D., L. Clarke, P. Della-Latta, G.A. Denys, S.D. Douglas, L.S. Garcia, K.C. Hazen, J.F. Hindler, and S.G. Jenkins. ASM Press, Washington DC.

Clinical Mycology. 2003. Anaissie, E.J., M.R. McGinnis, and M.A. Pfaller. Churchill Livingston, New York.

Clinical Mycology. 2003. Dismukes, W.E., P.G. Pappas, and J.D. Sobel. Oxford University Press, New York.

Color Atlas and Textbook of Diagnostic Microbiology. 1992. Koneman, E.W., S.D. Allen, W.M. Janda, P.C. Schreckenberger, and W.C. Winn, Jr., eds. J.B. Lippincott Co., Philadelphia, PA.

Diagnostic Microbiology. 1990. Finegold, S.M. and W.J. Martin. C.V. Mosby Co., St. Louis, MO.

Difco & BBL Manual: Dehydrated Culture Media and Reagents for Microbiology. 2003. Becton, Dickinson and Co., Sparks, MD.

Manual of Clinical Microbiology, eighth edition. 2003. Murray, P.R., E.J.Baron. J.H. Jorgensen, M.A. Pfaller, and R.H. Yolken. ASM Press, Washington, DC.

Media for Isolation–Cultivation–Identification–Maintenance of Medical Bacteria. 1985. McFaddin, J.F. Williams and Wilkins, Baltimore, MD.

Pocket Guide to Clinical Microbiology. 2004. Murray, P. R. ASM Press, Washington DC.

Practical Guide to Clinical Virology. 2002. Haaheim L.R., J.R. Pattison, and R.J. Whitley. John Wiley & Sons, Chichester, UK.

Principles and Practice of Clinical Virology. 2004. Zuckerman, A.J., J.E. Banatvala, P. Griffiths, J.R. Pattison, and B. Schoub. John Wiley & Sons, Chichester, UK.

Textbook of Diagnostic Microbiology. 2000. Mahon, C. Saunders, Philadelphia.

A 3 Agar

Composition per 202.4mL:

Agar base	140.0mL
Supplement solution	62.4mL

pH 6.0 ± 0.2 at 25°C

Agar Base:

Composition per liter:

Pancreatic digest of casein	17.0g
Ionagar no. 2	7.5g
NaCl	5.0g
Papaic digest of soybean meal	3.0g
K_2HPO_4	2.5g
Glucose	2.5g

Source: Ionagar no. 2 is available from Oxoid Unipath.

Preparation of Agar Base: Add components, except agar, to distilled/deionized water and bring volume to 1.0L. Adjust pH to 5.5. Add agar. Mix thoroughly. Gently heat and bring to boiling. Distribute into screw-capped bottles in 140.0mL volumes. Autoclave for 15 min at 15 psi pressure–121°C. Cool to 45°–50°C.

Supplement Solution:

Composition per 62.4mL:

Horse serum-urea solution	40.0mL
Fresh yeast extract solution	20.0mL
Penicillin solution	2.0mL
Phenol Red solution	0.4mL

Preparation of Supplement Solution: Aseptically combine components. Mix thoroughly.

Horse Serum-Urea Solution:

Composition per 40.0mL:

Urea	0.2g
Horse serum, unheated	40.0mL

Preparation of Horse Serum-Urea Solution: Add urea to 40.0mL of horse serum. Mix thoroughly. Filter sterilize.

Fresh Yeast Extract Solution:

Composition:

Baker's yeast, live, pressed, starch-free	25.0g

Preparation of Fresh Yeast Extract Solution: Add the live Baker's yeast to 100.0mL of distilled/deionized water. Autoclave for 90 min at 15 psi pressure–121°C. Allow to stand. Remove supernatant solution. Adjust pH to 6.6–6.8. Filter sterilize.

Penicillin Solution:

Composition per 10.0mL:

Penicillin G	1,000,000U

Preparation of Penicillin Solution: Add penicillin to distilled/deionized water and bring volume to 10.0mL. Mix thoroughly. Filter sterilize.

Phenol Red Solution:

Composition per 10.0mL:

Phenol Red	0.1g

Preparation of Phenol Red Solution: Add Phenol Red to distilled/deionized water and bring volume to 10.0mL. Mix thoroughly. Filter sterilize.

Preparation of Medium: Aseptically combine 140.0mL of cooled, sterile agar base and 62.4mL of sterile supplement solution. Mix thoroughly. Pour into sterile Petri dishes or distribute into sterile tubes.

Use: For the cultivation of *Ureaplasma urealyticum* from urine for the detction of urogenital infections. Also used for the cultivation of other *Ureaplasma* species.

A 3B Agar

Composition per 101.5mL:

Agar base	80.0mL
Supplement solution	21.5mL

pH 6.0 ± 0.2 at 25°C

Agar Base:

Composition per liter:

Pancreatic digest of casein	17.0g
Ionagar no. 2	7.5g
NaCl	5.0g
Papaic digest of soybean meal	3.0g
K_2HPO_4	2.5g
Glucose	2.5g

Source: Ionagar no. 2 is available from Oxoid Unipath.

Preparation of Agar Base: Add components, except agar, to distilled/deionized water and bring volume to 1.0L. Adjust pH to 5.5. Add agar. Mix thoroughly. Gently heat and bring to boiling. Distribute into screw-capped bottles in 80.0mL volumes. Autoclave for 15 min at 15 psi pressure–121°C. Cool to 45°–50°C.

Supplement Solution:

Composition per 21.5mL:

Horse serum-urea solution	20.0mL
Penicillin solution	1.0mL
L-Cysteine·HCl·H_2O solution	0.5mL

Preparation of Supplement Solution: Aseptically combine components. Mix thoroughly.

Horse Serum-Urea Solution:

Composition per 40.0mL:

Urea	0.2g
Horse serum, unheated	40.0mL

Preparation of Horse Serum-Urea Solution: Add urea to 40.0mL of horse serum. Mix thoroughly. Filter sterilize.

Penicillin Solution:

Composition per 10.0mL:

Penicillin G	1,000,000U

Preparation of Penicillin Solution: Add penicillin to distilled/deionized water and bring volume to 10.0mL. Mix thoroughly. Filter sterilize.

L-Cysteine·HCl·H_2O Solution:

Composition per 10.0mL:

L-Cysteine·HCl·H_2O	0.2g

Preparation of L-Cysteine·HCl·H_2O Solution: Add L-cysteine·HCl·H_2O to distilled/deionized water and bring volume to 10.0mL. Mix thoroughly. Filter sterilize.

Preparation of Medium: Aseptically combine 80.0mL of cooled, sterile agar base and 21.5mL of sterile supplement solution. Mix thoroughly.

Use: For the cultivation of *Ureaplasma urealyticum* from urine for the detction of urogenital infections. Also used for the cultivation of other *Ureaplasma* species.

A 7 Agar
(Shepard's Differential Agar)

Composition per 205.7mL:

Agar base	160.0mL
Supplement solution	45.7mL

pH 6.0 ± 0.2 at 25°C

Agar Base:

Composition per 165.0mL:

Pancreatic digest of casein	2.72g
Agar	2.1g
NaCl	0.8g
Papaic digest of soybean meal	0.48g
K_2HPO_4	0.4g
Glucose	0.4g
$MnSO_4·H_2O$	0.15g

Preparation of Agar Base: Add components, except agar, to distilled/deionized water and bring volume to 165.0mL. Adjust pH to 5.5. Add agar. Mix thoroughly. Autoclave for 15 min at 15 psi pressure–121°C. Cool to 45°–50°C.

Supplement Solution:

Composition per 45.72mL:

Horse serum, unheated	40.0mL
Fresh yeast extract solution	2.0mL
Penicillin solution	2.0mL
CVA enrichment	1.0mL
L-Cysteine·HCl·H_2O solution	0.5mL
Urea solution	0.22mL

Preparation of Supplement Solution: Aseptically combine components. Mix thoroughly.

Fresh Yeast Extract Solution:

Composition per 100.0mL:

Baker's yeast, live, pressed, starch-free	25.0g

Preparation of Fresh Yeast Extract Solution: Add the live Baker's yeast to 100.0mL of distilled/deionized water. Autoclave for 90 min at 15 psi pressure–121°C. Allow to stand. Remove supernatant solution. Adjust pH to 6.6–6.8. Filter sterilize.

Penicillin Solution:

Composition per 10.0mL:

Penicillin G	1,000,000U

Preparation of Penicillin Solution: Add penicillin to distilled/deionized water and bring volume to 10.0mL. Mix thoroughly. Filter sterilize.

CVA Enrichment:

Composition per liter:

Glucose	100.0g
L-Cysteine·HCl·H_2O	25.9g
L-Glutamine	10.0g
L-Cystine·2HCl	1.0g
Adenine	1.0g
Nicotinamide adenine dinucleotide	0.25g
Cocarboxylase	0.1g
Guanine·HCl	0.03g
$Fe(NO_3)_3$	0.02g
Vitamin B_{12}	0.01g
p-Aminobenzoic acid	0.013g
Thiamine·HCl	3.0mg

Preparation of CVA Enrichment: Add components to distilled/deionized water and bring volume to 1.0L. Mix thoroughly. Filter sterilize.

L-Cysteine·HCl·H_2O Solution:

Composition per 10.0mL:

L-Cysteine·HCl·H_2O	0.4g

Preparation of L-Cysteine·HCl·H_2O Solution: Add L-cysteine·HCl·H_2O solution to distilled/deionized water and bring volume to 10.0mL. Mix thoroughly. Filter sterilize.

Urea Solution:

Composition per 10.0mL:

Urea, ultrapure	1.0g

Preparation of Urea Solution: Add urea to distilled/deionized water and bring volume to 10.0mL. Mix thoroughly. Filter sterilize.

Preparation of Medium: Aseptically combine 160.0mL of cooled, sterile agar base and 45.9mL of sterile supplement solution. Mix thoroughly. Pour into sterile Petri dishes or distribute into sterile tubes.

Use: For the cultivation and differentiation of *Ureaplasma urealyticum* from urine based on its ability to produce ammonia from urea for the detction of urogenital infections. Bacteria that produce ammonia appear as golden to dark brown colonies. Also used for the cultivation of other *Ureaplasma* species.

A 7 Agar, Modified

Composition per 205.7mL:

Agar base	160.0mL
Supplement solution	45.7mL

pH 6.0 ± 0.2 at 25°C

Agar Base:

Composition per 165.0mL:

Agar	10.0g
Pancreatic digest of casein	2.72g
NaCl	0.8g
Papaic digest of soybean meal	0.48g
K_2HPO_4	0.4g
Glucose	0.4g
$MnSO_4 \cdot H_2O$	0.15g

Preparation of Agar Base: Add components, except agar, to distilled/deionized water and bring volume to 165.0mL. Adjust pH to 5.5. Add agar. Mix thoroughly. Autoclave for 15 min at 15 psi pressure–121°C. Cool to 45°–50°C.

Supplement Solution:

Composition per 45.72mL:

Horse serum, unheated	40.0mL
Fresh yeast extract solution	2.0mL
Penicillin solution	2.0mL
CVA enrichment	1.0mL
L-Cysteine·HCl·H_2O solution	0.5mL
Urea solution	0.22mL

Preparation of Supplement Solution: Aseptically combine components. Mix thoroughly.

Fresh Yeast Extract Solution:

Composition per 100.0mL:

Baker's yeast, live, pressed, starch-free	25.0g

Preparation of Fresh Yeast Extract Solution: Add the live Baker's yeast to 100.0mL of distilled/deionized water. Autoclave for 90 min at 15 psi pressure–121°C. Allow to stand. Remove supernatant solution. Adjust pH to 6.6–6.8. Filter sterilize.

Penicillin Solution:

Composition per 10.0mL:

Penicillin G	1,000,000U

Preparation of Penicillin Solution: Add penicillin to distilled/deionized water and bring volume to 10.0mL. Mix thoroughly. Filter sterilize.

CVA Enrichment:

Composition per liter:

Glucose	100.0g
L-Cysteine·HCl·H_2O	25.9g
L-Glutamine	10.0g
L-Cystine·2HCl	1.0g
Adenine	1.0g
Nicotinamide adenine dinucleotide	0.25g
Cocarboxylase	0.1g
Guanine·HCl	0.03g
$Fe(NO_3)_3$	0.02g
p-Aminobenzoic acid	0.013g
Vitamin B_{12}	0.01g
Thiamine·HCl	3.0mg

Preparation of CVA Enrichment: Add components to distilled/deionized water and bring volume to 1.0L. Mix thoroughly. Filter sterilize.

L-Cysteine·HCl·H_2O Solution:

Composition per 10.0mL:

L-Cysteine·HCl·H_2O	0.4g

Preparation of L-Cysteine·HCl·H_2O Solution: Add L-cysteine·HCl·H_2O solution to distilled/deionized water and bring volume to 10.0mL. Mix thoroughly. Filter sterilize.

Urea Solution:

Composition per 10.0mL:

Urea, ultrapure	1.0g

Preparation of Urea Solution: Add urea to distilled/deionized water and bring volume to 10.0mL. Mix thoroughly. Filter sterilize.

Preparation of Medium: Aseptically combine 160.0mL of cooled, sterile agar base and 45.9mL of sterile supplement solution. Mix thoroughly. Pour into sterile Petri dishes or distribute into sterile tubes.

Use: For the cultivation and differentiation of *Ureaplasma urealyticum* from urine based on its ability to produce ammonia from urea for the detction of urogenital infections. Bacteria that produce ammonia appear as golden to dark brown colonies. Also used for the cultivation of other *Ureaplasma* species.

A 7B Agar

Composition per 205.7mL:

Agar base	160.0mL
Supplement solution	45.7mL

pH 6.0 ± 0.2 at 25°C

Agar Base:

Composition per 165.0mL:

Pancreatic digest of casein	2.72g
Agar	2.1g
NaCl	0.8g
Papaic digest of soybean meal	0.48g
K_2HPO_4	0.4g
Glucose	0.4g
Putrescine·2HCl	0.33g
$MnSO_4 \cdot H_2O$	0.15g

Preparation of Agar Base: Add components, except agar, to distilled/deionized water and bring volume to 165.0mL. Adjust pH to 5.5. Add agar. Mix thoroughly. Autoclave for 15 min at 15 psi pressure–121°C. Cool to 45°–50°C.

Supplement Solution:

Composition per 45.72mL:

Horse serum, unheated	40.0mL
Fresh yeast extract solution	2.0mL
Penicillin solution	2.0mL
CVA enrichment	1.0mL
L-Cysteine·HCl·H_2O solution	0.5mL
Urea solution	0.22mL

Preparation of Supplement Solution: Aseptically combine components. Mix thoroughly.

Fresh Yeast Extract Solution:

Composition per 100.0mL:

Baker's yeast, live, pressed, starch-free	25.0g

Preparation of Fresh Yeast Extract Solution: Add the live Baker's yeast to 100.0mL of distilled/deionized water. Autoclave for 90 min at 15 psi pressure–121°C. Allow to stand. Remove supernatant solution. Adjust pH to 6.6–6.8. Filter sterilize.

Penicillin Solution:

Composition per 10.0mL:

Penicillin G	1,000,000U

Preparation of Penicillin Solution: Add penicillin to distilled/deionized water and bring volume to 10.0mL. Mix thoroughly. Filter sterilize.

CVA Enrichment:

Composition per liter:

Glucose	100.0g
L-Cysteine·HCl·H_2O	25.9g
L-Glutamine	10.0g
L-Cystine·2HCl	1.0g
Adenine	1.0g
Nicotinamide adenine dinucleotide	0.25g
Cocarboxylase	0.1g
Guanine·HCl	0.03g
$Fe(NO_3)_3$	0.02g
p-Aminobenzoic acid	0.013g
Vitamin B_{12}	0.01g
Thiamine·HCl	3.0mg

Preparation of CVA Enrichment: Add components to distilled/deionized water and bring volume to 1.0L. Mix thoroughly. Filter sterilize.

L-Cysteine·HCl·H_2O Solution:

Composition per 10.0mL:

L-Cysteine·HCl·H_2O	0.4g

Preparation of L-Cysteine·HCl·H_2O Solution: Add L-cysteine·HCl·H_2O solution to distilled/deionized water and bring volume to 10.0mL. Mix thoroughly. Filter sterilize.

Urea Solution:

Composition per 10.0mL:

Urea, ultrapure	1.0g

Preparation of Urea Solution: Add urea to distilled/deionized water and bring volume to 10.0mL. Mix thoroughly. Filter sterilize.

Preparation of Medium: Aseptically combine 160.0mL of cooled, sterile agar base and 45.9mL of sterile supplement solution. Mix thoroughly. Pour into sterile Petri dishes or distribute into sterile tubes.

Use: For the cultivation and differentiation of *Ureaplasma urealyticum* from urine based on its ability to produce ammonia from urea for the detction of urogenital infections. Bacteria that produce ammonia appear as golden to dark brown colonies. Also used for the cultivation of other *Ureaplasma* species.

A 8B Agar

Composition per 84.6mL:

Agar base	80.0mL
Supplement solution	4.6mL

pH 6.0 ± 0.2 at 25°C

Agar Base:

Composition per 165.0mL:

Pancreatic digest of casein	2.72g
Agar	2.1g
NaCl	0.8g
Papaic digest of soybean meal	0.48g
K_2HPO_4	0.4g
Glucose	0.4g
$MnSO_4$·H_2O	0.15g
$CaCl_2$·$2H_2O$	0.03g
Putrescine·2HCl	34.0mg

Preparation of Agar Base: Add components, except agar, to distilled/deionized water and bring volume to 165.0mL. Adjust pH to 5.5. Add agar. Mix thoroughly. Autoclave for 15 min at 15 psi pressure–121°C. Cool to 45°–50°C.

Supplement Solution:

Composition per 4.6mL:

Horse serum, unheated	1.0mL
Fresh yeast extract solution	1.0mL
Penicillin solution	1.0mL
Urea solution	1.0mL
L-Cysteine·HCl·H_2O solution	0.5mL
GHL tripeptide solution	0.1mL

Preparation of Supplement Solution: Aseptically combine components. Mix thoroughly.

Fresh Yeast Extract Solution:
Composition per 100.0mL:
Baker's yeast, live, pressed, starch-free............25.0g

Preparation of Fresh Yeast Extract Solution: Add the live Baker's yeast to 100.0mL of distilled/deionized water. Autoclave for 90 min at 15 psi pressure–121°C. Allow to stand. Remove supernatant solution. Adjust pH to 6.6–6.8. Filter sterilize.

Penicillin Solution:
Composition per 10.0mL:
Penicillin G 1,000,000U

Preparation of Penicillin Solution: Add penicillin to distilled/deionized water and bring volume to 10.0mL. Mix thoroughly. Filter sterilize.

GHL Tripeptide Solution:
Composition per 10.0mL:
GHL (Glycyl-L-histidyl-L-lysine
acetate) tripeptide 0.2g

Preparation of GHL Tripeptide Solution: Add GHL (Glycyl-L-histidyl-L-lysine acetate) tripeptide to distilled/deionized water and bring volume to 10.0mL. Mix thoroughly. Filter sterilize.

L-Cysteine·HCl·H_2O Solution:
Composition per 10.0mL:
L-Cysteine·HCl·H_2O..0.4g

Preparation of L-Cysteine·HCl·H_2O Solution: Add L-cysteine·HCl·H_2O solution to distilled/deionized water and bring volume to 10.0mL. Mix thoroughly. Filter sterilize.

Urea Solution:
Composition per 10.0mL:
Urea, ultrapure 1.0g

Preparation of Urea Solution: Add urea to distilled/deionized water and bring volume to 10.0mL. Mix thoroughly. Filter sterilize.

Preparation of Medium: Aseptically combine 80.0mL of cooled, sterile agar base and 4.6mL of sterile supplement solution. Mix thoroughly. Pour into sterile Petri dishes or distribute into sterile tubes.

Use: For the cultivation of *Ureaplasma urealyticum* from urine. Also used for the cultivation of other *Ureaplasma* species.

A 1 Broth

Composition per liter:
Pancreatic digest of casein...............................20.0g
Lactose 5.0g
NaCl 5.0g
Salicin 0.5g
Triton™ X-100 .... 1.0mL
pH 6.9 ± 0.1 at 25°C

Source: This medium is available as a premixed powder from BD Diagnostic Systems.

Preparation of Medium: Add components to distilled/deionized water and bring volume to 1.0L. Mix thoroughly. Gently heat and bring to boiling. Distribute into test tubes containing an inverted Durham tube. Autoclave for 10 min at 15 psi pressure–121°C.

Use: For the detection of fecal coliforms by a most-probable-number (MPN) method. Multiple dilutions of samples (3, 5, or 10 replicates per dilution) are added to tubes containing A 1 broth. After incubation, test tubes with gas accumulation in the Durham tubes are scored positive and those with no gas as negative. An MPN table is consulted to determine the most probable number of fecal coliforms.

Abeyta-Hunt Bark Agar

Composition per 1016.0mL:
Beef heart, infusion from 500.0g
Agar 15.0g
Tryptose 10.0g
NaCl 5.0g
Yeast extract...2.0g
Horse blood, lysed 50.0mL
Amphotericin B solution................................ 4.0mL
Cefoperazone solution 4.0mL
Rifampicin solution 4.0mL
Ferrous sulfate pyruvate
metabisulfite solution 4.0mL
pH 7.4 ± 0.2 at 25°C

Amphotericin B Solution:
Composition per 100.0mL:
Amphotericin B 0.05g

Preparation of Amphotericin B Solution: Add Amphotericin B to distilled/deionized water and bring volume to 100.0mL. Mix thoroughly. Filter sterilize.

Cefoperazone Solution:
Composition per 100.0mL:
Sodium cefoperazone......................................0.08g

Preparation of Cefoperazone Solution: Add sodium cefoperazone to distilled/deionized water and bring volume to 100.0mL. Mix thoroughly. Filter sterilize.

Rifampicin Solution:
Composition per 100.0mL:
Rifampicin 0.25g

Preparation of Rifampicin Solution: Add rifampicin to 70.0mL ethanol. Mix thoroughly. Add distilled/deionized water to bring volume to 100.0mL. Mix thoroughly. Filter sterilize.

Ferrous Sulfate, Pyruvate, Metabisulfite Solution:

Composition per 100.0mL:

$FeSO_4$	6.25g
Na-pyruvate	6.25g
Na-metabisulfite	6.25g

Preparation of Ferrous Sulfate, Pyruvate, Metabisulfite Solution: Add Na-pyruvate to 20mL distilled/deionized water. Mix thoroughly. Add Na-metabisulfite and $FeSO_4$. Bring volume to 100.0mL. Mix thoroughly. Filter sterilize.

Preparation of Medium: Add components, except cefoperazone solution, amphotericin B solution, rifampicin solution, ferrous sulfate pyruvate metabisulfide solution, and horse blood, to distilled/deionized water and bring volume to 950.0mL. Mix thoroughly. Gently heat and bring to boiling. Autoclave for 15 min at 15 psi pressure–121°C. Cool to 50°C. Aseptically add 4.0mL cefoperazone solution, 4.0mL amphotericin B solution, 4.0mL rifampicin solution, 4.0mL ferrous sulfate pyruvate metabisulfide solution, and 50.0mL lysed horse blood. Mix thoroughly. Pour into sterile Petri dishes or distribute into sterile tubes.

Use: For the isolation and cultivation of *Campylobacter* spp.

ABY Agar
(Acid Bismuth Yeast Agar)

Composition per liter:

Agar	20.0g
Glucose	20.0g
$Bi_2(SO_3)_2$	8.0g
$(NH_4)_2SO_4$	3.0g
KH_2PO_4	3.0g
$MgSO_4 \cdot 7H_2O$	0.25g
$CaCl_2 \cdot 2H_2O$	0.25g
Biotin	10.0μg

pH 7.2 ± 0.2 at 25°C

Preparation of Medium: Add components to distilled/deionized water and bring volume to 1.0L. Mix thoroughly. Gently heat and bring to boiling. Distribute into tubes or flasks. Autoclave for 15 min at 15 psi pressure–121°C. Cool tubes in a slanted position.

Use: For the selective isolation and differentiation of *Candida albicans* from other *Candida* species. *Candida albicans* and *Candida tropicalis* colonies appear as smooth, brownish-black round colonies. Other *Candida* species are differentially pigmented or produce diffusible pigments. Usually used in conjunction with BiGGY agar to differentiate further *Candida*; on BiGGY agar, *Candida albicans* appears as brown to black colonies with no pigment diffusion and no sheen, whereas *Candida tropicalis* appears as dark brown colonies with black centers, black pigment diffusion, and a sheen.

AC Agar
(AC Medium)

Composition per liter:

Proteose peptone no. 3	20.0g
Glucose	5.0g
Beef extract	3.0g
Yeast extract	3.0g
Malt extract	3.0g
Ascorbic acid	0.2g
Agar	1.0g

pH 7.2 ± 0.2 at 25°C

Source: This medium is available as a premixed powder from BD Diagnostic Systems.

Preparation of Medium: Add components to distilled/deionized water and bring volume to 1.0L. Mix thoroughly. Gently heat and bring to boiling. Distribute into tubes or flasks. Autoclave for 15 min at 15 psi pressure–121°C.

Use: For the cultivation and isolation of anaerobes, microaerophiles, and aerobes. Recommended for the sterility testing of solutions and other materials not containing mercurial preservatives.

AC Broth

Composition per liter:

Proteose peptone no. 3	20.0g
Glucose	5.0g
Beef extract	3.0g
Yeast extract	3.0g
Malt extract	3.0g
Ascorbic acid	0.2g

pH 7.2 ± 0.2 at 25°C

Source: This medium is available as a premixed powder from BD Diagnostic Systems.

Preparation of Medium: Add components to distilled/deionized water and bring volume to 1.0L. Mix thoroughly. Gently heat and bring to boiling. Distribute into tubes or flasks. Autoclave for 15 min at 15 psi pressure–121°C.

Use: For the cultivation and isolation of a wide variety of microorganisms, including anaerobes, microaerophiles, and aerobes. Recommended for the sterility testing of solutions and other materials not containing mercurial preservatives.

Acanthamoeba Medium

Composition per liter:

Proteose peptone	15.0g
Glucose	15.0g
KH_2PO_4	0.3g
L-Methionine	14.9mg
Thiamine	1.0mg
Biotin	0.2mg
Vitamin B_{12}	1.0μg
Salt solution	1.0mL

pH 5.5 ± 0.2 at 25°C

Salt Solution:

Composition per 100.0mL:

$MgSO_4 \cdot 7H_2O$	2.46g
$CaCl_2 \cdot 2H_2O$	0.15g
$FeCl_3$	0.02g

Preparation of Salt Solution: Add components to distilled/deionized water and bring volume to 100.0mL. Mix thoroughly.

Preparation of Medium: Add components to distilled/deionized water and bring volume to 1.0L. Mix thoroughly. Adjust pH to 5.5. Filter through Whatman paper to remove particles. Distribute into screw-capped tubes or flasks. Autoclave for 15 min at 15 psi pressure–121°C.

Use: For the cultivation of *Acanthamoeba* species.

ACC Medium

Composition per liter:

Proteose peptone	20.0g
Agar	12.0g
Glycerol	1.5g
K_2SO_4	1.5g
$MgSO_4 \cdot 7H_2O$	1.5g
Antibiotic solution	10.0mL

pH 7.2 ± 0.2 at 25°C

Antibiotic Solution:

Composition per 10.0mL:

Cycloheximide	0.075g
Ampicillin	0.05g
Chloramphenicol	0.0125g

Preparation of Antibiotic Solution: Add components to distilled/deionized water and bring volume to 10.0mL. Mix thoroughly. Filter sterilize.

Preparation of Medium: Add components, except antibiotic solution, to distilled/deionized water and bring volume to 990.0mL. Mix thoroughly. Gently heat and bring to boiling. Autoclave for 15 min at 15 psi pressure–121°C. Cool to 45°–50°C. Aseptically add sterile antibiotic solution. Mix thoroughly. Pour into sterile Petri dishes or distribute into sterile tubes.

Use: For the selective isolation and cultivation of fluorescent *Pseudomonas* species.

Acetamide Agar

Composition per liter:

Agar	15.0g
Acetamide	10.0g
NaCl	5.0g
K_2HPO_4	1.0g
$NH_4H_2PO_4$	1.0g
$MgSO_4 \cdot 7H_2O$	0.2g
Bromthymol Blue	0.08g

pH 6.9 ± 0.2 at 25°C

Preparation of Medium: Add components to distilled/deionized water and bring volume to 1.0L. Mix thoroughly. Gently heat and bring to boiling. Adjust pH. Distribute into tubes or flasks. Autoclave for 15 min at 15 psi pressure–121°C. Cool tubes in a slanted position to produce a long slant.

Use: For the differentiation of nonfermentative Gram-negative bacteria, especially *Pseudomonas aeruginosa.* Bacteria that deamidate acetamide turn the medium blue.

Acetamide Agar

Composition per liter:

Agar	15.0g
Acetamide	10.0g
NaCl	5.0g
K_2HPO_4	1.39g
KH_2PO_4	0.73g
$MgSO_4 \cdot 7H_2O$	0.5g
Phenol Red	0.012g

pH 6.9 ± 0.2 at 25°C

Source: This medium is available as a premixed powder from BD Diagnostic Systems.

Preparation of Medium: Add components to distilled/deionized water and bring volume to 1.0L. Mix thoroughly. Gently heat and bring to boiling. Adjust pH. Distribute into tubes or flasks. Autoclave for 15 min at 15 psi pressure–121°C. Cool tubes in a slanted position to produce a long slant.

Use: For the differentiation of nonfermentative Gram-negative bacteria, especially *Pseudomonas aeruginosa.* Bacteria that deamidate acetamide turn the medium blue.

Acetamide Broth

Composition per liter:

Acetamide	10.0g
NaCl	5.0g
K_2HPO_4	1.39g

KH_2PO_4 .... 0.73g
$MgSO_4 \cdot 7H_2O$ 0.5g
Phenol Red 0.012g

pH 6.9 ± 0.2 at 25°C

Preparation of Medium: Add components to distilled/deionized water and bring volume to 1.0L. Mix thoroughly. Adjust pH. Autoclave for 15 min at 15 psi pressure–121°C.

Use: For the differentiation of nonfermentative Gram-negative bacteria, especially *Pseudomonas aeruginosa.* Bacteria that deamidate acetamide turn the broth purplish red.

Acetamide Cetrimide Glycerol Mannitol Selective Medium

Composition per liter:

Agar 15.0g
K_2SO_4 .... 10.0g
D-Mannitol 5.0g
$MgCl_2 \cdot 6H_2O$ 1.4g
Cetrimide 0.3g
Peptone 0.2g
Acetamide solution 100.0mL
Glycerol 5.0mL

pH 7.0 ± 0.2 at 25°C

Acetamide Solution:

Composition per 100.0mL:

Acetamide 10.0g
Phenol Red 0.012g

Preparation of Acetamide Solution: Add components to distilled/deionized water and bring volume to 100.0mL. Mix thoroughly. Filter sterilize.

Preparation of Medium: Add components, except acetamide solution, to distilled/deionized water and bring volume to 900.0mL. Mix thoroughly. Adjust pH to 7.0. Gently heat and bring to boiling. Autoclave for 20 min at 15 psi pressure–121°C. Cool to 45°–50°C. Aseptically add sterile acetamide solution. Mix thoroughly. Pour into sterile Petri dishes.

Use: For the cultivation of *Pseudomonas aeruginosa, Pseudomonas fluorescens, Pseudomonas putida, Pseudomonas alcaligenes, Pseudomonas cepacia,* and *Pseudomonas pseudoalcaligenes.*

Acetamide Medium

Composition per liter:

Agar, noble 20.0g
Glucose 20.0g
KH_2PO_4 .... 15.0g
$CsCl_2$ solution 12.5mL
Acetamide solution 10.0mL
$CaCl_2 \cdot 2H_2O$ solution 4.1mL
$MgSO_4 \cdot 7H_2O$ solution 2.4mL
Trace elements solution 1.0mL

$CsCl_2$ Solution:

Composition per 100.0mL:

$CsCl_2$ 16.84g

Preparation of $CsCl_2$ Solution: Add $CsCl_2$ to distilled/deionized water and bring volume to 100.0mL. Mix thoroughly. Autoclave for 15 min at 15 psi pressure–121°C.

Acetamide Solution:

Composition per 100.0mL:

Acetamide 5.91g

Preparation of Acetamide Solution: Add acetamide to distilled/deionized water and bring volume to 100.0mL. Mix thoroughly. Autoclave for 15 min at 15 psi pressure–121°C.

$CaCl_2 \cdot 2H_2O$ Solution:

Composition per 100.0mL:

$CaCl_2 \cdot 2H_2O$ 14.7g

Preparation of $CaCl_2 \cdot 2H_2O$ Solution: Add $CaCl_2 \cdot 2H_2O$ to distilled/deionized water and bring volume to 100.0mL. Mix thoroughly. Autoclave for 15 min at 15 psi pressure–121°C.

$MgSO_4 \cdot 7H_2O$ Solution:

Composition per 100.0mL:

$MgSO_4 \cdot 7H_2O$ 24.65g

Preparation of $MgSO_4 \cdot 7H_2O$ Solution: Add $MgSO_4 \cdot 7H_2O$ to distilled/deionized water and bring volume to 100.0mL. Mix thoroughly. Autoclave for 15 min at 15 psi pressure–121°C.

Trace Elements Solution:

Composition per liter:

$FeSO_4 \cdot 7H_2O$ 5.0g
$CoCl_2 \cdot 6H_2O$ 3.7g
$MnSO_4 \cdot 1H_2O$ 1.6g
$ZnSO_4 \cdot 7H_2O$ 1.4g

Preparation of Trace Elements Solution: Add components to distilled/deionized water and bring volume to 1.0L. Mix thoroughly.

Preparation of Medium: Add components, except $CsCl_2$ solution, acetamide solution, $CaCl_2 \cdot 2H_2O$ solution, and $MgSO_4 \cdot 7H_2O$ solution, to distilled/deionized water and bring volume to 971.0mL. Mix thoroughly. Gently heat and bring to boiling. Autoclave for 15 min at 15 psi pressure–121°C. Cool to 50°–55°C. Aseptically add 12.5mL of sterile $CsCl_2$ solution, 10.0mL of sterile acetamide solution, 4.1mL of sterile $CaCl_2 \cdot 2H_2O$ solution, and 2.4mL of sterile $MgSO_4 \cdot 7H_2O$ solution. Mix thoroughly. Pour into sterile Petri dishes or distribute into sterile tubes.

Use: For the cultivation and maintenance of *Trichoderma longibrachiatum.*

Acetamide Medium

Composition per liter:

Stock basal solution	400.0mL
Stock acetamide solution	100.0mL

pH 6.9 ± 0.2 at 25°C

Stock Basal Solution:

Composition per 400.0mL:

Agar	0.5g
KH_2PO_4 solution, 0.5*M*	14.0mL
K_2HPO_4 solution, 0.5*M*	6.0mL

Preparation of Stock Basal Solution: Add components, except PR-CV solution, to distilled/deionized water and bring volume to 400.0mL. Mix thoroughly. Gently heat and bring to boiling with agitation to dissolve agar. Add 1.0mL PR-CV solution.

PR-CV Solution:

Composition per 200.0mL:

Phenol Red	2.0g
Crystal Violet	0.2g

Preparation of PR-CV Solution: Add components to distilled/deionized water and bring volume to 200.0mL. Mix thoroughly. Add 5N NaOH while stirring until components are dissolved.

Stock Acetamide Solution:

Composition per 100.0mL:

Acetamide	1.0g

Preparation of Stock Acetamide Solution: Add acetamide to distilled/deionized water and bring volume to 100.0mL. Mix thoroughly. Store over methylene chloride in a screw capped container. Can be stored indefinitely at room temperature.

Preparation of Medium: Combine 400.0mL stock basal solution and 100.0mL stock acetamide solution. Mix thoroughly. Distribute into tubes or flasks. Steam for 10 min at 100°C. Cool.

Use: For the differentiation of nonfermentative Gram-negative bacteria, especially *Pseudomonas aeruginosa.*

Acetate Differential Agar (Sodium Acetate Agar) (Simmons' Citrate Agar, Modified)

Composition per liter:

Agar	20.0g
NaCl	5.0g
Sodium acetate	2.0g
$(NH_4)H_2PO_4$	1.0g
K_2HPO_4	1.0g
$MgSO_4 \cdot 7H_2O$	0.2g
Bromthymol Blue	0.08g

pH 6.8 ± 0.2 at 25°C

Source: This medium is available as a premixed powder from BD Diagnostic Systems.

Preparation of Medium: Add components to cold distilled/deionized water and bring volume to 1.0L. Mix thoroughly. Gently heat and bring to boiling. Distribute into tubes to produce a 1 cm butt and 30 cm slant. Autoclave for 15 min at 15 psi pressure–121°C. Cool tubes in a slanted position.

Use: For the differentiation of *Shigella* species from *Escherichia coli* and also for the differentiation of nonfermenting Gram-negative bacteria. Bacteria that can utilize acetate as the sole carbon source turn the medium blue.

Acetobacteroides glycinophilus Medium

Composition per 1020.0mL:

Na_2HPO_4	5.8g
KH_2PO_4	3.0g
NH_4Cl	1.0g
$MgCl_2 \cdot 6H_2O$	0.2g
Resazurin	1.0mg
$CaCl_2 \cdot 2H_2O$	0.13g
Trace elements solution	10.0mL
Vitamin solution	5.0mL
Yeast extract solution	5.0mL
Glycine solution	5.0mL
$NaHCO_3$ solution	5.0mL
$Na_2S \cdot 9H_2O$ solution	5.0mL

pH 7.2–7.4 at 25°C

Trace Elements Solution:

Composition per liter:

Nitrilotriacetic acid	2.8g
NaCl	1.0g
$FeCl_3 \cdot 4H_2O$	0.2g
$CoCl_2 \cdot 6H_2O$	0.17g
$CaCl_2 \cdot 2H_2O$	0.1g
$MnCl_2 \cdot 4H_2O$	0.1g
$ZnCl_2$	0.1g
$NiCl_2 \cdot 6H_2O$	0.026g
$CuCl_2$	0.02g
$Na_2SeO \cdot 5H_2O$	0.02g
H_3BO_3	0.01g
$Na_2MoO \cdot 2H_2O$	0.01g

Preparation of Trace Elements Solution: Add nitrilotriacetic acid to 500.0mL of distilled/deionized water. Dissolve by adjusting pH to 6.5 with KOH. Add distilled/deionized water to 1.0L. Add remaining components. Mix thoroughly.

Vitamin Solution:
Composition per liter:

Pyridoxine·HCl	10.0mg
Calcium DL-pantothenate	5.0mg
Lipoic acid	5.0mg
Nicotinic acid	5.0mg
p-Aminobenzoic acid	5.0mg
Riboflavin	5.0mg
Thiamine·HCl	5.0mg
Biotin	2.0mg
Folic acid	2.0mg
Vitamin B_{12}	0.1mg

Preparation of Vitamin Solution: Add components to distilled/deionized water and bring volume to 1.0L. Mix thoroughly.

Yeast Extract Solution:
Composition per 5.0mL:

Yeast extract	0.5g

Preparation of Yeast Extract Solution: Add yeast extract to distilled/deionized water and bring volume to 5.0mL. Mix thoroughly. Sparge under 100% N_2 gas for 3 min. Autoclave for 15 min at 15 psi pressure–121°C. Store under N_2 gas.

Glycine Solution:
Composition per 5.0mL:

Glycine	1.5g

Preparation of Glycine Solution: Add glycine to distilled/deionized water and bring volume to 5.0mL. Mix thoroughly. Sparge under 100% N_2 gas for 3 min. Autoclave for 15 min at 15 psi pressure–121°C. Store under N_2 gas.

$NaHCO_3$ Solution:
Composition per 5.0mL:

$NaHCO_3$	0.5g

Preparation of $NaHCO_3$ Solution: Add $NaHCO_3$ to distilled/deionized water and bring volume to 5.0mL. Mix thoroughly. Sparge under 100% N_2 gas for 3 min. Autoclave for 15 min at 15 psi pressure–121°C. Store under N_2 gas.

$Na_2S{\cdot}9H_2O$ Solution:
Composition per 5.0mL:

$Na_2S{\cdot}9H_2O$	0.5g

Preparation of $Na_2S{\cdot}9H_2O$ Solution: Add $Na_2S{\cdot}9H_2O$ solution to distilled/deionized water and bring volume to 5.0mL. Mix thoroughly. Sparge under 100% N_2 gas for 3 min. Autoclave for 15 min at 15 psi pressure–121°C. Store under N_2 gas.

Preparation of Medium: Add components, except yeast extract solution, glycine solution, $NaHCO_3$ solution, and $Na_2S{\cdot}9H_2O$ solution, to distilled/deionized water and bring volume to 1.0L. Adjust pH to 7.2–7.4 with NaOH. Mix thoroughly. Gently heat and bring to boiling. Boil for a few minutes. Allow to cool to room temperature under 100% N_2. Distribute into tubes or flasks under 100% N_2. Autoclave for 15 min at 15 psi pressure–121°C. Cool to room temperature. Before inoculation, aseptically and anaerobically add yeast extract solution, glycine solution, $NaHCO_3$ solution, and $Na_2S{\cdot}9H_2O$ solution.

Use: For the cultivation and maintenance of *Acetobacteroides glycinophilus*.

Achromobacter Choline Medium

Composition per liter:

NaCl	30.0g
Agar	18.0g
Choline chloride	5.0g
K_2HPO_4	1.0g
$MgSO_4{\cdot}7H_2O$	0.5g
$FeSO_4{\cdot}7H_2O$	0.01g

Preparation of Medium: Add components to distilled/deionized water and bring volume to 1.0L. Mix well and warm gently until dissolved. Autoclave for 15 min at 15 psi pressure–121°C. Pour into sterile Petri dishes.

Use: For the cultivation and maintenance of *Achromobacter cholinophagum* and other bacteria that can utilize choline as a carbon source.

Achromobacter Choline Medium, Modified

Composition per liter:

NaCl	30.0g
Agar	15.0g
Choline chloride	5.0g
K_2HPO_4	1.0g
$MgSO_4{\cdot}7H_2O$	1.0g
$FeSO_4{\cdot}7H_2O$	0.018g

pH 7.4 ± 0.2 at 25°C

Preparation of Medium: Add agar, $MgSO_4{\cdot}7H_2O$, and $FeSO_4{\cdot}7H_2O$ to 500.0mL distilled/deionized water. Mix thoroughly. Bring volume to 1.0L with distilled/deionized water. Gently heat and bring to boiling. Add choline chloride. Mix thoroughly. Distribute into tubes or flasks. Autoclave for 15 min at 15 psi pressure–121°C. Pour into sterile Petri dishes or leave in tubes.

Use: For the cultivation and maintenance of *Achromobacter cholinophagum*.

Achromobacter Medium (ATCC Medium 457)

Composition per liter:

K_2HPO_4	7.32g
Ammonium tartrate	4.6g
KH_2PO_4	1.09g
$MgSO_4 \cdot 7H_2O$	0.04g
$FeSO_4 \cdot 7H_2O$	0.04g
$CaCl_2 \cdot 2H_2O$	0.014g
$MgSO_4 \cdot 7H_2O$	0.002g

pH 7.5 ± 0.2 at 25°C

Preparation of Medium: Add components to distilled/deionized water and bring volume to 1.0L. Mix well and warm gently until dissolved. Distribute into test tubes or flasks. Autoclave for 15 min at 15 psi pressure–121°C.

Use: For the cultivation and maintenance of *Achromobacter* species and *Alcaligenes* species.

Achromobacter Medium (ATCC Medium 589)

Composition per liter:

Agar	20.0g
K_2HPO_4	7.0g
Methionine	5.0g
KH_2PO_4	2.0g
$(NH_4)_2SO_4$	1.0g
Sodium citrate	0.4g
$MgSO_4 \cdot 7H_2O$	0.1g

Preparation of Medium: Add components to distilled/deionized water and bring volume to 1.0L. Mix thoroughly. Gently heat and bring to boiling. Autoclave for 15 min at 15 psi pressure–121°C. Pour into sterile Petri dishes.

Use: For the cultivation and maintenance of *Achromobacter* species.

Acid Egg Medium

Composition per 1640.0mL:

Potato starch	30.0g
KH_2PO_4	12.3g
Malachite Green	0.4g
$MgSO_4 \cdot 7H_2O$	0.3g
Penicillin G	100,000IU
Fresh egg mixture	1000.0mL
Glycerol	12.0mL

Source: This medium is available as a prepared medium from Oxoid Unipath.

Preparation of Medium: Add components to 1.0L of fresh egg mixture. Mix thoroughly. Gently heat and bring to boiling. Bring volume to 1640.0mL with distilled/deionized water. Distribute into tubes or flasks. Autoclave for 15 min at 15 psi pressure–121°C with tubes in an upright position.

Use: For the cultivation and maintenance of *Mycobacterium tuberculosis*.

Acid Tomato Broth

Composition per liter:

Glucose	10.0g
Peptone	10.0g
Yeast extract	5.0g
$MgSO_4 \cdot 7H_2O$	0.2g
$MnSO_4 \cdot 4H_2O$	0.05g
Tomato juice	250.0mL
L-Cysteine solution	0.5mL

L-Cysteine Solution:

Composition per 10.0mL:

L-Cysteine	0.1g

Preparation of L-Cysteine Solution: Add 0.1g of L-cysteine to distilled/deionized water and bring volume to 10.0mL. Mix thoroughly. Filter sterilize.

Preparation of Medium: Add components, except L-cysteine solution, to distilled/deionized water and bring volume to 999.5mL. Mix thoroughly. Adjust pH to 4.8. Autoclave for 15 min at 15 psi pressure–121°C. Aseptically add 0.5mL of sterile L-cysteine solution. Mix thoroughly. Aseptically distribute into sterile tubes or flasks.

Use: For the cultivation of a variety of fungi.

Actidione® Agar (Cycloheximide Agar)

Composition per liter:

Glucose	50.0g
Agar	15.0g
Pancreatic digest of casein	5.0g
Yeast extract	4.0g
KH_2PO_4	0.55g
KCl	0.425g
$CaCl_2 \cdot 2H_2O$	0.125g
$MgSO_4 \cdot 7H_2O$	0.125g
Bromocresol Green	22.0mg
Actidione (cycloheximide)	10.0mg
$FeCl_3$	2.5mg

pH 5.5 ± 0.2 at 25°C

Source: Actidione Agar is available as a prepared medium from Oxoid Unipath.

Preparation of Medium: Add components to distilled/deionized water and bring volume to 1.0L. Mix thoroughly. Gently heat and bring to boiling. Distribute into tubes or flasks. Autoclave for 15 min at 15 psi pressure–121°C. Pour into sterile Petri dishes or leave in tubes.

Use: For the enumeration and detection of bacteria in specimens containing large numbers of yeasts and molds.

Actinomyces Agar

Composition per liter:

Agar	20.0g
K_2HPO_4	13.0g
Heart muscle, solids from infusion	10.0g
Peptic digest of animal tissue	10.0g
Glucose	5.0g
Yeast extract	5.0g
NaCl	5.0g
Pancreatic digest of casein	4.0g
KH_2PO_4	2.0g
$(NH_4)_2SO_4$	1.0g
L-Cysteine·HCl·H_2O	1.0g
Soluble starch	1.0g
$MgSO_4 \cdot 7H_2O$	0.2g
$CaCl_2 \cdot 2H_2O$	0.01g

pH 6.9 ± 0.2 at 25°C

Preparation of Medium: Add components to distilled/deionized water and bring volume to 1.0L. If a semisolid medium is desired, add 7.0g of agar instead of 20.0g. Mix thoroughly. Gently heat and bring to boiling. Distribute into tubes or flasks. Autoclave for 10 min at 15 psi pressure–121°C. Pour into sterile Petri dishes or leave in tubes.

Use: For the maintenance or cultivation of a variety of anaerobic bacteria, including *Actinomyces* species, *Eubacterium* species, *Fusobacterium* species, *Propionibacterium* species, and others.

Actinomyces Broth

Composition per liter:

K_2HPO_4	13.0g
Heart muscle, solids from infusion	10.0g
Peptic digest of animal tissue	10.0g
Glucose	5.0g
Yeast extract	5.0g
NaCl	5.0g
Pancreatic digest of casein	4.0g
KH_2PO_4	2.0g
$(NH_4)_2SO_4$	1.0g
L-Cysteine·HCl·H_2O	1.0g
Soluble starch	1.0g
$MgSO_4 \cdot 7H_2O$	0.2g
$CaCl_2 \cdot 2H_2O$	0.01g

pH 6.9 ± 0.2 at 25°C

Source: This medium is available as a premixed powder from BD Diagnostic Systems.

Preparation of Medium: Add components to distilled/deionized water and bring volume to 1.0L. Mix thoroughly. Distribute into tubes or flasks. Autoclave for 10 min at 15 psi pressure–121°C.

Use: For the maintenance or cultivation of a variety of anaerobic bacteria including *Actinomyces* species, *Eubacterium* species, *Fusobacterium* species, *Propionibacterium* species and others.

Actinomyces Broth

Composition per liter:

Beef heart, infusion from	500.0g
KH_2PO_4	15.0g
Peptic digest of animal tissue	10.0g
Glucose	5.0g
Yeast extract	5.0g
NaCl	5.0g
Pancreatic digest of casein	4.0g
KH_2PO_4	2.0g
$(NH_4)_2SO_4$	1.0g
L-Cysteine·HCl·H_2O	1.0g
Soluble starch	1.0g
$MgSO_4 \cdot 7H_2O$	0.2g
$CaCl_2 \cdot 2H_2O$	0.02g

pH 7.2 ± 0.2 at 25°C

Source: This medium is available as a premixed powder from BD Diagnostic Systems.

Preparation of Medium: Add components to distilled/deionized water and bring volume to 1.0L. Mix thoroughly. Distribute into tubes or flasks. Autoclave for 10 min at 15 psi pressure–121°C.

Use: For the maintenance or cultivation of a variety of anaerobic bacteria, including *Actinomyces* species, *Eubacterium* species, *Fusobacterium* species, *Propionibacterium* species, and others.

Actinomyces Isolation Agar

Composition per liter:

Agar	15.0g
Glycerol	5.0g
Sodium propionate	4.0g
Sodium caseinate	2.0g
K_2HPO_4	0.5g
Asparagine	0.1g
$MgSO_4 \cdot 7H_2O$	0.1g
$FeSO_4 \cdot 7H_2O$	0.001g

Preparation of Medium: Add components to distilled/deionized water and bring volume to 1.0L. Mix thoroughly. Gently heat and bring to boiling. Distribute into tubes or flasks. Autoclave for 15 min at 15 psi pressure–121°C. Pour into sterile Petri dishes or leave in tubes.

Use: For the isolation and cultivation of *Actinomyces* species.

Actinomycete Growth Medium

Composition per liter:

Succinic acid	1.18g
L-Glutamine	0.29g
$CaCl_2 \cdot 2H_2O$	0.2g
KH_2PO_4	0.2g
$MgSO_4 \cdot 7H_2O$	0.2g
NaCl	0.1g
m-Inositol	0.09g
Ferric EDTA	0.037g
$MnSO_4 \cdot H_2O$	4.5mg
H_3BO_3	1.5mg
$ZnSO_4 \cdot 7H_2O$	1.5mg
Nicotonic acid	0.5mg
Pyridoxine-HCl	0.5mg
Thiamine-HCl	0.1mg
$CuSO_4 \cdot 5H_2O$	0.04mg
$Na_2MoO_4 \cdot 2H_2O$	0.025mg

pH 6.4 ± 0.2 at 25°C

Preparation of Medium: Add components to distilled/deionized water and bring volume to 1.0L. Mix thoroughly. Distribute into tubes or flasks. Autoclave for 15 min at 15 psi pressure–121°C.

Use: For the cultivation of actinomycetes.

Activated Carbon Medium (DSMZ Medium 811)

Composition per liter:

Agar	15.0g
$Na_2HPO_4 \cdot 12H_2O$	9.0g
Activated carbon	5.0g
KH_2PO_4	1.5g
NH_4Cl	1.5g
$MgSO_4 \cdot 7H_2O$	0.2g
$CaCl_2 \cdot 2H_2O$	20.0mg
NH_4-Fe-III-Citrate	1.2mg
Trace elements solution TS2	1.0mL

pH 7.5 ± 0.1 at 25°C

Trace Elements Solution TS2:

Composition per liter:

$Na_2MoO_4 \cdot 4H_2O$	900.0mg
H_3BO_3	300.0mg
$CoCl_2 \cdot 6H_2O$	200.0mg
$ZnSO_4 \cdot 7H_2O$	100.0mg
$MnCl_2 \cdot 4H_2O$	30.0mg
$NiCl_2 \cdot 6H_2O$	20.0mg
Na_2SeO_3	20.0mg
$CuCl_2 \cdot 2H_2O$	10.0mg

Preparation of Solution A: Add components to distilled/deionized water and bring volume to 1.0L. Mix thoroughly.

Preparation of Medium: Add components, except activated carbon, to distilled/deionized water and bring volume to 900.0mL. Mix thoroughly. After all components are dissolved add activated carbon. Bring volume to 1.0L with distilled/deionized water. Adjust pH to 7.5. Distribute into tubes or flasks. Autoclave for 15 min at 15 psi pressure–121°C. Pour into Petri dishes or leave in tubes.

Use: For the cultivation of *Streptomyces thermoautotrophicus.*

AE Sporulation Medium, Modified

Composition per 1079.2mL:

Polypeptone™	10.0g
Yeast extract	10.0g
Na_2HPO_4	4.36g
Ammonium acetate	1.5g
KH_2PO_4	0.25g
$MgSO_4 \cdot 7H_2O$	0.2g
Raffinose solution	39.6mL
Na_2CO_3 solution	13.2mL
$CoCl_2 \cdot 6H_2O$ solution	13.2mL
Sodium ascorbate solution	13.2mL

pH 7.8 ± 0.1 at 25°C

Raffinose Solution:

Composition per 100.0mL:

Raffinose	10.0g

Preparation of Raffinose Solution: Add raffinose to distilled/deionized water and bring volume to 100.0mL. Mix thoroughly. Filter sterilize.

Na_2CO_3 Solution:

Composition per 100.0mL:

Na_2CO_3	7.0g

Preparation of Na_2CO_3 Solution: Add Na_2CO_3 to distilled/deionized water and bring volume to 100.0mL. Mix thoroughly. Filter sterilize.

$CoCl_2$ Solution:

Composition per 100.0mL:

$CoCl_2 \cdot 6H_2O$	0.32g

Preparation of $CoCl_2$ Solution: Add $CoC_{l2} \cdot 6H_2O$ to distilled/deionized water and bring volume to 100.0mL. Mix thoroughly. Filter sterilize.

Sodium Ascorbate Solution:

Composition per 100.0mL:

Sodium ascorbate	1.5g

Preparation of Sodium Ascorbate Solution: Add sodium ascorbate to distilled/deionized water and bring volume to 100.0mL. Mix thoroughly. Filter sterilize. Use freshly prepared solution.

Preparation of Medium: Add components—except raffinose solution, Na_2CO_3 solution, $CoCl_2$ solution, and sodium ascorbate solution—to distilled/deionized water and bring volume to 1.0L. Mix thoroughly. Adjust pH to 7.5 using 2m sodium carbonate solution. Distribute into tubes in 15.0mL volumes. Autoclave for 15 min at 15 psi pressure–121°C. Aseptically add 0.6mL of sterile raffinose solution, 0.2mL of sterile Na_2CO_3 solution, and 0.2mL of sterile $CoCl_2$ solution to each tube. Mix thoroughly. Prior to inoculation, steam medium for 10 min. Cool to 25°C. Aseptically add 0.2mL of sterile sodium ascorbate solution to each tube.

Use: For the cultivation and sporulation of *Clostridium perfringens.*

Aeromonas Differential Agar (Dextrin Fuchsin Sulfite Agar)

Composition per liter:

Dextrin	15.0g
Agar	13.0g
Pancreatic digest of casein	10.0g
Na_2HPO_4	7.75g
NaCl	5.0g
Beef extract	3.0g
Na_2SO_3	1.6g
Acid Fuchsin solution	50.0mL

pH 7.5 ± 0.2 at 25°C

Acid Fuchsin Solution:

Composition per 50.0mL:

Acid Fuchsin	0.25g
Aquesou dioxan, 5%	50.0mL

Preparation of Acid Fuchsin Solution: Add Acid Fuchsin to 50.0mL of 5% aqueous dioxan. Mix well to dissolve.

Caution: Acid Fuchsin is a potential carcinogen and care must be taken to avoid inhalation of the powdered dye and contamination of the skin.

Preparation of Medium: Add components to distilled/deionized water and bring volume to 1.0L. Mix thoroughly. Gently heat while stirring and bring to boiling. Distribute into tubes or flasks. Autoclave for 15 min at 15 psi pressure–121°C. Pour into sterile Petri dishes or leave in tubes.

Use: For the isolation and differentiation of *Aeromonas* species from other Gram-negative rods such as *Pseudomonas* and Enterobacteriaceae. Specimens with low numbers of *Aeromonas* may first be enriched by growth in starch broth for 4–9 days. After 24h of growth on this agar, colonies are sprayed with Nadi reagent (1% solution of *N,N,N′,N′*-tetramethyl-*p*-phenylene-diammonium dichloride). A positive Nadi reaction (dextrin degradation) is indicated by a purple color at the periphery of the colony. Dextrin fermentation is also indicated by red colonies. *Aeromonas* species appear as large, convex, dark red colonies with a purple periphery.

Aeromonas hydrophila Medium

Composition per liter:

Inositol	10.0g
Pancreatic digest of casein	10.0g
L-Ornithine·HCl	5.0g
Proteose peptone	5.0g
Agar	3.0g
Yeast extract	3.0g
Mannitol	1.0g
Ferric ammonium citrate	0.5g
$Na_2S_2O_3 \cdot 5H_2O$	0.4g
Bromcresol Purple	0.02g

pH 6.7 ± 0.2 at 25°C

Preparation of Medium: Add components to distilled/deionized water and bring volume to 1.0L. Mix thoroughly. Gently heat until dissolved. Adjust pH to 6.7. Distribute into tubes in 5.0mL volumes. Autoclave for 12 min at 15 psi pressure–121°C.

Use: For the isolation and cultivation of *Aeromonas hydrophila.*

Aeromonas Medium (Ryan's *Aeromonas* Medium)

Composition per liter:

Agar	12.5g
$Na_2S_2O_3$	10.67g
Proteose peptone	5.0g
NaCl	5.0g
Xylose	3.75g
L-Lysine·HCl	3.5g
Yeast extract	3.0g
Sorbitol	3.0g
Bile salts no. 3	3.0g
Inositol	2.5g
L-Arginine·HCl	2.0g
Lactose	1.5g
Ferric ammonium citrate	0.8g
Bromthymol Blue	0.04g
Thymol Blue	0.04g

pH 8.0 ± 0.1 at 25°C

Source: This medium is available as a dehydrated powder from Oxoid Unipath.

Preparation of Medium: Add components to distilled/deionized water and bring volume to 1.0L. Mix thoroughly. Gently heat and bring to boiling. Do not autoclave. Cool to 50°C and aseptically add 5.0mg of ampicillin. Pour into sterile Petri dishes.

Use: For the isolation and selective differentiation of *Aeromonas hydrophila* and other *Aeromonas* species from clinical and nonclinical specimens. *Aeromonas* species appear as small (0.5–1.5mm), dark green colonies with darker centers.

AFPA
(*Aspergillus flavus/parasiticus* Agar)

Composition per liter:

Yeast extract	20.0g
Agar	15.0g
Peptone	10.0g
Ferric ammonium citrate	0.5g
Dichloran (Botran®)	2.0mg

pH 6.3 ± 0.2 at 25°C

Source: This medium is available as a dehydrated powder from Oxoid Unipath.

Preparation of Medium: Add components to distilled/deionized water and bring volume to 1.0L. Mix thoroughly. Gently heat while stirring and bring to boiling. Add 100.0mg of chloramphenicol. Autoclave for 15 min at 15 psi pressure–121°C. Pour into sterile Petri dishes.

Use: For the selective isolation and enumeration of *Aspergillus flavus* and *Aspergillus parasiticus*. Colonies of these fungi appear with dark yellow-orange color on the reverse side.

AH5 Medium

Composition per 205.9mL:

Agar base	160.0mL
Supplement solution	45.9mL

pH 6.0 ± 0.2 at 25°C

Agar Base:
Composition per 165.0mL:

Pancreatic digest of casein	2.72g
Agar	2.1g
NaCl	0.8g
Papaic digest of soybean meal	0.48g
K_2HPO_4	0.4g
Glucose	0.4g

Preparation of Agar Base: Add components, except agar, to distilled/deionized water and bring volume to 165.0mL. Adjust pH to 5.5. Add agar. Mix thoroughly. Autoclave for 15 min at 15 psi pressure–121°C. Cool to 45°–50°C.

Supplement Solution:
Composition per 45.9mL:

Horse serum, unheated	40.0mL
Fresh yeast extract solution	2.0mL
Penicillin solution	2.0mL
CVA enrichment	1.0mL
L-Cysteine·HCl·H_2O solution	0.5mL
Urea solution	0.4mL

Preparation of Supplement Solution: Aseptically combine components. Mix thoroughly.

Fresh Yeast Extract Solution:
Composition per 100.0mL:

Baker's yeast, live, pressed, starch-free	25.0g

Preparation of Fresh Yeast Extract Solution: Add the live Baker's yeast to 100.0mL of distilled/deionized water. Autoclave for 90 min at 15 psi pressure–121°C. Allow to stand. Remove supernatant solution. Adjust pH to 6.6–6.8.

Penicillin Solution:
Composition per 10.0mL:

Penicillin G	1,000,000U

Preparation of Penicillin Solution: Add penicillin to distilled/deionized water and bring volume to 10.0mL. Mix thoroughly. Filter sterilize.

CVA Enrichment:
Composition per liter:

Glucose	100.0g
L-Cysteine·HCl·H_2O	25.9g
L-Glutamine	10.0g
L-Cystine·2HCl	1.0g
Adenine	1.0g
Nicotinamide adenine dinucleotide	0.25g
Cocarboxylase	0.1g
Guanine·HCl	0.03g
$Fe(NO_3)_3$	0.02g
p-Aminobenzoic acid	0.013g
Vitamin B_{12}	0.01g
Thiamine·HCl	3.0mg

Preparation of CVA Enrichment: Add components to distilled/deionized water and bring volume to 1.0L. Mix thoroughly. Filter sterilize.

L-Cysteine·HCl·H_2O Solution:
Composition per 10.0mL:

L-Cysteine·HCl·H_2O	0.4g

Preparation of L-Cysteine·HCl·H_2O Solution: Add L-cysteine·HCl·H_2O solution to distilled/deionized water and bring volume to 10.0mL. Mix thoroughly. Filter sterilize.

Urea Solution:
Composition per 10.0mL:

Urea	1.0g

Preparation of Urea Solution: Add urea to distilled/deionized water and bring volume to 10.0mL. Mix thoroughly. Filter sterilize.

Preparation of Medium: Aseptically combine cooled, sterile components. Mix thoroughly. Pour into sterile Petri dishes or distribute into sterile tubes.

Use: For the cultivation of *Ureaplasma urealyticum* from urine and exudates and for the cultivation of other *Ureaplasma* species.

AKI Medium

Composition per liter:

Peptone	15.0g
NaCl	5.0g
Yeast extract	4.0g
Sodium bicarbonate solution	30.0mL

pH 7.2 ± 0.2 at 25°C

Sodium Bicarbonate Solution:
Composition per 100.0mL:

$NaHCO_3$	10.0g

Preparation of Sodium Bicarbonate Solution: Add sodium bicarbonate to distilled/deionized water and bring volume to 100.0mL. Mix thoroughly. Filter sterilize. Use freshly prepared solution.

Preparation of Medium: Add components, except sodium bicarbonate solution, to distilled/deionized water and bring volume to 970.0mL. Mix thoroughly. Autoclave for 15 min at 15 psi pressure–121°C. Cool to 45°–50°C. Aseptically add sterile sodium bicarbonate solution. Mix thoroughly. Aseptically distribute into sterile tubes or flasks. Prepare medium freshly.

Use: For the cultivation of *Vibrio cholerae* and other *Vibrio* species.

Alcaligenes Agar

Composition per liter:

Agar	10.0g
Peptone	5.0g
Ammonium lactate	3.0g
Meat extract	3.0g
Ferric citrate	0.2g

pH 7.0 ± 0.2 at 25°C

Preparation of Medium: Add ferric citrate to distilled/deionized water and bring volume to 100.0mL. In a separate flask, add remaining components to distilled/deionized water and bring volume to 900.0mL. Mix thoroughly. Adjust pH to 7.0. Steam the two solutions for 20 min on three consecutive days. Aseptically combine the two solutions. Pour into sterile Petri dishes or distribute into sterile tubes.

Use: For the cultivation of *Alcaligenes* species.

Alcaligenes Medium

Composition per liter:

Tris	6.06g
NaCl	4.68g
KCl	1.49g
NH_4Cl	1.07g
Na_2SO_4	0.43g
$Na_2HPO_4 \cdot 12H_2O$	0.23g
$MgCl_2 \cdot 6H_2O$	0.2g
$CaCl_2 \cdot 2H_2O$	0.03g
Ferric ammonium citrate	0.005g
Sodium succinate solution	10.0mL
$CuSO_4$ solution	2.5mL
Trace elements solution SL-7	1.0mL

Sodium Succinate Solution:
Composition per 100.0mL:

Sodium succinate	40.0g

Preparation of Sodium Succinate Solution: Add sodium succinate to distilled/deionized water and bring volume to 100.0mL. Mix thoroughly. Filter sterilize.

$CuSO_4$ Solution:
Composition per 100.0mL:

$CuSO_4$	16.0g

Preparation of $CuSO_4$ Solution: Add $CuSO_4$ to distilled/deionized water and bring volume to 100.0mL. Mix thoroughly. Filter sterilize.

Trace Elements Solution SL-7:
Composition per 1001.0mL:

$CoCl_2 \cdot 6H_2O$	200.0mg
$MnCl_2 \cdot 4H_2O$	100.0mg
$ZnCl_2$	70.0mg
H_3BO_3	60.0mg
$Na_2MoO_4 \cdot 2H_2O$	40.0mg
$CuCl_2 \cdot 2H_2O$	20.0mg
$NiCl_2 \cdot 6H_2O$	20.0mg
HCl (25%)	1.0mL

Preparation of Trace Elements Solution SL-7: Add components to distilled/deionized water and bring volume to 1.0L. Mix thoroughly.

Preparation of Medium: Add components, except $CuSO_4$ solution and sodium succinate solution, to distilled/deionized water and bring volume to 987.5mL. Mix thoroughly. Autoclave for 15 min at 15 psi pressure–121°C. Aseptically add 10.0mL of sterile $CuSO_4$ solution and 2.5mL of sterile sodium succinate solution. Mix thoroughly. Aseptically distribute into sterile tubes or flasks.

Use: For the cultivation of *Alcaligenes* species.

Alcaligenes N5 Medium

Composition per liter:

Sodium succinate·$2H_2O$	5.0g
KH_2PO_4	0.75g
NH_4Cl	0.67g
K_2HPO_4	0.61g
$MgSO_4 \cdot 7H_2O$	0.2g

$CaCl_2 \cdot 2H_2O$ 0.03g
$MnCl_2 \cdot 4H_2O$ 3.0mg
$FeCl_3$ 2.4mg
$Na_2MoO_4 \cdot 2H_2O$ 1.0mg

Preparation of Medium: Add components to distilled/deionized water and bring volume to 1.0L. Mix thoroughly. Gently heat while stirring and bring to boiling. Distribute into tubes or flasks. Autoclave for 15 min at 15 psi pressure–121°C.

Use: For the cultivation and maintenance of *Alcaligenes faecalis*.

Alcaligenes NA YE Medium (*Alcaligenes* Nutrient Agar Yeast Extract Medium)

Composition per liter:

Agar 15.0g
Pancreatic digest of gelatin 5.0g
Yeast extract 5.0g
Beef extract 3.0g
pH 7.0 ± 0.2 at 25°C

Preparation of Medium: Add components to distilled/deionized water and bring volume to 1.0L. Mix thoroughly. Gently heat while stirring and bring to boiling. Distribute into tubes or flasks. Autoclave for 15 min at 15 psi pressure–121°C. Pour into sterile Petri dishes or leave in tubes.

Use: For the cultivation and maintenance of *Alcaligenes* species.

Alcaligenes NB YE Agar (*Alcaligenes* Nutrient Broth Yeast Extract Agar)

Composition per liter:

Agar 15.0g
Pancreatic digest of gelatin 5.0g
Yeast extract 5.0g
Beef extract 3.0g

Preparation of Medium: Add components to distilled/deionized water and bring volume to 1.0L. Mix thoroughly. Gently heat while stirring and bring to boiling. Distribute into tubes or flasks. Autoclave for 15 min at 15 psi pressure–121°C. Pour into sterile Petri dishes or leave in tubes.

Use: For the cultivation and maintenance of *Alcaligenes faecalis*.

Alcaligenes NB YE Broth (*Alcaligenes* Nutrient Broth Yeast Extract Broth)

Composition per liter:

Pancreatic digest of gelatin 5.0g
Yeast extract 5.0g
Beef extract 3.0g

Preparation of Medium: Add components to distilled/deionized water and bring volume to 1.0L. Mix thoroughly. Gently heat while stirring and bring to boiling. Distribute into tubes or flasks. Autoclave for 15 min at 15 psi pressure–121°C.

Use: For the cultivation of *Alcaligenes faecalis*.

Alcaligenes NB YE Medium (*Alcaligenes* Nutrient Broth Yeast Extract Medium)

Composition per liter:

Pancreatic digest of gelatin 5.0g
Yeast extract 5.0g
Beef extract 3.0g
pH 7.0 ± 0.2 at 25°C

Preparation of Medium: Add components to distilled/deionized water and bring volume to 1.0L. Mix thoroughly. Distribute into tubes or flasks. Autoclave for 15 min at 15 psi pressure–121°C.

Use: For the cultivation and maintenance of *Alcaligenes* species.

Alginate Utilization Medium

Composition per liter:

Solution B 500.0mL
Solution A 400.0mL
Solution C 100.0mL

Solution A:
Composition per 400.0mL:

Marine salts 38.0g

Preparation of Solution A: Add marine salts to distilled/deionized water and bring volume to 400.0mL. Mix thoroughly. Autoclave for 15 min at 15 psi pressure–121°C.

Solution B:
Composition per 500.0mL:

Agar 20.0g
Sodium alginate 10.0g

Preparation of Solution B: Add components to distilled/deionized water and bring volume to 500.0mL. Mix thoroughly. Autoclave for 15 min at 15 psi pressure–121°C.

Solution C:
Composition per 100.0mL:

Tris·HCl buffer 0.067g
$NaNO_3$ 0.047g
Ferric EDTA 66.5mg
Sodium glycerophosphate 6.67mg

Thiamine·HCl 67.0μg
Vitamin B_{12} 1.3μg
Biotin 0.67μg

Preparation of Solution C: Add components to distilled/deionized water and bring volume to 100.0mL. Mix thoroughly. Filter sterilize.

Preparation of Medium: Aseptically combine solutions A, B, and C. For liquid medium, omit agar from solution B.

Use: For the cultivation of microorganisms that can utilize alginate as a carbon source. Growth on alginate (production of alginase) is a diagnostic test used in the differentiation of *Vibrio* species.

Alkaline Peptone Agar

Composition per liter:

NaCl 20.0g
Agar 15.0g
Peptone 10.0g

pH 8.5 ± 0.2 at 25°C

Preparation of Medium: Add components to distilled/deionized water and bring volume to 1.0L. Mix thoroughly. Gently heat and bring to boiling. Adjust pH to 8.5. Distribute into tubes. Autoclave for 15 min at 15 psi pressure–121°C. Allow tubes to cool in a slanted position.

Use: For the cultivation of *Vibrio cholerae* and other *Vibrio* species.

Alkaline Peptone Salt Broth (APS Broth)

Composition per liter:

NaCl 30.0g
Peptone 10.0g

Preparation of Medium: Add components to distilled/deionized water and bring volume to 1.0L. Mix thoroughly. Adjust pH to 8.5. Distribute into tubes in 10.0mL volumes. Autoclave for 10 min at 15 psi pressure–121°C.

Use: For the cultivation of *Vibrio cholerae* and other *Vibrio* species.

Alkaline Peptone Water

Composition per liter:

NaCl 10.0g
Peptone 10.0g

pH 8.5 ± 0.2 at 25°C

Preparation of Medium: Add components to distilled/deionized water and bring volume to 1.0L. Mix thoroughly. Adjust pH to 8.5. Distribute into tubes or flasks. Autoclave for 10 min at 15 psi pressure–121°C.

Use: For the cultivation and transport of *Vibrio cholerae* and other *Vibrio* species.

Alkaline Peptone Water

Composition per liter:

Peptone 10.0g
NaCl 5.0g

pH 9.0 ± 0.2 at 25°C

Preparation of Medium: Add components to distilled/deionized water and bring volume to 1.0L. Mix thoroughly. Adjust pH to 9.0. Distribute into tubes or flasks. Autoclave for 20 min at 15 psi pressure–121°C.

Use: For the cultivation of a variety of alkalophilic microorganisms, especially *Vibrio* species.

Alkvisco Medium

Composition per liter:

Agar 15.0g
Beef extract 10.0g
Peptone 10.0g
NaCl 5.0g
Acrylonitrile 0.5g
KCN 10.0mg

pH 6.5-8.0 at 25°C

Caution: Cyanide is toxic. Acrylonitrile is a carcinogen; use appropriate precautions.

Preparation of Medium: Add components, except acrylonitrile, to distilled/deionized water and bring volume to 980.0mL. Mix thoroughly. Gently heat and bring to boiling. Autoclave for 10 min at 15 psi pressure–121°C. Add acrylonitrile to 20.0mL of distilled/deionized water and filter sterilize. Add aseptically to the sterile basal medium.

Use: For the cultivation and maintenance of *Corynebacterium* species.

Allantoin Agar

Composition per liter:

Agar 15.0g
$Na_2HPO_4 \cdot 12H_2O$ 9.0g
NaCl 5.0g
KH_2PO_4 .... 1.5g
Meat extract 1.0g
Yeast extract 1.0g
$MgSO_4 \cdot 7H_2O$ 0.2g
$MnCl_2 \cdot 4H_2O$ 20.0mg
$CaCl_2$ 1.2mg
Glucose-allantoin solution 100.0mL

Glucose-Allantoin Solution:
Composition per 100.0mL:

Glucose	5.0g
Allantoin	1.0g

Preparation of Glucose-Allantoin Solution: Add components to distilled/deionized water and bring volume to 100.0mL. Mix thoroughly. Filter sterilize. Warm to 50°C.

Preparation of Medium: Add components, except glucose-allantoin solution, to distilled/deionized water and bring volume to 900.0mL. Mix thoroughly. Gently heat and bring to boiling. Autoclave for 15 min at 15 psi pressure–121°C. Cool to 50°–55°C. Aseptically add 100.0mL of sterile glucose-allantoin solution. Mix thoroughly. Pour into sterile Petri dishes or distribute into sterile tubes.

Use: For the cultivation and maintenance of *Bacillus* species.

Allantoin Broth

Composition per liter:

$Na_2HPO_4 \cdot 12H_2O$	9.0g
NaCl	5.0g
KH_2PO_4	1.5g
Meat extract	1.0g
Yeast extract	1.0g
$MgSO_4 \cdot 7H_2O$	0.2g
$MnCl_2 \cdot 4H_2O$	20.0mg
$CaCl_2$	1.2mg
Glucose-allantoin solution	100.0mL

Glucose-Allantoin Solution:
Composition per 100.0mL:

Glucose	5.0g
Allantoin	1.0g

Preparation of Glucose-Allantoin Solution: Add components to distilled/deionized water and bring volume to 100.0mL. Mix thoroughly. Filter sterilize.

Preparation of Medium: Add components, except glucose-allantoin solution, to distilled/deionized water and bring volume to 900.0mL. Mix thoroughly. Autoclave for 15 min at 15 psi pressure–121°C. Aseptically add 100.0mL of sterile glucose-allantoin solution. Mix thoroughly. Aseptically distribute into sterile tubes or flasks.

Use: For the cultivation and maintenance of *Bacillus* species.

ALOA Medium (Agar Listeria Ottavani & Agosti)

Composition per liter:

Agar	18.0g
Peptone	18.0g
LiCl	10.0g
Yeast extract	10.0g
Tryptone	6.0g
NaCl	5.0g
Na_2HPO_4	2.5g
Na-pyruvate	2.0g
Glucose	2.0g
Mg-glycerophosphate	1.0g
$MgSO_4$	0.5g
5-Bromo4-chloro-indolyl-β-D-glucopyranoside	0.05g
Phosphatidylinositol solution	50.0mL
Nalidixic acid solution	5.0mL
Ceftazidime solution	5.0mL
Cycloheximide solution	5.0mL
Polymyxin B solution	5.0mL

pH 7.2 ± 0.2 at 25°C

Nalidixic Acid Solution:
Composition per 5.0mL:

Nalidixic acid	0.02g

Preparation of Nalidixic Acid Solution: Add nalidixic acid to distilled/deionized water and bring volume to 5.0mL. Mix thoroughly. Filter sterilize.

Ceftazidime Solution:
Composition per 5.0mL:

Ceftazidime	1.5g

Preparation of Ceftazidime Solution: Add ceftazidime to distilled/deionized water and bring volume to 5.0mL. Mix thoroughly. Filter sterilize.

Cycloheximide Solution:
Composition per 5.0mL:

Cycloheximide	0.05g
Ethanol	2.5mL

Preparation of Cycloheximide Solution: Add cycloheximide to 2.5mL of ethanol. Mix thoroughly. Bring volume to 5.0mL with distilled/deionized water. Filter sterilize.

Polymyxin B Solution:
Composition per 5.0mL:

Polymyxin B	76700U

Preparation of Lioncomycin Solution: Add polymyxin B to distilled/deionized water and bring volume to 5.0mL. Mix thoroughly. Filter sterilize.

Phosphatidylinositol Solution:
Composition per 5.0mL:

Polymyxin B	76700U

Preparation of Phosphatidylinositol Solution: Add phosphatidylinositol to cold distilled/deionized water and bring volume to 50.0mL. Stir for 30 min so a homogeneous suspension is obtained.

Autoclave for 15 min at 15 psi pressure–121°C. Cool to 50°C.

Preparation of Medium: Add components, except phosphatidylinositol solution, nalidixic acid solution, cetazidime solution, cycloheximide solution, and polymyxin B solution, to distilled/deionized water and bring volume to 930.0mL. Mix thoroughly. Adjust the pH to 7.2. Gently heat and bring to boiling. Autoclave for 15 min at 15 psi pressure–121°C. Cool to 45°–50°C. Aseptically add 50.0mL sterile phosphatidylinositol solution, 5.0mL sterile nalidixic acid solution, 5.0mL sterile cetazidime solution, 5.0mL sterile cycloheximide solution, and 5.0mL sterile polymyxin B solution. Mix thoroughly. Pour into the Petri dishes or distribute into sterile tubes.

Use: For the isolaltion and cultivation of *Literia* spp.

ALP Basal Medium (Aerobic Low Peptone Basal Medium)

Composition per liter:

Agar	15.0g
$(NH_4)_2SO_4$	1.0g
Pancreatic digest of casein	0.5g
Yeast extract	0.5g
$MgSO_4 \cdot 7H_2O$	0.2g
KCl	0.2g
Phenol Red	0.02g
Substrate solution	50.0mL

pH 7.8 ± 0.2 at 25°C

Substrate Solution:

Composition per 50.0mL:

Substrate	0.1g

Preparation of Substrate Solution: Add substrate to distilled/deionized water and bring volume to 50.0mL. Use sugars, carbohydrates, *n*-butanol, other alcohols, or any acidogenic carbon source. Mix thoroughly. Filter sterilize.

Preparation of Medium: Add components, except substrate solution, to distilled/deionized water and bring volume to 950.0mL. Mix thoroughly. Gently heat and bring to boiling. Adjust pH to 7.8. Distribute into screw-capped tubes in 3.0mL volumes. Autoclave for 15 min at 15 psi pressure–121°C. Cool to 45°–50°C. Aseptically add 0.15mL of sterile substrate solution to each tube. Mix thoroughly. Allow tubes to cool in a slanted position.

Use: For the cultivation and differentiation of microorganisms based on their ability to utilize a variety of carbon sources such as carbohydrates, alcohols, and other acidogenic substrates.

ALP Basal Medium (Aerobic Low Peptone Basal Medium)

Composition per liter:

Agar	15.0g
$(NH_4)_2SO_4$	1.0g
Pancreatic digest of casein	0.5g
Yeast extract	0.5g
Glucose	0.2g
$MgSO_4 \cdot 7H_2O$	0.2g
KCl	0.2g
Phenol Red	0.02g
Substrate solution	50.0mL

pH 6.5 ± 0.2 at 25°C

Substrate Solution:

Composition per 50.0mL:

Substrate	0.1g

Preparation of Substrate Solution: Add substrate to distilled/deionized water and bring volume to 50.0mL. Use gelatin, aliphatic acids, or any alkalogenic carbon source. Mix thoroughly. Filter sterilize.

Preparation of Medium: Add components, except substrate solution, to distilled/deionized water and bring volume to 950.0mL. Mix thoroughly. Gently heat and bring to boiling. Adjust pH to 6.5. Distribute into screw-capped tubes in 3.0mL volumes. Autoclave for 15 min at 15 psi pressure–121°C. Cool to 45°–50°C. Aseptically add 0.15mL of sterile substrate solution to each tube. Mix thoroughly. Allow tubes to cool in a slanted position.

Use: For the cultivation and differentiation of microorganisms based on their ability to utilize a variety of carbon sources such as gelatin, aliphatic acids, and other alkalophilic substrates.

Amies Modified Transport Medium with Charcoal

Composition per liter:

Charcoal	10.0g
Agar	4.0g
NaCl	3.0g
Na_2HPO_4	1.15g
Sodium thioglycolate	1.0g
KCl	0.2g
KH_2PO_4	0.2g
$CaCl_2 \cdot 2H_2O$	0.1g
$MgCl_2 \cdot 6H_2O$	0.1g

pH 7.2 ± 0.2 at 25°C

Source: This medium is available as a premixed powder from BD Diagnostic Systems.

Preparation of Medium: Add components to distilled/deionized water and bring volume to 1.0L.

Mix thoroughly. Gently heat and bring to boiling. Distribute into flasks or tubes. Autoclave for 20 min at 15 psi pressure–121°C. While cooling, turn tubes to uniformly suspend charcoal.

Use: For the transport of swab specimens to prolong the survival of microorganisms, especially *Neisseria gonorrhoeae,* between collection and culturing. Addition of charcoal to this medium neutralizes metabolic products that may be toxic to *Neisseria gonorrhoeae.*

Amies Modified Transport Medium with Charcoal

Composition per liter:

Charcoal	10.0g
NaCl	8.0g
Agar	3.6g
Na_2HPO_4	1.15g
Sodium thioglycolate	1.0g
KCl	0.2g
KH_2PO_4	0.2g
$CaCl_2 \cdot 2H_2O$	0.1g
$MgCl_2 \cdot 6H_2O$	0.1g

pH 7.2 ± 0.2 at 25°C

Source: This medium is available as a premixed powder from BD Diagnostic Systems and Oxoid Unipath.

Preparation of Medium: Add components to distilled/deionized water and bring volume to 1.0L. Mix thoroughly. Gently heat and bring to boiling. Distribute into flasks or tubes. Autoclave for 20 min at 15 psi pressure–121°C. While cooling, turn tubes to uniformly suspend charcoal.

Use: For the transport of swab specimens to prolong the survival of microorganisms, especially *Neisseria gonorrhoeae,* between collection and culturing. Addition of charcoal to this medium neutralizes metabolic products that may be toxic to *Neisseria gonorrhoeae.*

Amies Transport Medium without Charcoal

Composition per liter:

Agar	4.0g
NaCl	3.0g
Na_2HPO_4	1.15g
Sodium thioglycolate	1.0g
KCl	0.2g
KH_2PO_4	0.2g
$CaCl_2 \cdot 2H_2O$	0.1g
$MgCl_2 \cdot 6H_2O$	0.1g

pH 7.2 ± 0.2 at 25°C

Source: This medium is available as a premixed powder from BD Diagnostic Systems.

Preparation of Medium: Add components to distilled/deionized water and bring volume to 1.0L. Mix thoroughly. Gently heat and bring to boiling. Distribute into flasks or tubes. Autoclave for 20 min at 15 psi pressure–121°C.

Use: For the transport of swab specimens to prolong the survival of microorganisms, especially *Neisseria gonorrhoeae,* between collection and culturing.

Amies Transport Medium without Charcoal

Composition per liter:

NaCl	8.0g
Agar	3.6g
Na_2HPO_4	1.15g
Sodium thioglycolate	1.0g
KCl	0.2g
KH_2PO_4	0.2g
$CaCl_2 \cdot 2H_2O$	0.1g
$MgCl_2 \cdot 6H_2O$	0.1g

pH 7.2 ± 0.2 at 25°C

Source: This medium is available as a premixed powder from BD Diagnostic Systems.

Preparation of Medium: Add components to distilled/deionized water and bring volume to 1.0L. Mix thoroughly. Gently heat and bring to boiling. Distribute into flasks or tubes. Autoclave for 20 min at 15 psi pressure–121°C.

Use: For the transport of swab specimens to prolong the survival of microorganisms, especially *Neisseria gonorrhoeae,* between collection and culturing.

AMO.1 Medium

Composition per 1012.0mL:

Yeast extract	10.0g
NaCl	5.8g
N-Methylhydantoin	5.64g
$NaHCO_3$	4.5g
L-Serine	2.0g
L-Threonine	2.0g
Pancreatic digest of casein	0.5g
L-Cysteine	0.5g
$MgCl_2 \cdot 6H_2O$	0.4g
KCl	0.3g
NH_4Cl	0.27g
KH_2PO_4	0.2g
$CaCl_2 \cdot 2H_2O$	0.15g
Wolfe's vitamin solution	10.0mL

Na_2HSeO_3 solution 1.0mL
Trace elements solution SL-10 .... 1.0mL

pH 8.3 ± 0.2 at 25°C

Wolfe's Vitamin Solution:
Composition per liter:

Pyridoxine·HCl 10.0mg
p-Aminobenzoic acid 5.0mg
Lipoic acid 5.0mg
Nicotinic acid 5.0mg
Riboflavin 5.0mg
Thiamine·HCl 5.0mg
Calcium DL-pantothenate 5.0mg
Biotin 2.0mg
Folic acid 2.0mg
Vitamin B_{12} 0.1mg

Preparation of Wolfe's Vitamin Solution: Add components to distilled/deionized water and bring volume to 1.0L. Mix thoroughly.

Na_2SeO_3 Solution:
Composition per liter:

$Na_2SeO_3 \cdot 5H_2O$ 0.2mg

Preparation of Na_2SeO_3 Solution: Add $Na_2SeO_3 \cdot 5H_2O$ to distilled/deionized water and bring volume to 1.0L. Mix thoroughly. Sparge with 100% N_2. Autoclave for 15 min at 15 psi pressure–121°C.

Trace Elements Solution SL-10:
Composition per liter:

$FeCl_2 \cdot 4H_2O$ 1.5g
$CoCl_2 \cdot 6H_2O$ 190.0mg
$MnCl_2 \cdot 4H_2O$ 100.0mg
$ZnCl_2$ 70.0mg
$Na_2MoO_4 \cdot 2H_2O$ 36.0mg
$NiCl_2 \cdot 6H_2O$ 24.0mg
H_3BO_3 .... 6.0mg
$CuCl_2 \cdot 2H_2O$ 2.0mg
HCl (25% solution) 10.0mL

Preparation of Trace Elements Solution SL-10: Add $FeCl_2 \cdot 4H_2O$ to 10.0mL of HCl solution. Mix thoroughly. Add distilled/deionized water and bring volume to 1.0L. Add remaining components. Mix thoroughly. Sparge with 100% N_2. Autoclave for 15 min at 15 psi pressure–121°C.

Preparation of Medium: Prepare anaerobically. Add components, except $NaHCO_3$ and L-cysteine, to distilled/deionized water and bring volume to 1.0L. Sparge with 100% N_2 for 30 min. Adjust pH to 8.3 with 10*N* NaOH. Add $NaHCO_3$ and L-cysteine (solid substances). Mix thoroughly. Sparge with 80% N_2 + 20% CO_2 mixture. Flush the headspace of the medium vessel with the 80% N_2 + 20% CO_2 mixture. Autoclave for 15 min at 15 psi pressure–121°C. Sparge with 80% N_2 + 20% CO_2 mixture for 30 min. Aseptically and anaerobically distribute into sterile tubes or bottles.

Use: For the cultivation of *Tissierella creatinini.*

Ampicillin Kanamycin Nutrient Agar

Composition per liter:

Agar 15.0g
Peptone 5.0g
NaCl 5.0g
Yeast extract 2.0g
Beef extract 1.0g
Ampicillin solution 10.0mL
Kanamycin solution 10.0mL

Ampicillin Solution:
Composition per 10.0mL:

Ampicillin 50.0mg

Preparation of Ampicillin Solution: Add ampicillin to distilled/deionized water and bring volume to 10.0mL. Mix thoroughly. Filter sterilize.

Kanamycin Solution:
Composition per 10.0mL:

Kanamycin 25.0mg

Preparation of Kanamycin Solution: Add kanamycin to distilled/deionized water and bring volume to 10.0mL. Mix thoroughly. Filter sterilize.

Preparation of Medium: Add components, except ampicllin solution and kanmycin solution, to distilled/deionized water and bring volume to 980.0mL. Mix thoroughly. Autoclave for 15 min at 15 psi pressure–121°C. Aseptically add 10.0mL of sterile ampicillin solution and 10.0mL of sterile kanamycin solution. Mix thoroughly. Aseptically distribute into sterile tubes or flasks.

Use: For the cultivation of fungi and various antibiotic-resitant bacteria, including *Escherichia coli.*

Ampicillin L Broth Medium

Composition per liter:

Pancreatic digest of casein 10.0g
NaCl 5.0g
Yeast extract 5.0g
Glucose 1.0g
Ampicillin solution 10.0mL

pH 7.0 ± 0.2 at 25°C

Ampicillin Solution:
Composition per 10.0mL:

Ampicillin 50.0mg

Preparation of Ampicillin Solution: Add ampicillin to distilled/deionized water and bring volume to 10.0mL. Mix thoroughly. Filter sterilize.

Preparation of Medium: Add components, except ampicillin solution, to distilled/deionized water and bring volume to 990.0mL. Mix thoroughly. Bring pH to 7.0. Autoclave for 15 min at 15 psi pressure–121°C. Aseptically add 10.0mL of sterile ampicillin solution. Mix thoroughly. Aseptically distribute into sterile tubes or flasks.

Use: For the cultivation of *Escherichia coli*.

Ampicillin TY Salt Medium

Composition per liter:

NaCl	10.0g
Pancreatic digest of casein	10.0g
Yeast extract	5.0g
Ampicillin solution	10.0mL

pH 7.0 ± 0.2 at 25°C

Ampicillin Solution:

Composition per 10.0mL:

Ampicillin	50.0mg

Preparation of Ampicillin Solution: Add ampicillin to distilled/deionized water and bring volume to 10.0mL. Mix thoroughly. Filter sterilize.

Preparation of Medium: Add components, except ampicillin solution, to distilled/deionized water and bring volume to 990.0mL. Mix thoroughly. Bring pH to 7.0. Autoclave for 15 min at 15 psi pressure–121°C. Aseptically add 10.0mL of sterile ampicillin solution. Mix thoroughly. Aseptically distribute into sterile tubes or flasks.

Use: For the cultivation of various antibiotic resistant bacteria, including *Escherichia coli*.

Amygdalin Medium

Composition per liter:

Peptone	10.0g
Beef extract	5.0g
NaCl	5.0g
Agar	3.0g
Amygdalin solution	200.0mL
Bromthymol Blue (0.05% solution)	5.0mL

pH 7.0 ± 0.2 at 25°C

Amygdalin Solution:

Composition per 200.0mL:

Amygdalin	10.0g

Preparation of Amygdalin Solution: Add amygdalin to distilled/deionized water and bring volume to 200.0mL. Mix thoroughly. Filter sterilize.

Preparation of Medium: Add components, except amygdalin solution, to distilled/deionized water and bring volume to 800.0mL. Mix thoroughly. Gently heat and bring to boiling. Adjust pH to 7.0. Autoclave for 20 min at 15 psi pressure–121°C. Cool to 45°–50°C. Aseptically add sterile amygdalin solution. Mix thoroughly. Aseptically distribute into sterile tubes with cotton plugs. Allow tubes to cool in a slanted position, forming a short slant.

Use: For the cultivation and differentiation of *Serratia* species based on their ability to produce acid and HCN from amygdalin.

Anaerobe Agar

Composition per 1001.5mL:

Agar	15.0g
Pancreatic digest of casein	15.0g
Pancreatic digest of soybean meal	5.0g
NaCl	5.0g
L-Cysteine·HCl·H_2O solution	5.0mL
Vitamin K_1 solution	1.0mL
Hemin solution	0.5mL

pH 7.0 ± 0.2 at 25°C

L-Cysteine·HCl·H_2O Solution:

Composition per 5.0mL:

L-Cysteine·HCl·H_2O	0.4g
NaOH (1*N* solution)	5.0mL

Preparation of L-Cysteine·HCl·H_2O Solution: Add L-cysteine·HCl·H_2O to 5.0mL 1*N* NaOH. Mix thoroughly. Filter sterilize.

Hemin Solution:

Composition per 100.0mL:

Hemin	1.0g

Preparation of Hemin Solution: Add hemin to distilled/deionized water and bring volume to 100.0mL. Mix thoroughly. Autoclave for 15 min at 15 psi pressure–121°C. Cool. Refrigerate at 4°C for storage.

Vitamin K_1 Solution:

Composition per 100.0mL:

Vitamin K_1	1.0g
Ethanol, absolute	20.0mL

Preparation of Vitamin K_1 Solution: Add vitamin K_1 to 100.0mL of 95% ethanol. Mix thoroughly. Solution may require 2–3 days with intermittent shaking to completely dissolve. Filter sterilize. Refrigerate at 4°C for storage.

Preparation of Medium: Add components, exhemin solution and vitamin K_1 solution, to distilled/deionized water and bring volume to 1.0L. Mix thoroughly. Gently heat and bring to boiling. Adjust pH to 7.0 ± 0.2 at 25°C. Autoclave for 15 min at 15 psi pressure–121°C. Cool to 45°–50°C. Aseptically add 0.5mL of sterile hemin solution and 1.0mL of sterile vitamin K_1 solution. Mix thoroughly. Pour into sterile Petri dishes or distribute into sterile tubes. Reduce medium for 24h by incubation in an anaerobic glove box of GasPak jar prior to use.

Use: For the cultivation of anaerobic bacteria such as *Brucella* and *Clostridium* spp.

Anaerobe Agar (LMG Medium 41)

Composition per liter:

Agar Base	800.0mL
Solution A	100.0mL
Solution B	100.0mL

pH 6.9 ± 0.2 at 25°C

Agar Base:

Composition per 800.0mL:

Agar	30.5g
Tryptone	5.0g
Yeast extract	5.0g
$(NH_4)_2SO_4$	0.5g
Sodium thioglycolate	0.5g
$MgSO_4 \cdot 7H_2O$	0.1g
$Fe(NH_4)_2(SO_4)_2 \cdot 6H_2O$	55.0mg
$Na_2MoO_4 \cdot 2H_2O$	2.4mg
$Na_2SeO_3 \cdot 5H_2O$	0.23mg

Preparation of Agar Base: Add components to distilled/deionized water and bring volume to 800.0mL. Mix thoroughly. Gently heat and bring to boiling. Autoclave for 15 min at 15 psi pressure–121°C. Cool to 45°–50°C.

Solution A:

Composition per 100.0mL:

Glucose	18.0g
K_2HPO_4	7.0g
KH_2PO_4	5.5g

Preparation of Solution A: Add components to 100.0mL of distilled/deionized water. Mix thoroughly. Filter sterilize.

Solution B:

Composition per 100.0mL:

$NaHCO_3$	10.0g

Preparation of Solution B: Add $NaHCO_3$ to 100.0mL of distilled/deionized water. Mix thoroughly. Filter sterilize.

Preparation of Medium: Aseptically add solutions A and B to the agar base. Adjust pH to 6.9. Mix thoroughly. Pour into sterile Petri dishes or distribute into sterile tubes. Incubate anaerobically under 100% CO_2 gas atmosphere.

Use: For the cultivation of anerobic bacteria.

Anaerobe Medium (LMG Medium 41)

Composition per liter:

Agar Base	800.0mL
Solution A	100.0mL
Solution B	100.0mL

pH 6.9 ± 0.2 at 25°C

Agar Base:

Composition per 800.0mL:

Tryptone	5.0g
Yeast extract	5.0g
$(NH_4)_2SO_4$	0.5g
Sodium thioglycolate	0.5g
Agar	0.5g
$MgSO_4 \cdot 7H_2O$	0.1g
$Fe(NH_4)_2(SO_4)_2 \cdot 6H_2O$	55.0mg
$Na_2MoO_4 \cdot 2H_2O$	2.4mg
$Na_2SeO_3 \cdot 5H_2O$	0.23mg

Preparation of Agar Base: Add components to distilled/deionized water and bring volume to 800.0mL. Mix thoroughly. Gently heat and bring to boiling. Autoclave for 15 min at 15 psi pressure–121°C. Cool to 45°–50°C.

Solution A:

Composition 100.0mL:

Glucose	18.0g
K_2HPO_4	7.0g
KH_2PO_4	5.5g

Preparation of Solution A: Add components to 100.0mL of distilled/deionized water. Mix thoroughly. Filter sterilize.

Solution B:

Composition 100.0mL:

$NaHCO_3$	10.0g

Preparation of Solution B: Add components to 100.0mL of distilled/deionized water. Mix thoroughly. Filter sterilize.

Preparation of Medium: Aseptically add solutions A and B to the sterile Agar Base. Adjust pH to 6.9. Mix thoroughly. Distribute into sterile tubes. Incubate anaerobically under 100% CO_2 gas atmosphere.

Use: For the cultivation of anerobic bacteria.

Anaerobic Agar

Composition per liter:

Agar	20.0g
Pancreatic digest of casein	20.0g
Glucose	10.0g
NaCl	5.0g
Sodium thioglycolate	2.0g

Sodium formaldehyde sulfoxylate 1.0g
Methylene Blue 2.0mg
pH 7.2 ± 0.2 at 25°C

Source: This medium is available as a premixed powder from BD Diagnostic Systems.

Preparation of Medium: Add components to distilled/deionized water and bring volume to 1.0L. Mix thoroughly. Gently heat and bring to boiling. Adjust pH to 7.2. Distribute into tubes until medium is 3 inches deep. Autoclave for 15 min at 15 psi pressure–121°C.

Use: For the cultivation of a variety of anaerobic microorganisms, especially *Clostridium* species.

Anaerobic Agar

Composition per liter:

Pancreatic digest of casein 17.5g
Agar 15.0g
Glucose 10.0g
Papaic digest of soybean meal 2.5g
NaCl 2.5g
Sodium thioglycolate 2.0g
Sodium formaldehyde sulfoxylate 1.0g
L-Cystine 0.4g
Methylene Blue 2.0mg
pH 7.2 ± 0.2 at 25°C

Source: This medium is available as a premixed powder from BD Diagnostic Systems.

Preparation of Medium: Add components to distilled/deionized water and bring volume to 1.0L. Mix thoroughly. Gently heat and bring to boiling. Autoclave for 15 min at 15 psi pressure–121°C. Use with Brewer anaerobic Petri dishes or in tubes or ordinary plates and incubate in anaerobic jars.

Use: For the cultivation of *Clostridium* species and for anaerobic microorganisms.

Anaerobic Broth

Composition per liter:

Pancreatic digest of casein 17.5g
Glucose 10.0g
NaCl 2.5g
Papaic digest of soybean meal 2.5g
Sodium thioglycolate 2.0g
Sodium formaldehyde sulfoxylate 1.0g
L-Cystine 0.4g
Methylene Blue 2.0mg
pH 7.2 ± 0.2 at 25°C

Preparation of Medium: Add components to distilled/deionized water and bring volume to 1.0L. Mix thoroughly. Gently heat and bring to boiling. Distribute into tubes or flasks. Autoclave for 15 min at 15 psi pressure–121°C.

Use: For the cultivation of a variety of anaerobic and microaerophilic microorganisms.

Anaerobic CNA Agar (Anaerobic Colistin Nalidixic Acid Agar)

Composition per liter:

Agar 13.0g
Pancreatic digest of casein 12.0g
Peptic digest of animal tissue 5.0g
NaCl 5.0g
Yeast extract 3.0g
Beef extract 3.0g
Cornstarch 1.0g
Glucose 1.0g
L-Cysteine·HCl·H_2O 0.5g
Vitamin K_1 .... 10.0mg
Hemin 10.0mg
Colistin 10.0mg
Nalidixic acid 10.0mg
Sheep blood, defibrinated 50.0mL

Source: This medium is available as a premixed powder from BD Diagnostic Systems.

Preparation of Medium: Add components, except sheep blood, to distilled/deionized water and bring volume to 950.0mL. Mix thoroughly. Gently heat and bring to boiling. Autoclave for 15 min at 15 psi pressure–121°C. Cool to 45°–50°C. Aseptically add 50.0mL of sterile, defibrinated sheep blood. Mix thoroughly. Pour into sterile Petri dishes.

Use: For the selective isolation of anaerobic streptococci.

Anaerobic Egg Yolk Agar

Composition per 1080.0mL:

Agar 20.0g
Proteose peptone 20.0g
NaCl 5.0g
Pancreatic digest of casein 5.0g
Yeast extract 5.0g
Egg yolk emulsion, 50% 80.0mL
pH 7.0 ± 0.2 at 25°C

Egg Yolk Emulsion, 50%:

Composition per 100.0mL:

Chicken egg yolks 11
Whole chicken egg 1
NaCl (0.9% solution) 50.0mL

Preparation of Egg Yolk Emulsion, 50%: Soak whole eggs with 1:100 dilution of saturated mercuric chloride solution for 1 min. Crack eggs and separate yolks from whites. Mix egg yolks with 1 chicken egg. Beat to form emultion. Measure 50.0mL of egg yolk emulsion and add to 50.0mL of

0.9% NaCl solution. Mix thoroughly. Filter sterilize. Warm to 45°–50°C.

Preparation of Medium: Add components, except egg yolk emulsion, 50%, to distilled/deionized water and bring volume to 1.0L. Mix thoroughly. Gently heat and bring to boiling. Autoclave for 15 min at 15 psi pressure–121°C. Cool to 45°–50°C. Aseptically add 80.0mL of sterile egg yolk emulsion, 50%. Mix thoroughly. Pour into sterile Petri dishes or distribute into sterile tubes. Allow plates to dry at 35°C for 24h.

Use: For the cultivation of *Clostridium* species.

Anaerobic Egg Yolk Agar

Composition per liter:

Agar	20.0g
Proteose peptone	20.0g
Pancreatic digest of casein	5.0g
NaCl	5.0g
Yeast extract	5.0g
Egg yolk emulsion, 50%	80.0mL

pH 7.0 ± 0.2 at 25°C

Egg Yolk Emulsion, 50%:
Composition per 80.0mL:

Chicken egg yolks	2 or more
NaCl (0.85% solution)	40.0mL

Preparation of Egg Yolk Emulsion, 50%: Wash fresh eggs wtih stiff brush and drain. Soak eggs in 70% ethanol for 1 hour. Crack eggs aseptically and separate yolks from whites. Drain contents of yolk sacs into sterile stoppered graduate cylinder and discard sacs. Measure 40.0mL of egg yolk emulsion and add 40.0mL of 0.85% NaCl solution. Mix thoroughly by inverting graduate cyliner. Warm to 45°–50°C.

Preparation of Medium: Add components, except egg yolk emulsion, to distilled/deionized water and bring volume to 1.0L. Mix thoroughly. Gently heat and bring to boiling. Autoclave for 15 min at 15 psi pressure–121°C. Cool to 45°–50°C. Aseptically add 80.0mL sterile egg yolk emulsion. Mix thoroughly. Pour into sterile Petri dishes. Allow plates to dry at ambient temperature for 2–3 days or at 35°C for 24h. Check plates for contamination before use.

Use: For the cultivation of *Yersinia enterocolitica.*

Anaerobic Egg Yolk Agar

Composition per liter:

Agar	20.0g
Proteose peptone	20.0g
Pancreatic digest of casein	5.0g
NaCl	5.0g
Yeast extract	5.0g
Egg yolk emulsion, 50%	20.0mL

pH 7.0 ± 0.2 at 25°C

Egg Yolk Emulsion, 50%:
Composition per 100.0mL:

Chicken egg yolks	2
NaCl (0.9% solution)	10.0mL

Preparation of Egg Yolk Emulsion, 50%: Soak eggs with 1:100 dilution of saturated mercuric chloride solution for 1 min. Crack eggs and separate yolks from whites. Beat to form emultion. Measure 10.0mL of egg yolk emulsion and add to 10.0mL of 0.9% NaCl solution. Mix thoroughly. Filter sterilize. Warm to 45°–50°C.

Preparation of Medium: Add components, except egg yolk emulsion, to distilled/deionized water and bring volume to 980.0mL. Mix thoroughly. Gently heat and bring to boiling. Autoclave for 15 min at 15 psi pressure–121°C. Cool to 45°–50°C. Aseptically add sterile egg yolk emulsion. Mix thoroughly. Pour into sterile Petri dishes. Allow plates to dry at 35°C for 24h.

Use: For the cultivation of *Yersinia enterocolitica.*

Anaerobic LKV Blood Agar

Composition per liter:

Agar	15.0g
Pancreatic digest of casein	13.0g
Peptic digest of animal tissue	10.0g
NaCl	5.0g
Yeast extract	2.0g
Glucose	1.0g
$NaHSO_3$	0.1g
Sheep blood, laked	50.0mL
Antibiotic solution	10.0mL
Hemin solution	1.0mL
Vitamin K_1 solution	1.0mL

pH 7.1–7.8 at 25°C

Source: This medium is available as a premixed powder from BD Diagnostic Systems.

Antibiotic Solution:
Composition per 10.0mL:

Kanamycin	0.075g
Vancomycin	7.5mg

Preparation of Antibiotic Solution: Add components to distilled/deionized water and bring volume to 10.0mL. Mix thoroughly. Filter sterilize.

Vitamin K_1 Solution:
Composition per 100.0mL:

Vitamin K_1	0.1g
Ethanol	99.0mL

Preparation of Vitamin K_1 Solution: Add vitamin K_1 to 99.0mL of absolute ethanol. Mix thoroughly.

Hemin Solution:

Composition per 100.0mL:

Hemin	0.01g
NaOH (1*N* solution)	20.0mL

Preparation of Hemin Solution: Add hemin to 20.0mL of 1*N* NaOH solution. Mix thoroughly. Bring volume to 100.0mL with distilled/deionized water.

Preparation of Medium: Add components—except sheep blood, antibiotic solution, and vitamin K_1 solution—to distilled/deionized water and bring volume to 939.0mL. Mix thoroughly. Gently heat and bring to boiling. Autoclave for 15 min at 15 psi pressure–121°C. Cool to 45°–50°C. Aseptically add 50.0mL of sterile sheep blood, 10.0mL of sterile antibiotic solution, and 1.0mL of sterile vitamin K_1 solution. Mix thoroughly. Pour into sterile Petri dishes or distribute into sterile tubes.

Use: For the isolation and cultivation of anaerobic Gram-negative microorganisms, expecially *Bacteroides* species.

Anaerobic Oxalate Medium

Composition per 1011.0mL:

Solution A	870.0mL
Solution C	100.0mL
Solution D	20.0mL
Solution E (Vitamin solution)	10.0mL
Solution F	10.0mL
Solution B (Trace elements solution SL-10)	1.0mL

pH 7.1–7.4 at 25°C

Solution A:

Composition per 870.0mL:

Na_2SO_4	3.0g
NaCl	1.0g
KCl	0.5g
$MgCl_2·6H_2O$	0.4g
NH_4Cl	0.3g
KH_2PO_4	0.2g
$CaCl_2·2H_2O$	0.15g
Resazurin	1.0mg

Preparation of Solution A: Add components to distilled/deionized water and bring volume to 870.0mL. Mix thoroughly. Gently heat and bring to boiling. Continue boiling for 3-4 min. Allow to cool to room temperature while gassing under 80% N_2 + 20% CO_2. Continue gassing until pH reaches below 6.0. Seal the flask under 80% N_2 + 20% CO_2. Autoclave for 15 min at 15 psi pressure–121°C.

Solution B (Trace Elements Solution SL-10):

Composition per liter:

$FeCl_2·4H_2O$	1.5g
$CoCl_2·6H_2O$	190.0mg
$MnCl_2·4H_2O$	100.0mg
$ZnCl_2$	70.0mg
$Na_2MoO_4·2H_2O$	36.0mg
$NiCl_2·6H_2O$	24.0mg
H_3BO_3	6.0mg
$CuCl_2·2H_2O$	2.0mg
HCl (25% solution)	10.0mL

Preparation of Solution B: Add $FeCl_2·4H_2O$ to 10.0mL of HCl solution. Mix thoroughly. Add distilled/deionized water and bring volume to 1.0L. Add remaining components. Mix thoroughly. Gas under 100% N_2. Autoclave for 15 min at 15 psi pressure–121°C.

Solution C:

Composition per 100.0mL:

$NaHCO_3$	5.0g

Preparation of Solution C: Add $NaHCO_3$ to distilled/deionized water and bring volume to 100.0mL. Mix thoroughly. Filter sterilize. Gas under 80% N_2 + 20% CO_2.

Solution D:

Composition per 20.0mL:

Ammonium oxalate	3.0g
Yeast extract	1.0g
Sodium acetate	0.41g

Preparation of Solution D: Add components to distilled/deionized water and bring volume to 20.0mL. Mix thoroughly. Gas under 100% N_2. Autoclave for 15 min at 15 psi pressure–121°C.

Solution E (Vitamin Solution):

Composition per liter:

Pyridoxine·HCl	10.0mg
Calcium DL-pantothenate	5.0mg
Lipoic acid	5.0mg
Nicotinic acid	5.0mg
p-Aminobenzoic acid	5.0mg
Riboflavin	5.0mg
Thiamine·HCl	5.0mg
Biotin	2.0mg
Folic acid	2.0mg
Vitamin B_{12}	0.1mg

Preparation of Solution E (Vitamin Solution): Add components to distilled/deionized water and bring volume to 1.0L. Mix thoroughly. Gas under 100% N_2. Autoclave for 15 min at 15 psi pressure–121°C.

Solution F:

Composition per 10.0mL:

$Na_2S \cdot 9H_2O$	0.4g

Preparation of Solution F: Add $Na_2S \cdot 9H_2O$ to distilled/deionized water and bring volume to 10.0mL. Mix thoroughly. Gas under 100% N_2. Autoclave for 15 min at 15 psi pressure–121°C.

Preparation of Medium: Aseptically and anaerobically combine solution A with solution B, solution C, solution D, solution E, and solution F, in that order. Mix thoroughly. Anaerobically distribute into sterile tubes or flasks under 80% N_2 + 20% CO_2.

Use: For the cultivation of *Clostridium oxalicum* and *Oxalobacter vibrioformis.*

Anaerobic Trypticase™ Soy Agar with Calf Blood (ATCC Medium 1664)

Composition per liter:

Pancreatic digest of casein	15.0g
Agar	15.0g
Papaic digest of soybean meal	5.0g
NaCl	5.0g
Calf blood, defibrinated	100.0mL

pH 7.3 ± 0.2 at 25°C

Preparation of Medium: Add components, except calf blood, to distilled/deionized water and bring volume to 900.0mL. Mix thoroughly. Prepare medium anaerobically with 80% N_2 + 10% CO_2 + 10% H_2. Gently heat while stirring and bring to boiling for 1 min. Autoclave for 15 min at 15 psi pressure–121°C. Do not overheat. Cool to 45°–50°C. Aseptically add 100.0mL sterile, defibrinated calf blood. Pour into sterile Petri dishes.

Use: For the isolation and cultivation of fastidious as well as nonfastidious microorganisms. For the differentiation of *Haemophilus* species.

Anaerobic TVLS Medium

Composition per liter:

Pancreatic digest of casein	17.0g
Beef extract	7.5g
Glucose	6.0g
Enzymatic hydrolysate of soybean meal	3.0g
Liver hydrolysate	3.0g
NaCl	2.5g
Na_2SO_3	0.7g
Sodium thioglycolate	0.5g
L-Cysteine·HCl·H_2O	0.25g
Agar	0.1g
Bovine serum	100.0mL

pH 7.3 ± 0.2 at 25°C

Preparation of Medium: Add components, except bovine serum, to distilled/deionized water and bring volume to 900.0mL. Mix thoroughly. Gently heat and bring to boiling. Autoclave for 15 min at 15 psi pressure–121°C. Cool to 45°–50°C. Aseptically add 100.0mL of bovine serum. Distribute into sterile tubes.

Use: For the isolation and cultivation of anaerobic microorganisms.

Andersen's Pork Pea Agar

Composition per 1685.0mL:

Agar	16.0g
Peptone	5.0g
Pancreatic digest of casein	1.6g
K_2HPO_4	1.25g
Soluble starch	1.0g
Sodium thioglycolate	0.5g
Pork infusion	800.0mL
Thioglycolate agar	660.0mL
Pea infusion	200.0mL
$NaHCO_3$ solution	25.0mL

pH 7.2 ± 0.2 at 25°C

Pork Infusion:

Composition per liter:

Pork, fresh lean ground	454.0g

Preparation of Pork Infusion: Add ground pork to distilled/deionized water and bring volume to 1.0L. Autoclave for 60 min at 0 psi pressure–100°C. Filter through two layers of cheesecloth. Cool to 4°C. Skim fat from surface. Warm to 25°C. Centrifuge at 5000 rpm for 10 min. Discard pellet.

Pea Infusion:

Composition per 450.0mL:

Green peas, fresh or frozen	454.0g
Diatomaceous earth (celite)	10.0g

Preparation of Pea Infusion: Add green peas to 450.0mL of distilled/deionized water. Blend until smooth. Autoclave for 60 min at 0 psi pressure–100°C. Centrifuge at 5000 rpm for 10 min. Discard pellet. Clarify supernatant solution with diatomaceous earth (celite). Filter through Whatman #4 filter paper. Use filtrate solution.

Thioglycolate Agar:

Composition per liter:

Agar	20.75g
Pancreatic digest of casein	15.0g
Glucose	5.5g
Yeast extract	5.0g
NaCl	2.5g
L-Cystine	0.5g

Sodium thioglycolate	0.5g
Resazurin	1.0mg

pH 7.1 ± 0.2 at 25°C

Preparation of Thioglycolate Agar: Add components to distilled/deionized water and bring volume to 1.0L. Mix thoroughly. Gently heat and bring to boiling. Autoclave for 15 min at 15 psi pressure–121°C. Cool to 45°–50°C.

$NaHCO_3$ Solution:

Composition per 100.0mL:

$NaHCO_3$	5.0g

Preparation of $NaHCO_3$ Solution: Add $NaHCO_3$ to distilled/deionized water and bring volume to 100.0mL. Mix thoroughly. Filter sterilize.

Preparation of Medium: Combine components, except $NaHCO_3$ solution and thioglycolate agar. Mix thoroughly. Adjust pH to 7.2. Autoclave for 5 min at 15 psi pressure–121°C. While medium is still hot, add 25.0g of celite. Filter through Whatman #4 filter paper with suction. Autoclave for 12 min at 15 psi pressure–121°C. Cool to 45°–50°C. Aseptically add 25.0mL of sterile $NaHCO_3$ solution. Mix thoroughly. Pour into sterile Petri dishes in 15.0mL volumes. Allow agar to solidify. Cover agar with 10.0mL of sterile, cooled thioglycolate agar.

Use: For the cultivation of mesophilic *Clostridium* species.

Andrade's Carbohydrate Broth and Indicator

Composition per liter:

Pancreatic digest of gelatin	10.0g
NaCl	10.0g
Beef extract	3.0g
Carbohydrate solution	100.0mL
Andrade's indicator	10.0mL

pH 7.2 ± 0.2 at 25°C

Source: This medium is available as a prepared medium from BD Diagnostic Systems, in tubes containing adonitol, arabinose, cellobiose, glucose, dulcitol, fructose, galactose, inositol, lactose, maltose, mannitol, raffinose, rhamnose, salicin, sorbitol, sucrose, trehalose, or xylose.

Andrade's Indicator

Composition per 26.0mL:

NaOH (1*N* solution)	16.0mL
Acid Fuchsin	0.21 g

Preparation of Andrade's Indicator: Add Acid Fuchsin to NaOH solution and bring volume to 26.0mL with distilled/deionized water.

Carbohydrate Solution:

Composition per 100.0mL:

Carbohydrate	5.0–10.0g

Preparation of Carbohydrate Solution: Add carbohydrate to distilled/deionized water and bring volume to 100.0mL. For glucose, lactose, sucrose, and mannitol, add 10.0g to distilled/deionized water and bring volume to 100.0mL. For dulcitol, salicin, and other carbohydrates, add 5.0g to distilled/deionized water and bring volume to 100.0mL. Mix thoroughly. Filter sterilize.

Preparation of Medium: Add components, except carbohydrate solution, to distilled/deionized water and bring volume to 1.0L. Mix thoroughly. Gently heat and bring to boiling. Cool. Aseptically add 100mL of sterile carbohydrate solution to 900mL of sterile medium. Mix thoroughly. Aseptically distribute into tubes or flasks. Alternately, prior to autoclaving, distribute 9.0mL volumes into test tubes containing inverted Durham tubes. Autoclave for 15 min at 15 psi pressure–121°C. Cool to 25°C. Add 1.0mL of sterile carbohydrate solution to each tube.

Caution: Acid Fuchsin is a potential carcinogen and care must be taken to avoid inhalation of the powdered dye and contact with the skin.

Use: For the determination of carbohydrate fermentation reactions of microorganisms, particularly members of the Enterobacteriaceae. A Durham tube is used to collect gas produced during the fermentation reaction. Acid production is indicated by a pink color.

Andrade's Broth

Composition per liter:

Pancreatic digest of gelatin	10.0g
NaCl	5.0g
Beef extract	3.0g
Andrade's indicator	10.0mL
Carbohydrate solution	50.0mL

pH 7.4 ± 0.2 at 25°C

Source: This medium is available as a prepared medium from BD Diagnostic Systems, in tubes containing adonitol, arabinose, cellobiose, dulcitol, fructose, galactose, glucose, inositol, lactose, maltose, mannitol, raffinose, rhamnose, salicin, sorbitol, sucrose, trehalose, or xylose.

Andrade's Indicator

Composition per 100.0mL:

NaOH (1*N* solution)	16.0mL
Acid Fuchsin	0.1 g

Preparation of Andrade's Indicator: Add Acid Fuchsin to NaOH solution and bring volume to 100.0mL with distilled/deionized water.

Carbohydrate Solution:

Composition per 100.0mL:

Carbohydrate 10.0g

Preparation of Carbohydrate Solution: Add carbohydrate to distilled/deionized water and bring volume to 100.0mL. Adonitol, arabinose, cellobiose, dulcitol, fructose, galactose, glucose, inositol, lactose, maltose, mannitol, raffinose, rhamnose, salicin, sorbitol, sucrose, trehalose, xylose, or other carbohydrates may be used. Mix thoroughly. Filter sterilize.

Preparation of Medium: Add components, except carbohydrate solution, to distilled/deionized water and bring volume to 1.0L. Mix thoroughly. Gently heat and bring to boiling. Distribute in 10.0mL volumes into test tubes containing inverted Durham tubes. Autoclave for 15 min at 15 psi pressure–121°C. Cool to 25°C. Add 0.5mL of sterile carbohydrate solution to each tube.

Caution: Acid Fuchsin is a potential carcinogen and care must be taken to avoid inhalation of the powdered dye and contact with the skin.

Use: For the determination of carbohydrate fermentation reactions of microorganisms, particularly members of the Enterobacteriaceae. A Durham tube is used to collect gas produced during the fermentation reaction. Acid production is indicated by a pink color.

Anthracis Chromogenic Agar

Composition per liter:

Proprietary.

Source: This medium is available as a premixed powder from BIOSYNTH International, Inc.

Preparation of Medium: Per manufacturers directions.

Use: For the rapid identification and isolation of *Bacillus anthracis* based on the detection of phosphatidylcholine-specific phospholipase C activity by 5-bromo-4-chloro-3-indoxyl-cholinphosphate hydrolysis. The medium incorporates chromogenic substrates for detecting specific enzyme activities in *Bacillus anthracis, B. cereus,* and *B. thuringiensis*. The enzymes targeted by the chromogenic medium are not present in other *Bacillus* species, allowing for specific isolation of these three *Bacillus* species. Inclusion of inhibitory compounds into the medium prevents the growth of environmental contaminants. The use of proprietary chromogenic substrates, X-IP and X-CP, allows for the differentiation of *Bacillus anthracis* from near-neighbors *B. cereus* and *B. thuringiensis*. Cream to pale teal-blue colored of *Bacillus anthracis* after 20-24h, teal-blue colonies of *Bacillus anthracis* after 36-48h at 35-37°C. Dark teal-blue colonies of *Bacillus cereus/Bacillus thuringiensis* after 20-24h at 35-37°C.

Antibiotic Medium 1 (Penassay Seed Agar) (Seed Agar)/(Agar Medium A)

Composition per liter:

Agar 15.0g
Pancreatic digest of gelatin 6.0g
Pancreatic digest of casein 4.0g
Yeast extract 3.0g
Beef extract 1.5g
Glucose 1.0g

pH 6.6 ± 0.1 at 25°C

Source: This medium is available as a premixed powder from BD Diagnostic Systems.

Preparation of Medium: Add components to distilled/deionized water and bring volume to 1.0L. Mix thoroughly. Gently heat and bring to boiling. Distribute into tubes or flasks. Autoclave for 15 min at 15 psi pressure–121°C. Pour into sterile Petri dishes or leave in tubes.

Use: For antibiotic assay testing, detection of antibiotics in milk, and determination of the antimicrobial effectiveness of antibiotics.

Antibiotic Medium 2 (Base Agar) (Penassay Base Agar)

Composition per liter:

Agar 15.0g
Pancreatic digest of gelatin 6.0g
Yeast extract 3.0g
Beef extract 1.5g

pH 6.6 ± 0.1 at 25°C

Source: This medium is available as a premixed powder from BD Diagnostic Systems and Oxoid Unipath.

Preparation of Medium: Add components to distilled/deionized water and bring volume to 1.0L. Mix thoroughly. Gently heat and bring to boiling. Distribute into tubes or flasks. Autoclave for 15 min at 15 psi pressure–121°C. Pour into sterile Petri dishes.

Use: For use as a base layer in antibiotic assay testing. Especially useful for the plate assay of bacitracin and penicillin G.

Antibiotic Medium 3 (Penassay Broth)

Composition per liter:

Pancreatic digest of gelatin 5.0g

K_2HPO_4 .... 3.68g
NaCl 3.5g
Yeast extract 1.5g
Beef extract 1.5g
KH_2PO_4 .... 1.32g
Glucose 1.0g
pH 7.0 ± 0.05 at 25°C

Source: This medium is available as a premixed powder from BD Diagnostic Systems and Oxoid Unipath.

Preparation of Medium: Add components to distilled/deionized water and bring volume to 1.0L. Mix thoroughly. Gently heat and bring to boiling. Distribute into tubes or flasks. Autoclave for 15 min at 15 psi pressure–121°C.

Use: For antibiotic assay testing. Used for the serial dilution assay of penicillins and other antibiotics. Used in the turbidimetric assay of penicillin and tetracycline with *Staphylococcus aureus*. For the cultivation and maintenance of *Staphylococcus aureus*.

Antibiotic Medium 3 Plus

Composition per liter:
Agar 15.0g
Peptone 5.0g
K_2HPO_4 .... 3.68g
NaCl 3.5g
Yeast extract 2.5g
Glucose 1.75g
Beef extract 1.5g
KH_2PO_4 .... 1.32g
pH 7.0 ± 0.05 at 25°C

Preparation of Medium: Add components to distilled/deionized water and bring volume to 1.0L. Mix thoroughly. Gently heat and bring to boiling. Distribute into tubes or flasks. Autoclave for 15 min at 15 psi pressure–121°C. Pour into sterile Petri dishes or leave in tubes.

Use: For antibiotic assay testing and for the cultivation of *Escherichia coli*.

Antibiotic Medium 4 (Yeast Beef Agar) (Agar Medium C)

Composition per liter:
Agar 15.0g
Pancreatic digest of gelatin 6.0g
Yeast extract 3.0g
Beef extract 1.5g
Glucose 1.0g
pH 6.6 ± 0.05 at 25°C

Source: This medium is available as a premixed powder from BD Diagnostic Systems.

Preparation of Medium: Add components to distilled/deionized water and bring volume to 1.0L. Mix thoroughly. Gently heat and bring to boiling. Distribute into tubes or flasks. Autoclave for 15 min at 15 psi pressure–121°C. Pour into sterile Petri dishes or leave in tubes.

Use: For antibiotic assay testing.

Antibiotic Medium 5 (Streptomycin Assay Agar with Yeast Extract)

Composition per liter:
Agar 15.0g
Pancreatic digest of gelatin 6.0g
Yeast extract 3.0g
Beef extract 1.5g
pH 7.9 ± 0.1 at 25°C

Source: This medium is available as a premixed powder from BD Diagnostic Systems and Oxoid Unipath.

Preparation of Medium: Add components to distilled/deionized water and bring volume to 1.0L. Mix thoroughly. Gently heat and bring to boiling. Distribute into tubes or flasks. Autoclave for 15 min at 15 psi pressure–121°C. Pour into sterile Petri dishes.

Use: For antibiotic assay testing. For the streptomycin assay using the cylinder plate technique and *Bacillus subtilis* as test organism.

Antibiotic Medium 6

Composition per liter:
Pancreatic digest of casein 17.0g
NaCl 5.0g
Papaic digest of soybean meal 3.0g
Glucose 2.5g
K_2HPO_4 .... 2.5g
$MnSO_4 \cdot H_2O$ 0.03g
pH 7.0 ± 0.1 at 25°C

Source: This medium is available as a premixed powder from BD Diagnostic Systems.

Preparation of Medium: Add components to distilled/deionized water and bring volume to 1.0L. Mix thoroughly. Gently heat and bring to boiling. Distribute into tubes or flasks. Autoclave for 15 min at 15 psi pressure–121°C. Pour into sterile Petri dishes.

Use: For antibiotic assay testing.

Antibiotic Medium 7

Composition per liter:
Agar 15.0g

Pancreatic digest of gelatin 6.0g
Yeast extract 3.0g
Beef extract 1.5g

pH 7.0 ± 0.1 at 25°C

Preparation of Medium: Add components to distilled/deionized water and bring volume to 1.0L. Mix thoroughly. Gently heat and bring to boiling. Adjust pH to 7.0. Distribute into tubes or flasks. Autoclave for 15 min at 15 psi pressure–121°C. Pour into sterile Petri dishes.

Use: For use as a base layer in antibiotic assay testing. Especially useful for the plate assay of bacitracin and penicillin G.

Antibiotic Medium 8 (Base Agar with Low pH)

Composition per liter:

Agar 15.0g
Pancreatic digest of gelatin 6.0g
Yeast extract 3.0g
Beef extract 1.5g

pH 5.9 ± 0.1 at 25°C

Source: This medium is available as a premixed powder from BD Diagnostic Systems.

Preparation of Medium: Add components to distilled/deionized water and bring volume to 1.0L. Mix thoroughly. Gently heat and bring to boiling. Distribute into tubes or flasks. Autoclave for 15 min at 15 psi pressure–121°C. Pour into sterile Petri dishes.

Use: For antibiotic assay testing. For use as the base agar and the seed agar in the plate assay of tetracycline. For use as the seed agar in the plate assay of vancomycin, mitomycin, and mithramycin.

Antibiotic Medium 9 (Polymyxin Base Agar)

Composition per liter:

Agar 20.0g
Pancreatic digest of casein 17.0g
NaCl 5.0g
Papaic digest of soybean meal 3.0g
K_2HPO_4 .... 2.5g
Glucose 2.5g

pH 7.2 ± 0.1 at 25°C

Source: This medium is available as a premixed powder from BD Diagnostic Systems.

Preparation of Medium: Add components to distilled/deionized water and bring volume to 1.0L. Mix thoroughly. Gently heat and bring to boiling. Distribute into tubes or flasks. Autoclave for 15 min at 15 psi pressure–121°C. Pour into sterile Petri dishes.

Use: For antibiotic assay testing. For base agar for the plate assay of carbenicillin, colistimethate, and polymyxin B.

Antibiotic Medium 10 (Polymyxin Seed Agar)

Composition per liter:

Pancreatic digest of casein 17.0g
Agar 12.0g
Polysorbate 80 .... 10.0g
NaCl 5.0g
Papaic digest of soybean meal 3.0g
K_2HPO_4 .... 2.5g
Glucose 2.5g

pH 7.3 ± 0.2 at 25°C

Source: This medium is available as a premixed powder from BD Diagnostic Systems.

Preparation of Medium: Add components to distilled/deionized water and bring volume to 1.0L. Mix thoroughly. Gently heat and bring to boiling. Distribute into tubes or flasks. Autoclave for 15 min at 15 psi pressure–121°C. Pour into sterile Petri dishes.

Use: For antibiotic assay testing. For seed agar for the plate assay of carbenicillin, colistimethate, and polymyxin B.

Antibiotic Medium 11 (Neomycin Assay Agar)

Composition per liter:

Agar 15.0g
Pancreatic digest of gelatin 6.0g
Pancreatic digest of casein 4.0g
Yeast extract 3.0g
Beef extract 1.5g
Glucose 1.0g

pH 8.0 ± 0.1 at 25°C

Source: This medium is available as a premixed powder from BD Diagnostic Systems and Oxoid Unipath.

Preparation of Medium: Add components to distilled/deionized water and bring volume to 1.0L. Mix thoroughly. Gently heat and bring to boiling. Distribute into tubes or flasks. Autoclave for 15 min at 15 psi pressure–121°C. Pour into sterile Petri dishes.

Use: For antibiotic assay testing. For base agar and seed agar for the plate assay to test the effectiveness of neomycin sulfate, amoxicillin, ampicillin, clindamycin, cyclacillin, erythromycin, gentamycin, neomycin, oleandomycin, and sisomycin.

Antibiotic Medium 12

Composition per liter:

Agar	25.0g
Peptone	10.0g
Glucose	10.0g
NaCl	10.0g
Yeast extract	5.0g
Beef extract	2.5g

pH 6.0 ± 0.1 at 25°C

Source: This medium is available as a premixed powder from BD Diagnostic Systems.

Preparation of Medium: Add components to distilled/deionized water and bring volume to 1.0L. Mix thoroughly. Gently heat and bring to boiling. Distribute into tubes or flasks. Autoclave for 15 min at 15 psi pressure–121°C. Pour into sterile Petri dishes.

Use: For antibiotic assay effectiveness testing.

Antibiotic Medium 13 (Sabouraud Liquid Broth, Modified) (Fluid Sabouraud Medium)

Composition per liter:

Glucose	20.0g
Pancreatic digest of casein	5.0g
Peptic digest of animal tissue	5.0g

pH 5.7 ± 0.1 at 25°C

Source: This medium is available as a premixed powder from BD Diagnostic Systems.

Preparation of Medium: Add components to distilled/deionized water and bring volume to 1.0L. Mix thoroughly. Gently heat and bring to boiling. Distribute into tubes or flasks. Autoclave for 15 min at 15 psi pressure–121°C. Pour into sterile Petri dishes.

Use: For testing the effectivness of antibiotics on yeast and molds.

Antibiotic Medium 19 (Nystatin Assay Agar)

Composition per liter:

Agar	23.5g
Glucose	10.0g
NaCl	10.0g
Pancreatic digest of gelatin	9.4g
Yeast extract	4.7g
Beef extract	2.4g

pH 6.1 ± 0.2 at 25°C

Source: This medium is available as a premixed powder from BD Diagnostic Systems.

Preparation of Medium: Add components to distilled/deionized water and bring volume to 1.0L. Mix thoroughly. Gently heat and bring to boiling. Distribute into tubes or flasks. Autoclave for 15 min at 15 psi pressure–121°C. Pour into sterile Petri dishes.

Use: For assaying the mycostatic activity of pharmaceutical preparations. For seed agar for the plate assay to test the effectiveness of nystatin, amphotericin B, and natamycin.

Antibiotic Medium 20

Composition per liter:

Glucose	11.0g
Pancreatic digest of casein	10.0g
Yeast extract	6.5g
Pancreatic digest of gelatin	5.0g
K_2HPO_4	3.68g
NaCl	3.5g
Beef extract	1.5g
KH_2PO_4	1.32g

pH 6.6 ± 0.2 at 25°C

Preparation of Medium: Add components to distilled/deionized water and bring volume to 1.0L. Mix thoroughly. Gently heat and bring to boiling. Distribute into tubes or flasks. Autoclave for 15 min at 15 psi pressure–121°C. Pour into sterile Petri dishes.

Use: For assaying the mycostatic activity of pharmaceutical preparations.

Antibiotic Medium 21

Composition per liter:

Glucose	11.0g
Pancreatic digest of gelatin	5.0g
K_2HPO_4	3.68g
NaCl	3.5g
Yeast extract	1.5g
Beef extract	1.5g
KH_2PO_4	1.32g

pH 6.6 ± 0.2 at 25°C

Preparation of Medium: Add components to distilled/deionized water and bring volume to 1.0L. Mix thoroughly. Gently heat and bring to boiling. Distribute into tubes or flasks. Autoclave for 15 min at 15 psi pressure–121°C. Pour into sterile Petri dishes.

Use: For assaying the mycostatic activity of pharmaceutical preparations.

Antibiotic Sulfonamide Sensitivity Test Agar (ASS Agar)

Composition per liter:

Agar	12.0g
Proteose peptone	10.0g
Beef extract	10.0g
NaCl	3.0g
Glucose	2.0g
Na_2HPO_4	2.0g
Sodium acetate	1.0g
Adenine	0.01g
Guanine	0.01g
Uracil	0.01g
Xanthine	0.01g

pH 7.4 ± 0.2 at 25°C

Preparation of Medium: Add components to distilled/deionized water and bring volume to 1.0L. Mix thoroughly. Gently heat and bring to boiling. Distribute into tubes or flasks. Autoclave for 15 min at 15 psi pressure–121°C. Pour into sterile Petri dishes or leave in tubes.

Use: For testing the antimicrobial effectiveness of antibiotics and sulfonamides. For detecting the presence of antimicrobial substances in urine.

AOAC Letheen Broth (Association of Official Analytical Chemists Letheen Broth)

Composition per liter:

Peptic digest of animal tissue	10.0g
Polysorbate 80	5.0g
NaCl	5.0g
Beef extract	5.0g
Lecithin	0.7g

pH 7.0 ± 0.2 at 25°C

Source: This medium is available as a premixed powder from BD Diagnostic Systems.

Preparation of Medium: Add components to distilled/deionized water and bring volume to 1.0L. Mix thoroughly. Gently heat and bring to boiling. Distribute into tubes in 10.0mL volumes. Autoclave for 15 min at 15 psi pressure–121°C.

Use: For the determination of phenol coefficients of disinfectant products containing cationic surface-active materials. Use according to *Official Methods of Analysis* of the *Association of Official Analytical Chemists* (AOAC).

Arenavirus Plaquing Medium

Composition per liter:

Eagle's basal medium	1.0L
Agarose solution	1.0L
Fetal calf serum, inactivated	100.0mL

pH 7.0 ± 0.2 at 25°C

Eagle's Basal Medium:

Composition per liter:

HEPES (*N*-2-Hydroxyethylpiperazine-*N*′-2-ethanesulfonic acid) buffer	9.53g
NaCl	6.8g
$NaHCO_3$	2.2g
Glucose	1.0g
KCl	0.4g
$CaCl_2·2H_2O$	0.2g
NaH_2PO_4	0.125g
$MgSO_4·7H_2O$	0.1g
L-Isoleucine	0.026g
L-Leucine	0.026g
L-Lysine	0.026g
L-Threonine	0.024g
L-Valine	0.0235g
L-Tyrosine	0.018g
L-Arginine	0.0174g
L-Phenylalanine	0.0165g
L-Cystine	0.012g
L-Histidine	8.0mg
L-Methionine	7.5mg
L-Tryptophan	4.0mg
Inositol	1.8mg
Biotin	1.0mg
Calcium pantothenate	1.0mg
Choline chloride	1.0mg
Folic acid	1.0mg
Nicotinamide	1.0mg
Pyridoxal·HCl	1.0mg
Thiamine·HCl	1.0mg
Riboflavin	0.1mg

pH 7.2–7.4 at 25°C

Source: This medium is available as a premixed powder from BD Diagnostic Systems.

Preparation of Eagle's Basal Medium: Add components to distilled/deionized water and bring volume to 1.0L. Mix thoroughly. Adjust pH to 7.0 with NaOH. Filter sterilize.

Agarose Solution:

Composition per liter:

Agarose	20.0g

Preparation of Agarose Solution: Add agarose to distilled/deionized water and bring volume to 1.0L. Mix thoroughly. Autoclave for 15 min at 15 psi pressure–121°C. Cool to 45°–50°C.

Preparation of Medium: To 1.0L of sterile Eagle's basal medium, aseptically add 1.0L of sterile, cooled agarose solution and 100.0mL of fetal calf serum. Mix thoroughly. Pour into sterile Petri dishes.

Use: For the cultivation of animal tissue culture cells used for the growth of arenaviruses.

Arginine Broth

Composition per liter:

L-Arginine	5.0g
Peptone or gelysate	5.0g
Yeast extract	3.0g
Glucose	1.0g
Bromcresol Purple	0.02g

pH 6.5 ± 0.2 at 25°C

Preparation of Medium: Add components to distilled/deionized water and bring volume to 1.0L. Mix thoroughly. Adjust pH so that is will be 6.5 ± 0.2 after sterilization. Distribute into 16 × 150mm screw-capped tubes in 5.0mL volumes. Autoclave medium with loosely capped tubes for 10 min at 15 psi pressure–121°C. Screw the caps on tightly for storage and after inoculation.

Use: For the cultivation and differentiation of bacteria based on their ability to decarboxylate the amino acid arginine. Bacteria that decarboxylate arginine turn the medium turbid purple.

Arginine Broth with NaCl

Composition per liter:

L-Arginine	5.0g
Peptone or gelysate	5.0g
Yeast extract	3.0g
Glucose	1.0g
Bromcresol Purple	0.02g

pH 6.5 ± 0.2 at 25°C

Preparation of Medium: Add components to distilled/deionized water and bring volume to 1.0L. Mix thoroughly. Adjust pH so that is will be 6.5 ± 0.2 after sterilization. Distribute into 16 × 150mm screw-capped tubes in 5.0mL volumes. Autoclave medium with loosely capped tubes for 10 min at 15 psi pressure–121°C. Screw the caps on tightly for storage and after inoculation.

Use: For the cultivation and differentiation of *Vibrio* spp. based on their ability to decarboxylate the amino acid arginine. Bacteria that decarboxylate arginine turn the medium turbid purple.

Arginine Glucose Slants (AGS)

Composition per liter:

NaCl	20.0g
Agar	13.5g
Pancreatic digest of casein	10.0g
L-Arginine·HCl	5.0g
Peptone	5.0g
Yeast extract	3.0g
Glucose	1.0g
Ferric ammonium citrate	0.5g
$Na_2S_2O_3 \cdot 5H_2O$	0.3g
Bromcresol Purple	0.02g

pH 6.8–7.0 at 25°C

Preparation of Medium: Add components to distilled/deionized water and bring volume to 1.0L. Mix thoroughly. Gently heat and bring to boiling. Distribute into tubes. Autoclave for 12 min at 15 psi pressure–121°C. Allow tubes to cool in a slanted position.

Use: For the cultivation and differentation of *Vibrio* species.

Armstrong *Fusarium* Medium

Composition per liter:

Glucose	20.0g
$Ca(NO_3)_2 \cdot 4H_2O$	8.4g
KH_2PO_4	1.09g
KCl	0.22g
$FeCl_3$	0.2μg
$MnSO_4$	0.2μg
$ZnSO_4$	0.2μg

Preparation of Medium: Add components to distilled/deionized water and bring volume to 1.0L. Mix thoroughly. Filter sterilize.

Use: For the cultivation of *Fusarium* species.

Arylsulfatase Agar (Wayne Sulfatase Agar)

Composition per liter:

Agar	15.0g
Na_2HPO_4	2.5g
L-Asparagine	1.0g
KH_2PO_4	1.0g
K_2HPO_4	1.0g
Trisodium phenolphthalein sulfate	0.65g
Pancreatic digest of casein	0.5g
Ferric ammonium citrate	0.05g
$MgSO_4 \cdot 7H_2O$	0.01g
$CaCl_2 \cdot 2H_2O$	0.5mg
$ZnSO_4 \cdot 7H_2O$	0.1mg
$CuSO_4$	0.1mg
Glycerol	10.0mL

pH 7.0 ± 0.2 at 25°C

Source: This medium is available as a premixed powder from BD Diagnostic Systems.

Preparation of Medium: Add glycerol to approximately 800.0mL of distilled/deionized water. Mix thoroughly. Add remaining components and bring volume to 1.0L with distilled/deionized water. Mix thoroughly. Gently heat and bring to boiling. Distribute into tubes. Autoclave for 15 min at 15 psi pressure–121°C. Cool tubes in an upright position.

Use: For the biochemical differentiation of species of *Mycobacterium*. Inoculate tubes with *Mycobacterium* cultures and incubate aerobically at 35°C for 3–14 days. Add 0.5–1.0mL of 2*N* Na_2CO_3 to each tube and observe color change within 30 min. Development of a pink color is indicative of *Mycobacterium fortuitum* or *Mycobacterium chelonae*. *Mycobacterium tuberculosis* gives a negative reaction.

Ascospore Agar

Composition per liter:

Potassium acetate	30.0g
Yeast extract	2.5g
Glucose	1.0g

pH 6.4 ± 0.2 at 25°C

Preparation of Medium: Add components to distilled/deionized water and bring volume to 1.0L. Mix thoroughly. Gently heat and bring to boiling. Distribute into tubes or flasks. Autoclave for 15 min at 15 psi pressure–121°C. Pour into sterile Petri dishes or leave in tubes.

Use: For the enrichment of ascosporogenous yeasts and their production of ascospores.

ASM Medium

Composition per 1001.2mL:

Na_2HPO_4	866.0mg
$NH_4 \cdot Cl$	535.0mg
KH_2PO_4	531.0mg
K_2SO_4	174.0mg
$MgSO_4 \cdot 7H_2O$	37.0mg
$CaCl_2 \cdot 2H_2O$	7.35mg
Trace elements solution	1.0mL
$FeSO_4$ solution	0.2mL

Trace Elements Solution:

Composition per liter:

$ZnSO_4 \cdot 7H_2O$	288.0mg
$MnSO_4 \cdot 4H_2O$	224.0mg
$CuSO_4 \cdot 5H_2O$	125.0mg
KI	83.0mg
H_3BO_3	61.8mg
$Na_2MoO_4 \cdot 2H_2O$	48.4mg
$CoCl_2 \cdot 6H_2O$	47.6mg
H_2SO_4, 1*M*	1.0mL

Preparation of Trace Elements Solution: Add components to distilled/deionized water and bring volume to 1.0L. Mix thoroughly. Autoclave for 15 min at 15 psi pressure–121°C.

$FeSO_4$ Solution:

Composition per liter:

$FeSO_4 \cdot 7H_2O$	278.0mg

Preparation of $FeSO_4$ Solution: Add $FeSO_4 \cdot 7H_2O$ to distilled/deionized water and bring volume to 10.0mL. Mix thoroughly. Filter sterilize.

Preparation of Medium: Add components, except trace elements solution and $FeSO_4$ solution, to distilled/deionized water and bring volume to 1.0L. Mix thoroughly. Autoclave for 15 min at 15 psi pressure–121°C. Aseptically add 1.0mL of trace elements solution and 0.2mL of sterile $FeSO_4$ solution. Mix thoroughly. Aseptically distribute into sterile tubes or flasks.

Use: For the cultivation of *Mycobacterium* species.

Asparaginate Glycerol Agar

Composition per liter:

Agar	15.0g
Sodium asparaginate	1.0g
K_2HPO_4	1.0g
Glycerol	10.0mL

pH 7.0 ± 0.2 at 25°C

Preparation of Medium: Add components to distilled/deionized water and bring volume to 1.0L. Mix thoroughly. Gently heat and bring to boiling. Distribute into tubes or flasks. Autoclave for 15 min at 15 psi pressure–121°C. Pour into sterile Petri dishes or leave in tubes.

Use: For the cultivation and maintenance of *Nocardia transvalensis*.

Asparagine Broth

Composition per liter:

DL-Asparagine	30.0g
K_2HPO_4	1.0g
$MgSO_4 \cdot 7H_2O$	0.5g

pH 6.9–7.2 at 25°C

Preparation of Medium: Add components to distilled/deionized water and bring volume to 1.0L. Mix well until dissolved. Adjust pH to between 6.9 and 7.2. Distribute into tubes or flasks. Autoclave for 15 min at 15 psi pressure–121°C.

Use: For a presumptive test medium in the differentiation of nonfermentative Gram-negative bacteria, especially *Pseudomonas aeruginosa*.

Aspergillus Differential Medium

Composition per liter:

Agar	15.0g
Pancreatic digest of casein	15.0g

Yeast extract...10.0g
Ferric citrate...0.5g

Preparation of Medium: Add components to distilled/deionized water and bring volume to 1.0L. Mix thoroughly. Gently heat and bring to boiling. Distribute into tubes in 7.0mL volumes. Autoclave for 15 min at 15 psi pressure–121°C. Allow tubes to cool in a slanted position.

Use: For the cultivation and differentiation of *Aspergillus flavus. Aspergillus flavus* appears as bright orange colonies.

Aspergillus Medium

Composition per liter:

Agar...15.0g
$NaNO_3$...6.0g
Casamino acids...1.0g
Peptone...1.0g
Yeast extract...1.0g
Adenine...0.15g
Vitamin solution...10.0mL
pH 6.0 ± 0.2 at 25°C

Vitamin Solution:

Composition per 100.0mL:

Biotin...0.01g
Nicotinic acid...0.01g
p-Aminobenzoic acid...0.01g
Pyridoxine·HCl...0.01g
Riboflavin...0.01g
Thiamine·HCl...0.01g

Preparation of Vitamin Solution: Add components to distilled/deionized water and bring volume to 100.0mL. Mix thoroughly. Autoclave for 10 min at 15 psi pressure–121°C.

Preparation of Medium: Add components, except vitamin solution, to distilled/deionized water and bring volume to 990.0mL. Mix thoroughly. Gently heat and bring to boiling. Adjust pH to 6.0. Autoclave for 15 min at 15 psi pressure–121°C. Cool to 50°C. Aseptically add 10.0mL of sterile vitamin solution. Mix thoroughly. Pour into sterile Petri dishes or distribute into sterile tubes.

Use: For the cultivation and maintenance of *Aspergillus* spp.

ATS Medium (American Trudeau Society Medium)

Composition per liter:

Potato...20.0g
Malachite Green...0.2g
Egg yolk emulsion...500.0mL
Glycerol...10.0mL
pH 6.5–7.0 at 25°C

Source: This medium is available as a prepared medium from BD Diagnostic Systems.

Egg Yolk Emulsion:

Composition:

Chicken egg yolks...11
Whole chicken egg...1

Preparation of Egg Yolk Emulsion: Soak eggs with 1:100 dilution of saturated mercuric chloride solution for 1 min. Crack eggs and separate yolks from whites. Mix egg yolks with 1 chicken egg.

Preparation of Medium: Add components to distilled/deionized water and bring volume to 1.0L. Distribute into tubes. Autoclave for 15 min at 15 psi pressure–121°C in a slanted position.

Use: For the isolation and cultivation of *Mycobacterium* species other than *Mycobacterium leprae.* Especially useful for the detection of *Mycobacterium tuberculosis* from clinical specimens such as cerebrospinal fluid, pleural fluid, and tissues.

AV Agar with Vitamins

Composition per liter:

Agar...15.0g
Glucose...1.0g
Glycerol...1.0g
L-Arginine...0.3g
K_2HPO_4...0.3g
NaCl...0.3g
$MgSO_4·7H_2O$...0.2g
Vitamin solution...100.0mL
Trace salts solution...1.0mL

Vitamin Solution:

Composition per 100.0mL:

p-Aminobenzoic acid...0.5mg
Calcium pantothenate...0.5mg
HCl...0.5mg
Inositol...0.5mg
Niacin...0.5mg
Pyridoxine...0.5mg
Ribovlavin...0.5mg
Thiamine·HCl...0.5mg
Biotin...0.25mg

Preparation of Vitamin Solution: Add components to distilled/deionized water and bring volume to 1.0L. Mix thoroughly. Filter sterilize.

Trace Salts Solution:

Composition per liter:

$FeSO_4·7H_2O$...10.0g
$CuSO_4·5H_2O$...1.0g

$MnSO_4 \cdot 7H_2O$ 1.0g
$ZnSO_4 \cdot 7H_2O$ 1.0g

Preparation of Trace Salts Solution: Add components to distilled/deionized water and bring volume to 1.0L. Mix thoroughly.

Preparation of Medium: Add components, except vitamin solution, to distilled/deionized water and bring volume to 900.0mL. Mix thoroughly. Gently heat and bring to boiling. Autoclave for 15 min at 15 psi pressure–121°C. Cool to 45°–50°C. Aseptically add 100.0mL of sterile vitamin solution. Mix thoroughly. Pour into sterile Petri dishes or distribute into sterile tubes.

Use: For the isolation and cultivation of *Actinomadura* species, *Actinopolyspora* species, *Excellospora* species, and *Microspora* species.

Avian *Mycoplasma* Agar

Composition per liter:

Agar, not inhibitory to mycoplasmas 10.0g
PPLO broth without Crystal Violet 700.0mL
Swine or horse serum, heat inactivated
at 56°C for 30 min. 150.0mL
Fresh yeast extract solution 100.0mL
Phenol Red solution 20.0mL
Glucose solution 10.0mL
Arginine solution 10.0mL
NAD solution 10.0mL

PPLO Broth without Crystal Violet:

Composition per 700.0mL:

Beef heart, infusion from 175.0g
Peptone 7.0g
NaCl 3.5g

Source: PPLO broth without Crystal Violet is available as a premixed powder from BD Diagnostic Systems.

Preparation of PPLO Broth without Crystal Violet: Add components to distilled/deionized water and bring volume to 700.0mL. Autoclave for 15 min at 15 psi pressure–121°C. Cool to 25°C. Beef heart for infusion may be substituted; 100.0g of beef heart for infusion is equivalent to 500.0g of fresh heart tissue.

Fresh Yeast Extract Solution:

Composition per 100.0mL:

Baker's yeast, live, pressed, starch-free 25.0g

Preparation of Fresh Yeast Extract Solution: Add the live Baker's yeast to 100.0mL of distilled/deionized water. Autoclave for 90 min at 15 psi pressure–121°C. Allow to stand. Remove supernatant solution. Adjust pH to 6.6–6.8.

Phenol Red Solution:

Composition per 20.0mL:

Phenol Red 0.02g

Preparation of Phenol Red Solution: Add Phenol Red to distilled/deionized water and bring volume to 20.0mL. Mix thoroughly. Filter sterilize.

Glucose Solution:

Composition per 10.0mL:

Glucose 1.0g

Preparation of Glucose Solution: Add glucose to distilled/deionized water and bring volume to 10.0mL. Mix thoroughly. Filter sterilize.

Arginine Solution:

Composition per 10.0mL:

Arginine 1.0g

Preparation of Arginine Solution: Add arginine to distilled/deionized water and bring volume to 10.0mL. Mix thoroughly. Filter sterilize.

NAD Solution:

Composition per 10.0mL:

NAD 0.1g

Preparation of NAD Solution: Add NAD to distilled/deionized water and bring volume to 10.0mL. Mix thoroughly. Filter sterilize.

Preparation of Medium: Add 10.0g of agar to 700.0mL of PPLO broth without Crystal Violet. Gently heat to boiling with frequent mixing. Autoclave for 15 min at 15 psi pressure–121°C. Cool to 50°–55°C. Warm other components to 50°–55°C using a water bath. Aseptically combine all components. Mix thoroughly. Pour into sterile Petri dishes or sterile tubes.

Use: For the cultivation and maintenance of *Mycoplasma* species.

Avian *Mycoplasma* Broth

Composition per liter:

PPLO broth without Crystal Violet 700.0mL
Swine or horse serum, heat inactivated
at 56°C for 30 min. 150.0mL
Fresh yeast extract solution 100.0mL
Phenol Red solution 20.0mL
Glucose solution 10.0mL
Arginine solution 10.0mL
NAD solution 10.0mL

PPLO Broth without Crystal Violet:

Composition per 700.0mL:

Beef heart, infusion from 175.0g
Peptone 7.0g
NaCl 3.5g

Source: PPLO broth without Crystal Violet is available as a premixed powder from BD Diagnostic Systems.

Preparation of PPLO Broth without Crystal Violet: Add components to distilled/deionized water and bring volume to 700.0mL. Autoclave for 15 min at 15 psi pressure–121°C. Cool to 25°C. Beef heart for infusion may be substituted; 100.0g of beef heart for infusion is equivalent to 500.0g of fresh heart tissue.

Fresh Yeast Extract Solution:
Composition per 100.0mL:
Baker's yeast, live, pressed, starch-free 25.0g

Preparation of Fresh Yeast Extract Solution: Add the live Baker's yeast to 100.0mL of distilled/deionized water. Autoclave for 90 min at 15 psi pressure–121°C. Allow to stand. Remove supernatant solution. Adjust pH to 6.6–6.8.

Phenol Red Solution:
Composition per 20.0mL:
Phenol Red 0.02g

Preparation of Phenol Red Solution: Add Phenol Red to distilled/deionized water and bring volume to 20.0mL. Mix thoroughly. Filter sterilize.

Glucose Solution:
Composition per 10.0mL:
Glucose 1.0g

Preparation of Glucose Solution: Add glucose to distilled/deionized water and bring volume to 10.0mL. Mix thoroughly. Filter sterilize.

Arginine Solution:
Composition per 10.0mL:
Arginine 1.0g

Preparation of Arginine Solution: Add arginine to distilled/deionized water and bring volume to 10.0mL. Mix thoroughly. Filter sterilize.

NAD Solution:
Composition per 10.0mL:
NAD 0.1g

Preparation of NAD Solution: Add NAD to distilled/deionized water and bring volume to 10.0mL. Mix thoroughly. Filter sterilize.

Preparation of Medium: Aseptically combine components. Distribute into sterile tubes or flasks.

Use: For the cultivation and maintenance of *Mycoplasma* species.

Azide Blood Agar

Composition per liter:
Agar 15.0g
Pancreatic digest of casein 5.0g
Peptic digest of animal tissue 5.0g
NaCl 5.0g
Beef extract 3.0g
NaN_3 .... 0.2g
Sheep blood, defibrinated 50.0mL
pH 7.2 ± 0.2 at 25°C

Source: This medium is available as a premixed powder from BD Diagnostic Systems and Oxoid Unipath.

Caution: Sodium azide is toxic. Azides also react with metals and disposal must be highly diluted.

Preparation of Medium: Add components, except sheep blood, to distilled/deionized water and bring volume to 950.0mL. Mix thoroughly. Gently heat and bring to boiling. Autoclave for 15 min at 15 psi pressure–121°C. Cool to 45–50°C. Aseptically add 50.0mL of sterile defibrinated sheep blood. Pour into sterile Petri dishes or distribute into sterile tubes. Allow tubes to cool in a slanted position.

Use: For the isolation and differentiation of streptococci and staphylococci from specimens containing mixed flora.

Azide Blood Agar with Crystal Violet (Packer's Agar)

Composition per liter:
Agar 15.0g
Pancreatic digest of casein 5.0g
Peptic digest of animal tissue 5.0g
NaCl 5.0g
Beef extract 3.0g
NaN_3 .... 0.9g
Crystal Violet 2.0mg
Sheep blood, defibrinated 50.0mL
pH 7.2 ± 0.2 at 25°C

Caution: Sodium azide is toxic. Azides also react with metals and disposal must be highly diluted.

Preparation of Medium: Add components, except sheep blood, to distilled/deionized water and bring volume to 950.0mL. Mix thoroughly. Gently heat and bring to boiling. Autoclave for 15 min at 15 psi pressure–121°C. Cool to 45°–50°C. Aseptically add 50.0mL of sterile defibrinated sheep blood. Pour into sterile Petri dishes or distribute into sterile tubes. Allow tubes to cool in a slanted position.

Use: For the isolation of *Streptococcus pneumoniae* and *Erysipelothrix rhusiopathiae*.

Azide Broth
(Azide Glucose Broth)
(Azide Dextrose Broth)

Composition per liter:

Pancreatic digest of casein	15.0g
Glucose	7.5g
NaCl	7.5g
Beef extract	4.5g
NaN_3	0.2g

pH 7.2 ± 0.2 at 25°C

Source: This medium is available as a premixed powder from BD Diagnostic Systems.

Caution: Sodium azide is toxic. Azides also react with metals and disposal must be highly diluted.

Preparation of Medium: Add components to distilled/deionized water and bring volume to 1.0L. Mix thoroughly. Gently heat and bring to boiling. Distribute into tubes or flasks. Autoclave for 15 min at 15 psi pressure–121°C. Prepare double-strength broth for samples larger than 1.0mL.

Use: Used in the multiple-tube technique as a presumptive test for the presence of fecal streptococci.

Azide Medium

Composition per liter:

Peptone	10.0g
K_2HPO_4	5.0g
Glucose	5.0g
NaCl	5.0g
Yeast extract	3.0g
KH_2PO_4	2.0g
NaN_3	0.25g
Bromcresol Purple solution	2.0mL

pH 7.2 ± 0.2 at 25°C

Bromcresol Purple Solution:
Composition per 10.0mL:

Bromcresol Purple	0.16g
Ethanol	10.0mL

Preparation of Bromcresol Purple Solution: Add Bromcresol Purple to ethanol and bring volume to 10.0mL. Mix thoroughly.

Caution: Sodium azide is toxic. Azides also react with metals and disposal must be highly diluted.

Preparation of Medium: Add components to distilled/deionized water and bring volume to 1.0L. Mix thoroughly. Distribute into tubes or flasks. Autoclave for 15 min at 15 psi pressure–121°C.

Use: For the cultivation of *Streptococcus* species and *Staphylococcus* species.

B Broth
(Medium for *Ureaplasma*)

Composition per 100.25mL:

Yeast extract	0.1g
GHL (Glycyl-L-histidyl-L-lysine)	2.0μg
PPLO broth without Crystal Violet	50.0mL
Horse serum, not inactivated	10.0mL
Bromthymol Blue (0.4% solution)	1.0mL
Urea solution	0.25mL

pH 6.0 ± 0.2 at 25°C

PPLO Broth without Crystal Violet:
Composition per 50.0mL:

Beef heart, infusion from	1.62g
Peptone	0.32g
NaCl	0.16g

Source: PPLO broth without Crystal Violet is available as a premixed powder from BD Diagnostic Systems.

Preparation of PPLO Broth without Crystal Violet: Add components to distilled/deionized water and bring volume to 50.0mL. Mix thoroughly.

Urea Solution:
Composition per 10.0mL:

Urea	1.0g

Preparation of Urea Solution: Add urea to distilled/deionized water and bring volume to 10.0mL. Mix thoroughly. Filter sterilize.

Preparation of Medium: Add components—except GHL, urea solution, and horse serum—to double glass-distilled water and bring volume to 90.0mL. Mix thoroughly. Gently heat and bring to boiling. Autoclave for 15 min at 15 psi pressure–121°C. Cool to 50°–55°C. To 90.0mL of the sterile medium, aseptically add 2.0μg of GHL, 10.0mL of horse serum, and 0.25mL of sterile urea solution. Mix thoroughly. Aseptically distribute into tubes or flasks.

Use: For the cultivation and maintenance of *Ureaplasma urealyticum* and other *Ureaplasma* species.

Bacillus cereus Medium
(BCM)

Composition per 110.0mL:

Agar	2.0g
D-Mannitol	1.0g
$(NH_4)_2PO_4$	0.1g
KCl	0.02g
$MgSO_4·7H_2O$	0.02g
Yeast extract	0.02g

Bromcresol Purple	4.0mg
Egg yolk emulsion, 20%	10.0mL

pH 7.0 ± 0.2 at 25°C

Egg Yolk Emulsion, 20%:

Composition per 100.0mL:

Chicken egg yolks	11
Whole chicken egg	1
NaCl (0.9% solution)	80.0mL

Preparation of Egg Yolk Emulsion, 20%: Soak eggs with 1:100 dilution of saturated mercuric chloride solution for 1 min. Crack eggs and separate yolks from whites. Mix egg yolks with 1 chicken egg. Measure 20.0mL of egg yolk emulsion and add to 80.0mL of 0.9% NaCl solution. Mix thoroughly. Filter sterilize. Warm to 45°–50°C.

Preparation of Medium: Add components—except egg yolk emulsion, 20%—to distilled/deionized water and bring volume to 100.0mL. Mix thoroughly. Gently heat and bring to boiling. Autoclave for 15 min at 15 psi pressure–121°C. Cool to 45°–50°C. Aseptically add 10.0mL of sterile egg yolk emulsion, 20%. Mix thoroughly. Pour into sterile Petri dishes or distribute into sterile tubes.

Use: For the cultivation of *Bacillus cereus.*

Bacillus cereus Selective Agar Base

Composition per liter:

Agar	15.0g
Sodium pyruvate	10.0g
Mannitol	10.0g
Na_2HPO_4	2.5g
NaCl	2.0g
Peptone	1.0g
KH_2PO_4	0.25g
Bromthymol Blue	0.12g
$MgSO_4 \cdot 7H_2O$	0.1g
Egg yolk emulsion	25.0mL
Polymyxin B solution	10.0mL

pH 7.2 ± 0.2 at 25°C

Source: This medium is available as a premixed powder from Oxoid Unipath.

Egg Yolk Emulsion:

Composition:

Chicken egg yolks	11
Whole chicken egg	1

Preparation of Egg Yolk Emulsion: Soak eggs with 1:100 dilution of saturated mercuric chloride solution for 1 min. Crack eggs and separate yolks from whites. Mix egg yolks with 1 chicken egg.

Polymyxin B Solution:

Composition per 10.0mL:

Polymyxin B	100,000U

Preparation of Polymyxin B Solution: Add Polymyxin B to distilled/deionized water and bring volume to 10.0mL. Mix thoroughly. Filter sterilize.

Preparation of Medium: Add components, except egg yolk emulsion and polymyxin B solution, to distilled/deionized water and bring volume to 965.0mL. Gently heat and bring to boiling. Distribute into tubes or flasks. Autoclave for 15 min at 15 psi pressure–121°C. Cool to 50°C. Aseptically add sterile polymyxin B and 25.0mL of sterile egg yolk emulsion. Mix thoroughly. Pour into sterile Petri dishes or leave in tubes.

Use: For the selection and presumptive identification of *Bacillus cereus.* Also for the isolation and enumeration of these bacteria. *Bacillus cereus* grows as moderate-sized (5mm) crenated colonies, which are turquoise, surrounded by a precipitate of egg yolk, which is also turquoise.

Bacillus Medium

Composition per liter:

Agar	25.0g
Peptone	6.0g
Pancreatic digest of casein	3.0g
Yeast extract	3.0g
Beef extract	1.5g
$MnSO_4 \cdot 4H_2O$	1.0μg

pH 7.0 ± 0.2 at 25°C

Preparation of Medium: Add components to distilled/deionized water and bring volume to 1.0L. Mix thoroughly. Gently heat and bring to boiling. Distribute into tubes or flasks. Autoclave for 15 min at 15 psi pressure–121°C. Pour into sterile Petri dishes or leave in tubes.

Use: For the cultivation of *Bacillus* species.

Bacillus Medium

Composition per liter:

$(NH_4)_2HPO_4$	1.0g
$MgSO_4 \cdot 7H_2O$	0.2g
KCl	0.2g
Yeast extract	0.2g
Glucose solution	50.0mL
Bromcresol Purple solution	15.0mL

pH 7.0 ± 0.2 at 25°C

Glucose Solution:

Composition per 100.0mL:

Glucose	10.0g

Preparation of Glucose Solution: Add glucose to distilled/deionized water and bring volume to 100.0mL. Mix thoroughly. Filter sterilize.

Preparation of Medium: Add components, except glucose solution, to distilled/deionized water and bring volume to 1.0L. Mix thoroughly. Gently heat and bring to boiling. Distribute 9.5mL volumes into test tubes that contain an inverted Durham tube. Autoclave for 20 min at 15 psi pressure–121°C. Cool to 25°C. Aseptically add 0.5mL of sterile glucose to each tube. Mix thoroughly.

Use: For cultivation and differentiation of *Bacillus* species based on acid and gas production from glucose.

Bacillus Medium (ATCC Medium 455)

Composition per liter:

Soluble starch	30.0g
Agar	20.0g
Polypeptone	5.0g
Yeast extract	5.0g

Preparation of Medium: Add components to distilled/deionized water and bring volume to 1.0L. Mix thoroughly. Gently heat and bring to boiling. Distribute into tubes or flasks. Autoclave for 15 min at 15 psi pressure–121°C. Swirl medium to resuspend starch. Pour into sterile Petri dishes or leave in tubes.

Use: For the cultivation of amylase-producing microorganisms.

Bacillus Medium (ATCC Medium 552)

Composition per liter:

Peptone	10.0g
Lactose	5.0g
NaCl	5.0g
Beef extract	3.0g
K_2HPO_4	2.0g

pH 7.2 ± 0.2 at 25°C

Preparation of Medium: Add components to distilled/deionized water and bring volume to 1.0L. Mix thoroughly. Gently heat and bring to boiling. Distribute into tubes or flasks. Autoclave for 15 min at 15 psi pressure–121°C.

Use: For the cultivation and maintenance of *Bacillus* species.

Bacteroides Bile Esculin Agar (BBE Agar)

Composition per liter:

Oxgall	20.0g
Pancreatic digest of casein	15.0g
Agar	15.0g
Papaic digest of soybean meal	5.0g
NaCl	5.0g
Esculin	1.0g
Ferric ammonium citrate	0.5g
Gentamicin solution	2.5mL
Hemin solution	2.5mL
Vitamin K_1 solution	1.0mL

pH 7.0 ± 0.2 at 25°C

Source: This medium is available as a premixed powder from BD Diagnostic Systems.

Gentamicin Solution:

Composition per 10.0mL:

Gentamicin	0.4mg

Preparation of Gentamicin Solution: Add gentamicin to 10.0mL of distilled/deionized water. Mix thoroughly. Filter sterilize.

Hemin Solution:

Composition per 100.0mL:

Hemin	0.5g
NaOH (1*N* solution)	10.0mL

Preparation of Hemin Solution: Add components to 100.0mL of distilled/deionized water. Mix thoroughly. Autoclave for 15 min at 15 psi pressure–121°C. Cool to 45°–50°C.

Vitamin K_1 Solution:

Composition per 100.0mL:

Vitamin K_1	1.0g
Ethanol	99.0mL

Preparation of Vitamin K_1 Solution: Add vitamin K_1 to 99.0mL of absolute ethanol. Mix thoroughly. Filter sterilize.

Preparation of Medium: Add components, except hemin solution, gentamicin solution, and vitamin K_1 solution, to distilled/deionized water and bring volume to 994.0mL. Mix thoroughly. Gently heat and bring to boiling. Autoclave for 15 min at 15 psi pressure–121°C. Cool to 45°–50°C. Aseptically add 2.5mL of sterile hemin solution, 2.5mL of sterile gentamicin solution, and 1.0mL of sterile vitamin K_1 solution.

Use: For the selection and presumptive identification of the *Bacteriodes fragilis* group. For the differentiation of *Bacteroides* species based on the hydrolysis of esculin and presence of catalase. After incubation for 48 hr, bacteria of the *Bacteroides fragilis* group appear as gray, circular, raised colonies larger than 1.0mm. Esculin hydrolysis is indicated by the presence of a blackened zone around the colonies.

BAF Agar

Composition per liter:

Glucose	30.0g
Agar	15.0g
Peptone	2.0g
KH_2PO_4	0.5g
$MgSO_4 \cdot 7H_2O$	0.5g
Yeast extract	0.2g
$CaCl_2 \cdot 2H_2O$	100.0mg
$FeCl_3 \cdot 6H_2O$	10.0mg
$MnSO_4$	5.0mg
$ZnSO_4 \cdot 7H_2O$	1.0mg
Folic acid	100.0μg
Inositol	50.0μg
Thiamine·HCl	50.0μg
Biotin	1.0μg

pH 5.8 ± 0.2 at 25°C

Preparation of Medium: Add components to distilled/deionized water and bring volume to 1.0L. Gently heat and bring to boiling. Adjust pH to 5.8. Distribute into tubes or flasks. Autoclave for 15 min at 15 psi pressure–121°C. Pour into sterile Petri dishes or leave in tubes.

Use: For the cultivation of a wide variety of bacteria.

BAGG Broth (Buffered Azide Glucose Glycerol Broth)

Composition per liter:

Pancreatic digest of casein	10.0g
Peptic digest of animal tissue	10.0g
Glucose	5.0g
NaCl	5.0g
K_2HPO_4	4.0g
KH_2PO_4	1.5g
NaN_3	0.5g
Bromcresol Purple	0.015g
Glycerol	5.0mL

pH 6.9 ± 0.2 at 25°C

Source: This medium is available as a premixed powder from BD Diagnostic Systems.

Caution: Sodium azide is toxic. Azides also react with metals and disposal must be highly diluted.

Preparation of Medium: Add 5.0mL of glycerol to 900.0mL of distilled/deionized water. Add remaining components and bring volume to 1.0L. Mix thoroughly. Gently heat and bring to boiling. Distribute into tubes in 10.0mL volumes. Autoclave for 15 min at 10 psi pressure–116°C.

Use: For the cultivation of fecal streptococci from a variety of clinical specimens. It is recommended for qualitative presumptive and confirmatory tests for fecal streptococci.

Baird-Parker Agar

Composition per liter:

Agar	17.0g
Glycine	12.0g
Sodium pyruvate	10.0g
Pancreatic digest of casein	10.0g
Beef extract	5.0g
LiCl	5.0g
Yeast extract	1.0g

pH 7.0 ± 0.2 at 25°C

Source: This medium is available as a premixed powder from Oxoid Unipath and BD Diagnostic Systems.

Preparation of Medium: Add components to distilled/deionized water and bring volume to 1.0L. Mix thoroughly. Gently heat and bring to boiling. Autoclave for 15 min at 15 psi pressure–121°C. Cool to 45°–50°C. Pour into sterile Petri dishes.

Use: Used as a base for the preparation of egg-tellurite-glycine-pyruvate agar for the selective isolation and enumeration of coagulase-positive staphylococci.

Baird-Parker Agar

Composition per liter:

Agar	17.0g
Glycine	12.0g
Sodium pyruvate	10.0g
Pancreatic digest of casein	10.0g
Beef extract	5.0g
LiCl	5.0g
Yeast extract	1.0g
Sulfamethazine solution	10.0mL

pH 7.0 ± 0.2 at 25°C

Sulfamethazine Solution:

Composition per 10.0mL:

Sulfamethazine	0.05g

Preparation of Sulfamethazine Solution: Add sulfamethazine to distilled/deionized water and bring volume to 10.0mL. Mix thoroughly. Filter sterilize.

Preparation of Medium: Add components, except sulfamethazine solution, to distilled/deionized water and bring volume to 990.0mL. Mix thoroughly. Gently heat and bring to boiling. Autoclave for 15 min at 15 psi pressure–121°C. Cool to 45°–50°C. Aseptically add sterile sulfamethazine solution. Mix thoroughly. Pour into sterile Petri dishes or distribute into sterile tubes.

Use: Used as a base for the preparation of egg-tellurite-glycine-pyruvate agar for the selective isolation and enumeration of coagulase-positive staphylococci.

Baird-Parker Agar, Supplemented

Composition per liter:

Agar	17.0g
Glycine	12.0g
Sodium pyruvate	10.0g
Pancreatic digest of casein	10.0g
Beef extract	5.0g
LiCl	5.0g
Yeast extract	1.0g
RPF supplement	100.0mL

pH 7.0 ± 0.2 at 25°C

RPF Supplement:

Composition per 100.0mL:

Bovine fibrinogen	3.75g
Trypsin inhibitor	25.0mg
K_2TeO_3	25.0mg
Rabbit plasma	25.0mL

Caution: Potassium tellurite is toxic.

Preparation of RPF Supplement: Add components to distilled/deionized water and bring volume to 100.0mL. Mix thoroughly. Filter sterilize.

Preparation of Medium: Add components, except RPF supplement, to distilled/deionized water and bring volume to 900.0mL. Mix thoroughly. Gently heat and bring to boiling. Autoclave for 15 min at 15 psi pressure–121°C. Cool to 45°–50°C. Aseptically add 100.0mL of filter-sterilized RPF supplement. Mix thoroughly but gently. Pour into sterile Petri dishes.

Use: For the selective isolation and enumeration of coagulase-positive staphylococci. For the differentiation and identification of staphylococci on the basis of their ability to coagulate plasma. Colonies surrounded by an opaque zone of coagulated plasma are diagnostic for *Staphylococcus aureus*.

Baird-Parker Medium

Composition per liter:

Agar	20.0g
Glycine	12.0g
Sodium pyruvate	10.0g
Pancreatic digest of casein	10.0g
Beef extract	5.0g
$LiCl·6H_2O$	5.0g
Yeast extract	1.0g
EY tellurite enrichment	50.0mL

pH 7.0 ± 0.2 at 25°C

EY Tellurite Enrichment:

Composition per 100.0mL:

Chicken egg yolks	10
K_2TeO_3	0.15g
NaCl (0.9% solution)	50.0mL

Preparation of EY Tellurite Enrichment: Soak eggs with 1:100 dilution of saturated mercuric chloride solution for 1 min. Crack 11 eggs and separate yolks from whites. Mix egg yolks. Measure 30.0mL of egg yolk emulsion and add to 70.0mL of 0.9% NaCl solution. Mix thoroughly. Add 0.15g K_2TeO_3. Filter sterilize. Warm to 45°–50°C.

Caution: Potassium tellurite is toxic.

Source: This medium is available as a premixed powder from BD Diagnostic Systems.

Preparation of Medium: Add components, except EY tellurite enrichment, to distilled/deionized water and bring volume to 950.0mL. Mix thoroughly. Gently heat and bring to boiling. Autoclave for 15 min at 15 psi pressure–121°C. Cool to 48°–50°C. Aseptically add 50.0mL of sterile EY tellurite enrichment. Mix thoroughly. Pour into sterile Petri dishes. The medium must be densely opqaue. Dry plates before use. Plates can be stored for up to 5 days at 20–25°C before use.

Use: For the selective isolation and enumeration of coagulase-positive staphylococci.

Balamuth Medium

Composition per 200.0mL:

Dehydrated egg yolk	36.0g
Dried liver concentrate	1.0g
Rice starch	0.2g
Potassium phosphate buffer, pH 7.5	125.0mL
NaCl solution	125.0mL

pH 7.3 ± 0.2 at 25°C

NaCl Solution

Composition per 200.0mL:

NaCl	1.6g

Preparation of NaCl Solution: Add NaCl to distilled/deionized water and bring volume to 200.0mL. Mix thoroughly.

Potassium Phosphate Buffer, 0.067*M*

Composition per 200.0mL:

K_2HPO_4 (1*M* solution)	8.6mL
KH_2PO_4 (1*M* solution)	4.66mL

Preparation of Potassium Phosphate Buffer: Combine the K_2HPO_4 and KH_2PO_4 solutions. Bring volume to 200.0mL with distilled/deionized water. Adjust pH to 7.5.

Preparation of Medium: Add dehydrated egg yolk to 36.0mL of distilled/deionized water. Add 125.0mL of 0.8% NaCl. Mix thoroughly in a blender. Heat in a covered, double boiler until infusion reaches 80°C and maintain at this temperature for 20 min. Add 20.0mL of distilled/deionized H_2O. Filter through a layer of cheesecloth. To 90–100.0mL of filtrate add 0.8% NaCl solution to bring volume to 125.0mL. Autoclave for 20 min at 15 psi pressure–121°C. Cool to 4°C. Filter. To filtrate, add an equal volume of 0.067*M* potassium phosphate buffer, pH 7.5. Add 1.0g of dried liver concentrate. Mix thoroughly. Distribute into tubes or flasks in 10.0mL volumes. Autoclave for 20 min at 15 psi pressure–121°C. Prior to inoculation, add 0.01g of rice starch to each tube.

Use: For the cultivation and maintenance of *Entamoeba histolytica.*

Basal Synthetic Medium

Composition per liter:

L-Glutamic acid	20.0g
$(NH_4)_2SO_4$	4.0g
K_2HPO_4	1.88g
KH_2PO_4	0.57g
$MgSO_4 \cdot 7H_2O$	0.2g
Salt solution	10.0mL

Salt Solution:

Composition per liter:

$FeCl_3 \cdot 6H_2O$	0.6g
$MnCl_2 \cdot 4H_2O$	0.6g
$ZnCl_2$	0.6g
$CuSO_4 \cdot 5H_2O$	0.6g
$CaCl_2 \cdot 2H_2O$	0.6g
NaCl	0.6g

Preparation of Salt Solution: Add components to 1.0L of distilled/deionized water. Mix thoroughly.

Preparation of Medium: Add components to distilled/deionized water and bring volume to 1.0L. Mix thoroughly. Gently heat and bring to boiling. Distribute into tubes or flasks. Autoclave for 15 min at 15 psi pressure–121°C.

Use: For the cultivation and maintenance of *Acinetobacter lwoffii.*

Basic Cultivation Medium

Composition per liter:

Yeast extract	10.0g
Glucose	5.0g
$(NH_4)_2PO_4$	1.5g
K_2HPO_4	1.0g
$MgSO_4 \cdot 7H_2O$	0.2g
$Fe_2(SO_4)_3 \cdot 5H_2O$	0.01g
$ZnSO_4 \cdot 7H_2O$	0.002g

pH 7.0 ± 0.2 at 25°C

Preparation of Medium: Add components to distilled/deionized water and bring volume to 1.0L. Mix thoroughly. Distribute into tubes or flasks. Autoclave for 15 min at 15 psi pressure–121°C.

Use: For the cultivation of a wide variety of microorganisms.

Basic Mineral Medium

Composition per liter:

NH_4NO_3	2.5g
$Na_2HPO_4 \cdot 2H_2O$	1.0g
$MgSO_4 \cdot 7H_2O$	0.5g
$Fe(SO_4)_3 \cdot 5H_2O$	0.01g
$Co(NO_3)_2 \cdot 6H_2O$	0.005g
$CaCl_2 \cdot 2H_2O$	1.0mg
KH_2PO_4	0.5mg
$MnSO_4 \cdot 2H_2O$	0.1mg
$(NH_4)_6Mo_7O_{24} \cdot 4H_2O$	0.1mg

Preparation of Medium: Add components to distilled/deionized water and bring volume to 1.0L. Mix thoroughly. Distribute into tubes or flasks. Autoclave for 15 min at 15 psi pressure–121°C.

Use: To supply the mineral nutrients necessary for the cultivation of a wide variety of microorganisms. Various carbon sources can be added as sterilized solutions for testing carbon utilization capabilities.

BC Motility Medium (*Bacillus cereus* Motility Medium)

Composition per liter:

Pancreatic digest of casein	10.0g
Glucose	5.0g
Agar	3.0g
Na_2HPO_4	2.5g
Yeast extract	2.5g

pH 7.4 ± 0.2 at 25°C

Preparation of Medium: Add components to distilled/deionized water and bring volume to 1.0L. Mix thoroughly. Gently heat and bring to boiling. Distribute into tubes in 2.0mL volumes. Autoclave for 15 min at 15 psi pressure–121°C.

Use: For the cultivation and observation of motility of *Bacillus cereus.*

BCM *Bacillus cereus* Group Plating Medium

Composition per liter:

Proprietary

Source: This medium is available from Biosynth International, Inc.

Use: For detection of *Bacillus cereus.* The medium contains 5-bromo-4-chloro-3-indoxyl-myoinositol-1-phosphate, which changes from colorless to turquoise upon enzymatic cleavage. *B. cereus*, *B. mycoides*, *B. thuringiensis*, and *B. weihenstephanensis* secrete phosphatidylinositol phospholipase C and so grow as turquoise colonies with species-specific morphologies.

BCM O157:H7(+) Plating Medium

Composition per liter:

Proprietary

Source: This medium is available from Biosynth International, Inc.

Use: For detection of this highly pathogenic EHEC serovar BCM *O157:H7(+)*.

BCP D Agar
(Bromcresol Purple Deoxycholate Agar)

Composition per liter:

Agar	25.0g
Lactose	10.0g
Sucrose	10.0g
Pancreatic digest of casein	7.5g
Thiopeptone	7.5g
NaCl	5.0g
Yeast extract	2.0g
Sodium citrate	2.0g
Sodium deoxycholate	1.0g
Bromcresol Purple	0.02g

pH 7.2 ± 0.2 at 25°C

Preparation of Medium: Add components to distilled/deionized water and bring volume to 1.0L. Mix thoroughly. Gently heat and bring to boiling. Pour into sterile Petri dishes without sterilization. Do not autoclave. Use the same day.

Use: For the isolation, cultivation, and differentiation of Gram-negative enteric bacilli. For the isolation, cultivation, and identification of microorganisms from fecal specimens. For the isolation and cultivation of *Salmonella, Shigella,* and other nonlactose- and nonsucrose-fermenting microorganisms. Nonlactose/nonsucrose fermenting microorganisms appear as colorless or blue colonies. Lactose/sucrose-fermenting microorganisms, such as coliform bacteria, appear as yellow-opaque white colonies surrounded by a zone of precipitated deoxycholate.

BCP DCLS Agar
(Bromcresol Purple Deoxycholate Citrate Lactose Sucrose Agar)

Composition per liter:

Agar	14.0g
Sodium citrate	10.0g
Lactose	7.5g
Sucrose	7.5g
Pancreatic digest of casein	7.5g
Peptone	7.5g
NaCl	5.0g
$Na_2S_2O_3 \cdot 5H_2O$	5.0g
Yeast extract	3.0g
Meat extract	3.0g
Sodium deoxycholate	2.5g
Bromcresol Purple	0.02g

pH 7.2 ± 0.2 at 25°C

Preparation of Medium: Add components to distilled/deionized water and bring volume to 1.0L. Mix thoroughly. Gently heat and bring to boiling. Pour into sterile Petri dishes without sterilization. Do not autoclave. Use the same day.

Use: For the differential isolation of Gram-negative enteric bacilli. For the isolation and identification of microorganisms from fecal specimens. For the isolation of *Salmonella*, *Shigella,* and other nonlactose- and nonsucrose-fermenting microorganisms. Nonlactose/nonsucrose-fermenting microorganisms appear as colorless or blue colonies. Lactose/sucrose-fermenting microorganisms, such as coliform bacteria, appear as yellow-opaque white colonies surrounded by a zone of precipitated deoxycholate.

BCYE Agar
(BCYE Alpha Base)
(Buffered Charcoal Yeast Extract Agar)

Composition per liter:

Agar	15.0g
Yeast extract	10.0g
ACES buffer (2-[(2-Amino-2-oxoethyl)-amino]-ethane sulfonic acid)	10.0g
Charcoal, activated	2.0g
α-Ketoglutarate	1.0g
L-Cysteine·HCl·H_2O	0.4g
$Fe_4(P_2O_7)_3 \cdot 9H_2O$	0.25g

pH 6.9 ± 0.2 at 25°C

Source: This medium is available as a premixed powder from BD Diagnostic Systems.

Preparation of Medium: Add components, except cysteine, to distilled/deionized water and bring volume to 1.0L. Mix thoroughly. Adjust medium to pH 6.9 with 1*N* KOH. Heat gently and bring to boil for 1 min. Autoclave for 15 min at 15 psi pressure–121°C. Cool to 50°–55°C. Add 4.0mL of a 10% solution of L-cysteine·HCl·H_2O that has been filter sterilized. Mix thoroughly. Pour into sterile Petri dishes with constant agitation to keep charcoal in suspension.

Use: For the isolation, cultivation, and maintenance of *Legionella pneumophila* and other *Legionella* species.

BCYE Differential Agar (Buffered Charcoal Yeast Extract Differential Agar)

Composition per liter:

Agar	15.0g
Yeast extract	10.0g
ACES buffer (2-[(2-Amino-2-oxoethyl)-amino]-ethane sulfonic acid)	10.0g
Charcoal, activated	2.0g
α-Ketoglutarate	1.0g
L-Cysteine·HCl·H_2O	0.4g
$Fe_4(P_2O_7)_3 \cdot 9H_2O$	0.25g
Bromcresol Purple	0.01g
Bromthymol Blue	0.01g

pH 6.9 ± 0.2 at 25°C

Source: This medium is available as a premixed powder from BD Diagnostic Systems.

Preparation of Medium: Add components, except L-cysteine·HCl·H_2O, to distilled/deionized water and bring volume to 1.0L. Mix thoroughly. Adjust medium to pH 6.9 with 1*N* KOH. Heat gently and bring to boil for 1 min. Autoclave for 15 min at 15 psi pressure–121°C. Cool to 50°–55°C. Add 4.0mL of a 10% solution of L-cysteine·HCl·H_2O that has been filter sterilized. Mix thoroughly. Pour into sterile Petri dishes with constant agitation to keep charcoal in suspension.

Use: For the isolation, cultivation, and maintenance of *Legionella pneumophila* and other *Legionella* species. For the presumptive differential identification of *Legionella* species based on colony color and morphology. *Legionella pneumophila* appears as light blue/green colonies. *Legionella micdadei* appears as blue/gray or dark blue colonies.

BCYE Medium, Diphasic Blood Culture (Buffered Charcoal Yeast Extract Medium, Diphasic Blood Culture)

Composition per liter:

Agar phase	1.0L
Broth phase	1.0L

pH 6.9 ± 0.2 at 25°C

Agar Phase:

Composition per liter:

Agar	20.0g
ACES buffer (2-[(2-Amino-2-oxoethyl)-amino]-ethane sulfonic acid)	10.0g
Yeast extract	10.0g
Charcoal, activated, acid washed	4.0g
KOH	2.8g
α-Ketoglutarate	1.0g
L-Cysteine·HCl·H_2O solution	10.0mL
$Fe_4(P_2O_7)_3 \cdot 9H_2O$ solution	10.0mL

L-Cysteine·HCl·H_2O Solution:

Composition per 10.0mL:

L-Cysteine·HCl·H_2O	0.4g

Preparation of L-Cysteine·HCl·H_2O Solution: Add L-cysteine·HCl·H_2O to distilled/deionized water and bring volume to 10.0mL. Mix thoroughly. Filter sterilize.

$Fe_4(P_2O_7)_3 \cdot 9H_2O$ Solution:

Composition per 10.0mL:

$Fe_4(P_2O_7)_3 \cdot 9H_2O$	0.25g

Preparation of $Fe_4(P_2O_7)_3 \cdot 9H_2O$ Solution: Add $Fe_4(P_2O_7)_3 \cdot 9H_2O$ to distilled/deionized water and bring volume to 10.0mL. Mix thoroughly. Filter sterilize.

Preparation of Agar Phase: Add components, except L-cysteine·HCl·H_2O solution and $Fe_4(P_2O_7)_3$ solution, to distilled/deionized water and bring volume to 980.0mL. Mix thoroughly. Adjust medium to pH 6.9 with 1*N* KOH. Heat gently and bring to boiling for 1 min. Autoclave for 15 min at 15 psi pressure–121°C. Cool to 50°–55°C. Aseptically add the L-cysteine·HCl·H_2O solution and $Fe_4(P_2O_7)_3 \cdot 9H_2O$ solution. Mix thoroughly.

Broth Phase:

Composition per liter:

ACES buffer (2-[(2-Amino-2-oxoethyl)-amino]-ethane sulfonic acid)	10.0g
Yeast extract	10.0g
Charcoal, activated, acid washed	4.0g
KOH	2.4g
α-Ketoglutarate	1.0g
Sodium polyaneolsulfonate	0.3g
L-Cysteine·HCl·H_2O solution	10.0mL
$Fe_4(P_2O_7)_3 \cdot 9H_2O$ solution	10.0mL

L-Cysteine·HCl·H_2O Solution:

Composition per 10.0mL:

L-Cysteine·HCl·H_2O	0.4g

Preparation of L-Cysteine·HCl·H_2O Solution: Add L-cysteine·HCl·H_2O to distilled/deionized water and bring volume to 10.0mL. Mix thoroughly. Filter sterilize.

$Fe_4(P_2O_7)_3 \cdot 9H_2O$ Solution:

Composition per 10.0mL:

$Fe_4(P_2O_7)_3 \cdot 9H_2O$	0.25g

Preparation of $Fe_4(P_2O_7)_3 \cdot 9H_2O$ Solution: Add $Fe_4(P_2O_7)_3 \cdot 9H_2O$ to distilled/deionized water and bring volume to 10.0mL. Mix thoroughly. Filter sterilize.

Preparation of Broth Phase: Add components, except L-cysteine·HCl·H_2O solution and $Fe_4(P_2O_7)_3$ solution, to distilled/deionized water and bring volume to 980.0mL. Mix thoroughly. Adjust medium to pH 6.9 with 1*N* KOH. Heat gently and bring to boiling for 1 min. Autoclave for 15 min at 15 psi pressure–121°C. Cool to 50–55°C. Aseptically add the cysteine·HCl·H_2O solution and $Fe_4(P_2O_7)_3$·$9H_2O$ solution. Mix thoroughly.

Preparation of Medium: Aseptically distribute cooled sterile agar phase into sterile blood culture bottles in 100.0mL volumes. Allow bottles to cool in a slanted position. Aseptically add 50.0mL of sterile broth phase to each blood culture bottle.

Use: For the isolation and cultivation of *Legionella pneumophila* and other *Legionella* species from blood samples.

BCYE Selective Agar with CCVC (Buffered Charcoal Yeast Extract Selective Agar with Cephalothin, Colistin, Vancomycin, and Cycloheximide)

Composition per 1014.0mL:

Agar	15.0g
Yeast extract	10.0g
ACES buffer (2-[(2-Amino-2-oxoethyl)-amino]-ethane sulfonic acid)	10.0g
Charcoal, activated	2.0g
α-Ketoglutarate	1.0g
$Fe_4(P_2O_7)_3$·$9H_2O$	0.25g
Antibiotic solution	10.0mL
L-Cysteine·HCl·H_2O solution	4.0mL

pH 6.9 ± 0.2 at 25°C

Source: This medium is available as a premixed powder from BD Diagnostic Systems.

L-Cysteine·HCl·H_2O Solution:

Composition per 10.0mL:

L-Cysteine·HCl·H_2O	1.0g

Preparation of L-Cysteine·HCl·H_2O Solution: Add L-cysteine·HCl·H_2O to distilled/deionized water and bring volume to 10.0mL. Mix thoroughly. Filter sterilize.

Antibiotic Solution:

Composition per 10.0mL:

Cycloheximide	80.0mg
Colistin	16.0mg
Cephalothin	4.0mg
Vancomycin	0.5mg

Preparation of Antibiotic Solution: Add components to distilled/deionized water and bring volume to 10.0mL. Mix thoroughly. Filter sterilize.

Preparation of Medium: Add components, except L-cysteine and antibiotic solutions, to distilled/deionized water and bring volume to 1.0L. Mix thoroughly. Adjust medium to pH 6.9 with 1*N* KOH. Heat gently and bring to boil for 1 min. Autoclave for 15 min at 15 psi pressure–121°C. Cool to 50°–55°C. Add 4.0mL of L-cysteine·HCl·H_2O solution and 10.0mL of sterile antibiotic solution. Mix thoroughly. Pour into sterile Petri dishes with constant agitation to keep charcoal in suspension.

Use: For the isolation, cultivation, and maintenance of *Legionella pneumophila* and other *Legionella* species from a variety of specimens.

BCYE Selective Agar with GPVA (Buffered Charcoal Yeast Extract Selective Agar with Glycine, Polymyxin B, Vancomycin, and Anisomycin)

Composition per 1014.0mL:

Agar	15.0g
Yeast extract	10.0g
ACES buffer (2-[(2-Amino-2-oxoethyl)-amino]-ethane sulfonic acid)	10.0g
Charcoal, activated	2.0g
α-Ketoglutarate	1.0g
$Fe_4(P_2O_7)_3$·$9H_2O$	0.25g
Antibiotic solution	10.0mL
L-Cysteine·HCl·H_2O solution	4.0mL

pH 6.9 ± 0.2 at 25°C

L-Cysteine·HCl·H_2O Solution:

Composition per 10.0mL:

L-Cysteine·HCl·H_2O	1.0g

Preparation of L-Cysteine·HCl·H_2O Solution: Add L-cysteine·HCl·H_2O to distilled/deionized water and bring volume to 10.0mL. Mix thoroughly. Filter sterilize.

Antibiotic Solution:

Composition per 10.0mL:

Glycine	3.0g
Anisomycin	0.08g
Vancomycin	5.0mg
Polymyxin B	100,000U

Preparation of Antibiotic Solution: Add components to distilled/deionized water and bring volume to 10.0mL. Mix thoroughly. Filter sterilize.

Preparation of Medium: Add components, except L-cysteine·HCl·H_2O solution and antibiotic solution, to distilled/deionized water and bring volume to 1.0L. Mix thoroughly. Adjust medium to pH 6.9

with 1*N* KOH. Heat gently and bring to boil for 1 min. Autoclave for 15 min at 15 psi pressure–121°C. Cool to 50°–55°C. Add 4.0mL of L-cysteine·HCl·H_2O solution and 10.0mL of sterile antibiotic solution. Mix thoroughly. Pour into sterile Petri dishes with constant agitation to keep charcoal in suspension.

Use: For the isolation, cultivation, and maintenance of *Legionella pneumophila* and other *Legionella* species.

BCYE Selective Agar with GVPC (Buffered Charcoal Yeast Extract Selective Agar with Glycine, Vancomycin, Polymyxin B, and Cycloheximide)

Composition per 1014.0mL:

Agar	15.0g
Yeast extract	10.0g
ACES buffer (2-[(2-Amino-2-oxoethyl)-amino]-ethane sulfonic acid)	10.0g
Charcoal, activated	2.0g
α-Ketoglutarate	1.0g
$Fe_4(P_2O_7)_3 \cdot 9H_2O$	0.25g
Antibiotic solution	10.0mL
L-Cysteine·HCl·H_2O solution	4.0mL

pH 6.9 ± 0.2 at 25°C

Source: This medium is available as a premixed powder from Oxoid Unipath.

L-Cysteine·HCl·H_2O Solution:

Composition per 10.0mL:

L-Cysteine·HCl·H_2O	1.0g

Preparation of L-Cysteine·HCl·H_2O Solution: Add L-cysteine·HCl·H_2O to distilled/deionized water and bring volume to 10.0mL. Mix thoroughly. Filter sterilize.

Antibiotic Solution:

Composition per 10.0mL:

Glycine	3.0g
Cycloheximide	0.08g
Vancomycin	1.0mg
Polymyxin B	79,200U

Preparation of Antibiotic Solution: Add components to distilled/deionized water and bring volume to 10.0mL. Mix thoroughly. Filter sterilize.

Preparation of Medium: Add components, except L-cysteine·HCl·H_2O solution and antibiotic solution, to distilled/deionized water and bring volume to 1.0L. Mix thoroughly. Adjust medium to pH 6.9 with 1*N* KOH. Heat gently and bring to boil for 1 min. Autoclave for 15 min at 15 psi pressure–121°C. Cool to 50°–55°C. Add 4.0mL of L-cysteine·HCl·H_2O solution and 10.0mL of sterile antibiotic solution. Mix thoroughly. Pour into sterile Petri dishes with constant agitation to keep charcoal in suspension.

Use: For the isolation, cultivation, and maintenance of *Legionella pneumophila* and other *Legionella* species.

BCYE Selective Agar with PAC (Buffered Charcoal Yeast Extract Selective Agar with Polymyxin B, Anisomycin, and Cefamandole)

Composition per 1014.0mL:

Agar	15.0g
Yeast extract	10.0g
ACES buffer (2-[(2-Amino-2-oxoethyl)-amino]-ethane sulfonic acid)	10.0g
Charcoal, activated	2.0g
α-Ketoglutarate	1.0g
$Fe_4(P_2O_7)_3 \cdot 9H_2O$	0.25g
Antibiotic solution	10.0mL
L-Cysteine·HCl·H_2O solution	4.0mL

pH 6.9 ± 0.2 at 25°C

Source: This medium is available as a premixed powder from BD Diagnostic Systems.

L-Cysteine·HCl·H_2O Solution:

Composition per 10.0mL:

L-Cysteine·HCl·H_2O	1.0g

Preparation of L-Cysteine·HCl·H_2O Solution: Add L-cysteine·HCl·H_2O to distilled/deionized water and bring volume to 10.0mL. Mix thoroughly. Filter sterilize.

Antibiotic Solution:

Composition per 10.0mL:

Polymyxin B	80,000 units
Anisomycin	80.0mg
Cefamandole	2.0mg

Preparation of Antibiotic Solution: Add components to distilled/deionized water and bring volume to 10.0mL. Mix thoroughly. Filter sterilize.

Preparation of Medium: Add components, except L-cysteine solution and antibiotic solution, to distilled/deionized water and bring volume to 1.0L. Mix thoroughly. Adjust medium to pH 6.9 with 1*N* KOH. Heat gently and bring to boil for 1 min. Autoclave for 15 min at 15 psi pressure–121°C. Cool to 50°–55°C. Add 4.0mL of L-cysteine solution and 10.0mL of sterile antibiotic solution. Mix thoroughly. Pour into sterile Petri dishes with constant agitation.

Use: For the isolation, cultivation, and maintenance of *Legionella pneumophila* and other *Legionella* species.

BCYE Selective Agar with PAV (Buffered Charcoal Yeast Extract Selective Agar with Polymyxin B, Anisomicin, and Vancomycin) (Wadowsky–Yee Medium)

Composition per 1014.0mL:

Agar	15.0g
Yeast extract	10.0g
ACES buffer (2-[(2-Amino-2-oxoethyl)-amino]-ethane sulfonic acid)	10.0g
Charcoal, activated	2.0g
α-Ketoglutarate	1.0g
$Fe_4(P_2O_7)_3 \cdot 9H_2O$	0.25g
Antibiotic solution	10.0mL
L-Cysteine·HCl·H_2O solution	4.0mL

pH 6.9 ± 0.2 at 25°C

Source: This medium is available as a premixed powder from BD Diagnostic Systems.

L-Cysteine·HCl·H_2O Solution:
Composition per 10.0mL:

L-Cysteine·HCl·H_2O	1.0g

Preparation of L-Cysteine·HCl·H_2O Solution: Add L-cysteine·HCl·H_2O to distilled/deionized water and bring volume to 10.0mL. Mix thoroughly. Filter sterilize.

Antibiotic Solution:
Composition per 10.0mL:

Polymyxin B	40,000 units
Anisomycin	80.0mg
Vancomycin	0.5mg

Preparation of Antibiotic Solution: Add components to distilled/deionized water and bring volume to 10.0mL. Mix thoroughly. Filter sterilize.

Preparation of Medium: Add components, except L-cysteine and antibiotic solution, to distilled/deionized water and bring volume to 1.0L. Mix thoroughly. Adjust medium to pH 6.9 with 1*N* KOH. Heat gently and bring to boil for 1 min. Autoclave for 15 min at 15 psi pressure–121°C. Cool to 50°–55°C. Add 4.0mL of L-cysteine·HCl·H_2O solution and 10.0mL of sterile antibiotic solution. Mix thoroughly. Pour into sterile Petri dishes with constant agitation to keep charcoal in suspension.

Use: For the isolation, cultivation, and maintenance of *Legionella pneumophila* and other *Legionella* species.

BCYEα with Alb (Buffered Charcoal Yeast Extract Agar with Albumin)

Composition per liter:

Agar	15.0g
Yeast extract	10.0g
ACES buffer (2-[(2-Amino-2-oxoethyl)-amino]-ethane sulfonic acid)	10.0g
Charcoal, activated	2.0g
α-Ketoglutarate	1.0g
Bovine serum albumin solution	10.0mL
L-Cysteine·HCl·H_2O solution	10.0mL
$Fe_4(P_2O_7)_3 \cdot 9H_2O$ solution	10.0mL

pH 6.9 ± 0.2 at 25°C

Bovine Serum Albumin Solution:
Composition per 10.0mL:

Bovine serum albumin	0.1g

Preparation of Bovine Serum Albumin Solution: Add bovine serum albumin to distilled/deionized water and bring volume to 10.0mL. Mix thoroughly. Filter sterilize.

L-Cysteine·HCl·H_2O Solution:
Composition per 10.0mL:

L-Cysteine·HCl·H_2O	0.4g

Preparation of L-Cysteine·HCl·H_2O Solution: Add L-cysteine·HCl·H_2O to distilled/deionized water and bring volume to 10.0mL. Mix thoroughly. Filter sterilize.

$Fe_4(P_2O_7)_3 \cdot 9H_2O$ Solution:
Composition per 10.0mL:

$Fe_4(P_2O_7)_3 \cdot 9H_2O$	0.25g

Preparation of $Fe_4(P_2O_7)_3 \cdot 9H_2O$ Solution: Add $Fe_4(P_2O_7)_3 \cdot 9H_2O$ to distilled/deionized water and bring volume to 10.0mL. Mix thoroughly. Filter sterilize.

Preparation of Medium: Add components—except L-cysteine·HCl·H_2O solution, $Fe_4(P_2O_7)_3 \cdot 9H_2O$ solution, and bovine serum albumin solution—to distilled/deionized water and bring volume to 970.0mL. Mix thoroughly. Adjust medium to pH 6.9 with 1*N* KOH. Heat gently and bring to boil for 1 min. Autoclave for 15 min at 15 psi pressure–121°C. Cool to 50°–55°C. Aseptically add the L-cysteine·HCl·H_2O solution, $Fe_4(P_2O_7)_3 \cdot 9H_2O$ solution, and 10.0mL of sterile bovine serum albumin solution. Mix thoroughly. Pour into sterile Petri dishes with constant agitation to keep charcoal in suspension.

Use: For the isolation, cultivation, and maintenance of *Legionella pneumophila* and other *Legionella* species.

BCYEα without L-Cysteine (Buffered Charcoal Yeast Extract Agar without L-Cysteine)

Composition per liter:

Agar	15.0g
Yeast extract	10.0g

ACES buffer (2-[(2-Amino-2-oxoethyl)-amino]-ethane sulfonic acid) 10.0g
Charcoal, activated 2.0g
α-Ketoglutarate 1.0g
$Fe_4(P_2O_7)_3 \cdot 9H_2O$ solution 10.0mL
pH 6.9 ± 0.2 at 25°C

$Fe_4(P_2O_7)_3 \cdot 9H_2O$ Solution:
Composition per 10.0mL:
$Fe_4(P_2O_7)_3 \cdot 9H_2O$ 0.25g

Preparation of $Fe_4(P_2O_7)_3 \cdot 9H_2O$ Solution: Add $Fe_4(P_2O_7)_3 \cdot 9H_2O$ to distilled/deionized water and bring volume to 10.0mL. Mix thoroughly. Filter sterilize.

Preparation of Medium: Add components, except $Fe_4(P_2O_7)_3 \cdot 9H_2O$ solution, to distilled/deionized water and bring volume to 990.0mL. Mix thoroughly. Adjust medium to pH 6.9 with 1*N* KOH. Heat gently and bring to boil for 1 min. Autoclave for 15 min at 15 psi pressure–121°C. Cool to 50°–55°C. Aseptically add 10.0mL of sterile $Fe_4(P_2O_7)_3 \cdot 9H_2O$ solution. Mix thoroughly. Pour into sterile Petri dishes with constant agitation to keep charcoal in suspension.

Use: For the isolation, cultivation, and maintenance of *Legionella pneumophila* and other *Legionella* species.

Beef Extract Agar

Composition per liter:
Agar 15.0g
Peptone 5.0g
Beef extract 3.0g
pH 7.4 ± 0.2 at 25°C

Preparation of Medium: Add components to distilled/deionized water and bring volume to 1.0L. Mix thoroughly. Heat gently and bring to boiling. Distribute into tubes or flasks. Autoclave for 15 min at 15 psi pressure–121°C. Pour into Petri dishes or leave in tubes.

Use: For the cultivation and maintenance of a wide variety of microorganisms.

Beef Extract Agar (ATCC Medium 225)

Composition per liter:
Agar 25.0g
Beef extract 10.0g
Peptone 10.0g
NaCl 5.0g
pH 7.2 ± 0.2 at 25°C

Preparation of Medium: Add components to distilled/deionized water and bring volume to 1.0L. Mix thoroughly. Heat gently and bring to boiling. Distribute into tubes or flasks. Autoclave for 15 min at 15 psi pressure–121°C. Pour into Petri dishes or leave in tubes.

Use: For the cultivation and maintenance of a wide variety of microorganisms, including *Alcaligenes* species and *Pseudomonas aeruginosa*.

Beef Extract Broth

Composition per liter:
Peptone 5.0g
Beef extract 3.0g
pH 7.4 ± 0.2 at 25°C

Preparation of Medium: Add components to distilled/deionized water and bring volume to 1.0L. Mix thoroughly. Heat gently and bring to boiling. Distribute into tubes or flasks. Autoclave for 15 min at 15 psi pressure–121°C.

Use: For the cultivation and maintenance of a wide variety of microorganisms.

Beef Extract Peptone Serum Medium

Composition per liter:
Agar 25.0g
Beef extract 10.0g
Peptone 10.0g
NaCl 1.0g
Bovine serum 50.0mL
pH 8.5 ± 0.2 at 25°C

Preparation of Medium: Add components, except bovine serum, to distilled/deionized water and bring volume to 950.0mL. Mix thoroughly. Adjust pH to 8.5. Heat gently and bring to boiling. Autoclave for 15 min at 15 psi pressure–121°C. Cool to 50°–55°C. Aseptically add 50.0mL of sterile bovine serum. Pour into sterile Petri dishes or leave in tubes.

Use: For the cultivation and maintenance of *Serratia marcescens*.

Beef Infusion Agar

Composition per liter:
Ground defatted beef 453.6g
Agar 20.0g
Peptone 10.0g
NaCl 5.0g
pH 7.6 ± 0.2 at 25°C

Preparation of Medium: Add ground beef to 1.0L of distilled/deionized water. Let stand overnight at 4°C. Gently heat and bring to 80°–90°C for 60 min. Let stand for 2h. Filter through muslin. To filtrate, add peptone and salt. Mix thoroughly. Adjust pH to 7.6 with 4% NaOH. Filter through Whatman #1 filter paper. Bring volume of filtrate to 1.0L. Add agar. Gently heat and bring to boiling. Distribute into tubes or flasks. Autoclave for 15 min at 15 psi pressure–121°C. Pour into sterile Petri dishes or leave in tubes.

Use: For the cultivation of a variety of microorganisms.

Beef Infusion Broth

Composition per liter:

Ground beef, defatted	453.6g
Peptone	10.0g
NaCl	5.0g

pH 7.6 ± 0.2 at 25°C

Preparation of Medium: Add ground beef to 1.0L of distilled/deionized water. Let stand overnight at 4°C. Gently heat and bring to 80°–90°C for 60 min. Let stand for 2h. Filter through muslin. To filtrate add peptone and salt. Mix thoroughly. Adjust pH to 7.6 with 4% NaOH. Filter through Whatman #1 filter paper. Bring volume of filtrate to 1.0L. Add agar. Gently heat and bring to boiling. Distribute into tubes or flasks. Autoclave for 15 min at 15 psi pressure–121°C.

Use: For the cultivation of a variety of microorganisms.

Beef Liver Medium for Anaerobes

Composition per liter:

Beef liver, minced	500.0g
Peptone	10.0g
K_2HPO_4	1.0g

pH 8.0 ± 0.2 at 25°C

Preparation of Medium: Add beef liver to 1.0L of tap water. Soak for 12–24h at 4°C. Skim fat off top. Autoclave for 10 min at 15 psi pressure–121°C. Filter through cheesecloth. Save meat. To filtrate, add peptone and K_2HPO_4. Adjust pH to 8.0. Filter through paper. Add tap water and bring volume to 1.0L. Add a small amount of $CaCO_3$ to a flask or test tube. Add 0.5 inch of reserved liver. Cover meat with 2 inches of broth. Cap tubes and autoclave for 15 min at 15 psi pressure–121°C.

Use: For the cultivation and maintenance of a variety of *Clostridium* species.

Bennett's Agar

Composition per liter:

Agar	15.0g
Glucose	10.0g
N-Z amine, type A	2.0g
Beef extract	1.0g
Yeast extract	1.0g

pH 7.3 ± 0.2 at 25°C

Preparation of Medium: Add components to distilled/deionized water and bring volume to 1.0L. Mix thoroughly. Adjust pH to 7.3. Gently heat and bring to boiling. Distribute into tubes or flasks. Autoclave for 15 min at 15 psi pressure–121°C. Pour into sterile Petri dishes or leave in tubes.

Use: For the cultivation and maintenance of *Actinomadura* spp., *Nocardia* spp., and *Streptomyces* species.

Bennett's Agar with Maltose

Composition per liter:

Agar	15.0g
Maltose, technical	10.0g
N-Z amine, type A	2.0g
Beef extract	1.0g
Yeast extract	1.0g

pH 7.3 ± 0.2 at 25°C

Preparation of Medium: Add components to distilled/deionized water and bring volume to 1.0L. Mix thoroughly. Adjust pH to 7.3. Gently heat and bring to boiling. Distribute into tubes or flasks. Autoclave for 15 min at 15 psi pressure–121°C. Pour into sterile Petri dishes or leave in tubes.

Use: For the cultivation and maintenance of *Streptomyces* species.

Bennett's Modified Agar Medium

Composition per liter:

Meer agar (washed agar)	20.0g
Dextrin	10.0g
Pancreatic digest of casein	2.0g
Yeast extract	1.0g
Beef extract	1.0g
$CoCl_2·6H_2O$	0.01g

pH 7.0 ± 0.2 at 25°C

Preparation of Medium: Add components to distilled/deionized water and bring volume to 1.0L. Mix thoroughly. Heat gently to boiling. Distribute into tubes or flasks. Autoclave for 15 min at 15 psi pressure–121°C. Pour into sterile Petri dishes or leave in tubes.

Use: For the cultivation and maintenance of *Streptomyces* species.

BG Sulfa Agar (Brilliant Green Sulfapyridine Agar)

Composition per liter:

Agar	20.0g
Proteose peptone no. 3	10.0g
Lactose	10.0g
Sucrose	10.0g

NaCl 5.0g
Yeast extract 3.0g
Sodium sulfapyridine 1.0g
Brilliant Green 0.125g
pH 6.9 ± 0.2 at 25°C

Source: This medium is available as a premixed powder from BD Diagnostic Systems.

Preparation of Medium: Add components to distilled/deionized water and bring volume to 1.0L. Mix thoroughly. Heat gently to boiling. Distribute into tubes or flasks. Autoclave for no longer than 15 min at 15 psi pressure–121°C. Pour into sterile Petri dishes if desired.

Use: For the selective isolation of *Salmonella* species other than *Salmonella typhi. Salmonella* appear as red, pink, or white colonies surrounded by zones of bright red.

BHI with Glucose (DSMZ Medium 215b)

Composition per liter:
Pancreatic digest of gelatin 14.5g
Glucose 8.0g
Brain heart, solids from infusion 6.0g
Peptic digest of animal tissue 6.0g
NaCl 5.0g
Na_2HPO_4 .... 2.5g
pH 7.4 ± 0.2 at 25°C

Preparation of Medium: Add components to distilled/deionized water and bring volume to 1.0L. Mix thoroughly. Distribute into tubes or flasks. Autoclave for 15 min at 15 psi pressure–121°C.

Use: For the cultivation of *Corynebacterium* spp., *Streptomyces flocculus, Mycobacterium spp., Nocardia* spp., *Rhodococcus* spp., *Dermatophilus congolensis,* and *Gordonia amicalis.*

BHI Glucose Medium

Composition per liter:
Agar 12.0g
Pancreatic digest of gelatin 7.25g
Glucose 6.5g
Brain heart, solids from infusion 3.0g
Peptic digest of animal tissue 3.0g
NaCl 2.5g
Na_2HPO_4 .... 1.25g
pH 7.4 ± 0.2 at 25°C

Preparation of Medium: Add components to distilled/deionized water and bring volume to 1.0L. Mix thoroughly. Gently heat and bring to boiling. Distribute into tubes or flasks. Autoclave for 15 min at 15 psi–121°C. Pour into sterile Petri dishes or leave in tubes.

Use: For the cultivation and maintenance of *Actinomadura* spp., *Dermatophilus* spp., *Mycobacterium* species, *Nocardia asteroides,* and *Streptococcus pyogenes.*

BHI with Glycerol and Reducing Agents (DSMZ Medium 215c)

Composition per liter:
Pancreatic digest of gelatin 14.5g
Brain heart, solids from infusion 6.0g
Peptic digest of animal tissue 6.0g
NaCl 5.0g
Glucose 3.0g
Na_2HPO_4 .... 2.5g
Glycerol solution 10.0mL
L-Cysteine·HCl–Na_2S solution 10.0mL
pH 7.4 ± 0.2 at 25°C

Glycerol Solution:
Composition per 100.0mL:
Glycerol 87.0g

Preparation of Glycerol Solution: Add glycerol to distilled/deionized water and bring volume to 100.0mL. Mix thoroughly.

L-Cysteine·HCl–Na_2S Solution:
Composition per 100.0mL:
L-Cysteine·HCl 2.5g
$Na_2S·9H_2O$ 2.5g

Preparation of L-Cysteine·HCl–Na_2S Solution: Add L-cysteine·HCl to distilled/deionized water and bring volume to 80.0mL. Mix thoroughly. Adjust pH to 11 with NaOH. Add $Na_2S·9H_2O$. Mix thoroughly. Bring volume to 100.0mL with distilled/deionized water. Gently heat and bring to boiling under 100% N_2. Cool to 25°C under 100% N_2. Autoclave for 15 min at 15 psi pressure–121°C.

Preparation of Medium: Add components, except L-cysteine·HCl–Na_2S solution, to distilled/deionized water and bring volume to 990.0mL. Mix thoroughly. Gently heat and bring to boiling under 100% N_2. Cool to 25°C under 100% N_2. Autoclave for 15 min at 15 psi pressure–121°C. Aseptically and anaerobically under 100% N_2 add 10.0mL of cysteine·HCl–Na_2S solution. Mix thoroughly. Aseptically under 100% N_2 distribute to tubes. Alternately distribute 10.0mL amounts of the medium without cysteine·HCl–Na_2S solution to tubes prior to autoclaving. Autoclave for 15 min at 15 psi pressure–121°C. Aseptically and anaerobically add 1.0mL of cysteine·HCl–Na_2S solution to each tube.

Use: For the cultivation of *Clostridium* sp.

BHI Medium (DSMZ Medium 215)

Composition per liter:

Pancreatic digest of gelatin	14.5g
Brain heart, solids from infusion	6.0g
Peptic digest of animal tissue	6.0g
NaCl	5.0g
Na_2HPO_4	2.5g

pH 7.4 ± 0.2 at 25°C

Preparation of Medium: Add components to distilled/deionized water and bring volume to 1.0L. Mix thoroughly. Distribute into tubes or flasks. Autoclave for 15 min at 15 psi pressure–121°C.

Use: For the cultivation of *Yersinia* spp., *Oligella urethralis=Moraxella urethralis, Moraxella (Branhamella) catarrhalis, Campylobacter sputorum,* and *Corynebacterium* spp.

BHI/1 Medium

Composition per liter:

Pancreatic digest of gelatin	14.5g
Casein hydrolysate	10.0g
Glucose	8.0g
Brain heart, solids from infusion	6.0g
Peptic digest of animal tissue	6.0g
NaCl	5.0g
Na_2HPO_4	2.5g

pH 7.4 ± 0.2 at 25°C

Preparation of Medium: Add components to distilled/deionized water and bring volume to 1.0L. Mix thoroughly. Distribute into tubes or flasks. Autoclave for 15 min at 15 psi–121°C.

Use: For the cultivation and maintenance of *Actinomyces israelii* and *Propionibacterium propionicus.*

BHI/2 Medium

Composition per liter:

Pancreatic digest of gelatin	14.5g
Casein hydrolysate	10.0g
Glucose	8.0g
Brain heart, solids from infusion	6.0g
Peptic digest of animal tissue	6.0g
NaCl	5.0g
Yeast extract	5.0g
Na_2HPO_4	2.5g

pH 7.4 ± 0.2 at 25°C

Preparation of Medium: Add components to distilled/deionized water and bring volume to 1.0L. Mix thoroughly. Distribute into tubes or flasks. Autoclave for 15 min at 15 psi pressure–121°C.

Use: For the cultivation and maintenance of *Actinomyces* spp.

BHI/3 Medium

Composition per liter:

Pancreatic digest of gelatin	14.5g
Casein hydrolysate	10.0g
Brain heart, solids from infusion	6.0g
Peptic digest of animal tissue	6.0g
Starch	5.0g
NaCl	5.0g
Glucose	3.0g
Na_2HPO_4	2.5g

pH 7.4 ± 0.2 at 25°C

Preparation of Medium: Add components to distilled/deionized water and bring volume to 1.0L. Mix thoroughly. Distribute into tubes or flasks. Autoclave for 15 min at 15 psi–121°C.

Use: For the cultivation and maintenance of *Actinomyces* species.

Bifidobacterium Medium

Composition per liter:

Glucose	20.0g
Pancreatic digest of casein	20.0g
Yeast extract	10.0g
Peptone	10.0g
Tomato juice	333.0mL
Tween™ 80	2.0mL

pH 6.8 ± 0.2 at 25°C

Preparation of Medium: Combine 333.0mL of tomato juice with 666.0mL of distilled/deionized water. Bring to boiling. Filter through paper. Add remaining components to filtrate. Mix thoroughly. Bring volume to 1.0L with distilled/deionized water. Distribute into tubes or flasks. Autoclave for 30 min at 15 psi pressure–110°C.

Use: For the cultivation of *Bifidobacterium infantis.*

Bifidobacterium Medium

Composition per liter:

Special peptone	23.0g
Agar	15.0g
NaCl	5.0g
Glucose	5.0g
Starch, soluble	1.0g
L-Cysteine·HCl	0.3g

Preparation of Medium: Add components to distilled/deionized water and bring volume to 1.0L. Mix thoroughly. Gently heat and bring to boiling. Distribute into tubes or flasks. Autoclave for 15 min at 15 psi pressure–121°C. Pour into sterile Petri dishes or leave in tubes.

Use: For the cultivation and maintenance of numerous *Bifidobacterium* species.

BiGGY Agar (Bismuth Sulfite Glucose Glycerin Yeast Extract Agar) (Nickerson Medium)

Composition per liter:

Agar	16.0g
Glucose	10.0g
Glycine	10.0g
Bismuth ammonium citrate	5.0g
Na_2SO_3	3.0g
Yeast extract	1.0g

pH 6.8 ± 0.2 at 25°C

Source: This medium is available as a premixed powder from Oxoid Unipath and BD Diagnostic Systems.

Preparation of Medium: Add components to distilled/deionized water and bring volume to 1.0L. Mix thoroughly and heat with frequent agitation until boiling. Distribute into tubes or flasks. Do not autoclave. Cool to approximately 45°–50°C. If desired, add 2mg/L of neomycin sulfate. Swirl to disperse the insoluble material and pour into sterile Petri dishes.

Use: For the detection, isolation, and presumptive identification of *Candida* species. Addition of neomycin helps inhibit bacterial species. *Candida albicans* appears as brown to black colonies with no pigment diffusion and no sheen. *Candida tropicalis* appears as dark brown colonies with black centers, black pigment diffusion, and a sheen. *Candida krusei* appears as shiny, wrinkled, brown to black colonies with yellow pigment diffusion. *Candida pseudotropicalis* appears as flat, shiny red to brown colonies with no pigment diffusion. *Candida parakrusei* appears as flat, shiny, wrinkled, dark reddish-brown colonies with light reddish-brown peripheries and a yellow fringe. *Candida stellatoidea* appears as flat dark brown colonies with a light fringe.

Bile Esculin Agar

Composition per liter:

Oxgall	20.0g
Agar	15.0g
Pancreatic digest of gelatin	5.0g
Beef extract	3.0g
Esculin	1.0g
Ferric citrate	0.5g
Horse serum	50.0mL

pH 6.8 ± 0.2 at 25°C

Source: This medium is available as a premixed powder from Oxoid Unipath and BD Diagnostic Systems.

Preparation of Medium: Add components, except horse serum, to distilled/deionized water and bring volume to 950.0L. Mix thoroughly and heat with frequent agitation until boiling. Autoclave for 15 min at 15 psi pressure–121°C. Cool to 45°–50°C. Aseptically add 50.0mL of filter sterilized horse serum. Distribute into sterile Petri dishes or test tubes. Cool tubes in a slanted position.

Use: For differentiation between group D streptococci and nongroup D streptococci. To differentiate members of the Enterobacteriaceae, particularly *Klebsiella, Enterobacter*, and *Serratia,* from other enteric bacteria. To differentiate *Listeria monocytogenes.* Bile tolerance and esculin hydrolysis (seen as a dark brown to black complex) are presumptive for enterococci (group D streptococci).

Bile Esculin Agar

Composition per liter:

Esculin	1.0g
Bile esculin agar base	1.0L

pH 6.6 ± 0.2 at 25°C

Bile Esculin Agar Base:

Composition per liter:

Oxgall	40.0g
Agar	15.0g
Peptone	5.0g
Beef extract	3.0g
Ferric citrate	0.5g

Source: This medium is available as a premixed powder from BD Diagnostic Systems.

Preparation of Bile Esculin Agar Base: Add components to distilled/deionized water and bring volume to 1.0L. Mix thoroughly.

Preparation of Medium: Add desired amount of esculin—typically 1.0g—to bile esculin agar base. Mix thoroughly and heat with frequent agitation until boiling. Autoclave for 15 min at 15 psi pressure–121°C. Cool to 45°–50°C. Distribute into sterile Petri dishes or test tubes. Cool tubes in a slanted position.

Use: For the isolation and presumptive identification of group D streptococci.

Bile Esculin Agar

Composition per liter:

Oxgall	40.0g
Agar	15.0g
Pancreatic digest of gelatin	5.0g
Beef extract	3.0g
Esculin	1.0g
Ferric citrate	0.5g

pH 6.6 ± 0.2 at 25°C

Preparation of Medium: Add components to distilled/deionized water and bring volume to 1.0L. Mix thoroughly and heat with frequent agitation until boiling. Autoclave for 15 min at 15 psi pressure–121°C. Distribute into sterile Petri dishes or test tubes. Cool tubes in a slanted position.

Use: For differentiation between group D streptococci and nongroup D streptococci. To differentiate members of the Enterobacteriaceae, particularly *Klebsiella, Enterobacter,* and *Serratia,* from other enteric bacteria. To differentiate *Listeria monocytogenes*. Bile tolerance and esculin hydrolysis (seen as a dark brown to black complex) are presumptive for enterococci (group D streptococci).

Bile Esculin Agar with Kanamycin

Composition per liter:

Oxgall	20.0g
Agar	15.0g
Beef extract	3.0g
Esculin	1.0g
Ferric citrate	0.5g
Hemin	10.0mg
Vitamin K_1	10.0mg
Horse serum	50.0mL
Kanamycin solution	10.0mL

pH 7.1 ± 0.2 at 25°C

Source: This medium is available as a premixed powder from BD Diagnostic Systems.

Kanamycin Solution:

Composition per 10.0mL:

Kanamycin	1.0g

Preparation of Kanamycin Solution: Add kanamycin to distilled/deionized water and bring volume to 10.0mL. Mix thoroughly. Filter sterilize.

Preparation of Medium: Add components to distilled/deionized water and bring volume to 1.0L. Mix thoroughly and heat with frequent agitation until boiling. Autoclave for 15 min at 15 psi pressure–121°C. Cool to 45°–50°C. Aseptically add 50.0mL of 5% filter-sterilized horse serum and 10.0mL of sterile kanamycin solution. Distribute into test tubes or flasks. Cool tubes in a slanted position.

Use: For the selective isolation and/or presumptive identification of bacteria of the *Bacteroides fragilis* group from specimens containing mixed flora. Examine colonies with a long-wavelength UV light. Pigmented colonies of the *Bacteroides* group will fluoresce red-orange. Growth on this medium with blackening of the medium is presumptive for *Bacteroides fragilis*.

Bile Esculin Azide Agar

Composition per liter:

Pancreatic digest of casein	17.0g
Agar	15.0g
Oxgall	10.0g
NaCl	5.0g
Yeast extract	5.0g
Proteose peptone no. 3	3.0g
Esculin	1.0g
Ferric ammonium citrate	0.5g
NaN_3	0.15g

pH 7.1 ± 0.2 at 25°C

Source: This medium is available as a premixed powder from BD Diagnostic Systems.

Caution: Sodium azide is toxic. Azides also react with metals and disposal must be highly diluted.

Preparation: Add components to distilled/deionized water and bring volume to 1.0L. Mix thoroughly and heat with frequent agitation until boiling. Distribute into tubes or flasks. Autoclave for 15 min at 15 psi pressure–121°C. Cool to 45°–50°C. Pour into sterile Petri dishes or leave in tubes. Cool tubes in a slanted position.

Use: For the isolation and presumptive identification of group D streptococci.

Bile Oxalate Sorbose Broth (BOS Broth)

Composition per liter:

Na_2HPO_4	9.14g
Sodium oxalate	5.0g
Bile salts	2.0g
NaCl	1.0g
$CaCl_2 \cdot 2H_2O$	0.01g
$MgSO_4 \cdot 7H_2O$	0.01g
Asparagine solution	100.0mL
Methionine solution	100.0mL
Sorbose solution	100.0mL
Yeast extract solution	10.0mL
Sodium pyruvate solution	10.0mL
Metanil Yellow solution	10.0mL
Sodium nitrofurantoin solution	10.0mL
Irgasan® solution	1.0mL

pH 7.6 ± 0.2 at 25°C

Asparagine Solution:

Composition per 100.0mL:

Asparagine	1.0g

Preparation of Asparagine Solution: Add asparagine to distilled/deionized water and bring volume to 100.0mL. Mix thoroughly. Filter sterilize.

Methionine Solution:
Composition per 100.0mL:

Methionine	1.0g

Preparation of Methionine Solution: Add methionine to distilled/deionized water and bring volume to 100.0mL. Mix thoroughly. Filter sterilize.

Sorbose Solution:
Composition per 100.0mL:

Sorbose	10.0g

Preparation of Sorbose Solution: Add sorbose to distilled/deionized water and bring volume to 100.0mL. Mix thoroughly. Filter sterilize.

Yeast Extract Solution:
Composition per 10.0mL:

Yeast extract	0.025g

Preparation of Yeast Extract Solution: Add yeast extract to distilled/deionized water and bring volume to 10.0mL. Mix thoroughly. Filter sterilize.

Sodium Pyruvate Solution:
Composition per 10.0mL:

Sodium pyruvate	0.05g

Preparation of Sodium Pyruvate Solution: Add sodium pyruvate to distilled/deionized water and bring volume to 10.0mL. Mix thoroughly. Filter sterilize.

Metanil Yellow Solution:
Composition per 10.0mL:

Metanil Yellow	0.025g

Preparation of Metanil Yellow Solution: Add Metanil Yellow to distilled/deionized water and bring volume to 10.0mL. Mix thoroughly. Filter sterilize.

Sodium Nitrofurantoin Solution:
Composition per 10.0mL:

Sodium nitrofurantoin	0.01g

Preparation of Sodium Nitrofurantoin Solution: Add sodium nitrofurantoin to distilled/deionized water and bring volume to 10.0mL. Mix thoroughly. Filter sterilize.

Irgasan Solution:
Composition per 10.0mL:

Irgasan	0.04g
Ethanol (95% solution)	10.0mL

Preparation of Irgasan Solution: Add Irgasan to 10.0mL of ethanol. Mix thoroughly. Filter sterilize.

Preparation of Medium: Add components, except asparagine solution, methionine solution, sorbose solution, yeast extract solution, sodium pyruvate solution, Metanil Yellow solution, sodium nitrofurantoin solution, and Irgasan solution, to distilled/deionized water and bring volume to 659.0mL. Mix thoroughly. Gently heat and bring to boiling. Autoclave for 15 min at 15 psi pressure–121°C. Cool to 45°–50°C. Aseptically add 100.0mL of sterile asparagine solution, 100.0mL of sterile methionine solution, 100.0mL of sterile sorbose solution, 10.0mL of sterile yeast extract solution, 10.0mL of sterile sodium pyruvate solution, 10.0mL of sterile Metanil Yellow solution, 10.0mL of sterile sodium nitrofurantoin solution, and 1.0mL of sterile Irgasan solution. Mix thoroughly. Pour into sterile Petri dishes or distribute into sterile tubes.

Use: For the isolation and cultivation of *Yersinia enterocolitica.*

Bile Salts Brilliant Green Starch Agar (BBGS Agar)

Composition per liter:

Agar	15.0g
Soluble starch	10.0g
Proteose peptone	10.0g
Beef extract	5.0g
Bile salts	5.0g
Brilliant Green (0.05% solution)	1.0mL

pH 7.2 ± 0.2 at 25°C

Preparation of Medium: Add components to distilled/deionized water and bring volume to 1.0L. Mix thoroughly. Gently heat while stirring and bring to boiling. Distribute into tubes or flasks. Autoclave for 15 min at 15 psi pressure–121°C. Pour into sterile Petri dishes or leave in tubes.

Use: For the isolation and cultivation of *Aeromonas hydrophila.*

Bile Salts Gelatin Agar

Composition per 100.0mL:

Gelatin	3.0g
Agar	1.5g
Pancreatic digest of casein	1.0g
NaCl	1.0g
Sodium taurocholate	0.5g
Na_2CO_3	0.1g
Water	100.0mL

pH 8.5 ± 0.2 at 25°C

Preparation of Medium: Add components to distilled/deionized water and bring volume to 1.0L. Mix thoroughly. Gently heat and bring to boiling. Distribute into tubes or flasks. Autoclave for 15 min at 15 psi pressure–121°C. Pour into sterile Petri dishes or leave in tubes.

Use: For the cultivation of *Vibrio cholerae.*

BIN Medium

Composition per liter:

Beef heart, infusion from	250.0g
Calf brains, infusion from	200.0g
Agar	15.0g
Proteose peptone	10.0g
NaCl	5.0g
Na_2HPO_4	2.5g
Glucose	2.0g
Irgasan solution	4.0mL
Crystal Violet solution	1.0mL
Sodium cholate solution	1.0mL
Sodium deoxycholate solution	1.0mL
Nystatin solution	1.0mL

pH 7.4 ± 0.2 at 25°C

Sodium Cholate Solution:

Composition per 100.0mL:

Sodium cholate	5.0g

Preparation of Sodium Cholate Solution: Add sodium cholate to distilled/deionized water and bring volume to 100.0mL. Mix thoroughly. Gently heat while stirring and bring to boiling. Autoclave for 15 min at 15 psi pressure–121°C. Cool to 25°C.

Sodium Deoxycholate Solution:

Composition per 100.0mL:

Sodium cholate	5.0g

Preparation of Sodium Deoxycholate Solution: Add sodium deoxycholate to distilled/deionized water and bring volume to 100.0mL. Mix thoroughly. Gently heat while stirring and bring to boiling. Autoclave for 15 min at 15 psi pressure–121°C. Cool 25°C.

Irgasam Solution:

Composition per 50.0mL:

Irgasan DP300	10.0mg
Ethanol, 90%	50.0mL

Preparation of Irgasam Solution: Add irgasam to 90% ethanol and bring volume to 50.0mL. Mix thoroughly.

Crystal Violet Solution:

Composition per 10.0mL:

Crystal Violet	10.0mg

Preparation of Sodium Cholate Solution: Add Crystal Violet to distilled/deionized water and bring volume to 10.0mL. Mix thoroughly. Gently heat while stirring and bring to boiling. Autoclave for 15 min at 15 psi pressure–121°C. Cool to 25°C.

Nystatin Solution:

Composition per 10.0mL:

Nystatin	2.5g

Preparation of Nystatin Solution: Add novobiocin to distilled/deionized water and bring volume to 10.0mL. Mix thoroughly. Filter sterilize.

Preparation of Medium: Add components, except irgasan solution, Crystal Violet solution, sodium cholate solution, sodium deoxycholate solution, and nystatin solution, to distilled/deionized water and bring volume to 992.0mL. Mix thoroughly. Gently heat while stirring and bring to boiling. Autoclave for 15 min at 15 psi pressure–121°C. Cool to 85°C. Aseptially add 4.0mL irgasam solution. Mix thoroughly to volatilize the ethanol. Cool to 50°C. Aseptially add 1.0mL each of Crystal Violet solution, sodium cholate solution, sodium deoxycholate solution, and nystatin solution. Mix thouroughly. Pour into sterile Petri dishes.

Use: For the efficient detection of *Yersinia pestis* from clinical and other specimens. The formulation of this medium is based on brain heart infusion agar, to which the selective agents irgasan, cholate salts, crystal violet, and nystatin are introduced to enhance efficiency of recovery of *Y. pestis.*

Biosynth Chromogenic Medium for *Listeria monocytogenes* (BCM for *Listeria monocytogenes)*

Composition per liter:

Proprietary

Source: This medium is available from Biosynth International, Inc.

Use: To differentiate *Listeria monocytogenes* and *L. ivanovii* from other *Listeria* spp. Supplements render the medium selective. Differential activity for all *Listeria* species is based upon a chromogenic substrate included in the medium. This is a complete test system with a fluorogenic selective enrichment broth and a chromogenic plating medium both detecting the virulence factor Phosphatidylinositol specific Phospholipase C (PI-PLC). The medium contains a substrate for phosphotidylinositol-specific phospholipase C (PlcA) enzymes. The selective enrichment broth is fluorogenic. The plating medium for rapid detection and enumeration of pathogenic *Listeria* combines cleavage of the chromogenic PI-PLC substrate with the additional production of a white precipitate surrounding the target colonies.

Biphasic Medium for *Neisseria*

Composition per liter:

Glucose starch agar	1.0L
Glucose starch broth	1.0L

pH 7.3 ± 0.2 at 25°C

Glucose Starch Agar:
Composition per liter:

Agar	20.0g
Gelatin	20.0g
Proteose peptone no. 3	15.0g
Soluble starch	10.0g
NaCl	5.0g
Na_2HPO_4	3.0g
Glucose	2.0g

Preparation of Glucose Starch Agar: Add components to distilled/deionized water and bring volume to 1.0L. Mix thoroughly. Gently heat and bring to boiling. Autoclave for 15 min at 15 psi pressure–121°C. Cool to 50°C.

Glucose Starch Broth:
Composition per liter:

Gelatin	20.0g
Proteose peptone no. 3	15.0g
Soluble starch	10.0g
NaCl	5.0g
Glucose	2.0g
Na_2HPO_4	3.0g

Preparation of Glucose Starch Broth: Add components to distilled/deionized water and bring volume to 1.0L. Mix thoroughly. Gently heat and bring to boiling. Autoclave for 15 min at 15 psi pressure–121°C. Cool to 25°C.

Preparation of Medium: Aseptically distibute glucose starch agar into flasks in 100–125mL volumes. Allow agar to solidify. Overlay agar with 25.0mL of sterile glucose starch broth.

Use: For selective isolation and cultivation of *Neisseria* species.

Bird Seed Agar (*Guizotia abyssinica* Creatinine Agar) (Niger Seed Agar)/(Staib Agar)

Composition per liter:

Agar	15.0g
Glucose	15.0g
Creatinine	5.0g
KH_2PO_4	3.0g
Biphenyl	1.0g
Chloramphenicol	0.5g
Guizotia abyssinica seed (niger seed) extract	1000.0mL

pH 6.7 ± 0.2 at 25°C

Preparation of Medium: Prepare seed extract by grinding 50.0g of *Guizotia abyssinica* seed in 1.0L of distilled/deionized water. Boil for 30 min. Filter through cheesecloth and filter paper. Add remaining components to seed filtrate. Mix thoroughly and heat with frequent agitation until boiling. Distribute into flasks or tubes. Autoclave for 25 min at 15 psi pressure–110°C.

Use: For the selective isolation and differentiation of *Cryptococcus neoformans* from other *Cryptococcus* species and other yeasts.

Bismuth Sulfite Agar

Composition per liter:

Agar	20.0g
$Bi_2(SO_3)_3$	8.0g
Pancreatic digest of casein	5.0g
Peptic digest of animal tissue	5.0g
Beef extract	5.0g
Glucose	5.0g
Na_2HPO_4	4.0g
$FeSO_4 \cdot 7H_2O$	0.3g

pH 7.5 ± 0.2 at 25°C

Source: This medium is available as a premixed powder from Oxoid Unipath and BD Diagnostic Systems.

Preparation of Medium: Add components to distilled/deionized water and bring volume to 1.0L. Mix thoroughly and heat with frequent agitation until boiling. Boil for 1 min. Do not autoclave. Cool to 45°–50°C. Pour into sterile Petri dishes while gently shaking flask to disperse precipitate. Use plates the same day as prepared.

Use: For the selective isolation and identification of *Salmonella typhi* and other enteric bacilli. *Salmonella typhi* appears as flat, black, "rabbit-eye" colonies surrounded by a zone of black with a metallic sheen.

Bismuth Sulfite Agar Wilson and Blair

Composition per liter:

Agar	20.0g
Pancreatic digest of casein	10.0g
$Bi_2(SO_3)_3$	8.0g
Beef extract	5.0g
Glucose	5.0g
Na_2HPO_4	4.0g
$FeSO_4 \cdot 7H_2O$	0.3g
Brilliant Green	0.025g

pH 7.7 ± 0.2 at 25°C

Preparation of Medium: Add components to distilled/deionized water and bring volume to 1.0L. Mix thoroughly and heat with frequent agitation until boiling. Boil for 1 min. Do not autoclave. Cool to 45°–50°C. Pour into sterile Petri dishes while gently shaking flask to disperse precipitate. Let plates dry for about 2h with lids partially removed. Use plates within one day of preparation; medium loses selectivity after 48h.

Use: For the selective isolation and identification of *Salmonella typhi* and other enteric bacilli. *Salmonella typhi* appears as flat, black, "rabbit-eye" colonies surrounded by a zone of black with a metallic sheen.

Bismuth Sulfite Broth (m-Bismuth Sulfite Broth)

Composition per liter:

$Bi_2(SO_3)_3$	16.0g
Pancreatic digest of casein	10.0g
Peptic digest of animal tissue	10.0g
Beef extract	10.0g
Glucose	10.0g
Na_2HPO_4	8.0g
$FeSO_4 \cdot 7H_2O$	0.6g

pH 7.7 ± 0.2 at 25°C

Preparation of Medium: Add components to distilled/deionized water and bring volume to 1.0L. Mix thoroughly and heat with frequent agitation until boiling. Boil for 1 min. Do not autoclave. Cool to 45°–50°C. Mix to disperse the precipitate and aseptically distribute into sterile tubes or flasks. Use 2.0–2.2mL of medium for each membrane filter.

Use: For the selective isolation of *Salmonella typhi* and other enteric bacilli and for the detection of *Salmonella* by the membrane filter method.

BL Agar (Glucose Blood Liver Agar)

Composition per liter:

Agar	15.0g
Glucose	10.0g
Proteose peptone no. 3	10.0g
Pancreatic digest of casein	5.0g
Yeast extract	5.0g
Meat extract	3.0g
Phytone™	3.0g
Tween 80	1.0g
Soluble starch	0.5g
Liver extract	150.0mL
Horse blood	50.0mL
L-Cysteine·HCl solution	10.0mL
Solution A	10.0mL
Solution B	5.0mL

pH 7.2 ± 0.2 at 25°C

Liver Extract:

Composition per 170.0mL:

Liver powder	10.0g

Preparation of Liver Extract: Add 10.0g of liver powder to 170mL of distilled/deionized water. Gently heat to 60°C. Maintain at 50°–60°C for 1h. Gently bring to boiling. Boil for 5 min. Adjust pH to 7.2. Filter through Whatman #2 filter paper.

L-Cysteine·HCl Solution:

Composition per 10.0mL:

L-Cysteine·HCl	0.5g

Preparation of L-Cysteine·HCl Solution: Dissolve 0.5g of L-cysteine·HCl in distilled/deionized water and bring volume to 10.0mL. Mix thoroughly. Filter sterilize. Warm to 50°C.

Solution A:

Composition per 100.0mL:

K_2HPO_4	10.0g
KH_2PO_4	10.0g

Preparation of Solution A: Add components to distilled/deionized water and bring volume to 100.0mL. Mix thoroughly. Autoclave for 15 min at 15 psi pressure–121°C. Cool to 50°–55°C.

Solution B:

Composition per 100.0mL:

$MgSO_4 \cdot 7H_2O$	4.0g
NaCl	0.2g
$FeSO_4 \cdot 7H_2O$	0.2g
$MnSO_4 \cdot H_2O$	0.2g

Preparation of Solution B: Add components to distilled/deionized water and bring volume to 100.0mL. Mix thoroughly. Autoclave for 15 min at 15 psi pressure–121°C. Cool to 50°–55°C.

Preparation of Medium: Add components, except liver extract, horse blood, L-cysteine·HCl solution, solution A, and solution B, to distilled/deionized water and bring volume to 775.0mL. Mix thoroughly. Gently heat and bring to boiling. Autoclave for 15 min at 15 psi pressure–121°C. Cool to 50°–55°C. Aseptically add 150.0mL of sterile liver extract, 50.0mL of sterile horse blood, 10.0mL of sterile L-cysteine·HCl solution, 10.0 mL of sterile solution A, and 5.0mL of sterile solution B. Mix thoroughly. Pour into sterile Petri dishes or distribute into sterile tubes.

Use: For the cultivation and maintenance of *Bacteroides* species.

Blood Agar

Composition per liter:

Agar	15.0g
Pancreatic digest of casein	15.0g
Papaic digest of soybean meal	5.0g
NaCl	5.0g
Sheep blood, defibrinated	50.0mL

pH 7.6 ± 0.2 at 25°C

Preparation of Medium: Add components, except sheep blood, to distilled/deionized water and bring volume to 950.0mL. Mix thoroughly. Gently heat and bring to boiling. Autoclave for 15 min at 15 psi pressure–121°C. Cool to 45°–50°C. Aseptically add 50.0mL of sterile sheep blood. Mix thoroughly. Pour into sterile Petri dishes in 20.0mL volumes.

Use: For the cultivation of fastidious microorganisms.

Blood Agar Base

Composition per liter:

Agar	15.0g
Beef extract	10.0g
Peptone	10.0g
NaCl	5.0g
Sheep blood, defibrinated	50.0mL

pH 7.3 ± 0.2 at 25°C

Source: This medium is available as a premixed powder from Oxoid Unipath.

Preparation of Medium: Add components, except sheep blood, to distilled/deionized water and bring volume to 950.0mL. Mix thoroughly. Heat with frequent agitation and boil for 1 min to completely dissolve. Autoclave for 15 min at 15 psi pressure–121°C. Cool to 45°–50°C. Aseptically add 50.0mL of sterile, defibrinated sheep blood. Mix thoroughly and pour into sterile Petri dishes.

Use: For the isolation, cultivation, and detection of hemolytic activity of streptococci and other fastidious microorganisms.

Blood Agar Base (ATCC Medium 368)

Composition per liter:

Beef heart, infusion from	500.0g
Agar	15.0g
Tryptose	10.0g
NaCl	5.0g

pH 6.8 ± 0.2 at 25°C

Source: This medium is available as a premixed powder from BD Diagnostic Systems.

Preparation of Medium: Add components to distilled/deionized water and bring volume to 1.0L. Mix thoroughly. Heat with frequent agitation and boil for 1 min to completely dissolve. Autoclave for 15 min at 15 psi pressure–121°C. Cool the basal medium to 45°–50°C. Aseptically add sterile, defibrinated blood to a final concentration of 5%. Mix thoroughly and pour into sterile Petri dishes.

Use: For the isolation, cultivation, and detection of hemolytic activity of staphylococci, streptococci, and other fastidious microorganisms.

Blood Agar Base (Infusion Agar)

Composition per liter:

Agar	15.0g
Pancreatic digest of casein	13.0g
NaCl	5.0g
Yeast extract	5.0g
Heart muscle, solids from infusion	2.0g
Sheep blood, defibrinated	50.0mL

pH 7.3 ± 0.2 at 25°C

Source: This medium is available as a premixed powder from BD Diagnostic Systems.

Preparation of Medium: Add components, except sheep blood, to distilled/deionized water and bring volume to 950.0mL. Mix thoroughly. Heat with frequent agitation and boil for 1 min to completely dissolve. Autoclave for 15 min at 15 psi pressure–121°C. Cool to 45°–50°C. Aseptically add 50.0mL of sterile, defibrinated sheep blood. Mix thoroughly and pour into sterile Petri dishes.

Use: For the isolation, cultivation, and detection of hemolytic activity of streptococci and other fastidious microorganisms.

Blood Agar Base (Infusion Agar) (FDA Medium M21)

Composition per liter:

Heart muscle, infusion from	375.0g
Agar	15.0g
Thiotone	10.0g
NaCl	5.0g

pH 7.3 ± 0.2 at 25°C

Preparation of Medium: Add components to distilled/deionized water and bring volume to 1.0L. Mix thoroughly. Gently heat and bring to boiling. Distribute into tubes or flasks. Autoclave for 20 min at 15 psi pressure–121°C. Pour into sterile Petri dishes or leave in tubes.

Use: For the cultivation of a variety of microorganisms. For the preparation of blood agar by the addition of sterile blood.

Blood Agar Base No. 2

Composition per 1004mL:

Proteose peptone	15.0g
Agar	12.0g
NaCl	5.0g
Yeast extract	5.0g
Liver digest	2.5g

Horse blood, defibrinated	50.0mL
FBP solution	4.0mL

pH 7.4 ± 0.2 at 25°C

FBP Solution:

Composition per 30.0mL:

$FeSO_4$	0.25g
$NaHSO_3$	0.25g
Sodium pyruvate	0.25g

Preparation of FBP Solution: Add components to distilled/deionized water and bring volume to 30.0mL. Mix thoroughly. Filter sterilize.

Preparation of Medium: Add components, except horse blood and FBP solution, to distilled/deionized water and bring volume to 950.0mL. Mix thoroughly. Gently heat and bring to boiling. Autoclave for 15 min at 15 psi pressure–121°C. Cool to 48°C. Aseptically add 50.0mL of sterile horse blood. Mix thoroughly. Aseptically add 4.0mL sterile FBP solution. Mix thoroughly. Pour into sterile Petri dishes in 20.0mL volumes.

Use: For the cultivation of *Brucella* spp. and other fastidious bacteria.

Blood Agar Base with Peptone

Composition per liter:

Agar	15.0g
Beef extract	10.0g
Peptone	10.0g
NaCl	5.0g

pH 7.3 ± 0.2 at 25°C

Preparation of Medium: Add components to distilled/deionized water and bring volume to 1.0L. Mix thoroughly. Gently heat and bring to boiling. Distribute into tubes or flasks. Autoclave for 15 min at 15 psi pressure–121°C. Pour into sterile Petri dishes or leave in tubes.

Use: For use as a base to which blood can be added; for the isolation, cultivation, and detection of hemolytic activity of streptococci and other fastidious microorganisms.

Blood Agar Base, Sheep

Composition per liter:

Pancreatic digest of casein	14.0g
Agar	12.5g
NaCl	5.0g
Peptone	4.5g
Yeast extract	4.5g
Sheep blood, defibrinated	70.0mL

ph 7.3 ± 0.2 at 25°C

Source: This medium is available as a premixed powder from Oxoid Unipath.

Preparation: Add components to distilled/deionized water and bring volume to 1.0L. Mix thoroughly. Autoclave for 15 min at 15 psi pressure–121°C. Cool the basal medium to 45°–50°C. Aseptically add 70.0mL of sterile, defibrinated sheep blood. Pour into sterile Petri dishes.

Use: For giving improved hemolytic reactions with sheep blood.

Blood Agar Base with Special Peptone

Composition per liter:

Agar	15.0g
Beef extract	10.0g
Special peptone	10.0g
NaCl	5.0g
Sheep blood, defibrinated	50.0mL

pH 7.3 ± 0.2 at 25°C

Source: Special peptone (L72) is available from Oxoid Unipath.

Preparation of Medium: Add components, except sheep blood, to distilled/deionized water and bring volume to 950.0mL. Mix thoroughly. Heat with frequent agitation and boil for 1 min to completely dissolve. Autoclave for 15 min at 15 psi pressure–121°C. Cool to 45°–50°C. Aseptically add 50.0mL of sterile, defibrinated sheep blood. Mix thoroughly and pour into sterile Petri dishes.

Use: For the isolation, cultivation, and detection of hemolytic activity of streptococci and other fastidious microorganisms.

Blood Agar, Diphasic

Composition per 800.0mL:

Lean beef, desiccated	25.0g
Agar	10.0g
Neopeptone	10.0g
NaCl	2.5g
Locke solution	200.0mL
Rabbit blood, defibrinated	100.0mL

pH 7.2–7.4 at 25°C

Locke Solution:

Composition per liter:

NaCl	8.0g
Glucose	2.5g
KH_2PO_4	0.3g
KCl	0.2g
$CaCl_2 \cdot 2H_2O$	0.2g

Preparation of Locke Solution: Add components to distilled/deionized water and bring volume to 1.0L. Mix thoroughly. Filter sterilize.

Preparation of Medium: Add beef to 500.0mL of distilled/deionized water. Let stand for 60 min.

Gently heat and bring to 80°C for 5 min. Filter through Whatman #1 filter paper. To filtrate, add remaining components, except Locke solution and rabbit blood. Mix thoroughly. Adjust pH to 7.2–7.4 with NaOH. Autoclave for 20 min at 15 psi pressure–121°C. Cool to 45°–50°C. Aseptically add sterile rabbit blood. Mix thoroughly. Aseptically distribute into sterile tubes in 5.0mL volumes. Allow tubes to cool in a slanted position. Immediately prior to inoculation, overlay agar in each tube with 2.0mL of sterile Locke solution.

Use: For the cultivation of *Trypanosoma* species and *Leishmania* species.

Blood Agar, Diphasic Base Medium

Composition per 750.0mL:

Beef	25.0g
Agar	10.0g
Neopeptone	10.0g
NaCl	2.5g

pH 7.2–7.4 at 25°C

Preparation of Medium: Trim beef to remove fat. Add 25.0g of lean beef to 250.0mL of distilled/deionized water. Gently heat and bring to boiling. Boil for 2–3 min. Filter through Whatman #2 filter paper. Add agar, neopeptone, and NaCl to filtrate. Bring volume to 750.0mL with distilled/deionized water. Mix thoroughly. Adjust pH to 7.2–7.4. Gently heat and bring to boiling. Autoclave for 15 min at 15 psi pressure–121°C. Pour into sterile Petri dishes or distribute into sterile tubes.

Use: For the cultivation of *Trypanosoma* species.

Blood Agar with Low pH

Composition per liter:

Beef heart, solids from infusion	500.0g
Agar	15.0g
Tryptose	10.0g
NaCl	5.0g
Sheep blood, defibrinated	50.0mL

pH 6. 8 ± 0.2 at 25°C

Source: This medium is available as a premixed powder from BD Diagnostic Systems.

Preparation of Medium: Add components, except sheep blood, to distilled/deionized water and bring volume to 950.0mL. Mix thoroughly. Heat with frequent agitation and boil for 1 min to completely dissolve. Autoclave for 15 min at 15 psi pressure–121°C. Cool to 45°–50°C. Aseptically add 50.0mL of sterile, defibrinated sheep blood. Mix thoroughly and pour into sterile Petri dishes.

Use: For the isolation and growth of a wide variety of microorganisms. For the detection of the hemolytic reactions of streptococci and other fastidious microorganisms. The slightly acid pH of this medium enhances distinct hemolytic reactions.

Blood Agar No. 2

Composition per liter:

Proteose peptone	15.0g
Agar	12.0g
NaCl	5.0g
Yeast extract	5.0g
Liver digest	2.5g

pH 7.4 ± 0.2 at 25°C

Source: This medium is available as a premixed powder from BD Diagnostic Systems and Oxoid Unipath.

Preparation of Medium: Add components to distilled/deionized water and bring volume to 1.0L. Mix thoroughly. Heat with frequent agitation and boil for 1 min to completely dissolve. Autoclave for 15 min at 15 psi pressure–121°C. Cool the basal medium to 45°–50°C. Aseptically add sterile, defibrinated blood to a final concentration of 7%. Pour into sterile Petri dishes.

Use: For the isolation, cultivation, and detection of hemolytic activity of streptococci, pneumococci, and other particularly fastidious microorganisms.

Blood Base Agar (LMG Medium 45)

Composition per liter:

Agar	15.0g
Lab-Lemco beef extract	10.0g
Special peptones	10.0g
NaCl	5.0g

pH 7.1 ± 0.2 at 25°C

Source: Special peptones is available as a premixed powder from Oxoid Unipath.

Preparation of Medium: Add components to distilled/deionized water and bring volume to 1.0L. Mix thoroughly. Gently heat and bring to boiling. Distribute into tubes or flasks. Autoclave for 15 min at 15 psi pressure–121°C. Pour into sterile Petri dishes or leave in tubes.

Use: For the cultivation and maintenance of heterotrophic bacteria.

Blood Base Agar with Charcoal (LMG Medium 46)

Composition per liter:

Agar	15.0g

Lab-Lemco beef extract	10.0g
Special peptones	10.0g
NaCl	5.0g
Charcoal	2.0g

pH 7.1 ± 0.2 at 25°C

Source: Special peptones is available as a premixed powder from Oxoid Unipath.

Preparation of Medium: Add components to distilled/deionized water and bring volume to 1.0L. Mix thoroughly. Gently heat and bring to boiling. Distribute into tubes or flasks. Autoclave for 15 min at 15 psi pressure–121°C. Pour into sterile Petri dishes or leave in tubes.

Use: For the cultivation and maintenance of various bacteria.

Blood Base Agar with Horse Blood (LMG Medium 47)

Composition per liter:

Agar	15.0g
Lab-Lemco beef extract	10.0g
Special peptones	10.0g
NaCl	5.0g
Horse blood, sterile defibrinated	50.0mL

pH 7.1 ± 0.2 at 25°C

Source: Special peptones is available as a premixed powder from Oxoid Unipath.

Preparation of Medium: Add components, except horse blood, to 950.0mL distilled/deionized water and bring volume to 1.0L. Mix thoroughly. Gently heat and bring to boiling. Autoclave for 15 min at 15 psi pressure–121°C. Cool to 45°–50°C. Aseptically add 50.0mL sterile horse blood. Mix thoroughly. Pour into sterile Petri dishes or distribute into sterile tubes.

Use: For the cultivation and maintenance of fastidious bacteria.

Blood Base Agar with Horse Blood, Fumarate and Formate (LMG Medium 48)

Composition per liter:

Agar	15.0g
Lab-Lemco beef extract	10.0g
Special peptones, Oxoid	10.0g
NaCl	5.0g
Sodium fumarate	3.0g
Sodium formate	2.0g
Horse blood, sterile defibrinated	50.0mL

pH 7.2 ± 0.2 at 25°C

Preparation of Medium: Add components, except horse blood, to 950.0mL distilled/deionized water and bring volume to 1.0L. Mix thoroughly. Gently heat and bring to boiling. Autoclave for 15 min at 15 psi pressure–121°C. Cool to 45°–50°C. Aseptically add 50.0mL sterile horse blood. Mix thoroughly. Pour into sterile Petri dishes or distribute into sterile tubes.

Use: For the cultivation and maintenance of fastidious bacteria.

Blood Glucose Cystine Agar

Composition per 100.0mL:

Nutrient agar	85.0mL
Glucose cystine solution	10.0mL
Human blood, fresh	5.0mL

pH 6.8 ± 0.2 at 25°C

Nutrient Agar:

Composition per liter:

Agar	15.0g
Pancreatic digest of gelatin	5.0g
Beef extract	3.0g

Source: Nutrient agar is available as a premixed powder from BD Diagnostic Systems.

Preparation of Nutrient Agar: Add components to distilled/deionized water and bring volume to 1.0L. Mix thoroughly. Gently heat while stirring and bring to boiling. Distribute into tubes or flasks. Autoclave for 15 min at 15 psi pressure–121°C. Cool to 45°–50°C.

Glucose Cystine Solution:

Composition per 50.0mL:

Glucose	12.5g
L-Cystine·HCl	0.5g

Preparation of Glucose Cystine Solution: Add components to distilled/deionized water and bring volume to 50.0mL. Mix thoroughly. Filter sterilize.

Preparation of Medium: To 85.0mL of cooled, sterile agar solution, aseptically add 10.0mL of sterile glucose cystine solution and 5.0mL of human blood. Mix thoroughly. Pour into sterile Petri dishes or distribute into sterile tubes.

Use: For the cultivation of *Francisella tularensis*.

BMPA-α Medium (Edelstein BMPA-α Medium)

Composition per liter:

Agar	13.0g
Yeast extract	10.0g
ACES buffer (2-[(2-Amino-2-oxoethyl)-amino]-ethane sulfonic acid)	2.0g
Charcoal, activated	2.0g
α-Ketoglutarate	0.2g

$Fe_4(P_2O_7)_3 \cdot 9H_2O$ 0.05g
Antibiotic inhibitor 10.0mL
L-Cysteine·HCl·H_2O solution 10.0mL
pH 6.9 ± 0.2 at 25°C

Source: This medium is available as premixed vials from Oxoid Unipath.

Antibiotic Inhibitor:
Composition per 10.0mL:
Anisomycin 0.08g
Cefamandole 4.0mg
Polymyxin B 80,000U

Preparation of Antibiotic Inhibitor: Add components to distilled/deionized water and bring volume to 10.0mL. Mix thoroughly. Filter sterilize.

L-Cysteine·HCl·H_2O Solution:
Composition per 10.0mL:
L-Cysteine·HCl·H_2O 0.08g

Preparation of L-Cysteine·HCl·H_2O Solution: Add L-cysteine·HCl·H_2O to distilled/deionized water and bring volume to 10.0mL. Mix thoroughly. Filter sterilize.

Preparation of Medium: Add components, except cysteine and antibiotic inhibitor, to distilled/deionized water and bring volume to 980.0mL. Mix thoroughly. Adjust medium to pH 6.9 with 1*N* KOH. Heat gently and bring to boiling for 1 min. Autoclave for 15 min at 15 psi pressure–121°C. Cool to 50°–55°C. Add 10.0mL of the sterile L-cysteine·HCl·H_2O solution and 10.0mL of the sterile antibiotic solution. Mix thoroughly. Pour into sterile Petri dishes with constant agitation to keep charcoal in suspension.

Use: For the selective isolation and cultivation of *Legionella pneumophila* and other *Legionella* species.

BMPA-α Medium (Semiselective Medium for *Legionella pneumophila*)

Composition per liter:
Agar 15.0g
Yeast extract 10.0g
ACES buffer (2-[(2-Amino-2-oxoethyl)-amino]-ethane sulfonic acid) 10.0g
Charcoal, activated 2.0g
α-Ketoglutarate 1.0g
$Fe_4(P_2O_7)_3 \cdot 9H_2O$ 0.25g
Antibiotic inhibitor 10.0mL
L-Cysteine·HCl·H_2O solution 10.0mL
pH 6.9 ± 0.2 at 25°C

Antibiotic Inhibitor:
Composition per 10.0mL:
Anisomycin 0.08g
Cefamandole 4.0mg
Polymyxin B 80,000U

Preparation of Antibiotic Inhibitor: Add components to distilled/deionized water and bring volume to 10.0mL. Mix thoroughly. Filter sterilize.

L-Cysteine·HCl·H_2O Solution:
Composition per 10.0mL:
L-Cysteine·HCl·H_2O 0.4g

Preparation of L-Cysteine·HCl·H_2O Solution: Add L-cysteine·HCl·H_2O to distilled/deionized water and bring volume to 10.0mL. Mix thoroughly. Filter sterilize.

Preparation of Medium: Add components, except cysteine and antibiotic inhibitor, to distilled/deionized water and bring volume to 980.0mL. Mix thoroughly. Adjust medium to pH 6.9 with 1*N* KOH. Heat gently and bring to boiling for 1 min. Autoclave for 15 min at 15 psi pressure–121°C. Cool to 50°–55°C. Add 10.0mL of the sterile L-cysteine·HCl·H_2O solution and 10.0mL of the sterile antibiotic solution. Mix thoroughly. Pour into sterile Petri dishes with constant agitation to keep charcoal in suspension.

Use: For the selective isolation and cultivation of *Legionella pneumophila* and other *Legionella* species.

Bonner-Addicott Medium

Composition per liter:
Agar 25.0g
Glucose 20.0g
$Ca(NO_3)_2 \cdot 4H_2O$ 0.236g
KNO_3 .... 0.081g
KCl 0.065g
$MgSO_4 \cdot 7H_2O$ 0.036g
KH_2PO_4 .... 0.012g
Ferric tartrate 1.0mg

Preparation of Medium: Add components to distilled/deionized water and bring volume to 1.0L. Mix thoroughly. Gently heat and bring to boiling. Distribute into tubes or flasks. Autoclave for 15 min at 15 psi pressure–121°C. Pour into sterile Petri dishes or leave in tubes.

Use: For the cultivation of a variety of fungi.

Bordetella pertussis Selective Medium with Bordet-Gengou Agar Base

Composition per 1210.0mL:
Bordet-Gengou agar base 1.0L
Horse blood, defibrinated 200.0mL
Cephalexin solution 10.0mL
pH 6.7± 0.2 at 25°C

Source: This medium is available as a premixed powder from Oxoid Unipath.

Bordet-Gengou Agar Base:

Composition per liter:

Agar	20.0g
NaCl	5.5g
Pancreatic digest of casein	5.0g
Peptic digest of animal tissue	5.0g

Preparation of Bordet-Gengou Agar Base: Add components to 1.0L of 1% glycerol solution. Autoclave for 15 min at 15 psi pressure–121°C. Cool to 50°C.

Cephalexin Solution:

Composition per 10.0mL:

Cephalexin	0.04g

Preparation of Cephalexin Solution: Add cephalexin to distilled/deionized water and bring volume to 10.0mL. Mix thoroughly. Filter sterilize.

Preparation of Medium: Aseptically add 10.0mL of sterile cephalexin solution and 200.0mL of defibrinated horse blood to 1.0L Bordet-Gengou agar base. Mix thoroughly and pour into sterile Petri dishes.

Use: For the selective isolation and presumptive identification of *Bordetella pertussis* and *Bordetella parapertussis. Bordetella pertussis* appears as small, nearly transparent, "bisected pearl-like" colonies.

Bordetella pertussis Selective Medium with Charcoal Agar Base

Composition per 1110.0mL:

Charcoal agar base	1.0L
Horse blood, defibrinated	100.0mL
Cephalexin solution	10.0mL

pH 6.7± 0.2 at 25°C

Source: This medium is available as a premixed powder from Oxoid Unipath.

Charcoal Agar Base:

Composition per liter:

Agar	12.0g
Beef extract	10.0g
Starch	10.0g
NaCl	5.0g
Pancreatic digest of casein	5.0g
Peptic digest of animal tissue	5.0g
Charcoal	4.0g
Nicotinic acid	1.0mg

Preparation of Charcoal Agar Base: Add components of charcoal agar base to distilled/deionized water and bring volume to 1.0L. Autoclave for 15 min at 15 psi pressure–121°C. Cool to 50°C.

Cephalexin Solution:

Composition per 10.0mL:

Cephalexin	0.04g

Preparation of Cephalexin Solution: Add cephalexin to distilled/deionized water and bring volume to 10.0mL. Mix thoroughly. Filter sterilize.

Preparation of Medium: Aseptically add 10.0mL of sterile cephalexin solution and 100.0mL of defibrinated horse blood to charcoal agar base. Mix thoroughly and pour into sterile Petri dishes.

Use: For the selective isolation and presumptive identification of *Bordetella pertussis* and *Bordetella parapertussis. Bordetella pertussis* appears as small, pale, shiny colonies.

Bordet-Gengou Agar

Composition per liter:

Agar	20.0g
Glycerol	10.0g
NaCl	5.5g
Pancreatic digest of casein	5.0g
Peptic digest of animal tissue	5.0g
Potato, solids from infusion	4.5g
Rabbit blood	200.0mL

pH 6.7± 0.2 at 25°C

Source: This medium is available as a premixed powder from Oxoid Unipath and BD Diagnostic Systems.

Preparation of Medium: Add 10.0g of glycerol to 980.0mL of distilled/deionized water. Add other components, except rabbit blood, to the glycerol solution. Mix thoroughly. Heat with occasional agitation of the medium. Boil for 1 min. Autoclave for 15 min at 15 psi pressure–121°C. Cool medium to 50°C. Aseptically add 200.0mL of rabbit blood (prewarmed to 35°C) to a concentration of 15–30%. 150.0–200.0mL of sterile, defibrinated horse blood may be used in place of rabbit blood. Mix thoroughly and pour plates or prepare slants.

Use: For the detection and isolation of *Bordetella pertussis* and *Bordetella parapertussis* from clinical specimens. The medium is rendered selective by the addition of methicillin. *Bordetella pertussis* appears as small (<1mm), smooth, pearl-like colonies surrounded by a narrow zone of hemolysis. *Bordetella parapertussis* appears as brown, nonshiny colonies with a green-black coloration on the reverse side. *Bordetella bronchiseptica* appears as brown, nonshiny, moderately sized colonies with a roughly pitted surface.

Bordet-Gengou Medium (ATCC Medium 35)

Composition per liter:

Agar	20.0g
Glycerol	10.0g
Proteose peptone	10.0g
NaCl	5.5g
Pancreatic digest of casein	5.0g
Peptic digest of animal tissue	5.0g
Potato, solids from infusion	4.5g
Rabbit blood	150.0mL

pH 6.7± 0.2 at 25°C

Source: This medium is available as a premixed powder from Oxoid Unipath and BD Diagnostic Systems.

Preparation of Medium: Add 10.0g of glycerol to 980.0mL of distilled/deionized water. Add other components, except rabbit blood, to the glycerol solution. Mix thoroughly. Heat with occasional agitation of the medium. Boil for 1 min. Autoclave for 15 min at 15 psi pressure–121°C. Cool medium to 50°C. Aseptically add 150.0mL of rabbit blood (prewarmed to 35°C). Mix thoroughly. Pour into sterile Petri dishes or distribute into sterile tubes. Allow tubes to cool in a slanted position.

Use: For the detection and isolation of *Bordetella pertussis* and *Bordetella parapertussis* from clinical specimens. The medium is rendered selective by the addition of methicillin. *Bordetella pertussis* appears as small (<1mm), smooth, pearl-like colonies surrounded by a narrow zone of hemolysis. *Bordetella parapertussis* appears as brown, nonshiny colonies with a green-black coloration on the reverse side. *Bordetella bronchiseptica* appears as brown, nonshiny, moderately sized colonies with a roughly pitted surface.

Bordet-Gengou Medium (LMG Medium 23)

Composition per liter:

Agar	20.0g
Proteose peptone	10.0g
NaCl	5.5g
Pancreatic digest of casein	5.0g
Peptic digest of animal tissue	5.0g
Potato, solids from 125g infusion	4.5g
Rabbit blood, sterile debrinated	150.0mL
Glycerol	10.0mL

pH 6.7± 0.2 at 25°C

Preparation of Medium: Add 10.0mL of glycerol to 850.0mL of distilled/deionized water. Add other components, except rabbit blood, to the glycerol solution. Mix thoroughly. Heat with occasional agitation of the medium. Boil for 1 min. Autoclave for 15 min at 15 psi pressure–121°C. Cool medium to 45–50°C. Aseptically add 150.0mL of rabbit blood (prewarmed to 35°C). Mix thoroughly. Pour into sterile Petri dishes or distribute into sterile tubes.

Use: For the detection and isolation of *Pseudomonas pertucinogena.*

Borrelia Medium

Composition per 370.0mL:

Solution 4	240.0mL
Solution 1	80.0mL
Solution 2	34.0mL
Rabbit serum, sterile	10.0mL
Solution 3	4.0mL
Solution 5	0.7mL

Solution 1:

Composition per liter:

$Na_2HPO_4 \cdot 7H_2O$	26.52g
Glucose	12.75g
Proteose peptone no.2	5.95g
Pancreatic digest of casein	2.55g
NaCl	1.2g
Sodium pyruvate	1.06g
$NaH_2PO_4 \cdot H_2O$	1.03g
KCl	0.85g
$MgCl_2 \cdot 6H_2O$	0.68g
N-acetylglucosamine	0.53g
Sodium citrate·$2H_2O$	0.47g

Preparation of Solution 1: Add components to distilled/deionized water and bring volume to 1.0L. Mix thoroughly. Store at –20° C.

Solution 2:

Composition per 100.0mL:

Bovine albumin fraction V	10.0g

Preparation of Solution 2: Add bovine albumin to distilled/deionized water and bring volume to 100.0mL. Mix thoroughly. Adjust pH to 7.8 with NaOH. Store at –20°C.

Solution 3:

Composition per 100.0mL:

$NaHCO_3$	4.5g

Preparation of Solution 3: Add $NaHCO_3$ to distilled/deionized water and bring volume to 100.0mL. Mix thoroughly. Prepare solution freshly.

Solution 4:

Composition per 100.0mL:

Gelatin	7.0g

Preparation of Solution 4: Add gelatin to distilled/deionized water and bring volume to 100.0mL.

Mix thoroughly. Autoclave for 15 min at 10 psi pressure–115°C. Store at 4°C.

Solution 5:

Composition per 100.0mL:

Phenol Red 0.5g

Preparation of Solution 5: Add Phenol Red to distilled/deionized water and bring volume to 100.0mL. Mix thoroughly. Store at 4°C.

Preparation of Medium: Combine 80.0mL of solution 1, 34.0mL of solution 2, 4.0mL of solution 3, 0.7mL of solution 5, and 1.3mL of distilled/deionized water. Mix thoroughly. Filter sterilize under pressure. Aseptically distribute into sterile borosilicate screw-capped tubes in 6.0mL volumes. Melt solution 4 by immersing tube in warm water. Add 2.0mL of solution 4 to each screw-capped tube. Add 0.5mL of sterile rabbit serum to each screw-capped tube.

Use: For the cultivation of *Borrelia hermsii*, *Borrelia turicatae*, and *Borrelia parkeri*.

Bouillon Medium

Composition per liter:

Peptone 15.0g
Meat extract 5.0g
NaCl 5.0g
K_2HPO_4 .... 5.0g

pH 7.0 ± 0.2 at 25°C

Preparation of Medium: Add components to distilled/deionized water and bring volume to 1.0L. Mix thoroughly. Heat with frequent agitation and bring to boiling. Distribute into tubes or flasks. Autoclave for 15 min at 15 psi pressure–121°C.

Use: For the general cultivation of heterotrophic microorganisms.

Bovine Albumin Tween 80 Medium, Ellinghausen and McCullough, Modified (Albumin Fatty Acid Broth, *Leptospira* Medium)

Composition per liter:

Basal medium 900.0mL
Albumin fatty acid supplement.................. 100.0mL

Basal Medium:

Composition per liter:

Na_2HPO_4, anhydrous 1.0g
NaCl 1.0g
KH_2PO_4, anhydrous 0.3g
NH_4Cl (25% solution).. 1.0mL
Glycerol (10% solution).. 1.0mL
Sodium pyruvate (10% solution)........................ 1.0mL
Thiamine·HCl (0.5% solution) 1.0mL

pH 7.4 ± 0.2 at 25°C

Preparation of Basal Medium: Add components to distilled/deionized water and bring volume to 1.0L. Mix thoroughly. Adjust pH to 7.4. Gently heat and bring to boiling. Autoclave for 15 min at 15 psi pressure–121°C. Cool to 25°C.

Albumin Fatty Acid Supplement:

Composition per 200.0mL:

Bovine albumin fraction V 20.0g
Polysorbate (Tween) 80 (10% solution) 25.0mL
$FeSO_4 \cdot 7H_2O$ (0.5% solution)........................ 20.0mL
$CaCl_2 \cdot 2H_2O$ (1.5% solution)........................ 2.0mL
$MgCl_2 \cdot 2H_2O$ (1.5% solution) 2.0mL
Vitamin B_{12} (0.2% solution)........................ 2.0mL
$ZnSO_4 \cdot 7H_2O$ (0.4% solution)........................ 2.0mL
$CuSO_4 \cdot 5H_2O$ (0.3% solution)........................ 0.2mL

Preparation of Albumin Fatty Acid Supplement: Add bovine albumin to 100.0mL of distilled/deionized water. Mix thoroughly. Add remaining components while stirring. Adjust pH to 7.4. Bring volume to 200.0mL with distilled/deionized water. Filter sterilize. Store at –20°C.

Preparation of Medium: Aseptically combine 100.0mL of sterile albumin fatty acid supplement and 900.0mL of sterile basal medium. Mix thoroughly. Aseptically distribute into sterile tubes or flasks.

Use: For the cultivation of *Leptospira* species.

Bovine Albumin Tween 80 Semisolid Medium, Ellinghausen and McCullough, Modified (Albumin Fatty Acid Semisolid Medium, Modified)

Composition per liter:

Basal medium 900.0mL
Albumin fatty acid supplement.................. 100.0mL

Basal Medium:

Composition per liter:

Agar 2.2g
Na_2HPO_4, anhydrous 1.0g
NaCl 1.0g
KH_2PO_4, anhydrous 0.3g
NH_4Cl (25% solution) 1.0mL
Glycerol (10% solution) 1.0mL
Sodium pyruvate (10% solution)........................ 1.0mL
Thiamine·HCl (0.5% solution) 1.0mL

pH 7.4 ± 0.2 at 25°C

Preparation of Basal Medium: Add components to distilled/deionized water and bring volume to 1.0L. Mix thoroughly. Adjust pH to 7.4. Gently heat and bring to boiling. Autoclave for 15 min at 15 psi pressure–121°C. Cool to 25°C.

Albumin Fatty Acid Supplement:
Composition per 200.0mL:

Bovine albumin fraction V	20.0g
Polysorbate (Tween) 80 (10% solution)	25.0mL
$FeSO_4 \cdot 7H_2O$ (0.5% solution)	20.0mL
$CaCl_2 \cdot 2H_2O$ (1.5% solution)	2.0mL
$MgCl_2 \cdot 2H_2O$ (1.5% solution)	2.0mL
Vitamin B_{12} (0.2% solution)	2.0mL
$ZnSO_4 \cdot 7H_2O$ (0.4% solution)	2.0mL
$CuSO_4 \cdot 5H_2O$ (0.3% solution)	0.2mL

Preparation of Albumin Fatty Acid Supplement: Add bovine albumin to 100.0mL of distilled/deionized water. Mix thoroughly. Add remaining components while stirring. Adjust pH to 7.4. Bring volume to 200.0mL with distilled/deionized water. Filter sterilize. Store at –20°C.

Preparation of Medium: Aseptically combine 100.0mL of sterile albumin fatty acid supplement and 900.0mL of sterile basal medium. Mix thoroughly. Aseptically distribute into sterile tubes or flasks.

Use: For the cultivation of *Leptospira* species.

Bovine Serum Albumin Tween 80 Agar (BSA Tween 80 Agar)

Composition per liter:

Basal medium	900.0mL
Albumin supplement	100.0mL

Basal Medium:
Composition per liter:

Agar	11.0g
Na_2HPO_4	1.0g
NaCl	1.0g
KH_2PO_4	0.3g
Glycerol (10% solution)	1.0mL
NH_4Cl (25% solution)	1.0mL
Sodium pyruvate (10% solution)	1.0mL
Thiamine (0.5% solution)	1.0mL

Preparation of Basal Medium: Add components to distilled/deionized water and bring volume to 1.0L. Mix thoroughly. Adjust pH to 7.4. Autoclave for 15 min at 15 psi pressure–121°C. Cool to 25°C.

Albumin Supplement:
Composition per 100.0mL:

Bovine albumin	10.0g
Tween 80 (10% solution)	12.5mL
$FeSO_4$ (0.5% solution)	10.0mL
$MgCl_2$–$CaCl_2$ solution	1.0mL
Cyanocobalamin (0.02% solution)	1.0mL
$ZnSO_4$ (0.4% solution)	1.0mL

Preparation of Albumin Supplement: Add components to distilled/deionized water and bring volume to 100.0mL. Mix thoroughly. Adjust pH to 7.4. Filter sterilize.

$MgCl_2$–$CaCl_2$ Solution:
Composition per 100.0mL:

$CaCl_2 \cdot 2H_2O$	1.5g
$MgCl_2 \cdot 6H_2O$	1.5g

Preparation of $MgCl_2$–$CaCl_2$ Solution: Add components to distilled/deionized water and bring volume to 100.0mL. Mix thoroughly.

Preparation of Medium: To 900.0mL of cooled, sterile basal medium, aseptically add 100.0mL of sterile albumin supplement. Mix thoroughly. Aseptically distribute into sterile tubes or flasks.

Use: For the cultivation and maintenance of *Leptospira* species.

Bovine Serum Albumin Tween 80 Broth (BSA Tween 80 Broth)

Composition per liter:

Basal medium	900.0mL
Albumin supplement	100.0mL

pH 7.4 ± 0.2 at 25°C

Basal Medium:
Composition per liter:

Na_2HPO_4	1.0g
NaCl	1.0g
KH_2PO_4	0.3g
Glycerol (10% solution)	1.0mL
NH_4Cl (25% solution)	1.0mL
Sodium pyruvate (10% solution)	1.0mL
Thiamine (0.5% solution)	1.0mL

Preparation of Basal Medium: Add components to distilled/deionized water and bring volume to 1.0L. Mix thoroughly. Adjust pH to 7.4. Autoclave for 15 min at 15 psi pressure–121°C. Cool to 25°C.

Albumin Supplement:
Composition per 100.0mL:

Bovine albumin	10.0g
Tween 80 (10% solution)	12.5mL
$FeSO_4$ (0.5% solution)	10.0mL
$MgCl_2$-$CaCl_2$ solution	1.0mL
Cyanocobalamin (0.02% solution)	1.0mL
$ZnSO_4$ (0.4% solution)	1.0mL

Preparation of Albumin Supplement: Add components to distilled/deionized water and bring volume to 100.0mL. Mix thoroughly. Adjust pH to 7.4. Filter sterilize.

$MgCl_2$–$CaCl_2$ Solution:
Composition per 100.0mL:

$CaCl_2 \cdot 2H_2O$	1.5g
$MgCl_2 \cdot 6H_2O$	1.5g

Preparation of $MgCl_2$–$CaCl_2$ Solution: Add components to distilled/deionized water and bring volume to 100.0mL. Mix thoroughly.

Preparation of Medium: To 900.0mL of cooled, sterile basal medium, aseptically add 100.0mL of sterile albumin supplement. Mix thoroughly. Aseptically distribute into sterile tubes or flasks.

Use: For the isolation and cultivation of *Leptospira* species.

Bovine Serum Albumin Tween 80 Soft Agar (BSA Tween 80 Soft Agar) (Semisolid BSA Tween 80 Medium)

Composition per liter:

Basal medium	900.0mL
Albumin supplement	100.0mL

Basal Medium:
Composition per liter:

Agar	2.0g
Na_2HPO_4	1.0g
NaCl	1.0g
KH_2PO_4	0.3g
Glycerol (10% solution)	1.0mL
NH_4Cl (25% solution)	1.0mL
Sodium pyruvate (10% solution)	1.0mL
Thiamine (0.5% solution)	1.0mL

Preparation of Basal Medium: Add components to distilled/deionized water and bring volume to 1.0L. Mix thoroughly. Adjust pH to 7.4. Autoclave for 15 min at 15 psi pressure–121°C. Cool to 25°C.

Albumin Supplement:
Composition per 100.0mL:

Bovine albumin	10.0g
Tween 80 (10% solution)	12.5mL
$FeSO_4$ (0.5% solution)	10.0mL
$CaCl_2$–$MgCl_2$ solution	1.0mL
Cyanocobalamin (0.02% solution)	1.0mL
$ZnSO_4$ (0.4% solution)	1.0mL

Preparation of Albumin Supplement: Add components to distilled/deionized water and bring volume to 100.0mL. Mix thoroughly. Adjust pH to 7.4. Filter sterilize.

$MgCl_2$–$CaCl_2$ Solution:
Composition per 100.0mL:

$CaCl_2 \cdot 2H_2O$	1.5g
$MgCl_2 \cdot 6H_2O$	1.5g

Preparation of $MgCl_2$–$CaCl_2$ Solution: Add components to distilled/deionized water and bring volume to 100.0mL. Mix thoroughly.

Preparation of Medium: To 900.0mL of cooled, sterile basal medium, aseptically add 100.0mL of sterile albumin supplement. Mix thoroughly. Aseptically distribute into sterile tubes or flasks.

Use: For the cultivation of *Leptospira* species.

Brain Heart CC Agar (Brain Heart Cycloheximide Chloramphenicol Agar)

Composition per liter:

Pancreatic digest of casein	16.0g
Agar	13.5g
Brain heart, solids from infusion	8.0g
Peptic digest of animal tissue	5.0g
NaCl	5.0g
Na_2HPO_4	2.5g
Glucose	2.0g
Cycloheximide	0.5g
Chloramphenicol	0.05g

pH 7.4 ± 0.2 at 25°C

Source: This medium is available as a premixed powder from BD Diagnostic Systems.

Preparation of Medium: Add components to distilled/deionized water and bring volume to 1.0L. Mix thoroughly. Distribute into tubes or flasks while shaking to distribute precipitate. Autoclave for 15 min at 15 psi pressure–118°C.

Use: For the selective isolation of fastidious pathogenic fungi such as *Histoplasma capsulatum* and *Blastomyces dermatiditis* from specimens heavily contaminated with bacteria and other fungi. It may also be used as a base supplemented with sheep blood and gentamicin for enrichment and additional selectivity.

Brain Heart Infusion (BHI)

Composition per liter:

Pancreatic digest of gelatin	14.5g
Brain heart, solids from infusion	6.0g
Peptic digest of animal tissue	6.0g
NaCl	5.0g
Glucose	3.0g
Na_2HPO_4	2.5g

pH 7.4 ± 0.2 at 25°C

Source: This medium is available as a premixed powder from Oxoid Unipath and BD Diagnostic Systems.

Preparation of Medium: Add components to distilled/deionized water and bring volume to 1.0L. Mix thoroughly. Distribute into tubes or flasks. Autoclave for 15 min at 15 psi pressure–121°C.

Use: For the cultivation of fastidious and nonfastidious microorganisms, including aerobic and anaerobic bacteria, from a variety of clinical and nonclinical specimens. It is particularly useful for culturing streptococci, pneumococci, and meningococci. It is also used for the preparation of inocula for use in antimicrobial susceptibility tests and as a base for blood culture.

Brain Heart Infusion Agar (BHI Agar)

Composition per liter:

Beef heart infusion	250.0g
Calf brain infusion	200.0g
Agar	13.5g
Proteose peptone	10.0g
NaCl	5.0g
$Na_2HPO_4 \cdot 12H_2O$	2.5g
Glucose	2.0g

pH 7.4 ± 0.2 at 25°C

Preparation of Medium: Add components to distilled/deionized water and bring volume to 1.0L. Mix thoroughly. Gently heat and bring to boiling. Distribute into tubes or flasks. Autoclave for 15 min at 15 psi pressure–121°C. Pour into sterile Petri dishes or leave in tubes.

Use: For the cultivation of a variety of fastidious and nonfastidious aerobic and anaerobic microorganisms.

Brain Heart Infusion Agar

Composition per liter:

Pancreatic digest of casein	16.0g
Agar	13.5g
Brain heart, solids from infusion	8.0g
Peptic digest of animal tissue	5.0g
NaCl	5.0g
Na_2HPO_4	2.5g
Glucose	2.0g

pH 7.4 ± 0.2 at 25°C

Source: This medium is available as a premixed powder from Oxoid Unipath and BD Diagnostic Systems.

Preparation of Medium: Add components to distilled/deionized water and bring volume to 1.0L. Mix thoroughly. Distribute into tubes or flasks while shaking to distribute precipitate. Autoclave for 15 min at 15 psi pressure–121°C.

Use: For the cultivation of a wide variety of fastidious microorganisms, including bacteria, yeasts, and molds. With the addition of 10% sheep blood, it is used for the isolation and cultivation of many fungal species, including systemic fungi, from clinical and nonclinical specimens. The addition of gentamicin and chloramphenicol with 10% sheep blood produces a selective medium used for the isolation of pathogenic fungi from specimens heavily contaminated with bacteria and saprophtic fungi. It is recommended for the isolation of *Histoplasma capsulatum* and other pathogenic fungi, including *Coccidioides immitis*.

Brain Heart Infusion Agar

Composition per liter:

Agar	15.0g
Pancreatic digest of gelatin	14.5g
Brain heart, solids from infusion	6.0g
Peptic digest of animal tissue	6.0g
NaCl	5.0g
Glucose	3.0g
Na_2HPO_4	2.5g

pH 7.4 ± 0.2 at 25°C

Source: This medium is available as a premixed powder from BD Diagnostic Systems.

Preparation of Medium: Add components to distilled/deionized water and bring volume to 1.0L. Mix thoroughly. Distribute into tubes or flasks while shaking to distribute precipitate. Autoclave for 15 min at 15 psi pressure–121°C. Mix thoroughly. Pour into sterile Petri dishes.

Use: For the cultivation of a wide variety of fastidious microorganisms, including bacteria, yeasts and molds.

Brain Heart Infusion Agar 0.7% (BHI Agar 0.7%)

Composition per liter:

Pancreatic digest of gelatin	14.5g
Agar	7.0g
Brain heart, solids from infusion	6.0g
Peptic digest of animal tissue	6.0g
NaCl	5.0g
Glucose	3.0g
Na_2HPO_4	2.5g

pH 5.3 ± 0.2 at 25°C

Source: This medium without agar is available as a premixed powder from BD Diagnostic Systems.

Preparation of Medium: Add components, except agar, to distilled/deionized water and bring volume to 1.0L. Mix thoroughly. Adjust pH to 5.3 with

1*N* HCl. Mix thoroughly. Add agar. Gently heat and bring to boiling. Distribute into tubes. Autoclave for 10 min at 15 psi pressure–121°C.

Use: For the detection of staphylococcal enterotoxin producing strains.

Brain Heart Infusion with 0.7% Agar

Composition per liter:

Beef heart infusion	250.0g
Calf brain infusion	200.0g
Proteose peptone	10.0g
Agar	7.0g
NaCl	5.0g
$Na_2HPO_4 \cdot 12H_2O$	2.5g
Glucose	2.0g

pH 5.3 ± 0.2 at 25°C

Preparation of Medium: Add components to distilled/deionized water and bring volume to 1.0L. Mix thoroughly. Gently heat and bring to boiling. Adjust pH to 5.3 with 1*N* HCl. Distribute into tubes in 25.0mL volumes. Autoclave for 10 min at 15 psi pressure–121°C.

Use: For the cultivation of *Staphylococcal* species for the production of enterotoxin.

Brain Heart Infusion Agar with Chloramphenicol

Composition per liter:

Pancreatic digest of casein	16.0g
Agar	13.5g
Brain heart, solids from infusion	8.0g
Peptic digest of animal tissue	5.0g
NaCl	5.0g
Na_2HPO_4	2.5g
Glucose	2.0g
Sheep blood, defibrinated	50.0mL
Chloramphenicol solution	10.0mL

pH 7.4 ± 0.2 at 25°C

Chloramphenicol Solution:
Composition per 10.0mL:

Chloramphenicol	0.05g

Preparation of Chloramphenicol Solution: Add chloramphenicol to distilled/deionized water and bring volume to 10.0mL. Mix thoroughly. Filter sterilize.

Preparation of Medium: Add components, except chloramphenicol solution and sheep blood, to distilled/deionized water and bring volume to 940.0mL. Mix thoroughly. Gently heat and bring to boiling. Autoclave for 15 min at 15 psi pressure–121°C. Cool to 50°C. Aseptically add sterile chloramphenicol solution and sheep blood. Mix thoroughly. Pour into sterile Petri dishes or distribute into sterile tubes.

Use: For the isolation and cultivation of a wide variety of fungal species, especially systemic fungi, from clinical and nonclinical specimens. For the selective isolation of pathogenic fungi from specimens heavily contaminated with bacteria and saprophytic fungi. For the maintenance of fungal species on slant cultures.

Brain Heart Infusion Agar with Kanamycin

Composition per liter:

Pancreatic digest of casein	16.0g
Agar	13.5g
Brain heart, solids from infusion	8.0g
Peptic digest of animal tissue	5.0g
NaCl	5.0g
Na_2HPO_4	2.5g
Glucose	2.0g
Sheep blood, defibrinated	50.0mL
Kanamycin solution	10.0mL

pH 7.4 ± 0.2 at 25°C

Kanamycin Solution:
Composition per 10.0mL:

Kanamycin	25.0mg

Preparation of Kanamycin Solution: Add kanamycin to distilled/deionized water and bring volume to 10.0mL. Mix thoroughly. Filter sterilize.

Preparation of Medium: Add components, except kanamycin solution and sheep blood, to distilled/deionized water and bring volume to 940.0mL. Mix thoroughly. Gently heat and bring to boiling. Autoclave for 15 min at 15 psi pressure–121°C. Cool to 45°–50°C. Aseptically add sterile kanamycin solution and sheep blood. Mix thoroughly. Pour into sterile Petri dishes or distribute into sterile tubes.

Use: For the cultivation of fastidious fungi.

Brain Heart Infusion Agar with Penicillin and Streptomycin

Composition per liter:

Pancreatic digest of casein	16.0g
Agar	13.5g
Brain heart, solids from infusion	8.0g
Peptic digest of animal tissue	5.0g
NaCl	5.0g
Na_2HPO_4	2.5g
Glucose	2.0g
Streptomycin	40.0mg
Penicillin	20,000U
Sheep blood, defibrinated	50.0mL

pH 7.4 ± 0.2 at 25°C

Preparation of Medium: Add components, except sheep blood, to distilled/deionized water and bring volume to 950.0mL. Mix thoroughly and while stirring bring to a boil for 1 min to completely dissolve. Autoclave for 15 min at 15 psi pressure–121°C. Cool to 50°C. Aseptically add 50.0mL of defibrinated sheep blood. Mix thoroughly. Pour into sterile Petri dishes while agitating gently to distribute the precipitate through the medium.

Use: For the isolation and cultivation of a wide variety of fungal species, especially systemic fungi, from clinical and nonclinical specimens. For the selective isolation of pathogenic fungi from specimens heavily contaminated with bacteria and saprophytic fungi. For the maintenance of fungal species on slant cultures.

Brain Heart Infusion Agar with 10% Sheep Blood, Gentamicin, and Chloramphenicol

Composition per liter:

Pancreatic digest of casein	16.0g
Agar	13.5g
Brain heart, solids from infusion	8.0g
Peptic digest of animal tissue	5.0g
NaCl	5.0g
Na_2HPO_4	2.5g
Glucose	2.0g
Sheep blood, defibrinated	100.0mL
Antibiotic solution	10.0mL

pH 7.4 ± 0.2 at 25°C

Antibiotic Solution:

Composition per 10.0mL:

Chloramphenicol	0.05g
Gentamicin	0.05g

Preparation of Antibiotic Solution: Add components to distilled/deionized water and bring volume to 10.0mL. Mix thoroughly. Filter sterilize.

Preparation of Medium: Add components, except antibiotic solution and sheep blood, to distilled/deionized water and bring volume to 890.0mL. Mix thoroughly. Gently heat and bring to boiling. Autoclave for 15 min at 15 psi pressure–121°C. Cool to 45°–50°C. Aseptically add sterile antibiotic solution and sheep blood. Mix thoroughly. Pour into sterile Petri dishes or distribute into sterile tubes.

Use: For the isolation and cultivation of a wide variety of fungal species, especially systemic fungi, from clinical and nonclinical specimens. For the selective isolation of pathogenic fungi from specimens heavily contaminated with bacteria and saprophytic fungi. For the maintenance of fungal species on slant cultures.

Brain Heart Infusion Agar with Tween 80 (ATCC Medium 1941)

Composition per liter:

Pancreatic digest of casein	16.0g
Agar	13.5g
Brain heart, solids from infusion	8.0g
Peptic digest of animal tissue	5.0g
NaCl	5.0g
Na_2HPO_4	2.5g
Glucose	2.0g
Tween 80	1.0g

pH 7.4 ± 0.2 at 25°C

Preparation of Medium: Add components to distilled/deionized water and bring volume to 1.0L. Mix thoroughly. Distribute into tubes or flasks while shaking to distribute precipitate. Autoclave for 15 min at 15 psi pressure–121°C. Pour into sterile Petri dishes or leave in tubes.

Use: For the cultivation of *Helicococcus kunzii.*

Brain Heart Infusion with Agar, Yeast Extract, NaCl, Inactivated Horse Serum, and Penicillin

Composition per liter:

NaCl	20.0g
Pancreatic digest of gelatin	14.5g
Agar	12.0g
Brain heart, solids from infusion	6.0g
Peptic digest of animal tissue	6.0g
Yeast extract	5.0g
NaCl	5.0g
Glucose	3.0g
Na_2HPO_4	2.5g
Horse serum, inactivated	100.0mL
Penicillin solution	10.0mL

pH 7.4 ± 0.2 at 25°C

Penicillin Solution:

Composition per 10.0mL:

Penicillin	1,000,000U

Preparation of Penicillin Solution: Add penicillin to distilled/deionized water and bring volume to 10.0mL. Mix thoroughly. Filter sterilize.

Preparation of Medium: Add components, except penicillin solution and inactivated horse serum, to distilled/deionized water and bring volume to 890.0mL. Mix thoroughly. Gently heat and bring to boiling. Autoclave for 15 min at 15 psi pressure–121°C. Cool to 45°–50°C. Aseptically add sterile penicillin solution and horse serum. Mix thoroughly. Pour into sterile Petri dishes or distribute into sterile tubes.

Use: For the cultivation of fastidious fungi.

Brain Heart Infusion with Agar, Yeast Extract, Sucrose, Horse Serum, and Penicillin

Composition per liter:

Sucrose	100.0g
Pancreatic digest of gelatin	14.5g
Agar	12.0g
Brain heart, solids from infusion	6.0g
Peptic digest of animal tissue	6.0g
Yeast extract	5.0g
NaCl	5.0g
Glucose	3.0g
Na_2HPO_4	2.5g
Horse serum	100.0mL
Penicillin solution	10.0mL

pH 7.4 ± 0.2 at 25°C

Penicillin Solution:

Composition per 10.0mL:

Penicillin	100,000U

Preparation of Penicillin Solution: Add penicillin to distilled/deionized water and bring volume to 10.0mL. Mix thoroughly. Filter sterilize.

Preparation of Medium: Add components, except penicillin solution and horse serum, to distilled/deionized water and bring volume to 890.0mL. Mix thoroughly. Gently heat and bring to boiling. Autoclave for 15 min at 15 psi pressure–121°C. Cool to 45°–50°C. Aseptically add sterile penicillin solution and horse serum. Mix thoroughly. Pour into sterile Petri dishes or distribute into sterile tubes.

Use: For the cultivation of fastidious fungi.

Brain Heart Infusion with Agar, Yeast Extract, Sucrose, Inactivated Horse Serum, and Penicillin

Composition per liter:

Sucrose	100.0g
Pancreatic digest of gelatin	14.5g
Agar	12.0g
Brain heart, solids from infusion	6.0g
Peptic digest of animal tissue	6.0g
Yeast extract	5.0g
NaCl	5.0g
Glucose	3.0g
Na_2HPO_4	2.5g
Horse serum, inactivated	100.0mL
Penicillin solution	10.0mL

pH 7.4 ± 0.2 at 25°C

Penicillin Solution:

Composition per 10.0mL:

Penicillin	1,000,000U

Preparation of Penicillin Solution: Add penicillin to distilled/deionized water and bring volume to 10.0mL. Mix thoroughly. Filter sterilize.

Preparation of Medium: Add components, except penicillin solution and inactivated horse serum, to distilled/deionized water and bring volume to 890.0mL. Mix thoroughly. Gently heat and bring to boiling. Autoclave for 15 min at 15 psi pressure–121°C. Cool to 45°–50°C. Aseptically add sterile penicillin solution and horse serum. Mix thoroughly. Pour into sterile Petri dishes or distribute into sterile tubes.

Use: For the cultivation of fastidious fungi.

Brain Heart Infusion Blood Agar

Composition per liter:

Blood agar	1.1L
Overlay solution	1.0L

pH 7.4 ± 0.2 at 25°C

Blood Agar:

Composition per liter:

Agar	18.0g
Pancreatic digest of gelatin	14.5g
Brain heart, solids from infusion	6.0g
Peptic digest of animal tissue	6.0g
NaCl	5.0g
Glucose	3.0g
Na_2HPO_4	2.5g
Rabbit blood, defibrinated	100.0mL

Preparation of Blood Agar: Add components, except rabbit blood, to distilled/deionized water and bring volume to 900.0mL. Mix thoroughly. Gently heat and bring to boiling. Adjust pH to 7.4. Distribute 5.0mL of agar solution into tubes. Autoclave for 25 min at 15 psi pressure–121°C. Cool to 50°C. Aseptically add 0.5mL of sterile rabbit blood to each tube. Mix thoroughly. Allow tubes to cool in a slanted position.

Overlay Solution:

Composition per liter:

Pancreatic digest of gelatin	14.5g
Brain heart, solids from infusion	6.0g
Peptic digest of animal tissue	6.0g
NaCl	5.0g
Glucose	3.0g
Na_2HPO_4	2.5g

Preparation of Overlay Solution: Add components to distilled/deionized water and bring volume to 1.0L. Mix thoroughly. Distribute into tubes or flasks. Autoclave for 15 min at 15 psi pressure–121°C.

Preparation of Medium: Prior to inoculation, add 0.5mL of overlay solution to the surface of solidified blood agar slants.

Use: For the cultivation of *Leishmania aethiopica, Leishmania amazonensis, Leishmania aristedesi, Leishmania braziliensis, Leishmania chagasi, Leishmania donovani, Leishmania enriettii, Leishmania garnhami, Leishmania gerbilli, Leishmania guyanensis, Leishmania hertigi, Leishmania infantum, Leishmania major, Leishmania mexicana, Leishmania panamensis, Leishmania pifanoi, Leishmania tropica*, and *Trypanosoma conorrhini.*

Brain Heart Infusion Broth (BHI Broth)

Composition per liter:

Beef heart infusion	250.0g
Calf brain infusion	200.0g
Proteose peptone	10.0g
NaCl	5.0g
$Na_2HPO_4 \cdot 12H_2O$	2.5g
Glucose	2.0g

pH 7.4 ± 0.2 at 25°C

Preparation of Medium: Add components to distilled/deionized water and bring volume to 1.0L. Mix thoroughly. Distribute into tubes or flasks. Autoclave for 15 min at 15 psi pressure–121°C.

Use: For the cultivation of a variety of fastidious and nonfastidious aerobic and anaerobic microorganisms.

Brain Heart Infusion Broth

Composition per liter:

Beef heart infusion	250.0g
Calf brain infusion	200.0g
Proteose peptone	10.0g
NaCl	5.0g
Na_2HPO_4	2.5g
Glucose	2.0g

pH 7.4 ± 0.2 at 25°C

Preparation of Medium: Add components to distilled/deionized water and bring volume to 1.0L. Mix thoroughly. Distribute into tubes or flasks while shaking to distribute precipitate. Autoclave for 15 min at 15 psi pressure–121°C.

Use: For the cultivation of a wide variety of microorganisms, including bacteria, yeasts, and molds, especially fastidious species.

Brain Heart Infusion Broth (BHI Broth)

Composition per liter:

Beef heart infusion	250.0g
Calf brain infusion	200.0g
Proteose peptone	10.0g
NaCl	5.0g
$Na_2HPO_4 \cdot 12H_2O$	2.5g
Glucose	2.0g

pH 7.4 ± 0.2 at 25°C

Preparation of Medium: Add components to distilled/deionized water and bring volume to 1.0L. Mix thoroughly. Distribute into tubes or flasks. Autoclave for 15 min at 15 psi pressure–121°C.

Use: For the cultivation of a variety of fastidious and nonfastidious aerobic and anaerobic microorganisms.

Brain Heart Infusion with Casein

Composition per liter:

Pancreatic digest of gelatin	14.5g
Brain heart, solids from infusion	6.0g
Peptic digest of animal tissue	6.0g
NaCl	5.0g
Casein	5.0g
Glucose	3.0g
Na_2HPO_4	2.5g

pH 7.4 ± 0.2 at 25°C

Preparation of Medium: Add components to distilled/deionized water and bring volume to 1.0L. Mix thoroughly. Distribute into tubes or flasks. Autoclave for 15 min at 15 psi pressure–121°C.

Use: For the cultivation of *Serratia marcescens.*

Brain Heart Infusion Casein Starch

Composition per liter:

Pancreatic digest of gelatin	14.5g
Casein hydrolysate	10.0g
Brain heart, solids from infusion	6.0g
Peptic digest of animal tissue	6.0g
NaCl	5.0g
Glucose	3.0g
Na_2HPO_4	2.5g
Soluble starch	1.0g

pH 7.4 ± 0.2 at 25°C

Preparation of Medium: Add components to distilled/deionized water and bring volume to 1.0L. Mix thoroughly. Distribute into tubes or flasks. Autoclave for 15 min at 15 psi pressure–121°C.

Use: For the cultivation of *Actinomyces odontolyticus.*

Brain Heart Infusion with Cystine

Composition per liter:

Pancreatic digest of gelatin	14.5g
Brain heart, solids from infusion	6.0g
Peptic digest of animal tissue	6.0g
NaCl	5.0g
Glucose	3.0g

Na_2HPO_4	2.5g
Cystine	1.0g

pH 7.4 ± 0.2 at 25°C

Preparation of Medium: Add components to distilled/deionized water and bring volume to 1.0L. Mix thoroughly. Distribute into tubes or flasks. Autoclave for 15 min at 15 psi pressure–121°C.

Use: For the cultivation of *Blastomyces dermatitidis, Histoplasma capsulatum, Mucor hiemalis, Paracoccidioides brasiliensis,* and *Sporothrix schenckii.*

Brain Heart Infusion with Glucose and Horse Serum (LMG Medium 183)

Composition per 1300.0mL:

Brain heart infusion agar	1.0L
Horse serum	200.0mL
Glucose solution	100.0mL

pH 7.4 ± 0.2 at 25°C

Brain Heart Infusion Agar:

Composition per liter:

Beef heart infusion	250.0g
Calf brain infusion	200.0g
Agar	13.5g
Proteose peptone	10.0g
NaCl	5.0g
$Na_2HPO_4 \cdot 12H_2O$	2.5g
Glucose	2.0g

Preparation of Brain Heart Infusion Agar: Add components to distilled/deionized water and bring volume to 1.0L. Mix thoroughly. Gently heat and bring to boiling. Distribute into tubes or flasks. Autoclave for 15 min at 15 psi pressure–121°C. Cool to 45°–50°C.

Glucose Solution:

Composition per 100.0mL:

Glucose	10.0g

Preparation of Glucose Solution: Add glucose to distilled/deionized water and bring volume to 100.0mL. Mix thoroughly. Filter sterilize.

Preparation of Medium: Aseptically add 200.0mL sterile horse serum and 100.0mL sterile glucose solution to 1.0L sterile brain heart infusion agar. Mix thoroughly. Pour into sterile Petri dishes or distribute into sterile tubes.

Use: For the cultivation of *Streptococcus pyogenes.*

Brain Heart Infusion with 5% NaCl

Composition per liter:

NaCl	50.0g
Pancreatic digest of gelatin	14.5g
Brain heart, solids from infusion	6.0g
Peptic digest of animal tissue	6.0g
Glucose	3.0g
Na_2HPO_4	2.5g

pH 7.4 ± 0.2 at 25°C

Preparation of Medium: Add components to distilled/deionized water and bring volume to 1.0L. Mix thoroughly. Distribute into tubes or flasks. Autoclave for 15 min at 15 psi pressure–121°C.

Use: For the cultivation of *Pediococcus halophilus.*

Brain Heart Infusion with PABA (Brain Heart Infusion with *p*-Aminobenzoic Acid)

Composition per liter:

Pancreatic digest of gelatin	14.5g
Brain heart, solids from infusion	6.0g
Peptic digest of animal tissue	6.0g
NaCl	5.0g
Glucose	3.0g
Na_2HPO_4	2.5g
p-Aminobenzoic acid	0.05g

pH 7.4 ± 0.2 at 25°C

Source: This medium is available as a premixed powder from BD Diagnostic Systems.

Preparation of Medium: Add components to distilled/deionized water and bring volume to 1.0L. Mix thoroughly. The addition of 1.0g agar to the medium enhances the growth of anaerobic and microaerophilic microorganisms. Heat with frequent agitation and boil for 1 min to dissolve. Distribute into tubes or flasks. Autoclave for 15 min at 15 psi pressure–121°C.

Use: For the detection of microorganisms in the blood of patients who have received sulfonamide therapy.

Brain Heart Infusion with PABA and Agar (Brain Heart Infusion with *p*-Aminobenzoic Acid and Agar)

Composition per liter:

Pancreatic digest of gelatin	14.5g
Brain heart, solids from infusion	6.0g
Peptic digest of animal tissue	6.0g
NaCl	5.0g
Glucose	3.0g
Na_2HPO_4	2.5g
Agar	1.0g
p-Aminobenzoic acid	0.05g

pH 7.4 ± 0.2 at 25°C

Source: This medium is available as a premixed powder from BD Diagnostic Systems.

Preparation of Medium: Add components to distilled/deionized water and bring volume to 1.0L. Mix thoroughly. The addition of 1.0g of agar to the medium enhances the growth of anaerobic and microaerophilic microorganisms. Heat with frequent agitation and boil for 1 min to dissolve. Distribute into tubes or flasks. Autoclave for 15 min at 15 psi pressure–121°C.

Use: For the detection of microorganisms in the blood of patients who have received sulfonamide therapy.

Brain Heart Infusion with Rabbit Serum

Composition per liter:

Yeast extract	20.0g
Pancreatic digest of casein	16.0g
Brain heart, solids from infusion	8.0g
Peptic digest of animal tissue	5.0g
NaCl	5.0g
Na_2HPO_4	2.5g
Glucose	2.0g
Rabbit serum, heat inactivated	50.0mL
Nicotinamide adenine dinucleotide solution	10.0mL

pH 7.2 ± 0.2 at 25°C

Nicotinamide Adenine Dinucleotide Solution:

Composition per 10.0mL:

Nicotinamide adenine dinucleotide	0.1g

Preparation of Nicotinamide Adenine Dinucleotide Solution: Add 0.1g of nicotinamide adenine dinucleotide to distilled/deinonized water and bring volume to 10.0mL. Filter sterilize.

Preparation of Medium: Add components, except nicotinamide adenine dinucleotide solution and rabbit serum, to distilled/deionized water and bring volume to 940.0mL. Autoclave for 15 min at 15 psi pressure–121°C. Cool to 50°–55°C. Aseptically add 10.0mL of a 0.1% filter-sterilized solution of nicotinamide adenine dinucleotide. Aseptically add 50.0mL of heat-inactivated rabbit serum. Distribute aseptically into tubes or flasks.

Use: For the cultivation and maintenance of *Actinobacillus* species.

Brain Heart Infusion with 3% Sodium Chloride

Composition per liter:

NaCl	30.0g
Pancreatic digest of gelatin	14.5g
Brain heart, solids from infusion	6.0g
Peptic digest of animal tissue	6.0g
Glucose	3.0g
Na_2HPO_4	2.5g

pH 7.4 ± 0.2 at 25°C

Preparation of Medium: Add components to distilled/deionized water and bring volume to 1.0L. Mix thoroughly. Distribute into tubes or flasks. Autoclave for 15 min at 15 psi pressure–121°C.

Use: For the cultivation of *Vibrio parahaemolyticus.*

Brain Heart Infusion Soil Extract Medium

Composition per liter:

Yeast extract	20.0g
Pancreatic digest of casein	16.0g
Brain heart, solids from infusion	8.0g
Peptic digest of animal tissue	5.0g
NaCl	5.0g
Na_2HPO_4	2.5g
Glucose	2.0g
Soil extract	250.0mL
Vitamin B_{12} solution	1.0mL

pH 7.2 ± 0.2 at 25°C

Soil Extract:

Composition per 400.0mL:

African Violet soil	1.0g
Na_2CO_3	1.0g

Preparation of Soil Extract: Autoclave for 60 min at 15 psi pressure–121°C. Filter through paper before using in medium.

Vitamin B_{12} Solution:

Composition per 1.0mL:

Vitamin B_{12}	2.0μg

Preparation of Vitamin B_{12} Solution: Add vitamin B_{12} to distilled/deionized water and bring volume to 1.0mL. Mix thoroughly. Filter sterilize.

Preparation of Medium: Add components, except glucose, yeast extract, and vitamin B_{12} solution, to tap water and bring volume to 799.0mL. Mix thoroughly. Autoclave for 15 min at 15 psi pressure–121°C. Add yeast extract and glucose to 200.0mL of tap water. Filter sterilize and add aseptically to cooled, sterile basal medium. Aseptically add 1.0mL of vitamin B_{12} solution. Mix thoroughly. Aseptically distribute into sterile tubes or flasks.

Use: For the cultivation of a wide variety of microorganisms, including bacteria, yeasts, and molds, especially fastidious species from soil. For the isolation of *Histoplasma capsulatum* and other pathogenic fungi, including *Coccidioides immitis.*

Brain Heart Infusion with Sucrose and Horse Serum

Composition per liter:

Sucrose	171.0g
Pancreatic digest of gelatin	14.5g
Brain heart, solids from infusion	6.0g
Peptic digest of animal tissue	6.0g
NaCl	5.0g
Glucose	3.0g
Na_2HPO_4	2.5g
Horse serum	100.0mL

pH 7.4 ± 0.2 at 25°C

Preparation of Medium: Add components, except horse serum, to distilled/deionized water and bring volume to 900.0mL. Mix thoroughly. Autoclave for 15 min at 15 psi pressure–121°C. Cool to 25°C. Aseptically add 100.0mL sterile horse serum. Mix thoroughly. Distribute into sterile tubes or flasks.

Use: For the cultivation of fastidious fungi.

Brain Liver Heart Semisolid Medium

Composition per liter:

Beef heart, infusion from	250.0g
Calf brains, infusion from	200.0g
Liver, infusion from	50.0g
Proteose peptone	10.0g
NaCl	5.0g
Neopeptone	3.25g
Pancreatic digest of casein	3.25g
Na_2HPO_4	2.5g
Glucose	2.0g
Agar	1.75g

pH 7.3 ± 0.2 at 25°C

Preparation of Medium: Add components to distilled/deionized water and bring volume to 1.0L. Mix thoroughly. Gently heat and bring to boiling. Distribute into tubes or flasks. Autoclave for 15 min at 15 psi pressure–121°C. Leave in tubes.

Use: For the cultivation of fastidious microorganisms. For the cultivation of *Actinomyces bovis*, *Actinomyces israelii*, and *Actinomyces naeslundii.*

Brewer Anaerobic Agar

Composition per liter:

Agar	20.0g
Proteose peptone no. 3	10.0g
Glucose	10.0g
Pancreatic digest of casein	5.0g
Yeast extract	5.0g
NaCl	5.0g
Sodium thioglycolate	2.0g
Sodium formaldehyde sulfoxylate	1.0g
Resazurin	2.0mg

pH 7.2 ± 0.2 at 25°C

Source: This medium is available as a premixed powder from BD Diagnostic Systems.

Preparation of Medium: Add components to distilled/deionized water and bring volume to 1.0L. Mix thoroughly. Distribute into tubes or flasks. Autoclave for 15 min at 15 psi pressure–121°C.

Use: For the cultivation and maintenance of anaerobic and microaerophilic microorganisms.

Brewer Thioglycolate Medium

Composition per liter:

Beef, infusion from	500.0g
Proteose peptone	10.0g
NaCl	5.0g
Glucose	5.0g
K_2HPO_4	2.0g
Sodium thioglycolate	0.5g
Agar	0.5g
Methylene Blue	2.0mg

pH 7.2 ± 0.2 at 25°C

Source: This medium is available as a premixed powder from BD Diagnostic Systems.

Preparation of Medium: Add components to distilled/deionized water and bring volume to 1.0L. Mix thoroughly. Gently heat to boiling. Distribute into tubes or flasks. Autoclave for 15 min at 15 psi pressure–121°C.

Use: For the cultivation and maintenance of anaerobic and microaerophilic microorganisms. For testing the sterility of biological products and materials.

Brewer Thioglycolate Medium, Modified

Composition per liter:

Tryptic digest of casein	17.0g
Glucose	10.0g
NaCl	5.0g
Enzymatic hydrolysate of soybean meal	3.0g
K_2HPO_4	2.0g
Sodium thioglycolate	1.0g
Agar	0.5g
Methylene Blue	2.0mg

pH 7.2 ± 0.2 at 25°C

Source: This medium is available as a premixed powder from BD Diagnostic Systems.

Preparation of Medium: Add components to distilled/deionized water and bring volume to 1.0L. Mix thoroughly. Gently heat to boiling. Distribute into tubes or flasks. Autoclave for 15 min at 15 psi pressure–121°C.

Use: For the cultivation and maintenance of anaerobic and microaerophilic microorganisms. For testing the sterility of biological products and materials.

BRILA MUG Broth (Brillant Green 2%-Bile MUG Broth)

Composition per liter:

Ox-bile (dried)	20.0g
Peptone	10.0g
Lactose	10.0g
L-Tryptophan	1.0g
Brillant Green	0.133g
4-Methylumbelliferyl-ß-D-glucuronide	0.1g

pH 7.2 ± 0.2 at 37°C

Source: This medium is available from Fluka, Sigma-Aldrich.

Preparation of Medium: Add components to distilled/deionized water and bring volume to 1.0L. Mix thoroughly. Distribute into test tubes that contain an inverted Durham tube in 10.0mL volumes. Autoclave for 15 min at 15 psi pressure–121°C.

Use: For the detection of *E. coli* and coliforms. Bile and brilliant green extensively inhibit the growth of accompanying flora, in particular gram-positive microorganisms. The presence of *E. coli* results in fluorescence in the UV. A positive indole test and possibly gas formation from lactose fermentation provide confirmation. β-D-glucoronidase, which is produced by *E. coli*, cleaves 4-Methylumbelliferyl-β-D-glucuronide to 4-methylumbelliferone and glucuronide. The fluorogen 4-methylumbelliferone can be detected under a long wavelength UV lamp.The broth can be used in conjunction with the MPN method for *E. coli* and coliform enumeration in the water of bathing areas.

Brilliant Green Agar

Composition per liter:

Agar	20.0g
Lactose	10.0g
Sucrose	10.0g
Peptic digest of animal tissue	5.0g
Pancreatic digest of casein	5.0g
NaCl	5.0g
Phenol Red	0.08g
Brilliant Green	0.0125g

pH 6.9 ± 0.2 at 25°C

Source: This medium is available as a premixed powder from Oxoid Unipath and BD Diagnostic Systems.

Preparation of Medium: Add components to distilled/deionized water and bring volume to1.0L. Mix thoroughly. Gently heat and bring to boiling. Distribute into tubes or flasks. Autoclave for 15 min at 15 psi pressure–121°C. Pour into sterile Petri dishes.

Use: For the selective isolation of *Salmonella* other than *Salmonella typhi* from feces and other specimens. *Salmonella* other than *Salmonella typhi* appear as red/pink/white colonies surrounded by a zone of red in the agar, indicating nonlactose/sucrose fermentation. *Proteus* or *Pseudomonas* species may appear as small red colonies. Lactose- or sucrose-fermenting bacteria appear as yellow-green colonies surrounded by a zone of yellow-green in the agar.

Brilliant Green Agar, Modified

Composition per liter:

Agar	12.0g
Lactose	10.0g
Sucrose	10.0g
Beef extract	5.0g
Peptone	5.0g
NaCl	5.0g
Yeast extract	3.0g
Na_2HPO_4	1.0g
NaH_2PO_4	0.6g
Phenol Red	0.09g
Brilliant Green	4.7mg

pH 6.9 ± 0.2 at 25°C

Source: This medium is available as a premixed powder from Oxoid Unipath.

Preparation of Medium: Add components to distilled/deionized water and bring volume to1.0L. Mix thoroughly. Gently heat and bring to boiling. Do not autoclave. Cool to 45°–50°C. Addition of 1.0g of sodium sulfacetamide and 250.0mg of sodium mandelate enhances inhibition of contaminating microorganisms. Pour into sterile Petri dishes.

Use: For the selective isolation of *Salmonella* other than *Salmonella typhi* from feces. *Salmonella* other than *Salmonella typhi* appear as red/pink/white colonies surrounded by a zone of red in the agar, indicating nonlactose/sucrose fermentation. *Proteus* or *Pseudomonas* species may appear as small red colonies. Lactose- or sucrose-fermenting bacteria appear as yellow-green colonies surrounded by a zone of yellow-green in the agar.

Brilliant Green 2%-Bile Broth, Fluorocult® (Fluorocult Brilliant Green 2%-Bile Broth) (BRILA)

Composition per liter:

Ox bile, dried	20.0g
Peptone	10.0g
Lactose	10.0g
L-Tryptophan	1.0g
4-Methylumbelliferyl-β-D-glucuronide	0.1g
Brilliant Green	0.0133g

pH 7.2 ± 0.2 at 25°C

Source: This medium is available from Merck.

Preparation of Medium: Add components to distilled/deionized water and bring volume to 1.0L. Mix thoroughly. Gently heat and bring to boiling. Cool. Distribute into test tubes containing inverted Durham tubes. Autoclave for 15 min at 15 psi pressure–121°C. Do not autoclave longer. The prepared broth is clear and green.

Use: For the cultivation of *Escherichia coli.* Bile and brilliant green almost completely inhibit the growth of undesired microbial flora, in particular Gram-positive microorganisms. *E. coli* shows a positive fluorescence under UV light (366 nm). A positive indole reaction and, if necessary gas formation due to fermenting lactose, confirm the findings.

Brilliant Green Lactose Bile Broth

Composition per liter:

Oxgall, dehydrated	20.0g
Lactose	10.0g
Pancreatic digest of gelatin	10.0g
Brilliant Green	0.0133g

pH 7.2 ± 0.1 at 25°C

Source: This medium is available as a premixed powder from BD Diagnostic Systems and Oxoid Unipath.

Preparation of Medium: Add lactose and pancreatic digest of gelatin to distilled/deionized water and bring volume to 500.0mL. Mix thoroughly. Add 20.0g oxgall dissolved in 200.0mL distilled/deionized water. The pH of this solution should be 7.0-7.5. Mix thoroughly. Bring volume to 975.0mL with distilled/deionized water. Adjust pH to 7.4. Add 13.3mL of 0.1% aqueous Brilliant Grenn in distilled/deionized water. Adjust volume to 1.0L with distilled/deionized water. Distribute into tubes containing inverted Durham tubes, in 10.0mL amounts for testing 1.0mL or less of sample. Make sure that the fluid level covers the inverted vials. Autoclave for 15 min at 15 psi pressure–121°C. After sterilization, cool the broth rapidly. Medium is sensitive to light.The final pH should be 7.2 ± 0.1 at 25°C.

Use: For the detection of coliform microorganisms. Turbidity in the broth and gas in the Durham tube are positive indications of *Escherichia coli.*

Brilliant Green Phenol Red Agar

Composition per liter:

Agar	15.0g
Lactose	15.0g
Peptone	10.0g
Meat extract	5.0g
NaCl	5.0g
Phenol Red	0.08g
Brilliant Green	0.0125g

pH 6.9 ± 0.2 at 25°C

Preparation of Medium: Add components to distilled/deionized water and bring volume to 1.0L. Mix thoroughly. Gently heat and bring to boiling. Distribute into tubes or flasks. Autoclave for 15 min at 15 psi pressure–121°C. Pour into sterile Petri dishes or leave in tubes.

Use: For the cultivation of *Salmonella* species.

BROLACIN MUG Agar (Bromthymol Blue Lactose Cystine MUG Agar) (C.L.E.D. MUG Agar)

Composition per liter:

Agar	12.0g
Lactose	10.0g
Universal peptone	4.0g
Casein peptone	4.0g
Meat extract	3.0g
L-Cystine	0.128g
4-Methylumbelliferyl-β-D-glucuronide	0.1g
Bromthymol Blue	0.02g

pH 7.3 ± 0.2 at 37°C

Source: This medium is available from Fluka, Sigma-Aldrich.

Preparation of Medium: Add components to distilled/deionized water and bring volume to 1.0L. Mix thoroughly. Gently heat while stirring and bring to boiling. Autoclave for 15 min at 15 psi pressure–121°C. Cool to 50°C. Pour into sterile Petri dishes.

Use: For the enumeration, isolation, and identification of microorganisms in urine. Growth of all urinary microorganisms is favored. Lactose catabolism produces a color change of Bromthymol Blue to yellow. Alkalization gives a color change to deep-blue.

β-D-glucoronidase, which is produced by *E. coli*, cleaves 4-methylumbelliferyl-β-D-glucuronide to 4-methylumbelliferone and glucuronide. The fluorogen 4-methylumbelliferone can be detected under a long wavelength UV lamp, permitting differentiation of *E. coli* colonies.

Bromcresol Purple Broth

Composition per liter:

Peptone	10.0g
NaCl	5.0g
Beef extract	3.0g
Bromcresol Purple	0.04g
Carbohydrate solution	10.0mL

pH 7.0 ± 0.2 at 25°C

Carbohydrate Solution:
Composition per 10.0mL:

Carbohydrate	5.0g

Preparation of Carbohydrate Solution: Add carbohydrate to distilled/deionized water and bring volume to 10.0mL. Mix thoroughly. Filter sterilize.

Preparation of Medium: Add components to distilled/deionized water and bring volume to 1.0L. Mix thoroughly. Gently heat and bring to boiling. Distribute into test tubes that contain an inverted Durham tube. Autoclave for 10 min at 15 psi pressure–121°C.

Use: For the differentiation of a variety of microorganisms based on their fermentation of specific carbohydrates. Bacteria that ferment the specific carbohydrate turn the medium yellow. When bacteria produce gas, the gas is trapped in the Durham tube.

Bromcresol Purple Broth

Composition per liter:

Peptone	10.0g
NaCl	5.0g
Beef extract	3.0g
Bromcresol Purple	0.04g
Carbohydrate solution	50.0mL

pH 7.0 ± 0.2 at 25°C

Carbohydrate Solution:
Composition per 50.0mL:

Carbohydrate	25.0g

Preparation of Carbohydrate Solution: Add carbohydrate to distilled/deionized water and bring volume to 50.0mL. Mix thoroughly. Filter sterilize.

Preparation of Medium: Add components, except carbohydrate solution, to distilled/deionized water and bring volume to 950.0mL. Mix thoroughly. Gently heat and bring to boiling. Autoclave for 10 min at 15 psi pressure–121°C. Cool to 25°C. Aseptically add 50.0mL of carbohydrate solution. Aseptically distribute into tubes or flasks. Alternately distribute the medium without the carbohydrate solution into test tubes that contain an inverted Durham tube prior to autoclaving. Then autoclave for 10 min at 15 psi pressure–121°C, cool to 25°C, and aseptically add the carbohydrate solution to each tube to yield a final carbohydrate concentration of 5%.

Use: For the differentiation of a variety of microorganisms based on their fermentation of specific carbohydrates. Bacteria that ferment the specific carbohydrate turn the medium yellow. When bacteria produce gas, the gas is trapped in the Durham tube.

Bromcresol Purple Dextrose Broth (BCP Broth)

Composition per liter:

Glucose	10.0g
Peptone	5.0g
Beef extract	3.0g
Bromcresol Purple solution	2.0mL

pH 7.0 ± 0.2 at 25°C

Bromcresol Purple Solution:
Composition per 10.0mL:

Bromcresol Purple	0.16g
Ethanol (95% solution)	10.0mL

Preparation of Bromcresol Purple Solution: Add Bromcresol Purple to 10.0mL of ethanol. Mix thoroughly.

Preparation of Medium: Add components to distilled/deionized water and bring volume to 1.0L. Mix thoroughly. Distribute into tubes in 12–15mL volumes. Autoclave for 15 min at 15 psi pressure–121°C.

Use: For the cultivation and differentiation of bacteria based on their ability to ferment glucose. Bacteria that ferment glucose turn the medium yellow.

Bromthymol Blue Agar

Composition per liter:

Agar	11.0g
Peptone	10.0g
NaCl	5.0g
Yeast extract	5.0g
Lactose (33% solution)	27.0mL
Bromthymol Blue (1% solution)	10.0mL
Sodium thiosulfate (50% solution)	2.0mL
Glucose (33% solution)	1.2mL
Maranil solution (5% solution)	1.0mL

pH 7.7–7.8 at 25°C

Preparation of Medium: Add agar, peptone, NaCl, and yeast extract to distilled/deionized water and bring volume to 1.0L. Mix thoroughly. Adjust pH to 8.0. Autoclave for 20 min at 15 psi pressure–121°C. Cool to 45°–50°C. Filter sterilize separately the lactose solution, Bromthymol Blue solution, sodium thiosulfate solution, glucose solution, and maranil solution. To the cooled, sterile agar solution aseptically add 27.0mL of sterile lactose solution, 10.0mL of sterile Bromthymol Blue solution, 2.0mL of sterile sodium thiosulfate solution, 1.2mL of sterile glucose solution, and 1.0mL of sterile maranil solution. Mix thoroughly. Adjust pH to 7.7–7.8. Pour into sterile Petri dishes or distribute into sterile tubes.

Use: For the selective isolation and cultivation of members of the Enterobacteriaceae.

Bromthymol Blue Broth

Composition per 101.45mL:

Pancreatic digest of casein	0.7g
NaCl	0.5g
Beef extract	0.3g
Yeast extract	0.3g
Beef heart, solids from infusion	0.2g
Horse serum	10.0mL
Bromthymol Blue solution	1.0mL
Ampicillin solution	1.0mL
Urea solution	0.25mL
Nystatin solution	0.1mL
Tripeptide solution	0.1mL

pH 6.0 ± 0.2 at 25°C

Bromthymol Blue Solution:

Composition per 50.0mL:

Bromthymol Blue	0.2g
NaOH (0.01*N* solution)	32.0mL

Preparation of Bromthymol Blue Solution: Add Bromthymol Blue to NaOH solution. Mix thoroughly. Bring volume to 50.0mL with distilled/deionized water. Autoclave for 15 min at 15 psi pressure–121°C. Store at 25°C.

Ampicillin Solution:

Composition per 10.0mL:

Ampicillin	1.0g

Preparation of Ampicillin Solution: Add ampicillin to distilled/deionized water and bring volume to 10.0mL. Mix thoroughly. Filter sterilize.

Urea Solution:

Composition per 100.0mL:

Urea	10.0g

Preparation of Urea Solution: Add urea to distilled/deionized water and bring volume to 100.0mL. Filter sterilize. Store at –20°C.

Nystatin Solution:

Composition per 1.0mL:

Nystatin	50,000U

Preparation of Nystatin Solution: Add nystatin to distilled/deionized water and bring volume to 1.0mL. Filter sterilize.

Tripeptide Solution:

Composition per 10.0mL:

Glycyl-L-histidyl-L-lysine acetate	0.2mg

Preparation of Tripeptide Solution: Add glycyl-L-histidyl-L-lysine acetate to distilled/deionized water and bring volume to 10.0mL. Mix thoroughly. Filter sterilize. Store at –20°C.

Preparation of Medium: Add components—except horse serum, ampicillin solution, urea solution, nystatin solution, and tripeptide solution—to distilled/deionized water and bring volume to 90.0mL. Mix thoroughly. Gently heat and bring to boiling. Autoclave for 15 min at 15 psi pressure–121°C. Cool to 45°–50°C. Aseptically add 10.0mL of sterile horse serum, 1.0mL of sterile ampicillin solution, 0.25mL of sterile urea solution, 0.1mL of sterile nystatin solution, and 0.1mL of sterile tripeptide solution. Mix thoroughly. Pour into sterile Petri dishes.

Use: For the cultivation of *Ureaplasma* species from clinical specimens.

Brucella Agar

Composition per liter:

Agar	15.0g
Pancreatic digest of casein	10.0g
Peptic digest of animal tissue	10.0g
NaCl	5.0g
Yeast extract	2.0g
Glucose	1.0g
$NaHSO_3$	0.1g
Horse blood, defibrinated	100.0mL

pH 7.0 ± 0.2 at 25°C

Source: This medium is available as a premixed powder from BD Diagnostic Systems and Oxoid Unipath.

Preparation of Medium: Add components to distilled/deionized water and bring volume to 900.0mL. Mix thoroughly. Heat gently with frequent mixing. Boil for 1 min. Autoclave for 15 min at 15 psi pressure–121°C. Cool to 45°–50°C. Add 100.0mL of sterile defibrinated horse blood. Mix gently and pour into sterile Petri dishes.

Use: For the cultivation and maintenance of *Brucella* species. For the isolation and cultivation of nonfastidious and fastidious microorganisms from a variety of clinical and nonclinical specimens.

Brucella Agar Base *Campylobacter* Medium

Composition per 1100.0mL:

Cycloheximide (actidione)	0.05g
Sodium cephazolin	0.015g
Novobiocin	5.0mg
Bacitracin	25,000U
Colistin sulfate	10,000U
Brucella agar base	1.0L
Horse blood, defibrinated	100.0mL

***Brucella* Agar Base**

Composition per liter:

Agar	15.0g
Pancreatic digest of casein	10.0g
Peptic digest of animal tissue	10.0g
NaCl	5.0g
Yeast extract	2.0g
Glucose	1.0g
$NaHSO_3$	0.1g

Preparation of *Brucella* Agar Base: Add components to distilled/deionized water and bring volume to 1.0L. Mix thoroughly.

Optional Supplement:

Composition per 10.0mL:

Sodium pyruvate	0.25g
$NaHSO_3$	0.25g
$FeSO_4 \cdot 7H_2O$	0.25g

Preparation of Optional Supplement: Add components to distilled/deionized water and bring volume to 10.0mL. Filter sterilize.

Preparation of Medium: Add components to 1.0L of prepared *Brucella* agar base. Mix thoroughly. Autoclave for 15 min at 15 psi pressure–121°C. Cool to 45°–50°C. Add 100.0mL of sterile, defibrinated horse blood. Addition of 10.0mL of optional supplement will improve growth. Mix thoroughly. Pour into sterile Petri dishes.

Use: For the selective isolation and cultivation of *Campylobacter jejuni* from fecal specimens or rectal swabs.

Brucella Agar with Vitamin K_1

Composition per liter:

Agar	17.5g
Pancreatic digest of casein	10.0g
Peptic digest of animal tissue	10.0g
NaCl	5.0g
Yeast extract	2.0g
Glucose	1.0g
$NaHSO_3$	0.1g
Sheep blood, defibrinated	50.0mL
Vitamin K_1 solution	1.0mL

pH 7.0 ± 0.2 at 25°C

Vitamin K_1 Solution:

Composition per 20.0mL:

Vitamin K_1	0.2g
Ethanol, absolute	20.0mL

Preparation of Vitamin K_1 Solution: Add vitamin K_1 to 20.0mL of ethanol. Mix thoroughly. Filter sterilize.

Preparation of Medium: Add components, except sheep blood and vitamin K_1 solution, to distilled/deionized water and bring volume to 949.0mL. Mix thoroughly. Gently heat and bring to boiling. Autoclave for 15 min at 15 psi pressure–121°C. Cool to 45°–50°C. Aseptically add 50.0mL of sterile sheep blood and 1.0mL of sterile vitamin K_1 solution. Mix thoroughly. Pour into sterile Petri dishes or distribute into sterile tubes.

Use: For the cultivation of *Brucella* species.

Brucella Albimi Broth with 0.16% Agar

Composition per liter:

Pancreatic digest of casein	10.0g
Peptic digest of animal tissue	10.0g
NaCl	5.0g
Yeast extract	2.0g
Agar	1.6g
Glucose	1.0g
$NaHSO_3$	0.1g
Horse blood, defibrinated	100.0mL

pH 7.0 ± 0.2 at 25°C

Preparation of Medium: Add components, except horse blood, to distilled/deionized water and bring volume to 900.0mL. Mix thoroughly. Heat gently with frequent mixing. Boil for 1 min. Autoclave for 15 min at 15 psi pressure–121°C. Cool to 45°–50°C. Aseptically add 100.0mL of sterile defibrinated horse blood. Mix thoroughly. Aseptically distribute into sterile tubes or flasks.

Use: For the cultivation and maintenance of *Campylobacter* species.

Brucella Albimi Broth with 0.16% Agar and 1% Glycine (ATCC Medium 2161)

Composition per liter:

Pancreatic digest of casein	10.0g
Peptic digest of animal tissue	10.0g
Glycine	10.0g
NaCl	5.0g
Yeast extract	2.0g

Agar 1.6g
Glucose 1.0g
$NaHSO_3$ 0.1g
Horse blood, defibrinated 100.0mL
pH 7.0 ± 0.2 at 25°C

Preparation of Medium: Add components, except horse blood, to distilled/deionized water and bring volume to 900.0mL. Mix thoroughly. Heat gently with frequent mixing. Boil for 1 min. Autoclave for 15 min at 15 psi pressure–121°C. Cool to 45°–50°C. Aseptically add 100.0mL of sterile defibrinated horse blood. Mix thoroughly. Aseptically distribute into sterile tubes or flasks.

Use: For the cultivation and maintenance of *Campylobacter* species.

Brucella Albimi Broth with Formate and Fumarate

Composition per 1050.0mL:
Pancreatic digest of casein 10.0g
Peptic digest of animal tissue 10.0g
NaCl 5.0g
Yeast extract 2.0g
Glucose 1.0g
$NaHSO_3$ 0.1g
Horse blood, defibrinated 100.0mL
Formate-fumarate solution 50.0mL
pH 7.0 ± 0.2 at 25°C

Formate-Fumarate Solution:
Composition per 100.0mL:
Sodium formate 6.0g
Fumaric acid 6.0g

Preparation of Formate-Fumarate Solution: Add components to distilled/deionized water and bring volume to 100.0mL. Mix thoroughly. Adjust pH to 7.0. Filter sterilize.

Preparation of Medium: Add components, except formate-fumarate solution and horse blood, to distilled/deionized water and bring volume to 900.0mL. Mix thoroughly. Heat gently with frequent mixing. Boil for 1 min. Autoclave for 15 min at 15 psi pressure–121°C. Cool to 45°–50°C. Add 100.0mL of sterile defibrinated horse blood. Mix gently and aseptically distribute into sterile tubes in 5.0mL volumes. Aseptically add 0.25mL of formate-fumarate solution to each tube containing 5.0mL of medium immediately prior to inoculation.

Use: For the cultivation and maintenance of *Campylobacter mucosalis*.

Brucella Albimi Broth with Sheep Blood

Composition per liter:
Pancreatic digest of casein 10.0g
Peptic digest of animal tissue 10.0g
NaCl 5.0g
Yeast extract 2.0g
Glucose 1.0g
$NaHSO_3$ 0.1g
Sheep blood, defibrinated 100.0mL
pH 7.0 ± 0.2 at 25°C

Preparation of Medium: Add components, except sheep blood, to distilled/deionized water and bring volume to 900.0mL. Mix thoroughly. Heat gently with frequent mixing. Boil for 1 min. Autoclave for 15 min at 15 psi pressure–121°C. Cool to 45°–50°C. Aseptically add 100.0mL of sterile defibrinated sheep blood. Mix thoroughly. Aseptically distribute into sterile tubes or flasks.

Use: For the cultivation and maintenance of *Helicobacter nemestrinae* and *Helicobacter pylori.*

Brucella Albimi Medium, Semisolid

Composition per liter:
Pancreatic digest of casein 10.0g
Peptic digest of animal tissue 10.0g
Glycine 10.0g
NaCl 8.5g
Yeast extract 2.0g
Agar 1.6g
Glucose 1.0g
L-Cysteine·HCl·H_2O 0.2g
$NaHSO_3$ 0.1g
pH 7.0 ± 0.2 at 25°C

Preparation of Medium: Add components to distilled/deionized water and bring volume to 1.0L. Mix thoroughly. Adjust pH to 7.0. Gently heat and bring to boiling. Distribute into tubes in 10.0mL volumes. Autoclave for 15 min at 15 psi pressure–121°C. Allow tubes to cool in an upright position.

Use: For the cultivation and identification of *Campylobacter* species.

Brucella Anaerobic Blood Agar

Composition per liter:
Vitamin K_1 .... 0.01g
Anaerobic agar base 1000.0mL
Sheep blood, sterile, defibrinated 50.0mL

Anaerobic Agar Base
Composition per liter:
Pancreatic digest of casein 17.5g
Agar 15.0g

Glucose 10.0g
Papaic digest of soybean meal 2.5g
NaCl 2.5g
Sodium thioglycolate 2.0g
Sodium formaldehyde sulfoxylate 1.0g
L-Cystine·HCl·H_2O 0.4g
Methylene Blue 0.002g
pH 7.0 ± 0.2 at 25°C

Preparation of Anaerobic Agar Base: Add components to distilled/deionized water and bring volume to 1.0L. Mix thoroughly. Autoclave for 15 min at 15 psi pressure–121°C. Cool to 45°–50°C.

Preparation of Medium: To 950.0mL of cooled, sterile anaerobic agar base, aseptically add 10.0mg of vitamin K_1 and 50.0mL of sterile, defibrinated sheep blood.

Use: For the isolation of anaerobes.

Brucella Blood Agar with Hemin and Vitamin K_1

Composition per liter:
Agar 15.0g
Pancreatic digest of casein 10.0g
Peptic digest of animal tissue 10.0g
NaCl 5.0g
Yeast extract 2.0g
Glucose 1.0g
$NaHSO_3$ 0.1g
Vitamin K_1 solution 1.0mL
Hemin solution 1.0mL
Sheep blood, defibrinated 50.0mL

Source: This medium is available as a prepared medium from BD Diagnostic Systems.

Vitamin K_1 Solution:
Composition per 100.0mL:
Vitamin K_1 .... 1.0g
Ethanol 99.0mL

Preparation of Vitamin K_1 Solution: Add vitamin K_1 to 99.0mL of absolute ethanol. Mix thoroughly. Filter sterilize.

Hemin Solution:
Composition per 100.0mL:
Hemin 1.0g
NaOH (1*N* solution) 20.0mL

Preparation of Hemin Solution: Add hemin to 20.0mL of 1*N* NaOH solution. Mix thoroughly. Bring volume to 100.0mL with distilled/deionized water.

Preparation of Medium: Add components, except vitamin K_1 solution and sheep blood, to distilled/deionized water and bring volume to 949.0mL. Mix thoroughly. Gently heat and bring to boiling. Autoclave for 15 min at 15 psi pressure–121°C. Cool to 50°C. Aseptically add 1.0mL of sterile vitamin K_1 solution and 50.0mL of sterile defibrinated sheep blood. Mix gently and pour into sterile Petri dishes.

Use: For the isolation and cultivation of anaerobic microorganisms from clinical specimens. After growth on agar plates, colonies should be examined under a dissecting microscope under long-wave UV light. Members of the pigmented *Bacteroides* group appear as red/orange fluorescent colonies.

Brucella Blood Culture Broth

Composition per liter:
Sucrose 100.0g
Hemin 0.5g
Sodium polyanetholsulfonate (SPS) 0.25g
Brucella broth base 1000.0mL
Vitamin K_1 solution 1.0mL
pH 7.0 ± 0.2 at 25°C

***Brucella* Broth Base:**
Composition per liter:
Pancreatic digest of casein 10.0g
Peptic digest of animal tissue 10.0g
NaCl 5.0g
Yeast extract 2.0g
Glucose 1.0g
$NaHSO_3$ 0.1g

Preparation of *Brucella* Broth Base: Add components to distilled/deionized water and bring volume to 1.0L. Mix thoroughly.

Vitamin K_1 Solution:
Composition per 100.0mL:
Vitamin K_1 .... 1.09g
Ethanol, absolute 99.0mL

Preparation of Vitamin K_1 Solution: Add vitamin K_1 to 99.0mL of absolute ethanol. Store in the dark at 4°C.

Preparation of Medium: Add components, except vitamin K_1 solution, to prepared *Brucella* broth base. Autoclave for 15 min at 15 psi pressure–121°C. Cool to 45°–50°C. Aseptically add 1.0mL of vitamin K_1 solution. Distribute into sterile tubes or flasks.

Use: For the isolation and cultivation of microorganisms from blood. Especially useful for the cultivation of anaerobes.

Brucella Broth (*Brucella* Albimi Broth)

Composition per liter:
Pancreatic digest of casein 10.0g
Peptic digest of animal tissue 10.0g

NaCl	5.0g
Yeast extract	2.0g
Glucose	1.0g
$NaHSO_3$	0.1g
Horse blood, defibrinated	50.0mL

pH 7.0 ± 0.2 at 25°C

Source: This medium is available as a premixed powder from BD Diagnostic Systems.

Preparation of Medium: Add components, except horse blood, to distilled/deionized water and bring volume to 950.0mL. Mix thoroughly. Heat gently with frequent mixing. Boil for 1 min. Autoclave for 15 min at 15 psi pressure–121°C. Cool to 45°–50°C. Aseptically add 50.0mL of sterile horse blood. Mix thoroughly. Aseptically distribute into sterile tubes or flasks.

Use: For the cultivation and maintenance of *Campylobacter coli, Campylobacter fecalis*, and *Brucella* species. Also used for the isolation and cultivation of a wide variety of fastidious and nonfastidious microorganisms.

Brucella Broth with Additives (ATCC Medium 489)

Composition per liter:

Pancreatic digest of casein	10.0g
Peptic digest of animal tissue	10.0g
NaCl	3.5g
Yeast extract	2.0g
Glucose	1.0g
$NaHSO_3$	0.1g
Horse serum, inactivated	100.0mL
Fresh yeast extract solution	50.0mL

pH 7.0 ± 0.2 at 25°C

Fresh Yeast Extract Solution:
Composition per 100.0mL:

Baker's yeast, live, pressed, starch-free	25.0g

Preparation of Fresh Yeast Extract Solution: Add the live Baker's yeast to 100.0mL of distilled/deionized water. Autoclave for 90 min at 15 psi pressure–121°C. Allow to stand. Remove supernatant solution. Adjust pH to 6.6–6.8.

Preparation of Medium: Add components, except horse serum and fresh yeast extract solution, to distilled/deionized water and bring volume to 850.0mL. Mix thoroughly. Gently heat and bring to boiling. Autoclave for 15 min at 15 psi pressure–121°C. Cool to 45°–50°C. Aseptically add 100.0mL of sterile horse serum and 50.0mL of sterile fresh yeast extract solution. Mix thoroughly. Aseptically distribute into sterile tubes or flasks.

Use: For the cultivation of *Corynebacterium* species.

Brucella Broth with Additives (ATCC Medium 490)

Composition per liter:

NaCl	30.0g
Pancreatic digest of casein	10.0g
Peptic digest of animal tissue	10.0g
Yeast extract	2.0g
Glucose	1.0g
$NaHSO_3$	0.1g
Horse serum, inactivated	100.0mL
Fresh yeast extract solution	50.0mL

pH 7.0 ± 0.2 at 25°C

Fresh Yeast Extract Solution:
Composition per 100.0mL:

Baker's yeast, live, pressed, starch-free	25.0g

Preparation of Fresh Yeast Extract Solution: Add the live Baker's yeast to 100.0mL of distilled/deionized water. Autoclave for 90 min at 15 psi pressure–121°C. Allow to stand. Remove supernatant solution. Adjust pH to 6.6–6.8.

Preparation of Medium: Add components, except horse serum and fresh yeast extract solution, to distilled/deionized water and bring volume to 850.0mL. Mix thoroughly. Gently heat and bring to boiling. Autoclave for 15 min at 15 psi pressure–121°C. Cool to 45°–50°C. Aseptically add 100.0mL of horse serum and 50.0mL of sterile fresh yeast extract solution. Mix thoroughly. Aseptically distribute into sterile tubes or flasks.

Use: For the cultivation of salt-tolerant *Corynebacterium* species.

Brucella Broth with 0.16% Agar (ATCC Medium 1116)

Composition per liter:

Pancreatic digest of casein	10.0g
Peptic digest of animal tissue	10.0g
NaCl	5.0g
Yeast extract	2.0g
Agar	1.6g
Glucose	1.0g
$NaHSO_3$	0.1g
Horse blood, defibrinated	50.0mL

pH 7.0 ± 0.2 at 25°C

Source: This medium without agar is available as a premixed powder from BD Diagnostic Systems.

Preparation of Medium: Add components, except horse blood, to distilled/deionized water and bring volume to 950.0mL. Mix thoroughly. Heat gently with frequent mixing. Boil for 1 min. Autoclave for 15 min at 15 psi pressure–121°C. Cool to 45°–50°C. Aseptically add 50.0mL of sterile horse blood.

Mix thoroughly. Aseptically distribute into sterile tubes or flasks.

Use: For the cultivation and maintenance of *Campylobacter fetus* subsp. *Fetus* and *Campylobacter jejuni* subsp. Jejuni.

Brucella Broth Base *Campylobacter* Medium

Composition per liter:

Cycloheximide (actidione)	50.0mg
Sodium cephazolin	15.0mg
Novobiocin	5.0mg
Bacitracin	25,000U
Colistin sulfate	10,000U
Brucella broth base	900.0mL
Horse blood, defibrinated	100.0mL

pH 7.0 ± 0.2 at 25°C

***Brucella* Broth Base:**
Composition per liter:

Pancreatic digest of casein	10.0g
Peptic digest of animal tissue	10.0g
NaCl	5.0g
Yeast extract	2.0g
Glucose	1.0g
$NaHSO_3$	0.1g

Preparation of *Brucella* Broth Base: Add components to distilled/deionized water and bring volume to 1.0L. Mix thoroughly.

Optional Supplement:
Composition per 10.0mL:

Sodium pyruvate	0.25g
$NaHSO_3$	0.25g
$FeSO_4·7H_2O$	0.25g

Preparation of Optional Supplement: Add components to distilled/deionized water and bring volume to 10.0mL. Filter sterilize.

Preparation of Medium: Add components, except horse blood, to 900.0mL of prepared *Brucella* broth base. Mix thoroughly. Autoclave for 15 min at 15 psi pressure–121°C. Cool to 45°–50°C. Aseptically add 100.0mL of sterile, defibrinated horse blood. Addition of 10.0mL of optional supplement will improve growth. Mix thoroughly. Pour into sterile Petri dishes.

Use: For the selective isolation and cultivation of *Campylobacter jejuni* from fecal specimens or rectal swabs. Addition of the optional supplement improves growth.

Brucella Broth, Modified

Composition per liter:

Pancreatic digest of casein	10.0g
Peptic digest of animal tissue	10.0g
NaCl	5.0g
$MgSO_4·7H_2O$	2.46g
Yeast extract	2.0g
$CaCl_2$	1.1g
Glucose	1.0g
$NaHSO_3$	0.1g
Horse blood, defibrinated	100.0mL

pH 7.0 ± 0.2 at 25°C

Preparation of Medium: Add components, except horse blood, to distilled/deionized water and bring volume to 900.0mL. Mix thoroughly. Gently heat and bring to boiling. Autoclave for 15 min at 15 psi pressure–121°C. Cool to 45°–50°C. Aseptically add sterile horse blood. Mix thoroughly. Aseptically distribute into sterile tubes or flasks.

Use: For the cultivation and maintenance of *Campylobacter coli* and *Campylobacter fecalis.*

Brucella FBP Agar

Composition per liter:

Agar	15.0g
Pancreatic digest of casein	10.0g
Peptic digest of animal tissue	10.0g
NaCl	5.0g
Yeast extract	2.0g
Glucose	1.0g
$NaHSO_3$	0.1g
FBP solution	30.0mL

pH 7.0 ± 0.2 at 25°C

FBP Solution:
Composition per 30.0mL:

$FeSO_4$	0.25g
$NaHSO_3$	0.25g
Sodium pyruvate	0.25g

Preparation of FBP Solution: Add components to distilled/deionized water and bring volume to 30.0mL. Mix thoroughly. Filter sterilize.

Preparation of Medium: Add components, except FBP solution, to distilled/deionized water and bring volume to 970.0mL. Mix thoroughly. Gently heat and bring to boiling. Autoclave for 15 min at 15 psi pressure–121°C. Cool to 45°–50°C. Aseptically add 30.0mL of sterile FBP solution. Mix thoroughly. Pour into sterile Petri dishes or distribute into sterile tubes.

Use: For the cultivation of *Brucella* species.

Brucella FBP Broth

Composition per liter:

Pancreatic digest of casein	10.0g
Peptic digest of animal tissue	10.0g

NaCl 5.0g
Yeast extract 2.0g
Glucose 1.0g
$NaHSO_3$ 0.1g
FBP solution 30.0mL

pH 7.0 ± 0.2 at 25°C

FBP Solution:

Composition per 30.0mL:

$FeSO_4$ 0.25g
Sodium metabisulfite, anhydrous 0.25g
Sodium pyruvate, anhydrous 0.25g

Preparation of FBP Solution: Add components to distilled/deionized water and bring volume to 30.0mL. Mix thoroughly. Filter sterilize.

Preparation of Medium: Add components, except FBP solution, to distilled/deionized water and bring volume to 970.0mL. Mix thoroughly. Gently heat and bring to boiling. Autoclave for 15 min at 15 psi pressure–121°C. Cool to 45°–50°C. Aseptically add 30.0mL of sterile FBP solution. Mix thoroughly. Aseptically distribute into sterile tubes.

Use: For the cultivation of *Brucella* species.

Brucella Medium Base

Composition per liter:

Agar 15.0g
Glucose 10.0g
Peptone 10.0g
Beef extract 5.0g
NaCl 5.0g

pH 7.5 ± 0.2 at 25°C

Preparation: Add components to distilled/deionized water and bring volume to 1.0L. Mix thoroughly. Heat gently and bring to boiling. Distribute into tubes or flasks. Autoclave for 15 min at 15 psi pressure–121°C. Pour into sterile Petri dishes or leave in tubes.

Use: For the isolation of *Campylobacter* species.

Brucella Medium, Selective

Composition per liter:

Agar 15.0g
Pancreatic digest of casein 10.0g
Peptic digest of animal tissue 10.0g
NaCl 5.0g
Yeast extract 2.0g
Glucose 1.0g
$NaHSO_3$ 0.1g
Horse serum 100.0mL
VCNF antibiotic solution 10.0mL

pH 7.0 ± 0.2 at 25°C

Preparation of VCNF Antibiotic Solution: Add components to distilled/deionized water and bring volume to 10.0mL. Mix thoroughly. Filter sterilize.

Preparation of Medium: Add components, except horse serum and VCNF antibiotic solution, to distilled/deionized water and bring volume to 890.0mL. Mix thoroughly. Gently heat and bring to boiling. Autoclave for 15 min at 15 psi pressure–121°C. Cool to 45°–50°C. Aseptically add 100.0mL of sterile horse serum and 10.0mL of VCNF antibiotic solution. Mix thoroughly. Pour into sterile Petri dishes or distribute into sterile tubes.

Use: For the selective isolation, cultivation, and maintenance of *Brucella* species.

Brucella Selective Medium

Composition per liter:

Beef heart, infusion from 500.0g
Agar 15.0g
Tryptose 10.0g
NaCl 5.0g
Glucose 2.5g
Gelatin 1.0g
Sheep blood 100.0mL
Antibiotic solution 10.0mL

pH 7.4 ± 0.2 at 25°C

Antibiotic Solution:

Composition per 10.0mL:

Cycloheximide 1.0g
Bacitracin 250,000U
Circulin 250,000U
Polymyxin B 100,000U

Preparation of Antibiotic Solution: Add components to distilled/deionized water and bring volume to 10.0mL. Mix thoroughly. Filter sterilize.

Preparation of Medium: Add components, except sheep blood and antibiotic solution, to distilled/deionized water and bring volume to 890.0mL. Mix thoroughly. Gently heat and bring to boiling. Autoclave for 15 min at 15 psi pressure–121°C. Cool to 45°–50°C. Aseptically add 100.0mL of sterile sheep blood and 10.0mL of sterile antibiotic solution. Mix thoroughly. Pour into sterile Petri dishes or distribute into sterile tubes.

Use: For the selective isolation and cultivation of *Brucella* species.

Brucella Semisolid Medium with Cysteine

Composition per liter:

Peptamin 10.0g
Pancreatic digest of casein 10.0g
Glycine 10.0g

NaCl	5.0g
Yeast extract	2.0g
Agar	1.8g
Glucose	1.0g
L-Cysteine·HCl·H_2O	0.2g
$NaHSO_3$	0.1g
Sodium citrate	0.1g
Neutral Red solution	10.0mL

pH 7.0 ± 0.2 at 25°C

Neutral Red Solution:

Composition per 100.0mL:

Neutral Red	0.2g
Ethanol	10.0mL

Preparation of Neutral Red Solution: Add Neutral Red to 10.0mL of ethanol. Bring volume to 100.0mL.

Preparation of Medium: Add components to distilled/deionized water and bring volume to 1.0L. Mix thoroughly. Gently heat and bring to boiling. Distribute into tubes in 10.0mL volumes. Autoclave for 15 min at 15 psi pressure–121°C.

Use: For the cultivation and differentiation of *Campylobacter* species based on H_2S production from cysteine.

Brucella Semisolid Medium with Glycine

Composition per liter:

Peptamine	10.0g
Pancreatic digest of casein	10.0g
Glycine	10.0g
NaCl	5.0g
Yeast extract	2.0g
Agar	1.8g
Glucose	1.0g
$NaHSO_3$	0.1g
Sodium citrate	0.1g
Neutral Red solution	10.0mL

pH 7.0 ± 0.2 at 25°C

Neutral Red Solution:

Composition per 100.0mL:

Neutral Red	0.2g
Ethanol	10.0mL

Preparation of Neutral Red Solution: Add Neutral Red to 10.0mL of ethanol. Bring volume to 100.0mL.

Preparation of Medium: Add components to distilled/deionized water and bring volume to 1.0L. Mix thoroughly. Gently heat and bring to boiling. Distribute into tubes in 10.0mL volumes. Autoclave for 15 min at 15 psi pressure–121°C.

Use: For the cultivation and differentiation of *Campylobacter* species based on glycine utilization.

Brucella Semisolid Medium with NaCl

Composition per liter:

NaCl	35.0g
Peptamin	10.0g
Pancreatic digest of casein	10.0g
Yeast extract	2.0g
Agar	1.8g
Glucose	1.0g
$NaHSO_3$	0.1g
Sodium citrate	0.1g
Neutral Red solution	10.0mL

pH 7.0 ± 0.2 at 25°C

Neutral Red Solution:

Composition per 100.0mL:

Neutral Red	0.2g
Ethanol	10.0mL

Preparation of Neutral Red Solution: Add Neutral Red to 10.0mL of ethanol. Bring volume to 100.0mL.

Preparation of Medium: Add components to distilled/deionized water and bring volume to 1.0L. Mix thoroughly. Gently heat and bring to boiling. Distribute into tubes in 10.0mL volumes. Autoclave for 15 min at 15 psi pressure–121°C.

Use: For the cultivation and differentiation of *Campylobacter* species based on glycine utilization.

Brucella Semisolid Medium with Nitrate

Composition per liter:

Peptamin	10.0g
Pancreatic digest of casein	10.0g
Glycine	10.0g
KNO_3	10.0g
NaCl	5.0g
Yeast extract	2.0g
Agar	1.8g
Glucose	1.0g
$NaHSO_3$	0.1g
Sodium citrate	0.1g

pH 7.0 ± 0.2 at 25°C

Preparation of Medium: Add components to distilled/deionized water and bring volume to 1.0L. Mix thoroughly. Gently heat and bring to boiling. Distribute into tubes in 10.0mL volumes. Autoclave for 15 min at 15 psi pressure–121°C.

Use: For the cultivation and differentiation of *Campylobacter* species based on nitrate reduction.

BSK Medium

Composition per 1260.0mL:

Bovine albumin fraction V	50.0g
HEPES (*N*-[2-Hydroxyethyl]piperazine-*N'*-2-ethanesulfonic acid) buffer	6.0g
Neopeptone	5.0g
Glucose	5.0g
$NaHCO_3$	2.2g
Sodium pyruvate	0.8g
Sodium citrate	0.7g
N-Acetylglucosamine	0.4g
Gelatin solution	200.0mL
CMRL 1066, without glutamine, without bicarbonate, 10X	100.0mL
Rabbit serum	72.0mL

pH 7.6-7.65 at 25°C

Gelatin Solution:

Composition per 200.0mL:

Gelatin	14.0g

Preparation of Gelatin Solution: Add gelatin to distilled/deionized water and bring volume to 200.0mL. Heat gently to boiling. Mix thoroughly. Filter sterilize.

CMRL 1066 Medium without Glutamine, without Bicarbonate, 10X:

Composition per liter:

NaCl	6.8g
D-Glucose	1.0g
KCl	0.4g
L-Cysteine·HCl·H_2O	0.26g
$CaCl_2$, anhydrous	0.2g
$MgSO_4$·$7H_2O$	0.2g
NaH_2PO_4·H_2O	0.14g
Sodium acetate·$3H_2O$	0.083g
L-Glutamic acid	0.075g
L-Arginine·HCl	0.07g
L-Lysine·HCl	0.07g
L-Leucine	0.06g
Glycine	0.05g
Ascorbic acid	0.05g
L-Proline	0.04g
L-Tyrosine	0.04g
L-Aspartic acid	0.03g
L-Threonine	0.03g
L-Alanine	0.025g
L-Phenylalanine	0.025g
L-Serine	0.025g
L-Valine	0.025g
L-Cystine	0.02g
L-Histidine·HCl·H_2O	0.02g
L-Isoleucine	0.02g
Phenol Red	0.02g
L-Methionine	0.015g
Deoxyadenosine	0.01g
Deoxycytidine	0.01g
Deoxyguanosine	0.01g
Glutathione, reduced	0.01g
Thymidine	0.01g
Hydroxy-L-proline	0.01g
L-Tryptophan	0.01g
Nicotinamide adenine dinucleotide	7.0mg
Tween 80	5.0mg
Sodium glucoronate·H_2O	4.2mg
Coenzyme A	2.5mg
Cocarboxylase	1.0mg
Flavin adenine dinucleotide	1.0mg
Nicotinamide adenine dinucleotide phosphate	1.0mg
Uridine triphosphate	1.0mg
Choline chloride	0.5mg
Cholesterol	0.2mg
5-Methyldeoxycytidine	0.1mg
Inositol	0.05mg
p-Aminobenzoic acid	0.05mg
Niacin	0.025mg
Niacinamide	0.025mg
Pyridoxine	0.025mg
Pyridoxal·HCl	0.025mg
Biotin	0.01mg
D-Calcium pantothenate	0.01mg
Folic acid	0.01mg
Riboflavin	0.01mg
Thiamine·HCl	0.01mg

pH 7.2 ± 0.2 at 25°C

Preparation of CMRL 1066 Medium without Glutamine, without Bicarbonate, 10X: Add components to distilled/deionized water and bring volume to 1.0L. Mix thoroughly. Adjust pH to 7.2. Filter sterilize.

Preparation of Medium: Add components, except gelatin solution and rabbit serum, to 628.0mL of glass-distilled water. Mix thoroughly. Adjust pH to 7.6–7.65. Add 200.0mL of 7% aqueous gelatin solution. Filter sterilize entire medium. Aseptically add 72.0mL of sterile rabbit serum.

Use: For the cultivation of a wide variety of microorganisms in a chemically defined medium. For the cultivation of *Borrelia* and *Spirochaeta* species.

BSK Medium, Modified

Composition per 1264.0mL:

Bovine serum albumin, fraction V	50.0g
HEPES (*N*-[2-Hydroxymethyl]piperazine-*N'* [ethane sulfonate]) buffer	6.0g
Neopeptone	5.0g
Glucose	5.0g
Yeastolate	2.54g
$NaHCO_3$	2.2g

Sodium pyruvate....0.8g
Sodium citrate....0.7g
$MgSO_4 \cdot 7H_2O$....0.6g
N-Acetylglucosamine....0.4g
$CaCl_2 \cdot 2H_2O$....0.07g
CMRL 1066, 10X
without glutamine or $NaHCO_3$....100.0mL
Rabbit serum, heat inactivated....64.0mL
pH 7.5 ± 0.2 at 25°C

CMRL 1066, 10X without Glutamine or $NaHCO_3$:
Composition per liter:
NaCl....6.8g
D-Glucose....1.0g
KCl....0.4g
L-Cysteine·HCl·H_2O....0.26g
$CaCl_2$, anhydrous....0.2g
$MgSO_4 \cdot 7H_2O$....0.2g
$NaH_2PO_4 \cdot H_2O$....0.14g
Sodium acetate·$3H_2O$....0.083g
L-Glutamic acid....0.075g
L-Arginine·HCl....0.070g
L-Lysine·HCl....0.070g
L-Leucine....0.060g
Glycine....0.050g
Ascorbic acid....0.050g
L-Proline....0.040g
L-Tyrosine....0.040g
L-Aspartic acid....0.030g
L-Threonine....0.030g
L-Alanine....0.025g
L-Phenylalanine....0.025g
L-Serine....0.025g
L-Valine....0.025g
L-Cystine....0.020g
L-Histidine·HCl·H_2O....0.020g
L-Isoleucine....0.020g
Phenol Red....0.020g
L-Methionine....0.015g
Deoxyadenosine....0.010g
Deoxycytidine....0.010g
Deoxyguanosine....0.010g
Glutathione, reduced....0.010g
Thymidine....0.010g
Hydroxy-L-proline....0.010g
L-Tryptophan....0.010g
Nicotinamide adenine dinucleotide....7.0mg
Tween 80....5.0mg
Sodium glucoronate·H_2O....4.2mg
Coenzyme A....2.5mg
Cocarboxylase....1.0mg
Flavin adenine dinucleotide....1.0mg
Nicotinamide adenine
dinucleotide phosphate....1.0mg
Uridine triphosphate....1.0mg
Choline chloride....0.50mg
Cholesterol....0.20mg
5-Methyldeoxycytidine....0.10mg
Inositol....0.05mg
p-Aminobenzoic acid....0.05mg
Niacin....0.025mg
Niacinamide....0.025mg
Pyridoxine....0.025mg
Pyridoxal·HCl....0.025mg
Biotin....0.01mg
D-Calcium pantothenate....0.01mg
Folic acid....0.01mg
Riboflavin....0.01mg
Thiamine·HCl....0.01mg

Preparation of CMRL 1066, 10X Without Glutamine or $NaHCO_3$: Add components to distilled/deionized water and bring volume to 1.0L. Mix thoroughly. Adjust pH to 7.2. Filter sterilize.

Preparation of Medium: Add components, except CMRL 1066, 10X without glutamine or $NaHCO_3$ and rabbit serum, to distilled/deionized water and bring volume to 1100.0mL. Mix thoroughly. Adjust pH to 7.5 with NaOH. Filter sterilize. Aseptically add 100.0mL of sterile CMRL 1066, 10X without glutamine or $NaHCO_3$ and 64.0mL of sterile rabbit serum. Mix thoroughly. Aseptically distribute 10.0mL volumes into sterile 16 × 125.0mm test tubes.

Use: For the cultivation of *Borrellia afzelii, Borrelia burgdorferi,* and *Borrelia gorinii.*

BSK Medium, Revised

Composition per 1164.0mL:
Bovine serum albumin fraction V....50.0g
HEPES (*N*-[2-Hydroxyethyl]piperazine-*N′*-2-ethanesulfonic acid) buffer....6.0g
Neopeptone....5.0g
Glucose....5.0g
TC-Yeastolate....2.54g
$NaHCO_3$....2.2g
Sodium pyruvate....0.8g
Sodium citrate....0.7g
N-Acetyl glucosamine....0.4g
CMRL 1066, without glutamine,
without bicarbonate, 10X....100.0mL
Rabbit serum....64.0mL
pH 7.6–7.65 at 25°C

CMRL 1066 Medium without Glutamine, without Bicarbonate, 10X:
Composition per liter:
NaCl....6.8g
D-Glucose....1.0g
KCl....0.4g

L-Cysteine·HCl·H_2O	0.26g
$CaCl_2$, anhydrous	0.2g
$MgSO_4 \cdot 7H_2O$	0.2g
$NaH_2PO_4 \cdot H_2O$	0.14g
Sodium acetate·$3H_2O$	0.083g
L-Glutamic acid	0.075g
L-Arginine·HCl	0.070g
L-Lysine·HCl	0.070g
L-Leucine	0.060g
Glycine	0.050g
Ascorbic acid	0.050g
L-Proline	0.040g
L-Tyrosine	0.040g
L-Aspartic acid	0.030g
L-Threonine	0.030g
L-Alanine	0.025g
L-Phenylalanine	0.025g
L-Serine	0.025g
L-Valine	0.025g
L-Cystine	0.020g
L-Histidine·HCl·H_2O	0.020g
L-Isoleucine	0.020g
Phenol Red	0.020g
L-Methionine	0.015g
Deoxyadenosine	0.010g
Deoxycytidine	0.010g
Deoxyguanosine	0.010g
Glutathione, reduced	0.010g
Thymidine	0.010g
Hydroxy-L-proline	0.010g
L-Tryptophan	0.010g
Nicotinamide adenine dinucleotide	7.0mg
Tween 80	5.0mg
Sodium glucoronate·H_2O	4.2mg
Coenzyme A	2.5mg
Cocarboxylase	1.0mg
Flavin adenine dinucleotide	1.0mg
Nicotinamide adenine dinucleotide phosphate	1.0mg
Uridine triphosphate	1.0mg
Choline chloride	0.50mg
Cholesterol	0.20mg
5-Methyldeoxycytidine	0.10mg
Inositol	0.05mg
p-Aminobenzoic acid	0.05mg
Niacin	0.025mg
Niacinamide	0.025mg
Pyridoxine	0.025mg
Pyridoxal·HCl	0.025mg
Biotin	0.01mg
Calcium DL-pantothenate	0.01mg
Folic acid	0.01mg
Riboflavin	0.01mg
Thiamine·HCl	0.01mg

pH 7.2 ± 0.2 at 25°C

Preparation of CMRL 1066 Medium without Glutamine, without Bicarbonate, 10X: Add components to distilled/deionized water and bring volume to 1.0L. Mix thoroughly. Adjust pH to 7.2. Filter sterilize.

Preparation of Medium: Add components, except rabbit serum and CMRL 1066, to 1.0L of glass-distilled/deionized water. Mix thoroughly. Adjust pH to 7.5 with NaOH. Filter sterilize. Aseptically add 100.0mL of sterile CMRL 1066 and 64.0mL of sterile rabbit serum. Adjust final pH to 7.5–7.6. Aseptically distribute into sterile tubes or flasks.

Use: For the cultivation of *Borrelia burgdorferi, Borrelia afzelii, Borrelia garinii, Borrelia anserina,* and *Borrelia japonica.*

BTB Lactose Agar (Bromthymol Blue Lactose Agar)

Composition per liter:

Agar	15.0g
Lactose	10.0g
Proteose peptone	5.0g
Beef extract	3.0g
Bromthymol Blue	0.17g

pH 8.7–7.2 at 25°C

Preparation of Medium: Add components to distilled/deionized water and bring volume to 1.0L. Mix thoroughly. Heat gently with frequent mixing. Bring to boiling. Distribute into tubes or flasks. Autoclave for 15 min at 15 psi pressure–121°C. Pour into sterile Petri dishes if desired.

Use: For the isolation and cultivation of pathogenic staphylococci.

BTB Teepol® Agar

Composition per liter:

NaCl	20.0g
Agar	15.0g
Peptone	10.0g
Sucrose	10.0g
Beef extract	5.0g
Bromthymol Blue	0.08g
Teepol	2.0mL

pH 7.8 ± 0.2 at 25°C

Preparation of Medium: Add components to distilled/deionized water and bring volume to 1.0L. Teepol may be substituted by 0.1mL of Tergitol™ 7. Mix thoroughly. Gently heat and bring to boiling. Adjust pH to 7.8. Autoclave for 15 min at 15 psi pressure–121°C. Pour into sterile Petri dishes.

Use: For the isolation and cultivation of *Vibrio anguillarum.*

BTU Medium

Composition per liter:

Ground meat, fat free	500.0g
Pancreatic digest of casein	30.0g
K_2HPO_4	5.0g
Yeast extract	5.0g
L-Cysteine·HCl	0.5g
Resazurin	1.0mg
NaOH (1*N* solution)	25.0mL
Formate-fumarate solution	4.26mL

pH 7.0 ± 0.2 at 25°C

Formate-Fumarate Solution:

Composition per 100.0mL:

Sodium formate	6.0g
Sodium fumarate	6.0g

Preparation of Formate-Fumarate Solution: Add components to distilled/deionized water and bring volume to 100.0mL. Mix thoroughly. Filter sterilize.

Preparation of Medium: Use lean beef or horse meat. Remove fat and connective tissue. Grind finely. Add ground meat and 25.0mL of NaOH solution to distilled/deionized water and bring volume to 1025.0mL. Gently heat and bring to boiling. Continue boiling for 15 min without stirring. Cool to room temperature. Remove fat from surface. Filter and retain both meat particles and filtrate. Adjust volume of filtrate to 1.0L with distilled/deionized water. Add casitone, K_2HPO_4, yeast extract, and resazurin. Gently heat and bring to boiling. Boil for 1–2 min. Add L-cysteine·HCl. Mix thoroughly. Distribute 7.0mL into tubes that contain meat particles (1 part meat particles to 5 parts fluid). Autoclave for 30 min at 15 psi pressure–121°C. Prior to inoculation, add 30.0μL of formate-fumarate solution for each milliliter of medium in the tubes.

Use: For the cultivation of *Bacteroides ureolyticus.*

Buffered Charcoal Yeast Extract Differential Agar (DIFF/BCYE)

Composition per 1014.0mL:

Agar	17.0g
ACES (2-[(2-Amino-2-oxoethyl)-amino]-ethane sulfonic acid) buffer	10.0g
Yeast extract	10.0g
Charcoal, activated	1.5g
$Fe_4(P_2O_7)_3 \cdot 9H_2O$	0.25g
Bromcresol Purple	0.01g
Bromthymol Blue	0.01g
Antibiotic solution	10.0mL
L-Cysteine·HCl·H_2O solution	4.0mL

pH 6.9 ± 0.2 at 25°C

Antibiotic Solution:

Composition per 10.0mL:

Vancomycin	1.0mg
Polymyxin B	50,000U

Preparation of Antibiotic Solution: Add components to distilled/deionized water and bring volume to 10.0mL. Mix thoroughly. Filter sterilize.

L-Cysteine·HCl·H_2O Solution:

Composition per 10.0mL:

L-Cysteine·HCl·H_2O	1.0g

Preparation of L-Cysteine·HCl·H_2O Solution: Add L-cysteine·HCl·H_2O to distilled/deionized water and bring volume to 10.0mL. Mix thoroughly. Filter sterilize.

Preparation of Medium: Add components, except L-cysteine·HCl·H_2O solution and antibiotic solution, to distilled/deionized water and bring volume to 1.0L. Mix thoroughly. Adjust medium to pH 6.9 with 1*N* KOH. Heat gently and bring to boil for 1 min. Autoclave for 15 min at 15 psi pressure–121°C. Cool to 50°–55°C. Add 4.0mL of sterile L-cysteine·HCl·H_2O solution and 10.0mL of sterile antibiotic solution. Mix thoroughly. Pour into sterile Petri dishes with constant agitation to keep charcoal in suspension.

Use: For the isolation, cultivation, and maintenance of *Legionella pneumophila* and other *Legionella* species from environmental and clinical specimens. For the selective recovery of *Legionella pneumophila* while reducing contaminating microorganisms from environmental water samples.

Buffered Enrichment Broth

Composition per liter:

Na_2HPO_4	9.6g
KH_2PO_4	1.35g
Pyruvic acid solution	11.1mL
Nalidixic acid solution	8.0mL
Cyclohiximide solution	5.0mL
Acriflavin solution	2.0mL

pH 7.3 ± 0.1 at 25°C

Nalidixic Acid Solution:

Composition per 10.0mL:

Nalidixic acid, sodium salt	0.05g

Preparation of Nalidixic Acid Solution: Add nalidixic acid to distilled/deionized water and bring volume to 10.0mL. Mix thoroughly. Filter sterilize.

Acriflavin Solution:

Composition per 10.0mL:

Acriflavin·HCl	0.05g

Preparation of Acriflavin Solution: Add acriflavin·HCl to distilled/deionized water and bring volume to 10.0mL. Mix thoroughly. Filter sterilize.

Cycloheximide Solution:
Composition per 10.0mL:

Cycloheximide	0.1g
Ethanol, 40%	10.0mL

Preparation of Cycloheximide Solution: Add cycloheximide to 40% ethanol and bring volume to 10.0mL. Mix thoroughly. Filter sterilize.

Pyruvate Solution:
Composition per 20.0mL:

Na-pyruvate	2.0g

Preparation of Pyruvate Solution: Add Na-pyruvate to distilled/deionized water and bring volume to 20.0mL. Mix thoroughly. Filter sterilize.

Preparation of Medium: Add components, except pyruvate solution, nalidixic acid solution, acriflavin solution, and cycloheximide solution, to distilled/deionized water and bring volume to 973.9.0L. Mix thoroughly. Gently heat and bring to boiling. Autoclave for 15 min at 15 psi pressure–121°C. Cool to 25°C. Aseptically add 11.1mL sterile pyruvate solution. Mix thoroughly. Aseptically add 8.0mL sterile nalidixic acid solution, 5.0mL sterile cycloheximide solution, and 2.0mL sterile acriflavin solution. Mix thoroughly. Aseptically distribute into sterile tubes or flasks.

Use: For the cultivation of of *Listeria* spp.

Buffered Peptone Water

Composition per liter:

Pancreatic digest of gelatin	10.0g
NaCl	5.0g
Na_2HPO_4	3.5g
KH_2PO_4	1.5g

pH 7.2 ± 0.2 at 25°C

Source: This medium is available as a premixed powder from BD Diagnostic Systems and Oxoid Unipath.

Preparation of Medium: Add components to distilled/deionized water and bring volume to 1.0L. Mix thoroughly. Distribute into tubes or flasks. Autoclave for 15 min at 15 psi pressure–121°C.

Use: Used as a preenrichment medium for the isolation of *Salmonella*, especially injured microorganisms.

Burkholderia cepacia Agar

Composition per liter:

Agar	12.0g
Sodium pyruvate	7.0g
Peptone	5.0g
KH_2PO_4	4.4g
Yeast Extract	4.0g
Bile salts	1.5g
Na_2HPO_4	1.4g
$(NH_4)_2SO_4$	1.0g
$MgSO_4$	0.2g
Phenol Red	0.02g
$Fe(NH_4)_2(SO_4)_2 \cdot 6H_2O$	0.01g
Crystal Violet	0.001g
Selective supplement solution	10.0mL

pH 6.2 ± 0.2 at 25°C

Source: This medium is available as a premixed powder from Oxoid Unipath.

Selective Supplement Solution:
Composition per 10.0mL:

Polymyxin B	150,000IU
Ticarcillin	100.0mg
Gentamicin	5.0mg

Preparation of Selective Supplement Solution: Add components to distilled/deionized water and bring volume to 10.0mL. Mix thoroughly. Filter sterilize.

Preparation of Medium: Add components, except selective supplement solution, to distilled/deionized water and bring volume to 990.0mL. Mix thoroughly. Gently heat while stirring and bring to boiling. Autoclave for 15 min at 15 psi pressure–121°C. Cool to 50°C. Aseptially add 10.0mL selective supplement solution. Mix thoroughly. Pour into sterile Petri dishes.

Use: For the selective isolation of *Burkholderia cepacia* from the respiratory secretions of patients with cystic fibrosis and for routine testing of nonsterile inorganic salt solutions containing preservative.Slow growing *B. cepacia* can be missed on conventional media such as blood or MacConkey Agar due to overgrowth caused by other faster growing organisms found in the respiratory tract of CF patients such as mucoid *Klebsiella* species, *Pseudomonas aeruginosa,* and *Staphylococcus* species. This may lead to the infection being missed or wrongly diagnosed.

BYE Agar

Composition per liter:

Pancreatic digest of casein	16.0g
Agar	13.5g
Brain heart, solids from infusion	8.0g
Peptic digest of animal tissue	5.0g
NaCl	5.0g
Na_2HPO_4	2.5g
Glucose	2.0g

Yeast extract 2.0g
Blood, human or animal, sterile 150.0mL
pH 7.8–8.0 ± 0.2 at 25°C

Preparation of Medium: Add components, except blood, to distilled/deionized water and bring volume to 850.0mL. Mix thoroughly. Autoclave for 15 min at 15 psi pressure–121°C. Cool to 45°–50°C. Aseptically add 150.0mL of sterile blood. Outdated, citrated, or heparinized blood (blood from a blood bank is acceptable). Pour into sterile Petri dishes.

Use: For the isolation and cultivation of *Mycoplasma* species and L-forms of bacteria. For the detection of *Mycoplasma* species in tissue culture and cell lines.

BYEB (Buffered Yeast Extract Broth)

Composition per liter:

ACES buffer (2-[(2-Amino-2-oxoethyl)-amino]-ethane sulfonic acid) 10.0g
Yeast extract 10.0g
α-Ketoglutarate 1.0g
L-Cysteine·HCl·H_2O 0.4g
$Fe_4(P_2O_7)_3 \cdot 9H_2O$ 0.25g
pH 6.9 ± 0.2 at 25°C

Preparation of Medium: Add components to distilled/deionized water and bring volume to 1.0L. Mix thoroughly. Adjust pH to 6.9. Filter sterilize. Aseptically distribute into sterile tubes or flasks.

Use: For the cultivation of *Legionella pneumophila*.

Caffeic Acid Ferric Citrate Test Medium (CAFC Test Medium) (Caffeic Acid Agar)

Composition per liter:

Agar 20.0g
$(NH_4)_2SO_4$ 5.0g
Glucose 5.0g
Yeast extract 2.0g
K_2HPO_4 .... 0.8g
$MgSO_4 \cdot 3H_2O$ 0.7g
Caffeic acid·1/2H_2O 0.18g
Chloramphenicol 0.05g
Ferric citrate solution 4.0mL
pH 6.5 ± 0.2 at 25°C

Ferric Citrate Solution:

Composition per 20.0mL:

Agar 100.0mg

Preparation of Ferric Citrate Solution: Add ferric citrate to 20.0mL of distilled/deionized water. Mix thoroughly.

Preparation of Medium: Add components, except chloramphenicol, to distilled/deionized water and bring volume to 1.0L. Mix thoroughly. Heat to boiling. Autoclave for 15 min at 15 psi pressure–121°C. Cool to 45°–50°C. Aseptically add 0.05g of chloramphenicol. Mix thoroughly. Pour into sterile Petri dishes.

Use: For the isolation and presumptive identification of *Cryptococcus neoformans*. *Cryptococcus neoformans* appears as dark brown colonies. All other *Cryptococcus* species appear as light brown or nonpigmented colonies.

Caffeine Medium

Composition per liter:

Agar 15.0g
Solution A 400.0mL
Solution B 400.0mL
Solution C 200.0mL
pH 5.0 ± 0.2 at 25°C

Solution A:

Composition per 400.0mL:

Na_2HPO_4 .... 7.8g
KH_2PO_4 .... 3.0g
Caffeine 1.0g
NaCl 0.58g

Preparation of Solution A: Add components to distilled/deionized water and bring volume to 400.0mL. Mix thoroughly. Adjust pH to 5.0.

Solution B:

Composition per 400.0mL:

$MgSO_4 \cdot 7H_2O$ 0.12g
$CaCl_2 \cdot 2H_2O$ 11.0mg

Preparation of Solution B: Add components to distilled/deionized water and bring volume to 400.0mL. Mix thoroughly.

Solution C:

Composition per 200.0mL:

$FeCl_3$ 16.0mg

Preparation of Solution C: Add $FeCl_3$ to distilled/deionized water and bring volume to 200.0mL. Mix thoroughly.

Preparation of Medium: To 400.0mL of solution A, add 400.0mL of solution B and 200.0mL of solution C. Adjust pH to 5.0. Add agar. Mix thoroughly. Gently heat and bring to boiling. Autoclave for 15 min at 15 psi pressure–121°C. Pour into sterile Petri dishes or distribute into sterile tubes.

Use: For the cultivation of *Pseudomonas* species.

CAL Agar
(Cellobiose Arginine Lysine Agar)
(*Yersinia* Isolation Agar)

Composition per liter:

Agar	20.0g
L-Arginine·HCl	6.5g
L-Lysine·HCl	6.5g
NaCl	5.0g
Cellobiose	3.5g
Yeast extract	3.0g
Sodium deoxycholate	1.5g
Neutral Red	0.03g

Preparation of Medium: Add components to distilled/deionized water and bring volume to 1.0L. Mix thoroughly. Heat to boiling. Do not autoclave. Pour into sterile Petri dishes.

Use: For the isolation and characterization of *Yersinia enterocolitica* from fecal specimens and enumeration of *Yersinia enterocolitica* from water and other liquid specimens.

CAL Broth
(Cellobiose Arginine Lysine Broth)

Composition per liter:

L-Arginine·HCl	6.5g
L-Lysine·HCl	6.5g
NaCl	5.0g
Cellobiose	3.5g
Yeast extract	3.0g
Sodium deoxycholate	1.5g
Neutral Red	0.03g

Preparation of Medium: Add components to distilled/deionized water and bring volume to 1.0L. Mix thoroughly. Heat to boiling. Do not autoclave. Distribute into sterile tubes in 6.0–8.0mL volumes.

Use: For the isolation and characterization of *Yersinia enterocolitica* from fecal specimens and enumeration of *Yersinia enterocolitica* from water and other liquid specimens.

Calymmatobacterium granulomatis Semidefined Medium

Composition per liter:

Papaic digest of soybean meal	20.0g
NaCl	2.5g
K_2HPO_4	1.5g
Sodium thioglycolate	0.6g
L-Cystine	0.4g

pH 7.2 ± 0.2 at 25°C

Preparation of Medium: Add components to distilled/deionized water and bring volume to 1.0L. Mix thoroughly. Adjust pH to 7.2. Distribute into screw-capped tubes in 20–22mL volumes. Autoclave for 15 min at 15 psi pressure–121°C. Tighten screw caps.

Use: For the cultivation of *Calymmatobacterium granulomatis.*

CAMG Broth

Composition per liter:

K_2HPO_4	5.0g
Pancreatic digest of casein	5.0g
Yeast extract	5.0g
Glucose	2.0g
Tween 80	0.5mL

pH 7.0 ± 0.1 at 25°C

Preparation of Medium: Prepare and dispense anaerobically under an atmosphere of 80% N_2 + 10% CO_2 + 10% H_2. Add components to distilled/deionized water and bring volume to 1.0L. Mix thoroughly. Sparge with a gas mixture of 80% N_2 + 10% CO_2 + 10% H_2. Adjust pH to 7.5. Distribute into tubes or flasks. Autoclave for 15 min at 15 psi pressure–121°C.

Use: For the cultivation of *Actinomyces naeslundii.*

Campy THIO Medium

Composition per liter:

Pancreatic digest of casein	20.0g
Agar	15.0g
NaCl	2.5g
K_2HPO_4	1.5g
Sodium thioglycolate	0.6g
L-Cystine	0.4g
Na_2SO_3	0.2g
Antibiotic supplement	10.0mL

Antibiotic Supplement:

Composition per 10.0mL:

Cephalothin	15.0mg
Vancomycin	10.0mg
Trimethoprim	5.0mg
Amphotericin B	2.0mg
Polymyxin B	2500U

Preparation of Antibiotic Supplement: Add components to 10.0mL of distilled/deionized water. Filter sterilize.

Preparation of Medium: Add components, except antibiotic solution, to distilled/deionized water and bring volume to 990.0mL. Mix thoroughly. Gently heat and bring to boiling. Autoclave for 15 min at 15 psi pressure–121°C. Cool to 45°–50°C. Aseptically add 10.0mL of sterile antibiotic solution. Mix thoroughly. Aseptically distribute into sterile screw-capped tubes in 3.0mL volumes for 1.5cm swabs or 5.0mL volumes for 3.0cm swabs.

Use: For the maintenance—as a holding or transport medium—of *Campylobacter* species isolated from clinical specimens on swabs.

Campylobacter Agar

Composition per liter:

Agar	15.0g
Polypeptone	10.0g
Meat extract	5.0g
NaCl	5.0g
Yeast extract	5.0g
Sodium L-glutamate	2.0g
Sodium succinate·$6H_2O$	2.0g
$MgCl_2·6H_2O$	1.0g
Sheep blood, defibrinated	50.0mL

pH 7.0 ± 0.2 at 25°C

Preparation of Medium: Add components, except sheep blood, to distilled/deionized water and bring volume to 950.0mL. Mix thoroughly. Gently heat and bring to boiling. Adjust pH to 7.0. Autoclave for 15 min at 15 psi pressure–121°C. Cool to 50°–55°C. Aseptically add 50.0mL of sterile defibrinated sheep blood. Mix thoroughly. Pour into sterile Petri dishes or distribute into sterile tubes.

Use: For the cultivation of *Moraxella lincolnii.*

Campylobacter Agar with 5 Antimicrobics and 10% Sheep Blood

Composition per liter:

Agar	15.0g
Pancreatic digest of casein	10.0g
Peptic digest of animal tissue	10.0g
NaCl	5.0g
Yeast extract	2.0g
Glucose	1.0g
$NaHSO_3$	0.1g
Sheep blood, defibrinated	100.0mL
Antibiotic supplement	10.0mL

pH 7.2 ± 0.2 at 25°C

Source: This medium is available as a prepared medium from BD Diagnostic Systems.

Antibiotic Supplement:

Composition per 10.0mL:

Cephalothin	0.015g
Vancomycin	0.01g
Trimethoprim	5.0mg
Amphotericin B	2.0mg
Polymyxin B	2500U

Preparation of Antibiotic Supplement: Add components to 10.0mL of distilled/deionized water. Filter sterilize.

Preparation of Medium: Add components, except sheep blood and antibiotic solution, to distilled/deionized water and bring volume to 890.0mL. Mix thoroughly. Gently heat and bring to boiling. Autoclave for 15 min at 15 psi pressure–121°C. Cool to 45°–50°C. Aseptically add 100.0mL of sterile sheep blood and 10.0mL of sterile antibiotic solution. Mix thoroughly. Pour into sterile Petri dishes or distribute into sterile tubes.

Use: For the primary selective isolation and cultivation of *Campylobacter jejuni* from human fecal specimens.

Campylobacter Agar, Blaser's (Blaser's *Campylobacter* Agar)

Composition per liter:

Campylobacter agar base	990.0mL
Supplement B	10.0mL

pH 7.4 ± 0.2 at 25°C

Campylobacter Agar Base:

Composition per liter:

Proteose peptone	15.0g
Agar	12.0g
NaCl	5.0g
Yeast extract	5.0g
Liver digest	2.5g

Source: *Campylobacter* agar base and *Campylobacter* antimicrobic supplement B are available as a premixed powder from BD Diagnostic Systems.

Preparation of *Campylobacter* Agar Base: Add components to distilled/deionized water and bring volume to 990.0mL. Mix thoroughly. Gently heat and bring to boiling. Autoclave for 15 min at 15 psi pressure–121°C. Cool to 45°–50°C.

Supplement B:

Composition per 10.0mL:

Cephalothin	15.0mg
Vancomycin	10.0mg
Trimethoprim	5.0mg
Amphotericin B	2.0mg
Polymyxin B	2500U

Preparation of Supplement B: Add components to 10.0mL of distilled/deionized water. Filter sterilize.

Preparation of Medium: Prepare 990.0mL of *Campylobacter* agar base. Autoclave and cool to 45°–50°C. Aseptically add 10.0mL of sterile supplement B. Mix thoroughly. Pour into sterile Petri dishes.

Use: For the selective isolation of *Campylobacter jejuni* from fecal specimens.

Campylobacter Agar, Skirrow's (Skirrow's *Campylobacter* Agar)

Composition per liter:

Campylobacter agar base	990.0mL
Supplement S	10.0mL

pH 7.4 ± 0.2 at 25°C

***Campylobacter* Agar Base:**

Composition per liter:

Proteose peptone	15.0g
Agar	12.0g
NaCl	5.0g
Yeast extract	5.0g
Liver digest	2.5g

Source: *Campylobacter* agar base and *Campylobacter* antimicrobic supplement S are available as a premixed powder from BD Diagnostic Systems.

Preparation of *Campylobacter* Agar Base: Add components to distilled/deionized water and bring volume to 990.0mL. Mix thoroughly. Gently heat and bring to boiling. Autoclave for 15 min at 15 psi pressure–121°C. Cool to 45°–50°C.

Supplement S:

Composition per 10.0mL:

Vancomycin	10.0mg
Trimethoprim	5.0mg
Polymyxin B	2500U

Preparation of Supplement S: Add components to 10.0mL of distilled/deionized water. Filter sterilize.

Preparation of Medium: Prepare 990.0mL of *Campylobacter* agar base. Autoclave for 15 min at 15 psi pressure–121°C. Cool to 45°–50°C. Aseptically add 10.0mL of sterile supplement S. Mix thoroughly. Pour into sterile Petri dishes.

Use: For the selective isolation of *Campylobacter jejuni* from fecal specimens.

Campylobacter Blood-Free Agar Base Modified (CCDA, Modified)

Composition per liter:

Peptone	20.0g
Agar	12.0g
NaCl	5.0g
Activated charcoal	4.0g
Casein hydrolysate	3.0g
Na-desoxycholate	1.0g
Na-pyruvate	0.25g
$FeSO_4$	0.25g
Cefaperazone-amhotericin B solution	10.0mL

pH 7.4 ± 0.2 at 25°C

Cefoperazone-Amphotericin B Solution:

Composition per 10.0mL:

Cefoperazone	0.016g
Amphotericn B	0.005g

Preparation of Cefoperazone-Amphotericin B Solution: Add cefoperazone and amphotericin B to distilled/deionized water and bring volume to 10.0mL. Mix thoroughly. Filter sterilize.

Preparation of Medium: Add components except cefaperazone-amphotericin B solution to distilled/deionized water and bring volume to 990.0mL. Mix thoroughly. Autoclave for 15 min at 15 psi pressure–121°C. Cool to 45°–50°C. Aseptically add 10.0mL of cefoperazone-amphotericin B solution. Mix thoroughly. Pour into sterile Petri dishes or aseptically distribute into tubes or flasks.

Use: For the isolation of *Campylobacter* spp. Amphotericin largely reduces the growth of yeasts and molds. Cefoperazone especially inhibits Enterobacteriaceae.

Campylobacter Blood-Free Selective Agar

Composition per liter:

Agar	12.0g
Beef extract	10.0g
Peptone	10.0g
Charcoal	4.0g
Casein hydrolysate	3.0g
Sodium deoxycholate	1.0g
$Fe_2SO_4 \cdot H_2O$	0.25g
Sodium pyruvate	0.25g
Cefoperazone solution	10.0mL

pH 7.4 ± 0.2 at 25°C

Cefoperazone Solution:

Composition per 10.0mL:

Sodium cefoperazone	0.032g

Preparation of Cefoperazone Solution: Add sodium cefoperazone to distilled/deionized water and bring volume to 10.0mL. Mix thoroughly. Filter sterilize.

Preparation of Medium: Add components, except cefoperazone solution, to distilled/deionized water and bring volume to 990.0mL. Mix thoroughly. Heat with frequent agitation and boil for 1 min to completely dissolve. Autoclave for 15 min at 15 psi pressure–121°C. Cool to 50°–55°C. Add 10.0mL of sterile cefoperazone solution. Addition of 10.0mg/L of amphotericin B improves the selectivity of the medium. Mix thoroughly. Pour into sterile Petri dishes.

Use: For the selective isolation of *Campylobacter* species, especially *Campylobacter jejuni, Campylobacter coli*, and *Campylobacter laridis*.

Campylobacter Charcoal Differential Agar (CCDA) (Preston Blood-Free Medium)

Composition per liter:

Agar	12.0g
Beef extract	10.0g
Peptone	10.0g
NaCl	5.0g
Charcoal	4.0g
Casein hydrolysate	3.0g
Sodium deoxycholate	1.0g
$FeSO_4$	0.25g
Sodium pyruvate	0.25g
Cefoperazone solution	10.0mL

pH 7.5 ± 0.2 at 25°C

Cefoperazone Solution:

Composition per 10.0mL:

Sodium cefoperazone	0.032g

Preparation of Cefoperazone Solution: Add sodium cefoperazone to distilled/deionized water and bring volume to 10.0mL. Mix thoroughly. Filter sterilize.

Preparation of Medium: Add components, except cefoperazone solution, to distilled/deionized water and bring volume to 990.0mL. Mix thoroughly. Gently heat and bring to boiling. Autoclave for 15 min at 15 psi pressure–121°C. Cool to 45°–50°C. Aseptically add 10.0mL of sterile cefoperazone solution. Mix thoroughly. Pour into sterile Petri dishes or distribute into sterile tubes.

Use: For the cultivation of *Campylobacter* species.

Campylobacter Enrichment Broth (FDA Medium M29)

Composition per 1024.0mL:

Basal medium	950.0mL
Horse blood, lysed	50.0mL
Cefoperazone solution	8.0mL
FBP solution	4.0mL
Trimethoprim lactate solution	4.0mL
Vancomycin solution	4.0mL
Cycloheximide solution	4.0mL

pH 7.5 ± 0.2 at 25°C

Basal Medium:

Composition per 950.0mL:

Beef extract	10.0g
Peptone	10.0g
Yeast extract	6.0g
NaCl	5.0g

Preparation of Basal Medium: Add components to distilled/deionized water and bring volume to 950.0mL. Mix thoroughly. Autoclave for 15 min at 15 psi pressure–121°C. Cool to 45°–50°C.

FBP Solution:

Composition per 100.0mL:

$FeSO_4 \cdot 7H_2O$	6.25g
$Na_2S_2O_5$	6.25g
Sodium pyruvate	6.25g

Preparation of FBP: Add components to distilled/deionized water and bring volume to 100.0mL. Mix thoroughly. Filter sterilize.

Cefoperazone Solution:

Composition per 10.0mL:

Cefoperazone	0.037g

Preparation of Cefoperazone Solution: Add cefoperazone to distilled/deionized water and bring volume to 10.0mL. Mix thoroughly. Filter sterilize.

Trimethoprim Lactate Solution:

Composition per 10.0mL:

Trimethoprim lactate	0.031g

Preparation of Trimethoprim Lactate Solution: Add trimethoprim lactate to distilled/deionized water and bring volume to 10.0mL. Mix thoroughly. Filter sterilize.

Vancomycin Solution:

Composition per 10.0mL:

Vancomycin	0.025g

Preparation of Vancomycin Solution: Add vancomycin to distilled/deionized water and bring volume to 10.0mL. Mix thoroughly. Filter sterilize.

Cycloheximide Solution:

Composition per 10.0mL:

Cycloheximide	0.025g

Preparation of Cycloheximide Solution: Add cycloheximide to distilled/deionized water and bring volume to 10.0mL. Mix thoroughly. Filter sterilize.

Preparation of Medium: To 950.0mL of cooled sterile basal medium, aseptically add 50.0mL of lysed (fresh, frozen, and thawed) horse blood, 4.0mL of sterile FBP solution, 8.0mL of sterile cefoperazone solution, 4.0mL of sterile trimethoprim lactate solution, 4.0mL of sterile vancomycin solution, and 4.0mL of sterile cycloheximide solution. Mix thoroughly. Aseptically distribute into sterile screw-capped tubes or bottles. Close caps tightly to reduce O_2 absorbtion. Use within 2 weeks.

Use: For the selective isolation and cultivation of *Campylobacter* species.

Campylobacter Enrichment Broth (FDA Medium M29)

Composition per 1020.0mL:

Basal medium	950.0mL
Horse blood, lysed	50.0mL
FBP solution	4.0mL
Cefoperazone solution	4.0mL
Trimethoprim lactate solution	4.0mL
Vancomycin solution	4.0mL
Cycloheximide solution	4.0mL

pH 7.5 ± 0.2 at 25°C

Basal Medium:

Composition per 950.0mL:

Beef extract	10.0g
Peptone	10.0g
Yeast extract	6.0g
NaCl	5.0g

Preparation of Basal Medium: Add components to distilled/deionized water and bring volume to 950.0mL. Mix thoroughly. Autoclave for 15 min at 15 psi pressure–121°C. Cool to 45°–50°C.

FBP Solution:

Composition per 100.0mL:

$FeSO_4 \cdot 7H_2O$	6.25g
$Na_2S_2O_5$	6.25g
Sodium pyruvate	6.25g

Preparation of FBP: Add components to distilled/deionized water and bring volume to 100.0mL. Mix thoroughly. Filter sterilize.

Cefoperazone Solution:

Composition per 10.0mL:

Cefoperazone	0.037g

Preparation of Cefoperazone Solution: Add cefoperazone to distilled/deionized water and bring volume to 10.0mL. Mix thoroughly. Filter sterilize.

Trimethoprim Lactate Solution:

Composition per 10.0mL:

Trimethoprim lactate	0.031g

Preparation of Trimethoprim Lactate Solution: Add trimethoprim lactate to distilled/deionized water and bring volume to 10.0mL. Mix thoroughly. Filter sterilize.

Vancomycin Solution:

Composition per 10.0mL:

Vancomycin	0.025g

Preparation of Vancomycin Solution: Add vancomycin to distilled/deionized water and bring volume to 10.0mL. Mix thoroughly. Filter sterilize.

Cycloheximide Solution:

Composition per 10.0mL:

Cycloheximide	0.025g

Preparation of Cycloheximide Solution: Add cycloheximide to distilled/deionized water and bring volume to 10.0mL. Mix thoroughly. Filter sterilize.

Preparation of Medium: To 950.0mL of cooled, sterile basal medium, aseptically add 50.0mL of lysed (fresh, frozen, and thawed) horse blood, 4.0mL of sterile FBP solution, 4.0mL of sterile cefoperazone solution, 4.0mL of sterile trimethoprim lactate solution, 4.0mL of sterile vancomycin solution, and 4.0mL of sterile cycloheximide solution. Mix thoroughly. Aseptically distribute into sterile screw-capped tubes or bottles. Close caps tightly to reduce O_2 absorption. Use within 2 weeks.

Use: For the selective isolation and cultivation of *Campylobacter* species.

Campylobacter Enrichment Broth

Composition per 1016.0mL:

Peptone	10.0g
Yeast extract	5.0g
Lactalbumin hydrolysate	5.0g
NaCl	5.0g
α-Ketoglutamic acid	1.0g
Na_2CO_3	0.6g
Sodium pyruvate	0.5g
$Na_2S_2O_5$	0.5g
Haemin	0.01g
Cefoperazone solution	4.0mL
Trimethoprim lactate solution	4.0mL
Vancomycin solution	4.0mL
Cycloheximide or amphotericin B solution	4.0mL

pH 7.4 ± 0.2 at 25°C

Cefoperazone Solution:

Composition per 10.0mL:

Cefoperazone	0.05g

Preparation of Cefoperazone Solution: Add cefoperazone to distilled/deionized water and bring volume to 10.0mL. Mix thoroughly. Filter sterilize. Can be stored for 5 days at 4°C, 14 days at –20°C, and 5 months at –70°C.

Trimethoprim Solution:

Composition per 10.0mL:

Trimethoprim lactate	0.066g

Preparation of Trimethoprim Solution: Add trimethoprim lactate to distilled/deionized water and bring volume to 10.0mL. Mix thoroughly. Filter sterilize. Alternately add 0.05g tremethoprim hydrochloride to 3.0mL 0.05N HCl. Heat to 50°C. Stir until dissolved. Add distilled/deionized water and bring volume to 10.0mL. Mix thoroughly. Filter sterilize. Can be stored for 1 year at 4°C.

Vancomycin Solution:

Composition per 10.0mL:

Vancomycin	0.05g

Preparation of Vancomycin Solution: Add vancomycin to distilled/deionized water and bring volume to 10.0mL. Mix thoroughly. Filter sterilize. Can be stored for 2 months at 4°C.

Cycloheximide Solution:

Composition per 10.0mL:

Cycloheximide	0.025g
Ethanol	2.0mL

Preparation of Cycloheximide Solution: Add cycloheximide to 2.0mL ethanol to dissolve. Mix thoroughly. Add distilled/deionized water and bring volume to 10.0mL. Mix thoroughly. Filter sterilize. Can be stored for 1 year at 4°C.

Amphotericin B Solution:

Composition per 10.0mL:

Amphotericin B	0.005g

Preparation of Amphotericin B Solution: Add amphotericin B to distilled/deionized water and bring volume to 10.0mL. Mix thoroughly. Filter sterilize. Can be stored for 1 year at –20°C.

Preparation of Medium: Add components, except antimicrobic solutions, to distilled/deionized water and bring volume to 1.0L. Mix thoroughly. Autoclave for 15 min at 15 psi pressure–121°C. Cool to 25°C. Aseptically add 4.0mL of sterile cefoperazone solution, 4.0mL of sterile trimethoprim lactate solution, 4.0mL of sterile vancomycin solution, and either 4.0mL of sterile cycloheximide solution of 4.0mL of sterile amphotericin B solution. Mix thoroughly. Aseptically distribute into sterile screw-capped tubes or bottles. Close caps tightly to reduce O_2 absorbtion. Use within 2 weeks.

Use: For the selective isolation and cultivation of *Campylobacter* species.

Campylobacter fecalis Medium

Composition per 1133.6mL:

Starch, soluble	2.4g
Yeast extract	2.4g
Blood agar base	1.0L
Bovine blood	120.0mL
Antibiotic solution	10.0mL
Sodium lactate (60% syrup)	3.6mL

pH 6.8 ± 0.2 at 25°C

Blood Agar Base:

Composition per liter:

Beef heart, solids from infusion	500.0g
Agar	15.0g
Tryptose	10.0g
NaCl	5.0g

Preparation of Blood Agar Base: Add components to distilled/deionized water and bring volume to 1.0L. Mix thoroughly. Gently heat while stirring and bring to boiling.

Antibiotic Solution:

Composition per 10.0mL:

Cycloheximide	0.12g
Albamycin	6.0mg
Bacitracin	6000U

Preparation of Antibiotic Solution: Add components to distilled/deionized water and bring volume to 10.0mL. Mix thoroughly. Filter sterilize.

Preparation of Medium: To 1.0L of blood agar base, add soluble starch, yeast extract, and sodium lactate. Mix thoroughly. Autoclave for 15 min at 15 psi pressure–121°C. Cool to 45°–50°C. Aseptically add sterile bovine blood and antibiotic solution. Mix thoroughly. Pour into sterile Petri dishes or distribute into sterile tubes.

Use: For the cultivation and isolation of *Campylobacter fecalis.*

Campylobacter fetus Medium

Composition per liter:

Proteose peptone	10.0g
NaCl	5.0g
Beef extract	3.0g
Bovine blood	50.0mL
Antibiotic solution	10.0mL

pH 7.2–7.4 at 25°C

Antibiotic Solution:

Composition per 10.0mL:

Novobiocin	2.0mg
Bacitracin	2000U

Preparation of Antibiotic Solution: Add components to distilled/deionized water and bring volume to 10.0mL. Mix thoroughly. Filter sterilize.

Preparation of Medium: Add components, except bovine blood and antibiotic solution, to distilled/deionized water and bring volume to 940.0mL. Mix thoroughly. Gently heat and bring to boiling. Autoclave for 15 min at 15 psi pressure–121°C. Cool to 45°–50°C. Aseptically add sterile bovine blood and antibiotic solution. Mix thoroughly. Aseptically distribute into sterile tubes or flasks.

Use: For the isolation and cultivation of *Campylobacter fetus*.

Campylobacter fetus Medium

Composition per 1160.0mL:

Fluid thioglycolate agar	1.0L
Sheep blood, defibrinated	150.0mL
Antibiotic solution	10.0mL

pH 7.1 ± 0.2 at 25°C

Fluid Thioglycolate Agar:
Composition per liter:

Agar	15.0g
Pancreatic digest of casein	15.0g
Glucose	5.5g
Yeast extract	5.0g
NaCl	2.5g
Agar	0.75g
L-Cystine	0.5g
Sodium thioglycolate	0.5g
Resazurin	1.0mg

Preparation of Fluid Thioglycolate Agar: Add components to distilled/deionized water and bring volume to 1.0L. Mix thoroughly. Gently heat and bring to boiling. Autoclave for 15 min at 15 psi pressure–121°C. Cool to 25°C.

Antibiotic Solution:
Composition per 10.0mL:

Cycloheximide	0.05g
Novobiocin	5.0mg
Bacitracin	25,000U
Polymyxin B sulfate	10,000U

Preparation of Antibiotic Solution: Add components to distilled/deionized water and bring volume to 10.0mL. Mix thoroughly. Filter sterilize.

Preparation of Medium: To 1.0L of cooled, sterile, fluid thioglycolate agar, aseptically add 150.0mL of sterile sheep blood and 10.0mL of sterile antibiotic solution.

Use: For the isolation and cultivation of *Campylobacter fetus* from human specimens.

Campylobacter fetus Selective Medium

Composition per liter:

Fluid thioglycolate agar	1.0L
Sheep blood, defibrinated	150.0mL
Antibiotic solution	10.0mL

Fluid Thioglycolate Agar:
Composition per liter:

Agar	15.0g
Pancreatic digest of casein	15.0g
Glucose	5.5g
Yeast extract	5.0g
NaCl	2.5g
Agar	0.75g
L-Cystine	0.5g
Sodium thioglycolate	0.5g
Resazurin	1.0mg

Preparation of Fluid Thioglycolate Agar: Add components to distilled/deionized water and bring volume to 1.0L. Mix thoroughly. Gently heat and bring to boiling. Autoclave for 15 min at 15 psi pressure–121°C. Cool to 25°C.

Antibiotic Solution:
Composition per 10.0mL:

Cycloheximide	0.05g
Cephalothin	0.02g
Novobiocin	5.0mg
Bacitracin	25,000U
Colistin	10,000U

Preparation of Antibiotic Solution: Add components to distilled/deionized water and bring volume to 10.0mL. Mix thoroughly. Filter sterilize.

Preparation of Medium: To 1.0L of cooled, sterile, fluid thioglycolate agar, aseptically add 150.0mL of sterile sheep blood and 10.0mL of sterile antibiotic solution.

Use: For the isolation and cultivation of *Campylobacter fetus*.

Campylobacter Isolation Agar A

Composition per liter:

Agar	12.0g
Beef extract	10.0g
Peptone	10.0g
NaCl	5.0g
Charcoal	4.0g
Casein hydrolysate	3.0g
Yeast extract	2.0g
Sodium deoxycholate	1.0g
$FeSO_4$	0.25g
Sodium pyruvate	0.25g
Antibiotic solution	10.0mL

pH 7.4 ± 0.2 at 25°C

Antibiotic Solution:
Composition per 10.0mL:

Cycloheximide	0.1g
Sodium cefoperazone	0.03g

Preparation of Antibiotic Solution: Add components to distilled/deionized water and bring volume to 10.0mL. Mix thoroughly. Filter sterilize.

Preparation of Medium: Add components, except antibiotic solution, to distilled/deionized water and bring volume to 990.0mL. Mix thoroughly. Gently heat and bring to boiling. Autoclave for 15 min at

15 psi pressure–121°C. Cool to 45°–50°C. Aseptically add sterile antibiotic solution. Mix thoroughly. Pour into sterile Petri dishes. Swirl flask while pouring to distribute charcoal.

Use: For the isolation and cultivation of *Campylobacter* species.

Campylobacter Isolation Agar B (*Campy* Cefex Agar)

Composition per liter:

Agar	15.0g
Pancreatic digest of casein	10.0g
Peptic digest of animal tissue	10.0g
NaCl	5.0g
Yeast extract	2.0g
Glucose	1.0g
$FeSO_4$	0.5g
Sodium pyruvate	0.5g
$NaHSO_3$	0.35g
Horse blood, laked	50.0mL
Antibiotic solution	10.0mL

pH 7.0 ± 0.2 at 25°C

Horse Blood, Laked:
Composition per 50.0mL:

Horse blood, fresh	50.0mL

Preparation of Horse Blood, Laked: Add blood to a sterile polypropylene bottle. Freeze overnight at –20°C. Thaw at 8°C. Refreeze at –20°C. Thaw again at 8°C.

Antibiotic Solution:
Composition per 10.0mL:

Cycloheximide	0.1g
Sodium cefoperazone	0.033g

Preparation of Antibiotic Solution: Add components to distilled/deionized water and bring volume to 10.0mL. Mix thoroughly. Filter sterilize.

Preparation of Medium: Add components, except horse blood and antibiotic solution, to distilled/deionized water and bring volume to 940.0mL. Mix thoroughly. Gently heat and bring to boiling. Autoclave for 15 min at 15 psi pressure–121°C. Cool to 45°–50°C. Aseptically add sterile horse blood and antibiotic solution. Mix thoroughly. Pour into sterile Petri dishes or distribute into sterile tubes.

Use: For the isolation and cultivation of *Campylobacter* species.

Campylobacter Medium

Composition per liter:

Sodium aspartate	10.0g
$MgSO_4 \cdot 7H_2O$	1.0g
K_2HPO_4	0.75g
Yeast extract	0.2g
$CaCl_2 \cdot 2H_2O$	28.0mg
Resazurin	1.0mg
Phosphate-cysteine solution	100.0mL
Trace elements solution SL-10	1.0mL

pH 7.0 ± 0.2 at 25°C

Phosphate-Cysteine Solution:
Composition per 100.0mL:

NaH_2PO_4	0.25g
L-Cysteine·HCl	0.25g

Preparation of Phosphate-Cysteine Solution: Add components to distilled/deionized water and bring volume to 100.0mL. Mix thoroughly. Autoclave for 15 min at 15 psi pressure–121°C.

Trace Elements Solution SL-10:
Composition per liter:

$FeCl_2 \cdot 4H_2O$	1.5g
$CoCl_2 \cdot 6H_2O$	190.0mg
$MnCl_2 \cdot 4H_2O$	100.0mg
$ZnCl_2$	70.0mg
$Na_2MoO_4 \cdot 2H_2O$	36.0mg
$NiCl_2 \cdot 6H_2O$	24.0mg
H_3BO_3	6.0mg
$CuCl_2 \cdot 2H_2O$	2.0mg
HCl (25% solution)	10.0mL

Preparation of Trace Elements Solution SL-10: Add $FeCl_2 \cdot 4H_2O$ to 10.0mL of HCl solution. Mix thoroughly. Add distilled/deionized water and bring volume to 1.0L. Add remaining components. Mix thoroughly.

Preparation of Medium: Add components, except phosphate-cysteine solution, to distilled/deionized water and bring volume to 900.0mL. Mix thoroughly. Autoclave for 15 min at 15 psi pressure–121°C. Aseptically add 100.0mL of sterile phosphate-cysteine solution. Mix thoroughly. Aseptically distribute into sterile tubes or flasks.

Use: For the cultivation and maintenance of *Campylobacter* species, *Actinobacillus ureae, Erysipelothrix rhusiopathiae, Helicobacter pylori, Moraxella bovis, Moraxella nonliquefaciens, Moraxella osloensis, Pasteurella haemolytica,* and *Pasteurella multocida.*

Campylobacter mucosalis Medium

Composition per liter:

Agar	15.0g
Beef extract	10.0g
Special peptone	10.0g
NaCl	5.0g
Sodium fumarate	3.0g

Sodium formate	2.0g
Horse blood, defibrinated	50.0mL

pH 7.2 ± 0.2 at 25°C

Source: Special peptone (L72) is available from Oxoid Unipath.

Preparation of Medium: Add components, except horse blood, to distilled/deionized water and bring volume to 950.0mL. Mix thoroughly. Heat with frequent agitation and boil for 1 min to completely dissolve. Autoclave for 15 min at 15 psi pressure–121°C. Cool to 45°–50°C. Aseptically add 50.0mL of sterile, defibrinated horse blood. Mix thoroughly and pour into sterile Petri dishes.

Use: For the cultivation and maintenance of *Campylobacter mucosalis*.

Campylobacter rectus Medium

Composition per liter:

Yeast extract	11.0g
Pancreatic digest of casein	9.0g
Beef extract	3.0g
NaCl	2.0g
Na_2HPO_4	0.4g
Na_2CO_3	0.25g
Resazurin	1.0mg
Formate-fumarate solution	100.0mL
Hemin solution	10.0mL

pH 7.5 ± 0.2 at 25°C

Formate-Fumarate Solution:

Composition per 100.0mL:

Sodium fumarate	3.0g
Sodium formate	2.0g

Preparation of Formate-Fumarate Solution: Add components to distilled/deionized water and bring volume to 100.0mL. Mix thoroughly. Filter sterilize. Sparge with 100% N_2 for 3–4 min.

Hemin Solution:

Composition per 10.0mL:

Hemin	5.0mg
NaOH (1*N* solution)	0.1mL

Preparation of Hemin Solution: Add hemin to NaOH solution to dissolve. Add distilled/deionized water and bring volume to 10.0mL. Mix thoroughly. Sparge with 100% N_2 for 3–4 min. Autoclave under 100% N_2 for 15 min at 15 psi pressure–121°C.

Preparation of Medium: Add components, except formate-fumarate solution and hemin solution, to distilled/deionized water and bring volume to 890.0mL. Sparge with 100% N_2 for 10 min. Autoclave under 100% N_2 for 15 min at 15 psi pressure–121°C. Aseptically and anaerobically add 100.0mL of sterile formate-fumarate solution and 10.0mL of sterile hemin solution under 100% N_2. Mix thoroughly. Aseptically and anaerobically distribute into sterile anaerobic tubes.

Use: For the cultivation and maintenance of *Campylobacter curvus* and *Campylobacter rectus*.

Campylobacter Selective Medium, Blaser-Wang (Blaser-Wang *Campylobacter* Medium) (Blaser's Agar) (Campy BAP Medium)

Composition per liter:

Brucella agar base	890.0mL
Sheep blood	100.0mL
Antibiotic supplement	10.0mL

***Brucella* Agar Base:**

Composition per 890.0mL:

Agar	15.0g
Glucose	10.0g
Pancreatic digest of casein	10.0g
NaCl	5.0g
Peptic digest of animal tissue	5.0g

pH 7.5 ± 0.2 at 25°C

Preparation of *Brucella* Agar Base: Add components to distilled/deionized water and bring volume to 890.0mL. Mix thoroughly. Gently heat and bring to boiling. Autoclave for 15 min at 15 psi pressure–121°C. Cool to 45°–50°C.

Antibiotic Supplement:

Composition per 10.0mL:

Cephalothin	15.0mg
Vancomycin	10.0mg
Trimethoprim	5.0mg
Amphotericin B	2.0mg
Polymyxin B	2500U

Preparation of Antibiotic Supplement: Add components to 10.0mL of distilled/deionized water. Filter sterilize.

Preparation of Medium: Prepare 890.0mL of *Brucella* agar base. Sterilize as directed. Cool to 50°–55°C and add 100.0mL of sheep blood or 50.0–70.0mL of laked horse blood. Laked blood is prepared by freezing whole blood overnight and thawing to room temperature. Aseptically add 10.0mL of sterile antibiotic supplement. Mix thoroughly. Pour into sterile Petri dishes.

Use: For the selective isolation of *Campylobacter* species.

Campylobacter Selective Medium, Blaser-Wang (Blaser-Wang *Campylobacter* Medium)

Composition per liter:

Columbia agar base	890.0mL
Sheep blood	100.0mL
Antibiotic supplement	10.0mL

pH 7.3 ± 0.2 at 25°C

Columbia Agar Base:

Composition per liter:

Special peptone	25.0g
Agar	10.0g
NaCl	5.0g
Starch	1.0g

Preparation of Columbia Agar Base: Add components to distilled/deionized water and bring volume to 890.0mL. Mix thoroughly. Gently heat and bring to boiling. Autoclave for 15 min at 15 psi pressure–121°C. Cool to 45°–50°C.

Antibiotic Supplement:

Composition per 10.0mL:

Cephalothin	15.0mg
Vancomycin	10.0mg
Trimethoprim	5.0mg
Amphotericin B	2.0mg
Polymyxin B	2,500U

Preparation of Antibiotic Supplement: Add components to 10.0mL of distilled/deionized water. Filter sterilize.

Preparation of Medium: To 890.0mL of cooled, sterile Columbia agar base, aseptically add 100.0mL of sheep blood or 50.0–70.0mL of laked horse blood. Laked blood is prepared by freezing whole blood overnight and thawing to room temperature. Aseptically add 10.0mL of sterile antibiotic supplement. Mix thoroughly. Pour into sterile Petri dishes.

Use: For the selective isolation of *Campylobacter* species.

Campylobacter Selective Medium, Butzler's (Butzler's *Campylobacter* Medium)

Composition per liter:

Brucella agar base	940.0mL
Sheep or horse blood, defibrinated	50.0mL
Antibiotic supplement	10.0mL

pH 7.5 ± 0.2 at 25°C

***Brucella* Agar Base:**

Composition per liter:

Agar	15.0g
Glucose	10.0g
Pancreatic digest of casein	10.0g
NaCl	5.0g
Peptic digest of animal tissue	5.0g

Preparation of *Brucella* Agar Base: Add components to distilled/deionized water and bring volume to 940.0mL. Mix thoroughly. Gently heat and bring to boiling. Autoclave for 15 min at 15 psi pressure–121°C. Cool to 45°–50°C.

Antibiotic Supplement:

Composition per 10.0mL:

Cycloheximide	50.0mg
Cephazolin	15.0mg
Novobiocin	5.0mg
Bacitracin	25,000U
Colistin sulfate	10,000U

Preparation of Antibiotic Supplement: Add components to 10.0mL of distilled/deionized water. Filter sterilize.

Preparation of Medium: To 940.0mL of cooled, sterile *Brucella* agar base, aseptically add 50.0mL of defibrinated sheep or horse blood and 10.0mL of sterile antibiotic supplement. Mix thoroughly. For enhanced growth, medium may also be supplemented with 0.25g of $Fe_2SO_4 \cdot H_2O$, 0.25g of sodium metabisulfite, and 0.25g of sodium pyruvate. Pour into sterile Petri dishes.

Use: For the selective isolation of *Campylobacter* species.

Campylobacter Selective Medium, Butzler's (Butzler's *Campylobacter* Medium)

Composition per liter:

Columbia agar base	940.0mL
Blood, horse or sheep	50.0mL
Antibiotic supplement	10.0mL

pH 7.3 ± 0.2 at 25°C

Columbia Agar Base:

Composition per liter:

Peptone	25.0g
Agar	10.0g
NaCl	5.0g
Starch	1.0g

Preparation of Columbia Agar Base: Add components to distilled/deionized water and bring volume to 940.0mL. Mix thoroughly. Gently heat and bring to boiling. Autoclave for 15 min at 15 psi pressure–121°C. Cool to 45°–50°C.

Antibiotic Supplement:

Composition per 10.0mL:

Cycloheximide	50.0mg

Cephazolin	15.0mg
Novobiocin	5.0mg
Bacitracin	25,000U
Colistin sulfate	10,000U

Preparation of Antibiotic Supplement: Add components to 10.0mL of distilled/deionized water. Filter sterilize.

Preparation of Medium: To 940.0mL of cooled, sterile Columbia agar base, aseptically add 50.0mL of defibrinated sheep or horse blood and 10.0mL of sterile antibiotic supplement. Mix thoroughly. The medium may also be supplemented with 0.25g of $Fe_2SO_4·H_2O$, 0.25g of sodium metabisulfite, and 0.25g of sodium pyruvate. Pour into sterile Petri dishes.

Use: For the selective isolation of *Campylobacter* species.

Campylobacter Selective Medium, Karmali's (Karmali's *Campylobacter* Medium)

Composition per liter:

Activated charcoal	4.0g
Columbia agar base	990.0mL
Antibiotic supplement	10.0mL

pH 7.4 ± 0.2 at 25°C

Source: This medium is available as a premixed powder from Oxoid Unipath.

Columbia Agar Base:

Composition per 990.0mL:

Peptone	25.0g
Agar	10.0g
NaCl	5.0g
Starch	1.0g

Preparation of Columbia Agar Base: Add components to distilled/deionized water and bring volume to 990.0mL. Mix thoroughly. Gently heat and bring to boiling. Autoclave for 15 min at 15 psi pressure–121°C. Cool to 45°–50°C.

Antibiotic Supplement:

Composition per 10.0mL:

Sodium pyruvate	0.05g
Cycloheximide	0.05g
Cefoperazone	0.016g
Hemin	0.016g
Vancomycin	0.01g

Preparation of Antibiotic Supplement: Add components to 10.0mL of distilled/deionized water. Filter sterilize.

Preparation of Medium: Prepare 990.0mL of Columbia agar base. Sterilize as directed. Cool to 50°–55°C. Add defibrinated sheep or horse blood to a final concentration of 5–7%. Add 10.0mL of sterile antibiotic supplement. Mix thoroughly. For enhanced growth, medium may also be supplemented with 0.25g of $Fe_2SO_4·H_2O$, 0.25g of sodium metabisulfite, and 0.25g of sodium pyruvate. Pour into sterile Petri dishes. Swirl while pouring to keep charcoal in suspension.

Use: For the selective isolation of *Campylobacter* species.

Campylobacter Selective Medium, Preston's (Preston's *Campylobacter* Medium)

Composition per liter:

Campylobacter agar base	940.0mL
Horse blood, lysed	50.0mL
Antibiotic supplement	10.0mL

pH 7.5 ± 0.2 at 25°C

***Campylobacter* Agar Base:**

Composition per liter:

Agar	12.0g
Beef extract	10.0g
Peptone	10.0g
NaCl	5.0g

Preparation of *Campylobacter* Agar Base: Add components to distilled/deionized water and bring volume to 940.0mL. Mix thoroughly. Gently heat and bring to boiling. Autoclave for 15 min at 15 psi pressure–121°C. Cool to 45°–50°C.

Antibiotic Supplement:

Composition per 10.0mL:

Cycloheximide	0.1g
Rifampicin	0.01g
Trimethoprim lactate	0.01g
Polmyxin B	5000U

Preparation of Antibiotic Supplement: Add components to 10.0mL of 50:50 acetone:distilled/deionized water. Filter sterilize.

Preparation of Medium: To 940.0mL of cooled, sterile *Campylobacter* agar base, aseptically add 50.0mL of lysed horse blood and 10.0mL of sterile antibiotic supplement. Mix thoroughly. Pour into sterile Petri dishes.

Use: For the selective isolation of *Campylobacter* species.

Campylobacter sputorum Subspecies *bubulus* Medium

Composition per liter:

Agar	1.5g
Brilliant Green	0.01g
Ethyl Violet	1.25mg

Brucella broth base 1.0L
Antibiotic solution 10.0mL
pH 7.0 ± 0.2 at 25°C

***Brucella* Broth Base:**
Composition per liter:
Pancreatic digest of casein 10.0g
Peptic digest of animal tissue 10.0g
NaCl 5.0g
Yeast extract 2.0g
Glucose 1.0g
$NaHSO_3$ 0.1g

Preparation of *Brucella* Broth Base: Add components to distilled/deionized water and bring volume to 1.0L. Mix thoroughly.

Antibiotic Solution:
Composition per 10.0mL:
Cycloheximide 0.1g
Bacitracin 20,000U

Preparation of Antibiotic Solution: Add components to distilled/deionized water and bring volume to 10.0mL. Mix thoroughly. Filter sterilize.

Preparation of Medium: To 1.0L of *Brucella* broth base, add agar, Brilliant Green, and Ethyl Violet. Mix thoroughly. Autoclave for 15 min at 15 psi pressure–121°C. Cool to 45°–50°C. Aseptically add sterile antibiotic solution. Mix thoroughly. Aseptically distribute into sterile tubes or flasks.

Use: For the cultivation and isolation of *Campylobacter sputorum* subspecies *bubulus.*

Campylobacter sputorum Subspecies *mucosalis* Medium

Composition per liter:
Yeast extract 2.8g
KNO_3 .... 1.0g
Fluid thioglycolate broth without glucose 1.0L

Fluid Thioglycolate Broth without Glucose:
Composition per liter:
Pancreatic digest of casein 15.0g
Yeast extract 5.0g
NaCl 2.5g
Agar 0.75g
L-Cystine 0.5g
Sodium thioglycolate 0.5g
Resazurin 1.0mg

Preparation of Fluid Thioglycolate Broth without Glucose: Add components to distilled/deionized water and bring volume to 1.0L. Mix thoroughly.

Preparation of Medium: Combine components. Mix thoroughly. Gently heat and bring to boiling. Autoclave for 15 min at 15 psi pressure–121°C. Cool to 25°C. Aseptically distribute into sterile tubes or flasks.

Use: For the cultivation and isolation of *Campylobacter sputorum* subspecies *mucosalis.*

Campylobacter Thioglycolate Medium with 5 Antimicrobics

Composition per liter:
Pancreatic digest of casein 17.0g
Glucose 6.0g
Papaic digest of soybean meal 3.0g
NaCl 2.5g
Agar 1.6g
Sodium thioglycolate 0.5g
Na_2SO_3 .... 0.1g
Antibiotic supplement solution 10.0mL
pH 7.0 ± 0.2 at 25°C

Antibiotic Supplement Solution:
Composition per 10.0mL:
Cephalothin 0.015g
Vancomycin 0.01g
Trimethoprim 5.0mg
Amphotericin B 2.0mg
Polymyxin B 2500U

Preparation of Antibiotic Supplement Solution: Add components to 10.0mL of distilled/deionized water. Mix thoroughly. Filter sterilize.

Preparation of Medium: Add components, except cephalothin, vancomycin, trimethoprim, amphotericin, and polymyxin B, to distilled deionized water and bring volume to 990.0mL. Mix thoroughly. Gently heat and bring to boiling. Autoclave for 15 min at 15 psi pressure–121°C. Cool to 45°–50°C. Add 10.0mL of sterile antibiotic supplement. Mix thoroughly. Pour into sterile Petri dishes.

Use: For the maintenence—as a holding medium or transport medium— of fecal specimens or swabs suspected of containing *Campylobacter jejuni* or other *Campylobacter* species when immediate inoculation of *Campylobacter* growth medium is unavailable.

Candida BCG Agar Base (*Candida* Bromcresol Green Agar Base)

Composition per liter:
Glucose 40.0g
Agar 15.0g
Peptone 10.0g
Yeast extract 1.0g
Bromcresol Green 0.02g
Neomycin solution 10.0mL
pH 6.1 ± 0.1 at 25°C

Source: This medium is available as a premixed powder from BD Diagnostic Systems.

Neomycin Solution:
Composition per 10.0mL:

Neomycin	0.5g

Preparation of Neomycin Solution: Add neomycin to distilled/deionized water and bring volume to 10.0mL. Mix thoroughly. Filter sterilize.

Preparation of Medium: Add components, except neomycin solution, to distilled/deionized water and bring volume to 1.0L. Mix thoroughly and heat gently until boiling. Autoclave for 15 min at 15 psi pressure–121°C. Cool to 50°–55°C. Aseptically add 10.0mL of sterile neomycin solution. Mix thoroughly. Pour into sterile Petri dishes or leave in tubes.

Use: For the selective isolation and identification of *Candida* species. It is a highly differential medium that is used for demonstrating morphological and biochemical reactions characterizing different *Candida* species. *Candida albicans* appears as blunt conical colonies with smooth edges and yellow to blue-green color. *Candida stellatoidea* appears as convex colonies with smooth edges and yellow to green color. *Candida tropicalis* appears as convex colonies with wavy edges and yellow-green to green color with a dark blue-green base. *Candida pseudotropicalis* appears as convex, shiny colonies with smooth edges and green color with a light green edge. *Candida krusei* appears as low conical colonies with spreading edges and blue-green color. *Candida stellatoidea* appears as convex colonies with smooth edges and yellow to green color.

Candida Isolation Agar

Composition per liter:

Agar	20.0g
Glucose	10.0g
Peptone	5.0g
Yeast extract	3.0g
Malt extract	3.0g
Aniline Blue	0.1g

pH 5.9 ± 0.5 at 25°C

Source: This medium is available as a premixed powder from BD Diagnostic Systems.

Preparation of Medium: Add components to distilled/deionized water and bring volume to 1.0L. Mix thoroughly. Gently heat and bring to boiling. Distribute into tubes or flasks. Autoclave for 15 min at 15 psi pressure–121°C. Pour into sterile Petri dishes or leave in tubes.

Use: For the isolation and differentiation of *Candida albicans*. *Candida albicans* turns the medium blue.

Capnocytophaga Medium

Composition per liter:

Pancreatic digest of casein	17.0g
KNO_3	3.0g
NaCl	3.0g
Yeast extract	3.0g
Hemin	3.0mg
Glucose solution	20.0mL

pH 7.0 ± 0.2 at 25°C

Glucose Solution:
Composition per 20.0mL:

D-Glucose	3.0g

Preparation of Glucose Solution: Add glucose to distilled/deionized water and bring volume to 20.0mL. Mix thoroughly. Autoclave for 15 min at 15 psi pressure–121°C.

Preparation of Medium: Add components, except glucose solution, to distilled/deionized water and bring volume to 990.0mL. Mix thoroughly. Autoclave for 15 min at 15 psi pressure–121°C. Aseptically add 10.0mL of sterile glucose solution. Mix thoroughly. Aseptically distribute into sterile tubes or flasks.

Use: For the cultivation and maintenance of *Capnocytophaga* species.

Capnocytophaga II Medium (DSMZ Medium 779)

Composition per liter:

Proteose peptone no. 3	10.0g
Yeast extract	5.0g
Na_2HPO_4	4.0g
Lab-Lemco meat extract	2.4g
Glucose	1.5g
Starch, soluble	0.5g
Cysteine-HCl·H_2O	0.5g
Horse blood	50.0mL

pH 6.8 ± 0.2 at 25°C

Preparation of Medium: Add components, except horse blood, to distilled/deionized water and bring volume to 950.0L. Mix thoroughly. Adjust pH to 7.6-7.8. Autoclave for 15 min at 15 psi pressure–121°C. Cool to room temperature. Aseptically add 50.0mL horse blood. Mix thoroughly. Aseptically distribute into tubes or flasks. Incubate under 95% air + 5% CO_2 or anaerobically under 95% N_2 + 5% CO_2.

Use: For the cultivation of *Capnocytophaga haemolytica* and *Capnocytophaga granulosa*.

Carbohydrate Fermentation Broth

Composition per liter:

Peptone	10.0g

NaCl	5.0g
Meat extract	3.0g
Carbohydrate solution	50.0mL
Andrade's indicator	10.0mL

pH 7.1 ± 0.2 at 25°C

Andrade's Indicator:
Composition per 100.0mL:

Acid Fuchsin	0.1 g
NaOH (1*N* solution)	16.0mL

Preparation of Andrade's Indicator: Add components to distilled/deionized water and bring volume to 100.0mL. Mix thoroughly.

Carbohydrate Solution:
Composition per 100.0mL:

Carbohydrate	10.0g

Preparation of Carbohydrate Solution: Add carbohydrate to distilled/deionized water and bring volume to 100.0mL. Adonitol, arabinose, cellobiose, glucose, dulcitol, fructose, galactose, inositol, lactose, maltose, mannitol, raffinose, rhamnose, salicin, sorbitol, sucrose, trehalose, xylose, or other carbohydrates may be used. Mix thoroughly. Filter sterilize.

Caution: Acid Fuchsin is a potential carcinogen and care must be taken to avoid inhalation of the powdered dye and contact with the skin.

Preparation of Medium: Add components, except carbohydrate solution, to distilled/deionized water and bring volume to 1.0L. Mix thoroughly. Gently heat and bring to boiling. Distribute in 10.0mL volumes into test tubes containing inverted Durham tubes. Autoclave for 15 min at 15 psi pressure–121°C. Cool to 25°C. Add 0.5mL of sterile carbohydrate solution to each tube.

Use: For the determination of carbohydrate fermentation reactions of microorganisms, particularly members of the Enterobacteriaceae. A Durham tube is used to collect gas produced during the fermentation reaction. Acid production is indicated by a pink reaction.

Carbon Assimilation Medium

Composition per liter:

Agar solution	500.0mL
Mineral base medium	500.0mL

pH 6.5 ± 0.1 at 25°C

Agar Solution:
Composition per liter:

Agar	32.0g

Preparation of Agar Solution: Add agar to distilled/deionized water and bring volume to 1.0L. Mix thoroughly. Gently heat and bring to boiling. Autoclave for 15 min at 15 psi pressure–121°C. Cool to 45°–50°C.

Mineral Base Medium:
Composition per 500.0mL:

Carbohydrate	10.0g
NaCl	5.0g
NH_4HPO_4	1.0g
K_2HPO_4	1.0g
$MgSO_4 \cdot 7H_2O$, anhydrous	0.1g

Preparation of Mineral Base Medium: Add components to distilled/deionized water and bring volume to 500.0mL. Mix thoroughly. Gently heat until dissolved. Filter sterilize. Warm to 45°–50°C.

Preparation of Medium: Combine 500.0mL of cooled, sterile agar solution and 500.0mL of sterile mineral base medium. Mix thoroughly. Aseptically distribute into sterile tubes. Allow tubes to cool in a slanted position.

Use: For the cultivation and differentiation of microorganisms based on their ability to utilize a particular carbon source.

Carbon Assimilation Medium, Auxanographic Method for Yeast Identification

Composition per liter:

Noble agar	20.0g
$(NH_4)_2SO_4$	0.5g
KH_2PO_4	0.1g
$MgSO_4 \cdot 7H_2O$	0.05g
NaCl	0.01g
$CaCl_2 \cdot 2H_2O$	0.01g
DL-Methionine	2.0mg
DL-Tryptophan	2.0mg
L-Histidine·HCl	1.0mg
Inositol	0.2mg
H_3BO_3	0.05mg
$ZnSO_4 \cdot 7H_2O$	0.04mg
$MnSO_4 \cdot 4H_2O$	0.04mg
Thiamine·HCl	0.04mg
Pyroxidine·HCl	0.04mg
Niacin	0.04mg
Calcium pantothenate	0.04mg
p-Aminobenzoic acid	0.02mg
Riboflavin	0.02mg
$FeCl_3$	0.02mg
$Na_2MoO_4 \cdot 4H_2O$	0.02mg
KI	0.01mg
$CuSO_4 \cdot 5H_2O$	4.0μg
Folic acid	0.2μg
Biotin	0.2μg

pH 4.5 ± 0.2 at 25°C

Preparation of Medium: Add components to distilled/deionized water and bring volume to 1.0L. Mix thoroughly. Gently heat and bring to boiling. Distribute into screw-capped tubes in 20.0mL volumes. Autoclave for 15 min at 15 psi pressure–121°C.

Use: For carbohydrate assimilation tests by the auxanographic method for the identification of yeasts.

Carbon Utilization Test

Composition per liter:

Ionagar	10.0g
NH_4Cl	1.0g
$MgSO_4·7H_2O$	0.5g
Ferric ammonium citrate	0.05g
$CaCl_2$	0.5mg
Sodium potassium phosphate buffer (0.33*M* solution, pH 6.8)	1.0L
Carbon source	10.0mL

pH 6.8 ± 0.2 at 25°C

Carbon Source:

Composition per 10.0mL:

Carbon source	1.0g

Preparation of Carbon Source: Add carbon source to distilled/deionized water and bring volume to 10.0mL. Mix thoroughly. Filter sterilize.

Preparation of Medium: Add components, except carbon source, to distilled/deionized water and bring volume to 990.0mL. Mix thoroughly. Gently heat and bring to boiling. Autoclave for 15 min at 15 psi pressure–121°C. Cool to 45°–50°C. Aseptically add sterile carbon source. Mix thoroughly. Pour into sterile Petri dishes or distribute into sterile tubes.

Use: For the cultivation and differentiation of *Pseudomonas* species based on their ability to utilize a specific carbon source.

Cardiobacterium hominis Medium

Composition per liter:

Glucose	5.0g
Leucine	0.43g
Threonine	0.28g
Glutamic acid	0.2g
Valine	0.19g
Glycine	0.18g
Arginine	0.16g
Histidine	0.13g
Proline	0.1g
Tyrosine	0.04g
Buffered salts solution	100.0mL
Vitamin solution	10.0mL

pH 7.0 ± 0.2 at 25°C

Buffered Salts Solution:

Composition per liter:

Na_2PHO_4	284.0.g
KH_2PO_4	272.0.g
NaCl	5.0g
$FeSO_4·7H_2O$	4.0g
$MgSO_4·7H_2O$	4.0g
$ZnSO_4·7H_2O$	0.4g
$MnSO_4·H_2O$	0.3g
$CuSO_4·5H_2O$	0.05g

Preparation of Buffered Salts Solution: Add components to distilled/deionized water and bring volume to 1.0L. Mix thoroughly.

Vitamin Solution:

Composition per liter:

Pyridoxine·HCl	2.0mg
Calcium pantothenate	1.0mg
Nicotinamide	1.0mg
Thiamine·HCl	1.0mg
Biotin	0.1mg

Preparation of Vitamin Solution: Add components to distilled/deionized water and bring volume to 1.0L. Mix thoroughly.

Preparation of Medium: Add components to distilled/deionized water and bring volume to 1.0L. Mix thoroughly. Adjust pH to 7.0. Filter sterilize.

Use: For the isolation and cultivation of *Cardiobacterium hominis*.

Cardiobacterium hominis Medium

Composition per liter:

K_2HPO_4	7.0g
Yeast extract	5.0g
KH_2PO_4	3.0g
$(NH_4)_2SO_4$	0.1g
$MgSO_4·7H_2O$	0.01g

pH 7.0 ± 0.2 at 25°C

Preparation of Medium: Add components to distilled/deionized water and bring volume to 1.0L. Mix thoroughly. Distribute into tubes or flasks. Autoclave for 15 min at 15 psi pressure–121°C.

Use: For the cultivation of *Cardiobacterium hominis*.

Carnitine Chloride Medium

Composition per liter:

Noble agar	15.0g
DL-Carnitine chloride	10.0g
Na_2HPO_4	10.0g
KH_2PO_4	5.5g
$(NH_4)_2HPO_4$	2.0g
$NH_4H_2PO_4$	1.5g
$MgSO_4·7H_2O$	0.2g

Yeast extract..0.05g
$CaCl_2$..0.015g
$Fe_2(SO_4)_3$..0.6mg
$CuSO_4 \cdot 5H_2O$..0.2mg
$MnSO_4 \cdot H_2O$...0.2mg
$ZnSO_4 \cdot 7H_2O$..0.2mg
pH 7.0 ± 0.1 at 25°C

Preparation of Medium: Add components to distilled/deionized water and bring volume to 1.0L. Adjust pH to 7.0 with NaOH. Mix thoroughly. Heat gently until boiling. Distribute into tubes or flasks. Autoclave for 15 min at 15 psi pressure–121°C. Pour into sterile Petri dishes or leave in tubes.

Use: For the cultivation and maintenance of bacteria that can use carnitine as a carbon source.

Carrot Potato Dextrose Agar (ATCC Medium 1829)

Composition per liter:
Agar 25.0g
Glucose 20.0g
Pancreatic digest of casein.................................2.5g
Yeast extract..0.5g
$MgSO_4 \cdot 7H_2O$..0.3g
$CaCO_3$..0.2g
Potatoes, infusion from 500.0mL
Carrot juice (any commercial brand)........... 15.0 ml
pH 5.6 ± 0.2 at 25°C

Source: Potato dextrose agar, without carrot juice is available as a premixed powder from BD Diagnostic Systems.

Potato Infusion:
Composition per 500.0mL:
Potatoes...300.0g

Preparation of Potato Infusion: Peel and dice potatoes. Add 500.0mL of distilled/deionized water. Gently heat and bring to boiling. Continue boiling for 30 min. Filter through cheesecloth. Reserve filtrate.

Preparation of Medium: Add components to distilled/deionized water and bring volume to 1.0L. Mix thoroughly. Gently heat and bring to boiling. Distribute into tubes or flasks. Autoclave for 15 min at 15 psi pressure–121°C. Pour into sterile Petri dishes or leave in tubes.

Use: For the cultivation of yeasts and molds. Also used to induce sporulation in many fungi.

Cary and Blair Transport Medium

Composition per liter:
Agar 5.0g
NaCl..5.0g
Sodium thioglycolate 1.5g
Na_2HPO_4..1.1g
$CaCl_2$ solution 9.0mL
pH 8.0 ± 0.5 at 25°C

Source: This medium is available as a premixed powder from BD Diagnostic Systems and Oxoid Unipath.

$CaCl_2$ Solution:
Composition per 10.0mL:
$CaCl_2$...0.1g

Preparation of $CaCl_2$ Solution: Add $CaCl_2$ to distilled/deionized water and bring volume to 10.0mL. Mix thoroughly. Filter sterilize.

Preparation of Medium: Add components to distilled/deionized water and bring volume to 1.0L. Mix thoroughly and heat gently until boiling. Cool to 50°C. Add 9.0mL of a 1% $CaCl_2$ solution. Adjust the pH to 8.4. Distribute into screw-capped tubes in 7.0mL volumes. Sterilize under flowing steam for 15 min. After sterilization, tighten the screwcaps.

Use: For the maintenance—as a holding medium or transport medium—of clinical specimens during collection or shipment.

Cary and Blair Transport Medium, Modified

Composition per liter:
Agar 5.0g
NaCl..5.0g
Sodium thioglycolate 1.5g
L-Cysteine·HCl·H_2O..0.5g
$CaCl_2 \cdot 2H_2O$..0.1g
Na_2HPO_4..0.1g
$NaHSO_3$..0.1g
Resazurin solution 4.0mL
pH 8.4 ± 0.2 at 25°C

Resazurin Solution:
Composition per 380.0mL:
Resazurin 0.05g
Ethanol (95% solution)............................. 200.0mL

Preparation of Resazurin Solution: Add resazurin to 200.0mL of ethanol. Mix thoroughly. Bring volume to 380.0mL with distilled/deionized water.

Preparation of Medium: Add components, except L-cysteine·HCl·H_2O, to distilled/deionized water and bring volume to 1.0L. Mix thoroughly. Gas the solution with 100% CO_2 for 10–15 min. Add the L-cysteine·HCl·H_2O. Mix thoroughly. Adjust pH to 8.4. Anaerobically distribute into tubes under 100% N_2. Cap tubes with butyl rubber stoppers. Autoclave for 15 min at 0 psi pressure–100°C on three consecutive days.

Use: For the maintenance—as a holding medium—of clinical specimens during collection or shipment.

Casamino Acids Yeast Extract Lincomycin Medium

Composition per liter:

Casamino acids	20.0g
K_2HPO_4	8.71g
Yeast extract	6.0g
NaCl	2.5g
Lincomycin solution	5.0mL
Trace salts solution	1.0mL

pH 8.5 ± 0.2 at 25°C

Lincomycin Solution:

Composition per 5.0mL:

Lincomycin	45.0mg

Preparation of Lincomycin Solution: Add lincomycin to distilled/deionized water and bring volume to 5.0mL. Mix thoroughly. Filter sterilize.

Trace Salts Solution:

Composition per liter:

$MgSO_4 \cdot 7H_2O$	50.0g
$MnCl_2 \cdot 4H_2O$	5.0g
$FeCl_2$	5.0g

Preparation of Trace Salts Solution: Add components to distilled/deionized water and bring volume to 1.0L. Add sufficient 0.1*N* H_2SO_4 to dissolve components. Mix thoroughly. Filter sterilize.

Preparation of Medium: Add components, except trace salts solution and lincomycin solution, to distilled/deionized water and bring volume to 994.0mL. Mix thoroughly. Adjust pH to 8.5. Autoclave for 15 min at 15 psi pressure–121°C. Cool to 25°C. Aseptically add 1.0mL of sterile trace salts solution and 5.0mL of sterile lincomycin solution. Mix thoroughly. Aseptically distribute into sterile tubes or flasks.

Use: For the cultivation of heat-labile, toxin-producing enterotoxigenic *Escherichia coli.*

Casamino Acids Yeast Extract Salts Broth, Gorbach (CA YE Broth)

Composition per liter:

Casamino acids	20.0g
K_2HPO_4	8.71g
Yeast extract	6.0g
NaCl	2.5g
Trace salts solution	1.0mL

pH 8.5 ± 0.2 at 25°C

Trace Salts Solution:

Composition per liter:

$MgSO_4 \cdot 7H_2O$	50.0g
$MnCl_2 \cdot 4H_2O$	5.0g
$FeCl_2$	5.0g

Preparation of Trace Salts Solution: Add components to distilled/deionized water and bring volume to 1.0L. Add sufficient 0.1*N* H_2SO_4 to dissolve components. Mix thoroughly. Filter sterilize.

Preparation of Medium: Add components, except trace salts solution, to distilled/deionized water and bring volume to 999.0mL. Mix thoroughly. Adjust pH to 8.5. Autoclave for 15 min at 15 psi pressure–121°C. Cool to 25°C. Aseptically add 1.0mL of sterile trace salts solution. Mix thoroughly. Aseptically distribute into sterile tubes or flasks.

Use: For the cultivation of enterotoxigenic *Escherichia coli.*

Casein Agar

Composition per liter:

Agar	10.0g
Skim milk	50.0mL

Preparation of Medium: Add components to distilled/deionized water and bring volume to 1.0L. Mix thoroughly. Gently heat and bring to boiling. Distribute into tubes or flasks. Autoclave for 15 min at 15 psi pressure–121°C. Pour into sterile Petri dishes or leave in tubes.

Use: For the cultivation and differentiation of aerobic actinomycetes based on casein utilization. Bacteria that utilize casein, such as *Streptomyces* and *Actinomadura* species, appear as colonies surrounded by a clear zone. *Nocardia asteroides, Nocardia caviae,* and *Mycobacterium fortuitum* do not utilize casein.

Casein Yeast Extract Glucose Agar (CYG Agar)

Composition per liter:

Agar	20.0g
Glucose	5.0g
Casein hydrolysate	5.0g
Yeast extract	5.0g

pH 7.0 ± 0.2 at 25°C

Preparation of Medium: Add components to distilled/deionized water and bring volume to 1.0L. Mix thoroughly. Gently heat and bring to boiling. Distribute into tubes or flasks. Autoclave for 15 min at 15 psi pressure–121°C.

Use: For agar dilution susceptibility tests with imidazole antifungal agents.

Casein Yeast Extract Glucose Broth (CYG Broth)

Composition per liter:

Casein hydrolysate	5.0g
Glucose	5.0g
Yeast extract	5.0g

pH 7.0 ± 0.2 at 25°C

Preparation of Medium: Add components to distilled/deionized water and bring volume to 1.0L. Mix thoroughly. Gently heat and bring to boiling. Distribute into tubes or flasks. Autoclave for 15 min at 15 psi pressure–121°C.

Use: For agar dilution susceptibility tests with imidazole antifungal agents.

Casman Agar Base

Composition per liter:

Noble agar	14.0g
Proteose peptone no. 3	10.0g
Tryptose	10.0g
NaCl	5.0g
Beef extract	3.0g
Cornstarch	1.0g
Glucose	0.5g
p-Aminobenzoic acid	0.05g
Nicotinamide	0.05g
Blood	50.0mL
Water-lysed blood solution	1.5mL

pH 7.3 ± 0.2 at 25°C

Water-Lysed Blood Solution:
Composition per 8.0mL:

Blood	2.0mL

Preparation of Water-Lysed Blood Solution: Add blood to distilled/deionized water and bring volume to 8.0mL. Mix thoroughly. Filter sterilize.

Preparation of Medium: Add components, except blood and water-lysed blood solution, to distilled/deionized water and bring volume to 948.5mL. Mix thoroughly. Gently heat to boiling. Autoclave for 15 min at 15 psi pressure–121°C. Cool to 50°C. Aseptically add 50.0mL of sterile blood and 1.5mL of sterile water-lysed blood solution (one part blood to three parts water). Water-lysed blood may be omitted if sterile blood is partially lysed due to storage. Mix thoroughly. Pour into sterile Petri dishes or distribute into sterile tubes.

Use: For the isolation of fastidious bacteria from clinical specimens. For the cultivation under reduced oxygen tension of fastidious microorganisms such as *Haemophilus influenzae*, *Neisseria meningitidis*, and *Neisseria gonorrhoeae*.

Casman Agar Base with Rabbit Blood (Casman-Medium) (DSMZ Medium 439)

Composition per liter:

Noble agar	14.0g
Proteose peptone no. 3	10.0g
Tryptose	10.0g
NaCl	5.0g
Beef extract	3.0g
Cornstarch	1.0g
Glucose	0.5g
p-Aminobenzoic acid	0.05g
Nicotinamide	0.05g
Rabbit blood	50.0mL
Water-lysed blood solution	1.5mL

pH 7.3 ± 0.2 at 25°C

Source: Casman agar base is available as a premixed powder from BD Diagnostic Systems.

Water-Lysed Blood Solution:
Composition per 8.0mL:

Rabbit blood	2.0mL

Preparation of Water-Lysed Blood Solution: Add blood to distilled/deionized water and bring volume to 8.0mL. Mix thoroughly. Filter sterilize.

Preparation of Medium: Add components, except rabbit blood and water-lysed blood solution, to distilled/deionized water and bring volume to 950.0L. Mix thoroughly. Gently heat to boiling. Autoclave for 15 min at 15 psi pressure–121°C. Cool to 50°C. Aseptically add 50.0mL of sterile rabbit blood and 1.5mL of sterile water-lysed blood solution. Water-lysed blood may be omitted if sterile blood is partially lysed due to storage. Mix thoroughly. Pour into sterile Petri dishes or distribute into sterile tubes.

Use: For the cultivation and maintenance of *Gardnerella vaginalis*.

CASO MUG Agar

Composition per liter:

Casein peptone	16.0g
Agar	13.0g
NaCl	6.0g
Soy peptone	5.0g
Tryptophan	1.0g
4-Methylumbelliferyl-β-D-glucuronide	0.07g

pH 7.3 ± 0.2 at 25°C

Source: This medium is available from Fluka, Sigma-Aldrich.

Preparation of Medium: Add components to distilled/deionized water and bring volume to 1.0L. Mix thoroughly. Gently heat while stirring and bring to boiling. Autoclave for 15 min at 15 psi pressure–121°C. Cool to 50°C. Pour into sterile Petri dishes.

Use: This universal medium without indicator or inhibitor is intended for a broad range of application including enumeration and cultivation of a wide variety of microorganisms. It is also suitable for the cultivation of more fastidious microorganisms. β-D-glucoronidase, which is produced by *E. coli*, cleaves 4-methylumbelliferyl-β-D-glucuronide to 4-methylumbelliferone and glucuronide. The fluorogen 4-methylumbelliferone can be detected under a long wavelength UV lamp. A positive indole reaction provides confirmation.

CCY Modified Medium

Composition per liter:

Yeast extract	30.0g
Casamino acids	20.0g
Na_2HPO4	2.48g
KH_2PO_4	0.41g
$MgSO_4 \cdot 7H_2O$	20.0mg
$MnSO_4 \cdot H_2O$	7.5mg
Citric acid	6.4mg
$FeSO_4 \cdot 7H_2O$	6.4mg
Sodium pyruvate solution	100.00mL

pH 7.3 ± 0.2 at 25°C

Sodium Pyruvate Solution:
Composition per 100.0mL:

Sodium pyruvate	23.2g

Preparation of Sodium Pyruvate Solution: Add sodium pyruvate to distilled/deionized water and bring volume to 100.0mL. Mix thoroughly. Filter sterilize.

Preparation of Medium: Add components, except sodium pyruvate solution, to distilled/deionized water and bring volume to 900.0mL. Mix thoroughly. Adjust pH to 7.3. Autoclave for 15 min at 15 psi pressure–121°C. Aseptically add 100.0mL of sterile sodium pyruvate solution. Mix thoroughly. Aseptically distribute into sterile tubes or flasks.

Use: For the cultivation of *Staphylococcus aureus.*

CDC Anaerobe Blood Agar

Composition per liter:

Agar	20.0g
Pancreatic digest of casein	15.0g
Papaic digest of soybean meal	5.0g
NaCl	5.0g
Yeast extract	5.0g
L-Cystine	0.4g
Sheep blood, defibrinated	50.0mL
Vitamin K_1 solution	1.0mL
Hemin solution	0.5mL

pH 7.5 ± 0.2 at 25°C

Source: This medium is available as a prepared medium from BD Diagnostic Systems.

Vitamin K_1 Solution:
Composition per 100.0mL:

Vitamin K_1	1.0g
Ethanol	99.0mL

Preparation of Vitamin K_1 Solution: Add vitamin K_1 to 99.0mL of absolute ethanol. Mix thoroughly. Filter sterilize.

Hemin Solution:
Composition per 100.0mL:

Hemin	1.0g
NaOH (1*N* solution)	20.0mL

Preparation of Hemin Solution: Add hemin to 20.0mL of 1*N* NaOH solution. Mix thoroughly. Bring volume to 100.0mL with distilled/deionized water.

Preparation of Medium: Add components, except vitamin K_1 and sheep blood, to distilled/deionized water and bring volume to 949.0mL. Mix thoroughly. Heat gently and bring to boiling for 1 min. Autoclave for 15 min at 15 psi pressure–121°C. Cool to 50°–55°C. Aseptically add 1.0mL of vitamin K_1 solution and 50.0mL of sterile, defibrinated sheep blood. Mix thoroughly. Pour into sterile Petri dishes.

Use: For the isolation and cultivation of fastidious and slow-growing, obligate anaerobic bacteria from a variety of clinical and nonclinical specimens. For the isolation and cultivation of *Actinomyces israelii, Bacteroides melaninogenicus, Bacteroides thetaiotaomicron, Clostridium haemolyticum,* and *Fusobacterium necrophorum.*

CDC Anaerobe Blood Agar with Kanamycin and Vancomycin

Composition per liter:

Agar	20.0g
Pancreatic digest of casein	15.0g
NaCl	5.0g
Papaic digest of soybean meal	5.0g
Yeast extract	5.0g
L-Cystine	0.4g
Sheep blood, defibrinated	50.0mL
Antibiotic solution	10.0mL
Vitamin K_1 solution	1.0mL
Hemin solution	0.5mL

pH 7.5 ± 0.2 at 25°C

Source: This medium is available as a prepared medium from BD Diagnostic Systems.

Antibiotic Solution:
Composition per 10.0mL:

Kanamycin	0.1g
Vancomycin	7.5mg

Preparation of Antibiotic Solution: Add components to distilled/deionized water and bring volume to 10.0mL. Mix thoroughly. Filter sterilize.

Vitamin K_1 Solution:
Composition per 100.0mL:

Vitamin K_1	1.0g
Ethanol	99.0mL

Preparation of Vitamin K_1 Solution: Add vitamin K_1 to 99.0mL of absolute ethanol. Mix thoroughly. Filter sterilize.

Hemin Solution:
Composition per 100.0mL:

Hemin	1.0g
NaOH (1*N* solution)	20.0mL

Preparation of Hemin Solution: Add hemin to 20.0mL of 1*N* NaOH solution. Mix thoroughly. Bring volume to 100.0mL with distilled/deionized water.

Preparation of Medium: Add components, except vitamin K_1 solution and sheep blood, to distilled/deionized water and bring volume to 949.0mL. Mix thoroughly. Heat gently and bring to boiling for 1 min. Autoclave for 15 min at 15 psi pressure–121°C. Cool to 50°–55°C. Aseptically add 1.0mL of sterile vitamin K_1 solution and 50.0mL of sterile, defibrinated sheep blood. Mix thoroughly. Pour into sterile Petri dishes.

Use: For the selective isolation of fastidious and slow-growing, obligate anaerobic Gram-negative bacteria, especially *Bacteroides* species, from a variety of clinical and nonclinical specimens.

CDC Anaerobe Blood Agar with Phenylethyl Alcohol (CDC Anaerobe Blood Agar with PEA)

Composition per liter:

Agar	20.0g
Pancreatic digest of casein	15.0g
NaCl	5.0g
Papaic digest of soybean meal	5.0g
Yeast extract	5.0g
L-Cystine	0.4g
Sheep blood, defibrinated	50.0mL
Vitamin K_1 solution	10.0mL
Hemin solution	0.5mL

pH 7.5 ± 0.2 at 25°C

Source: This medium is available as a prepared medium from BD Diagnostic Systems.

Vitamin K_1 Solution:
Composition per 100.0mL:

Vitamin K_1	0.1g
Phenylethyl alcohol	25.0g
Ethanol	74.0mL

Preparation of Vitamin K_1 Solution: Add components to 74.0mL of absolute ethanol. Mix thoroughly. Filter sterilize.

Hemin Solution:
Composition per 100.0mL:

Hemin	1.0g
NaOH (1*N* solution)	20.0mL

Preparation of Hemin Solution: Add hemin to 20.0mL of 1*N* NaOH solution. Mix thoroughly. Bring volume to 100.0mL with distilled/deionized water.

Preparation of Medium: Add components, except vitamin K_1 solution and sheep blood, to distilled/deionized water and bring volume to 940.0mL. Mix thoroughly. Heat gently and bring to boiling for 1 min. Autoclave for 15 min at 15 psi pressure–121°C. Cool to 50°–55°C. Aseptically add 1.0mL of vitamin K_1 solution and 50.0mL of sterile, defibrinated sheep blood. Mix thoroughly. Pour into sterile Petri dishes.

Use: For the selective isolation of fastidious and slow-growing, obligate anaerobic bacteria from a variety of clinical and nonclinical specimens.

CDC Anaerobe Laked Blood Agar with Kanamycin and Vancomycin (CDC Anaerobe Laked Blood Agar with KV)

Composition per liter:

Agar	20.0g
Pancreatic digest of casein	15.0g
Papaic digest of soybean meal	5.0g
Yeast extract	5.0g
L-Cystine	0.4g
Sheep blood, defibrinated, laked	50.0mL
Antibiotic solution	10.0mL
Vitamin K_1 solution	1.0mL
Hemin solution	0.5mL

pH 7.5 ± 0.2 at 25°C

Source: This medium is available as a prepared medium from BD Diagnostic Systems.

Antibiotic Solution:
Composition per 10.0mL:

Kanamycin	0.1g
Vancomycin	7.5mg

Preparation of Antibiotic Solution: Add components to distilled/deionized water and bring volume to 10.0mL. Mix thoroughly. Filter sterilize.

Vitamin K_1 Solution:

Composition per 100.0mL:

Vitamin K_1	1.0g
Ethanol	99.0mL

Preparation of Vitamin K_1 Solution: Add vitamin K_1 to 99.0mL of absolute ethanol. Mix thoroughly. Filter sterilize.

Hemin Solution:

Composition per 100.0mL:

Hemin	1.0g
NaOH (1*N* solution)	20.0mL

Preparation of Hemin Solution: Add hemin to 20.0mL of 1*N* NaOH solution. Mix thoroughly. Bring volume to 100.0mL with distilled/deionized water.

Preparation of Medium: Add components, except antibiotic solution, vitamin K_1, and laked sheep blood, to distilled/deionized water and bring volume to 939.0mL. Mix thoroughly. Heat gently and bring to boiling for 1 min. Autoclave for 15 min at 15 psi pressure–121°C. Cool to 50°–55°C. Aseptically add the 1.0mL of sterile vitamin K_1 solution and 10.0mL of sterile antibiotic solution. Mix thoroughly. Aseptically add 50.0mL of sterile, defibrinated, laked sheep blood. Laked blood is prepared by freezing whole blood overnight and thawing to room temperature. Mix thoroughly. Pour into sterile Petri dishes.

Use: For the selective isolation of fastidious and slow-growing, obligate anaerobic bacteria from a variety of clinical and nonclinical specimens.

Cefiximine Rhamnose Sorbitol MacConkey Agar (CR-SMAC Agar Base)

Composition per liter:

Peptone	20.0g
Agar	15.0g
Sorbitol	10.0g
NaCl	5.0g
Rhamnose	5.0g
Bile salts no. 3	1.5g
Neutral Red	0.03g
Crystal Violet	0.001g
Selective supplement solution	10.0mL

pH 7.1 ± 0.2 at 25°C

Source: This medium is available as a premixed powder from Oxoid Unipath.

Selective Supplement Solution:

Composition per 10.0mL:

Cefiximine	0.05mg

Preparation of Selective Supplement Solution: Add cefiximine to distilled/deionized water and bring volume to 10.0mL. Mix thoroughly. Filter sterilize.

Preparation of Medium: Add components, except selective supplement solution, to distilled/deionized water and bring volume to 990.0mL. Mix thoroughly. Gently heat while stirring and bring to boiling. Autoclave for 15 min at 15 psi pressure–121°C. Cool to 50°C. Aseptially add selective supplement solution. Mix thoroughly. Pour into sterile Petri dishes.

Use: For the detection of *Escherichia coli* O157:H7. This is selective, differential medium based on Sorbitol MacConkey Agar with added rhamnose and cefixime. This medium provides a selective base with improved differentiation of *E. coli* O157. The addition of rhamnose aids in the differentiation of *Escherichia coli* O157 from background flora. Cefixime reduces the level of competing flora, particularly *Proteus* spp., that often account for large numbers of non-sorbitol fermenting colonies *E. coli* O157 do not usually ferment sorbitol or rhamnose, so will appear as straw colored colonies. However, rhamnose is fermented by most sorbitol negative *E. coli* of other serogroups. These colonies will be pink/red and will not be counted as presumptive *E. coli* O157 colonies.

Cefsulodin Irgasan Novobiocin Agar (CIN Agar) (*Yersinia* Selective Agar)

Composition per 1008.0mL:

Basal medium	757.0mL
Desoxycholate solution	200.0mL
Cefsulodin solution	10.0mL
Novobiocin solution	10.0mL
Crystal Violet solution	10.0mL
Strontium chloride solution	10.0mL
Neutral Red solution	10.0mL
NaOH, 5*N*	1.0mL
Irgasan solution	1.0mL

pH 7.4 ± 0.2 at 25°C

Basal Medium:

Composition per 757.0mL:

Mannitol	20.0g
Special peptone	20.0g
Agar	12.0g
Sodium pyruvate	2.0g
Yeast extract	2.0g

NaCl 1.0g
Magnesium sulfate solution 1.0mL

Preparation of Basal Medium: Add components to distilled/deionized water and bring volume to 757.0mL. Mix thoroughly. Gently heat and bring to boiling with stirring. Cool to about 80°C by placing in a 50°C water bath for about 10 min.

Magnesium Sulfate Solution:
Composition per 10mL:
$MgSO_4 \cdot 7H_2O$ 0.1g

Preparation of Magnesium Sulfate Solution: Add $MgSO_4 \cdot 7H_2O$ to distilled/deionized water and bring volume to 10.0mL. Mix thoroughly.

Irgasan Solution:
Composition per 10mL:
Irgasan (triclosan) 0.04g

Preparation of Irgasan Solution: Add irgasan to 95% ethanol and bring volume to 10.0mL. Mix thoroughly. Can be stored for 4 weeks at –20°C.

Desoxycholate Solution:
Composition per 200.0mL:
Na-desoxycholate 0.5g

Preparation of Desoxycholate Solution: Add desoxycholate to distilled/deionized water and bring volume to 200.0mL. Mix thoroughly. Gently heat and bring to boiling with stirring. Cool to 50–55°C.

Neutral Red Solution:
Composition per 10.0mL:
Neutral Red 30.0mg

Preparation of Neutral Red Solution: Add Neutral Red to 10.0mL of distilled/deionized water. Mix thoroughly. Autoclave for 15 min at 15 psi pressure–121°C. Cool to 25°C.

Crystal Violet Solution:
Composition per 10.0mL:
Crystal Violet 1.0mg

Preparation of Crystal Violet Solution: Add Crystal Violet to 10.0mL of distilled/deionized water. Mix thoroughly. Autoclave for 15 min at 15 psi pressure–121°C. Cool to 25°C.

Cefsulodin Solution:
Composition per 10.0mL:
Cefsulodin 15.0mg

Preparation of Cefsulodin Solution: Add cefsulodin to 10.0mL of distilled/deionized water. Mix thoroughly. Filter sterilize.

Novobiocin Solution:
Composition per 10.0mL:
Novobiocin 2.5mg

Preparation of Novobiocin Solution: Add novobiocin to 10.0mL of distilled/deionized water. Mix thoroughly. Filter sterilize.

Strontium Chloride Solution:
Composition per 10.0mL:
$SrCl_2 \cdot 6H_2O$ 1.0g

Preparation of Strontium Chloride: Add strontium chloride to 10.0mL of distilled/deionized water. Mix thoroughly. Filter sterilize.

Preparation of Medium: Add 1.0mL irgasan solution to 757.0mL basal medium. Mix thoroughly. Cool to 50–55°C. Add 200.0mL desoxychlolate solution. Mix thoroughly. Solution should remain clear. Aseptically add 1.0mL 5*N* NaOH, 10.0mL Neutral Red solution, 10.0mL Crystal Violet solution, 10.0mL cefsulodin solution, and 10.0mL novobiocin solution. Mix thoroughly. Slowly add 10.0mL strontium chloride solution while continuously stirring. Adjust pH to 7.4 with 5N NaOH. Pour into sterile Petri dishes or distribute into sterile tubes.

Use: For the selective isolation and differentiation of *Yersinia enterocolitica* based on mannitol fermentation. *Yersinia enterocolitica* appears as "bull's eye" colonies with deep red centers surrounded by a transparent periphery.

Cell Growth Medium

Composition per 2250.0mL:
Eagle's minimal essential medium
with Hanks' salts (MEMH) 1.0L
L 15 medium, modified Leibovitz 1.0L
Fetal calf serum 200.0mL
$NaHCO_3$ solution 50.0mL
pH 7.5 at 25°C

MEMH:
Composition per liter:
NaCl 8.0g
Glucose 1.0g
KCl 0.4g
$CaCl_2 \cdot 2H_2O$ 0.14g
$MgSO_4 \cdot 7H_2O$ 0.1g
KH_2PO_4 .... 0.06g
Na_2HPO_4 .... 0.05g
L-Isoleucine 0.026g
L-Leucine 0.026g
L-Lysine 0.026g
L-Threonine 0.024g
L-Valine 0.0235g
L-Tyrosine 0.018g
L-Arginine 0.0174g
L-Phenylalanine 0.0165g
L-Cystine 0.012g
L-Histidine 8.0mg

L-Methionine	7.5mg
Phenol Red	5.0mg
L-Tryptophan	4.0mg
Inositol	1.8mg
Biotin	1.0mg
Folic acid	1.0mg
Calcium pantothenate	1.0mg
Choline chloride	1.0mg
Nicotinamide	1.0mg
Pyridoxal·HCl	1.0mg
Thiamine·HCl	1.0mg
Riboflavin	0.1mg

Preparation of MEMH: Add components to distilled/deionized water and bring volume to 1.0L. Mix thoroughly. Filter sterilize.

L 15 Medium, Modified Leibovitz:
Composition per liter:

NaCl	8.0g
DL-Threonine	0.6g
Sodium pyruvate	0.6g
DL-Alanine	0.5g
L-Arginine, free base	0.5g
KCl	0.4g
L-Asparagine·H_2O	0.3g
L-Histidine, free base	0.3g
L-Glutamine	0.3g
L-Isoleucine	0.3g
L-Phenylalanine	0.3g
L-Tyrosine	0.3g
DL-Methionine	0.2g
DL-Valine	0.2g
Glycine	0.2g
L-Serine	0.2g
Na_2HPO_4, anhydrous	0.2g
$CaCl_2$, anhydrous	0.1g
L-Cysteine, free base	0.1g
L-Leucine·HCl	0.1g
Streptomycin	0.1g
$MgCl_2$, anhydrous	0.094g
D-Galactose	0.09g
KH_2PO_4	0.06g
Gentamicin	0.05g
L-Tryptophan	0.02g
Phenol Red	0.01g
i-Inositol	2.0mg
Choline chloride	1.0mg
D-Calcium pantothenate	1.0mg
Folic acid	1.0mg
Nicotinamide	1.0mg
Pyridoxine·HCl	1.0mg
Thiamine monophosphate·$2H_2O$	1.0mg
Riboflavin-5-phosphate	0.1mg
Penicillin G	100,000U

pH 7.5 ± 0.2 at 25°C

Preparation of L 15 Medium, Modified Leibovitz: Add components to distilled/deionized water and bring volume to 1.0L. Mix thoroughly. Filter sterilize. Store at 5°C.

$NaHCO_3$ Solution:
Composition per 100.0mL:

$NaHCO_3$	7.5g

Preparation of $NaHCO_3$ Solution: Add $NaHCO_3$ to distilled/deionized water and bring volume to 100.0mL. Mix thoroughly. Filter sterilize.

Preparation of Medium: Aseptically combine 1.0L of sterile Eagle's minimal essential medium with Hanks' salts (MEMH) and 1.0L of sterile L 15 medium, modified Leibovitz. Mix thoroughly. Immediately prior to use, aseptically add 200.0mL of fetal calf serum and 50.0mL of $NaHCO_3$ solution. Mix thoroughly. Aseptically distribute into sterile containers.

Use: For the cultivation of mammalian HeLa or Vero tissue culture cells to test the cytopathic effects of *Escherichia coli.*

Cellobiose Polymyxin B Colistin Agar, Modified

Composition per liter:

Solution 1	900.0mL
Solution 2	100.0mL

pH 7.6 ± 0.2 at 25°C

Solution 1:
Composition per 900.0mL:

NaCl	20.0g
Agar	15.0g
Peptone	10.0g
Beef extract	5.0g
1000× dye stock solution	1.0mL

Preparation of Solution 1: Add components to distilled/deionized water and bring volume to 900.0mL. Mix thoroughly. Adjust pH to 7.6. Gently heat and bring to boiling. Do not autoclave. Cool to 48°–55°C.

1000X Dye Stock Solution:
Composition per 100.0mL:

Bromthymol Blue	4.0g
Cresol Red	4.0g
Ethanol (95% solution)	100.0mL

Preparation of 1000X Dye Stock Solution: Add Bromthymol Blue and Cresol Red to 100.0mL of ethanol. Mix thoroughly.

Solution 2:
Composition per 100.0mL:

Cellobiose	10.0g

Colistin..400,000U
Polymyxin B.....................................100,000U

Preparation of Solution 2: Add cellobiose to distilled/deionized water and bring volume to 100.0mL. Mix thoroughly. Gently heat until dissolved. Cool to 25°C. Add colistin and polymyxin B. Mix thoroughly.

Preparation of Medium: Combine cooled solution 1 and solution 2. Mix thoroughly. Do not autoclave. Pour into sterile Petri dishes.

Use: For the cultivation of *Vibrio* species.

Cellobiose Polymyxin Colistin Agar (CPC Agar)

Composition per liter:

Solution A.....................................900.0mL
Solution B.....................................100.0mL

pH 7.6 ± 0.2 at 25°C

Solution A:

Composition per 900.0mL:

NaCl...20.0g
Agar...15.0g
Peptone......................................10.0g
Beef extract.................................5.0g
Bromthymol Blue..............................0.04g
Cresol Red...................................0.04g

Preparation of Solution A: Add components to distilled/deionized water and bring volume to 900.0mL. Mix thoroughly. Adjust pH to 7.6. Gently heat and bring to boiling. Autoclave for 15 min at 15 psi pressure–121°C. Cool to 50°–55°C.

Solution B:

Composition per 100.0mL:

Cellobiose...................................15.0g
Colistin.....................................1,360,000U
Polymyxin B..................................100,000U

Preparation of Solution B: Add components to distilled/deionized water and bring volume to 100.0mL. Mix thoroughly. Filter sterilize.

Preparation of Medium: Aseptically combine 900.0mL of cooled, sterile solution A and 100.0mL of sterile solution B. Mix thoroughly. Pour into sterile Petri dishes. Use within 7 days.

Use: For the cultivation and identification of *Vibrio* species.

Cereal Agar

Composition per liter:

Cereal, precooked mixed......................100.0g
Agar...15.0g

Preparation of Medium: Add components to distilled/deionized water and bring volume to 1.0L. Mix thoroughly. Gently heat and bring to boiling. Autoclave for 15 min at 15 psi pressure–121°C. Pour into sterile Petri dishes or distribute into sterile tubes. Allow tubes to cool in a slanted position.

Use: For the cultivation and sporulation of fungi.

Cetrimide Agar, Non-USP

Composition per liter:

Beef heart, solids from infusion.............500.0g
Agar...15.0g
Tryptose.....................................10.0g
NaCl...5.0g
Cetrimide....................................0.9g

pH 7.2 ± 0.2 at 25°C

Preparation of Medium: Add components to distilled/deionized water and bring volume to 1.0L. Mix thoroughly. Gently heat and bring to boiling. Distribute into tubes or flasks. Autoclave for 15 min at 13 psi pressure–118°C. Pour into sterile Petri dishes or leave in tubes.

Use: For the selective isolation, cultivation, and identification of *Pseudomonas aeruginosa* and other Gram-negative, nonfermentative bacteria.

Cetrimide Agar, USP (Pseudosel® Agar)

Composition per liter:

Pancreatic digest of gelatin.................20.0g
Agar...13.6g
K_2SO_4....................................10.0g
$MgCl_2$.....................................1.4g
Cetrimide....................................0.3g
Glycerol.....................................10.0mL

pH 7.2 ± 0.2 at 25°C

Source: This medium is available as a premixed powder from BD Diagnostic Systems.

Preparation of Medium: Add components to distilled/deionized water and bring volume to 1.0L. Mix thoroughly. Gently heat and bring to boiling. Distribute into tubes or flasks. Autoclave for 15 min at 13 psi pressure–118°C. Pour into sterile Petri dishes or leave in tubes.

Use: For the selective isolation, cultivation, and identification of *Pseudomonas aeruginosa* and other Gram-negative, nonfermentative bacteria.

CF Assay Medium (Citrovorum Factor Assay Medium)

Composition per liter:

Glucose......................................50.0g

Sodium acetate	40.0g
Vitamin assay casamino acids	10.0g
NH_4Cl	6.0g
K_2HPO_4	1.2g
KH_2PO_4	1.2g
$MgSO_4 \cdot 7H_2O$	0.4g
DL-Alanine	0.2g
DL-Tryptophan	0.2g
L-Cystine	0.2g
L-Cysteine·HCl	0.2g
$MgSO_4 \cdot 7H_2O$	0.04g
Adenine sulfate	0.02g
$FeSO_4$	0.02g
Glycine	0.02g
Guanine·HCl	0.02g
NaCl	0.02g
Uracil	0.02g
Xanthine	0.02g
Pyridoxamine·HCl	6.0mg
Nicotinic acid	2.0mg
Pyridoxine·HCl	2.0mg
Calcium pantothenate	1.0mg
Riboflavin	1.0mg
Thiamine·HCl	1.0mg
Pyridoxal·HCl	600.0μg
p-Aminobenzoic acid	200.0μg
Folic acid	20.0μg
Biotin	2.0μg

pH 6.7 ± 0.2 at 25°C

Preparation of Medium: Add components to distilled/deionized water and bring volume to 1.0L. Mix thoroughly. Gently heat and bring to boiling. Continue boiling for 2–3 min. Allow precipitate to settle out. Distribute supernatant into tubes in 5.0mL volumes. Add standard solution or test solutions to each tube. Adjust the volume of each tube to 10.0mL with distilled/deionized water. Autoclave for 10 min at 15 psi pressure–121°C.

Use: For the microbiological assay of citrovorum factor using *Pediococcus acidilactici.*

CFAT Medium (Cadmium Fluoride Acriflavin Tellurite Medium)

Composition per liter:

Pancreatic digest of casein	17.0g
Agar	15.0g
Glucose	7.5g
NaCl	5.0g
Papaic digest of soybean meal	3.0g
K_2HPO_4	2.5g
NaF	0.8g
$CdSO_4$	0.013g
K_2TeO_3	2.5mg
Neutral acriflavin	1.2mg
Basic Fuchsin	0.25mg
Sheep blood, defibrinated	50.0mL

Caution: Potassium tellurite is toxic.

Preparation of Medium: Add components, except sheep blood, to distilled/deionized water and bring volume to 950.0mL. Mix thoroughly. Gently heat and bring to boiling. Autoclave for 15 min at 15 psi pressure–121°C. Cool to 45°–50°C. Add 50.0mL of sterile, defibrinated sheep blood. Mix thoroughly. Pour into sterile Petri dishes or leave in tubes.

Use: For the isolation, cultivation, and enumeration of *Actinomyces viscosus* and *Actinomyces naeslundii* from clinical specimens, especially dental plaque.

CGY Agar

Composition per liter:

Agar	20.0g
Pancreatic digest of casein	5.0g
Yeast extract	1.0g
Glycerol	10.0mL

pH 7.2 ± 0.2 at 25°C

Preparation of Medium: Add components to distilled/deionized water and bring volume to 1.0L. Mix thoroughly. Gently heat and bring to boiling. Distribute into tubes or flasks. Autoclave for 15 min at 15 psi pressure–121°C. Pour into sterile Petri dishes or leave in tubes.

Use: For the cultivation and maintenance of *Bacillus pseudogordonae.*

CH 1 Medium

Composition per liter:

NaCl	250.0g
Tris(hydroxymethyl)amino methane buffer	12.0g
Glycerol	10.0g
Hy-Case SF	5.0g
Yeast extract	5.0g
Solution 1	50.0mL

Solution 1:

Composition per liter:

$MgCl_2 \cdot 6H_2O$	40.0g
KCl	4.0g
$CaCl_2 \cdot 2H_2O$	0.4g

pH 7.4 ± 0.2 at 25°C

Preparation of Solution 1: Add components to distilled/deionized water and bring volume to 1.0L. Mix thoroughly.

Preparation of Medium: Add components to distilled/deionized water and bring volume to 1.0L. Mix

thoroughly. Adjust pH to 7.4. Distribute into tubes or flasks. Autoclave for 15 min at 15 psi pressure–121°C.

Use: For the cultivation of *Haloarcula vallismortis*.

Chapman Stone Agar

Composition per liter:

$(NH_4)_2SO_4$	75.0g
NaCl	55.0g
Gelatin	30.0g
Agar	15.0g
D-Mannitol	10.0g
Pancreatic digest of casein	10.0g
K_2HPO_4	5.0g
Yeast extract	2.0g

pH 7.0 ± 0.2 at 25°C

Source: This medium is available as a premixed powder from BD Diagnostic Systems.

Preparation of Medium: Add components to distilled/deionized water and bring volume to 1.0L. Mix thoroughly. Autoclave for 10 min at 15 psi pressure–121°C. Pour into sterile Petri dishes while the medium is still hot. Add 25.0mL of medium per Petri dish.

Use: For the isolation of staphylococci from a variety of specimens.

Charcoal Agar

Composition per liter:

Beef heart, solids from infusion	500.0g
Agar	18.0g
Peptone	10.0g
Soluble starch	10.0g
NaCl	5.0g
Charcoal, activated, acid washed	4.0g
Yeast extract	3.5g

pH 7.3 ± 0.2 at 25°C

Source: This medium is available as a premixed powder from BD Diagnostic Systems.

Preparation of Medium: Add components to distilled/deionized water and bring volume to 1.0L. Mix thoroughly. Gently heat and bring to boiling with frequent stirring. Autoclave for 15 min at 15 psi pressure–121°C. Cool to 45°–50°C. Pour into sterile Petri dishes or leave in tubes. Shake flask while dispensing to keep charcoal in suspension. Allow tubes to cool in a slanted position.

Use: For the cultivation and maintenance of fastidious microorganisms.

Charcoal Agar

Composition per liter:

Agar	12.0g
Beef extract	10.0g
Peptone	10.0g
Starch	10.0g
NaCl	5.0g
Charcoal	4.0g
Nicotinic acid	0.001g

pH 7.4 ± 0.2 at 25°C

Source: This medium is available as a premixed powder from Oxoid Unipath.

Preparation of Medium: Add components to distilled/deionized water and bring volume to 1.0L. Mix thoroughly. Gently heat and bring to boiling with frequent stirring. Autoclave for 15 min at 15 psi pressure–121°C. Cool to 45°–50°C. This medium may be enriched by the addition of blood. Pour into sterile Petri dishes or distribute into tubes. Shake flask while dispensing to keep charcoal in suspension.

Use: For the cultivation and isolation of various bacteria; with the addition of blood, for the cultivation of fastidious bacteria.

Charcoal Agar with Horse Blood

Composition per liter:

Agar	12.0g
Beef extract	10.0g
Peptone	10.0g
Starch	10.0g
NaCl	5.0g
Charcoal, bacteriological	4.0g
Nicotinic acid	1.0mg
Horse blood, defibrinated	100.0mL

pH 7.4 ± 0.2 at 25°C

Preparation of Medium: Add components to distilled/deionized water and bring volume to 900.0L. Mix thoroughly. Gently heat and bring to boiling with frequent stirring. Autoclave for 15 min at 15 psi pressure–121°C. Cool to 80°C. Aseptically add 100.0mL of sterile, defibrinated horse blood. Maintain at 80°C for 10 min to form chocolate agar. Pour into sterile Petri dishes or distribute into tubes. Shake flask while dispensing to keep charcoal in suspension.

Use: For the cultivation and isolation of *Haemophilus influenzae*.

Charcoal Agar with Horse Blood and Cepahalexin

Composition per liter:

Agar	12.0g
Beef extract	10.0g
Peptone	10.0g
Starch	10.0g

NaCl 5.0g
Charcoal 4.0g
Nicotinic acid 1.0mg
Horse blood, defibrinated 100.0mL
Cephalexin solution 10.0mL
pH 7.4 ± 0.2 at 25°C

Cephalexin Solution:
Composition per 10.0mL:
Cephalexin 0.04g

Preparation of Cephalexin Solution: Add cephalexin to distilled/deionized water and bring volume to 10.0mL. Mix thoroughly. Filter sterilize.

Preparation of Medium: Add components, except cephalexin solution and horse blood, to distilled/deionized water and bring volume to 890.0L. Mix thoroughly. Gently heat and bring to boiling with frequent stirring. Autoclave for 15 min at 15 psi pressure–121°C. Cool to 45°–50°C. Aseptically add 100.0mL of sterile, defibrinated horse blood and 10.0mL of sterile cephalexin solution. Pour into sterile Petri dishes or distribute into tubes. Shake flask while dispensing to keep charcoal in suspension.

Use: For the cultivation and isolation of *Bordetella pertussis*.

Charcoal Blood Medium

Composition per liter:
Beef heart, solids from infusion 500.0g
Agar 18.0g
Peptone 10.0g
Soluble starch 10.0g
NaCl 5.0g
Charcoal, activated, acid washed 4.0g
Yeast extract 3.5g
Horse or sheep blood, defibrinated 100.0mL
Cephalexin solution 10.0mL
pH 7.4 ± 0.2 at 25°C

Cephalexin Solution:
Composition per 10.0mL:
Cephalexin 0.04g

Preparation of Cephalexin Solution: Add cephalexin to distilled/deionized water and bring volume to 10.0mL. Mix thoroughly. Filter sterilize.

Preparation of Medium: Add components, except blood and cephalexin solution, to distilled/deionized water and bring volume to 890.0mL. Mix thoroughly. Gently heat and bring to boiling. Autoclave for 15 min at 15 psi pressure–121°C. Cool to 45°–50°C. Aseptically add sterile blood and cephalexin solution. Mix thoroughly. Pour into sterile Petri dishes or distribute into sterile tubes.

Use: For the cultivation of *Haemophilus influenzae*.

Chlamydia Growth Medium

Composition per 500.0mL:
Eagle minimum essential medium
with Earle salts, 10X 50.0mL
Fetal calf serum 50.0mL
L-Glutamine solution 5.0mL
pH 7.4 ± 0.2 at 25°C

Eagle Minimum Essential Medium with Earle Salts, 10X:
Composition per liter:
NaCl 6.8g
Glucose 1.0g
KCl 0.4g
$CaCl_2 \cdot 2H_2O$ 0.2g
$MgCl_2 \cdot 6H_2O$ 0.2g
NaH_2PO_4 .... 0.15g
L-Arginine 0.1g
L-Lysine 0.06g
L-Isoleucine 0.05g
L-Leucine 0.05g
L-Threonine 0.05g
L-Valine 0.05g
L-Tyrosine 0.04g
L-Phenylalanine 0.03g
L-Histidine 0.03g
L-Cystine 0.02g
L-Methionine 0.02g
L-Tryptophan 0.01g
i-Inositol 2.0mg
Calcium pantothenate 1.0mg
Choline chloride 1.0mg
Folic acid 1.0mg
Nicotinamide 1.0mg
Pyridoxal 1.0mg
Thiamine·HCl 1.0mg
Riboflavin 0.1mg

Preparation of Eagle Minimum Essential Medium With Earle Salts, 10X: Add components to distilled/deionized water and bring volume to 1.0L. Mix thoroughly. Adjust pH to 7.4 with 7.5% Na_2CO_3 solution. Filter sterilize.

Glutamine Solution:
Composition per 100.0mL:
L-Glutamine 2.92g
NaCl (0.85% solution) 100.0mL

Preparation of Glutamine Solution: Add the glutamine to the 0.85% NaCl solution. Mix thoroughly. Filter sterilize.

Preparation of Medium: Aseptically combine 50.0mL of sterile Eagle minimum essential medium-with Earle salts, 10X, 50.0mL of fetal calf serum, and 5.0mL of sterile glutamine solution. Bring volume to 500.0mL with sterile distilled/deionized water. Mix thoroughly. Aseptically distribute into sterile tubes or flasks.

Use: For the cultivation of *Chlamydia* species.

Chlamydia Isolation Medium

Composition per 500.0mL:

Eagle minimum essential medium with Earle salts, 10X	50.0mL
Fetal calf serum	50.0mL
Selective supplement	10.0mL
L-Glutamine solution	5.0mL

pH 7.4 ± 0.2 at 25°C

Eagle Minimum Essential Medium with Earle Salts, 10X:

Composition per liter:

NaCl	6.8g
Glucose	1.0g
KCl	0.4g
$CaCl_2 \cdot 2H_2O$	0.2g
$MgCl_2 \cdot 6H_2O$	0.2g
NaH_2PO_4	0.15g
L-Arginine	0.1g
L-Lysine	0.06g
L-Isoleucine	0.05g
L-Leucine	0.05g
L-Threonine	0.05g
L-Valine	0.05g
L-Tyrosine	0.04g
L-Phenylalanine	0.03g
L-Histidine	0.03g
L-Cystine	0.02g
L-Methionine	0.02g
L-Tryptophan	0.01g
i-Inositol	2.0mg
Calcium pantothenate	1.0mg
Choline chloride	1.0mg
Folic acid	1.0mg
Nicotinamide	1.0mg
Pyridoxal	1.0mg
Thiamine·HCl	1.0mg
Riboflavin	0.1mg

Preparation of Eagle Minimum Essential Medium With Earle Salts, 10X: Add components to distilled/deionized water and bring volume to 1.0L. Mix thoroughly. Adjust pH to 7.4 with 7.5% Na_2CO_3 solution. Filter sterilize.

Selective Supplement:

Composition per 10.0mL:

Glucose	0.594g
Vancomycin	0.05g
Gentamicin	0.01g
Amphotericin B	2.0mg
Cycloheximide	2.0mg

Preparation of Selective Supplement: Add components to distilled/deionized water and bring volume to 10.0mL. Mix thoroughly. Filter sterilize.

Glutamine Solution:

Composition per 100.0mL:

L-Glutamine	2.92g
NaCl (0.85% solution)	100.0mL

Preparation of Glutamine Solution: Add the glutamine to the 0.85% NaCl solution. Mix thoroughly. Filter sterilize.

Preparation of Medium: Aseptically combine 50.0mL of sterile Eagle minimum essential medium with Earle salts, 10X, 50.0mL of fetal calf serum, 10.0mL of selective supplement, and 5.0mL of sterile glutamine solution. Bring volume to 500.0mL with sterile distilled/deionized water. Mix thoroughly. Aseptically distribute into sterile tubes or flasks.

Use: For the isolation and cultivation of *Chlamydia* species.

Chlamydospore Agar

Composition per liter:

Purified polysaccharide	20.0g
Agar	15.0g
KH_2PO_4	1.0g
$(NH_4)_2SO_4$	1.0g
Trypan Blue	0.1g
Biotin	5.0μg

pH 5.1 ± 0.2 at 25°C

Source: This medium is available as a premixed powder from BD Diagnostic Systems.

Preparation of Medium: Add components to distilled/deionized water and bring volume to 1.0L. Mix thoroughly. Gently heat and bring to boiling. Distribute into tubes or flasks. Autoclave for 15 min at 15 psi pressure–121°C. Pour into sterile Petri dishes or leave in tubes.

Use: For differentiating *Candida albicans* from other *Candida* species on the basis of chlamydospore formation.

Chloramphenicol Ampicillin LB Medium

Composition per liter:

NaCl	10.0g
Pancreatic digest of casein	10.0g
Yeast extract	5.0g
Ampicillin solution	10.0mL
Chloramphenicol solution	10.0mL

pH 7.0 ± 0.2 at 25°C

Ampicillin Solution:
Composition per 10.0mL:

Ampicillin	40.0mg

Preparation of Ampicillin Solution: Add ampicillin to distilled/deionized water and bring volume to 10.0mL. Mix thoroughly. Filter sterilize.

Chloramphenicol Solution:
Composition per 10.0mL:

Chloramphenicol	5.0mg

Preparation of Chloramphenicol Solution: Add chloramphenicol to distilled/deionized water and bring volume to 10.0mL. Mix thoroughly. Filter sterilize.

Preparation of Medium: Add components, except ampicillin solution and chloramphenicol solution, to distilled/deionized water and bring volume to 980.0mL. Mix thoroughly. Adjust pH to 7.0. Autoclave for 15 min at 15 psi pressure–121°C. Aseptically add 10.0mL of sterile ampicillin solution and 10.0mL of sterile chloramphenicol solution. Mix thoroughly. Aseptically distribute into sterile tubes or flasks.

Use: For the cultivation and maintenance of *Escherichia coli.*

Chloramphenicol Erythromycin LB Medium

Composition per liter:

NaCl	10.0g
Pancreatic digest of casein	10.0g
Yeast extract	5.0g
Erythromycin	10.0mL
Chloramphenicol	10.0mL

pH 7.0 ± 0.2 at 25°C

Erythromycin Solution:
Composition per 10.0mL:

Erythromycin	10mg

Preparation of Erythromycin Solution: Add erythromycin to distilled/deionized water and bring volume to 10.0mL. Mix thoroughly. Filter sterilize.

Chloramphenicol Solution:
Composition per 10.0mL:

Chloramphenicol	5.0mg

Preparation of Chloramphenicol Solution: Add chloramphenicol to distilled/deionized water and bring volume to 10.0mL. Mix thoroughly. Filter sterilize.

Preparation of Medium: Add components, except erythromycin solution and chloramphenicol solution, to distilled/deionized water and bring volume to 980.0mL. Mix thoroughly. Adjust pH to 7.0. Autoclave for 15 min at 15 psi pressure–121°C. Aseptically add 10.0mL of sterile ampicillin solution and 10.0mL of sterile chloramphenicol solution. Mix thoroughly. Aseptically distribute into sterile tubes or flasks.

Use: For the cultivation and maintenance of *Escherichia coli.*

Chloramphenicol L Broth Medium No. 1

Composition per liter:

Pancreatic digest of casein	10.0g
NaCl	5.0g
Yeast extract	5.0g
Glucose	1.0g
Chloramphenicol	10.0mL

pH 7.0 ± 0.2 at 25°C

Chloramphenicol Solution:
Composition per 10.0mL:

Chloramphenicol	5.0mg

Preparation of Chloramphenicol Solution: Add chloramphenicol to distilled/deionized water and bring volume to 10.0mL. Mix thoroughly. Filter sterilize.

Preparation of Medium: Add components, except chloramphenicol solution, to distilled/deionized water and bring volume to 990.0mL. Mix thoroughly. Adjust pH to 7.0. Autoclave for 15 min at 15 psi pressure–121°C. Aseptically add 10.0mL of sterile ampicillin solution and 10.0mL of sterile chloramphenicol solution. Mix thoroughly. Aseptically distribute into sterile tubes or flasks.

Use: For the cultivation and maintenance of *Escherichia coli.*

Chloramphenicol L Broth Medium No. 2

Composition per liter:

Pancreatic digest of casein	10.0g
NaCl	5.0g
Yeast extract	5.0g
Glucose	1.0g
Chloramphenicol	10.0mL

pH 7.0 ± 0.2 at 25°C

Chloramphenicol Solution:
Composition per 10.0mL:

Chloramphenicol	12.5mg

Preparation of Chloramphenicol Solution: Add chloramphenicol to distilled/deionized water and bring volume to 10.0mL. Mix thoroughly. Filter sterilize.

Preparation of Medium: Add components, except chloramphenicol solution, to distilled/deionized water and bring volume to 990.0mL. Mix thoroughly.

Adjust pH to 7.0. Autoclave for 15 min at 15 psi pressure–121°C. Aseptically add 10.0mL of sterile ampicillin solution and 10.0mL of sterile chloramphenicol solution. Mix thoroughly. Aseptically distribute into sterile tubes or flasks.

Use: For the cultivation and maintenance of *Escherichia coli.*

Chloramphenicol L Broth Medium No. 3

Composition per liter:

Pancreatic digest of casein	10.0g
NaCl	5.0g
Yeast extract	5.0g
Glucose	1.0g
Chloramphenicol	10.0mL

pH 7.0 ± 0.2 at 25°C

Chloramphenicol Solution:
Composition per 10.0mL:

Chloramphenicol	50.0mg

Preparation of Chloramphenicol Solution: Add chloramphenicol to distilled/deionized water and bring volume to 10.0mL. Mix thoroughly. Filter sterilize.

Preparation of Medium: Add components, except chloramphenicol solution, to distilled/deionized water and bring volume to 990.0mL. Mix thoroughly. Adjust pH to 7.0. Autoclave for 15 min at 15 psi pressure–121°C. Aseptically add 10.0mL of sterile ampicillin solution and 10.0mL of sterile chloramphenicol solution. Mix thoroughly. Aseptically distribute into sterile tubes or flasks.

Use: For the cultivation and maintenance of *Escherichia coli.*

Chloramphenicol LB Medium No.1

Composition per liter:

NaCl	10.0g
Pancreatic digest of casein	10.0g
Yeast extract	5.0g
Chloramphenicol	50.0μg

Preparation of Medium: Adjust pH to 7.0. Autoclave at 15 psi pressure–121°C for 15 min. For solid medium, add 15.0g of agar.

Use: For the cultivation of and maintenance of *Escherichia coli.*

Chloramphenicol LB Medium No. 2

Composition per liter:

NaCl	10.0g
Pancreatic digest of casein	10.0g
Yeast extract	5.0g
Chloramphenicol	10.0mL

pH 7.0 ± 0.2 at 25°C

Chloramphenicol Solution:
Composition per 10.0mL:

Chloramphenicol	5.0mg

Preparation of Chloramphenicol Solution: Add chloramphenicol to distilled/deionized water and bring volume to 10.0mL. Mix thoroughly. Filter sterilize.

Preparation of Medium: Add components, except chloramphenicol solution, to distilled/deionized water and bring volume to 990.0mL. Mix thoroughly. Adjust pH to 7.0. Autoclave for 15 min at 15 psi pressure–121°C. Aseptically add 10.0mL of sterile ampicillin solution and 10.0mL of sterile chloramphenicol solution. Mix thoroughly. Aseptically distribute into sterile tubes or flasks.

Use: For the cultivation and maintenance of *Escherichia coli.*

CHO Medium Base (Carbohydrate Medium Base)

Composition per liter:

Pancreatic digest of casein	15.0g
Yeast extract	7.0g
NaCl	2.5g
Agar	0.75g
Sodium thioglycolate	0.5g
L-Cystine	0.25g
Ascorbic acid	0.1g
Bromthymol Blue	0.01g

pH 7.0 ± 0.2 at 25°C

Preparation of Medium: Add components to distilled/deionized water and bring volume to 1.0L. Mix thoroughly. Gently heat and bring to boiling. Distribute into tubes or flasks. Autoclave for 15 min at 15 psi pressure–121°C. Cool to 45°–50°C.

Use: Used as a basal medium to which carbohydrates are added for fermentation studies of anaerobic bacteria. Generally, 6.25mL of a 10% filter-sterilized solution of carbohydrate is added to the sterile basal medium.

Chocolate Agar

Composition per liter:

Agar	15.0g
Pantone	10.0g
Bitone	10.0g
NaCl	5.0g
Tryptic digest of beef heart	3.0g
Cornstarch	1.0g

Sheep blood, defibrinated 100.0mL
Supplement B... 10.0mL

pH 7.3 ± 0.2 at 25°C

Supplement B:
Composition per 10.0mL:

Cephalothin... 15.0mg
Vancomycin 10.0mg
Trimethoprim 5.0mg
Amphotericin B.. 2.0mg
Polymyxin B 2500U

Preparation of Supplement B: Add components to 10.0mL of distilled/deionized water. Mix thoroughly. Filter sterilize.

Source: Supplemement B is available from BD Diagnostic Systems.

Preparation of Medium: Add components, except hemoglobin solution and sheep blood, to distilled/deionized water and bring volume to 890.0mL. Mix thoroughly. Gently heat until boiling. Autoclave for 15 min at 15 psi pressure–121°C. Cool to 45°–50°C. Aseptically add 100.0mL of sterile, defibrinated sheep blood. Gently heat while stirring and bring to 85°C for 5–10 min. Cool to 50°C. Aseptically add 10.0mL of sterile supplement B. Mix thoroughly. Pour into sterile Petri dishes or distribute into sterile tubes.

Use: For the isolation and cultivation of a variety of fastidious microorganisms.

Chocolate Agar

Composition per liter:

Proteose peptone no. 3 .... 15.0g
Agar 10.0g
NaCl 5.0g
K_2HPO_4 .... 4.0g
Cornstarch 1.0g
KH_2PO_4 .... 1.0g
Hemoglobin solution................................. 100.0mL
Supplement B.. 10.0mL

pH 7.0 ± 0.2 at 25°C

Source: This medium is available from BD Diagnostic Systems.

Supplement B:
Composition per 10.0mL:

Cephalothin... 15.0mg
Vancomycin 10.0mg
Trimethoprim 5.0mg
Amphotericin B.. 2.0mg
Polymyxin B 2500U

Preparation of Supplement B: Add components to distilled/deionized water and bring volume to 10.0mL. Mix thoroughly. Filter sterilize.

Hemoglobin Solution:
Composition per 100.0mL:

Hemoglobin 10.0g

Preparation of Hemoglobin Solution: Add hemoglobin to distilled/deionized water and bring volume to 100.0mL. Mix thoroughly. Filter sterilize.

Preparation of Medium: Add components, except hemoglobin solution and supplement B, to distilled/deionized water and bring volume to 990.0mL. Mix thoroughly. Gently heat and bring to boiling. Autoclave for 15 min at 15 psi pressure–121°C. Cool to 45°–50°C. Aseptically add 100.0mL of sterile hemoglobin solution. Gently heat while stirring and bring to 85°C for 5–10 min. Cool to 50°C. Aseptically add 10.0mL of sterile supplement B. Mix thoroughly. Pour into sterile Petri dishes or distribute into sterile tubes.

Use: For the isolation and cultivation of fastidious microorganisms.

Chocolate Agar, Enriched

Composition per liter:

GC medium base... 740.0mL
Hemoglobin solution 250.0mL
Supplement B... 10.0mL

pH 7.3 ± 0.2 at 25°C

Source: This medium is available from BD Diagnostic Systems.

GC Medium Base:
Composition per 740.0mL:

Agar 20.0g
Proteose peptone no. 3 .... 15.0g
NaCl 5.0g
K_2HPO_4 .... 4.0g
Glucose 1.5g
Cornstarch 1.0g
KH_2PO_4 .... 1.0g

pH 7.2 ± 0.2 at 25°C

Preparation of GC Medium Base: Add components to distilled/deionized water and bring volume to 740.0mL. Mix thoroughly. Gently heat until boiling. Autoclave for 15 min at 15 psi pressure–121°C. Cool to 45°–50°C.

Hemoglobin Solution:
Composition per 250.0mL:

Hemoglobin 10.0g

Preparation of Hemoglobin Solution: Add hemoglobin to distilled/deionized water and bring volume to 250.0mL. Mix thoroughly. Autoclave for 15 min at 15 psi pressure–121°C. Cool to 45°–50°C.

Supplement B:
Composition per 10.0mL:

Cephalothin	15.0mg
Vancomycin	10.0mg
Trimethoprim	5.0mg
Amphotericin B	2.0mg
Polymyxin B	2500U

Preparation of Supplement B: Add components to distilled/deionized water and bring volume to 10.0mL. Mix thoroughly. Filter sterilize.

Preparation of Medium: To 740.0mL of cooled sterile GC medium base, aseptically add 250.0mL of sterile hemoglobin solution and 10.0mL of sterile supplement B. Mix thoroughly. Pour into sterile Petri dishes or distribute into sterile tubes.

Use: For the cultivation of fastidious microorganisms, especially *Neisseria* species.

Chocolate Agar, Enriched

Composition per liter:

GC medium base	740.0mL
Hemoglobin solution	250.0mL
Supplement VX	10.0mL

pH 7.3 ± 0.2 at 25°C

Source: This medium is available from BD Diagnostic Systems.

GC Medium Base:
Composition per 740.0mL:

Agar	20.0g
Proteose peptone no. 3	15.0g
NaCl	5.0g
K_2HPO_4	4.0g
Glucose	1.5g
Cornstarch	1.0g
KH_2PO_4	1.0g

pH 7.2 ± 0.2 at 25°C

Preparation of GC Medium Base: Add components to distilled/deionized water and bring volume to 740.0mL. Mix thoroughly. Gently heat until boiling. Autoclave for 15 min at 15 psi pressure–121°C. Cool to 45°–50°C.

Hemoglobin Solution:
Composition per 250.0mL:

Hemoglobin	10.0g

Preparation of Hemoglobin Solution: Add hemoglobin to distilled/deionized water and bring volume to 250.0mL. Mix thoroughly. Autoclave for 15 min at 15 psi pressure–121°C. Cool to 45°–50°C.

Supplement VX:
Composition per 10.0mL:
Supplement VX contains essential growth factors.

Preparation of Supplement VX: Add components to distilled/deionized water and bring volume to 10.0mL. Mix thoroughly. Filter sterilize.

Preparation of Medium: To 740.0mL of cooled sterile GC medium base, aseptically add 250.0mL of sterile hemoglobin solution and 10.0mL of sterile supplement B. Mix thoroughly. Pour into sterile Petri dishes or distribute into sterile tubes.

Use: For the cultivation of fastidious microorganisms, especially *Neisseria* species.

Chocolate Agar-*Bartonella* C-29 (ATCC Medium 2119)

Composition per 1010.0mL:

GC agar base solution	500.0 ml
Hemoglobin solution	500.0 ml
IsoVitaleX® enrichment	10.0mL

IsoVitaleX Enrichment:
Composition per liter:

Glucose	100.0g
L-Cysteine·HCl	25.9g
L-Glutamine	10.0g
L-Cystine	1.1g
Adenine	1.0g
Nicotinamide adenine dinucleotide	0.25g
Vitamin B_{12}	0.1g
Thiamine pyrophosphate	0.1g
Guanine·HCl	0.03g
$Fe(NO_3)_3 \cdot 6H_2O$	0.02g
p-Aminobenzoic acid	0.013g
Thiamine·HCl	3.0mg

Preparation of IsoVitaleX: Add components to distilled/deionized water and bring volume to 1.0L. Mix thoroughly. Filter sterilize.

GC Agar Base Solution:
Composition per 500.0mL:

Agar	10.0g
Pancreatic digest of casein	7.5g
Peptic digest of animal tissue	7.5g
NaCl	5.0g
K_2HPO_4	4.0g
Cornstarch	1.0g
KH_2PO_4	1.0g

Preparation of GC Agar Base: Add components to distilled/deionized water and bring volume to 500.0mL. Mix thoroughly. Gently heat until boiling. Autoclave for 15 min at 15 psi pressure–121°C. Cool to 45°–50°C.

Hemoglobin Solution:
Composition per 500.0mL:

Hemoglobin	10.0g

Preparation of Hemoglobin Solution: Add hemoglobin to distilled/deionized water and bring volume to 500.0mL. Mix thoroughly. Gently heat until boiling. Autoclave for 15 min at 15 psi pressure–121°C. Cool to 45°–50°C.

Preparation of Medium: Aseptically combine 500.0mL sterile, cooled GC agar base solution, and 500.0mL cooled sterile hemoglobin solution. Aseptically add 10.0mL of sterile IsoVitaleX enrichment. Mix thoroughly. Pour into sterile Petri dishes or distribute into sterile tubes.

Use: For the isolation and cultivation of fastidious microorganisms, especially *Neisseria* and *Haemophilus* species, from a variety of clinical specimens.

Chocolate II Agar

Composition per liter:

Agar	12.0g
Hemoglobin	10.0g
Pancreatic digest of casein	7.5g
Selected meat peptone	7.5g
NaCl	5.0g
K_2HPO_4	4.0g
Cornstarch	1.0g
KH_2PO_4	1.0g

Preparation of Medium: Add components to distilled/deionized water and bring volume to 1.0L. Mix thoroughly. Gently heat to boiling. Autoclave for 15 min at 15 psi pressure–121°C. Pour into sterile Petri dishes or leave in tubes.

Use: For the isolation and cultivation of fastidious microorganisms.

Chocolate II Agar with Hemoglobin and IsoVitaleX (GCII Agar with Hemoglobin and IsoVitaleX)

Composition per liter:

GCII agar base	990.0mL
IsoVitaleX enrichment	10.0mL

pH 7.3 ± 0.2 at 25°C

Source: This medium is available as a prepared medium from BD Diagnostic Systems.

GCII Agar Base:

Composition per liter:

Agar	12.0g
Hemoglobin	10.0g
Pancreatic digest of casein	7.5g
Selected meat peptone	7.5g
NaCl	5.0g
K_2HPO_4	4.0g
Cornstarch	1.0g
KH_2PO_4	1.0g

Preparation of GCII Agar Base: Add components to distilled/deionized water and bring volume to 1.0L. Mix thoroughly. Gently heat to boiling. Autoclave for 15 min at 15 psi pressure–121°C. Cool to 45°–50°C.

IsoVitaleX Enrichment:

Composition per liter:

Glucose	100.0g
L-Cysteine·HCl	25.9g
L-Glutamine	10.0g
L-Cystine	1.1g
Adenine	1.0g
Nicotinamide adenine dinucleotide	0.25g
Vitamin B_{12}	0.1g
Thiamine pyrophosphate	0.1g
Guanine·HCl	0.03g
$Fe(NO_3)_3 \cdot 6H_2O$	0.02g
p-Aminobenzoic acid	0.013g
Thiamine·HCl	3.0mg

Preparation of IsoVitaleX: Add components to distilled/deionized water and bring volume to 1.0L. Mix thoroughly. Filter sterilize.

Preparation of Medium: Aseptically add 10.0mL of sterile IsoVitaleX enrichment to 990.0L of sterile, cooled GCII agar base. Mix thoroughly. Pour into sterile Petri dishes or distribute into sterile tubes.

Use: For the isolation and cultivation of fastidious microorganisms, especially *Neisseria* and *Haemophilus* species, from a variety of clinical specimens.

Chocolate Tellurite Agar (Tellurite Blood Agar)

Composition per liter:

Agar	10.0g
Casein/meat (50/50) peptone	10.0g
Hemoglobin	10.0g
NaCl	5.0g
K_2HPO_4	4.0g
Cornstarch	1.0g
KH_2PO_4	1.0g
K_2TeO_3	0.1g
Bio-X enrichment	10.0mL

Bio-X Enrichment:

Composition per liter:

Glucose	100.0g
L-Cysteine·HCl	25.9g
L-Glutamate	10.0g
L-Cystine	1.1g
Adenine	1.0g

Cocarboxylase 0.1g
Guanine·HCl 0.03g
$FeNO_3$ 0.02g
p-Aminobenzoic acid 0.013g
Vitamin B_{12} 0.01g
NAD
(nicotinamide adenine dinucleotide) 250.0mg
Thiamine·HCl 3.0mg
pH 7.2 ± 0.2 at 25°C

Preparation of Bio-X Enrichment: Add components to distilled/deionized water and bring volume to 1.0L. Mix thoroughly. Filter sterilize.

Caution: Potassium tellurite is toxic.

Preparation of Medium: Add components, except Bio-X enrichment, to distilled/deionized water and bring volume to 990.0mL. Mix thoroughly. Gently heat and bring to boiling. Autoclave for 15 min at 15 psi pressure–121°C. Cool to 45°–50°C. Aseptically add filter-sterilized Bio-X enrichment. Mix thoroughly. Pour into sterile Petri dishes or distribute into sterile tubes.

Use: For the selective isolation and cultivation of *Corynebacterium* species. *Corynebacterium diphtheriae* appears as gray-black colonies.

Cholera Medium TCBS

Composition per liter:

Sucrose 20.0g
Agar 14.0g
Peptone 10.0g
NaCl 10.0g
Sodium citrate 10.0g
$Na_2S_2O_3 \cdot 5H_2O$ 10.0g
Ox bile 8.0g
Yeast extract 5.0g
Ferric citrate 1.0g
Bromthymol Blue 0.04g
Thymol Blue 0.04g
pH 8.6 ± 0.2 at 25°C

Source: This medium is available as a premixed powder from Oxoid Unipath.

Preparation of Medium: Add components to distilled/deionized water and bring volume to 1.0L. Mix thoroughly. Gently heat and bring to boiling. Do not autoclave. Pour into sterile Petri dishes. Dry agar plates before using.

Use: For the growth of *Vibrio cholerae, Vibrio parahaemolyticus,* and other *Vibrio* species.

Cholic Acid Medium

Composition per liter:

Noble agar 15.0g
K_2HPO_4 .... 3.5g
Cholic acid 2.0g
$(NH_4)_2SO_4$ 2.0g
KH_2PO_4 .... 1.5g
$MgSO_4 \cdot 7H_2O$ 0.1g
$CaCl_2 \cdot 2H_2O$ 0.01g
$FeSO_4 \cdot 7H_2O$ 0.5mg
pH 7.0 ± 0.2 at 25°C

Preparation of Medium: Add components to distilled/deionized water and bring volume to 1.0L. Mix thoroughly. Adjust pH to 7.0. Gently heat and bring to boiling. Autoclave for 15 min at 15 psi pressure–121°C. Pour into sterile Petri dishes or distribute into sterile tubes.

Use: For the cultivation and maintenance of *Nocardia* species and other bacteria that can utilize cholic acid as a carbon source.

Cholera Medium TCBS

Composition per liter:

Sucrose 20.0g
Agar 14.0g
Peptone 10.0g
$Na_2S_2O_3$ 10.0g
Sodium citrate 10.0g
NaCl 10.0g
Ox bile 8.0g
Yeast extract 5.0g
Ferric citrate 1.0g
Bromthymol Blue 0.04g
Thymol Blue 0.04g
pH 8.6 ± 0.2 at 25°C

Source: This medium is available as a premixed powder from Oxoid Unipath.

Preparation of Medium: Add components to distilled/deionized water and bring volume to 1.0mL Mix thoroughly. Gently heat while stirring and bring to boiling. Do not autoclave. Cool to 45°C. Pour into sterile Petri dishes.

Use: For the isolation of pathogenic vibrios, especially *Vibrio cholerae.* This medium is suitable for the growth of *Vibrio cholerae, Vibrio parahaemolyticus,* and most other *Vibrios.* Most of the Enterobacteriaceae encountered in faeces are totally suppressed for at least 24 hours. Slight growth of *Proteus* species and *Enterococcus faecalis* may occur but the colonies are easily distinguished from vibrio colonies. Whilst inhibiting non-vibrios, it promotes rapid growth of pathogenic vibrios after overnight incubation at 35°C. *Vibrio cholerae* El Tor biotype forms yellow colonies, *Vibrio parahaemolyticus* forms blue-green colonies, *Vibrio alginolyticus* forms yellow colonies, *Vibrio metschnikovii* forms yellow colonies, *Vibrio fluvialis* forms yellow colonies, *Vibrio vulnificus* forms blue-green colonies, *Vibrio mimicus* forms blue-green colonies, *Enterococcus* species form yellow colonies, *Proteus* species form yellow-green colonies, *Pseudomonas* species form blue-green colonies and some strains of *Aeromonas hydrophila* produce yellow colonies, but *Plesimonas shigelloides* does not usually grow well on this medium.

Chopped Meat Agar

Composition per liter:

Ground meat, fat free	500.0g
Pancreatic digest of casein	30.0g
Agar	15.0g
K_2HPO_4	5.0g
Yeast extract	5.0g
L-Cysteine·HCl	0.5g
Resazurin	1.0mg
NaOH (1*N* solution)	25.0mL

pH 7.0 ± 0.2 at 25°C

Preparation of Medium: Use lean beef or horse meat. Remove fat and connective tissue. Grind finely. Add ground meat and 25.0mL of NaOH solution to distilled/deionized water and bring volume to 1025.0mL. Gently heat and bring to boiling. Continue boiling for 15 min without stirring. Cool to room temperature. Remove fat from surface. Filter and retain both meat particles and filtrate. Adjust volume of filtrate to 1.0L with distilled/deionized water. Add pancreatic digest of casein, agar, K_2HPO_4, yeast extract, and resazurin. Gently heat and bring to boiling. Boil for 1–2 min. Add L-cysteine·HCl. Mix thoroughly. Distribute 7.0mL into tubes that contain meat particles (1 part meat particles to 5 parts fluid). Autoclave for 30 min at 15 psi pressure–121°C.

Use: For the cultivation of various anaerobes.

Chopped Meat Broth

Composition per liter:

Ground meat, fat free	500.0g
Pancreatic digest of casein	30.0g
K_2HPO_4	5.0g
Yeast extract	5.0g
L-Cysteine·HCl	0.5g
Resazurin	1.0mg
NaOH (1*N* solution)	25.0mL

pH 7.0 ± 0.2 at 25°C

Preparation of Medium: Use lean beef or horse meat. Remove fat and connective tissue. Grind finely. Add ground meat and 25.0mL of NaOH solution to distilled/deionized water and bring volume to 1025.0mL. Gently heat and bring to boiling. Continue boiling for 15 min without stirring. Cool to room temperature. Remove fat from surface. Filter and retain both meat particles and filtrate. Adjust volume of filtrate to 1.0L with distilled/deionized water. Add pancreatic digest of casein, K_2HPO_4, yeast extract, and resazurin. Gently heat and bring to boiling. Boil for 1–2 min. Add L-cysteine·HCl. Mix thoroughly. Distribute 7.0mL into tubes that contain meat particles (1 part meat particles to 5 parts fluid). Autoclave for 30 min at 15 psi pressure–121°C.

Use: For the cultivation of various anaerobes.

Chopped Meat Broth with Carbohydrates (DSMZ Medium 110)

Composition per liter:

Ground meat, fat free	500.0g
Pancreatic digest of casein	30.0g
K_2HPO_4	5.0g
Yeast extract	5.0g
Glucose	4.0g
Cellobiose	1.0g
Maltose	1.0g
Starch, soluble	1.0g
L-Cysteine·HCl	0.5g
Resazurin	1.0mg
NaOH (1*N* solution)	25.0mL

pH 7.0 ± 0.2 at 25°C

Preparation of Medium: Use lean beef or horse meat. Remove fat and connective tissue. Grind finely. Add ground meat and 25.0mL of NaOH solution to distilled/deionized water and bring volume to 1025.0mL. Gently heat and bring to boiling. Continue boiling for 15 min. without stirring. Cool to room temperature. Remove fat from surface. Filter and retain both meat particles and filtrate. Adjust volume of filtrate to 1.0L with distilled/deionized water. Add pancreatic digest of casein, K_2HPO_4, yeast extract, and resazurin. Gently heat and bring to boiling. Boil for 1–2 min. Add glucose, cellobiose, maltose, and soluble starch. Add L-cysteine·HCl. Mix thoroughly. Distribute 7.0mL into tubes that contain meat particles (1 part meat particles to 5 parts fluid). Autoclave for 30 min at 15 psi pressure–121°C.

Use: For the cultivation of numerous anaerobes.

Chopped Meat Broth with Formate and Fumarate (LMG Medium 69)

Composition per liter:

Ground meat, fat free	500.0g
Pancreatic digest of casein	30.0g
K_2HPO_4	5.0g

Yeast extract	5.0g
L-Cysteine·HCl	0.5g
Resazurin	1.0mg
NaOH (1*N* solution)	25.0mL
Formate-fumarate solution	7.7mL

pH 7.0 ± 0.2 at 25°C

Formate-Fumarate Solution:
Composition per 100.0mL:

Sodium formate	6.0g
Fumaric acid	6.0g

Preparation of Formate-Fumarate Solution: Add components to distilled/deionized water and bring volume to 100.0mL. Adjust pH to 7.0. Filter sterilize.

Preparation of Medium: Use lean beef or horse meat. Remove fat and connective tissue. Grind finely. Add ground meat and 25.0mL of NaOH solution to distilled/deionized water and bring volume to 1025mL. Gently heat and bring to boiling. Continue boiling for 15 min without stirring. Cool to room temperature. Remove fat from surface. Filter and retain both meat particles and filtrate. Adjust volume of filtrate to 1.0L with distilled/deionized water. Add pancreatic digest of casein, agar, K_2HPO_4, yeast extract, and resazurin. Gently heat and bring to boiling. Boil for 1–2 min. Add L-cysteine·HCl. Mix thoroughly. Distribute 6.5mL into tubes that contain meat particles (1 part meat particles to 5 parts fluid). Autoclave for 30 min at 15 psi pressure–121°C. Cool to 25°C. Aseptically add 50 µL of sterile formate/fumarate solution to each tube prior to inoculation.

Use: For the cultivation of various *Clostridium* spp.

Chopped Meat Broth with Vitamin K_1 (LMG Medium 70)

Composition per liter:

Ground meat, fat free	500.0g
Pancreatic digest of casein	30.0g
K_2HPO_4	5.0g
Yeast extract	5.0g
L-Cysteine·HCl	0.5g
Resazurin	1.0mg
NaOH (1*N* solution)	25.0mL
Vitamin K_1 solution	7.7mL

pH 7.0 ± 0.2 at 25°C

Vitamin K_1 Solution:
Composition per 50.0mL:

Ethanol (20% solution)	50.0mL
Vitamin K_1	0.7gL

Preparation of Vitamin K_1 Solution: Mix components. Filter sterilize. Store solution protected from light at 5°C. Discard after one month.

Preparation of Medium: Use lean beef or horse meat. Remove fat and connective tissue. Grind finely. Add ground meat and 25.0mL of NaOH solution to distilled/deionized water and bring volume to 1025.0mL. Gently heat and bring to boiling. Continue boiling for 15 min without stirring. Cool to room temperature. Remove fat from surface. Filter and retain both meat particles and filtrate. Adjust volume of filtrate to 1.0L with distilled/deionized water. Add pancreatic digest of casein, agar, K_2HPO_4, yeast extract, and resazurin. Gently heat and bring to boiling. Boil for 1–2 min. Add L-cysteine·HCl. Mix thoroughly. Distribute 6.5mL into tubes that contain meat particles (1 part meat particles to 5 parts fluid). Autoclave for 30 min at 15 psi pressure–121°C. Cool to 25°C. Aseptically add 50.0µL of sterile vitamin K_1 solution to each tube prior to inoculation.

Use: For the cultivation of various *Clostridium* spp.

Chopped Meat Carbohydrate Medium

Composition per 1240.0mL:

Peptone	30.0g
K_2HPO_4	5.0g
Yeast extract	5.0g
Cellobiose	1.0g
Maltose	1.0g
Starch	1.0g
L-Cysteine·HCl·H_2O	0.5g
Chopped meat extract filtrate	1.0L
Chopped meat extract solids	200.0mL
Resazurin (0.025% solution)	4.0mL

pH 7.0 ± 0.2 at 25°C

Chopped Meat Extract:
Composition per liter:

Beef or horse meat	500.0g
NaOH (1*N* solution)	25.0mL

Preparation of Chopped Meat Extract: Use lean beef or horse meat. Remove fat and connective tissue. Grind. Add meat and NaOH to distilled/deionized water and bring volume to 1.0L. Gently heat and bring to boiling while stirring. Cool to 25°C. Remove fat from surface. Filter. Reserve ground meat particles and filtrate. Add distilled/deionized water to filtrate and bring volume to 1.0L.

Preparation of Medium: To 1.0L of chopped meat extract filtrate, add the remaining components, except the L-cysteine·HCl·H_2O and chopped meat solids. Mix thoroughly. Gently heat to boiling. Cool to room temperature. Add the L-cysteine·HCl·H_2O. Adjust pH to 7.0. Distribute 1 part chopped meat solids (by volume) and 5 parts of liquid (by volume) into tubes under O_2-free 97% N_2 + 3% H_2. Cap with rubber stoppers and place tubes in a press. Autoclave for 15 min at 15 psi pressure–121°C with fast exhaust.

Use: For the cultivation of anaerobic bacteria, including *Clostridium* species, *Eubacterium* species, and *Gemmiger formicilis*.

Chopped Meat Glucose Agar (LMG Medium 68)

Composition per liter:

Ground meat, fat free	500.0g
Pancreatic digest of casein	30.0g
Agar	15.0g
Glucose	10.0g
K_2HPO_4	5.0g
Yeast extract	5.0g
L-Cysteine·HCl	0.5g
Resazurin	1.0mg
NaOH (1*N* solution)	25.0mL

pH 7.0 ± 0.2 at 25°C

Preparation of Medium: Use lean beef or horse meat. Remove fat and connective tissue. Grind finely. Add ground meat and 25.0mL of NaOH solution to distilled/deionized water and bring volume to 1025.0mL. Gently heat and bring to boiling. Continue boiling for 15 min without stirring. Cool to room temperature. Remove fat from surface. Filter and retain both meat particles and filtrate. Adjust volume of filtrate to 1.0L with distilled/deionized water. Add pancreatic digest of casein, agar, K_2HPO_4, yeast extract, and resazurin. Gently heat and bring to boiling. Boil for 1–2 min. Add L-cysteine·HCl. Mix thoroughly. Distribute 7.0mL into tubes that contain meat particles (1 part meat particles to 5 parts fluid). Autoclave for 30 min at 15 psi pressure–121°C.

Use: For the cultivation of various *Clostridium* spp.

Chopped Meat Glucose Broth (LMG Medium 68)

Composition per liter:

Ground meat, fat free	500.0g
Pancreatic digest of casein	30.0g
Glucose	10.0g
K_2HPO_4	5.0g
Yeast extract	5.0g
L-Cysteine·HCl	0.5g
Resazurin	1.0mg
NaOH (1*N* solution)	25.0mL

pH 7.0 ± 0.2 at 25°C

Preparation of Medium: Use lean beef or horse meat. Remove fat and connective tissue. Grind finely. Add ground meat and 25.0mL of NaOH solution to distilled/deionized water and bring volume to 1025.0mL. Gently heat and bring to boiling. Continue boiling for 15 min without stirring. Cool to room temperature. Remove fat from surface. Filter and retain both meat particles and filtrate. Adjust volume of filtrate to 1.0L with distilled/deionized water. Add pancreatic digest of casein, agar, K_2HPO_4, yeast extract, and resazurin. Gently heat and bring to boiling. Boil for 1–2 min. Add L-cysteine·HCl. Mix thoroughly. Distribute 7.0mL into tubes that contain meat particles (1 part meat particles to 5 parts fluid). Autoclave for 30 min at 15 psi pressure–121°C.

Use: For the cultivation of various *Clostridium* spp.

Chopped Meat Glucose Medium

Composition per 1240.0mL:

Peptone	30.0g
K_2HPO_4	5.0g
Yeast extract	5.0g
Glucose	5.0g
L-Cysteine·HCl·H_2O	0.5g
Chopped meat extract filtrate	1.0L
Chopped meat extract solids	200.0mL
Resazurin (0.025% solution)	4.0mL

pH 7.0 ± 0.2 at 25°C

Chopped Meat Extract:

Composition per liter:

Beef or horse meat	500.0g
NaOH (1*N* solution)	25.0mL

Preparation of Chopped Meat Extract: Use lean beef or horse meat. Remove fat and connective tissue. Grind. Add meat and NaOH to distilled/deionized water and bring volume to 1.0L. Gently heat and bring to boiling while stirring. Cool to 25°C. Remove fat from surface. Filter. Reserve ground meat particles and filtrate. Add distilled/deionized water to filtrate and bring volume to 1.0L.

Preparation of Medium: To 1.0L of chopped meat extract filtrate, add the remaining components, except L-cysteine·HCl·H_2O and chopped meat solids. Mix thoroughly. Gently heat to boiling. Cool to room temperature. Add the L-cysteine·HCl·H_2O. Adjust pH to 7.0. Distribute 1 part chopped meat solids (by volume) and 5 parts of liquid (by volume) into tubes under O_2-free 97% N_2 + 3% H_2. Cap with rubber stoppers and place tubes in a press. Autoclave for 15 min at 15 psi pressure–121°C with fast exhaust.

Use: For the cultivation of *Clostridium* species and *Selenomonas noxia*.

Chopped Meat Medium

Composition per 1205.0mL:

Peptone	30.0g
K_2HPO_4	5.0g
Yeast extract	5.0g

L-Cysteine·HCl·H_2O 0.5g
Chopped meat extract filtrate 1.0L
Chopped meat extract solids 200.0mL
Resazurin (0.025% solution) 4.0mL
pH 7.0 ± 0.2 at 25°C

Chopped Meat Extract:
Composition per liter:
Beef or horse meat 500.0g
NaOH (1*N* solution) 25.0mL

Preparation of Chopped Meat Extract: Use lean beef or horse meat. Remove fat and connective tissue. Grind. Add meat and NaOH to distilled/deionized water and bring volume to 1.0L. Gently heat and bring to boiling while stirring. Cool to 25°C. Remove fat from surface. Filter. Reserve ground meat particles and filtrate. Add distilled/deionized water to filtrate and bring volume to 1.0L.

Preparation of Medium: To 1.0L of chopped meat extract filtrate, add the remaining components, except the L-cysteine·HCl·H_2O and chopped meat solids. Mix thoroughly. Gently heat to boiling. Cool to room temperature. Add the L-cysteine·HCl·H_2O. Adjust pH to 7.0. Distribute 1 part chopped meat solids (by volume) and 5 parts of liquid (by volume) into tubes under O_2-free 97% N_2 + 3% H_2. Cap with rubber stoppers and place tubes in a press. Autoclave for 15 min at 15 psi pressure–121°C with fast exhaust.

Use: For the cultivation and maintenance of a variety of anaerobic bacteria, including *Bacteroides* species, *Bifidobacterium* species, *Capnocytophaga* species, *Clostridium* species, *Eubacterium* species, *Fusobacterium* species, and others.

Chopped Meat Medium with Formate and Fumarate

Composition per 1230.0mL:
Peptone 30.0g
K_2HPO_4 .... 5.0g
Yeast extract 5.0g
L-Cysteine·HCl·H_2O 0.5g
Chopped meat extract filtrate 1.0L
Chopped meat extract solids 200.0mL
Resazurin (0.025% solution) 4.0mL
Formate-fumarate solution 0.05mL
pH 7.0 ± 0.2 at 25°C

Chopped Meat Extract:
Composition per liter:
Beef or horse meat 500.0g
NaOH (1*N* solution) 25.0mL

Preparation of Chopped Meat Extract: Use lean beef or horse meat. Remove fat and connective tissue. Grind. Add meat and NaOH to distilled/deionized water and bring volume to 1.0L. Gently heat and bring to boiling while stirring. Cool to 25°C. Remove fat from surface. Filter. Reserve ground meat particles and filtrate. Add distilled/deionized water to filtrate and bring volume to 1.0L.

Formate-Fumarate Solution:
Composition per 100.0mL:
Sodium formate 6.0g
Fumaric acid 6.0g

Preparation of Formate-Fumarate Solution: Add components to distilled/deionized water and bring volume to 100.0mL. Adjust pH to 7.0. Filter sterilize.

Preparation of Medium: To 1.0L of chopped meat extract filtrate, add the remaining components, except the L-cysteine·HCl·H_2O, formate-fumarate solution, and chopped meat solids. Mix thoroughly. Gently heat to boiling. Cool to room temperature. Add the L-cysteine·HCl·H_2O. Adjust pH to 7.0. Distribute 1 part chopped meat solids (by volume) and 5 parts of liquid (by volume) into tubes under O_2-free 97% N_2 + 3% H_2. Cap with rubber stoppers and place tubes in a press. Autoclave for 15 min at 15 psi pressure–121°C with fast exhaust. Prior to inoculation, add 0.05mL of formate-fumarate solution to each tube containing approximately 6.5mL of chopped meat medium.

Use: For the cultivation and maintenance of *Bacteroides ureolyticus* and *Wolinella* species.

Chopped Meat Medium, Modified

Composition per 1230.0mL:
Pancreatic digest of casein 30.0g
Peptone 30.0g
Agar 20.0g
K_2HPO_4 .... 5.0g
Yeast extract 5.0g
L-Cysteine·HCl·H_2O 0.5g
Chopped meat extract filtrate 1.0L
Chopped meat extract solids 200.0mL
Hemin solution 10.0mL
Resazurin (0.025% solution) 4.0mL
Vitamin K_1 solution 0.2mL
pH 7.0 ± 0.2 at 25°C

Chopped Meat Extract:
Composition per liter:
Beef or horse meat 500.0g
NaOH (1*N* solution) 25.0mL

Preparation of Chopped Meat Extract: Use lean beef or horse meat. Remove fat and connective tissue. Grind. Add meat and NaOH to distilled/deionized water and bring volume to 1.0L. Gently heat and bring to boiling while stirring. Cool to 25°C. Remove fat from surface. Filter. Reserve ground meat particles and filtrate. Add distilled/deionized water to filtrate and bring volume to 1.0L.

Hemin Solution:
Composition per 100.0mL:

Hemin	0.05g
NaOH (1*N* solution)	1.0mL

Preparation of Hemin Solution: Add components to distilled/deionized water and bring volume to 100.0mL. Mix thoroughly.

Vitamin K_1 Solution:
Composition per 30.0mL:

Ethanol (95% solution)	30.0mL
Vitamin K_1	0.15mL

Preparation of Vitamin K_1 Solution: Mix components. Store solution protected from light at 5°C. Discard after 1 month.

Preparation of Medium: To 1.0L of chopped meat extract filtrate, add the remaining components, except the L-cysteine·HCl·H_2O, hemin solution, vitamin K_1 solution, and chopped meat solids. Mix thoroughly. Gently heat to boiling. Cool to room temperature. Add the L-cysteine·HCl·H_2O, hemin solution, and vitamin K_1 solution. Adjust pH to 7.0. Distribute 1 part chopped meat solids (by volume) and 5 parts of liquid (by volume) into tubes under O_2-free 97% N_2 + 3% H_2. Cap with rubber stoppers and place tubes in a press. Autoclave for 15 min at 15 psi pressure–121°C with fast exhaust.

Use: For the cultivation and maintenance of a variety of anaerobic bacteria, including *Actinomyces* species, *Bacteroides* species, *Clostridium* species, *Eubacterium* species, *Fusobacterium* species, *Peptostreptococcus* species, *Porphyromonas* species, *Prevotella* species, *Propionibacterium* species, *Selenomonas* species, and others.

Chopped Meat Medium, Modified with Formate and Fumarate

Composition per 1230.0mL:

Pancreatic digest of casein	30.0g
Peptone	30.0g
Agar	20.0g
K_2HPO_4	5.0g
Yeast extract	5.0g
L-Cysteine·HCl·H_2O	0.5g
Chopped meat extract filtrate	1.0L
Chopped meat extract solids	200.0mL
Hemin solution	10.0mL
Resazurin (0.025% solution)	4.0mL
Formate-fumarate solution	0.25mL
Vitamin K_1 solution	0.2mL

pH 7.0 ± 0.2 at 25°C

Chopped Meat Extract:
Composition per liter:

Beef or horse meat	500.0g
NaOH (1*N* solution)	25.0mL

Preparation of Chopped Meat Extract: Use lean beef or horse meat. Remove fat and connective tissue. Grind. Add meat and NaOH to distilled/deionized water and bring volume to 1.0L. Gently heat and bring to boiling while stirring. Cool to 25°C. Remove fat from surface. Filter. Reserve ground meat particles and filtrate. Add distilled/deionized water to filtrate and bring volume to 1.0L.

Hemin Solution:
Composition per 100.0mL:

Hemin	0.05g
NaOH (1*N* solution)	1.0mL

Preparation of Hemin Solution: Add components to distilled/deionized water and bring volume to 100.0mL. Mix thoroughly.

Formate-Fumarate Solution:
Composition per 100.0mL:

Sodium formate	6.0g
Fumaric acid	6.0g

Preparation of Formate-Fumarate Solution: Add components to distilled/deionized water and bring volume to 100.0mL. Adjust pH to 7.0. Filter sterilize.

Vitamin K_1 Solution:
Composition per 30.0mL:

Ethanol (95% solution)	30.0mL
Vitamin K_1	0.15mL

Preparation of Vitamin K_1 Solution: Mix components. Store solution protected from light at 5°C. Discard after one month.

Preparation of Medium: To 1.0L of chopped meat extract filtrate, add the remaining components, except the L-cysteine·HCl·H_2O, hemin solution, vitamin K_1 solution, formate-fumarate solution, and chopped meat solids. Mix thoroughly. Gently heat to boiling. Cool to room temperature. Add the L-cysteine·HCl·H_2O, hemin solution, and vitamin K_1 solution. Adjust pH to 7.0. Distribute 1 part chopped meat solids (by volume) and 5 parts of liquid (by volume) into tubes under O_2-free 97% N_2 + 3% H_2. Cap with rubber stoppers and place tubes in a press. Autoclave for 15 min at 15 psi pressure–121°C with fast exhaust. Prior to inoculation, add 0.25mL of formate-fumarate solution to each tube containing approximately 5.0mL of chopped meat medium, modified.

Use: For the cultivation and maintenance of *Bacteroides gracilis, Bacteroides ureolyticus, Campylobacter mucosalis,* and *Wolinella succinogenes.*

Christensen Agar

Composition per liter:

Agar	15.0g
NaCl	5.0g
Sodium citrate	3.0g
KH_2PO_4	1.0g
L-Cysteine·HCl·H_2O	0.1g
Phenol Red	12.0mg

pH 6.9 ± 0.2 at 25°C

Preparation of Medium: Add components to distilled/deionized water and bring volume to 1.0L. Mix thoroughly. Gently heat and bring to boiling. Dispense into tubes or flasks. Autoclave for 15 min at 15 psi pressure–121°C. Pour into sterile Petri dishes or leave in tubes. Allow tubes to cool in a slanted position.

Use: For the differentiation of enteric pathogens, especially members of the Enterobacteriaceae, and coliforms based on their ability to utilize citrate as a carbon source. Bacteria that can utilize citrate turn the medium pink-red.

Christensen Agar

Composition per liter:

Agar	15.0g
NaCl	5.0g
Sodium citrate	3.0g
KH_2PO_4	1.0g
Yeast extract	0.5g
Glucose	0.2g
L-Cysteine·HCl·H_2O	0.1g
Phenol Red	12.0mg

pH 6.9 ± 0.2 at 25°C

Source: This medium is available as a premixed powder from BD Diagnostic Systems.

Preparation of Medium: Add components to distilled/deionized water and bring volume to 1.0L. Mix thoroughly. Gently heat and bring to boiling. Dispense into tubes or flasks. Autoclave for 15 min at 15 psi pressure–121°C. Pour into sterile Petri dishes or leave in tubes. Allow tubes to cool in a slanted position.

Use: For the differentiation of enteric pathogens, especially members of the Enterobacteriaceae, and coliforms based on their ability to utilize citrate as a carbon source. Bacteria that can utilize citrate turn the medium pink-red.

Christensen Citrate Agar

Composition per liter:

Agar	15.0g
NaCl	5.0g
Sodium citrate	3.0g
KH_2PO_4	1.0g
Yeast extract	0.5g
Ferric ammonium citrate	0.4g
L-Cysteine·HCl·H_2O	0.1g
$Na_2S_2O_5$	0.08g
Phenol Red	12.0mg

pH 6.9 ± 0.2 at 25°C

Preparation of Medium: Add components to distilled/deionized water and bring volume to 1.0L. Mix thoroughly. Gently heat and bring to boiling. Dispense into tubes or flasks. Autoclave for 15 min at 15 psi pressure–121°C. Pour into sterile Petri dishes or leave in tubes. Allow tubes to cool in a slanted position.

Use: For the differentiation of enteric pathogens, especially members of the Enterobacteriaceae, and coliforms based on their ability to utilize citrate as a carbon source. Bacteria that can utilize citrate turn the medium pink-red.

Christensen Citrate Agar, Modified (Citrate Agar)

Composition per liter:

Agar	12.0g
NaCl	5.0g
Sodium citrate	3.8g
KH_2PO_4	1.0g
Yeast extract	0.5g
Glucose	0.2g
L-Cysteine·HCl·H_2O	0.1g
Phenol Red	0.02g

pH 6.7 ± 0.2 at 25°C

Preparation of Medium: Add components to distilled/deionized water and bring volume to 1.0L. Mix thoroughly. Gently heat and bring to boiling. Dispense into tubes or flasks. Autoclave for 15 min at 15 psi pressure–121°C. Pour into sterile Petri dishes or leave in tubes. Allow tubes to cool in a slanted position.

Use: For the differentiation of enteric pathogens, especially members of the Enterobacteriaceae, and coliforms based on their ability to utilize citrate as a carbon source. Bacteria that can utilize citrate turn the medium pink-red.

Christensen Citrate Sulfide Medium

Composition per liter:

Agar	15.0g
NaCl	5.0g
Sodium citrate·$2H_2O$	3.0g
KH_2HPO_4	1.0g
Yeast extract	0.5g
Ferric citrate	0.2g
Ammonium citrate	0.2g
Glucose	0.2g
L-Cysteine·HCl·H_2O	0.1g

$Na_2S_2O_3 \cdot 5H_2O$ 0.08g
Phenol Red 0.012g

pH 6.7± 0.2 at 25°C

Preparation of Medium: Add components to distilled/deionized water and bring volume to 1.0L. Mix thoroughly. Gently heat and bring to boiling. Dispense into tubes or flasks. Autoclave for 15 min at 15 psi pressure–121°C. Pour into sterile Petri dishes or leave in tubes. Allow tubes to cool in a slanted position.

Use: For the differentiation of enteric pathogens, especially members of the Enterobacteriaceae, and coliforms based on their ability to utilize citrate as a carbon source and production of H_2S. Bacteria that can utilize citrate turn the medium pink-red. H_2S production appears as a blackening of the butt of the tube.

Christensen's Urea Agar

Composition per liter:

Agar 15.0g
NaCl 5.0g
KH_2PO_4 .... 2.0g
Peptone 1.0g
Glucose 1.0g
Phenol Red 0.012g
Urea solution 100.0mL

pH 6.8 ± 0.1 at 25°C

Urea Solution:

Composition per 100.0mL:

Urea 20.0g

Preparation of Urea: Add urea to 100.0mL of distilled/deionized water. Mix thoroughly. Filter sterilize.

Preparation of Medium: Add components, except urea solution, to distilled/deionized water and bring volume to 900.0mL. Mix thoroughly. Gently heat and bring to boiling. Autoclave for 15 min at 15 psi pressure–121°C. Cool to 50–55°C. Aseptically add 100.0mL of sterile urea solution. Mix thoroughly. Pour into Petri dishes or distribute into sterile tubes. Allow tubes to solidify in a slanted position.

Use: For the differentiation of a variety of microorganisms, especially members of the Enterobacteriaceae, aerobic actinomycetes, streptococci, and nonfermenting Gram-negative bacteria, on the basis of urease production.

CHROMagar *Candida*

Composition per liter:

Glucose 20.0g
Agar 15.0g
Peptone 10.0g
Chromogenic mix 2.0g
Chloramphenicol 0.5g

Source: CHROMagar *Candida* is available from CHROMagar Microbiology.

Preparation of Medium: Add components to distilled/deionized water and bring volume to 1.0L. Mix thoroughly. Gently heat in a boiling water bath or steam bath. Shake periodically during heating to dissolve components. Heat long enough with shaking every 5 min to ensure complete dissolution. Do not overheat. Cool to 45–50°C. Pour into sterile Petri dishes.

Use: For the differentiation of *Candida* spp. Specific *Candida* spp. give characteristic color reactions, e.g., *Candida albicans* produce distinctive green colonies and *Candida tropicalis* produce distinctive dark blue-gray colonies.

CHROMagar *E. coli*

Composition per liter:

Proprietary

Source: CHROMagar *E. coli* is available from CHROMagar Microbiology.

Preparation of Medium: Add components to distilled/deionized water and bring volume to 1.0L. Mix thoroughly. Gently heat in a boiling water bath or steam bath. Shake periodically during heating to dissolve components. Heat long enough with shaking every 5 min to ensure complete dissolution. Do not overheat. Adding tellurite can increase specificity. Cool to 45–50°C. Pour into sterile Petri dishes.

Use: For the differentiation and presumptive identification of *Eshcrichia coli* which forms blue colonies.

CHROMagar ECC

Composition per liter:

Proprietary

Source: CHROMagar EEC is available from CHROMagar Microbiology.

Preparation of Medium: Add components to distilled/deionized water and bring volume to 1.0L. Mix thoroughly. Gently heat in a boiling water bath or steam bath. Shake periodically during heating to dissolve components. Heat long enough with shaking every 5 min to ensure complete dissolution. Do not overheat. Adding tellurite can increase specificity. Cool to 45–50°C. Pour into sterile Petri dishes.

Use: For the differentiation and presumptive identification of *Eshcrichia coli* and other coliform bacteria which form red colonies.

CHROMagar *Listeria*

Composition per liter:

Proprietary

Source: CHROMagar *Listeria* is available from CHROMagar Microbiology.

Preparation of Medium: Add components to distilled/deionized water and bring volume to 1.0L. Mix thoroughly. Gently heat in a boiling water bath or steam bath. Shake periodically during heating to dissolve components. Heat long enough with shaking every 5 min to ensure complete dissolution. Do not overheat. Adding tellurite can increase specificity. Cool to 45–50°C. Pour into sterile Petri dishes.

Use: For the differentiation and presumptive identification of *Listeria monocytogenes* which form blue colonies surrounded by white halos.

CHROMagar O157

Composition per liter:

Proprietary

Source: CHROMagar O157 is available from CHROMagar Microbiology.

Preparation of Medium: Add components to distilled/deionized water and bring volume to 1.0L. Mix thoroughly. Gently heat in a boiling water bath or steam bath. Shake periodically during heating to dissolve components. Heat long enough with shaking every 5 min to ensure complete dissolution. Do not overheat. Adding tellurite can increase specificity. Cool to 45–50°C. Pour into sterile Petri dishes.

Use: For the differentiation and presumptive identification of *Eshcrichia coli* O157.

CHROMagar Orientation

Composition per liter:

Peptone	16.0g
Meat extract	16.0g
Peptone	16.0g
Yeast extract	16.0g
Agar	15.0g
Chromogenic mix	2.0g

pH 7.0 ± 0.2 at 25°C

Source: CHROMagar Orientation is available from CHROMagar Microbiology.

Preparation of Medium: Add components to distilled/deionized water and bring volume to 1.0L. Mix thoroughly. Gently heat in a boiling water bath or steam bath. Shake periodically during heating to dissolve components. Heat long enough with shaking every 5 min to ensure complete dissolution. Do not overheat. Cool to 50°C. Pour into sterile Petri dishes.

Use: For the differentiation and presumptive identification of Gram-negative bacteria and *Enterococcus* spp. For use in identifying urinary tract pathogens. Isolates produce characteristic diagnostic colors, e.g., *Escherichia coli* produces pinto red colonies.

CHROMagar *Salmonella*

Composition per liter:

Proprietary

Source: CHROMagar *Salmonella* is available from CHROMagar Microbiology.

Preparation of Medium: Add components to distilled/deionized water and bring volume to 1.0L. Mix thoroughly. Gently heat in a boiling water bath or steam bath. Shake periodically during heating to dissolve components. Heat long enough with shaking every 5 min to ensure complete dissolution. Do not overheat. Cool to 50°C. Pour into sterile Petri dishes.

Use: For the differentiation and presumptive identification of *Salmonella* spp.

CHROMagar *Staph. aureus*

Composition per liter:

Proprietary

Source: CHROMagar *Staph. aureus* is available from CHROMagar Microbiology.

Preparation of Medium: Add components to distilled/deionized water and bring volume to 1.0L. Mix thoroughly. Gently heat in a boiling water bath or steam bath. Shake periodically during heating to dissolve components. Heat long enough with shaking every 5 min to ensure complete dissolution. Do not overheat. Adding an antibiotic such as tobramycin or methicillin can be used to identify resistant strains. Cool to 45–50°C. Pour into sterile Petri dishes.

Use: For the differentiation and presumptive identification of *Staphylococcus aureus.*

CHROMagar *Vibrio*

Composition per liter:

Proprietary

Preparation of Medium: Add components to distilled/deionized water and bring volume to 1.0L. Mix thoroughly. Gently heat in a boiling water bath or steam bath. Shake periodically during heating to dissolve components. Heat long enough with shaking every 5 min to ensure complete dissolution. Do not overheat. Adding tellurite can increase specificity. Cool to 45–50°C. Pour into sterile Petri dishes.

Source: CHROMagar *Vibrio* is available from CHROMagar Microbiology.

Use: For the differentiation and presumptive identification of *Vibrio parahaemolyiticus* which form mauve colonies; *Vibrio cholerae* form turquoise blue colonies and *Vibrio alginolyitcus* colonies are colorless.

Chromogenic *Candida* Agar

Composition per liter:

Chromogenic mix	13.6g
Agar	13.6g
Peptone	4.0g
Selective supplement solution	10.0mL

pH 6.0 ± 0.2 at 25°C

Source: This medium is available as a premixed powder from Oxoid Unipath.

Selective Supplement Solution:
Composition per 10.0mL:

Chloramphenicol	500.0mg

Preparation of Selective Supplement Solution: Add chloramphenicol to distilled/deionized water and bring volume to 10.0mL. Mix thoroughly. Filter sterilize.

Preparation of Medium: Add components to distilled/deionized water and bring volume to 1.0mL Mix thoroughly. Gently heat while stirring and bring to boiling. Do not autoclave. Cool to 45°C. Pour into sterile Petri dishes.

Use: For the rapid isolation and identification of clinically important *Candida* species. The medium incorporates two chromogens that indicate the presence of the target enzymes: X-NAG (5-bromo-4-chloro-3-indolyl N acetyl ß-D-glucosaminide) detects the activity of hexosaminidase. BCIP (5-bromo-6-chloro-3-indolyl phosphate p-toluidine salt) detects alkaline phosphatase activity. An opaque agent has been incorporated into the formulation to improve the color definition on the agar. The broad-spectrum antibacterial agent chloramphenicol is added to the agar to inhibit bacterial growth on the plates.

Chromogenic *Listeria* Agar

Composition per liter:

Peptone	18.5g
LiCl	15.0g
Agar	14.0g
NaCl	9.5g
Yeast extract	4.0g
Maltose	4.0g
Sodium pyruvate	2.0g
X-glucoside chromogenic mix	0.2g
Differential lecithin solution	40.0mL
Selective supplement solution	20.0mL

pH 7.2 ± 0.2 at 25°C

Source: This medium is available as a premixed powder from Oxoid Unipath.

Differential Lecithin Solution:
Composition per 40.0mL:

Lecithin	Proprietary

Preparation of Differential Lecithin Solution: Available as premixed solution.

Selective Supplement Solution:
Composition per 20.0mL:

Nalidixic acid	26.0mg
Polymyxin B	10.0mg
Ceftazidime	6.0mg
Amphotericin	10.0mg

Preparation of Selective Supplement Solution: Add components to distilled/deionized water and bring volume to 20.0mL. Mix thoroughly. Filter sterilize.

Preparation of Medium: Add components, except differential lecithin solution and selective supplement solution, to distilled/deionized water and bring volume to 940.0mL. Mix thoroughly. Gently heat while stirring and bring to boiling. Autoclave for 15 min at 15 psi pressure–121°C. Cool to 46°C. Aseptially add differential lecithin solution and selective supplement solution. Mix thouroughly. Pour into sterile Petri dishes.

Use: For the isolation, enumeration and presumptive identification of *Listeria* spp. and *Listeria monocytogenes*. This selective medium contains the substrate lecithin, which permits differentiation of *L. monocytogenes* and *L. Ivanovii* from other *Listeria* species. Differential activity for all *Listeria* species is due to the addition of a chromogenic substrate.

Chromogenic MRSA Agar

Composition per liter:
Proprietary

Source: This medium is available as a premixed powder from Oxoid Unipath.

Use: For the detection of methicillin resistant *Staphylococcus aureus*. The chromogen in this medium detects phosphatase activity, which is present in MRSA strains, producing distinct, easily visible, denim-blue colonies. Antimicrobial compounds within the medium, including cefoxitin, inhibit the growth of competitor organisms to allow accurate diagnosis of MRSA. The medium may be inoculated directly from swabs, or from isolates or culture suspensions. The plate can then be examined for MRSA colonies after a short, 18-hour incubation.

Chromogenic *Salmonella* Esterase Agar (CSE Agar)

Composition per liter:

Agar	12.0g
Lactose	14.6g
Peptone	4.0g
Tryptone	4.0g
Tween 20	3.0g
Lab Lemco	3.0g
Na_3-citrate dihydrate	0.5g
L-cysteine	0.128g
Tris	0.06g
SLA-octonoate solution	50.0mL
Novobiocin solution	10.0mL
Ethyl 4-dimethylaminobenzoate solution	10.0mL

pH 7.0 ± 0.2 at 25°C

Novobiocin Solution:

Composition per 10.0mL:

Novobiocin	70.0mg

Preparation of Novobiocin Solution: Add novobiocin to distilled/deionized water and bring volume to 10.0mL. Mix thoroughly. Filter sterilize.

Ethyl 4-dimethylaminobenzoate Solution:

Composition per 10.0mL:

Ethyl 4-dimethylaminobenzoate	0.35g
Methanol	8.0mL

Preparation of Ethyl 4-dimethylaminobenzoate Solution: Add ethyl 4-dimethylaminobenzoate to 8.0mL methanol. Mix thoroughly. Bring volume to 10.0mL with distilled/deionized water. Mix thoroughly. Filter sterilize.

SLA-Octonoate Solution:

Composition per 50.0mL:

4-[2-(4-octanoyloxy-3,5-dimethoxyphenyl)-vinyl]-quinolinium-1-(propan-3-yl carboxylic acid) bromide (SLPA-octanoate; bromide form)	0.3223g

Preparation of SLA-Octonoate Solution: Add SLA-octonoate to distilled/deionized water and bring volume to 50.0mL. Mix thoroughly. Filter sterilize.

Preparation of Medium: Add components, except novobiocin solution, SLA-octonoate solution, and ethyl 4-dimethylaminobenzoate solution, to distilled/deionized water and bring volume to 920.0mL. Mix thoroughly. Gently heat and bring to boiling. Autoclave for 15 min at 15 psi pressure–121°C. Cool to 50°C. Aseptically add 10.0mL novobiocin solution, 50.0mL SLA-octonoate solution, and 10.0mL ethyl 4-dimethylaminobenzoate solution. Mix thoroughly. Pour into sterile Petri dishes.

Use: For the detection of *Salmonella* spp. in clinical specimens. For the differentiation of *Salmonella* spp.

Chromogenic Substrate Broth

Composition per liter:

NaCl	10.0g
HEPES (*N*-[2-Hydroxyethyl] piperazine-*N*′-[2-ethanesulfonic acid]) buffer	6.9g
$(NH_4)_2SO_4$	5.0g
o-Nitrophenyl-β-D-galactopyranoside	0.5g
Solanium	0.5g
$MgSO_4$	0.1g
4-Methylumbelliferyl-β-D-glucuronide	0.075g
$CaCl_2$	0.05g
Na_2SO_3	0.04g
Amphotericin B	1.0mg
$MnSO_4$	0.5mg
$ZnSO_4$	0.5mg

Preparation of Medium: Add components to distilled/deionized water and bring volume to 1.0L. Mix thoroughly. Distribute into tubes or flasks. Autoclave for 15 min at 15 psi pressure–121°C.

Use: For the detection of coliform bacteria based on their hydrolysis of chromogenic substrates by production of β-D-galactopyranosidase. Bacteria that produce β-D-galactopyranosidase turn the medium yellow.

Chromogenic Urinary Tract Infection (UTI) Medium

Composition per liter:

Chromogenic mix	26.3g
Peptone	15.0g
Agar	15.0g

pH 6.8 ± 0.2 at 25°C

Source: This medium is available as a premixed powder from Oxoid Unipath.

Preparation of Medium: Add components to distilled/deionized water and bring volume to 1.0L. Mix thoroughly. Gently heat while stirring and bring to boiling. Autoclave for 15 min at 15 psi pressure–121°C. Cool to 50°C. Pour into sterile Petri dishes.

Use: For the presumptive identification and differentiation of all the main microorganisms that cause urinary tract infections (UTIs). The medium contains two specific chromogenic substrates which are cleaved by enzymes produced by *Enterococcus* spp., *Escherichia coli,* and coliforms. In addition, it contains phenylalanine and tryptophan which provide an indication of tryptophan deaminase activity, indicating the presence of *Proteus* spp., *Morganella* spp., and *Providencia* spp. It is based on electrolyte deficient CLED Medium which provides a valuable noninhibitory diagnostic agar for plate culture of other urinary organisms, whilst preventing the swarming of *Proteus* spp. One chromogen, X-Gluc, is targeted towards β-glucosidase, and allows the specific detection of enterococci through the formation of blue colonies. The other chromogen, Red-Gal, is cleaved by the enzyme β-galactosidase which is produced by Escherichia coli, resulting in pink colonies. Cleavage of both chromogens occurs in the presence of coliforms, resulting in purple colonies. The medium also contains tryptophan which acts as an indicator of tryptophan deaminase activity, resulting in colonies of *Proteus, Morganella,* and *Providencia* spp. appearing brown.

Chromogenic UTI Medium, Clear

Composition per liter:

Peptone	15.0g
Agar	15.0g
Chromogenic mix	13.0g

pH 7.0 ± 0.2 at 25°C

Source: This medium is available as a premixed powder from Oxoid Unipath.

Preparation of Medium: Add components to distilled/deionized water and bring volume to 1.0L. Mix thoroughly. Gently heat while stirring and bring to boiling. Autoclave for 15 min at 15 psi pressure–121°C. Cool to 50°C. Pour into sterile Petri dishes.

Use: For the presumptive identification and differentiation of all the main microorganisms that cause urinary tract infections (UTIs). This medium uses the same chromogenic substrates as the existing opaque Chromogenc UTI Medium but has a clear background to make multiple sample testing easier. The medium contains two specific chromogenic substrates which are cleaved by enzymes produced by *Enterococcus* spp., *E. coli,* and coliforms. In addition, it contains tryptophan which indicates tryptophan deaminase activity (TDA), indicating the presence of *Proteus* spp. It is based on Cystine Lactose Electrolyte Deficient (CLED) Medium which provides a valuable non-inhibitory diagnostic agar for plate culture of other urinary organisms, whilst preventing the swarming of *Proteus* spp. The chromogen, X-glucoside, is targeted towards ß-glucosidase enzyme activity, and allows the specific detection of enterococci through the formation of blue colonies. The other chromogen, Red-Galactoside, is cleaved by the enzyme ß-galactosidase which is produced by *E. coli,* resulting in pink colonies. Cleavage of both the chromogens by members of the coliform group, results in purple colonies. The medium also contains tryptophan which acts as an indicator of tryptophan deaminase activity (TDA), resulting in halos around the colonies of *Proteus, Morganella,* and *Providencia* spp.

CIN Agar
(*Yersinia* Selective Agar)
(Cefsulodin Irgasan Novobiocin Agar)

Composition per liter:

Mannitol	20.0g
Agar	12.0g
Pancreatic digest of gelatin	10.0g
Beef extract	5.0g
Peptic digest of animal tissue	5.0g
Sodium pyruvate	2.0g
Yeast extract	2.0g
NaCl	1.0g
Sodium deoxycholate	0.5g
Neutral Red	0.03g
Cefsulodin	0.015g
Irgasan (triclosan)	4.0mg
Novobiocin	2.5mg
Crystal Violet	1.0mg

pH 7.4 ± 0.2 at 25°C

Source: This medium is available as a premixed powder from BD Diagnostic Systems.

Preparation of Medium: Add components, except cefsulodin and novobiocin, to distilled/deionized water and bring volume to 1.0L. Heat, mixing continuously, until boiling. Do not autoclave. Cool to 45°–50°C. Aseptically add cefsulodin and novobiocin. Mix thoroughly. Pour into sterile Petri dishes or distribute into sterile tubes.

Use: For the selective isolation and differentiation of *Yersinia enterocolitica* from a variety of clinical and nonclinical specimens based on mannitol fermentation. *Yersinia enterocolitica* appears as "bull's eye" colonies with deep red centers surrounded by a transparent periphery.

Citrate Medium, Koser's Modified

Composition per liter:

NaCl	5.0g
Citric acid	2.0g
$(NH_4)H_2PO_4$	1.0g
K_2HPO_4	1.0g
$MgSO_4 \cdot 7H_2O$	0.2g

pH 6.8 ± 0.2 at 25°C

Preparation of Medium: Add components to distilled/deionized water and bring volume to 1.0L. Mix thoroughly. Adjust pH to 6.8. Distribute into tubes in 5.0mL volumes. Autoclave for 15 min at 15 psi pressure–121°C.

Use: For the cultivation and differentiation of bacteria based on their ability to utilize citrate as a carbon source.

CLED Agar (Cystine Lactose Electrolyte Deficient Agar) (Brolacin Agar)

Composition per liter:

Agar	15.0g
Lactose	10.0g
Pancreatic digest of casein	4.0g
Pancreatic digest of gelatin	4.0g
Beef extract	3.0g
L-Cystine	0.128g
Bromthymol Blue	0.02g

pH 7.3 ± 0.2 at 25°C

Preparation of Medium: Add components to distilled/deionized water and bring volume to 1.0L. Mix thoroughly. Gently heat while stirring and bring to boiling. Autoclave for 15 min at 15 psi pressure–121°C. Cool to 50°–55°C. Pour into sterile Petri dishes or distribute into sterile tubes.

Use: For the isolation, enumeration, and presumptive identification of microorganisms from urine.

CLED Agar with Andrade Indicator (Cystine Lactose Electrolyte Deficient Agar with Andrade Indicator)

Composition per liter:

Agar	15.0g
Pancreatic digest of casein	10.0g
Peptone	4.0g
Beef extract	3.0g
L-Cystine	0.128g
Bromthymol Blue	0.02g
Andrade indicator	10.0mL

pH 7.5 ± 0.2 at 25°C

Source: This medium is available as a premixed powder from Oxoid Unipath.

Caution: Acid Fuchsin is a potential carcinogen and care must be taken to avoid inhalation of the powdered dye and contamination of the skin.

Andrade's Indicator:
Composition per 100.0mL:

NaOH (1*N* solution)	16.0mL
Acid Fuchsin	0.1g

Preparation of Andrade's Indicator: Add Acid Fuchsin to NaOH solution and bring volume to 100.0mL with distilled/deionized water.

Preparation of Medium: Add components to distilled/deionized water and bring volume to 1.0L. Mix thoroughly. Gently heat while stirring and bring to boiling. Autoclave for 15 min at 15 psi pressure–121°C. Cool to 50°–55°C. Pour into sterile Petri dishes or distribute into sterile tubes.

Use: For the differentiation of microorganisms based on colony characteristics.

Clostridia Medium

Composition per liter:

Sodium L-lactate	10.0g
Sodium acetate	8.0g
K_2HPO_4	0.5g
$(NH_4)_2·7H_2O$	0.5g
Sodium thioglycolate	0.5g
Yeast extract	0.5g
$MgSO_4·7H_2O$	0.1g
$FeSO_4·7H_2O$	0.02g
p-Aminobenzoate	100.0μg
Biotin	0.1μg

pH 6.0–7.0 at 25°C

Preparation of Medium: Add components to distilled/deionized water and bring volume to 1.0L. Mix thoroughly. Adjust pH to 6.0–7.0. Distribute into tubes or flasks. Autoclave for 20 min at 15 psi pressure–121°C.

Use: For the isolation and cultivation of *Clostridium* species that ferment lactate and acetate.

Clostridium botulinum Isolation Agar (CBI Agar)

Composition per 1033.0mL:

Egg yolk agar base	900.0mL
Egg yolk emulsion, 50%	100.0mL
Cycloserine solution	25.0mL
Sulfamethoxazole solution	4.0mL
Trimethoprim solution	4.0mL

pH7.4 ± 0.2 at 25°C

Egg Yolk Agar Base:
Composition per 900.0mL:

Pancreatic digest of casein	40.0g
Agar	20.0g
Na_2HPO_4	5.0g
Yeast extract	5.0g
Glucose	2.0g
NaCl	2.0g
$MgSO_4·7H_2O$ solution	0.2mL

Preparation of Egg Yolk Agar Base: Add components to distilled/deionized water and bring volume to 900.0mL. Mix thoroughly. Gently heat to boiling. Autoclave for 15 min at 15 psi pressure–121°C. Cool to 45°–50°C.

$MgSO_4·7H_2O$ Solution:
Composition per 100.0mL:

$MgSO_4·7H_2O$	5.0g

Preparation of $MgSO_4 \cdot 7H_2O$ Solution: Add $MgSO_4 \cdot 7H_2O$ to distilled/deionized water and bring volume to 100.0mL. Mix thoroughly.

Cycloserine Solution:

Composition per 100.0mL:

Cycloserine	1.0g

Preparation of Cycloserine Solution: Add cycloserine to distilled/deionized water and bring volume to 100.0mL. Mix thoroughly. Filter sterilize.

Sulfamethoxazole Solution:

Composition per 100.0mL:

Sulfamethoxazole	1.9g

Preparation of Sulfamethoxazole Solution: Add sulfamethoxazole to distilled/deionized water and bring volume to 50.0mL. Add sufficient 10% NaOH to dissolve. Bring volume to 100.0mL with distilled/deionized water. Mix thoroughly. Filter sterilize.

Trimethoprim Solution:

Composition per 100.0mL:

Trimethoprim	0.1g

Preparation of Trimethoprim Solution: Add trimethoprim to distilled/deionized water and bring volume to 50.0mL. Gently heat to 55°C. Add sufficient 0.05*N* HCl to dissolve. Bring volume to 100.0mL with distilled/deionized water. Mix thoroughly. Filter sterilize.

Egg Yolk Emulsion, 50%:

Composition per 100.0mL:

Chicken egg yolks	11
Whole chicken egg	1
NaCl (0.9% solution)	50.0mL

Preparation of Egg Yolk Emulsion, 50%: Soak eggs with 1:100 dilution of saturated mercuric chloride solution for 1 min. Crack eggs and separate yolks from whites. Mix egg yolks with 1 chicken egg. Beat to form emulsion. Measure 50.0mL of egg yolk emulsion and add to 50.0mL of 0.9% NaCl solution. Mix thoroughly. Filter sterilize. Warm to 45°–50°C.

Preparation of Medium: Aseptically add warmed, sterile egg yolk emulsion, 50%, and sterile cycloserine solution, sterile sulfamethoxazole solution, and sterile trimethoprim solution to cooled, sterile egg yolk agar base. Mix thoroughly. Pour into sterile Petri dishes.

Use: For isolation, cultivation, and differentiation based on lipase activity of *Clostridium botulinum* types A, B, and F from fecal specimens associated with foodborne and infant botulism. *Clostridium botulinum* types A, B, and F appear as raised colonies surrounded by an opaque zone. Other *Clostridium* species and *Clostridium botulinum* type G appear as pinpoint colonies with no opaque zone.

Clostridium difficile Agar

Composition per liter:

Clostridum difficile agar base	920.0mL
Horse blood, defibrinated	70.0mL
Clostridium difficile selective supplement	10.0mL

pH 7.4 ± 0.2 at 25°C

Source: This medium is available as a premixed powder from Oxoid Unipath.

***Clostridum difficile* Agar Base:**

Composition per 920.0mL:

Proteose peptone	40.0g
Agar	15.0g
Fructose	6.0g
Na_2HPO_4	5.0g
NaCl	2.0g
KH_2PO_4	1.0g
$MgSO_4 \cdot 7H_2O$	0.1g

Preparation of *Clostridum difficile* Agar Base: Add components to distilled/deionized water and bring volume to 920.0mL. Mix thoroughly. Gently heat to boiling. Autoclave for 15 min at 15 psi pressure–121°C. Cool to 45°–50°C.

***Clostridium difficile* Selective Supplement:**

Composition per 10.0mL:

D-Cycloserine	500.0mg
Cefoxitin	16.0mg

Preparation of *Clostridium difficile* Selective Supplement: Add components to distilled/deionized water and bring volume to 10.0mL. Mix thoroughly. Filter sterilize.

Preparation of Medium: Add 10.0mL of sterile *Clostridium difficile* selective supplement and 70.0mL of sterile, defibrinated horse blood to 920.0mL of cooled, sterile *Clostridum difficile* agar base. Mix thoroughly. Pour into sterile Petri dishes or distribute into sterile tubes.

Use: For the selective isolation and cultivation of *Clostridium difficile* from clinical and nonclinical specimens.

Clostridium difficile Agar (Cycloserine Cefoxitin Fructose Agar) (CCFA)

Composition per liter:

Peptic digest of animal tissue	32.0g
Agar	20.0g
Fructose	6.0g

Na_2HPO_4	5.0g
NaCl	2.0g
KH_2PO_4	1.0g
Cycloserine	0.25g
$MgSO_4$	0.1g
Neutral Red	0.03g
Cefoxitin solution	10.0mL

pH 7.2 ± 0.2 at 25°C

Source: This medium is available as a premixed powder from BD Diagnostic Systems.

Cefoxitin Solution:
Composition per 10.0mL:

Cefoxitin	16.0mg

Preparation of Cefoxitin Solution: Add cefoxitin to distilled/deionized water and bring volume to 10.0mL. Mix thoroughly. Filter sterilize.

Preparation of Medium: Add components to distilled/deionized water and bring volume to 990.0mL. Mix thoroughly. Gently heat to boiling. Autoclave for 15 min at 15 psi pressure–121°C. Cool to 45°–50°C. Aseptically add 10.0mL of sterile cefoxitin solution. Mix thoroughly. Pour into sterile Petri dishes or distribute into sterile tubes.

Use: For the selective isolation and cultivation of *Clostridium difficile* from clinical and nonclinical specimens.

Clostridium Medium

Composition per liter:

Sodium L-glutamate	10.0g
Sodium thioglycolate	0.5g
Yeast extract	0.5g
K_2HPO_4	0.2g
$MgSO_4 \cdot 7H_2O$	0.1g

pH 7.6 ± 0.2 at 25°C

Preparation of Medium: Add components to distilled/deionized water and bring volume to 1.0L. Mix thoroughly. Distribute into tubes or flasks. Autoclave for 15 min at 15 psi pressure–121°C.

Use: For the enrichment and isolation of glutamate-fermenting *Clostridium* species.

Clostridium Medium

Composition per liter:

Uric acid	2.0g
Yeast extract	1.2g
$MgSO_4 \cdot 7H_2O$	0.05g
$CaCl_2 \cdot 2H_2O$	5.0mg
$FeSO_4 \cdot 7H_2O$	2.0mg
Resazurin	1.0mg
KOH (10*N* solution)	3.0mL
$K_2HPO_4 \cdot 3H_2O$ (70% solution)	1.5mL
Mercaptoacetic acid	1.5mL

pH 7.2 ± 0.2 at 25°C

Preparation of Medium: Add KOH solution and $K_2HPO_4 \cdot 3H_2O$ solution to distilled/deionized water and bring volume to 500.0mL. Gently heat and bring to boiling. Mix thoroughly. Add uric acid slowly. Cool to 45°–50°C. Add remaining components. Add mercaptoacetic acid immediately prior to sterilization. Bring volume to 1.0L with distilled/deionized water. Mix thoroughly. Autoclave for 15 min at 15 psi pressure–121°C. Adjust pH to 7.2 with sterile 60% K_2CO_3 solution.

Use: For the isolation and cultivation of purine-fermenting *Clostridium* species.

Clostridium Medium (ATCC Medium 39)

Composition per liter:

K_2HPO_4	7.0g
γ-Aminobutyric acid	5.0g
Yeast extract	3.0g
Agar	1.5g
KH_2PO_4	1.3g
$MgCl_2 \cdot 6H_2O$	0.2g
$CaCl_2 \cdot 2H_2O$	0.01g
$FeCl_3 \cdot 6H_2O$	0.01g
Methylene Blue	2.0mg
$MnSO_4$	1.0mg
Na_2MoO_4	1.0mg
$Na_2S \cdot 9H_2O$ solution	10.0mL

$Na_2S \cdot 9H_2O$ Solution:
Composition per 20.0mL:

$Na_2S \cdot 9H_2O$	0.6g

Preparation of $Na_2S \cdot 9H_2O$ Solution: Add $Na_2S \cdot 9H_2O$ to distilled/deionized water and bring volume to 20.0mL. Autoclave for 15 min at 15 psi pressure–121°C. Use freshly prepared solution.

Preparation of Medium: Add components, except $Na_2S \cdot 9H_2O$, to distilled/deionized water and bring volume to 1.0L. Mix thoroughly. Gently heat to boiling. Autoclave for 15 min at 15 psi pressure–121°C. Cool to 45°–50°C. Distribute anaerobically into sterile tubes. Aseptically add 0.1mL of sterile 1.5% $Na_2S \cdot 9H_2O$ solution to each 5.0mL of the medium. Cap with rubber stoppers.

Use: For the cultivation and maintenance of a variety of *Clostridium* species.

Clostridium Medium (ATCC Medium 40)

Composition per liter:

K_2HPO_4	7.0g

δ-Aminovaleric acid·HCl (neutralized) 5.0g
Agar 1.5g
KH_2PO_4 .... 1.3g
Yeast extract 1.0g
$MgCl_2 \cdot 6H_2O$ 0.2g
$CaCl_2 \cdot 2H_2O$ 0.01g
$FeCl_3 \cdot 6H_2O$ 0.01g
Methylene Blue 2.0mg
$MnSO_4$ 1.0mg
Na_2MoO_4 .... 1.0mg
$Na_2S \cdot 9H_2O$ solution 20.0mL

$Na_2S \cdot 9H_2O$ Solution:
Composition per 100.0mL:
$Na_2S \cdot 9H_2O$ 1.5g

Preparation of $Na_2S \cdot 9H_2O$ Solution: Add $Na_2S \cdot 9H_2O$ to distilled/deionized water and bring volume to 100.0mL. Autoclave for 15 min at 15 psi pressure–121°C. Use freshly prepared solution.

Preparation of Medium: Add components, except $Na_2S \cdot 9H_2O$ solution, to distilled/deionized water and bring volume to 1.0L. Mix thoroughly. Gently heat to boiling. Autoclave for 15 min at 15 psi pressure–121°C. Cool to 45°–50°C. Distribute anaerobically into sterile tubes. Aseptically add 0.1mL of sterile $Na_2S \cdot 9H_2O$ solution to each 5.0mL of the medium. Cap with rubber stoppers.

Use: For the cultivation and maintenance of a variety of *Clostridium* species.

Clostridium Medium (ATCC Medium 43)

Composition per liter:
Agar 15.0g
Yeast extract 5.0g
L-Arginine·HCl 2.0g
L-Lysine·HCl 2.0g
NH_4Cl 2.0g
Sodium formate 2.0g
K_2HPO_4 .... 1.75g
$MgSO_4 \cdot 7H_2O$ 0.2g
$CaCl_2 \cdot 2H_20$ 0.01g
$FeSO_4 \cdot 7H_2O$ 0.01g
Methylene Blue 2.0mg
$Na_2S \cdot 9H_2O$ solution 30.0mL

$Na_2S \cdot 9H_2O$ Solution:
Composition per 100.0mL:
$Na_2S \cdot 9H_2O$ 1.0g

Preparation of $Na_2S \cdot 9H_2O$ Solution: Add $Na_2S \cdot 9H_2O$ to distilled/deionized water and bring volume to 100.0mL. Autoclave for 15 min at 15 psi pressure–121°C. Use freshly prepared solution.

Preparation of Medium: Add components, except $Na_2S \cdot 9H_2O$ solution, to tap water and bring volume to 1.0L. Mix thoroughly. Gently heat to boiling. Autoclave for 15 min at 15 psi pressure–121°C. Cool to 45°–50°C. Distribute anaerobically into sterile tubes. Aseptically add 0.15mL of sterile $Na_2S \cdot 9H_2O$ solution to each 5.0mL of the medium. Cap with rubber stoppers.

Use: For the cultivation and maintenance of a variety of *Clostridium* species.

Clostridium Medium (ATCC Medium 163)

Composition per liter:
Agar 20.0g
Sodium glutamate 17.0g
Yeast extract 6.0g
Sodium thioglycolate 0.5g
Phosphate buffer (1.0*M*, pH 7.4) 40.0mL
$MgSO_4$ (2.0*M* solution) 0.5mL
$FeSO_4$ (0.2*M* solution) 0.2mL
$CaCl_2$ (1.0*M* solution) 0.1mL
$CoCl_2$ (0.1*M* solution) 0.1mL
$MnCl_2$ (0.1*M* solution) 0.1mL
Na_2MoO_4 (0.1*M* solution) 0.1mL

Preparation of Medium: Add components to distilled/deionized water and bring volume to 1.0L. Mix thoroughly. Gently heat to boiling. Autoclave for 15 min at 15 psi pressure–121°C. Pour into sterile Petri dishes or distribute into sterile tubes.

Use: For the cultivation and maintenance of a variety of *Clostridium* species.

Clostridium Medium (ATCC Medium 511)

Composition per liter:
Yeast extract 4.0g
Alanine 3.0g
Peptone 3.0g
L-Cysteine 0.2g
$MgSO_4$ 0.05g
$FeSO_4$ 0.01g
Potassium phosphate
buffer (1.0*M*, pH 7.1) 5.0mL
$CaSO_4$ (saturated solution) 2.5mL

Preparation of Medium: Add components to distilled/deionized water and bring volume to 1.0L. Mix thoroughly. Distribute into tubes or flasks. Autoclave for 15 min at 15 psi pressure–121°C.

Use: For the cultivation of a variety of *Clostridium* species.

Clostridium Medium (ATCC Medium 568)

Composition per liter:

Na_2CO_3	10.0g
Fructose	3.0g
K_2HPO_4	2.0g
Yeast extract	2.0g
$(NH_4)_2SO_4$	1.0g
$MgSO_4 \cdot 7H_2O$	0.5g
Sodium thioglycolate	0.05g
$CaSO_4$	0.015g
$FeSO_4 \cdot 7H_2O$	2.5mg
$MnSO_4 \cdot H_2O$	0.5mg
$Na_2MoO_4 \cdot 2H_2O$	0.5mg

pH 7.8 ± 0.2 at 25°C

Preparation of Medium: Add components to distilled/deionized water and bring volume to 1.0L. Mix thoroughly. Distribute into tubes or flasks. Autoclave for 15 min at 15 psi pressure–121°C.

Use: For the cultivation of a variety of *Clostridium* species.

Clostridium Medium (ATCC Medium 591)

Composition per liter:

Solution 1	600.0mL
Solution 2	400.0mL

pH 8.0 ± 0.2 at 25°C

Solution 1:

Composition per 600.0mL:

Peptone	5.0g

Preparation of Solution 1: Add component to distilled/deionized water and bring volume to 600.0mL. Mix thoroughly. Autoclave for 15 min at 15 psi pressure–121°C.

Solution 2:

Composition per 400.0mL:

$NaHCO_3$	20.0g
Fructose	10.0g
K_2HPO_4	10.0g
Sodium thioglycolate	0.75g
Vitamin solution	14.0mL
Trace elements solution	10.0mL

Preparation of Solution 2: Add components, except sodium thioglycolate, to distilled/deionized water and bring volume to 400.0mL. Mix thoroughly. Gas with 100% CO_2. Add sodium thioglycolate. Adjust pH to 8.0. Filter sterilize.

Vitamin Solution:

Composition per 100.0mL:

Thiamine	0.1g
Nicotinic acid	0.05g
Pyridoxine	0.05g
Pantothenic acid	0.025g
p-Aminobenzoic acid	5.0mg
Vitamin B_{12}	2.0mg
Biotin	1.0mg

Preparation of Vitamin Solution: Add components to distilled/deionized water and bring volume to 100.0mL. Mix thoroughly.

Trace Elements Solution:

Composition per liter:

EDTA	0.5g
$FeSO_4 \cdot 7H_2O$	0.2g
H_3BO_3	0.03g
$CoCl_2 \cdot 6H_2O$	0.02g
$ZnSO_4 \cdot 7H_2O$	0.01g
$MnCl_2 \cdot 4H_2O$	3.0mg
$Na_2MoO_4 \cdot 2H_2O$	3.0mg
$NiCl_2 \cdot 6H_2O$	2.0mg
$CuCl_2 \cdot 2H_2O$	1.0mg

Preparation of Trace Elements Solution: Add components to distilled/deionized water and bring volume to 1.0L. Mix thoroughly.

Preparation of Medium: Aseptically combine 600.0mL of sterile solution 1 and 400.0mL of sterile solution 2. Distribute into sterile tubes or flasks.

Use: For the cultivation and maintenance of a variety of *Clostridium* species.

Clostridium novyi Blood Agar

Composition per 100.0mL:

Agar	2.0g
Glucose	1.0g
Neopeptone	1.0g
Proteolyzed liver	0.5g
Yeast extract	0.5g
Horse blood, defibrinated	10.0mL
Reducing solution	0.75mL
Salts solution	0.5mL

pH 7.6–7.8 at 25°C

Salts Solution:

Composition per 100.0mL:

$MgSO_4 \cdot 7H_2O$	4.0g
$MnSO_4 \cdot 4H_2O$	0.2g
HCl	0.05g
$FeCl_3$	0.04g

Preparation of Salts Solution: Add components to distilled/deionized water and bring volume to 100.0mL. Mix thoroughly.

Reducing Solution:

Composition per 10.0mL:

L-Cysteine·HCl·H_2O	0.12g

Dithiothreitol..0.12g
Glutamine...0.06g

Preparation of Reducing Solution: Add components to distilled/deionized water and bring volume to 10.0mL. Mix thoroughly. Adjust pH to 7.6–7.8. Filter sterilize.

Preparation of Medium: Add agar to distilled/deionized water and bring volume to 50.0mL. Mix thoroughly. Gently heat and bring to boiling. In another flask, add neopeptone, yeast extract, liver extract, and salts solution to distilled/deionized water and bring volume to 50.0mL. Mix thoroughly. Gently heat until dissolved. Combine the two solutions. Distribute into screw-capped bottles in 18.0mL volumes. Autoclave for 10 min at 10 psi pressure–115°C. Cool to 45°–50°C. Medium may be stored at 4°C at this point. Immediately prior to inoculation, aseptically add 2.0mL of horse blood and 0.15mL of sterile reducing solution to each tube of melted agar at 50°C. Mix thoroughly. Pour the contents of each tube into a sterile Petri dish.

Use: For the cultivation of *Clostridium novyi.*

Clostridium perfringens Agar, OPSP (Perfringens Agar, OPSP)

Composition per liter:

Pancreatic digest of casein...............................15.0g
Agar..10.0g
Liver extract...7.0g
Papaic digest of soybean meal...........................5.0g
Yeast extract...5.0g
Tris(hydroxymethyl)aminomethane buffer.........1.5g
Ferric ammonium citrate...................................1.0g
$Na_2S_2O_5$ 1.0g
Antibiotic inhibitor 10.0mL

pH 7.3 ± 0.2 at 25°C

Source: This medium is available as a premixed powder from Oxoid Unipath.

Antibiotic Inhibitor:

Composition per 10.0mL:

Sodium sulfadiazine..0.1g
Oleandomycin phosphate................................0.5mg
Polymyxin B 10,000U

Preparation of Antibiotic Inhibitor: Add components to distilled/deionized water and bring volume to 10.0mL. Mix thoroughly. Filter sterilize.

Preparation of Medium: Add components, except antibiotic inhibitor, to distilled/deionized water and bring volume to 990.0mL. Mix thoroughly. Gently heat and bring to boiling. Autoclave for 15 min at 15 psi pressure–121°C. Cool to 45°–50°C. Aseptically add sterile antibiotic inhibitor. Mix thoroughly. Pour into sterile Petri dishes or distribute into sterile tubes.

Use: For the presumptive identification and enumeration of *Clostridium perfringens.*

Clostridium Selective Agar (Clostrisel Agar)

Composition per liter:

Pancreatic digest of casein...............................17.0g
Agar..14.0g
Glucose...6.0g
Papaic digest of soybean meal...........................3.0g
NaCl...2.5g
Sodium thioglycolate...1.8g
Sodium formaldehyde sulfoxylate......................1.0g
L-Cystine...0.25g
NaN_3 ....0.15g
Neomycin sulfate...0.15g

pH 7.0 ± 0.2 at 25°C

Source: This medium is available as a premixed powder from BD Diagnostic Systems.

Preparation of Medium: Add components to distilled/deionized water and bring volume to 1.0L. Mix thoroughly. Gently heat while stirring and bring to boiling. Distribute into tubes or flasks. Autoclave for 15 min at 15 psi pressure–118°C. Pour into sterile Petri dishes or leave in tubes.

Caution: Sodium azide is toxic. Azides also react with metals and disposal must be highly diluted.

Use: For the selective isolation of pathogenic *Clostridium* species from specimens containing mixed flora, e.g., from wounds, fecal specimens, and other specimens.

CM3 Medium

Composition per liter:

3-(*N*-Morpholino)propanesulfonic
acid (MOPS) buffer...................................20.0g
Cellobiose (or Cellulose MN 300)....................10.0g
K_2HPO_4..4.4g
Urea..1.5g
L-Cysteine·HCl·H_2O 1.0g
$MgSO_4·7H_2O$ 0.5g
$(NH_4)_2SO_4$ 0.4g
$CaCl_2·2H_2O$ 0.05g
$FeSO_4·7H_2O$...1.0mg

pH 7.1 ± 0.2 at 25°C

Preparation of Medium: Add components to distilled/deionized water and bring volume to 1.0L. Mix thoroughly. Distribute into tubes or flasks. Autoclave for 15 min at 15 psi pressure–121°C.

Use: For the cultivation of *Clostridium* species.

CN Screen Medium (*Cryptococcus neoformans* Screen Medium)

Composition per liter:

Agar	15.0g
K_2HPO_4	4.0g
$MgSO_4 \cdot 7H_2O$	2.5g
Glucose	1.25g
Asparagine	1.0g
Glutamine	1.0g
Glycine	1.0g
Thiamine·HCl	1.0g
Tryptophan	1.0g
EDTA	0.6g
Biotin	0.51g
Dihydroxyphenylalanine (Dopa)	0.2g
Phenol Red	0.2g

pH 5.5–5.6 ± 0.2 at 25°C

Preparation of Medium: Add components to distilled/deionized water and bring volume to 1.0L. Mix thoroughly. Gently heat until boiling. Distribute into tubes or flasks. Autoclave for 15 min at 15 psi pressure–121°C.

Use: For the screening of yeast isolates for the presumptive identification of *Cryptococcus neoformans*. *Cryptococcus neoformans* forms black colonies.

Coagulase Agar Base

Composition per liter:

Agar	25.0g
Brain heart infusion	10.5g
Pancreatic digest of casein	10.5g
D-Mannitol	10.0g
Brain heart infusion	5.0g
NaCl	3.5g
Papaic digest of soybean meal	3.5g
Bromcresol Purple	0.02g
Rabbit plasma	100.0mL

pH 7.4 ± 0.2 at 25°C

Preparation of Medium: Add components, except rabbit plasma, to distilled/deionized water and bring volume to 1.0L. Mix thoroughly. Gently heat, while stirring, until boiling. Distribute into tubes or flasks. Autoclave for 15 min at 15 psi pressure–121°C. Cool to 45°–50°C. Add rabbit plasma to a final concentration of 7–15%. Mix thoroughly. Pour into sterile Petri dishes in 18.0mL volume per plate.

Use: For the cultivation and differentiation of *Staphylococcus aureus* from other *Staphylococcus* species based on coagulase production.

Coagulase Mannitol Agar

Composition per liter:

Agar	14.5g
Pancreatic digest of casein	10.5g
D-Mannitol	10.0g
Brain heart infusion	5.0g
NaCl	3.5g
Papaic digest of soybean meal	3.5g
Bromcresol Purple	0.02g
Rabbit plasma with 0.15% EDTA	100.0mL

pH 7.3 ± 0.2 at 25°C

Source: This medium is available as a premixed powder from BD Diagnostic Systems.

Preparation of Medium: Add components, except rabbit plasma, to distilled/deionized water and bring volume to 1.0L. Mix thoroughly. Gently heat while stirring until boiling. Distribute into tubes or flasks. Autoclave for 15 min at 15 psi pressure–121°C. Cool to 45°–50°C. Add rabbit plasma with 0.15% EDTA to a final concentration of 7–15%. Mix thoroughly. Pour into sterile Petri dishes in 18.0mL volume per plate.

Use: For the cultivation and differentiation of *Staphylococcus aureus* from other *Staphylococcus* species based on coagulase production and mannitol fermentation.

COBA (Colistin Oxolinic Acid Blood Agar)

Composition per liter:

Columbia agar base	930.0mL
Horse blood, defibrinated, sterile	50.0mL
Colistin sulfate solution	10.0mL
Oxolinic acid solution	10.0mL

pH 7.3 ± 0.2 at 25°C

Columbia Agar Base:
Composition per 930.0mL:

Agar	13.5g
Pancreatic digest of casein	10.0g
Peptic digest of animal tissue	10.0g
NaCl	5.0g
Beef extract	3.0g
Yeast extract	3.0g
Cornstarch	1.0g

Preparation of Columbia Agar Base: Add components to distilled/deionized water and bring volume to 930.0mL. Mix thoroughly. Gently heat until boiling. Autoclave for 15 min at 15 psi pressure–121°C. Cool to 45°–50°C.

Colistin Sulfate Solution:
Composition per 10.0mL:

Colistin sulfate	10.0mg

Preparation of Colistin Sulfate Solution: Add colistin sulfate to distilled/deionized water and bring volume to 10.0mL. Mix thoroughly. Filter sterilize.

Oxolinic Acid Solution:

Composition per 10.0mL:

Oxolinic acid	5.0–10.0mg

Preparation of Oxolinic Acid Solution: Add oxolinic acid to distilled/deionized water and bring volume to 10.0mL. Mix thoroughly. Filter sterilize.

Preparation of Medium: To 930.0mL of sterile, cooled Columbia agar base, add sterile colistin sulfate, sterile oxolinic acid, and sterile, defibrinated horse blood. Mix thoroughly. Pour into sterile Petri dishes.

Use: For the isolation and cultivation of streptococci in pure culture from mixed flora in clinical specimens.

Coletsos Medium

Composition per 1625mL:

Potato starch	10.0g
Gelatin	4.0g
Asparagine	2.25g
KH_2PO_4	1.5g
Na-glutamate	1.0g
Na-pyruvate	1.0g
Mg-citrate	0.375g
Litmus	0.25g
Malachite green	0.25g
$MgSO_4$	0.15g
Activated carbon	0.1g
Oligonucleotide mixture	3.0mg
Egg mixture	625.0mL
Glycerol	7.5mL

Egg Mixture:

Composition per liter:

Whole eggs	18–24

Preparation of Egg Mixture: Use fresh eggs, less than 1 week old. Scrub the shells with soap. Let stand in a soap solution for 30 min. Rinse in running water. Soak eggs in 70% ethanol for 15 min. Break the eggs into a sterile container. Separate egg whites from egg yolks. Combine 8 parts egg white with 2 parts egg yolk. Homogenize by shaking. Filter through four layers of sterile cheesecloth into a sterile graduated cylinder. Bring volume to 1.0L distilled/deionized water.

Preparation of Medium: Add glycerol to 600.0mL of distilled/deionized water. Mix thoroughly. Add remaining components, except egg mixture. Bring volume to 1.0L. Mix thoroughly. Gently heat while stirring and bring to boiling. Autoclave for 15 min at 15 psi pressure–121°C. Cool to 50°C. Aseptically add 625.0mL of egg mixture. Mix thoroughly. Distribute into sterile screw-capped tubes. Place tubes in a slanted position. Inspissate at 85°C (moist heat) for 45 min.

Use: For the cultivation of *Mycobacterium tuberculosis*.

Coletsos Selective Medium

Composition per 1625mL:

Potato starch	10.0g
Gelatin	4.0g
Asparagine	2.25g
KH_2PO_4	1.5g
Na-glutamate	1.0g
Na-pyruvate	1.0g
Mg-citrate	0.375g
Litmus	0.25g
Malachite green	0.25g
$MgSO_4$	0.15g
Activated carbon	0.1g
Oligonucleotide mixture	3.0mg
Homogenized egg	625.0mL
Glycerol	7.5mL
Nalidixic acid solution	1.0mL
Lincomycin solution	1.0mL
Cycloheximide solution	1.0mL

Nalidixic Acid Solution:

Composition per 100.0mL:

Nalidixic acid	0.5g

Preparation of Nalidixic Acid Solution: Add nalidixic acid to distilled/deionized water and bring volume to 100.0mL. Mix thoroughly. Filter sterilize.

Cycloheximide Solution:

Composition per 100.0mL:

Cycloheximide	1.5g
Ethanol	40.0mL

Preparation of Cycloheximide Solution: Add cycloheximide to 40.0mL of ethanol. Mix thoroughly. Bring volume to 100.0mL with distilled/deionized water. Filter sterilize.

Lincomycin Solution:

Composition per 100.0mL:

Lincomycin	0.5g

Preparation of Lioncomycin Solution: Add lincomycin to distilled/deionized water and bring volume to 100.0mL. Mix thoroughly. Filter sterilize.

Egg Mixture Solution:

Composition per liter:

Whole eggs	18–24

Preparation of Egg Mixture Solution: Use fresh eggs, less than 1 week old. Scrub the shells with soap. Let stand in a soap solution for 30 min. Rinse in running water. Soak eggs in 70% ethanol for 15 min. Break the eggs into a sterile container. Separate egg whites from egg yolks. Combine 8 parts egg white with 2 parts egg yolk. Homogenize by shaking. Filter through four layers of sterile cheesecloth into a sterile graduated cylinder. Bring volume to 1.0L distilled/deionized water.

Preparation of Medium: Add glycerol to 600.0mL of distilled/deionized water. Mix thoroughly. Add remaining components, except egg mixture, lincomycin solution, cycloheximide solution, and nalidixic acid solution. Mix thoroughly. Bring volume to 1.0L. Gently heat while stirring and bring to boiling. Autoclave for 15 min at 15 psi pressure–121°C. Cool to 50°C. Aseptically add 625.0mL of egg mixture. Mix thoroughly. Aseptically add 1.0mL cycloheximide solution, 1.0mL lincomycin solution, and 1.0mL nalidixic acid solution. Distribute into sterile screw-capped tubes. Place tubes in a slanted position. Inspissate at 85°C (moist heat) for 45 min.

Use: For the isolation and cultivation of *Mycobacterium tuberculosis.*

Coliform Agar, Chromocult® (Chromocult Coliform Agar)

Composition per liter:

Agar	10.0g
NaCl	5.0g
Peptone	3.0g
Na_2HPO_4	2.7g
NaH_2PO_4	2.2g
Tryptophan	1.0g
Na-pyruvatge	1.0g
Chromogenic mixture	0.4g
Tergiotol 7	0.15g

pH 7.0 ± 0.2 at 25°C

Source: This medium is available from Merck.

Preparation of Medium: Add components to distilled/deionized water and bring volume to 1.0L. Mix well and warm gently until dissolved. Autoclave for 15 min at 15 psi pressure–121°C. Pour into sterile Petri dishes. Some turbidity may occur, but this does not effect the performance

Use: For the detection of *E. coli* and coliform bacteria. The interaction of selected peptones, pyruvate, sorbitol, and phosphate buffer guarantees rapid colony growth, even for sublethally injured coliforms. The growth of Gram-positive bacteria as well as some Gram-negative bacteria is largely inhibited by the content of Tergitol 7 which has no negative effect on the growth of the coliform bacteria. A combination of two chromogenic substrates which allow for the simultaneous detection of total coliforms and *E. coli.* The characteristic enzyme for coliforms, β-D-galactosidase cleaves the Salmon-GAL substrate and causes a salmon to red color of the coliform colonies. The substrate X-glucuronide is used for the identification of β-D-glucuronidase, which is characteristic for *E. coli. E. coli* cleaves both Salmon-GAL and X-glucuronide, so that positive colonies take on a dark-blue to violet color. These are easily distinguished from other coliform colonies which have a salmon to red color. As part of an additional confirmation of *E. coli*, the inclusion of tryptophan improves the indole reaction, thereby increasing detection reliability when it is used in combination with the Salmon-GAL and X-glucuronide reaction.

Coliform Agar ES, Chromocult (Chromocult Coliform Agar ES) (Chromocult Enhanced Selectivity Agar)

Composition per liter:

Agar	10.0g
MOPS	10.0g
KCl	7.5g
Peptone	5.0g
Bile salts	1.15g
Na-propionate	0.5g
6-Chloro-3-indoxyl-β-D-galactopyranoside	0.15g
5-Bromo-4-chloro-3-indoxyl-β-D-glucuronic acid	0.1g
Isopropyl-β-D-thiogalactopyranoside	0.1g

pH 7.0 ± 0.2 at 25°C

Source: This medium is available from Merck.

Preparation of Medium: Add components to distilled/deionized water and bring volume to 1.0L. Mix thoroughly and heat with frequent agitation until components are completely dissolved (approximately 45 min.). Do not autoclave. Cool to 45°–50°C. Pour into sterile Petri dishes. The plates should be clear and colorless.

Use: For the detection of *E.coli* and total coliforms. The combination of suitable peptones and the buffering using MOPS allow rapid growth of coliforms and an optimal transformation of the chromogenic substrates. The amount of bile salts and propionate largely inhibit growth of Gram-positive and Gram-negative accompanying flora. The simultaneous detection of total coliforms and *E.coli* is achieved using the combination of two chromogrenic substrates. The substrate Salmon™--β-D-GAL is split by β-D-galactosidase, characteristic for coliforms, resulting in a salmon to red coloration of coliform colonies. The detection of the β-D-glucuronidase, characteristic for *E.coli,* is cleaved via the substrate X-β-D-glucuronide, causing a blue coloration of positive colonies. As *E.coli* splits Salmon-β-D-GAL as well as X-β-D-glucuronide, the colonies turn to a dark violet color and can be easily differentiated from the other coliforms being salmon-red.

Columbia Agar

Composition per liter:

Columbia agar base	950.0mL
Sheep blood	50.0mL

pH 7.3 ± 0.2 at 25°C

Columbia Agar Base:

Composition per liter:

Agar	13.5g
Pancreatic digest of casein	12.0g
NaCl	5.0g
Peptic digest of animal tissue	5.0g
Beef extract	3.0g
Yeast extract	3.0g
Cornstarch	1.0g

Preparation of Columbia Agar Base: Add components to distilled/deionized water and bring volume to 1.0L. Mix thoroughly. Gently heat until boiling. Autoclave for 15 min at 15 psi pressure–121°C. Cool to 45°–50°C.

Preparation of Medium: To 950.0mL of cooled, sterile Columbia agar base, aseptically add 50.0mL of sterile, defibrinated sheep blood. Mix thoroughly. Pour into sterile Petri dishes or distribute into sterile tubes.

Use: For the isolation and cultivation of nonfastidious and fastidious microorganisms from a variety of clinical and nonclinical specimens.

Columbia Blood Agar

Composition per liter:

Columbia blood agar base	950.0mL
Sheep blood	50.0mL

pH 7.3 ± 0.2 at 25°C

Columbia Blood Agar Base:

Composition per liter:

Agar	15.0g
Pantone	10.0g
Bitone	10.0g
NaCl	5.0g
Tryptic digest of beef heart	3.0g
Cornstarch	1.0g

Source: Columbia blood agar base is available as a premixed powder from BD Diagnostic Systems.

Preparation of Columbia Blood Agar Base: Add components to distilled/deionized water and bring volume to 1.0L. Mix thoroughly. Gently heat until boiling. Autoclave for 15 min at 15 psi pressure–121°C. Cool to 45°–50°C.

Preparation of Medium: To 950.0mL of cooled, sterile Columbia blood agar base, aseptically add 50.0mL of sterile, defibrinated sheep blood. Mix thoroughly. Pour into sterile Petri dishes or distribute into sterile tubes.

Use: With the addition of blood or other enrichments, used for the isolation and cultivation of fastidious microorganisms.

Columbia Blood Agar (DSMZ Medium 693)

Composition per liter:

Columbia blood agar base	950.0mL
Sheep blood	50.0mL

pH 7.3 ± 0.2 at 25°C

Source: This medium is available as a premixed powder from Oxoid Unipath.

Columbia Blood Agar Base:

Composition per liter:

Special peptone	23.0g
Agar	10.0g
NaCl	5.0g
Starch	1.0g

Preparation of Columbia Blood Agar Base: Add components to distilled/deionized water and bring volume to 1.0L. Mix thoroughly. Gently heat until boiling. Autoclave for 15 min at 15 psi pressure–121°C. Cool to 45°–50°C.

Preparation of Medium: To 950.0mL of cooled, sterile Columbia blood agar base, aseptically add 50.0mL of sterile, defibrinated sheep blood. Mix thoroughly. Pour into sterile Petri dishes or distribute into sterile tubes.

Use: For the cultivation of *Corynebacterium* spp., *Actinomyces* spp., *Arcanobacterium* spp., *Streptococcus pneumoniae, Lactobacillus iners, Isobaculum melis, Nocardia paucivorans,* and a variety of fastidious microorganisms.

Columbia Blood Agar Base

Composition per liter:

Agar	15.0g
Pantone	10.0g
Bitone	10.0g
NaCl	5.0g
Tryptic digest of beef heart	3.0g
Cornstarch	1.0g

Source: Columbia blood agar base is available as a premixed powder from BD Diagnostic Systems.

Preparation of Medium: Add components to distilled/deionized water and bring volume to 1.0L. Mix thoroughly. Gently heat until boiling. Autoclave for 15 min at 15 psi pressure–121°C. Pour into sterile Petri dishes or distribute into sterile tubes.

Use: For the cultivation of *Balneatrix alpica.*

Columbia Blood Agar Base with Horse Blood (LMG Medium 151)

Composition per liter:

Columbia blood agar base	950.0mL
Horse blood	50.0mL

pH 7.3 ± 0.2 at 25°C

Columbia Blood Agar Base:

Composition per liter:

Special peptone	23.0g
Agar	10.0g
NaCl	5.0g
Starch	1.0g

Preparation of Columbia Blood Agar Base: Add components to distilled/deionized water and bring volume to 1.0L. Mix thoroughly. Gently heat until boiling. Autoclave for 15 min at 15 psi pressure–121°C. Cool to 45°–50°C.

Preparation of Medium: To 950.0mL of cooled, sterile Columbia blood agar base, aseptically add 50.0mL of sterile, defibrinated horse blood. Mix thoroughly. Pour into sterile Petri dishes or distribute into sterile tubes.

Use: For the cultivation of *Arcanobacterium* spp., *Paenibacillus* spp., *Corynebacterium* spp., *Lactobacillus iners, Enterococcus spp.* and other bacteria.

Columbia Blood Agar Base with Horse Blood (LMG Medium 210)

Composition per liter:

Columbia blood agar base	950.0mL
Horse blood	50.0mL

pH 7.3 ± 0.2 at 25°C

Columbia Blood Agar Base:

Composition per liter:

Agar	15.0g
Pancreatic digest of casein	10.0g
Proteose peptone no. 3	5.0g
Yeast extract	5.0g
NaCl	5.0g
Beef heart digest	3.0g
Corn starch	1.0g

Preparation of Columbia Blood Agar Base: Add components to distilled/deionized water and bring volume to 1.0L. Mix thoroughly. Gently heat until boiling. Autoclave for 15 min at 15 psi pressure–121°C. Cool to 45°–50°C.

Preparation of Medium: To 950.0mL of cooled, sterile Columbia blood agar base, aseptically add 50.0mL of sterile, defibrinated horse blood. Mix thoroughly. Pour into sterile Petri dishes or distribute into sterile tubes.

Use: For the cultivation of *Actinomyces* spp., *Streptococcus dysgalactiae,* and *Actinobaculum* spp.

Columbia Blood Agar Base with Horse Blood and Charcoal (DSMZ Medium 429a)

Composition per liter:

Columbia blood agar base with charcoal	960.0mL
Horse blood	40.0mL

pH 7.3 ± 0.2 at 25°C

Columbia Blood Agar Base with Charcoal:

Composition per liter:

Agar	15.0g
Pancreatic digest of casein	10.0g
Proteose peptone no. 3	5.0g
Yeast extract	5.0g
NaCl	5.0g
Beef heart digest	3.0g
Charcoal	2.0g
Corn starch	1.0g

Preparation of Columbia Blood Agar Base with Charcoal: Add components to distilled/deionized water and bring volume to 1.0L. Mix thoroughly. Gently heat until boiling. Autoclave for 15 min at 15 psi pressure–121°C. Cool to 45°–50°C.

Preparation of Medium: To 960.0mL of cooled, sterile Columbia blood agar base, aseptically add 40.0mL of sterile, defibrinated horse blood. Mix thoroughly. Pour into sterile Petri dishes or distribute into sterile tubes.

Use: For the cultivation of *Neisseria gonorrhoeae.*

Columbia Broth

Composition per liter:

Bitone	10.0g
Pancreatic digest of casein	5.0g
Peptic digest of animal tissue	5.0g
NaCl	5.0g
Tryptic digest of beef heart	3.0g
Tris(hydroxymethyl)aminomethane·HCl	2.86g
Glucose	2.5g
Tris(hydroxymethyl)aminomethane	0.83g
Na_2CO_3	0.6g
L-Cysteine·HCl	0.1g

$MgSO_4$, anhydrous....0.1g
$FeSO_4$....0.02g

pH 7.5 ± 0.2 at 25°C

Source: This medium is available as a premixed powder from BD Diagnostic Systems.

Preparation of Medium: Add components to distilled/deionized water and bring volume to 1.0L. Mix thoroughly. Gently heat until boiling. Distribute into tubes or flasks. Autoclave for 15 min at 15 psi pressure–121°C.

Use: For the cultivation and isolation of fastidious bacteria from clinical specimens or as a general purpose broth.

Columbia Broth

Composition per liter:

Pancreatic digest of casein....10.0g
Peptic digest of animal tissue....8.0g
NaCl....5.0g
Yeast extract....5.0g
Tris(hydroxymethyl) aminomethane·HCl buffer....2.86g
Glucose....2.5g
Tris(hydroxymethyl) aminomethane buffer....0.83g
L-Cysteine·HCl·H_2O....0.1g
$MgSO_4$·$7H_2O$....0.05g
$FeSO_4$....0.012g

pH 7.4 ± 0.2 at 25°C

Source: This medium is available as a premixed powder from BD Diagnostic Systems.

Preparation of Medium: Add components to distilled/deionized water and bring volume to 1.0L. Mix thoroughly. Gently heat until boiling. Distribute into tubes or flasks. Autoclave for 15 min at 15 psi pressure–121°C.

Use: For the cultivation of a wide variety of microorganisms. Used as a general purpose medium.

Columbia CNA Agar (Columbia Colistin Nalidixic Acid Agar)

Composition per liter:

Columbia blood agar base....950.0L
Sheep blood....50.0mL

pH 7.3 ± 0.2 at 25°C

Source: This medium is available as a premixed powder from BD Diagnostic Systems.

Columbia Blood Agar Base:

Composition per liter:

Agar....13.5g
Pancreatic digest of casein....12.0g
NaCl....5.0g
Peptic digest of animal tissue....5.0g
Beef extract....3.0g
Yeast extract....3.0g
Cornstarch....1.0g
Nalidixic acid....15.0mg
Colistin....10.0mg

Preparation of Columbia Blood Agar Base: Add components to distilled/deionized water and bring volume to 1.0L. Mix thoroughly. Gently heat until boiling. Autoclave for 15 min at 15 psi pressure–121°C. Cool to 45°–50°C.

Preparation of Medium: To 950.0mL of cooled, sterile Columbia blood agar base, aseptically add 50.0mL of sterile, defibrinated sheep blood. Mix thoroughly. Pour into sterile Petri dishes or distribute into sterile tubes.

Use: For the selective isolation, cultivation, and differentiation of Gram-positive cocci from clinical and nonclinical specimens.

Columbia CNA Agar, Modified with Sheep Blood

Composition per liter:

Columbia blood agar base....950.0mL
Sheep blood, defribinated....50.0mL

pH 7.3 ± 0.2 at 25°C

Source: This medium is available as a premixed powder from BD Diagnostic Systems.

Columbia Blood Agar Base:

Composition per liter:

Agar....13.5g
Pancreatic digest of casein....12.0g
NaCl....5.0g
Peptic digest of animal tissue....5.0g
Beef extract....3.0g
Yeast extract....3.0g
Cornstarch....1.0g
Colistin....10.0mg
Nalidixic acid....5.0mg

Preparation of Columbia Blood Agar Base: Add components to distilled/deionized water and bring volume to 1.0L. Mix thoroughly. Gently heat until boiling. Autoclave for 15 min at 15 psi pressure–121°C. Cool to 45°–50°C.

Preparation of Medium: To 950.0L of cooled, sterile Columbia blood agar base, aseptically add 50.0mL of sterile, defibrinated sheep blood. Mix thoroughly. Pour into sterile Petri dishes or distribute into sterile tubes.

Use: For the selective isolation, cultivation, and differentiation of Gram-positive cocci from clinical and nonclinical materials.

Complex Medium

Composition per liter:

NaCl	250.0g
$MgSO_4 \cdot 7H_2O$	20.0g
Yeast extract	10.0g
Casamino acids	7.5g
Trisodium citrate	3.0g
KCl	2.0g

pH 7.5–7.8 at 25°C

Preparation of Medium: Add components to distilled/deionized water and bring volume to 1.0L. Mix thoroughly. Autoclave for 5 min at 15 psi pressure–121°C. Filter through Whatman #1 filter paper. Adjust pH of filtrate to 7.4. Distribute into tubes or flasks. Autoclave for 15 min at 15 psi pressure–121°C.

Use: For the isolation and cultivation of *Actinomadura* species, *Actinopolyspora* species, *Excellospora* species, and *Microspora* species.

Congo Red Acid Morpholinepropanesulfonic Acid Pigmentation Agar (CRAMP Agar)

Composition per liter:

Agarose	14.0g
Morpholinepropanesulfonic acid	8.4g
NaCl	2.9g
Casamino acids	2.0g
Galactose	2.0g
Tricine (*n*-Tris-hydroxymethyl-methylglycine) buffer	1.8g
$Na_2S_2O_3 \cdot 5H_2O$	0.6g
NH_4Cl	0.5g
K_2HPO_4	0.24g
$MgSO_4 \cdot 7H_2O$	0.1g
Congo Red	5.0mg

pH 5.3 ± 0.2 at 25°C

Preparation of Medium: Add components to distilled/deionized water and bring volume to 1.0L. Mix thoroughly. Gently heat and bring to boiling. Adjust pH to 5.3. Distribute into tubes or flasks. Autoclave for 15 min at 15 psi pressure–121°C. Pour into sterile Petri dishes or leave in tubes.

Use: For the cultivation of *Yersinia* species with plasmids.

Congo Red Agar (CR Agar)

Composition per liter:

GC agar base	890.0mL
Hemoglobin solution	100.0mL
Supplement solution	10.0mL
Congo Red (0.01% solution)	0.1mL

pH 7.2 ± 0.2 at 25°C

GC Agar Base:

Composition per 890.0mL:

Agar	10.0g
Pancreatic digest of casein	7.5g
Peptic digest of animal tissue	7.5g
NaCl	5.0g
K_2HPO_4	4.0g
Cornstarch	1.0g
KH_2PO_4	1.0g

Preparation of GC Agar Base: Add components to distilled/deionized water and bring volume to 890.0mL. Mix thoroughly. Gently heat until boiling. Autoclave for 15 min at 15 psi pressure–121°C. Cool to 45°–50°C.

Hemoglobin Solution:

Composition per 100.0mL:

Hemoglobin	2.0g

Preparation of Hemoglobin Solution: Add hemoglobin to distilled/deionized water and bring volume to 100.0mL. Mix thoroughly. Autoclave for 15 min at 15 psi pressure–121°C. Cool to 50°C.

Congo Red Solution:

Composition per 100.0mL:

Congo Red	0.01g

Preparation of Congo Red Solution: Add Congo Red to 100.0mL of distilled/deionized water. Mix thoroughly. Autoclave for 15 min at 15 psi pressure–121°C.

Supplement Solution:

Composition per liter:

Glucose	100.0g
L-Cysteine·HCl	25.9g
L-Glutamine	10.0g
L-Cystine	1.1g
Adenine	1.0g
Nicotinamide adenine dinucleotide	0.25g
Vitamin B_{12}	0.1g
Thiamine pyrophosphate	0.1g
Guanine·HCl	0.03g
$Fe(NO_3)_3 \cdot 6H_2O$	0.02g
p-Aminobenzoic acid	0.013g
Thiamine·HCl	3.0mg

Source: The supplement solution IsoVitaleX enrichment is available from BD Diagnostic Systems.

This enrichment may be replaced by supplement VX from BD Diagnostic Systems.

Preparation of Supplement Solution: Add components to distilled/deionized water and bring volume to 1.0L. Mix thoroughly. Filter sterilize.

Preparation of Medium: To 890.0mL of sterile, cooled GC agar base aseptically add 100.0mL of sterile, cooled hemoglobin solution, 10.0mL of sterile supplement solution, and 0.1mL of sterile Congo Red solution. Mix thoroughly. Pour into sterile Petri dishes.

Use: For the isolation and differentiation of virulent and avirulent strains of *Shigella, Vibrio cholerae, Escherichia coli,* and *Neisseria meningitidis.* Used for the detection and differentiation of "iron-responsive" avirulent mutants. Used in the preparation of live vaccines. Used for the differentiation of sensitive *Neisseria gonorrhoeae* (no growth) from other *Neisseria* species (growth) that are resistant to Congo Red.

Congo Red Agar (CR Agar)

Composition per liter:

Soybean-casein digest agar	890.0mL
Hemoglobin solution	100.0mL
Supplement solution	10.0mL
Congo Red (0.01% solution)	0.1mL

pH 7.3 ± 0.2 at 25°C

Soybean-Casein Digest Agar:
Composition per 890.0mL:

Pancreatic digest of casein	17.0g
Agar	15.0g
NaCl	5.0g
Papaic digest of soybean meal	3.0g
Glucose	2.5g
K_2HPO_4	2.5g

Preparation of Soybean-Casein Digest Agar: Add components to distilled/deionized water and bring volume to 890.0mL. Mix thoroughly. Gently heat until boiling. Autoclave for 15 min at 15 psi pressure–121°C. Cool to 45°–50°C.

Hemoglobin Solution:
Composition per 100.0mL:

Hemoglobin	2.0g

Preparation of Hemoglobin Solution: Add hemoglobin to distilled/deionized water and bring volume to 100.0mL. Mix thoroughly. Autoclave for 15 min at 15 psi pressure–121°C. Cool to 50°C.

Congo Red Solution:
Composition per 100.0mL:

Congo Red	0.01g

Preparation of Congo Red Solution: Add Congo Red to 100.0mL of distilled/deionized water. Mix thoroughly. Autoclave for 15 min at 15 psi pressure–121°C.

Supplement Solution:
Composition per liter:

Glucose	100.0g
L-Cysteine·HCl	25.9g
L-Glutamine	10.0g
L-Cystine	1.1g
Adenine	1.0g
Nicotinamide adenine dinucleotide	0.25g
Vitamin B_{12}	0.1g
Thiamine pyrophosphate	0.1g
Guanine·HCl	0.03g
$Fe(NO_3)_3 \cdot 6H_2O$	0.02g
p-Aminobenzoic acid	0.013g
Thiamine·HCl	3.0mg

Preparation of Supplement Solution: Add components to distilled/deionized water and bring volume to 1.0L. Mix thoroughly. Filter sterilize.

Source: The supplement solution IsoVitaleX enrichment is available from BD Diagnostic Systems. This enrichment may be replaced by supplement VX from BD Diagnostic Systems.

Preparation of Medium: To 890.0mL of sterile, cooled soybean-casein digest agar, aseptically add 100.0mL of sterile, cooled hemoglobin solution, 10.0mL of sterile supplement solution, and 0.1mL of sterile Congo Red solution. Mix thoroughly. Pour into sterile Petri dishes.

Use: For the isolation and differentiation of virulent and avirulent strains of *Shigella, Vibrio cholerae, Escherichia coli,* and *Neisseria meningitidis.* Used for the detection and differentiation of "iron-responsive" avirulent mutants. Used in the preparation of live vaccines. Used for the differentiation of sensitive *Neisseria gonorrhoeae* (no growth) from other *Neisseria* species (growth) that are resistant to Congo Red.

Congo Red BHI Agarose Medium

Composition per liter:

Agarose	15.0g
Pancreatic digest of gelatin	14.5g
Brain heart, solids from infusion	6.0g
Peptic digest of animal tissue	6.0g
NaCl	5.0g
Glucose	3.0g
Na_2HPO_4	2.5g
Congo Red	0.075g

pH 7.4 ± 0.2 at 25°C

Preparation of Medium: Add components to distilled/deionized water and bring volume to 1.0L. Mix thoroughly. Gently heat and bring to boiling. Distribute into tubes or flasks. Autoclave for 15 min at 15 psi pressure–121°C. Pour into sterile Petri dishes in 20.0mL volumes.

Use: For the isolation, cultivation, and detection of virulent strains of *Yersinia enterocolitica.*

Congo Red BHI Agarose Medium (CRBHO Medium)

Composition per liter:

Agarose	12.0g
Pancreatic digest of gelatin	14.5g
Brain heart, solids from infusion	6.0g
Peptic digest of animal tissue	6.0g
NaCl	5.0g
Glucose	3.0g
Na_2HPO_4	2.5g
$MgCl_2$	1.0g
Congo Red solution	20.0mL

pH 7.4 ± 0.2 at 25°C

Preparation of Medium: Add components to distilled/deionized water and bring volume to 1.0L. Mix thoroughly. Gently heat and bring to boiling. Distribute into tubes or flasks. Autoclave for 15 min at 15 psi pressure–121°C. Pour into sterile Petri dishes in 20.0mL volumes.

Congo Red Solution:
Composition per 100.0mL:

Congo Red	375.0mg

Preparation of Congo Red Solution: Add Congo Red to 100.0mL of distilled/deionized water. Mix thoroughly. Autoclave for 15 min at 15 psi pressure–121°C. Cool to 25°C.

Use: For the isolation, cultivation, and detection of virulent strains of *Yersinia enterocolitica.*

Congo Red Magnesium Oxalate Agar (CRMOX Agar)

Composition per liter:

Solution 1	825.0mL
Solution 2	80.0mL
Solution 3	80.0mL
Solution 4	10.0mL
Solution 5	5.0mL

pH 7.3 ± 0.2 at 25°C

Solution 1:
Composition per 825.0mL:

Pancreatic digest of casein	15.0g
Agar	15.0g
Papaic digest of soybean meal	5.0g
NaCl	5.0g

pH 7.3 ± 0.2 at 25°C

Preparation of Solution 1: Add components to distilled/deionized water and bring volume to 825.0mL. Mix thoroughly. Gently heat and bring to boiling. Autoclave for 15 min at 15 psi pressure–121°C. Do not overheat.

Solution 2:
Composition per liter:

$MgCl_2 \cdot 6H_2O$	50.8g

Preparation of Solution 2: Add $MgCl_2 \cdot 6H_2O$ to distilled/deionized water and bring volume to 1.0L. Mix thoroughly. Autoclave for 15 min at 15 psi pressure–121°C.

Solution 3:
Composition per liter:

Sodium oxalate	33.2g

Preparation of Solution 3: Add sodium oxalate to distilled/deionized water and bring volume to 1.0L. Mix thoroughly. Autoclave for 15 min at 15 psi pressure–121°C.

Solution 4:
Composition per 100.0mL:

D-Galactose	20.0g

Preparation of Solution 4: Add D-galactose to distilled/deionized water and bring volume to 100.0mL. Mix thoroughly. Filter sterilize.

Solution 5:
Composition per 10.0mL:

Congo Red	0.1g

Preparation of Solution 5: Add Congo Red to distilled/deionized water and bring volume to 10.0mL. Mix thoroughly. Autoclave for 15 min at 15 psi pressure–121°C.

Preparation of Medium: Aseptically combine 80.0mL of sterile solution 2, 80.0mL of sterile solution 3, 10.0mL of sterile solution 4, and 5.0mL of sterile solution 5. Mix thoroughly. Warm to 50°C. Add this mixture to 825.0mL of cooled, sterile solution 1. Mix thoroughly. Pour into sterile Petri dishes.

Use: For the cultivation and identification of pathogenic serotypes of *Yersinia enterocolitica*. For the determination of whether *Yersinia* strains contain the *Yersinia* virulence plasmid.

Conradi Drigalski Agar

Composition per liter:

Agar	15.0g
Casein	10.0g

Lactose	10.0g
Peptone	10.0g
NaCl	5.0g
Bromcresol Purple	0.03g
Crystal Violet	4.0mg

pH 6.8 ± 0.2 at 25°C

Preparation of Medium: Add components to distilled/deionized water and bring volume to 1.0L. Mix thoroughly. Gently heat until boiling. Distribute into tubes or flasks. Autoclave for 15 min at 15 psi pressure–121°C. Pour into sterile Petri dishes or leave in tubes.

Use: For the isolation and cultivation of Gram-negative enteric bacilli.

Converse Liquid Medium, Levine Modification

Composition per liter:

Ionagar no. 2 or Noble agar	10.0g
Glucose	4.0g
Ammonium acetate	1.23g
K_2HPO_4	0.52g
Tamol	0.5g
$MgSO_4 \cdot 7H_2O$	0.4g
KH_2PO_4	0.4g
NaCl	0.014g
Na_2CO_3	0.012g
$CaCl_2 \cdot 2H_2O$	0.002g
$ZnSO_4 \cdot 7H_2O$	0.002g

Preparation of Medium: Add components to distilled/deionized water and bring volume to 1.0L. Mix thoroughly. Gently heat and bring to boiling. Autoclave for 15 min at 15 psi pressure–121°C. Pour into sterile Petri dishes in 15.0mL volumes.

Use: For the cultivation and induction of spherules of *Coccidioides immitis*.

Cooke Rose Bengal Agar

Composition per liter:

Agar	20.0g
Glucose	10.0g
Enzymatic hydrolysate of soybean meal	5.0g
KH_2PO_4	1.0g
$MgSO_4 \cdot 7H_2O$	0.5g
Rose Bengal	35.0mg

pH 6.0 ± 0.2 at 25°C

Source: This medium is available as a premixed powder from BD Diagnostic Systems.

Preparation of Medium: Add components to distilled/deionized water and bring volume to 1.0L. Mix thoroughly. Gently heat until boiling. Distribute into tubes or flasks. Autoclave for 15 min at 15 psi pressure–121°C. Pour into sterile Petri dishes or leave in tubes.

Use: For the isolation of fungi.

Cooked Meat Medium (LMG Medium 140)

Composition per liter:

Heart muscle	454.0g
Peptone	40.0g
Beef extract	10.0g
NaCl	5.0g
Yeast extract	5.0g
K_2HPO_4	5.0g
Glucose	2.0g
Resazurin solution	4.0mL

pH 7.0 ± 0.2 at 25°C

Resazurin Solution:

Composition per 100.0mL:

Resazurin	0.025g

Preparation of Resazurin Solution: Add resazurin to distilled/deionized water and bring volume to 100.0mL. Mix thoroughly.

Preparation of Medium: Finely chop beef heart. Add approximately 1.5g of heart particles to test tubes. Add remaining components to distilled/deionized water and bring volume to 1.0L. Mix thoroughly. Distribute into tubes in 10.0mL volumes. Autoclave for 15 min at 15 psi pressure–121°C. Slowly cool tubes to prevent expulsion of meat particles.

Use: For the cultivation and maintenance of *Peptostreptococcus magnus*.

Cooked Meat Medium

Composition per liter:

Beef heart	454.0g
Proteose peptone	20.0g
NaCl	5.0g
Glucose	2.0g

pH 7.2 ± 0.2 at 25°C

Source: This medium is available as a premixed powder from BD Diagnostic Systems.

Preparation of Medium: Finely chop beef heart. Add approximately 1.5g of heart particles to test tubes. Add remaining components to distilled/deionized water and bring volume to 1.0L. Mix thoroughly. Distribute into tubes in 10.0mL volumes. Autoclave for 15 min at 15 psi pressure–121°C. Slowly cool tubes to prevent expulsion of meat particles.

Use: For the cultivation and maintenance of anaerobic microorganisms.

Cooked Meat Medium

Composition per liter:

Heart muscle	454.0g
Beef extract	10.0g
Peptone	10.0g
NaCl	5.0g
Glucose	2.0g

pH 7.2 ± 0.2 at 25°C

Source: This medium is available as a premixed powder from Oxoid Unipath.

Preparation of Medium: Finely chop beef heart. Add approximately 1.5g of heart particles to test tubes. Add remaining components to distilled/deionized water and bring volume to 1.0L. Mix thoroughly. Distribute into tubes in 10.0mL volumes. Autoclave for 15 min at 15 psi pressure–121°C. Slowly cool tubes to prevent expulsion of meat particles.

Use: For the cultivation and maintenance of aerobic and anaerobic microorganisms. For the cultivation of anaerobes, especially pathogenic clostridia.

Cooked Meat Medium

Composition per liter:

Heart tissue granules	98.0g
Peptic digest of animal tissue	20.0g
NaCl	5.0g
Glucose	2.0g

pH 7.2 ± 0.2 at 25°C

Source: This medium is available as a premixed powder from BD Diagnostic Systems.

Preparation of Medium: Add approximately 1.0g of heart tissue granules to test tubes. Add remaining components to distilled/deionized water and bring volume to 1.0L. Mix thoroughly. Distribute into tubes in 10.0mL volumes. Autoclave for 15 min at 15 psi pressure–121°C. Slowly cool tubes to prevent expulsion of meat particles.

Use: For the cultivation of anaerobes, especially pathogenic clostridia.

Cooked Meat Medium with Glucose, Hemin, and Vitamin K

Composition per liter:

Heart tissue granules	98.0g
Peptic digest of animal tissue	20.0g
NaCl	5.0g
Glucose	5.0g
Yeast extract	5.0g
Hemin	5.0mg
Vitamin K	1.0mg

pH 7.2 ± 0.2 at 25°C

Source: This medium is available as a premixed powder from BD Diagnostic Systems.

Preparation of Medium: Add approximately 1.0g of heart tissue granules to test tubes. Add remaining components to distilled/deionized water and bring volume to 1.0L. Mix thoroughly. Distribute into tubes in 10.0mL volumes. Autoclave for 15 min at 15 psi pressure–121°C. Slowly cool tubes to prevent expulsion of meat particles.

Use: For the cultivation of anaerobes, especially pathogenic clostridia.

Cooked Meat Medium, Modified

Composition per tube:

Cooked meat medium	1.0g
Diluent	1.0L

pH 6.8 ± 0.2 at 25°C

Cooked Meat Medium:

Composition per 481g:

Beef heart	454.0g
Proteose peptone	20.0g
NaCl	5.0g
Glucose	2.0g

Source: Cooked meat medium is available in dehydrated form from BD Diagnostic Systems.

Diluent:

Composition per liter:

Pancreatic digest of casein	10.0g
Glucose	2.0g
Soluble starch	1.0g
Sodium thioglycolate	1.0g
Neutral Red (1% aqueous)	5.0mL

Preparation of Diluent: Add components to distilled/deionized water and bring volume to 1.0L. Mix thoroughly. Gently heat until dissolved.

Preparation of Medium: Add 1.0g of dehydrated cooked meat medium and 15.0mL diluent to 20 × 150mm test tubes. Let meat particles rehydrate. Gently heat and bring to boiling. Autoclave for 15 min at 15 psi pressure–121°C.

Use: For the cultivation of anaerobic bateria.

Cooked Meat Medium, Modified

Composition per liter:

Cooked meat medium	66.0g
Solution A	1.0L

pH 6.8 ± 0.2 at 25°C

Solution A:

Composition per liter:

Pancreatic digest of casein	10.0g
Glucose	2.0g
Soluble starch	1.0g

Sodium thioglycolate 1.0g
Neutral Red (1% aqueous) 5.0mL

Preparation of Solution A: Add components to distilled/deionized water and bring volume to 1.0L. Mix thoroughly. Gently heat until dissolved.

Preparation of Medium: Add 1.0g of cooked meat medium to each of 66 test tubes. Add 15.0mL of solution A to each test tube. Allow meat particles to rehydrate. Autoclave for 15 min at 15 psi pressure–121°C.

Use: For the cultivation of a variety of anaerobic microorganisms.

Cooked Meat Medium with Peptone and Yeast Extract

Composition per liter:

Heart muscle 454.0g
Peptone 40.0g
Beef extract 10.0g
NaCl 5.0g
Yeast extract 5.0g
Glucose 2.0g
pH 7.2 ± 0.2 at 25°C

Preparation of Medium: Finely chop beef heart. Add approximately 1.5g of heart particles to test tubes. Add remaining components to distilled/deionized water and bring volume to 1.0L. Mix thoroughly. Distribute into tubes in 10.0mL volumes. Autoclave for 15 min at 15 psi pressure–121°C. Slowly cool tubes to prevent expulsion of meat particles.

Use: For the cultivation and maintenance of *Peptostreptococcus magnus.*

Cornmeal Agar (ATCC Medium 307)

Composition per liter:

Cornmeal 50.0g
Agar 7.5g

Preparation of Medium: Add cornmeal to distilled/deionized water and bring volume to 800.0mL. Leave overnight in refrigerator. Heat to 60°C for 1h. Bring volume to 1.0L with distilled/deionized water. Add agar. Gently heat and bring to boiling. Autoclave for 15 min at 15 psi pressure–121°C. Pour into sterile Petri dishes or distribute into sterile tubes.

Use: For the cultivation and maintenance of numerous fungi.

Cornmeal Agar (CMA)

Composition per liter:

Agar 20.0g
Cornmeal polenta 15.0g
pH 7.0 ± 0.2 at 25°C

Preparation of Medium: Add cornmeal polenta to distilled/deionized water and bring volume to 1.0L. Mix thoroughly. Gently heat and bring to boiling. Continue boiling for 30 min. Filter through Whatman #1 filter paper. Add agar to filtrate. Gently heat and bring to boiling. Distribute into tubes or flasks. Autoclave for 10 min at 15 psi pressure–121°C. Pour into sterile Petri dishes or leave in tubes.

Use: For the cultivation and maintenance of many filamentous fungi.

Cornmeal Agar

Composition per liter:

Agar 15.0g
Cornmeal, solids from infusion 2.0g
pH 5.6–6.0 at 25°C

Source: This medium is available as a premixed powder from BD Diagnostic Systems and Oxoid Unipath.

Preparation of Medium: Add components to distilled/deionized water and bring volume to 1.0L. Mix thoroughly. Gently heat until boiling. Distribute into tubes or flasks. Autoclave for 15 min at 15 psi pressure–121°C. Pour into sterile Petri dishes or leave in tubes.

Use: For the cultivation and maintenance of fungi.

Cornmeal Agar with Strep100 and Tet100 (ATCC Medium 2285)

Composition per liter:

Agar 15.0g
Cornmeal, solids from infusion 2.0g
Antibiotic solution 10.0mL
pH 5.6–6.0 at 25°C

Source: This medium without antibiotics is available as a premixed powder from BD Diagnostic Systems and Oxoid Unipath.

Preparation of Medium: Add components except antibiotic solution to 990.0mL distilled/deionized water. Mix thoroughly. Gently heat until boiling. Distribute into tubes or flasks. Autoclave for 15 min at 15 psi pressure–121°C. Cool to 45-50°C. Aseptically add 10.0mL sterile antibiotic solution. Mix thoroughly. Pour into sterile Petri dishes or leave in tubes.

Antibiotic Solution:

Composition per 10.0mL:

Tetracycline 0.1g
Streptomycin sulfate 0.1g

Preparation of Antibiotic Solution: Add components to distilled/deionized water and bring volume to 10.0mL. Mix thoroughly. Filter sterilize.

Use: For the cultivation and maintenance of fungi.

Cornmeal Yeast Glucose Agar (CMYG)

Composition per liter:

Agar	15.0g
Cornmeal, solids from infusion	2.0g
Glucose	2.0g
Yeast extract	1.0g

pH 5.6–6.0 at 25°C

Preparation of Medium: Add components to distilled/deionized water and bring volume to 1.0L. Mix thoroughly. Gently heat until boiling. Distribute into tubes or flasks. Autoclave for 15 min at 15 psi pressure–121°C. Pour into sterile Petri dishes or leave in tubes.

Use: For the cultivation of numerous filamentous fungi.

CREA (Creatine Agar)

Composition per liter:

Sucrose	30.0g
Agar	15.0g
Creatin·H_2O	3.0g
$K_3PO_4 \cdot 7H_2O$	1.6g
Bromcresol Purple	50.0mg
Minerals solution	10.0mL
Trace minerals solution	1.0mL

Minerals Solution:

Composition per 100.0mL:

KCl	5.0g
$MgSO_4 \cdot 7H_2O$	5.0g
$FeSO_4 \cdot 7H_2O$	0.1g

Preparation of Minerals Solution: Add components to distilled/deionized water and bring volume to 100.0mL. Mix thoroughly.

Trace Minerals Solution:

Composition per 100.0mL:

$ZnSO_4 \cdot 7H_2O$	1.0g
$CuSO_4 \cdot 5H_2O$	0.5g

Preparation of Trace Minerals Solution: Add components to distilled/deionized water and bring volume to 100.0mL. Mix thoroughly.

Preparation of Medium: Add components to distilled/deionized water and bring volume to 1.0L. Mix thoroughly. Gently heat and bring to boiling. Distribute into tubes or flasks. Autoclave for 15 min at 15 psi pressure–121°C. Pour into sterile Petri dishes or leave in tubes.

Use: For the cultivation and maintenance of *Penicillium* species.

Creatinine Agar

Composition per liter:

Agar	15.0g
$Na_2HPO_4 \cdot 12H_2O$	9.0g
NaCl	5.0g
KH_2PO_4	1.5g
Creatinine	1.0g
Meat extract	1.0g
Yeast extract	1.0g
$MgSO_4 \cdot 7H_2O$	0.2g
$MnCl_2 \cdot 4H_2O$	20.0mg
$CaCl_2$	1.2mg
Glucose solution	100.0mL

Glucose Solution:

Composition per 100.0mL:

Glucose	5.0g

Preparation of Glucose Solution: Add glucose to distilled/deionized water and bring volume to 100.0mL. Mix thoroughly. Filter sterilize. Warm to 50°C.

Preparation of Medium: Add components, except glucose solution, to distilled/deionized water and bring volume to 900.0mL. Mix thoroughly. Gently heat and bring to boiling. Autoclave for 15 min at 15 psi pressure–121°C. Cool to 50°–55°C. Aseptically add 100.0mL of sterile glucose solution. Mix thoroughly. Pour into sterile Petri dishes or distribute into sterile tubes.

Use: For the cultivation and maintenance of *Pseudomonas* species and other bacteria that can utilize creatinine.

Creatinine Medium

Composition per liter:

Creatinine	5.0g
Agar	2.0g
Fumaric acid	2.0g
K_2HPO_4	2.0g
Yeast extract	1.0g
Salt solution	10.0mL

pH 6.8 ± 0.2 at 25°C

Salt Solution:

Composition per liter:

$MgSO_4$	12.2g
$FeSO_4 \cdot 7H_2O$	2.8g
$MnSO_4 \cdot H_2O$	1.7g
$CaCl_2 \cdot 2H_2O$	0.76g
NaCl	0.6g
$Na_2MoO_4 \cdot 2H_2O$	0.1g
$ZnSO_4 \cdot 7H_2O$	0.06g
HCl (0.1*N* solution)	1.0L

Preparation of Salt Solution: Dissolve salts in 1.0L of 0.1*N* HCl solution. Mix thoroughly.

Preparation of Medium: Add components to distilled/deionized water and bring volume to 1.0L. Mix thoroughly. Adjust pH to 6.8 with NaOH or KOH. Gently heat until boiling. Distribute into tubes or flasks. Autoclave for 15 min at 15 psi pressure–121°C.

Use: For the cultivation and maintenance of *Pseudomonas* species.

CreDm1 Medium

Composition per 1002.0mL:

Solution A	980.0mL
Solution D (Vitamin solution)	10.0mL
Solution E	10.0mL
Solution B (Trace elements solution SL-10)	1.0mL
Solution C (Selentite-tungstate solution)	1.0mL

pH 6.7–6.9 at 25°C

Solution A:

Composition per 980.0mL:

KH_2PO_4	1.4g
NH_4Cl	0.5g
$MgCl_2 \cdot 6H_2O$	0.2g
$CaCl_2 \cdot 2H_2O$	0.15g
Yeast extract	50.0mg

Preparation of Solution A: Add components to distilled/deionized water and bring volume to 980.0mL. Mix thoroughly. Autoclave for 15 min at 15 psi pressure–121°C.

Solution B (Trace Elements Solution SL-10):

Composition per liter:

$FeCl_2 \cdot 4H_2O$	1.5g
$CoCl_2 \cdot 6H_2O$	190.0mg
$MnCl_2 \cdot 4H_2O$	100.0mg
$ZnCl_2$	70.0mg
$Na_2MoO_4 \cdot 2H_2O$	36.0mg
$NiCl_2 \cdot 6H_2O$	24.0mg
H_3BO_3	6.0mg
$CuCl_2 \cdot 2H_2O$	2.0mg
HCl (25% solution)	10.0mL

Preparation of Solution B (Trace Elements Solution SL-10): Add $FeCl_2 \cdot 4H_2O$ to 10.0mL of HCl solution. Mix thoroughly. Add distilled/deionized water and bring volume to 1.0L. Add remaining components. Mix thoroughly. Autoclave for 15 min at 15 psi pressure–121°C.

Solution C (Selenite-Tungstate Solution):

Composition per liter:

NaOH	0.5g
$Na_2WO_4 \cdot 2H_2O$	4.0mg
$Na_2SeO_3 \cdot 5H_2O$	3.0mg

Preparation of Solution C (Selenite-Tungstate Solution): Add components to distilled/deionized water and bring volume to 1.0L. Mix thoroughly. Autoclave for 15 min at 15 psi pressure–121°C.

Solution D (Vitamin Solution):

Composition per liter:

Pyridoxine·HCl	10.0mg
Calcium DL-pantothenate	5.0mg
Lipoic acid	5.0mg
Nicotinic acid	5.0mg
p-Aminobenzoic acid	5.0mg
Riboflavin	5.0mg
Thiamine·HCl	5.0mg
Biotin	2.0mg
Folic acid	2.0mg
Vitamin B_{12}	0.1mg

Preparation of Solution D (Vitamin Solution): Add components to distilled/deionized water and bring volume to 1.0L. Mix thoroughly. Filter sterilize.

Solution E:

Composition per 10.0mL:

Disodium-DL-malate	1.6g

Preparation of Solution E: Add disodium-DL-malate to distilled/deionized water and bring volume to 10.0mL. Mix thoroughly. Autoclave for 15 min at 15 psi pressure–121°C.

Preparation of Medium: Aseptically combine 980.0mL of sterile solution A with 1.0mL of sterile solution B, 1.0mL of sterile solution C, 10.0mL of sterile solution D, and 10.0mL of sterile solution E, in that order. Mix thoroughly. Adjust pH to 6.7–6.9. Aseptically distribute into sterile tubes or flasks.

Use: For the cultivation of *Campylobacter* species.

Crystal Violet Agar

Composition per liter:

Agar	15.0g
Lactose	10.0g
Proteose peptone	5.0g
Beef extract	3.0g
Crystal Violet	3.3mg

pH 6.8 ± 0.1 at 25°C

Preparation of Medium: Add components to distilled/deionized water and bring volume to 1.0L. Mix thoroughly. Gently heat until boiling. Distribute into tubes or flasks. Autoclave for 15 min at 15 psi pressure–121°C. Pour into sterile Petri dishes or leave in tubes.

Use: For the differentiation of pathogenic staphylococci from non-pathogenic staphylococci. Hemolytic and coagulating strains of *Staphylococcus aureus* appear as purple or yellow colonies. Nonhemolytic and noncoagulating strains of *Staphylococcus* species appear as white colonies.

Crystal Violet Azide Esculin Agar

Composition per liter:

Agar	15.0g
Glucose	5.0g
NaCl	5.0g
Proteose peptone	5.0g
Pancreatic digest of casein	5.0g
Meat extract	3.0g
Esculin	1.0g
NaN_3	1.0g
Crystal Violet	0.1g
Bovine blood, citrated	100.0mL

pH 7.5 ± 0.2 at 25°C

Caution: Sodium azide is toxic. Azides also react with metals and disposal must be highly diluted.

Preparation of Medium: Add components, except citrated bovine blood, to distilled/deionized water and bring volume to 900.0mL. Mix thoroughly. Gently heat and bring to boiling. Autoclave for 15 min at 15 psi pressure–121°C. Cool to 45°–50°C. Aseptically add sterile, citrated bovine blood. Mix thoroughly. Pour into sterile Petri dishes or distribute into sterile tubes.

Use: For the cultivation of *Erysipelothrix rhusiopathiae.*

Crystal Violet Esculin Agar

Composition per liter:

Agar	15.0g
Glucose	5.0g
NaCl	5.0g
Proteose peptone	5.0g
Pancreatic digest of casein	5.0g
Meat extract	3.0g
Esculin	1.0g
Crystal Violet	2.0mg
Blood, citrated	100.0mL

pH 7.5 ± 0.2 at 25°C

Preparation of Medium: Add components, except citrated blood, to distilled/deionized water and bring volume to 900.0mL. Mix thoroughly. Gently heat and bring to boiling. Autoclave for 15 min at 15 psi pressure–121°C. Cool to 45°–50°C. Aseptically add citrated blood. Mix thoroughly. Pour into sterile Petri dishes or distribute into sterile tubes.

Use: For the cultivation of *Erysipelothrix rhusiopathiae.*

CT Agar (Caprylate Thallous Agar)

Composition per liter:

Solution A	500.0mL
Solution B	500.0mL

pH 7.2 ± 0.2 at 25°C

Solution A:

Composition per 500.0mL:

K_2HPO_4	2.61g
KH_2PO_4	0.68g
Thallous sulfate	0.25g
$MgSO_4 \cdot 7H_2O$	0.12g
$CaCl_2 \cdot 2H_2O$	0.016g
Trace elements solution	10.0mL
Yeast extract	2.0mL
Caprylic acid	1.1mL

Preparation of Solution A: Add components to distilled/deionized water and bring volume to 500.0mL. Mix thoroughly. Adjust pH to 7.2 with NaOH. Autoclave for 20 min at 10 psi pressure–115°C.

Trace Elements Solution:

Composition per liter:

H_3PO_4	1.96g
$FeSO_4 \cdot 7H_2O$	0.056g
$ZnSO_4 \cdot 4H_2O$	0.029g
$CuSO_4 \cdot 5H_2O$	0.025g
$MnSO_4 \cdot 4H_2O$	0.022g
H_3BO_3	6.2mg
$Co(NO_3)_2 \cdot 6H_2O$	3.0mg

Preparation of Trace Elements Solution: Add components to distilled/deionized water and bring volume to 1.0L. Mix thoroughly. Store at 4°C.

Solution B:

Composition per liter:

Agar	15.0g
NaCl	7.0g
$(NH_4)_2SO_4$	1.0g

Preparation of Solution B: Add components to distilled/deionized water and bring volume to 500.0mL. Mix thoroughly. Gently heat and bring to boiling. Adjust pH to 7.2. Autoclave for 20 min at 10 psi pressure–115°C.

Preparation of Medium: Aseptically combine 500.0mL of sterile solution A and 500.0mL of sterile solution B. Mix thoroughly. Pour into sterile Petri dishes in 25.0–30.0mL volumes.

Use: For the isolation and cultivation of the *Serratia* species.

CTA Agar
(Cystine Trypticase Agar)

Composition per liter:

Pancreatic digest of casein	20.0g
Agar	14.0g
NaCl	5.0g
L-Cystine	0.5g
Na_2SO_3	0.5g
Phenol Red	0.017g

pH 7.3 ± 0.2 at 25°C

Preparation of Medium: Add components to distilled/deionized water and bring volume to 1.0L. Mix thoroughly. Gently heat until boiling. Distribute into tubes or flasks. Autoclave for 15 min at 15 psi pressure–118°C. Pour into sterile Petri dishes or leave in tubes. Two drops of sterile rabbit serum added per tube prior to solidification enhance the recovery of *Corynebacterium diphtheriae.*

Use: For the cultivation and maintenance of a variety of fastidious microorganisms, including *Corynebacterium diphtheriae.* For carbohydrate fermentation tests in the differentiation of *Neisseria* species.

CTA Medium
(Cystine Trypticase Agar Medium)
(Cystine Tryptic Agar)

Composition per liter:

Pancreatic digest of casein	20.0g
NaCl	5.0g
Carbohydrate	5.0g
Agar	2.5g
L-Cystine	0.5g
Na_2SO_3	0.5g
Phenol Red	0.017g

pH 7.3 ± 0.2 at 25°C

Source: The medium is available as a premixed powder from BD Diagnostic Systems.

Preparation of Medium: Add components to distilled/deionized water and bring volume to 1.0L. Mix thoroughly. Adjust pH to 7.3. Gently heat until boiling. Distribute into tubes or flasks. Autoclave for 15 min at 15 psi pressure–118°C. Cool tubes in an upright position. Store at room temperature.

Use: For the cultivation and maintenance of a variety of fastidious microorganisms. For the detection of bacterial motility. Used, with added specific carbohydrate, for fermentation reactions of fastidious microorganisms, especially *Neisseria* species, pneumococci, streptococci, and nonspore-forming anaerobes.

CTA Medium with Yeast Extract and Rabbit Serum
(Cystine Trypticase Agar Medium with Yeast Extract and Rabbit Serum)

Composition per liter:

Yeast extract	50.0g
Pancreatic digest of casein	20.0g
NaCl	5.0g
Carbohydrate	5.0g
Agar	2.5g
L-Cystine	0.5g
Na_2SO_3	0.5g
Phenol Red	0.017g
Rabbit serum	250.0mL

pH 7.3 ± 0.2 at 25°C

Preparation of Medium: Add components, except rabbit serum, to distilled/deionized water and bring volume to 750.0mL. Mix thoroughly. Adjust pH to 7.3. Gently heat until boiling. Autoclave for 15 min at 15 psi pressure–118°C. Cool to 50°C. Aseptically add sterile rabbit serum. Mix thoroughly. Distribute into sterile tubes. Store at room temperature. Do not refrigerate.

Use: For the cultivation and maintenance of fastidious microorganisms, especially mycoplasmas and related microorganisms.

CTLM Medium

Composition per 1100.0mL:

Beef liver, infusion from	125.0g
Tryptose	25.0g
Proteose peptone	2.5g
L-Cysteine·HCl	1.75g
Maltose	1.25g
NaCl	1.25g
Agar	1.15g
L-Ascorbic acid	0.25g
$NaHCO_3$	0.075g
Horse serum, heat inactivated	100.0mL
Ringer's salt solution, 10×	75.0mL

pH 6.0 ± 0.2 at 25°C

Ringer's Salt Solution, 10×:
Composition per 100.0mL:

NaCl	9.0g
KCl	0.42g
$CaCl_2$	0.24g

Preparation of Ringer's Salt Solution, 10×: Add components to distilled/deionized water and bring volume to 100.0mL. Mix thoroughly.

Preparation of Medium: Add components, except horse serum, to distilled/deionized water and bring volume to 1.0L. Mix thoroughly. Adjust pH to

6.0. Gently heat and bring to boiling. Autoclave for 25 min at 15 psi pressure–121°C. Cool to 25°C. Aseptically add 100.0mL of sterile, heat-inactivated horse serum. Mix thoroughly. Aseptically distribute into sterile, screw-capped tubes or flasks.

Use: For the cultivation of *Trichomonas vaginalis*.

CVA Medium (Cefoperazone Vancomycin Amphotericin Medium)

Composition per liter:

Agar	15.0g
Casein peptone	10.0g
Meat peptone	10.0g
NaCl	5.0g
Yeast autolysate	2.0g
Glucose	1.0g
$NaHSO_3$	0.1g
Sheep blood, defibrinated	50.0mL
CVA antibiotic solution	10.0mL

pH 7.0 ± 0.2 at 25°C

CVA Antibiotic Solution:

Composition per 10.0mL:

Cefoperazone	20.0mg
Vancomycin	10.0mg
Amphotericin B	2.0mg

Preparation of CVA Antibiotic Solution: Add components to distilled/deionized water and bring volume to 10.0mL. Mix thoroughly. Filter sterilize.

Preparation of Medium: Add components, except CVA antibiotic solution and sheep blood, to distilled/deionized water and bring volume to 940.0mL. Mix thoroughly. Gently heat until boiling. Autoclave for 15 min at 15 psi pressure–121°C. Cool to 45°–50°C. Aseptically add sterile CVA antibiotic solution and sterile, defibrinated sheep blood. Mix thoroughly. Pour into sterile Petri dishes.

Use: For the isolation and cultivation of *Campylobacter* species from clinical specimens.

Cycloserine Cefoxitin Egg Yolk Fructose Agar

Composition per liter:

Proteose peptone No. 2	40.0g
Agar	25.0g
Fructose	6.0g
Na_2HPO_4	5.0g
NaCl	2.0g
KH_2PO_4	1.0g
$MgSO_4 \cdot 7H_2O$	0.1g
Egg yolk emulsion	100.0mL
Antibiotic solution	10.0mL
Neutral Red solution	3.0mL
Hemin solution	1.0mL

Egg Yolk Emulsion:

Composition:

Chicken egg yolks	11
Whole chicken egg	1

Preparation of Egg Yolk Emulsion: Soak eggs with 1:100 dilution of saturated mercuric chloride solution for 1 min. Crack eggs. Separate yolks from whites for 11 eggs. Mix egg yolks with 1 chicken egg.

Antibiotic Solution:

Composition per 10.0mL:

Cycloserine	0.5g
Cefoxitin	0.016g

Preparation of Antibiotic Solution: Add components to distilled/deionized water and bring volume to 10.0mL. Mix thoroughly. Filter sterilize.

Neutral Red Solution:

Composition per 10.0mL:

Neutral Red	0.1g
Ethanol	10.0mL

Preparation of Neutral Red Solution: Add Neutral Red to 10.0mL of ethanol. Mix thoroughly.

Hemin Solution:

Composition per 100.0mL:

Hemin	0.5g
NaOH (1*N* solution)	10.0mL

Preparation of Hemin Solution: Add hemin to 10.0mL of 1*N* NaOH solution. Mix thoroughly. Bring volume to 100.0mL with distilled/deionized water.

Preparation of Medium: Add components, except egg yolk emulsion and antibiotic solution, to distilled/deionized water and bring volume to 890.0mL. Mix thoroughly. Gently heat and bring to boiling. Autoclave for 15 min at 15 psi pressure–121°C. Cool to 45°–50°C. Aseptically add sterile egg yolk emulsion and antibiotic solution. Mix thoroughly. Pour into sterile Petri dishes.

Use: For the selective isolation and cultivation of *Clostridium difficile* from feces.

CYE Agar (Charcoal Yeast Extract Agar)

Composition per liter:

Agar	17.0g
Yeast extract	10.0g
Charcoal, activated, acid-washed	2.0g
L-Cysteine·HCl·H_2O solution	10.0mL
$Fe_4(P_2O_7)_3$ solution	10.0mL

pH 6.9 ± 0.5 at 50°C

L-Cysteine·HCl·H_2O Solution:
Composition per 10.0mL:

L-Cysteine·HCl·H_2O	0.4g

Preparation of L-Cysteine·HCl·H_2O solution: Add L-cysteine·HCl·H_2O to distilled/deionized water and bring volume to 10.0mL. Mix thoroughly. Filter sterilize.

$Fe_4(P_2O_7)_3$ Solution:
Composition per liter:

$Fe_4(P_2O_7)_3$	0.25g

Preparation of $Fe_4(P_2O_7)_3$ Solution: Add soluble $Fe_4(P_2O_7)_3$ to distilled/deionized water and bring volume to 10.0mL. Mix thoroughly. Filter sterilize. The soluble $Fe_4(P_2O_7)_3$ must be kept dry and in the dark. Do not use if brown or yellow. Prepare solutions freshly. Do not heat over 60°C to dissolve. The mixture dissolves readily in a 50°C water bath.

Preparation of Medium: Add components, except L-cysteine·HCl·H_2O solution and $Fe_4(P_2O_7)_3$ solution, to distilled/deionized water and bring volume to 980.0mL. Mix thoroughly. Gently heat to boiling. Autoclave for 15 min at 15 psi pressure–121°C. Cool to 50°C. Add 10.0mL of sterile L-cysteine·HCl·H_2O solution and 10.0mL of sterile $Fe_4(P_2O_7)_3$ solution. Adjust pH to 6.9 at 50°C by adding 4.0–4.5mL of 1.0*N* KOH. This is a critical step. Mix thoroughly. Pour in 20.0mL volumes into sterile Petri dishes. Swirl medium while pouring to keep charcoal in suspension.

Use: For the cultivation and maintenance of *Legionella* species.

CYE Agar, Buffered
(Charcoal Yeast Extract Agar, Buffered)

Composition per liter:

Agar	17.0g
ACES buffer (*N*-2-acetamido-2-aminoethane sulfonic acid)	10.0g
Yeast extract	10.0g
Charcoal, activated, acid-washed	2.0g
L-Cysteine·HCl·H_2O solution	10.0mL
$Fe_4(P_2O_7)_3$ solution	10.0mL

pH 6.9 ± .05 at 50°C

L-Cysteine·HCl·H_2O Solution:
Composition per 10.0mL:

L-Cysteine·HCl·H_2O	0.4g

Preparation of L-Cysteine·HCl·H_2O Solution: Add L-cysteine·HCl·H_2O to distilled/deionized water and bring volume to 10.0mL. Mix thoroughly. Filter sterilize.

$Fe_4(P_2O_7)_3$ Solution:
Composition per liter:

$Fe_4(P_2O_7)_3$	0.25g

Preparation of $Fe_4(P_2O_7)_3$ Solution: Add soluble $Fe_4(P_2O_7)_3$ to distilled/deionized water and bring volume to 10.0mL. Mix thoroughly. Filter sterilize. The soluble $Fe_4(P_2O_7)_3$ must be kept dry and in the dark. Do not use if brown or yellow. Prepare solutions freshly. Do not heat over 60°C to dissolve. The mixture dissolves readily in a 50°C water bath.

Preparation of Medium: Add components, except L-cysteine·HCl·H_2O solution and $Fe_4(P_2O_7)_3$ solution, to distilled/deionized water and bring volume to 980.0mL. Mix thoroughly. Gently heat to boiling. Autoclave for 15 min at 15 psi pressure–121°C. Cool to 50°C. Add 10.0mL of sterile L-cysteine·HCl·H_2O solution and 10.0mL of sterile $Fe_4(P_2O_7)_3$ solution. Adjust pH to 6.9 at 50°C by adding 4.0–4.5mL of 1.0 *N* KOH. This is a critical step. Mix thoroughly. Pour in 20.0mL volumes into sterile Petri dishes. Swirl medium while pouring to keep charcoal in suspension.

Use: For the cultivation and maintenance of *Legionella* species.

Cystine Heart Agar

Composition per liter:

Beef heart, solids from infusion	500.0g
Agar	15.0g
Glucose	10.0g
Proteose peptone	10.0g
NaCl	5.0g
L-Cystine	1.0g
Hemoglobin solution	100.0mL

pH 6.8 ± 0.2 at 25°C

Source: This medium is available as a premixed powder from BD Diagnostic Systems.

Hemoglobin Solution:
Composition per 100.0mL:

Hemoglobin	2.0g

Preparation of Hemoglobin Solution: Add hemoglobin to cold distilled/deionized water and bring volume to 100.0mL. Mix thoroughly by shaking for 10–15 min. Autoclave for 15 min at 15 psi pressure–121°C. Cool to 50°–60°C.

Preparation of Medium: Add components, except hemoglobin solution, to distilled/deionized water and bring volume to 900.0mL. Mix thoroughly. Gently heat until boiling. Autoclave for 15 min at 15 psi pressure–121°C. Cool to 50–60°C. Aseptically add 100.0mL of sterile cooled hemoglobin solution. Mix thoroughly. Pour into sterile Petri dishes or distribute into sterile tubes.

Use: For the cultivation and maintenance of *Francisella tularensis* and *Francisella philomiragia*. Without the hemoglobin enrichment, it supports excellent growth of Gram-negative cocci and other pathogenic microorganisms.

Cystine Heart Agar with Rabbit Blood

Composition per liter:

Beef heart, solids from infusion	500.0g
Agar	15.0g
Glucose	10.0g
Proteose peptone	10.0g
NaCl	5.0g
L-Cystine	1.0g
Rabbit blood, defibrinated	50.0mL

pH 6.8 ± 0.2 at 25°C

Source: This medium is available as a premixed powder from BD Diagnostic Systems.

Preparation of Medium: Add components, except rabbit blood, to distilled/deionized water and bring volume to 950.0mL. Mix thoroughly. Gently heat until boiling. Autoclave for 15 min at 15 psi pressure–121°C. Cool to 50°–60°C. Aseptically add 50.0mL of sterile, defibrinated rabbit blood. Mix thoroughly. Pour into sterile Petri dishes or distribute into sterile tubes.

Use: For the cultivation and maintenance of *Francisella tularensis* and *Francisella philomiragia.* Without the hemoglobin enrichment, it supports excellent growth of Gram-negative cocci and other pathogenic microorganisms.

Cystine Tellurite Blood Agar

Composition per liter:

Heart infusion agar	900.0mL
K_2TeO_3 solution	75.0mL
Rabbit blood	25.0mL
L-Cystine	22.0mg

pH 7.4 ± 0.2 at 25°C

Heart Infusion Agar:

Composition per 900.0mL:

Beef heart, solids from infusion	500.0g
Agar	20.0g
Tryptose	10.0g
Yeast extract	5.0g
NaCl	5.0g

Preparation of Heart Infusion Agar: Add components to distilled/deionized water and bring volume to 900.0mL. Mix thoroughly. Autoclave for 15 min at 15 psi pressure–121°C. Cool to 45°–50°C.

K_2TeO_3 Solution:

Composition per 100.0mL:

K_2TeO_3	0.3g

Preparation of K_2TeO_3 Solution: Add K_2TeO_3 to distilled/deionized water and bring volume to 100.0mL. Mix thoroughly. Autoclave for 15 min at 15 psi pressure–121°C.

Caution: Potassium tellurite is toxic.

Preparation of Medium: Add sterile K_2TeO_3 solution, sterile rabbit blood, and sterile, solid L-cystine to sterile, cooled heart infusion agar. Mix thoroughly. Pour into sterile Petri dishes or distribute into sterile tubes.

Use: For the isolation, differentiation, and cultivation of *Corynebacterium diphtheriae. Corynebacterium diphtheriae* appears as dark gray to black colonies.

Cystine Tellurite Blood Agar

Composition per 120.0mL:

Heart infusion agar	100.0mL
K_2TeO_3 solution	15.0mL
Sheep blood	5.0mL
L-Cystine	5.0mg

pH 7.4 ± 0.2 at 25°C

Heart Infusion Agar:

Composition per liter:

Beef heart, infusion from	500.0g
Agar	20.0g
Tryptose	10.0g
Yeast extract	5.0g
NaCl	5.0g

Preparation of Heart Infusion Agar: Add components to distilled/deionized water and bring volume to 1.0L. Mix thoroughly. Autoclave for 15 min at 15 psi pressure–121°C. Cool to 45°–50°C.

K_2TeO_3 Solution:

Composition per 100.0mL:

K_2TeO_3	0.3g

Preparation of K_2TeO_3 Solution: Add K_2TeO_3 to distilled/deionized water and bring volume to 100.0mL. Mix thoroughly. Autoclave for 15 min at 15 psi pressure–121°C.

Caution: Potassium tellurite is toxic.

Preparation of Medium: Add sterile K_2TeO_3 solution, sterile, defibrinated sheep blood, and sterile, solid L-cystine to sterile, cooled heart infusion agar. Mix thoroughly. Pour into sterile Petri dishes or distribute into sterile tubes.

Use: For the isolation, differentiation, and cultivation of *Corynebacterium diphtheriae. Corynebacterium diphtheriae* appears as dark gray to black colonies.

Czapek Dox Agar, Modified

Composition per liter:

Sucrose	30.0g
Agar	12.0g
$NaNO_3$	2.0g

Magnesium glycerophosphate 0.5g
KCl...0.5g
K_2SO_4..0.35g
$FeSO_4$..0.01g

pH 6.8 ± 0.2 at 25°C

Source: This medium is available as a premixed powder from Oxoid Unipath.

Preparation of Medium: Add components to distilled/deionized water and bring volume to 1.0L. Mix thoroughly. Distribute into tubes or flasks. Autoclave for 15 min at 15 psi pressure–121°C. Pour into sterile Petri dishes or leave in tubes.

Use: For the cultivation and maintenance of numerous fungal species. For chlamydospore production by *Candida albicans*.

Czapek Dox Broth

Composition per liter:

Sucrose..30.0g
$NaNO_3$..3.0g
K_2HPO_4..1.0g
$MgSO_4 \cdot 7H_2O$...0.5g
KCl...0.5g
$FeSO_4 \cdot 7H_2O$..0.01g

pH 7.3 ± 0.2 at 25°C

Source: This medium is available as a premixed powder from BD Diagnostic Systems.

Preparation of Medium: Add components to distilled/deionized water and bring volume to 1.0L. Mix thoroughly. Distribute into tubes or flasks. Autoclave for 15 min at 15 psi pressure–121°C.

Use: For the cultivation and maintenance of a variety of fungal and bacterial species that can use nitrate as sole nitrogen source.

Czapek Dox Liquid Medium, Modified

Composition per liter:

Sucrose..30.0g
$NaNO_3$..2.0g
Magnesium glycerophosphate 0.5g
KCl...0.5g
K_2SO_4..0.35g
$FeSO_4$..0.01g

pH 6.8 ± 0.2 at 25°C

Source: This medium is available as a premixed powder from Oxoid Unipath.

Preparation of Medium: Add components to distilled/deionized water and bring volume to 1.0L. Mix thoroughly. Distribute into tubes or flasks. Autoclave for 15 min at 15 psi pressure–121°C.

Use: For the cultivation of fungi and bacteria capable of utilizing sodium nitrate as the sole source of nitrogen.

Czapek Solution Agar

Composition per liter:

Sucrose..30.0g
Agar...15.0g
$NaNO_3$..2.0g
K_2HPO_4..1.0g
KCl...0.5g
$MgSO_4 \cdot 7H_20$...0.5g
$FeSO_4 \cdot 7H_2O$..0.01g

pH 7.3 ± 0.2 at 25°C

Source: This medium is available as a premixed powder from BD Diagnostic Systems.

Preparation of Medium: Add components to distilled/deionized water and bring volume to 1.0L. Mix thoroughly. Distribute into tubes or flasks. Autoclave for 15 min at 15 psi pressure–121°C. Pour into sterile Petri dishes or leave in tubes.

Use: For the cultivation of *Aspergillus, Penicillium*, and other fungi. For the cultivation and maintenance of microorganisms that can utilize nitrate as sole nitrogen source.

Czapek Yeast Extract Agar

Composition per liter:

Sucrose..30.0g
Agar...15.0g
Yeast extract..5.0g
$NaNO_3$..3.0g
K_2HPO_4..1.0g
$MgSO_4 \cdot 7H_2O$...0.5g
KCl...0.5g
$FeSO_4 \cdot 7H_2O$..0.01g
Trace metal solution..1.0mL

pH 7.3 ± 0.2 at 25°C

Trace Metal Solution:

Composition per 100.0mL:

$ZnSO_4 \cdot 7H_2O$..1.0g
$CuSO_4 \cdot 5H_2O$..0.5g

Preparation of Trace Metal Solution: Add components to 100.0mL distilled/deionized water. Mix thoroughly.

Preparation of Medium: Add components to distilled/deionized water and bring volume to 1.0L. Mix thoroughly. Adjust pH to 6.2. Gently heat and bring to boiling. Distribute into tubes or flasks. Autoclave for 15 min at 15 psi pressure–121°C. Pour into sterile Petri dishes or leave in tubes.

Use: For the cultivation and maintenance of a variety of fungal and bacterial species that can use nitrate as sole nitrogen source.

DCLS Agar (Deoxycholate Citrate Lactose Sucrose Agar)

Composition per liter:

Agar	12.0g
Sodium citrate·3H_2O	10.5g
Lactose	5.0g
$Na_2S_2O_3$	5.0g
Sucrose	5.0g
Pancreatic digest of casein	3.5g
Peptic digest of animal tissue	3.5g
Beef extract	3.0g
Sodium deoxycholate	2.5g
Neutral Red	0.03g

pH 7.2 ± 0.1 at 25°C

Source: This medium is available as a premixed powder from BD Diagnostic Systems and Oxoid.

Preparation of Medium: Add components to distilled/deionized water and bring volume to 1.0L. Mix thoroughly. Gently heat while stirring and bring to boiling. Do not overheat. Do not autoclave. Pour into sterile Petri dishes in 20.0mL volumes.

Use: For the selective isolation of *Salmonella* species, *Shigella* species, and *Vibrio* species from fecal specimens.

Decarboxylase Base, Møller

Composition per liter:

Amino acid	10.0g
Beef extract	5.0g
Peptone	5.0g
Glucose	0.5g
Bromcresol Purple	0.01g
Cresol Red	5.0mg
Pyridoxal	5.0mg
Mineral oil	200.0mL

pH 6.0 ± 0.2 at 25°C

Source: This medium is available as a premixed powder from BD Diagnostic Systems.

Preparation of Medium: Add components, except mineral oil, to distilled/deionized water and bring volume to 1.0L. For amino acid, use L-arginine, L-lysine, or L-ornithine. Mix thoroughly. Distribute into screw-capped tubes in 5.0mL volumes. Autoclave medium and mineral oil separately for 15 min at 15 psi pressure–121°C. After inoculation, overlay medium with 1.0mL of sterile mineral oil per tube.

Use: For the cultivation and differentiation of bacteria based on their ability to decarboxylate the amino acid. Bacteria that decarboxylate arginine, lysine, or ornithine turn the medium turbid purple.

Decarboxylase Medium Base, Falkow

Composition per liter:

Amino acid	5.0g
Peptone	5.0g
Yeast extract	3.0g
Glucose	1.0g
Bromcresol Purple	0.02g
Mineral oil	200.0mL

pH 6.8 ± 0.2 at 25°C

Source: This medium is available as a premixed powder from BD Diagnostic Systems.

Preparation of Medium: Add components, except mineral oil, to distilled/deionized water and bring volume to 1.0L. For amino acid, use L-arginine, L-lysine, or L-ornithine. Mix thoroughly. Distribute into screw-capped tubes in 5.0mL volumes. Autoclave medium and mineral oil separately for 15 min at 15 psi pressure–121°C. After inoculation, overlay medium with 1.0mL of sterile mineral oil per tube.

Use: For the cultivation and differentiation of bacteria based on their ability to decarboxylate the amino acid. Bacteria that decarboxylate arginine, lysine, or ornithine turn the medium turbid purple.

Decarboxylase Medium, Ornithine Modified

Composition per liter:

L-Ornithine	10.0g
Meat peptone	5.0g
Yeast extract	3.0g
Bromcresol Purple solution	5.0mL

pH 5.5 ± 0.2 at 25°C

Bromcresol Purple Solution:

Composition per 100.0mL:

Bromcresol Purple	0.2g
Ethanol	50.0mL

Preparation of Bromcresol Purple Solution: Add Bromcresol Purple to ethanol. Mix thoroughly. Bring volume to 100.0mL with distilled/deionized water. Mix thoroughly. Filter sterilize.

Preparation of Medium: Add components to distilled/deionized water and bring volume to 1.0L. Mix thoroughly. Gently heat until dissolved. Adjust pH to 5.5 with HCl or NaOH. Distribute into screw-capped tubes. Autoclave for 15 min at 15 psi pressure–121°C.

Use: For the cultivation and differentiation of bacteria based on their ability to decarboxylate ornithine.

Bacteria that decarboxylate ornithine turn the medium turbid purple.

Demi-Fraser Broth

Composition per liter:

NaCl	20.0g
Tryptose	10.0g
Na_2HPO_4	9.6g
Beef extract	5.0g
Yeast extract	5.0g
LiCl	3.0g
KH_2PO_4	1.35g
Esculin	1.0g
Acriflavin·HCl	12.5mg
Nalidixic acid	10.0mg
Ferric ammonium citrate supplement	10.0mL

pH 7.2 ± 0.2 at 25°C

Source: This medium is available as a premixed powder and supplement from BD Diagnostic Systems.

Ferric Ammonium Citrate Supplement:
Composition per 10.0mL:

Ferric ammonium citrate	0.5g

Preparation of Ferric Ammonium Citrate Supplement: Add ferric ammonium citrate to distilled/deionized water and bring volume to 10.0mL. Mix thoroughly. Filter sterilize.

Preparation of Medium: Add components, except ferric ammonium citrate supplement, to distilled/deionized water and bring volume to 990.0mL. Mix thoroughly. Autoclave for 15 min at 15 psi pressure–121°C. Aseptically add 10.0mL of sterile ferric ammonium citrate supplement. Mix thoroughly. Aseptically distribute into sterile tubes or flasks.

Use: For the cultivation of *Listeria* species.

Deoxycholate Agar

Composition per liter:

Agar	16.0g
Lactose	10.0g
NaCl	5.0g
Pancreatic digest of casein	5.0g
Peptic digest of animal tissue	5.0g
K_2HPO_4	2.0g
Ferric citrate	1.0g
Sodium citrate	1.0g
Sodium deoxycholate	1.0g
Neutral Red	0.033g

pH 7.3 ± 0.2 at 25°C

Source: This medium is available as a premixed powder from BD Diagnostic Systems.

Preparation of Medium: Add components to distilled/deionized water and bring volume to 1.0L. Mix thoroughly. Gently heat and bring to boiling. Do not autoclave. Cool to 45°–50°C. Pour into sterile Petri dishes.

Use: For the selective isolation, cultivation, enumeration, and differentiation of Gram-negative enteric microorganisms from a variety of clinical and nonclinical specimens. *Escherichia coli* appears as large, flat, rose-red colonies. *Enterobacter* and *Klebsiella* species appear as large, mucoid, pale colonies with a pink center. *Proteus* and *Salmonella* species appear as large, colorless to tan colonies. *Shigella* species appear as colorless to pink colonies. *Pseudomonas* species appear as irregular colorless to brown colonies.

Deoxycholate Agar (Desoxycholate Agar)

Composition per liter:

Agar	15.0g
Lactose	10.0g
Peptone	10.0g
NaCl	5.0g
K_2HPO_4	2.0g
Ferric citrate	1.0g
Sodium citrate	1.0g
Sodium deoxycholate	1.0g
Neutral Red	0.03g

pH 7.3 ± 0.2 at 25°C

Source: This medium is available as a premixed powder from Oxoid Unipath and BD Diagnostic Systems.

Preparation of Medium: Add components to distilled/deionized water and bring volume to 1.0L. Mix thoroughly. Gently heat and bring to boiling. Do not autoclave. Cool to 50°C. Pour into sterile Petri dishes.

Use: For the selective isolation, cultivation, enumeration, and differentiation of Gram-negative enteric microorganisms from a variety of clinical and nonclinical specimens. *Escherichia coli* appears as large, flat, rose-red colonies. *Enterobacter* and *Klebsiella* species appear as large, mucoid, pale colonies with a pink center. *Proteus* and *Salmonella* species appear as large, colorless to tan colonies. *Shigella* species appear as colorless to pink colonies. *Pseudomonas* species appear as irregular colorless to brown colonies.

Deoxycholate Citrate Agar

Composition per liter:

Sodium citrate	50.0g
Agar	15.0g
Lactose	10.0g
Beef extract	5.0g
Peptone	5.0g
$Na_2S_2O_3·5H_2O$	5.0g
Sodium deoxycholate	2.5g

Ferric citrate 1.0g
Neutral Red 0.025g

pH 7.3 ± 0.2 at 25°C

Source: This medium is available as a premixed powder from Oxoid Unipath.

Preparation of Medium: Add components to distilled/deionized water and bring volume to 1.0L. Mix thoroughly. Gently heat and bring to boiling. Do not autoclave. Cool to 45°–50°C. Pour into sterile Petri dishes. Dry the agar surface before use.

Use: For the selective isolation and cultivation of enteric pathogens, especially *Salmonella* and *Shigella* species.

Deoxycholate Citrate Agar

Composition per liter:

Sodium citrate 20.0g
Agar 17.0g
Lactose 10.0g
Meat, solids from infusion 10.0g
Peptic digest of animal tissue 10.0g
Sodium deoxycholate 5.0g
Ferric citrate 1.0g
Neutral Red 0.02g

pH 7.3 ± 0.2 at 25°C

Source: This medium is available as a premixed powder from BD Diagnostic Systems.

Preparation of Medium: Add components to distilled/deionized water and bring volume to 1.0L. Mix thoroughly. Gently heat and bring to boiling. Do not autoclave. Cool to 45°–50°C. Pour into sterile Petri dishes. Dry the agar surface before use.

Use: For the selective isolation and cultivation of enteric pathogens, especially *Salmonella* and *Shigella* species.

Deoxycholate Citrate Agar (Desoxycholate Citrate Agar)

Composition per liter:

Pork infusion 330.0g
Sodium citrate 20.0g
Agar 13.5g
Lactose 10.0g
Proteose peptone no. 3 .... 10.0g
Sodium deoxycholate 5.0g
Ferric ammonium citrate 2.0g
Neutral Red 0.02g

pH 7.5 ± 0.2 at 25°C

Source: This medium is available as a premixed powder from BD Diagnostic Systems.

Preparation of Medium: Add components to distilled/deionized water and bring volume to 1.0L. Mix thoroughly. Gently heat and bring to boiling. Do not autoclave. Cool to 45°–50°C. Pour into sterile Petri dishes. Dry the agar surface before use.

Use: For the selective isolation and cultivation of enteric pathogens, especially *Salmonella* and *Shigella* species.

Deoxycholate Citrate Agar, Hynes

Composition per liter:

Agar 12.0g
Lactose 10.0g
Sodium citrate 8.5g
$Na_2S_2O_3 \cdot 5H_2O$ 5.4g
Beef extract powder 5.0g
Peptone 5.0g
Sodium deoxycholate 5.0g
Ferric citrate 1.0g
Neutral Red 0.02g

pH 7.3 ± 0.2 at 25°C

Source: This medium is available as a premixed powder from Oxoid Unipath.

Preparation of Medium: Add components to distilled/deionized water and bring volume to 1.0L. Mix thoroughly. Gently heat and bring to boiling. Do not autoclave. Cool to 45°–50°C. Pour into sterile Petri dishes. Dry the agar surface before use.

Use: For the selective isolation, cultivation, and differentiation of enteric pathogens, especially *Salmonella* and *Shigella* species. Lactose-fermenting bacteria appear as pink colonies that may or may not be surrounded by a zone of precipitated deoxycholate. Nonlactose-fermenting bacteria appear as colorless colonies that are surrounded by a clear orange-yellow zone.

Deoxycholate Lactose Agar

Composition per liter:

Agar 15.0g
Lactose 10.0g
NaCl 5.0g
Pancreatic digest of casein 5.0g
Peptic digest of animal tissue 5.0g
Sodium citrate 2.0g
Sodium deoxycholate 0.5g
Neutral Red 0.033g

pH 7.1 ± 0.2 at 25°C

Source: This medium is available as a premixed powder from BD Diagnostic Systems.

Preparation of Medium: Add components to distilled/deionized water and bring volume to 1.0L. Mix

thoroughly. Gently heat and bring to boiling. Do not autoclave. Cool to 45°–50°C. Pour into sterile Petri dishes. Dry the agar surface before use.

Use: For the selective isolation, cultivation, and differentiation of enteric pathogens, especially *Salmonella* and *Shigella* species. Lactose-fermenting bacteria appear as pink colonies that may or may not be surrounded by a zone of precipitated deoxycholate. Nonlactose-fermenting bacteria appear as colorless colonies that are surrounded by a clear orange-yellow zone. Also used for the enumeration of coliform bacteria from water, milk, and dairy products.

Dermasel Agar Base

Composition per liter:

Glucose	20.0g
Agar	14.5g
Papaic digest of soybean meal	10.0g
Antibiotic inhibitor	10.0mL

pH 6.8–7.0 at 25°C

Source: This medium is available as a premixed powder from Oxoid Unipath.

Antibiotic Inhibitor:

Composition per 10.0mL:

Cycloheximide	0.4g
Chloramphenicol	0.05g
Acetone	10.0mL

Preparation of Antibiotic Inhibitor: Add cycloheximide and chloramphenicol to 10.0mL of acetone. Mix thoroughly.

Preparation of Medium: Add components to distilled/deionized water and bring volume to 990.0mL. Mix thoroughly. Gently heat and bring to boiling. Do not overheat. Add antibiotic inhibitor. Mix thoroughly. Autoclave for 10 min at 15 psi pressure–121°C. Pour into sterile Petri dishes.

Use: For the isolation and cultivation of dermatophytic fungi isolated from hair, nails, or skin scrapings.

Dermatophyte Test Medium Base

Composition per liter:

Agar	20.0g
Glucose	10.0g
Papaic digest of soybean meal	10.0g
Cycloheximide	0.5g
Phenol Red	0.2g
Gentamycin sulfate	0.1g
Chlortetracycline	0.1g

pH 5.5 ± 0.2 at 25°C

Source: This medium is available as a premixed powder from BD Diagnostic Systems.

Preparation of Medium: Add components, except gentamycin sulfate and chlortetracycline, to distilled/deionized water and bring volume to 1.0L. Mix thoroughly. Gently heat while stirring and bring to boiling. Autoclave for 15 min at 15 psi pressure–121°C. Cool to 45°–50°C. Aseptically add gentamycin sulfate and chlortetracycline. Mix thoroughly. Pour into sterile Petri dishes.

Use: For the selective isolation and cultivation of pathogenic fungi from cutaneous sources.

DEV Lactose Peptone MUG Broth

Composition per liter:

Lactose	10.0g
Meat peptone	10.0g
NaCl	5.0g
Tryptophan	1.0g
4-Methylumbelliferyl-β-D-glucuronide	0.1g
Bromocresol Purple	0.01g

pH 7.2 ± 0.2 at 37°C

Source: This medium is available from Fluka, Sigma-Aldrich.

Preparation of Medium: Add components to distilled/deionized water and bring volume to 1.0L. Mix thoroughly. Distribute into test tubes that contain an inverted Durham tube in 10.0mL volumes. Autoclave for 15 min at 15 psi pressure–121°C.

Use: For the enrichment and titre determination of coliform bacteria. The presence of *E. coli* can be demonstrated by fluorescence in the UV and a positive indole test.

Dextrose Agar

Composition per liter:

Agar	15.0g
Glucose	10.0g
NaCl	5.0g
Pancreatic digest of casein	5.0g
Peptic digest of animal tissue	5.0g
Beef extract	3.0g

pH 6.9 ± 0.2 at 25°C

Source: This medium is available as a premixed powder from BD Diagnostic Systems.

Preparation of Medium: Add components to distilled/deionized water and bring volume to 1.0L. Mix thoroughly. Gently heat and bring to boiling. Distribute into tubes or flasks. Autoclave for 15 min at 15 psi pressure–121°C. Pour into sterile Petri dishes or leave in tubes.

Use: For use as a base for the preparation of blood agar.

Dextrose Agar

Composition per liter:

Agar	15.0g
Glucose	10.0g
Tryptose	10.0g
NaCl	5.0g
Beef extract	3.0g

pH 7.3 ± 0.2 at 25°C

Source: This medium is available as a premixed powder from BD Diagnostic Systems.

Preparation of Medium: Add components to distilled/deionized water and bring volume to 1.0L. Mix thoroughly. Gently heat and bring to boiling. Distribute into tubes or flasks. Autoclave for 15 min at 15 psi pressure–121°C. Pour into sterile Petri dishes or leave in tubes.

Use: For the cultivation of a wide variety of microorganisms. For use as a base for the preparation of blood agar and for general laboratory procedures.

Dextrose Ascitic Fluid Semisolid Agar

Composition per liter:

Pancreatic digest of casein	2.66g
NaCl	1.33g
Agar	0.5g
Phenol Red	4.8mg
Ascitic fluid	50.0mL
Glucose solution	15.0mL

pH 7.4 ± 0.2 at 25°C

Glucose Solution:

Composition per 15.0mL:

Glucose	3.0g

Preparation of Glucose Solution: Add glucose to distilled/deionized water and bring volume to 15.0mL. Mix thoroughly. Filter sterilize.

Preparation of Medium: Add components, except ascitic fluid and glucose solution, to distilled/deionized water and bring volume to 935.0mL. Mix thoroughly. Gently heat and bring to boiling. Autoclave for 15 min at 15 psi pressure–121°C. Cool to 45°–50°C. Aseptically add sterile ascitic fluid and glucose solution. Mix thoroughly. Aseptically distribute into sterile tubes.

Use: For the isolation and cultivation of microorganisms from spinal fluid.

Dextrose Broth

Composition per liter:

Tryptose	10.0g
Glucose	5.0g
NaCl	5.0g
Beef extract	3.0g

pH 7.2 ± 0.2 at 25°C

Source: This medium is available as a premixed powder from BD Diagnostic Systems and Oxoid Unipath.

Preparation of Medium: Add components to distilled/deionized water and bring volume to 1.0L. Mix thoroughly. Distribute into tubes or flasks. Autoclave for 15 min at 15 psi pressure–121°C.

Use: For the isolation and enrichment of fastidious or damaged microorganisms.

Dextrose Broth

Composition per liter:

Pancreatic digest of casein	10.0g
Glucose	5.0g
NaCl	5.0g

pH 7.3 ± 0.2 at 25°C

Source: This medium is available as a premixed powder from BD Diagnostic Systems.

Preparation of Medium: Add components to distilled/deionized water and bring volume to 1.0L. Mix thoroughly. Distribute into tubes or flasks. Autoclave for 15 min at 15 psi pressure–121°C.

Use: For the cultivation and differentiation of microorganisms based on their ability to ferment glucose. If desired, a Durham tube may be added to the test tubes to determine gas production.

Dextrose Proteose No. 3 Agar

Composition per liter:

Proteose peptone no. 3	20.0g
Agar	13.0g
NaCl	5.0g
Glucose	2.0g
Tellurite blood solution	50.0mL

pH 7.4 ± 0.2 at 25°C

Tellurite Blood Solution:

Composition per 60.0mL:

Sheep blood, defibrinated	50.0mL
Chapman tellurite solution	10.0mL

Preparation of Tellurite Blood Solution: Aseptically combine 10.0mL of Chapman tellurite solution with 50.0mL of sterile, defibrinated sheep blood. Mix thoroughly.

Chapman Tellurite Solution:

Composition per 100.0mL:

K_2TeO_3	1.0g

Preparation of Chapman Tellurite Solution: Add K_2TeO_3 to distilled/deionized water and bring volume to 100.0mL. Mix thoroughly. Filter sterilize.

Caution: Potassium tellurite is toxic.

Preparation of Medium: Add components, except tellurite blood solution, to distilled/deionized water and bring volume to 940.0mL. Mix thoroughly. Gently heat and bring to boiling. Autoclave for 15 min at 15 psi pressure–121°C. Cool to 75°–80°C. Aseptically add 50.0mL of sterile tellurite blood solution. Mix thoroughly. Maintain at 75°–80°C for 10–15 min or until the agar becomes chocolatized. Cool slowly to 50°C. Pour into sterile Petri dishes or distribute into sterile tubes.

Use: For propagating pure cultures of *Neisseria gonorrhoeae* and other fastidious microorganisms.

Dextrose Starch Agar

Composition per liter:

Gelatin	20.0g
Proteose peptone	15.0g
Agar	10.0g
Starch	10.0g
Glucose	5.0g
NaCl	5.0g
Na_2HPO_4	3.0g

pH 7.3 ± 0.2 at 25°C

Source: This medium is available as a premixed powder from BD Diagnostic Systems.

Preparation of Medium: Add components to distilled/deionized water and bring volume to 1.0L. Mix thoroughly. Gently heat while stirring and bring to boiling. Distribute into tubes or flasks. Autoclave for 15 min at 15 psi pressure–121°C. Pour into sterile Petri dishes or leave in tubes.

Use: For the cultivation and maintenance of *Neisseria gonorrhoeae*, *Neisseria animalis*, and other fastidious microorganisms.

Diagnostic Sensitivity Test Agar (DST AGAR)

Composition per liter:

Agar	12.0g
Proteose peptone	10.0g
Veal infusion solids	10.0g
NaCl	3.0g
Glucose	2.0g
Na_2HPO_4	2.0g
Sodium acetate	1.0g
Adenine sulfate	0.01g
Guanine hydrochloride	0.01g
Uracil	0.01g
Xanthine	0.01g
Aneurine	0.00002g
Horse blood, sterile lysed or defibrinated	70.0mL

pH 7.4 ± 0.2 at 25°C

Source: This medium is available as a premixed powder from Oxoid Unipath.

Preparation of Medium: Add components, except horse blood, to 930.0mL distilled/deionized water in a 2.5L flask. Mix thoroughly. Gently heat and bring to boiling. Autoclave for 15 min at 15 psi pressure–121°C. Cool to 50°C. Aseptically add 70.0mL sterile horse blood. Mix thoroughly to ensure adequate aeration of the blood. Pour into sterile Petri dishes or distribute into sterile tubes.

Use: For antimicrobial testing of a various pathogenic microorganisms. DSTA is primarily used for susceptibility tests rather than the primary isolation of organisms from clinical samples. An essential requirement for satisfactory antimicrobial susceptibility media is that the reactive levels of thymidine and thymine must be sufficiently reduced to avoid antagonism of trimethoprim and sulphonamides. DSTA meets this requirement and in the presence of lysed horse blood (or defibrinated horse blood if the plates are stored long enough to allow some lysis of the erythrocytes) the level of thymidine will be further reduced. This is caused by the action of the enzyme thymidine phosphorylase which is released from lysed horse erythrocytes. Thymidine is an essential growth factor for thymidine-dependent organisms and they will not grow in its absence or they will grow poorly in media containing reduced levels.

Diagnostic Sensitivity Test Agar (DST Agar)

Composition per liter:

Agar	12.0g
Proteose peptone	10.0g
Veal infusion solids	10.0g
NaCl	3.0g
Na_2HPO_4	2.0g
Glucose	2.0g
Sodium acetate	1.0g
Adenine sulfate	0.01g
Guanine·HCl	0.01g
Uracil	0.01g
Xanthine	0.01g
Thiamine	0.02mg
Horse blood, defibrinated	70.0mL

pH 7.4 ± 0.2 at 25°C

Source: This medium is available as a premixed powder from Oxoid Unipath.

Preparation of Medium: Add components, except horse blood, to distilled/deionized water and bring volume to 930.0mL. Mix thoroughly. Gently heat and bring to boiling. Autoclave for 15 min at 15 psi pressure–121°C. Cool to 45°–50°C. Aseptically add sterile horse blood. Mix thoroughly. Pour into sterile Petri dishes or distribute into sterile tubes.

Use: For the cultivation of microorganisms for antimicrobial sensitivity testing.

Diamonds Medium, Modified

Composition per liter:

Pancreatic digest of casein	20.0g
Yeast extract	1.0g
L-Cysteine·HCl·H_2O	0.5g
Maltose	0.5g
L-Ascorbic acid	0.02g
Horse serum, inactivated	100.0mL
Antibiotic inhibitor	10.0mL

pH 6.5 ± 0.2 at 25°C

Antibiotic Inhibitor:

Composition per 10.0mL:

Streptomycin sulfate	0.15g
Amphotericin B	0.2mg
Penicillin G	100,000U

Preparation of Antibiotic Inhibitor: Add components to distilled/deionized water and bring volume to 10.0mL. Mix thoroughly. Filter sterilize.

Preparation of Medium: Add components, except antibiotic inhibitor and horse serum, to distilled/deionized water and bring volume to 890.0mL. Mix thoroughly. Gently heat and bring to boiling. Autoclave for 15 min at 15 psi pressure–121°C. Cool to 25°C. Aseptically add sterile antibiotic inhibitor and horse serum. Mix thoroughly. Aseptically distribute into sterile tubes in 5.0mL volumes.

Use: For the cultivation of *Trichomonas* species.

Diethyl Phosphonate Agar

Composition per liter:

Agar	12.0g
Tris(hydroxymethyl)methylamine	6.0g
p-Hydroxybenzoate, Na salt	0.75g
KCl	0.2g
$MgSO_4 \cdot 7H_2O$	0.2g
NH_4Cl	0.2g
Diethyl phosphonate solution	100.0mL

pH 7.4 ± 0.2 at 25°C

Diethyl Phosphonate Solution:

Composition per 100.0mL:

Diethyl phosphonate	0.015g

Source: Diethyl phosphonate is available from Eastman Organic Chemical Division, Rochester, NY.

Preparation of Diethyl Phosphonate Solution: Add diethyl phosphonate to distilled/deionized water and bring volume to 100.0mL. Mix thoroughly. Filter sterilize.

Preparation of Medium: Add components, except diethyl phosphonate solution, to distilled/deionized water and bring volume to 900.0mL. Mix thoroughly. Gently heat and bring to boiling. Autoclave for 15 min at 15 psi pressure–121°C. Cool to 25°C. Aseptically add 100.0mL diethyl phosphate solution. Mix. Pour into sterile Petri dishes or distribute into sterile tubes.

Use: For the cultivation and maintenance of *Comamonas acidovorans.*

Differential Agar Medium A8 for *Ureaplasma urealyticum*

Composition per 103.1mL:

Basal agar	80.0mL
Horse serum, unheated	20.0mL
Fresh yeast extract solution	1.0mL
Urea solution	1.0mL
CVA enrichment	0.5mL
L-Cysteine·HCl·H_2O solution	0.5mL
GHL tripeptide solution	0.1mL

pH 5.5 ± 0.2 at 25°C

Basal Agar:

Composition per 80.0mL:

Tryptic soy broth	2.4g
Noble agar	1.05g
Putrescine·2HCl	0.17g
$CaCl_2 \cdot 2H_2O$	0.015g

Preparation of Basal Agar: Add components to distilled/deionized water and bring volume to 80.0mL. Mix thoroughly. Adjust pH to 5.5 with 2*N* HCl. Gently heat and bring to boiling. Autoclave for 15 min at 15 psi pressure–121°C. Cool to 50°–55°C.

Fresh Yeast Extract Solution:

Composition per 100.0mL:

Baker's yeast, live, pressed, starch-free	25.0g

Preparation of Fresh Yeast Extract Solution: Add the live Baker's yeast to 100.0mL of distilled/deionized water. Autoclave for 90 min at 15 psi pressure–121°C. Allow to stand. Remove supernatant solution. Adjust pH to 6.6–6.8. Filter sterilize.

Urea Solution:

Composition per 30.0mL:

Urea	3.0g

Preparation of Urea Solution: Add urea to distilled/deionized water and bring volume to 30.0mL. Mix thoroughly. Filter sterilize.

CVA Enrichment:

Composition per liter:

Glucose	100.0g

L-Cysteine·HCl·H_2O 25.9g
L-Glutamine 10.0g
Adenine 1.0g
L-Cystine·2HCl 1.0g
Nicotinamide adenine dinucleotide 0.25g
Cocarboxylase 0.1g
Guanine·HCl 0.03g
$Fe(NO_3)_3$ 0.02g
Vitamin B_{12} 0.01g
p-Aminobenzoic acid 0.013g
Thiamine·HCl 3.0mg

Preparation of CVA Enrichment: Add components to distilled/deionized water and bring volume to 1.0L. Mix thoroughly. Filter sterilize.

L-Cysteine·HCl·H_2O Solution:
Composition per 50.0mL:
L-Cysteine·HCl·H_2O 1.0g

Preparation of L-Cysteine·HCl·H_2O Solution: Add L-cysteine·HCl·H_2O to distilled/deionized water and bring volume to 50.0mL. Mix thoroughly. Filter sterilize.

GHL Tripeptide Solution:
Composition per 10.0mL:
GHL tripeptide 0.2mg

Preparation of GHL Tripeptide Solution: Add GHL tripeptide (glycyl-L-histidyl-L-lysine acetate) to distilled/deionized water and bring volume to 10.0mL. Mix thoroughly. Filter sterilize.

Preparation of Medium: To 80.0mL of cooled, sterile basal agar, aseptically add 20.0mL of sterile horse serum, 1.0mL of sterile fresh yeast extract solution, 1.0mL of sterile urea solution, 0.5mL of sterile CVA enrichment, 0.5mL of sterile L-cysteine·HCl·H_2O solution, and 0.1mL of sterile GHL tripeptide solution. Mix thoroughly. Pour into sterile Petri dishes in 20.0mL volumes.

Use: For the cultivation and maintenance of *Ureaplasma urealyticum*.

Dilute Peptone Water

Composition per liter:
NaCl 1.0g
Peptone 1.0g
pH 7.0 ± 0.2 at 25°C

Preparation of Medium: Add components to distilled/deionized water and bring volume to 1.0L. Mix thoroughly. Distribute into tubes or flasks. Autoclave for 15 min at 15 psi pressure–121°C.

Use: For the cultivation of various heterotrophic bacteria.

Dilute Potato Medium (DSMZ Medium 789)

Composition per liter:
Glucose 1.0g
Na_2HPO_4 .... 0.12g
$Ca(NO_3)_2·4H_2O$ 0.05g
Peptone 0.05g
Potato decoction 100.0mL
pH 7.3 ± 0.2 at 25°C

Potato Decoction:
Diced potato 20.0g

Preparation of Potato Decoction: Add diced potatoes to distilled/deionized water and bring volume to 1.0L. Boil for 30 min. Filter to remove solid potatoes. Bring volume to 1.0L with distilled/deionized water.

Preparation of Medium: Add components to distilled/deionized water and bring volume to 1.0L. Mix thoroughly. Distribute into tubes or flasks. Autoclave for 15 min at 15 psi pressure–121°C.

Use: For the cultivation of various fungi.

Diphasic Blood Agar Base Medium (ATCC Medium 449)

Composition per 500.0mL:
Beef 25.0g
Agar 10.0g
Neopeptone 10.0g
NaCl 2.5g

Preparation of Diphasic Blood Agar Base Medium: Trim beef to remove fat. Add 25.0g of lean beef to 250.0mL of distilled/deionized water. Gently heat and bring to boiling. Boil for 2-3 min. Filter through Whatman #2 filter paper. Add agar, neopeptone, and NaCl to filtrate. Bring volume to 500.0mL with distilled/deionized water. Mix thoroughly. Adjust pH to 7.2–7.4. Gently heat and bring to boiling. Autoclave for 20 min at 15 psi pressure–121°C. Cool to 50°–55°C. Add as required to other diphasic blood agars.

Use: As the base medium for diphasic blood agars.

Diphasic Blood Agar Medium with 10% Blood

Composition per 1120.0mL:
Blood agar, diphasic base medium 630.0mL
Locke's solution 420.0mL
Rabbit blood, defibrinated 70.0mL
pH 7.2–7.4 at 25°C

Blood Agar, Diphasic Base Medium:
Composition per 750.0mL:
Beef 25.0g
Agar 10.0g

Neopeptone 10.0g
NaCl 2.5g

Preparation of Blood Agar, Diphasic Base Medium: Trim beef to remove fat. Add 25.0g of lean beef to 250.0mL of distilled/deionized water. Gently heat and bring to boiling. Boil for 2-3 min. Filter through Whatman #2 filter paper. Add agar, neopeptone, and NaCl to filtrate. Bring volume to 750.0mL with distilled/deionized water. Mix thoroughly. Adjust pH to 7.2–7.4. Gently heat and bring to boiling. Autoclave for 15 min at 15 psi pressure–121°C. Cool to 50°–55°C.

Locke's Solution:
Composition per liter:
NaCl 8.0g
Glucose 2.5g
KH_2PO_4 .... 0.3g
$CaCl_2$ 0.2g
KCl 0.2g

Preparation of Locke's Solution: Add components to distilled/deionized water and bring volume to 1.0L. Mix thoroughly. Autoclave for 15 min at 15 psi pressure–121°C. Cool to 50°–55°C.

Preparation of Medium: Aseptically combine 630.0mL of sterile blood agar, diphasic base medium, with 70.0mL of sterile defibrinated rabbit blood warmed to 50°–55°C. Mix thoroughly. Aseptically distribute 5.0mL volumes into16 × 125mm screw-capped test tubes. Allow to cool in a slanted position. Overlay the agar in each tube with 3.0mL of sterile Locke's solution.

Use: For the cultivation of *Leishmania braziliensis, Leishmania enriettii, Leishmania tropica, Trypanosoma conorrhini, Trypanosoma cruzi,* and *Trypanosoma rangeli.*

Diphasic Blood Agar Medium with 30% Blood

Composition per 1450.0mL:
Blood agar, diphasic base medium 700.0mL
Locke's solution 450.0mL
Rabbit blood, defibrinated 300.0mL
pH 7.2–7.4 at 25°C

Blood Agar, Diphasic Base Medium:
Composition per 750.0mL:
Beef 25.0g
Agar 10.0g
Neopeptone 10.0g
NaCl 2.5g

Preparation of Blood Agar, Diphasic Base Medium: Trim beef to remove fat. Add 25.0g of lean beef to 250.0mL of distilled/deionized water. Gently heat and bring to boiling. Boil for 2-3 min. Filter through Whatman #2 filter paper. Add agar, neopeptone, and NaCl to filtrate. Bring volume to 750.0mL with distilled/deionized water. Mix thoroughly. Adjust pH to 7.2–7.4. Gently heat and bring to boiling. Autoclave for 15 min at 15 psi pressure–121°C. Cool to 50°–55°C.

Locke's Solution:
Composition per liter:
NaCl 8.0g
Glucose 2.5g
KH_2PO_4 .... 0.3g
$CaCl_2$ 0.2g
KCl 0.2g

Preparation of Locke's Solution: Add components to distilled/deionized water and bring volume to 1.0L. Mix thoroughly. Autoclave for 15 min at 15 psi pressure–121°C. Cool to 50°–55°C.

Preparation of Medium: Aseptically combine 700.0mL of sterile blood agar, diphasic base medium, with 300.0mL of sterile defibrinated rabbit blood warmed to 50°–55°C. Mix thoroughly. Aseptically distribute 5.0mL volumes into 16 × 125mm screw-capped test tubes. Allow to cool in a slanted position. Overlay the agar in each tube with 3.0mL of sterile Locke's solution.

Use: For the cultivation and maintenance of *Leishmania* spp. and *Trypanosoma* spp.

Diphasic Medium for Amoeba (Charcoal Agar Slants)

Composition per liter:
Agar slants 1.0L
Buffered saline overlay 1.0L
pH 7.4 ± 0.2 at 25°C

Agar Slants:
Composition per liter:
Agar 10.0g
Charcoal, activated 10.0g
Pancreatic digest of casein 5.0g
KH_2PO_4 .... 4.0g
Na_2HPO_4 .... 3.0g
Asparagine 2.0g
Sodium citrate 1.0g
Ferric ammonium citrate 0.1g
$MgSO_4 \cdot 7H_2O$ 0.1g
Cholesterol solution 25.0mL
Glycerol 10.0mL

Cholesterol Solution:
Composition per 25.0mL:
Cholesterol 0.25g
Acetone 25.0mL

Preparation of Cholesterol Solution: Add cholesterol to 25.0mL of acetone. Mix thoroughly.

Preparation of Agar Slants: Add components, except agar, charcoal, and cholesterol solution, to distilled/deionized water and bring volume to 1.0L. Mix thoroughly. Gently heat to dissolve. Do not boil. Add agar, charcoal, and cholesterol solution. Mix thoroughly. Gently heat and bring to boiling. Distribute into tubes in 3.0mL volumes. Autoclave for 15 min at 15 psi pressure–121°C. Resuspend charcoal. Allow tubes to cool in a slanted position with short butts or no butts.

Buffered Saline Overlay:
Composition per liter:

NaCl	5.0g
Solution B	810.0mL
Solution A	190.0mL

Solution A:
Composition per liter:

KH_2PO_4, anhydrous	9.07g

Preparation of Solution A: Add KH_2PO_4 to distilled/deionized water and bring volume to 1.0L. Mix thoroughly.

Solution B:
Composition per liter:

Na_2HPO_4, anhydrous	9.46g

Preparation of Solution B: Add Na_2HPO_4 to distilled/deionized water and bring volume to 1.0L. Mix thoroughly.

Preparation of Buffered Saline Overlay: Combine 810.0mL of solution A and 190.0mL of solution B. Add the NaCl. Mix thoroughly. Autoclave for 15 min at 15 psi pressure–121°C. Cool to 25°C. Store at 4°C.

Preparation of Medium: To each agar slant, aseptically add 3.0mL of sterile, buffered saline overlay.

Use: For the cultivation and maintenance of *Amoebae* species.

Dixon Agar

Composition per liter:

Malt extract	30.0g
Oxbile	20.0g
Agar	15.0g
Mycological peptone	5.0g
Glycerol mono-oleate	2.50g
Tween 40	10.0mL

pH 5.4 ± 0.2 at 25°C

Preparation of Medium: Add components to distilled/deionized water and bring volume to 1.0L. Mix thoroughly. Gently heat while stirring until boiling. Distribute into tubes or flasks. Autoclave for 10 min at 15 psi pressure–115°C. Do not overheat or agar will not harden. If a lower pH (3.5) is desired, cool medium to 55°C and aseptically add 100.0mL of sterile lactic acid. Pour into sterile Petri dishes or distribute into sterile tubes.

Use: For the cultivation and maintenance of *Malassezia* species.

DNase Agar

Composition per liter:

Tryptose	20.0g
Agar	12.0g
NaCl	5.0g
Deoxyribonucleic acid	2.0g

pH 7.3 ± 0.2 at 25°C

Source: This medium is available as a premixed powder from Oxoid Unipath.

Preparation of Medium: Add components to distilled/deionized water and bring volume to 1.0L. Mix thoroughly. Gently heat and bring to boiling. Distribute into tubes or flasks. Autoclave for 15 min at 15 psi pressure–121°C. Pour into sterile Petri dishes or leave in tubes.

Use: For the differentiation of microorganisms, especially *Staphylococcus* species and *Serratia marcescens*, based on their production of deoxyribo-nuclease.

DNase Medium

Composition per liter:

Agar	15.0g
Pancreatic digest of casein	10.0g
Peptic digest of animal tissue	10.0g
L-Arabinose	10.0g
NaCl	5.0g
Deoxyribonucleic acid	2.0g
Methyl Green	0.09g
Phenol Red	0.05g
Antibiotic solution	10.0mL

pH 7.3 ± 0.2 at 25°C

Antibiotic Solution:
Composition per 10.0mL:

Cephalothin	0.01g
Ampicillin	5.0mg
Colistimethate	5.0mg
Amphotericin B	2.5mg

Preparation of Antibiotic Solution: Add components to distilled/deionized water and bring volume to 10.0mL. Mix thoroughly. Filter sterilize.

Preparation of Medium: Add components, except antibiotic solution, to distilled/deionized water and bring volume to 990.0mL. Mix thoroughly. Gently heat and bring to boiling. Autoclave for 15 min at 15 psi pressure–121°C. Cool to 45°–50°C. Aseptically add sterile components. Mix thoroughly. Pour into sterile Petri dishes or distribute into sterile tubes.

Use: For the isolation and cultivation of *Serratia marcescens*.

DNase Test Agar

Composition per liter:

Agar 15.0g
Pancreatic digest of casein 15.0g
NaCl 5.0g
Papaic digest of soybean meal 5.0g
Deoxyribonucleic acid 2.0g

pH 7.3 ± 0.2 at 25°C

Source: This medium is available as a premixed powder from BD Diagnostic Systems.

Preparation of Medium: Add components to distilled/deionized water and bring volume to 1.0L. Mix thoroughly. Gently heat while stirring and bring to boiling. Distribute into tubes or flasks. Autoclave for 15 min at 13 psi pressure–118°C. Pour into sterile Petri dishes or leave in tubes.

Use: For the differentiation of microorganisms, especially *Staphylococcus* species and *Serratia marcescens*, based on their production of deoxyribonuclease.

DNase Test Agar with Methyl Green

Composition per liter:

Agar 15.0g
Pancreatic digest of casein 10.0g
Peptic digest of animal tissue 10.0g
NaCl 5.0g
Deoxyribonucleic acid 2.0g
Methyl Green 0.05g

pH 7.3 ± 0.2 at 25°C

Source: This medium is available as a premixed powder from BD Diagnostic Systems.

Preparation of Medium: Add components to distilled/deionized water and bring volume to 1.0L. Mix thoroughly. Gently heat while stirring and bring to boiling. Distribute into tubes or flasks. Autoclave for 15 min at 13 psi pressure–118°C. Pour into sterile Petri dishes or leave in tubes.

Use: For the differentiation of microorganisms, especially *Staphylococcus* species and *Serratia marcescens*, based on their production of deoxyribonuclease.

DNase Test Agar with Toluidine Blue

Composition per liter:

Agar 15.0g
Pancreatic digest of casein 10.0g
Peptic digest of animal tissue 10.0g
NaCl 5.0g
Deoxyribonucleic acid 2.0g
Toluidine Blue 0.1g

pH 7.3 ± 0.2 at 25°C

Preparation of Medium: Add components to distilled/deionized water and bring volume to 1.0L. Mix thoroughly. Gently heat while stirring and bring to boiling. Distribute into tubes or flasks. Autoclave for 15 min at 13 psi pressure–118°C. Pour into sterile Petri dishes or leave in tubes.

Use: For the differentiation of microorganisms, especially *Staphylococcus* species and *Serratia marcescens*, based on their production of deoxyribonuclease.

Dorset Egg Medium

Composition per liter:

Homogenized whole egg 950.0mL
Glycerol 50.0mL

pH 6.8–7.4 at 25°C

Source: This medium is available as a prepared medium from BD Diagnostic Systems.

Homogenized Whole Egg:
Composition per liter:

Whole eggs 18–24

Preparation of Homogenized Whole Egg: Use fresh eggs, less than 1 week old. Scrub the shells with soap. Let stand in a soap solution for 30 min. Rinse in running water. Soak eggs in 70% ethanol for 15 min. Break the eggs into a sterile container. Homogenize by shaking. Filter through four layers of sterile cheesecloth into a sterile graduated cylinder. Measure out 1.0L.

Preparation of Medium: Filter sterilize glycerol. Combine glycerol and homogenized whole egg. Mix thoroughly. Distribute into sterile screw-capped tubes. Place tubes in a slanted position. Inspissate at 85°C (moist heat) for 45 min.

Use: For the maintenance of *Mycobacterium* species.

DTC Agar

Composition per liter:

Agar 20.0g
Pancreatic digest of casein 15.0g
NaCl 5.0g
Papaic digest of soybean meal 5.0g
Deoxyribonucleic acid 2.0g
Toluidine Blue O 0.1g
Cephalothin solution 10.0mL

Cephalothin Solution:

Composition per 10.0mL:

Cephalothin	1.0g

Preparation of Cephalothin Solution: Add cephalothin to distilled/deionized water and bring volume to 10.0mL. Mix thoroughly. Filter sterilize.

Preparation of Medium: Add components, except cephalothin solution, to distilled/deionized water and bring volume to 990.0mL. Mix thoroughly. Gently heat and bring to boiling. Autoclave for 15 min at 15 psi pressure–121°C. Cool to 45°–50°C. Aseptically add sterile cephalothin solution. Mix thoroughly. Pour into sterile Petri dishes.

Use: For the isolation and cultivation of *Serratia* species. *Serratia* appear as colonies with red halos.

DTM Agar (Dermatophyte Test Medium Agar)

Composition per liter:

Agar	20.0g
Enzymatic digest of soybean meal	10.0g
Glucose	10.0g
Cycloheximide	0.5g
Phenol Red	0.2g
Chlortetracycline	0.1g
Gentamicin	0.1g

pH 7.3 ± 0.2 at 25°C

Source: Available as a prepared medium from BD Diagnostic Systems.

Preparation of Medium: Add components to distilled/deionized water and bring volume to 1.0L. Mix thoroughly. Gently heat and bring to boiling. Distribute into tubes or flasks. Autoclave for 15 min at 15 psi pressure–121°C. Pour into sterile Petri dishes or leave in tubes.

Use: For the isolation and cultivation of dermatophytic fungi.

Dulaney Slants

Composition per liter:

Egg yolks	50.0mL
Locke solution	50.0mL

Locke Solution:

Composition per 100.0mL:

NaCl	0.9g
Glucose	0.25g
KCl	0.042g
$CaCl_2 \cdot 2H_2O$	0.024g
Na_2CO_3	0.02g

Preparation of Locke Solution: Add components to distilled/deionized water and bring volume to 100.0mL. Mix thoroughly. Filter sterilize.

Preparation of Medium: Aseptically remove the yolks from 5–8 day old hen egg embryos. Add an equal volume of sterile Locke solution containing sterile glass beads. Mix thoroughly to homogenize. Aseptically distribute into sterile tubes. Inspissate tubes in a slanted position at 80°C (moist heat) for 15 min.

Use: For the cultivation of *Calymmatobacter granulomatis* from clinical specimens.

Duncan-Strong Sporulation Medium, Modified (DS Sporulation Medium, Modified) (Sporulation Medium, Modified)

Composition per liter:

Proteose peptone	15.0g
$Na_2HPO_4 \cdot 7H_2O$	10.0g
Raffinose	4.0g
Yeast extract	4.0g
Sodium thioglycolate	1.0g

pH 7.8 ± 0.2 at 25°C

Preparation of Medium: Add components to distilled/deionized water and bring volume to 1.0L. Mix thoroughly. Gently heat and bring to boiling. Distribute into tubes or flasks. Autoclave for 15 min at 15 psi pressure–121°C. Adjust pH to 7.8 with filter-sterilized 0.66*M* Na_2CO_3. Pour into sterile Petri dishes or leave in tubes.

Use: For the cultivation and induction of sporulation of *Clostridium perfringens.*

Dunkelberg Carbohydrate Medium, Modified

Composition per 100.0mL:

Proteose peptone no. 3	1.5g
Carbohydrate	1.0g
$Na_2HPO_4 \cdot 2H_2O$	0.207g
Phenol Red	0.055g
$NaH_2PO_4 \cdot H_2O$	0.038g
Horse serum	5.0mL

pH 7.4 ± 0.2 at 25°C

Preparation of Medium: Add components, except horse serum, to distilled/deionized water and bring volume to 95.0mL. For carbohydrate, use glucose, maltose, or starch. Mix thoroughly. Filter sterilize. Aseptically add sterile horse serum. Mix thoroughly. Aseptically distribute into sterile tubes or flasks.

Use: For the cultivation and differentiation of *Gardnerella vaginalis* based on its ability to ferment glucose, maltose, or starch.

Dunkelberg Maintenance Medium

Composition per liter:

Proteose peptone no. 3	20.0g
Soluble starch	10.0g
Agar	8.0g
Glucose	2.0g
Na_2HPO_4	1.0g
NaH_2PO_4	1.0g

pH 6.8 ± 0.2 at 25°C

Preparation of Medium: Add starch to approximately 100.0mL of cold distilled/deionized water. Mix thoroughly. Add starch solution to 400.0mL of boiling distilled/deionized water. Add remaining components. Mix thoroughly. Bring volume to 1.0L with distilled/deionized water. Distribute into screw-capped tubes. Autoclave for 12 min at 8 psi pressure–112°C.

Use: For the cultivation and maintenance of *Gardnerella vaginalis.*

Dunkelberg Semisolid Carbohydrate Fermentation Medium

Composition per liter:

Proteose peptone no. 3	20.0g
Carbohydrate	10.0g
Agar	5.0g
Bromcresol Purple solution	1.0mL

pH 7.4 ± 0.2 at 25°C

Bromcresol Purple Solution:
Composition per 10.0mL:

Bromcresol Purple	0.16g
Ethanol (95% solution)	10.0mL

Preparation of Bromcresol Purple Solution: Add Bromcresol Purple to 10.0mL of ethanol. Mix thoroughly. Filter sterilize.

Preparation of Medium: Add components to distilled/deionized water and bring volume to 1.0L. For carbohydrate, use glucose, maltose, or starch. Mix thoroughly. Gently heat and bring to boiling. Filter sterilize. Aseptically distribute into sterile tubes or flasks.

Use: For the cultivation and differentiation of *Gardnerella vaginalis* based on its ability to ferment glucose, maltose, or starch.

E. coli O157:H7 MUG Agar

Composition per liter:

Casein peptone	20.0g
Agar	13.0g
Sorbitol	10.0g
NaCl	5.0g
Meat extract	2.0g
$Na_2S_2O_3$	2.0g
Na-deoxycholate	1.12g
Yeast extract	1.0g
Ammonium ferric citrate	0.5g
4-methylumbelliferyl-β-D-glucuronide	0.1g
Bromthymol Blue	0.025g

pH 7.4 ± 0.2 at 37°C

Source: This medium is available from Fluka, Sigma-Aldrich.

Preparation of Medium: Add components to distilled/deionized water and bring volume to 1.0L. Mix thoroughly. Gently heat while stirring and bring to boiling. Autoclave for 15 min at 15 psi pressure–121°C. Cool to 50°C. Pour into sterile Petri dishes.

Use: For the isolation and differentiation of enterohaemorrhagic (EHEC) *E. coli* O157:H7-strains from clinical specimens.

Eagle Medium

Composition per 99.1mL:

Eagle MEM in Hanks BSS	87.0mL
Fetal bovine serum	10.0mL
$NaHCO_3$ (7.5% solution)	1.0mL
Penicillin-streptomycin solution	1.0mL
Amphotericin B solution	0.1mL

pH 7.2–7.4 at 25°C

Eagle MEM in Hanks BSS:
Composition per liter:

NaCl	8.0g
Glucose	1.0g
KCl	0.4g
$CaCl_2 \cdot 2H_2O$	0.14g
$MgSO_4 \cdot 7H_2O$	0.1g
KH_2PO_4	0.06g
Na_2HPO_4	0.05g
L-Isoleucine	0.026g
L-Leucine	0.026g
L-Lysine	0.026g
L-Threonine	0.024g
L-Valine	0.0235g
L-Tyrosine	0.018g
L-Arginine	0.0174g
L-Phenylalanine	0.0165g
L-Cystine	0.012g
L-Histidine	8.0mg
L-Methionine	7.5mg
Phenol Red	5.0mg
L-Tryptophan	4.0mg
Inositol	1.8mg
Biotin	1.0mg

Folic acid	1.0mg
Calcium pantothenate	1.0mg
Choline chloride	1.0mg
Nicotinamide	1.0mg
Pyridoxal·HCl	1.0mg
Thiamine·HCl	1.0mg
Riboflavin	0.1mg

Preparation of Eagle MEM in Hanks BSS: Add components to distilled/deionized water and bring volume to 1.0L. Mix thoroughly.

Penicillin-Streptomycin Solution:
Composition per 1.0mL:

Streptomycin	0.01g
Penicillin	10,000U

Preparation of Penicillin-Streptomycin Solution: Add components to distilled/deionized water and bring volume to 1.0mL. Mix thoroughly.

Amphotericin B Solution:
Composition per 1.0mL:

Amphotericin B	1.0mg

Preparation of Amphotericin B Solution: Add amphotericin B to distilled/deionized water and bring volume to 1.0mL. Mix thoroughly.

Preparation of Medium: Combine components. Mix thoroughly. Filter sterilize.

Use: For the cultivation of animal tissue culture cell lines.

Eagle Medium

Composition per liter:

Hanks balanced salt solution (10X)	100.0mL
Calf serum	50.0mL
$NaHCO_3$ (7.5% solution)	29.6mL
Tissue culture amino acids (50X)	20.0mL
Tissue culture vitamins (100X)	10.0mL
Glutamine solution	10.0mL
Phenol Red (0.5% solution)	4.0mL
Penicillin solution	1.0mL
Streptomycin solution	0.4mL

pH 7.0 ± 0.2 at 25°C

Hanks Balanced Salt Solution (10X):
Composition per 100.0mL:

NaCl	8.0g
Glucose	1.0g
KCl	0.4g
$NaHCO_3$	0.35g
$CaCl_2 \cdot 2H_2O$	0.14g
$MgCl_2 \cdot 6H_2O$	0.1g
$MgSO_4 \cdot 7H_2O$	0.1g
Na_2HPO_4	0.06g
KH_2PO_4	0.06g
Phenol Red	0.02g

Preparation of Hanks Balanced Salt Solution (10X): Add components to distilled/deionized water and bring volume to 100.0mL. Mix thoroughly.

Tissue Culture Amino Acids (50X):
Composition per liter:

L-Arginine	0.1g
L-Lysine	0.058g
L-Isoleucine	0.052g
L-Leucine	0.052g
L-Threonine	0.048g
L-Valine	0.046g
L-Tyrosine	0.036g
L-Phenylalanine	0.032g
L-Histidine	0.031g
L-Cystine	0.024g
L-Methionine	0.015g
L-Tryptophan	0.01g

Preparation of Tissue Culture Amino Acids, Minimal Eagle 50X: Add components to distilled/deionized water and bring volume to 1.0L. Mix thoroughly.

Tissue Culture Vitamins (100X):
Composition per liter:

Inositol	2.0mg
Calcium pantothenate	1.0mg
Choline chloride	1.0mg
Folic acid	1.0mg
Nicotinamide	1.0mg
Pyridoxal	1.0mg
Thiamine·HCl	1.0mg
Riboflavin	0.1mg

Preparation of TC Vitamins, Minimal Eagle 100X: Add components to distilled/deionized water and bring volume to 1.0L. Mix thoroughly.

Glutamine Solution:
Composition per 100.0mL:

L-Glutamine	2.9g

Preparation of Glutamine Solution: Add glutamine to distilled/deionized water and bring volume to 100.0mL. Mix thoroughly.

Penicillin Solution:
Composition per 1.0mL:

Penicillin	200,000U

Preparation of Penicillin Solution: Add penicillin to distilled/deionized water and bring volume to 1.0mL. Mix thoroughly.

Streptomycin Solution:

Composition per 1.0mL:

Streptomycin	0.5g

Preparation of Streptomycin Solution: Add streptomycin to distilled/deionized water and bring volume to 1.0mL. Mix thoroughly.

Preparation of Medium: Combine components. Mix thoroughly. Adjust pH to 7.0 with 1*N* NaOH. Filter sterilize.

Use: For the cultivation of animal tissue culture cell lines, especially for use with rhinoviruses.

Eagle Medium

Composition per 100.1mL:

Eagle MEM in Earle BSS	94.0mL
$NaHCO_3$ (7.5% solution)	3.0mL
Fetal bovine serum, inactivated	2.0mL
Penicillin-streptomycin solution	1.0mL
Amphotericin B solution	0.1mL

pH 7.2–7.4 at 25°C

Eagle MEM in Earle BSS:

Composition per liter:

NaCl	6.8g
Glucose	1.0g
KCl	0.4g
$CaCl_2 \cdot 2H_2O$	0.2g
$MgCl_2 \cdot 6H_2O$	0.2g
NaH_2PO_4	0.15g
L-Arginine	0.1g
L-Lysine	0.06g
L-Isoleucine	0.05g
L-Leucine	0.05g
L-Threonine	0.05g
L-Valine	0.05g
L-Tyrosine	0.04g
L-Phenylalanine	0.03g
L-Histidine	0.03g
L-Cystine	0.02g
L-Methionine	0.02g
L-Tryptophan	0.01g
i-Inositol	2.0mg
Calcium pantothenate	1.0mg
Choline chloride	1.0mg
Folic acid	1.0mg
Nicotinamide	1.0mg
Pyridoxal	1.0mg
Thiamine·HCl	1.0mg
Riboflavin	0.1mg

Preparation of Eagle MEM in Earle BSS: Add components to distilled/deionized water and bring volume to 1.0L. Mix thoroughly.

Penicillin-Streptomycin Solution:

Composition per 1.0mL:

Streptomycin	0.01g
Penicillin	10,000U

Preparation of Penicillin-Streptomycin Solution: Add components to distilled/deionized water and bring volume to 1.0mL. Mix thoroughly.

Amphotericin B Solution:

Composition per 1.0mL:

Amphotericin B	1.0mg

Preparation of Amphotericin B Solution: Add amphotericin B to distilled/deionized water and bring volume to 1.0mL. Mix thoroughly.

Preparation of Medium: Combine components. Mix thoroughly. Filter sterilize.

Use: For the cultivation of animal tissue culture cell lines.

Eagle Medium, Modified

Composition per liter:

Eagle MEM (10X)	100.0mL
Fetal bovine serum	100.0mL
Glucose solution	20.0mL
HEPES (*N*-2-Hydroxyethyl piperazine-*N*′-2-ethanesulfonic acid) buffer, 1*M*, pH 7.2	20.0mL
Glutamine solution	10.0mL
$NaHCO_3$ (7.5% solution)	7.5mL
Gentamicin sulfate solution	0.2mL

pH 7.2 ± 0.2 at 25°C

Eagle MEM (10X):

Composition per 100.0mL:

Sterile salt solution	97.0mL
TC amino acids, minimal Eagle 50X	2.0mL
TC vitamins, minimal Eagle 100X	1.0mL

Preparation of Eagle MEM (10X): Combine components. Mix thoroughly. Filter sterilize.

Sterile Salt Solution:

Composition per 100.0mL:

NaCl	6.8g
Glucose	1.0g
KCl	0.4g
$CaCl_2$	0.2g
$MgCl_2$	0.2g
NaH_2PO_4	0.15g

Preparation of Sterile Salt Solution: Add components to distilled/deionized water and bring volume to 100.0mL. Mix thoroughly. Filter sterilize.

Tissue Culture Amino Acids, Minimal Eagle 50X:

Composition per liter:

L-Arginine	0.1g
L-Lysine	0.06g
L-Isoleucine	0.05g
L-Leucine	0.05g
L-Threonine	0.05g
L-Valine	0.05g
L-Tyrosine	0.04g
L-Phenylalanine	0.03g
L-Histidine	0.03g
L-Cystine	0.02g
L-Methionine	0.02g
L-Tryptophan	0.01g

Preparation of Tissue Culture Amino Acids, Minimal Eagle 50X: Add components to distilled/deionized water and bring volume to 1.0L. Mix thoroughly. Adjust pH to 7.2–7.4. Filter sterilize.

TC Vitamins, Minimal Eagle 100X:

Composition per liter:

Inositol	2.0mg
Calcium pantothenate	1.0mg
Choline chloride	1.0mg
Folic acid	1.0mg
Nicotinamide	1.0mg
Pyridoxal	1.0mg
Thiamine·HCl	1.0mg
Riboflavin	0.1mg

Preparation of TC Vitamins, Minimal Eagle 100X: Add components to distilled/deionized water and bring volume to 1.0L. Mix thoroughly. Filter sterilize.

Glucose Solution:

Composition per 100.0mL:

Glucose	27.0g

Preparation of Glucose Solution: Add glucose to distilled/deionized water and bring volume to 100.0mL. Mix thoroughly. Filter sterilize.

Glutamine Solution:

Composition per 10.0mL:

L-Glutamine	5.0g

Preparation of Glutamine Solution: Add glutamine to distilled/deionized water and bring volume to 10.0mL. Mix thoroughly. Filter sterilize.

Gentamicin Solution:

Composition per 1.0mL:

Gentamicin sulfate	0.05g

Preparation of Gentamicin Solution: Add gentamicin sulfate to distilled/deionized water and bring volume to 1.0mL. Mix thoroughly. Filter sterilize.

Preparation of Medium: Combine components. Mix thoroughly. Filter sterilize.

Use: For the cultivation of animal tissue culture cell lines, especially for McCoy cells.

Eagle's Minimal Essential Medium with Earle's Salts and Nonessential Amino Acids (MEM with Earle's Salts and Nonessential Amino Acids)

Composition per liter:

NaCl	6.8g
$NaHCO_3$	2.2g
Glucose	1.0g
KCl	0.4g
$CaCl_2 \cdot 2H_2O$	0.265g
$MgSO_4 \cdot 7H_2O$	0.2g
L-Arginine·H_2O	0.15g
$NaH_2PO_4 \cdot H_2O$	0.14g
L-Arginine·HCl	0.126g
L-Lysine·HCl	72.5mg
L-Tyrosine, disodium salt	52.1mg
L-Leucine	52.0mg
L-Threonine	48.0mg
L-Valine	46.0mg
L-Histidine·HCl·H_2O	42.0mg
D-Phenylalanine	32.0mg
L-Cysteine·2HCl	31.29mg
L-Methionine	15.0mg
L-Glutamic acid	14.7mg
L-Aspartic acid	13.3mg
L-Proline	11.5mg
L-Serine	10.5mg
L-Tryptophan	10.0mg
Phenol Red	10.0mg
L-Alanine	8.9mg
L-Glycine	7.5mg
i-Inositol	2.0mg
D-Calcium pantothenate	1.0mg
Choline chloride	1.0mg
Folic acid	1.0mg
Nicotinamide	1.0mg
Pyridoxal·HCl	1.0mg
Thiamine·HCl	1.0mg
Riboflavin	0.1mg

pH 7.2 ± 0.2 at 25°C

Preparation of Medium: Add components to 1.0L of distilled/deionized water. Mix thoroughly. Filter sterilize.

Use: For the cultivation of animal cells in tissue culture, for example cells for viral detection and identification by characteristic cytopathic effects.

Earle's Balanced Salts, Phenol Red-Free

Composition per liter:

NaCl	6.8g
$NaHCO_3$	2.2g
Glucose	1.0g
KCl	0.4g
$CaCl_2 \cdot 2H_2O$	0.265g
$MgSO_4 \cdot 7H_2O$	0.2g
$NaH_2PO_4 \cdot H_2O$	0.14g

pH 7.2 ± 0.2 at 25°C

Preparation of Medium: Add components to distilled/deionized water and bring volume to 1.0L. Mix thoroughly. Filter sterilize.

Use: For the preparation of tissue culture media where Phenol Red is not desired.

EB Motility Medium

Composition per liter:

Peptone or gelysate	10.0g
NaCl	5.0g
Agar	4.0g
Beef extract	3.0g

pH 7.4 ± 0.2 at 25°C

Preparation of Medium: Add components to distilled/deionized water and bring volume to 1.0L. Mix thoroughly. Gently heat and bring to boiling. Distribute into tubes in 8.0mL volumes. Autoclave for 15 min at 15 psi pressure–121°C.

Use: For the cultivation and differentiation of bacteria based on motility.

EC Broth (*Escherichia coli* Broth) (EC Medium)

Composition per liter:

Pancreatic digest of casein	20.0g
Lactose	5.0g
NaCl	5.0g
K_2HPO_4	4.0g
Bile salts mixture	1.5g
KH_2PO_4	1.5g

pH 6.9 ± 0.2 at 25°C

Source: This medium is available as a premixed powder from BD Diagnostic Systems.

Preparation of Medium: Add components to distilled/deionized water and bring volume to 1.0L. Mix thoroughly. Distribute into test tubes that contain an inverted Durham tube. Autoclave for 12 min at 15 psi pressure–121°C. Cool broth as quickly as possible.

Use: For the cultivation and differentiation of coliform bacteria at 37°C and of *Escherichia coli* at 45.5°C.

EC Broth with MUG

Composition per liter:

Pancreatic digest of casein	20.0g
Lactose	5.0g
NaCl	5.0g
K_2HPO_4	4.0g
Bile salts mixture	1.5g
KH_2PO_4	1.5g
4-Methylumbeliferyl-β-D-glucuronide (MUG)	0.05g

pH 6.9 ± 0.2 at 25°C

Source: This medium is available as a premixed powder from BD Diagnostic Systems.

Preparation of Medium: Add components to distilled/deionized water and bring volume to 1.0L. Mix thoroughly. Distribute into test tubes that contain an inverted Durham tube in 10.0mL volumes. Autoclave for 15 min at 15 psi pressure–121°C.

Use: For the detection of *Escherichia coli* by a fluorogenic procedure.

EC Medium, Modified with Novobiocin

Composition per liter:

Tryptone	20.0g
NaCl	5.0g
Lactose	5.0g
K_2HPO_4	4.0g
KH_2PO_4	1.5g
Bile salts	1.12g
Novobiocin supplement	10.0mL

pH 6.9 ± 0.2 at 25°C

Source: This medium is available as a premixed powder and supplement from BD Diagnostic Systems.

Novobiocin Supplement:
Composition per 10.0mL:

Sodium novobiocin	20.0mg

Preparation of Novobiocin Supplement: Add sodium novobiocin to distilled/deionized water and bring volume to 10.0mL. Mix thoroughly. Filter sterilize.

Preparation of Medium: Add components, except novobiocin supplement, to distilled/deionized water and bring volume to 990.0mL. Mix thoroughly.

Autoclave for 15 min at 15 psi pressure–121°C. Aseptically add 10.0mL of sterile novobiocin supplement. Mix thoroughly. Aseptically distribute into sterile tubes or flasks.

Use: For the cultivation of *Escherichia coli* O157:H7.

ECD MUG Agar

Composition per liter:

Casein peptone	20.0g
Agar	15.0g
NaCl	5.0g
Lactose	5.0g
K_2HPO_4	4.0g
Bile salt mixture	1.5g
KH_2PO_4	1.5g
Tryptophan	1.0g
4-Methylumbelliferyl-β-D-glucuronide	0.07g

pH 7.0 ± 0.2 at 37°C

Source: This medium is available from Fluka, Sigma-Aldrich.

Preparation of Medium: Add components to distilled/deionized water and bring volume to 1.0L. Mix thoroughly. Gently heat while stirring and bring to boiling. Autoclave for 15 min at 15 psi pressure–121°C. Cool to 50°C. Pour into sterile Petri dishes.

Use: For detection of *Escherichia coli* in a variety of specimens. The bile-salt mixture in this *E. coli* direct agar extensively inhibits the non-obligatory intestinal accompanying flora. Fluorescence in the UV and a positive indole test demonstrate the presence of *E. coli* in the colonies.

Egg Yolk Agar

Composition per liter:

Proteose peptone no. 2	40.0g
Agar	25.0g
Na_2HPO_4	5.0g
Glucose	2.0g
NaCl	2.0g
KH_2PO_4	1.0g
$MgSO_4·7H_2O$	0.1g
Egg yolk emulsion	100.0mL
Hemin solution	1.0mL

pH 7.6 ± 0.2 at 25°C

Hemin Solution:

Composition per 100.0mL:

Hemin	0.5g
NaOH (1*N* solution)	20.0mL

Preparation of Hemin Solution: Add hemin to 20.0mL of 1*N* NaOH solution. Mix thoroughly. Bring volume to 100.0mL with distilled/deionized water.

Egg Yolk Emulsion:

Composition:

Chicken egg yolks	11
Whole chicken egg	1

Preparation of Egg Yolk Emulsion: Soak eggs with 1:100 dilution of saturated mercuric chloride solution for 1 min. Crack eggs and separate yolks from whites. Mix egg yolks with 1 chicken egg.

Preparation of Medium: Add components, except egg yolk emulsion, to distilled/deionized water and bring volume to 900.0mL. Mix thoroughly. Gently heat and bring to boiling. Autoclave for 15 min at 15 psi pressure–121°C. Cool to 45°–50°C. Aseptically add sterile egg yolk emulsion. Mix thoroughly. Pour into sterile Petri dishes.

Use: For the isolation, cultivation, and differentiation of *Clostridium* species and some other anaerobic bacteria.

Egg Yolk Agar, Modified

Composition per liter:

Agar	20.0g
Pancreatic digest of casein	15.0g
Vitamin K_1	10.0g
NaCl	5.0g
Papaic digest of soybean meal	5.0g
Yeast extract	5.0g
L-Cystine	0.4g
Hemin	5.0mg
Egg yolk emulsion	100.0mL

Source: This medium is available as a prepared medium from BD Diagnostic Systems.

Egg Yolk Emulsion:

Composition:

Chicken egg yolks	11
Whole chicken egg	1

Preparation of Egg Yolk Emulsion: Soak eggs with 1:100 dilution of saturated mercuric chloride solution for 1 min. Crack eggs and separate yolks from whites. Mix egg yolks with 1 chicken egg.

Preparation of Medium: Add components, except egg yolk emulsion, to distilled/deionized water and bring volume to 900.0mL. Mix thoroughly. Gently heat and bring to boiling. Autoclave for 15 min at 15 psi pressure–121°C. Cool to 45°–50°C. Aseptically add sterile egg yolk emulsion. Mix thoroughly. Pour into sterile Petri dishes.

Use: For the isolation, cultivation, and differentiation of *Clostridium* species and some other anaerobic bacteria.

Egg Yolk Emulsion

Composition per 100mL:

Sterile saline	70.0mL
Egg yolk	30.0mL

Source: Sterile egg yolk emulsion is available from Fluka, Sigma-Aldrich.

Preparation of Medium: Use fresh eggs, less than 1 week old. Scrub the shells with soap. Let stand in a soap solution for 30 min. Rinse in running water. Soak eggs in 70% ethanol for 15 min. or soak eggs with 1:100 dilution of saturated mercuric chloride solution for 1 min. Crack eggs and separate yolks from whites, placing egg yolks into a sterile container. Use enough egs to produce at least 30.0mL egg yolk. Homogenize by shaking. Add 0.9g NaCl to distilled/deionized water and bring volume to 100.0mL. Sterilze the saline solution by filtration or by autoclaving for 15 min at 15 psi pressure–121°C. If autoclaving is used, cool to 25°C. Aseptically add 30.0mL homogenized egg yolks to 70.0mL of sterile saline solution. Mix thoroughly.

Use: Sterile stabilized emulsion of egg yolk recommended for use in various culture media.

Egg Yolk Emulsion

Composition per 100mL:

NaCl	0.45g
Egg yolk	50.0mL

Preparation of Medium: Use fresh eggs, less than 1 week old. Scrub the shells with soap. Let stand in a soap solution for 30 min. Rinse in running water. Soak eggs in 70% ethanol for 15 min. or soak eggs with 1:100 dilution of saturated mercuric chloride solution for 1 min. Crack eggs and separate yolks from whites, placing egg yolks into a sterile container. Use enough egs to produce at least 50.0mL egg yolk. Homogenize by shaking. Add 0.45g NaCl to distilled/deionized water and bring volume to 50.0mL. Sterilze the saline solution by filtration or by autoclaving for 15 min at 15 psi pressure–121°C. If autoclaving is used, cool to 25°C. Aseptically add 50.0mL homogenized egg yolks to 50.0mL of the sterile NaCl solution. Mix thoroughly.

Use: Sterile stabilized emulsion of egg yolk recommended for use in various culture media.

Egg Yolk Emulsion, 50%

Composition per 100.0mL:

Chicken egg yolks	variable
NaCl (0.85% solution)	40.0mL

Preparation of Egg Yolk Emulsion: Wash fresh eggs wtih stiff brush and drain. Soak eggs in 70% ethanol for 1h. Crack eggs aseptically and separate yolks from whites. Remove egg yolks with a sterile syringe or a wide-mouth pipet. Place 50.0mL of egg yolks into a sterile container. Add 50.0mL sterile 0.85% saline

Use: For use in media requiring egg yolk emulsion.

Egg-yolk Tellurite Emulsion 20%

Composition per 100mL:

NaCl	0.425g
K_2TeO_3	0.21g
Egg yolk	20.0mL

Preparation of Medium: Use fresh eggs, less than 1 week old. Scrub the shells with soap. Let stand in a soap solution for 30 min. Rinse in running water. Soak eggs in 70% ethanol for 15 min. or soak eggs with 1:100 dilution of saturated mercuric chloride solution for 1 min. Crack eggs and separate yolks from whites, placing egg yolks into a sterile container. Use enough eggs to produce at least 20.0mL egg yolk. Homogenize by shaking. Add 0.45g NaCl and 0.21g K_2TeO_3 to distilled/deionized water and bring volume to 80.0mL. Sterilze the saline-tellurite solution by filtration or by autoclaving for 15 min at 15 psi pressure–121°C. If autoclaving is used, cool to 25°C. Aseptically add 20.0mL homogenized egg yolks to 80.0mL of the sterile saline-tellurite solution. Mix thoroughly.

Use: For use in various culture media. It may be added directly to nutrient media for the identification of *Clostridium, Bacillus,* and *Staphylococcus* species by their lipase activity.

Eijkman Lactose Medium

Composition per liter:

Pancreatic digest of casein	15.0g
K_2HPO_4	10.0g
KH_2PO_4	4.0g
Lactose	3.0g
NaCl	2.5g

pH 6.8 ± 0.1 at 25°C

Preparation of Medium: Add components to distilled/deionized water and bring volume to 1.0L. Mix thoroughly. Distribute into test tubes that contain an inverted Durham tube. Autoclave for 15 min at 15 psi pressure–121°C.

Use: For the cultivation and differentiation of *Escherichia coli* from other coliform organisms based on their ability to ferment lactose and produce gas.

Eijkman Lactose Medium

Composition per liter:

Tryptose	15.0g

NaCl 5.0g
K_2HPO_4 .... 4.0g
Lactose 3.0g
KH_2PO_4 .... 1.5g
pH 6.8 ± 0.1 at 25°C

Preparation of Medium: Add components to distilled/deionized water and bring volume to 1.0L. Mix thoroughly. Distribute into test tubes that contain an inverted Durham tube. Autoclave for 15 min at 15 psi pressure–121°C.

Use: For the cultivation and differentiation of *Escherichia coli* from other coliform organisms based on their ability to ferment lactose and produce gas.

EMB Agar (Eosin Methylene Blue Agar)

Composition per liter:
Agar 13.5g
Pancreatic digest of casein 10.0g
Lactose 5.0g
Sucrose 5.0g
K_2HPO_4 .... 2.0g
Eosin Y 0.4g
Methylene Blue 0.065g
pH 7.2 ± 0.2 at 25°C

Source: This medium is available as a premixed powder from BD Diagnostic Systems.

Preparation of Medium: Add components to distilled/deionized water and bring volume to 1.0L. Mix thoroughly. Gently heat and bring to boiling. Distribute into tubes or flasks. Autoclave for 15 min at 15 psi pressure–121°C. Pour into sterile Petri dishes.

Use: For the isolation, cultivation, and differentiation of Gram-negative enteric bacteria based on lactose fermentation. Bacteria that ferment lactose, especially the coliform bacterium *Escherichia coli*, appear as colonies with a green metallic sheen or blue-black to brown color. Bacteria that do not ferment lactose appear as colorless or transparent, light purple colonies.

EMB Agar Base

Composition per liter:
Agar 15.0g
Peptone 10.0g
K_2HPO_4 .... 2.0g
Eosin Y 0.4g
Methylene Blue 0.065g
pH 7.3 ± 0.2 at 25°C

Preparation of Medium: Add components to distilled/deionized water and bring volume to 1.0L. Mix thoroughly. Gently heat and bring to boiling. Distribute into tubes or flasks. Autoclave for 15 min at 15 psi pressure–121°C. Pour into sterile Petri dishes.

Use: For the isolation, cultivation, and differentiation of Gram-negative enteric bacteria based on lactose fermentation. Bacteria that ferment lactose, especially the coliform bacterium *Escherichia coli*, appear as colonies with a green metallic sheen or blue-black to brown color. Bacteria that do not ferment lactose appear as colorless or transparent, light purple colonies.

EMB Agar, Modified (Eosin Methylene Blue Agar, Modified)

Composition per liter:
Agar 15.0g
Lactose 10.0g
Pancreatic digest of gelatin 10.0g
K_2HPO_4 .... 2.0g
Eosin Y 0.4g
Methylene Blue 0.065g
pH 6.8 ± 0.2 at 25°C

Source: This medium is available as a premixed powder from Oxoid.

Preparation of Medium: Add components to distilled/deionized water and bring volume to 1.0L. Mix thoroughly. Gently heat and bring to boiling. Distribute into tubes or flasks. Autoclave for 15 min at 15 psi pressure–121°C. Cool to 60°C. Shake medium to oxidize methylene blue. Pour into sterile Petri dishes. Swirl flask while pouring plates to distribute precipitate.

Use: For the isolation, cultivation, and differentiation of Gram-negative enteric bacteria based on lactose fermentation. Bacteria that ferment lactose, especially the coliform bacterium *Escherichia coli*, appear as colonies with a green metallic sheen or blue-black to brown color. Bacteria that do not ferment lactose appear as colorless or transparent, light purple colonies.

Endo Agar

Composition per liter:
Agar 15.0g
Lactose 10.0g
Peptic digest of animal tissue 10.0g

K_2HPO_4 .... 3.5g
Na_2SO_3 .... 2.5g
Basic Fuchsin 0.5g
pH 7.4 ± 0.2 at 25°C

Source: This medium is available as a premixed powder from BD Diagnostic Systems.

Caution: Basic Fuchsin is a potential carcinogen and care must be taken to avoid inhalation of the powdered dye and contact with the skin.

Preparation of Medium: Add components to distilled/deionized water and bring volume to 1.0L. Mix thoroughly. Gently heat and bring to boiling. Autoclave for 15 min at 15 psi pressure–121°C. Cool to 45°–50°C. Pour into sterile Petri dishes. Swirl flask while pouring plates to keep precipitate in suspension. Protect from the light.

Use: For the selective isolation, cultivation, and differentiation of coliform and other enteric microorganisms based on their ability to ferment lactose. Lactose-fermenting bacteria appear as dark red colonies with a gold metallic sheen. Lactose-nonfermenting bacteria appear as colorless or translucent colonies.

Endo Agar

Composition per liter:
Agar 10.0g
Lactose 10.0g
Peptic digest of animal tissue 10.0g
K_2HPO_4 .... 3.5g
Na_2SO_3 .... 2.5g
Basic Fuchsin solution 4.0mL
pH 7.5 ± 0.2 at 25°C

Source: This medium is available as a premixed powder from Oxoid.

Basic Fuchsin Solution:
Composition per 10.0mL:
Basic Fuchsin 1.0g
Ethanol (95% solution) 10.0mL

Preparation of Basic Fuchsin Solution: Add Basic Fuchsin to 10.0mL of ethanol. Mix thoroughly.

Caution: Basic Fuchsin is a potential carcinogen and care must be taken to avoid inhalation of the powdered dye and contact with the skin.

Preparation of Medium: Add components to distilled/deionized water and bring volume to 1.0L. Mix thoroughly. Gently heat and bring to boiling. Autoclave for 15 min at 15 psi pressure–121°C. Cool to 45°–50°C. Pour into sterile Petri dishes. Swirl flask while pouring plates to keep precipitate in suspension. Protect from the light.

Use: For the selective isolation, cultivation, and differentiation of coliform and other enteric microorganisms based on their ability to ferment lactose. Lactose-fermenting bacteria appear as dark red colonies with a gold metallic sheen. Lactose-nonfermenting bacteria appear as colorless or translucent colonies.

Endo Agar, LES (Endo Agar, Laurance Experimental Station) (m-Endo Agar, LES) (m-LES, Endo Agar)

Composition per liter:
Agar 14.0g
Lactose 9.4g
Peptones (pancreatic digest of casein 65% and yeast extract 35%) 7.5g
NaCl 3.7g
Pancreatic digest of casein 3.7g
Peptic digest of animal tissue 3.7g
K_2HPO_4 .... 3.3g
Na_2SO_3 .... 1.6g
Yeast extract 1.2g
KH_2PO_4 .... 1.0g
Basic Fuchsin 0.8g
Sodium lauryl sulfate 0.05g
Ethanol 20.0mL
pH 7.2 ± 0.2 at 25°C

Source: This medium is available as a premixed powder from BD Diagnostic Systems.

Caution: Basic Fuchsin is a potential carcinogen and care must be taken to avoid inhalation of the powdered dye and contact with the skin.

Preparation of Medium: Add ethanol to approximately 900.0mL of distilled/deionized water. Add remaining components. Bring volume to 1.0L with distilled/deionized water. Mix thoroughly. Gently heat and bring to boiling. Autoclave for 15 min at 15 psi pressure–121°C. Pour into sterile 60mm Petri dishes in 4.0mL volumes. Protect from the light.

Use: For the cultivation and enumeration of coliform bacteria by the membrane filter method.

Endo Agar, LES (m-Endo Agar, LES)

Composition per liter:
Agar 10.0g
Lactose 9.4g

Tryptose 7.5g
NaCl 3.7g
Peptone 3.7g
Pancreatic digest of casein 3.7g
K_2HPO_4 .... 3.3g
Na_2SO_3 .... 1.6g
Yeast extract 1.2g
KH_2PO_4 .... 1.0g
Sodium deoxycholate 0.1g
Sodium lauryl sulfate 0.05g
Basic Fuchsin solution 8.0mL

pH 7.2 ± 0.2 at 25°C

Basic Fuchsin Solution:
Composition per 10.0mL:

Basic Fuchsin 1.0g
Ethanol (95% solution) 10.0mL

Caution: Basic Fuchsin is a potential carcinogen and care must be taken to avoid inhalation of the powdered dye and contact with the skin.

Preparation of Basic Fuchsin Solution: Add Basic Fuchsin to 10.0mL of ethanol. Mix thoroughly.

Preparation of Medium: Add components to distilled/deionized water and bring volume to 1.0L. Mix thoroughly. Gently heat and bring to boiling. Autoclave for 15 min at 15 psi pressure–121°C. Cool to 45°–50°C. Pour into sterile Petri dishes. Swirl flask while pouring plates to keep precipitate in suspension. Protect from the light.

Use: For the cultivation and enumeration of coliform bacteria by the membrane filter method.

Endo Broth (m-Endo Broth)

Composition per liter:

Lactose 12.5g
Peptone 10.0g
NaCl 5.0g
Pancreatic digest of casein 5.0g
Peptic digest of animal tissue 5.0g
K_2HPO_4 .... 4.375g
Na_2SO_3 .... 2.1g
Yeast extract 1.5g
KH_2PO_4 .... 1.375g
Basic Fuchsin 1.05g
Sodium deoxycholate 0.1g
Ethanol (95% solution) 20.0mL

pH 7.2 ± 0.1 at 25°C

Source: This medium is available as a premixed powder from BD Diagnostic Systems.

Caution: Basic Fuchsin is a potential carcinogen and care must be taken to avoid inhalation of the powdered dye and contact with the skin.

Preparation of Medium: Add ethanol to approximately 900.0mL of distilled/deionized water. Add remaining components. Bring volume to 1.0L with distilled/deionized water. Mix thoroughly. Gently heat and bring to boiling. Rapidly cool broth below 45°C. Do not autoclave. Use 1.8–2.0mL for each filter pad. Protect from the light. Prepare broth freshly.

Use: For the cultivation and enumeration of coliform bacteria by the membrane filter method.

Enriched Nutrient Broth

Composition per liter:

Beef heart, infusion from 300.0g
Tryptose 7.5g
NaCl 3.0g
Yeast extract 3.0g
Peptone 2.5g
NaCl 2.5g
Beef extract 0.5g

Preparation of Medium: Add components to distilled/deionized water and bring volume to 1.0L. Mix thoroughly. Distribute into tubes or flasks. Autoclave for 15 min at 15 psi pressure–121°C.

Use: For the cultivation of fastidious bacteria.

Enrichment Broth for *Aeromonas hydrophila*

Composition per liter:

NaCl 5.0g
Maltose 3.5g
Yeast extract 3.0g
Bile salts no. 3 .... 1.0g
L-Cysteine·HCl·H_2O 0.3g
Bromthymol Blue 0.03g
Novobiocin 5.0mg

pH 7.0 ± 0.2 at 25°C

Preparation of Medium: Add components to distilled/deionized water and bring volume to 1.0L. Mix thoroughly. Distribute into tubes or flasks. Autoclave for 15 min at 15 psi pressure–121°C.

Use: For the cultivation and enrichment of *Aeromonas hydrophila*.

Entamoeba Medium (Endamoeba Medium)

Composition per liter:

Liver infusion 272.0g

Rice powder	14.2g
Agar	11.0g
Proteose peptone	5.5g
Sodium glycerophosphate	3.0g
NaCl	2.7g
Horse serum	50.0mL

pH 7.0 ± 0.2 at 25°C

Source: This medium is available as a premixed powder from BD Diagnostic Systems.

Rice Powder:
Composition per 15.0g:

Rice powder	15.0g

Preparation of Rice Powder: Sterilize rice powder at 160°C for 60 min. Do not overheat or rice powder will scorch.

Preparation of Medium: Add components, except horse serum and rice powder, to distilled/deionized water and bring volume to 994.0mL. Mix thoroughly. Gently heat and bring to boiling. Distribute into tubes in 7.0mL volumes. Autoclave for 15 min at 15 psi pressure–121°C. Allow tubes to cool in a slanted position. Aseptically add enough sterile horse serum to each tube to cover about half the slant. Aseptically add 0.1g of sterile rice powder to each tube.

Use: For the cultivation of *Entamoeba histolytica.*

Enteric Fermentation Base (Fermentation Base for *Campylobacter*)

Composition per liter:

Peptic digest of animal tissue	10.0g
NaCl	5.0g
Beef extract	3.0g
Carbohydrate solution	100.0mL
Andrade's indicator	10.0mL

pH 7.2 ± 0.1 at 25°C

Source: This medium is available as a premixed powder from BD Diagnostic Systems.

Carbohydrate Solution:
Composition per 100.0mL:

Carbohydrate	10.0g

Preparation of Carbohydrate Solution: Add carbohydrate to distilled/deionized water and bring volume to 100.0mL. Mix thoroughly. Filter sterilize. Glucose, lactose, mannitol, sucrose, adonitol, arabinose, cellobiose, dulcitol, glycerol, inositol, salicin, xylose, or other carbohydrates may be used. For the preparation of expensive carbohydrate solutions (adonitol, arabinose, cellobiose, dulcitol, glycerol, inositol, salicin, or xylose), 5.0g of carbohydrate per 100.0mL of distilled/deionized water may be used.

Andrade's Indicator:
Composition per 100.0mL:

NaOH (1*N* solution)	16.0mL
Acid Fuchsin	0.1g

Caution: Acid Fuchsin is a potential carcinogen and care must be taken to avoid inhalation of the powdered dye and contact with the skin.

Preparation of Andrade's Indicator: Add components to distilled/deionized water and bring volume to 100.0mL. Mix thoroughly.

Preparation of Medium: Add components, except carbohydrate solution, to distilled/deionized water and bring volume to 900.0mL. Mix thoroughly. Gently heat and bring to boiling. Distribute into tubes that contain an inverted Durham tube in 9.0mL volumes. Autoclave for 15 min at 15 psi pressure–121°C. Cool to 25°C. Aseptically add 1.0mL of sterile carbohydrate solution per tube. Mix thoroughly.

Use: For the cultivation and differentiation of a variety of bacteria based on their ability to ferment different carbohydrates. Bacteria that produce acid from carbohydrate fermentation turn the medium dark pink to red. Bacteria that produce gas have a bubble trapped in the Durham tube.

Enterobacter Medium

Composition per 800.0mL:

Casein hydrolysate	2.0g
K_2HPO_4	1.4g
K_2SO_4	1.0g
Yeast extract	1.0g
KH_2PO_4	0.6g
$MgSO_4$	0.5g
Glycerol	20.0mL

Preparation of Medium: Add components to distilled/deionized water and bring volume to 800.0mL. Mix thoroughly. Distribute into tubes or flasks. Autoclave for 15 min at 15 psi pressure–121°C.

Use: For the cultivation and maintenance of *Enterobacter* species and *Klebsiella pneumoniae.*

Enterococci Broth, Chromocult (Chromocult Enterococci Broth)

Composition per liter:

Peptone	8.6g
NaCl	6.4g
Tween 80	2.2g

NaN_3 .... 0.6g
5-Bromo-4-chloro-3-indolyl-β-D-glucopyranoside (X-GLU) 0.04

pH 7.5 ± 0.2 at 25°C

Source: This medium is available from Merck.

Preparation of Medium: Add components to distilled/deionized water and bring volume to 1.0L. Mix well. Distribute into tubes. Autoclave for 15 min at 15 psi pressure–121°C. The prepared broth is clear and yellowish.

Use: For the detection of enterococci. The sodium-azide present in this medium largely inhibits the growth of the accompanying, and especially the Gram-negative microbial flora while sparing the enterococci. The substrate X-GLU (5-bromo-4-chloro-3-indolyl-β-D-glucopyranoside) is cleaved, stimulated by selected peptones, by the enzyme β-D-glucosidase which is characteristic for enterococci. This results in an intensive blue-green color of the broth. Azide, at the same time, prevents a false positive result by most other β-D-glucosidase positive bacteria. Therefore, the color-change of the broth largely confirms the presence of enterococci and group D-streptococci.

Enterococcosel™ Agar

Composition per liter:

Pancreatic digest of casein	17.0g
Agar	13.5g
Oxgall	10.0g
NaCl	5.0g
Yeast extract	5.0g
Peptic digest of animal tissue	3.0g
Esculin	1.0g
Sodium citrate	1.0g
Ferric ammonium citrate	0.5g
NaN_3	0.25g

pH 7.1 ± 0.2 at 25°C

Source: This medium is available as a premixed powder from BD Diagnostic Systems.

Caution: Sodium azide is toxic. Azides also react with metals and disposal must be highly diluted.

Preparation of Medium: Add components to distilled/deionized water and bring volume to 1.0L. Mix thoroughly. Gently heat while stirring and bring to boiling. Distribute into tubes or flasks. Autoclave for 15 min at 15 psi pressure–121°C. Pour into sterile Petri dishes or leave in tubes.

Use: For the rapid, selective isolation, cultivation, and enumeration of fecal group D streptococci (enterococci). For the cultivation of staphylococci and *Listeria monocytogenes.*

Enterococcosel Broth

Composition per liter:

Pancreatic digest of casein	17.0g
Oxgall	10.0g
NaCl	5.0g
Yeast extract	5.0g
Peptic digest of animal tissue	3.0g
Esculin	1.0g
Sodium citrate	1.0g
Ferric ammonium citrate	0.5g
NaN_3	0.25g

pH 7.1 ± 0.2 at 25°C

Source: This medium is available as a premixed powder from BD Diagnostic Systems.

Caution: Sodium azide is toxic. Azides also react with metals and disposal must be highly diluted.

Preparation of Medium: Add components to distilled/deionized water and bring volume to 1.0L. Mix thoroughly. Gently heat while stirring until dissolved. Distribute into tubes or flasks. Autoclave for 15 min at 15 psi pressure–121°C.

Use: For the cultivation and differentiation of group D streptococci (enterococci).

Enterococcus Agar (m-*Enterococcus* Agar) (Azide Agar)

Composition per liter:

Pancreatic digest of casein	15.0g
Agar	10.0g
Papaic digest of soybean meal	5.0g
Yeast extract	5.0g
KH_2PO_4	4.0g
Glucose	2.0g
NaN_3	0.4g
Triphenyltetrazolium chloride	0.1g

pH 7.2 ± 0.2 at 25°C

Source: This medium is available as a premixed powder from BD Diagnostic Systems.

Caution: Sodium azide is toxic. Azides also react with metals and disposal must be highly diluted.

Preparation of Medium: Add components to distilled/deionized water and bring volume to 1.0L. Mix thoroughly. Gently heat and bring to boiling. Cool to 45°–50°C. Do not autoclave. Pour into sterile Petri dishes.

Use: For the direct plating of specimens for the detection and enumeration of fecal streptococci.

Enterococcus faecium Medium

Composition per liter:

Sucrose	97.3g
Brain heart, solids from infusion	37.0g
Agar	13.3g
NaCl	9.3g
Yeast extract	5.0g
$MgSO_4$	0.25g

pH 7.4 ± 0.2 at 25°C

Preparation of Medium: Add components to distilled/deionized water and bring volume to 1.0L. Mix thoroughly. Gently heat and bring to boiling. Distribute into tubes or flasks. Autoclave for 15 min at 15 psi pressure–121°C. Pour into sterile Petri dishes or leave in tubes.

Use: For the cultivation and maintenance of *Enterococcus faecium.*

Erythritol Agar

Composition per liter:

Agar	15.0g
Erythritol	2.0g
K_2HPO_4	1.15g
NH_4NO_3	1.0g
KH_2PO_4	0.625g
$MgSO_4 \cdot 7H_2O$	0.02g

Preparation of Medium: Add components to distilled/deionized water and bring volume to 1.0L. Mix thoroughly. Gently heat and bring to boiling. Distribute into tubes or flasks. Autoclave for 15 min at 15 psi pressure–121°C. Pour into sterile Petri dishes or leave in tubes.

Use: For the cultivation and maintenance of *Klebsiella pneumoniae.*

Erythritol Broth

Composition per liter:

Erythritol	2.0g
K_2HPO_4	1.15g
NH_4NO_3	1.0g
Yeast extract	1.0g
KH_2PO_4	0.625g
$MgSO_4 \cdot 7H_2O$	0.02g

Preparation of Medium: Add components to distilled/deionized water and bring volume to 1.0L. Mix thoroughly. Distribute into tubes or flasks. Autoclave for 15 min at 15 psi pressure–121°C.

Use: For the cultivation of *Klebsiella pneumoniae.*

Esculin Agar

Composition per liter:

Agar	15.0g
Pancreatic digest of casein	13.0g
NaCl	5.0g
Yeast extract	5.0g
Heart muscle, solids from infusion	2.0g
Esculin	1.0g
Ferric citrate	0.5g

pH 7.3 ± 0.2 at 25°C

Preparation of Medium: Add components to distilled/deionized water and bring volume to 1.0L. Mix thoroughly. Gently heat and bring to boiling. Distribute into screw-capped tubes in 3.0mL volumes. Autoclave for 15 min at 15 psi pressure–121°C. Allow tubes to cool in a slanted position.

Use: For the cultivation and differentiation of bacteria based on their ability to hydrolyze esculin and produce H_2S. Bacteria that hydrolyze esculin appear as colonies surrounded by a reddish-brown to dark brown zone. Bacteria that produce H_2S appear as black colonies.

Esculin Agar, Modified CDC

Composition per liter:

Heart muscle, infusion from	375.0g
Agar	15.0g
Thiotone	10.0g
NaCl	5.0g
Esculin	1.0g
Ferric citrate	0.5g

pH 7.0 ± 0.2 at 25°C

Preparation of Medium: Add components to distilled/deionized water and bring volume to 1.0L. Mix thoroughly. Gently heat and bring to boiling. Cool to 55°C. Adjust pH to 7.0. Distribute into tubes or leave in flask. Autoclave for 20 min at 15 psi pressure–121°C. Pour into sterile Petri dishes or leave in tubes. Allow tubes to cool in inclined position to produce slants

Use: For the differentiation of *Enterobacter* spp.

Esculin Broth

Composition per liter:

Beef heart, solids from infusion	500.0g
Tryptose	10.0g
NaCl	5.0g
Agar	1.0g
Esculin	1.0g

pH 7.0 ± 0.2 at 25°C

Preparation of Medium: Add components to distilled/deionized water and bring volume to 1.0L.

Mix thoroughly. Gently heat and bring to boiling. Distribute into screw-capped tubes in 7.0mL volumes. Autoclave for 15 min at 15 psi pressure–121°C.

Use: For the cultivation and differentiation of bacteria based on their ability to hydrolyze esculin. Bacteria that hydrolyze esculin turn the medium brown-black to black.

Esculin Iron Agar

Composition per liter:

Agar	15.0g
Esculin	1.0g
Ferric ammonium citrate	0.5g

pH 7.1 ± 0.2 at 25°C

Source: This medium is available as a premixed powder from BD Diagnostic Systems.

Preparation of Medium: Add components to distilled/deionized water and bring volume to 1.0L. Mix thoroughly. Gently heat and bring to boiling. Distribute into tubes or flasks. Autoclave for 15 min at 15 psi pressure–121°C. Pour into sterile Petri dishes.

Use: For the cultivation and identification of enterococci based on their ability to hydrolyze esculin. Used in conjunction with E agar and the membrane filter method.

Esculin Mannitol Agar

Composition per liter:

Agar	13.5g
Polypeptone	10.0g
D-Mannitol	10.0g
Pancreatic digest of casein	5.0g
Yeast extract	5.0g
NaCl	5.0g
Heart peptone	3.0g
Cornstarch	1.0g
Esculin	1.0g
Ferric ammonium citrate	0.5g
Phenol Red	0.025g
Nalidixic acid solution	10.0mL
Colistin solution	10.0mL

pH 7.3 ± 0.2 at 25°C

Nalidixic Acid Solution:

Composition per 10.0mL:

Nalidixic acid	0.015g

Preparation of Nalidixic Acid Solution: Add nalidixic acid to distilled/deionized water and bring volume to 10.0mL. Mix thoroughly. Filter sterilize.

Colistin Solution:

Composition per 10.0mL:

Colistin	0.01g

Preparation of Colistin Solution: Add colistin to distilled/deionized water and bring volume to 10.0mL. Mix thoroughly. Filter sterilize.

Preparation of Medium: Add components, except nalidixic acid solution and colistin solution, to distilled/deionized water and bring volume to 980.0mL. Mix thoroughly. Gently heat and bring to boiling. Autoclave for 15 min at 15 psi pressure–121°C. Cool to 45°–50°C. Aseptically add sterile nalidixic acid solution and colistin solution. Mix thoroughly. Pour into sterile Petri dishes or distribute into sterile tubes.

Use: For the selective isolation, cultivation, and differentiation of *Staphylococcus aureus* and group D streptococci based on mannitol fermentation and hydrolysis of esculin. Bacteria that ferment mannitol appear as yellow colonies surrounded by a yellow zone. Bacteria that hydrolyze esculin appear as dark brown to black colonies surrounded by a dark brown to black zone.

ETGPA
(Egg Tellurite Glycine Pyruvate Agar)

Composition per liter:

Agar	17.0g
Glycine	12.0g
Sodium pyruvate	10.0g
Pancreatic digest of casein	10.0g
Beef extract	5.0g
LiCl	5.0g
Yeast extract	1.0g
Egg yolk emulsion	50.0mL
K_2TeO_3 solution	10.0mL

pH 7.0 ± 0.2 at 25°C

Source: This medium is available as a premixed powder from BD Diagnostic Systems.

Egg Yolk Emulsion:

Composition:

Chicken egg yolks	11
Whole chicken egg	1

Preparation of Egg Yolk Emulsion: Soak egg with 1:100 dilution of saturated mercuric chloride solution for 1 min. Crack eggs and separate yolks from whites. Mix egg yolks with 1 chicken egg.

K_2TeO_3 Solution:

Composition per 100.0mL:

K_2TeO_3	1.0g

Preparation of K_2TeO_3 Solution: Add K_2TeO_3 to distilled/deionized water and bring volume to 100.0mL. Mix thoroughly. Filter sterilize.

Caution: Potassium tellurite is toxic.

Preparation of Medium: Add components to distilled/deionized water and bring volume to 940.0mL. Mix thoroughly. Gently heat and bring to boiling. Autoclave for 15 min at 15 psi pressure–121°C. Cool to 45°–50°C. Add 10.0mL of sterile 1% tellurite solution and 50.0mL of sterile egg yolk emulsion. If desired, add sulfamethazine to a final concentration of 50.0mg/mL. Mix thoroughly but gently and pour into sterile Petri dishes.

Use: For the selective isolation and enumeration of coagulase-positive staphylococci from skin. For the differentiation and identification of staphylococci on the basis of their ability to clear egg yolk. Addition of sulfamethazine inhibits the growth of *Proteus*. Gray-black colonies surrounded by a clear zone are diagnostic for *Staphylococcus aureus*.

Ethyl Violet Azide Broth (EVA Broth)

Composition per liter:

Pancreatic digest of casein	13.5g
Yeast extract	6.5g
Glucose	5.0g
NaCl	5.0g
K_2HPO_4	2.7g
KH_2PO_4	2.7g
NaN_3	0.4g
Ethyl Violet	0.83mg

pH 7.0 ± 0.2 at 25°C

Source: This medium is available as a premixed powder from BD Diagnostic Systems.

Caution: Sodium azide is toxic. Azides also react with metals and disposal must be highly diluted.

Preparation of Medium: Add components to distilled/deionized water and bring volume to 1.0L. Mix thoroughly. Gently heat and bring to boiling. Distribute into tubes in 10.0mL volumes. Autoclave for 15 min at 15 psi pressure–121°C.

Use: For the isolation, cultivation, and enumeration of enterococci from water and other specimens. Fecal enterococci turn the medium turbid with a purple sediment on the bottom of the tube.

Ethyl Violet Azide Broth (EVA Broth)

Composition per liter:

Tryptose	20.0g
Glucose	5.0g
NaCl	5.0g
K_2HPO_4	2.7g
KH_2PO_4	2.7g
NaN_3	0.4g
Ethyl Violet	0.83mg

pH 7.0 ± 0.2 at 25°C

Source: This medium is available as a premixed powder from BD Diagnostic Systems.

Caution: Sodium azide is toxic. Azides also react with metals and disposal must be highly diluted.

Preparation of Medium: Add components to distilled/deionized water and bring volume to 1.0L. Mix thoroughly. Gently heat and bring to boiling. Distribute into tubes in 10.0mL volumes. Autoclave for 15 min at 15 psi pressure–121°C.

Use: For the isolation, cultivation, and enumeration of enterococci from water and other specimens. Fecal enterococci turn the medium turbid with a purple sediment on the bottom of the tube.

Ethyl Violet Azide Broth (EVA Broth)

Composition per liter:

Tryptose	20.0g
Glucose	5.0g
NaCl	5.0g
K_2HPO_4	2.7g
KH_2PO_4	2.7g
NaN_3	0.3g
Ethyl Violet	0.5mg

pH 6.8 ± 0.2 at 25°C

Source: This medium is available as a premixed powder from Oxoid Unipath.

Preparation of Medium: Add components to distilled/deionized water and bring volume to 1.0L. Mix thoroughly. Gently heat and bring to boiling. Distribute into tubes in 10.0mL volumes. Autoclave for 15 min at 15 psi pressure–121°C.

Caution: Sodium azide is toxic. Azides also react with metals and disposal must be highly diluted.

Use: For the isolation, cultivation, and enumeration of enterococci from water and other specimens. Fecal enterococci turn the medium turbid with a purple sediment on the bottom of the tube.

Eubacterium Medium

Composition per liter:

Pancreatic digest of casein	20.0g
Agar	15.0g
Meat extract	15.0g
Glucose	5.0g
$Na_2HPO_4 \cdot 12H_2O$	4.0g
L-Cysteine·HCl	0.5g

pH 7.4 ± 0.2 at 25°C

Preparation of Medium: Add components to distilled/deionized water and bring volume to 1.0L. Mix thoroughly. Gently heat and bring to boiling. Distribute into tubes or flasks. Autoclave for 15 min at 15 psi pressure–121°C. Pour into sterile Petri dishes or leave in tubes. Use freshly prepared medium.

Use: For the cultivation of *Eubacterium* species.

Eubacterium Medium

Composition per liter:

Beef brain powder	33.33g
Pancreatic digest of casein	15.0g
Yeast extract	10.0g
Glucose	5.5g
Yeast extract	5.0g
NaCl	2.5g
Sodium thioglycolate	1.8g
L-Cystine	0.5g

pH 7.0 ± 0.2 at 25°C

Preparation of Medium: Add components, except beef brain powder, to distilled/deionized water and bring volume to 1.0L. Mix thoroughly. Gently heat and bring to boiling under 97% N_2 + 3% H_2. Continue boiling for 15–20 min. Adjust pH to 7.0. Cool to 25°C under 97% N_2 + 3% H_2. Anaerobically distribute into tubes in 9.0mL volumes. Add 0.3g of beef brain powder to each tube. Cap tubes with rubber stoppers. Place tubes in a press. Autoclave for 15 min at 15 psi pressure–121°C with fast exhaust.

Use: For the cultivation of *Eubacterium* species.

Fastidious Anaerobe Agar (FAA)

Composition per liter:

Peptone	23.0g
Agar	12.0g
NaCl	5.0g
Glucose	1.0g
L-Arginine	1.0g
Sodium pyruvate	1.0g
Soluble starch	1.0g
L-Cysteine·HCl·H_2O	0.5g
Sodium succinate	0.5g
$NaHCO_3$	0.4g
$Na_4P_2O_7 \cdot 10H_2O$	0.25g
Sheep blood, defibrinated	50.0mL
Hemin solution	1.0mL
Vitamin K_1 solution	0.1mL

pH 7.2 ± 0.2 at 25°C

Vitamin K_1 Solution:
Composition per 100.0mL:

Vitamin K_1	1.0g
Ethanol	99.0mL

Preparation of Vitamin K_1 Solution: Add vitamin K_1 to 99.0mL of absolute ethanol. Mix thoroughly.

Hemin Solution:
Composition per 100.0mL:

Hemin	1.0g
NaOH (1*N* solution)	20.0mL

Preparation of Hemin Solution: Add hemin to 20.0mL of 1*N* NaOH solution. Mix thoroughly. Bring volume to 100.0mL with distilled/deionized water.

Preparation of Medium: Add components, except defibrinated sheep blood, to distilled/deionized water and bring volume to 950.0mL. Mix thoroughly. Gently heat and bring to boiling. Autoclave for 15 min at 15 psi pressure–121°C. Cool to 45°–50°C. Aseptically add 50.0mL of sterile defibrinated sheep blood. Mix thoroughly. Pour into sterile Petri dishes or distribute into sterile tubes.

Use: For the cultivation of a variety of fastidious anaerobes from clinical and nonclinical specimens.

Fastidious Anaerobe Agar, Alternative Selective (FAA Alternative Selective)

Composition per liter:

Peptone	23.0g
Agar	12.0g
NaCl	5.0g
Glucose	1.0g
L-Arginine	1.0g
Sodium pyruvate	1.0g
Soluble starch	1.0g
L-Cysteine·HCl·H_2O	0.5g
Sodium succinate	0.5g
$NaHCO_3$	0.4g
$Na_4P_2O_7 \cdot 10H_2O$	0.25g
Sheep blood, defibrinated	50.0mL
Hemin solution	1.0mL
Vitamin K_1 solution	0.1mL

pH 7.2 ± 0.2 at 25°C

Vitamin K_1 Solution:
Composition per 100.0mL:

Vitamin K_1	1.0g
Ethanol	99.0mL

Preparation of Vitamin K_1 Solution: Add vitamin K_1 to 99.0mL of absolute ethanol. Mix thoroughly.

Hemin Solution:

Composition per 100.0mL:

Hemin	1.0g
NaOH (1*N* solution)	20.0mL

Preparation of Hemin Solution: Add hemin to 20.0mL of 1*N* NaOH solution. Mix thoroughly. Bring volume to 100.0mL with distilled/deionized water.

Preparation of Medium: Add components, except defibrinated sheep blood, to distilled/deionized water and bring volume to 950.0mL. Mix thoroughly. Gently heat and bring to boiling. Autoclave for 15 min at 15 psi pressure–121°C. Cool to 45°–50°C. Aseptically add 50.0mL of sterile defibrinated sheep blood. Mix thoroughly. Pour into sterile Petri dishes or distribute into sterile tubes.

Use: For the cultivation of a variety of fastidious anaerobes from clinical and nonclinical specimens.

Fastidious Anaerobe Agar, Alternative Selective with Neomycin, Vancomycin, and Josamycin (FAA Alternative Selective Medium with Neomycin, Vancomycin, and Josamycin)

Composition per liter:

Peptone	23.0g
Agar	12.0g
NaCl	5.0g
Glucose	1.0g
L-Arginine	1.0g
Sodium pyruvate	1.0g
Soluble starch	1.0g
L-Cysteine·HCl·H_2O	0.5g
Sodium succinate	0.5g
$NaHCO_3$	0.4g
$Na_4P_2O_7 \cdot 10H_2O$	0.25g
Neomycin	0.1g
Sheep blood, defibrinated	50.0mL
Vancomycin solution	10.0mL
Josamycin	10.0mL
Hemin solution	1.0mL
Vitamin K_1	0.1mL

pH 7.2 ± 0.2 at 25°C

Vitamin K_1 Solution:

Composition per 100.0mL:

Vitamin K_1	1.0g
Ethanol	99.0mL

Preparation of Vitamin K_1 Solution: Add vitamin K_1 to 99.0mL of absolute ethanol. Mix thoroughly.

Hemin Solution:

Composition per 100.0mL:

Hemin	1.0g
NaOH (1*N* solution)	20.0mL

Preparation of Hemin Solution: Add hemin to 20.0mL of 1*N* NaOH solution. Mix thoroughly. Bring volume to 100.0mL with distilled/deionized water.

Vancomycin Solution:

Composition per 10.0mL:

Vancomycin	5.0mg

Preparation of Vancomycin Solution: Add vancomycin to distilled/deionized water and bring volume to 10.0mL. Mix thoroughly. Filter sterilize.

Josamycin Solution:

Composition per 10.0mL:

Josamycin	3.0mg

Preparation of Josamycin Solution: Add josamycin to distilled/deionized water and bring volume to 10.0mL. Mix thoroughly. Filter sterilize.

Preparation of Medium: Add components, except defibrinated sheep blood, vancomycin solution, and josamycin solution, to distilled/deionized water and bring volume to 930.0mL. Mix thoroughly. Gently heat and bring to boiling. Autoclave for 15 min at 15 psi pressure–121°C. Cool to 45°–50°C. Aseptically add 50.0mL of sterile defibrinated sheep blood, 10.0mL vancomycin solution, and 10.0mL of josamycin solution. Mix thoroughly. Pour into sterile Petri dishes or distribute into sterile tubes.

Use: For the selective cultivation of *Fusobacterium* species from clinical and nonclinical specimens.

Fastidious Anaerobe Agar, Selective (FAA Selective)

Composition per liter:

Peptone	23.0g
Agar	12.0g
NaCl	5.0g
Glucose	1.0g
L-Arginine	1.0g
Sodium pyruvate	1.0g
Soluble starch	1.0g
L-Cysteine·HCl·H_2O	0.5g
Sodium succinate	0.5g
$NaHCO_3$	0.4g
$Na_4P_2O_7 \cdot 10H_2O$	0.25g
Sheep blood, defibrinated	50.0mL
Hemin solution	1.0mL
Vitamin K_1 solution	0.1mL

pH 7.2 ± 0.2 at 25°C

Vitamin K_1 Solution:

Composition per 100.0mL:

Vitamin K_1	1.0g
Ethanol	99.0mL

Preparation of Vitamin K_1 Solution: Add vitamin K_1 to 99.0mL of absolute ethanol. Mix thoroughly.

Hemin Solution:

Composition per 100.0mL:

Hemin	1.0g
NaOH (1*N* solution)	20.0mL

Preparation of Hemin Solution: Add hemin to 20.0mL of 1*N* NaOH solution. Mix thoroughly. Bring volume to 100.0mL with distilled/deionized water.

Preparation of Medium: Add components, except defibrinated sheep blood, to distilled/deionized water and bring volume to 950.0mL. Mix thoroughly. Gently heat and bring to boiling. Autoclave for 15 min at 15 psi pressure–121°C. Cool to 45°–50°C. Aseptically add 50.0mL of sterile defibrinated sheep blood. Mix thoroughly. Pour into sterile Petri dishes or distribute into sterile tubes.

Use: For the cultivation of a variety of fastidious anaerobes from clinical and nonclinical specimens.

Fastidious Anaerobe Agar, Selective with Neomycin and Vancomycin (FAA Selective with Neomycin and Vancomycin)

Composition per liter:

Peptone	23.0g
Agar	12.0g
NaCl	5.0g
Glucose	1.0g
L-Arginine	1.0g
Sodium pyruvate	1.0g
Soluble starch	1.0g
L-Cysteine·HCl·H_2O	0.5g
Sodium succinate	0.5g
$NaHCO_3$	0.4g
$Na_4P_2O_7·10H_2O$	0.25g
Neomycin	0.1g
Sheep blood, defibrinated	50.0mL
Vancomycin solution	10.0mL
Hemin solution	1.0mL
Vitamin K_1 solution	0.1mL

pH 7.2 ± 0.2 at 25°C

Vitamin K_1 Solution:

Composition per 100.0mL:

Vitamin K_1	1.0g
Ethanol	99.0mL

Preparation of Vitamin K_1 Solution: Add vitamin K_1 to 99.0mL of absolute ethanol. Mix thoroughly.

Hemin Solution:

Composition per 100.0mL:

Hemin	1.0g
NaOH (1*N* solution)	20.0mL

Preparation of Hemin Solution: Add hemin to 20.0mL of 1*N* NaOH solution. Mix thoroughly. Bring volume to 100.0mL with distilled/deionized water.

Vancomycin Solution:

Composition per 10.0mL:

Vancomycin	7.5mg

Preparation of Vancomycin Solution: Add vancomycin to distilled/deionized water and bring volume to 10.0mL. Mix thoroughly. Filter sterilize.

Preparation of Medium: Add components, except defibrinated sheep blood and vancomycin solution, to distilled/deionized water and bring volume to 940.0mL. Mix thoroughly. Gently heat and bring to boiling. Autoclave for 15 min at 15 psi pressure–121°C. Cool to 45°–50°C. Aseptically add 50.0mL of sterile defibrinated sheep blood and 10.0mL of vancomycin solution. Mix thoroughly. Pour into sterile Petri dishes or distribute into sterile tubes.

Use: For the selective cultivation of *Fusobacterium* species from clinical and nonclinical specimens.

Fay and Barry Medium

Composition per liter:

Amino acid	10.0g
Peptone	5.0g
Yeast extract	3.0g
Bromcresol Purple solution	5.0mL

pH 5.5 ± 0.2 at 25°C

Bromcresol Purple Solution:

Composition per 100.0mL:

Bromcresol Purple	0.2g
Ethanol	50.0mL

Preparation of Bromcresol Purple Solution: Add Bromcresol Purple to 50.0mL of absolute ethanol. Add distilled/deionized water and bring volume to 100.0mL. Mix thoroughly.

Preparation of Medium: Add components to distilled/deionized water and bring volume to 1.0L. The amino acid may be L-arginine, L-ornithine, or L-lysine, depending on which amino acid decarboxylase activity is being measured. Mix thoroughly. Distribute into tubes or flasks. Autoclave for 15 min at 15 psi pressure–121°C.

Use: For the determination of decarboxylase activities of *Aeromonas* species.

FDA Agar (ATCC Medium 182) (AATCC Bacteriostasis Agar) (American Association of Textile Chemists and Colorists Bacteriostasis Agar)

Composition per liter:

Agar	15.0g
Peptic digest of animal tissue	10.0g
Beef extract	5.0g
NaCl	5.0g

pH 6.8 ± 0.1 at 25°C

Source: This medium is available as a premixed powder from BD Diagnostic Systems.

Preparation of Medium: Add components to distilled/deionized water and bring volume to 1.0L. Mix thoroughly. Gently heat and bring to boiling. Distribute into tubes or flasks. Autoclave for 15 min at 15 psi pressure–121°C. Pour into sterile Petri dishes or leave in tubes.

Use: For testing the antibacterial activities of antiseptics and disinfectants.

FDA Broth (AATCC Bacteriostasis Broth) (American Association of Textile Chemists and Colorists Bacteriostasis Broth)

Composition per liter:

Peptic digest of animal tissue	10.0g
Beef extract	5.0g
NaCl	5.0g

pH 6.8 ± 0.1 at 25°C

Source: This medium is available as a premixed powder from BD Diagnostic Systems.

Preparation of Medium: Add components to distilled/deionized water and bring volume to 1.0L. Mix thoroughly. Distribute into tubes or flasks. Autoclave for 15 min at 15 psi pressure–121°C.

Use: For testing the antibacterial activities of antiseptics and disinfectants.

Fecal Coliform Agar, Modified (m-Fecal Coliform Agar, Modified) (FCIC)

Composition per liter:

Agar	15.0g
Inositol	10.0g
Tryptose	10.0g
Proteose peptone no. 3	5.0g
NaCl	5.0g
Yeast extract	3.0g
Bile salts no. 3	1.5g
Aniline Blue	0.1g

pH 7.4 ± 0.2 at 25°C

Preparation of Medium: Add components and bring volume to 1.0L. Mix thoroughly. Gently heat and bring to boiling. Do not autoclave. Cool to 50°C. Adjust pH to 7.4. Pour into sterile Petri dishes in 20.0mL volumes. Allow surface of plates to dry before using.

Use: For the isolation, cultivation, and enumeration of *Klebsiella* species using the membrane filter method.

Fecal Coliform Agar, Modified

Composition per liter:

Agar	15.0g
Lactose	12.5g
Tryptose	10.0g
Proteose peptone no. 3	5.0g
NaCl	5.0g
Yeast extract	3.0g
Bile salts no. 3	1.5g
Aniline Blue	0.1g

pH 7.4 ± 0.2 at 25°C

Preparation of Medium: Add components and bring volume to 1.0L. Mix thoroughly. Gently heat and bring to boiling. Do not autoclave. Cool to 50°C. Adjust pH to 7.4. Pour into sterile Petri dishes in 20.0mL volumes. Allow surface of plates to dry before using.

Use: For the isolation, cultivation, and identification of stressed fecal coliform microorganisms based on their ability to ferment lactose. Lactose-fermenting bacteria turn the medium blue.

Fermentation Broth (CHO Medium)

Composition per liter:

Pancreatic digest of casein	15.0g
Yeast extract	7.0g
NaCl	2.5g
Agar	0.75g
Sodium thioglycolate	0.5g
L-Cystine	0.25g
Ascorbic acid	0.1g
Bromthymol Blue	0.01g
Carbohydrate or starch solution	100.0mL

pH 7.0 ± 0.1 at 25°C

Source: This medium is available as a premixed powder from BD Diagnostic Systems.

Carbohydrate Solution:
Composition per 100.0mL:
Carbohydrate 6.0g

Preparation of Carbohydrate Solution: Add carbohydrate to distilled/deionized water and bring volume to 10.0mL. Mix thoroughly. Filter sterilize.

Starch Solution:
Composition per 100.0mL:
Starch 2.5g

Preparation of Starch Solution: Add starch to distilled/deionized water and bring volume to 100.0mL. Mix thoroughly. Filter sterilize.

Preparation of Medium: Add components, except carbohydrate solution, to distilled/deionized water and bring volume to 900.0mL. Mix thoroughly. Distribute into tubes or flasks. Autoclave for 15 min at 15 psi pressure–121°C. Cool to 45°–50°C. Aseptically add 100.0mL of sterile carbohydrate solution. Mix thoroughly. Aseptically distribute into sterile tubes or flasks. Loosen caps on tubes. Place in an anaerobic chamber under an atmosphere of 85% N_2, 10% H_2, and 5% CO_2. Fasten the caps securely or maintain in an anaerobic chamber.

Use: For the differentiation of anaerobic bacteria based upon carbohydrate fermentation. Bacteria that ferment carbohydrates turn the medium yellow.

Fermentation Medium

Composition per liter:
Glucose or mannitol 10.0g
Pancreatic digest of casein 10.0g
Agar 2.2g
Yeast extract 1.0g
Bromcresol Purple 0.04g
pH 7.0 ± 0.2 at 25°C

Preparation of Medium: Add components to distilled/deionized water and bring volume to 1.0L. Mix thoroughly. Gently heat and bring to boiling. Distribute into tubes or flasks. Autoclave for 10 min at 15 psi pressure–121°C. Pour into sterile Petri dishes or leave in tubes.

Use: For differentiating *Staphylococcus* and *Micrococcus* species based upon the fermentation of glucose and mannitol.

F-G Agar (Feeley-Gorman Agar)

Composition per liter:
Casein, acid hydrolyzed 17.5g
Agar 17.0g
Beef extract 3.0g
Starch 1.5g
L-Cysteine solution 10.0mL
$Fe_4(P_2O_7)_3$ solution 10.0mL
pH 6.9 ± 0.05 at 25°C

L-Cysteine Solution:
Composition per 10.0mL:
L-Cysteine·HCl·H_2O 0.4g

Preparation of L-Cysteine Solution: Add L-cysteine·HCl·H_2O to distilled/deionized water and bring volume to 10.0mL. Mix thoroughly. Filter sterilize.

$Fe_4(P_2O_7)_3$ Solution:
Composition per 10.0mL:
$Fe_4(P_2O_7)_3$ 0.25g

Preparation of $Fe_4(P_2O_7)_3$ Solution: Add $Fe_4(P_2O_7)_3$ to distilled/deionized water and bring volume to 10.0mL. Mix thoroughly. Filter sterilize.

Preparation of Medium: Add components, except L-cysteine solution and $Fe_4(P_2O_7)_3$ solution, to distilled/deionized water and bring volume to 980.0mL. Mix thoroughly. Gently heat and bring to boiling. Autoclave for 15 min at 15 psi pressure–121°C. Cool to 45°–50°C. Aseptically add 10.0mL of L-cysteine solution. Mix thoroughly. Aseptically add 10.0mL of $Fe_4(P_2O_7)_3$ solution. Mix thoroughly. Adjust pH to 6.9. Pour into sterile Petri dishes or distribute into sterile tubes.

Use: For the isolation and cultivation of *Legionella pneumophila*.

F-G Agar with Selenium (Feeley-Gorman Agar with Selenium)

Composition per liter:
Casein, acid hydrolyzed 17.5g
Agar 17.0g
Beef extract 3.0g
Starch 1.5g
L-Cysteine solution 10.0mL
$Fe_4(P_2O_7)_3$ solution 10.0mL
$Na_2SeO_3·5H_2O$ solution 10.0mL
pH 6.9 ± 0.05 at 25°C

L-Cysteine Solution:
Composition per 10.0mL:
L-Cysteine·HCl·H_2O 0.4g

Preparation of L-Cysteine Solution: Add L-cysteine·HCl·H_2O to distilled/deionized water and bring volume to 10.0mL. Mix thoroughly. Filter sterilize.

$Fe_4(P_2O_7)_3$ Solution:
Composition per 10.0mL:
$Fe_4(P_2O_7)_3$ 0.25g

Preparation of $Fe_4(P_2O_7)_3$ Solution: Add $Fe_4(P_2O_7)_3$ to distilled/deionized water and bring volume to 10.0mL. Mix thoroughly. Filter sterilize.

$Na_2SeO_3 \cdot 5H_2O$ Solution:
Composition per 10.0mL:
$Na_2SeO_3 \cdot 5H_2O$ 0.01g

Preparation of $Na_2SeO_3 \cdot 5H_2O$ Solution: Add $Na_2SeO_3 \cdot 5H_2O$ to distilled/deionized water and bring volume to 10.0mL. Mix thoroughly. Filter sterilize.

Preparation of Medium: Add components—except L-cysteine solution, $Fe_4(P_2O_7)_3$ solution, and $Na_2SeO_3 \cdot 5H_2O$ solution—to distilled/deionized water and bring volume to 970.0mL. Mix thoroughly. Gently heat and bring to boiling. Autoclave for 15 min at 15 psi pressure–121°C. Cool to 45°–50°C. Aseptically add 10.0mL of sterile L-cysteine solution. Mix thoroughly. Aseptically add 10.0mL of sterile $Fe_4(P_2O_7)_3$ solution and 10.0mL of sterile $Na_2SeO_3 \cdot 5H_2O$ solution. Mix thoroughly. Adjust pH to 6.9. Pour into sterile Petri dishes or distribute into sterile tubes.

Use: For the isolation and cultivation of *Legionella pneumophila*.

F-G Broth (Feeley-Gorman Broth)

Composition per liter:
Casein, acid hydrolyzed 17.5g
Beef extract 3.0g
Starch 1.5g
L-Cysteine solution 10.0mL
$Fe_4(P_2O_7)_3$ solution 10.0mL
pH 6.9 ± 0.05 at 25°C

L-Cysteine Solution:
Composition per 10.0mL:
L-Cysteine·HCl·H_2O 0.4g

Preparation of L-Cysteine Solution: Add L-cysteine·HCl·H_2O to distilled/deionized water and bring volume to 10.0mL. Mix thoroughly. Filter sterilize.

$Fe_4(P_2O_7)_3$ Solution:
Composition per 10.0mL:
$Fe_4(P_2O_7)_3$ 0.25g

Preparation of $Fe_4(P_2O_7)_3$ Solution: Add $Fe_4(P_2O_7)_3$ to distilled/deionized water and bring volume to 10.0mL. Mix thoroughly. Filter sterilize.

Preparation of Medium: Add components, except L-cysteine solution and $Fe_4(P_2O_7)_3$ solution, to distilled/deionized water and bring volume to 980.0mL. Mix thoroughly. Gently heat and bring to boiling. Autoclave for 15 min at 15 psi pressure–121°C. Cool to 45°–50°C. Aseptically add 10.0mL of L-cysteine solution. Mix thoroughly. Aseptically add 10.0mL of $Fe_4(P_2O_7)_3$ solution. Mix thoroughly. Adjust pH to 6.9. Aseptically distribute into sterile tubes or flasks.

Use: For the cultivation of *Legionella pneumophila*.

FGTC Agar

Composition per liter:
Pancreatic digest of casein 15.0g
Agar 15.0g
Papaic digest of soybean meal 5.0g
NaCl 5.0g
KH_2PO_4 .... 5.0g
Amylose Azure 3.0g
Galactose 1.0g
Thallous acetate 0.5g
MUG (4-Methylumbelliferyl- α-D-galactoside) .. 0.1g
$NaHCO_3$ solution 20.0mL
Gentamicin solution 2.5mL
Tween 80 .... 0.75mL
pH 7.3 ± 0.2 at 25°C

Gentamicin Solution:
Composition per 10.0mL:
Gentamicin 0.01g

Preparation of Gentamicin Solution: Add gentamicin to distilled/deionized water and bring volume to 10.0mL. Mix thoroughly.

$NaHCO_3$ Solution:
Composition per 20.0mL:
$NaHCO_3$ 2.0g

Preparation of $NaHCO_3$ Solution: Add the $NaHCO_3$ to distilled/deionized water and bring volume to 20.0mL. Mix thoroughly. Filter sterilize. Use freshly prepared solution.

Preparation of Medium: Add components, except $NaHCO_3$ solution, to distilled/deionized water and bring volume to 980.0mL. Mix thoroughly. Gently heat and bring to boiling. Autoclave for 15 min at 15 psi pressure–121°C. Cool to 50°C. Aseptically add sterile $NaHCO_3$ solution. Mix thoroughly. Pour into sterile Petri dishes.

Use: For the cultivation, differentiation, and enumeration of *Enterococcus* species based on starch hydrolysis and production of fluorescence. Bacteria that hydrolyze starch, such as *Streptococcus bovis*, appear as colonies surrounded by a clear zone. Bacteria that produce fluorescence, such as *Streptococcus bovis* and *Enterococcus faecium*, appear as colonies surrounded by a zone of bright bluish fluorescence when viewed under a long-wave UV lamp. Other bacteria, such as *Enterococcus faecalis*, *Enterococcus avium*, or *Streptococcus equinus*, do not hydrolyze starch or produce fluorescence.

Fildes Enrichment Agar

Composition per liter:

Agar	15.0g
Peptone	5.0g
Beef extract	3.0g
Fildes enrichment solution	50.0mL

Fildes Enrichment Solution:

Composition per 206.0mL:

Pepsin	1.0g
NaCl (0.85% solution)	150.0mL
Sheep blood, defibrinated	50.0mL
HCl	6.0mL

pH 7.0–7.2 at 25°C

Source: Fildes enrichment solution is available from BD Diagnostic Systems.

Preparation of Fildes Enrichment Solution: Combine components. Mix thoroughly. Incubate at 56°C for 4h. Bring pH to 7.0 with 20% NaOH. Adjust pH to 7.2 with HCl. Do not autoclave. Add 0.25 mL of chloroform and store at 4°C. Before use, heat to 56°C to remove chloroform.

Preparation of Medium: Add components, except Fildes enrichment solution, to distilled/deionized water and bring volume to 950.0mL. Mix thoroughly. Gently heat and bring to boiling. Autoclave for 15 min at 15 psi pressure–121°C. Cool to 56°C. Aseptically add 50.0mL of sterile Fildes enrichment solution. Mix thoroughly. Pour into sterile Petri dishes or distribute into sterile tubes.

Use: For the isolation and cultivation of *Haemophilus influenzae*.

Fish Peptone Agar

Composition per liter:

Agar	5.0g
Maltose	5.0g
NaCl	5.0g
Peptone	5.0g
Pancreatic digest of casein	5.0g
Yeast extract	5.0g
Trout tissue extract solution	50.0mL

pH 7.0 ± 0.2 at 25°C

Trout Tissue Extract Solution:

Composition per liter:

Fish (brook trout)	500.0g
Pepsin	1.0g
HCl, concentrated	15.0mL

Preparation of Trout Tissue Extract: Add 1.0L of distilled/deionized water to brook trout and blend for 20–30 min. Add 1.0g of pepsin and 15.0mL of concentrated HCl to digest the trout proteins. Incubate for 12 hr at 45°C. Adjust pH to 7.0. Allow solids to settle. Filter sterilize. Do not autoclave. Store at 5°C.

Preparation of Medium: Add components, except trout tissue extract, to distilled/deionized water and bring volume to 950.0L. Mix thoroughly. Gently heat and bring to boiling. Autoclave for 15 min at 13 psi pressure–118°C. Cool to 45°–50°C. Aseptically add 50.0mL of sterile trout tissue extract. Mix thoroughly. Pour into sterile Petri dishes or distribute into sterile tubes.

Use: For the cultivation and maintenance of *Aeromonas salmonicida*.

Fish Peptone Broth

Composition per liter:

Maltose	5.0g
NaCl	5.0g
Peptone	5.0g
Pancreatic digest of casein	5.0g
Yeast extract	5.0g
Trout tissue extract solution	50.0mL

pH 7.0 ± 0.2 at 25°C

Trout Tissue Extract Solution:

Composition per liter:

Fish (brook trout)	500.0g
Pepsin	1.0g
HCl, concentrated	15.0mL

Preparation of Trout Tissue Extract: Add 1.0L of distilled/deionized water to brook trout and blend for 20–30 min. Add 1.0g of pepsin and 15.0mL of concentrated HCl to digest the trout proteins. Incubate for 12 hr at 45°C. Adjust pH to 7.0. Allow solids to settle. Filter sterilize. Do not autoclave. Store at 5°C.

Preparation of Medium: Add components, except trout tissue extract, to distilled/deionized water and bring volume to 950.0L. Mix thoroughly. Gently heat and bring to boiling. Autoclave for 15 min at 10 psi pressure–118°C. Cool to 45°–50°C. Aseptically add 50.0mL of sterile trout tissue extract. Mix thoroughly. Aseptically distribute into sterile tubes or flasks.

Use: For the cultivation of *Aeromonas salmonicida*.

Five g Agar (5g Agar)

Composition per liter:

Glycerol	50.0g
Agar	15.0g
Yeast extract	5.0g
$CaCO_3$	1.0g

Preparation of Medium: Add components to distilled/deionized water and bring volume to 1.0L. Mix thoroughly. Gently heat and bring to boiling. Distribute into tubes or flasks. Autoclave for 15 min at 15 psi pressure–121°C. Pour into sterile Petri dishes or leave in tubes.

Use: For the cultivation and maintenance of *Dermatophilus congolensis* and *Geodermatophilus obscurus.*

Fletcher Medium

Composition per liter:

Agar	1.5g
NaCl	0.5g
Peptone	0.3g
Beef extract	0.2g
Rabbit serum	50.0mL

pH 7.9 ± 0.1 at 25°C

Source: This medium is available as a premixed powder from BD Diagnostic Systems.

Preparation of Medium: Add components, except rabbit serum, to distilled/deionized water and bring volume to 950.0mL. Mix thoroughly. Gently heat and bring to boiling. Autoclave for 15 min at 15 psi pressure–121°C. Cool to 50°–55°C. Aseptically add 50.0mL of sterile rabbit serum. Mix thoroughly. Aseptically distribute into sterile tubes or flasks.

Use: For the isolation, cultivation, and maintenance of cultures of *Leptospira* species.

Fletcher Medium with Fluorouracil (Fluorouracil *Leptospira* Medium)

Composition per liter:

Agar	1.5g
NaCl	0.5g
Peptone	0.3g
Beef extract	0.2g
Rabbit serum	50.0mL
Fluorouracil solution	20.0mL

pH 7.9 ± 0.1 at 25°C

Fluorouracil Solution:

Composition per 100.0mL:

Fluorouracil	10.0g

Preparation of Fluorouracil Solution: Add fluorouracil to 50.0mL of distilled/deionized water. Add 1.0mL of 2*N* NaOH and bring volume to 100.0mL. Gently heat to 56°C for 2h. Adjust pH to 7.4–7.6 with NaOH. Mix thoroughly. Filter sterilize.

Preparation of Medium: Add components, except rabbit serum and fluorouracil solution, to distilled/deionized water and bring volume to 930.0mL. Mix thoroughly. Gently heat and bring to boiling. Autoclave for 15 min at 15 psi pressure–121°C. Cool to 50°–55°C. Aseptically add 80.0mL of sterile rabbit serum. Mix thoroughly. Aseptically distribute into sterile tubes or flasks. Immediately prior to use, add 0.1mL of fluorouracil solution per 5.0mL of medium.

Use: For the isolation, cultivation, and maintenance of cultures of *Leptospira* species.

Fletcher's Semisolid Medium

Composition per 2120.0mL:

Agar	1.5g
NaCl	0.5g
Peptone	0.3g
Beef extract	0.2g
Rabbit serum	240.0mL

pH 7.9 ± 0.1 at 25°C

Preparation of Medium: Add components, except rabbit serum, to distilled/deionized water and bring volume to 1880.0mL. Mix thoroughly. Gently heat and bring to boiling. Autoclave for 15 min at 15 psi pressure–121°C. Cool to 50°–55°C. Aseptically add 240.0mL of sterile rabbit serum. Mix thoroughly. Aseptically distribute into sterile tubes or flasks.

Use: For the isolation, cultivation, and maintenance of cultures of *Leptospira* species.

FLN Medium (Fluorescence Lactose Nitrate Medium)

Composition per liter:

Lactose	20.0g
Agar	15.0g
Proteose peptone no. 3	10.0g
KNO_3	2.0g
K_2HPO_4	1.5g
$MgSO_4 \cdot 7H_2O$	1.5g
$NaNO_2$	0.5g
Phenol Red	0.02g

pH 7.2 ±0.2 at 25°C

Preparation of Medium: Add components to distilled/deionized water and bring volume to 1.0L. Mix thoroughly. Gently heat and bring to boiling. Distribute into tubes or flasks. Autoclave for 15 min at 15 psi pressure–121°C. Pour into sterile Petri dishes or leave in tubes.

Use: For the differentiation of pseudomonads from other nonfermentative bacilli. Lactose fermentation is indicated by the medium turning yellow. *Pseudomonas cepacia* often produces acid from lactose. Denitrification from nitrate or nitrite is indicated by the formation of gas bubbles in the solid medium. *Pseudomonas aeruginosa*, *Pseudomonas mendocina*, and *Pseudomonas denitrificans* are positive for denitrification. Fluorescein production is indicated by fluorescence under UV light. *Pseudomonas aeruginosa* is positive for fluorescein production; *Pseudomonas denitrificans* does not produce fluorescein.

Fluid Sabouraud Medium

Composition per liter:

Glucose	20.0g
Pancreatic digest of casein	5.0g
Peptic digest of animal tissue	5.0g

pH5.7 ± 0.2 at 25°C

Source: This medium is available as a premixed powder from BD Diagnostic Systems.

Preparation of Medium: Add components to distilled/deionized water and bring volume to 1.0L. Mix thoroughly. Distribute into tubes or flasks. Autoclave for 15 min at 15 psi pressure–121°C.

Use: For cultivation of yeasts, molds, and aciduric microorganisms.

Fluid Thioglycolate Agar

Composition per liter:

Pancreatic digest of casein	15.0g
Glucose	5.0g
Yeast extract	5.0g
NaCl	2.5g
Agar	0.75g
L-Cystine	0.5g
Sodium thioglycolate	0.5g
Resazurin	1.0mg

pH 7.1 ± 0.2 at 25°C

Preparation of Medium: Add components to distilled/deionized water and bring volume to 1.0L. Mix thoroughly. Gently heat and bring to boiling. Distribute into tubes or flasks. Autoclave for 15 min at 15 psi pressure–121°C. Pour into sterile Petri dishes or leave in tubes.

Use: For the cultivation of anaerobic bacteria. For the cultivation of *Campylobacter fetus, Campylobacter jejuni, Leptotrichia buccalis,* and *Streptococcus* species.

Fluid Thioglycolate Agar with Calcium Carbonate

Composition per liter:

Agar	75.0g
Pancreatic digest of casein	15.0g
$CaCO_3$	10.0g
Glucose	5.0g
Yeast extract	5.0g
NaCl	2.5g
L-Cystine	0.5g
Sodium thioglycolate	0.5g
Resazurin	0.001g

pH 7.1 ± 0.2 at 25°C

Preparation of Medium: Add components, except $CaCO_3$, to distilled/deionized water and bring volume to 1.0L. Mix thoroughly. Gently heat and bring to boiling. Distribute into tubes or flasks. Autoclave for 15 min at 15 psi pressure–121°C. Sterilize $CaCO_3$ by autoclaving for 15 min at 15 psi pressure–121°C. Aseptically add sterile $CaCO_3$ to sterile tubes or plates—0.1g of sterile $CaCO_3$ per 10.0mL of medium to be added. Pour medium into sterile Petri dishes or distribute into sterile tubes.

Use: For the rapid cultivation of anaerobic bacteria. For the cultivation and maintenance of *Campylobacter fetus, Campylobacter jejuni, Leptotrichia buccalis,* and *Streptococcus* species.

Fluid Thioglycolate Medium

Composition per liter:

Pancreatic digest of casein	15.0g
Glucose	5.5g
Yeast extract	5.0g
NaCl	2.5g
Agar	0.75g
L-Cystine	0.5g
Sodium thioglycolate	0.5g
Resazurin	1.0mg

pH 7.1 ± 0.2 at 25°C

Source: This medium is available as a premixed powder from BD Diagnostic Systems.

Preparation of Medium: Add components to distilled/deionized water and bring volume to 1.0L. Mix thoroughly. Gently heat and bring to boiling. Distribute into tubes or flasks. Autoclave for 15 min at 15 psi pressure–121°C. If medium becomes oxidized before use (resazurin turns red), heat in a boiling water bath to expel absorbed O_2. Cool to 25°C.

Use: For the cultivation of anaerobic, microaerophilic, and aerobic microorganisms. For use in sterility testing of a variety of specimens.

Fluid Thioglycolate Medium with Beef Extract

Composition per liter:

Pancreatic digest of casein	15.0g
Glucose	5.5g
Yeast extract	5.0g
Beef extract	5.0g
NaCl	2.5g
Agar	0.75g
L-Cystine	0.5g
Sodium thioglycolate	0.5g
Resazurin	1.0mg

pH 7.2 ± 0.2 at 25°C

Source: This medium is available as a premixed powder from BD Diagnostic Systems.

Preparation of Medium: Add components to distilled/deionized water and bring volume to 1.0L. Mix thoroughly. Gently heat and bring to boiling. Distribute into tubes or flasks. Autoclave for 15 min at 15 psi pressure–121°C. If medium becomes oxidized before use (resazurin turns red), heat in a boiling water bath to expel absorbed O_2. Cool to 25°C.

Use: For the cultivation of anaerobic, microaerophilic, and aerobic microorganisms. For use in sterility testing of a variety of specimens.

Fluid Thioglycolate Medium without Glucose or E_h Indicator

Composition per liter:

Pancreatic digest of casein	20.0g
NaCl	2.5g
Agar	0.75g
L-Cystine	0.5g
Sodium thioglycolate	0.5g

pH 7.1 ± 0.2 at 25°C

Source: This medium is available as a premixed powder from BD Diagnostic Systems.

Preparation of Medium: Add components to distilled/deionized water and bring volume to 1.0L. Mix thoroughly. Gently heat and bring to boiling. Distribute into tubes or flasks. Autoclave for 15 min at 15 psi pressure–121°C. Aseptically distribute into sterile screw-capped tubes or flasks.

Use: For cultivation of anaerobic bacteria. For use as a basal medium in carbohydrate fermentation tests for differentiating anaerobic bacteria. For carbohydrate fermentation tests, 1.0 mL of a 10% filter-sterilized carbohydrate solution is aseptically added to 9.0mL of fluid thioglycolate medium.

Fluid Thioglycolate Medium with K Agar

Composition per liter:

Pancreatic digest of casein	15.0g
Glucose	5.0g
Yeast extract	5.0g
KCl	2.5g
L-Cystine	0.5g
K agar	0.45g
Resazurin	1.0mg
Thioglycolic acid	0.3mL

pH 7.2 ± 0.2 at 25°C

Source: This medium is available as a premixed powder from BD Diagnostic Systems.

Preparation of Medium: Add components to distilled/deionized water and bring volume to 1.0L. Mix thoroughly. Gently heat and bring to boiling. Distribute into tubes or flasks. Autoclave for 15 min at 15 psi pressure–121°C. If medium becomes oxidized before use (resazurin turns red), heat in a boiling water bath to expel absorbed O_2. Cool to 25°C.

Use: For the cultivation of anaerobic, microaerophilic, and aerobic microorganisms. For use in sterility testing of a variety of specimens.

Fluid Thioglycolate Medium with Rabbit Serum

Composition per liter:

Pancreatic digest of casein	15.0g
Glucose	5.5g
Yeast extract	5.0g
NaCl	2.5g
Agar	0.75g
L-Cystine	0.5g
Sodium thioglycolate	0.5g
Resazurin	1.0mg
Rabbit serum	100.0mL

pH 7.1 ± 0.2 at 25°C

Preparation of Medium: Add components, except rabbit serum, to distilled/deionized water and bring volume to 900.0mL. Mix thoroughly. Gently heat and bring to boiling. Distribute into tubes in 9.0mL volumes. Autoclave for 15 min at 15 psi pressure–121°C. If medium becomes oxidized before use (resazurin turns red), heat in a boiling water bath to expel absorbed O_2. Cool to 25°C. Immediately prior to inoculation, aseptically and anerobically add 1.0mL of sterile rabbit serum to each tube. Mix thoroughly.

Use: For the cultivation of anaerobic, microaerophilic, and aerobic microorganisms. For use in the sterility testing of a variety of specimens.

Fluorocult *E. coli* 0157:H7 Agar (*E. coli* 0157:H7 Agar, Fluorocult)

Composition per liter:

Peptone from casein	20.0g
Agar	13.0g
Sorbitol	10.0g
NaCl	5.0g
Meat extract	2.0g
$Na_2S_2O_3$	2.0g
Sodium deoxycholate	1.12g
Yeast extract	1.0g
Ammonium ferric citrate	0.5g
4-Methylumbelliferyl-β-D-glucuronide	0.1g
Bromthymol Blue	0.025g

pH 7.4 ± 0.2 at 25°C

Source: This medium is available from Merck.

Preparation of Medium: Add components to distilled/deionized water and bring volume to 1.0L. Mix thoroughly. Autoclave for 15 min at 15 psi pressure–121°C. Cool to 45°–50°C. Pour into sterile Petri dishes.

Use: For the isolation and differentiation of enterohemorrhagic (EHEC) *Escherichia coli* 0157:H7-strains. In contrast to most other *E. coli* strains, *E. coli* 0157:H7 shows the following characteristics: no sorbitol-cleavage capacity within 48h. and no formation of glucuronidase (MUG-negative/no fluorescence). Sodium deoxycholate inhibits the growth of the Gram-positive accompanying flora for the greater part. Sorbitol serves, together with the pH indicator Bromthymol Blue, to determine the degradation of sorbitol which, in the case of sorbitol-positive microorganisms, results in the colonies turning yellow in color. Sorbitol-negative strains, on the other hand, do not lead to any change in the color of the culture medium and thus proliferate as greenish colonies. Sodium thiosulfate and ammonium iron(III) citrate result in black-brown discoloration of the agar for colonies, in the presence of hydrogen-sulfide-forming pathogens, precipitating iron sulfide. *Proteus mirabilis* in particular, which displays biochemical properties similar to those of *E. coli* 0157:H7, can thus be very easily differentiated from *E. coli* 0157:H7 on account of the brownish discoloration. 4-methylumbelliferyl-β-D-glucuronide (MUG) is converted into 4-methylumbelliferone by β-D-glucuronidase-forming pathogens; 4-methylumbelliferone fluoresces under UV light. The activity of β-D-glucuronidase is a highly specific characteristic of *E. coli.* In contrast to most *E. coli* strains, *E. coli* 0157:H7 is not capable of forming β-D-glucoronidase. When irradiated with long-wave UV light, no fluorescence is formed.

FN Medium (Fluorescence Denitrification Medium)

Composition per liter:

Agar	15.0g
Proteose peptone no. 3	10.0g
KNO_3	2.0g
K_2HPO_4	1.5g
$MgSO_4 \cdot 7H_2O$	1.5g
$NaNO_2$	0.5g

pH 7.2 ±0.2 at 25°C

Preparation of Medium: Add components to distilled/deionized water and bring volume to 1.0L. Mix thoroughly. Gently heat and bring to boiling. Distribute into tubes or flasks. Autoclave for 15 min at 15 psi pressure–121°C. Pour into sterile Petri dishes or leave in tubes.

Use: For the differentiation of pseudomonads from other nonfermentative bacilli. Denitrification from nitrate or nitrite is indicated by the formation of gas bubbles in the solid medium. *Pseudomonas aeruginosa, Pseudomonas mendocina,* and *Pseudomonas denitrificans* are positive for denitrification. Fluorescein production is indicated by fluorescence under UV light. *Pseudomonas aeruginosa* is positive for fluorescein production.

Folic Acid Agar

Composition per liter:

Noble agar	15.0g
$K_2HPO_4 \cdot 3H_2O$	1.2g
Folic acid	1.0g
KH_2PO_4	0.5g
Salts A	5.0mL
Salts B	1.5mL

Salts A:

Composition per 100.0mL:

$MgSO_4 \cdot 7H_2O$	1.0g
$CaCl_2 \cdot 2H_2O$	0.1g
$FeSO_4 \cdot 7H_2O$	0.1g

Preparation of Salts A Solution: Add components to distilled/deionized water and bring volume to 100.0mL. Mix thoroughly. Maintain for 3 days at 25°C to dissolve. Filter sterilize the supernatant.

Salts B:

Composition per 100.0mL:

$MnSO_4$	0.1g
Na_2MoO_4	0.1g

Preparation of Salts B Solution: Add components to distilled/deionized water and bring volume to 100.0mL. Mix thoroughly. Filter sterilize.

Preparation of Medium: Add components, except salts A solution and salts B solution, to distilled/

deionized water and bring volume to 994.5mL. Mix thoroughly. Gently heat and bring to boiling. Autoclave for 15 min at 15 psi pressure–121°C. Cool to 45°–50°C. Aseptically add 5.0mL of sterile salts A solution and 1.5mL of sterile salts B solution. Mix thoroughly. Pour into sterile Petri dishes or distribute into sterile tubes.

Use: For the cultivation and maintenance of *Pseudomonas* species.

Formate Fumarate Medium

Composition per 900.0mL:

Agar	20.0g
Sodium fumarate	16.6g
Sodium formate	6.8g
Yeast extract	4.0g
Sodium thioglycolate solution	100.0mL

pH 7.3 ± 0.2 at 25°C

Sodium Thioglycolate Solution:

Composition per 100.0mL:

Sodium thioglycolate	0.5g

Preparation of Sodium Thioglycolate Solution: Add sodium thioglycolate to distilled/deionized water and bring volume to 100.0mL. Mix thoroughly. Filter sterilize. Sparge with 100% N_2 gas. Warm to 50°–55°C.

Preparation of Medium: Add components, except sodium thioglycolate solution, to distilled/deionized water and bring volume to 900.0mL. Mix thoroughly. Gently heat and bring to boiling. Continue boiling for 5 min. Cool to 50°–55°C while sparging with 80% N_2 + 20% CO_2. Anaerobically distribute 4.5mL volumes into anaerobic tubes. Autoclave for 15 min at 15 psi pressure–121°C. Cool to 50°–55°C. Prior to inoculation, add 0.5mL of sterile sodium thioglycolate solution to each tube.

Use: For the cultivation and identification of *Bacteroides gracilis, Bacteroides ureolyticus, Campylobacter* species, and *Wolinella* species.

FRAG Agar (Fragilis Agar)

Composition per 1025.0mL:

L-Cysteine·HCl·H_2O	0.5g
Basal solution	995.0mL
Glucuronic acid solution	25.0mL
Gentamicin solution	1.0mL
Hemin-vitamin K_1 solution	1.0mL
Ferric sulfate solution	1.0mL
Mineral solution	1.0mL
Phenol Red (1% solution)	1.0mL
Vitamin B_{12} solution	0.05mL

pH 7.0 ± 0.1 at 25°C

Basal Solution:

Composition per 995.0mL:

Oxgall	20.0g
Agar	15.4g
K_2HPO_4	2.26g
Yeast extract	2.0g
Pancreatic digest of casein	1.4g
$(NH_4)_2HPO_4$	1.0g
K_2HPO_4	0.9g
Papaic digest of soybean meal	0.12g
NaCl	0.12g

Preparation of Basal Solution: Add components to distilled/deionized water and bring volume to 995.0mL. Mix thoroughly. Gently heat and bring to boiling. Autoclave for 15 min at 15 psi pressure–121°C. Cool to 45°–50°C.

Glucuronic Acid Solution:

Composition per 100.0mL:

D-Glucuronic acid	40.0g

Preparation of Glucuronic Acid Solution: Add glucuronic acid to distilled/deionized water and bring volume to 100.0mL. Mix thoroughly. Filter sterilize.

Gentamicin Solution:

Composition per 10.0mL:

Gentamicin	0.1mg

Preparation of Gentamicin Solution: Add gentamicin to distilled/deionized water and bring volume to 10.0mL. Mix thoroughly. Filter sterilize.

Vitamin K_1-Hemin Solution:

Composition per liter:

Vitamin K_1 solution	10.0mL
Hemin solution	10.0mL

Preparation of Vitamin K_1-Hemin Solution: Add components to distilled/deionized water and bring volume to 1.0L. Mix thoroughly. Filter sterilize.

Vitamin K_1 Solution:

Composition per 100.0mL:

Vitamin K_1	1.0g
Ethanol	99.0mL

Preparation of Vitamin K_1 Solution: Add vitamin K_1 to 99.0mL of absolute ethanol. Mix thoroughly.

Hemin Solution:

Composition per 10.0mL:

Hemin	0.5g
NaOH	0.4g

Preparation of Hemin Solution: Add hemin and NaOH to distilled/deionized water and bring volume to 10.0mL. Mix thoroughly.

Ferric Sulfate Solution:
Composition per 100.0mL:

$FeSO_4 \cdot 9H_2O$	0.04g

Preparation of Ferric Sulfate Solution: Add $FeSO_4 \cdot 9H_2O$ to distilled/deionized water and bring volume to 100.0mL. Mix thoroughly. Filter sterilize.

Mineral Solution:
Composition per 100.0mL:

NaCl	9.0g
$CaCl_2 \cdot 2H_2O$	0.27g
$MgCl_2 \cdot 6H_2O$	0.2g
$CoCl_2 \cdot 6H_2O$	0.1g
$MnCl_2 \cdot 4H_2O$	0.1g

Preparation of Mineral Solution: Add components to distilled/deionized water and bring volume to 100.0mL. Mix thoroughly. Filter sterilize.

Vitamin B_{12} Solution:
Composition per 10.0mL:

Vitamin B_{12}	0.1mg

Preparation of Vitamin B_{12} Solution: Add vitamin B_{12} to distilled/deionized water and bring volume to 10.0mL. Mix thoroughly. Filter sterilize.

Preparation of Medium: To 995.0mL of cooled, sterile basal solution, aseptically add 0.5g of L-cysteine·HCl·H_2O, 25.0mL of sterile glucuronic acid solution, 1.0mL of sterile gentamicin solution, 1.0mL of sterile hemin-vitamin K_1 solution, 1.0mL of sterile ferric sulfate solution, 1.0mL of sterile mineral solution, 1.0mL of Phenol Red solution, and 0.05mL of sterile vitamin B_{12} solution. Mix thoroughly. Pour into sterile Petri dishes or distribute into sterile tubes.

Use: For the isolation, cultivation, and differentiation of the *Bacteroides fragilis* group (*Bacteroides fragilis, Bacteroides thetaiotamicron, Bacteroides vulgatus, Bacteroides distasonis, Bacteriodes ovatus,* and *Bacteroides uniformis*) from clinical specimens.

Francisella tularensis Isolation Medium

Composition per liter:

Agar	10.0g
Glucose	10.0g
Pancreatic digest of casein	10.0g
Peptic digest of animal tissue	10.0g
L-Cysteine·HCl·H_2O	5.0g
NaCl	5.0g
Sodium thioglycolate	2.0g
Glucose	1.0g
Thiamine·HCl	5.0mg

pH 7.2 ± 0.2 at 25°C

Preparation of Medium: Add components, except agar, to distilled/deionized water and bring volume to 1.0L. Mix thoroughly. Adjust pH to 7.2. Add agar. Gently heat and bring to boiling. Autoclave for 20 min at 15 psi pressure–121°C. Cool to 45°–50°C. Pour into sterile Petri dishes.

Use: For the isolation and cultivation of *Francisella tularensis.*

Freezing Medium

Composition per 11.5mL:

Bolton broth base	9.5mL
Fetal bovine serum	1.0mL
Glycerol solution	1.0mL

Bolton Broth Base:
Composition per liter:

Peptone	10.0g
Lactalbumin hydrolysate	5.0g
Yeast extract	5.0g
NaCl	5.0g
α-ketoglutarate	1.0g
Na-pyruvate	0.5g
Na-metabisulfite	0.5g
Na_2CO_3	0.6g
Hemin	0.01g

Preparation of Bolton Broth Base: Add components to distilled/deionized water and bring volume to 1.0L. Mix thoroughly. Autoclave for 15 min at 15 psi pressure–121°C. Cool to room temperature.

Glycerol Solution:
Composition per 100.0mL:

Glycerol	10.0g

Preparation of Glycerol Solution: Add glycerol to distilled/deionized water and bring volume to 100.0mL. Mix thoroughly. Autoclave for 15 min at 15 psi pressure–121°C. Cool to room temperature.

Preparation of Medium: Aseptically combine 9.5mL Bolton broth base, 1.0mL sterile glycerol solution, and 1.0mL filter sterilized fetal bovine serum. Mix thoroughly.

Use: For the preservation of *Campylobacter* spp.

G Medium

Composition per liter:

$(NH_4)_2SO_4$	2.0g
Yeast extract	2.0g
Glucose	1.0g
K_2HPO_4	0.6g
KH_2PO_4	0.4g
$MgSO_4 \cdot 7H_2O$	0.2g
$CaCl_2$	0.08g
$MnSO_4 \cdot H_2O$	0.05g

$CuSO_4 \cdot 5H_2O$ 5.0mg
$ZnSO_4 \cdot 7H_2O$ 5.0mg
$FeSO_4 \cdot 7H_2O$ 0.5mg

pH 7.8 ± 0.2 at 25°C

Preparation of Medium: Add components to distilled/deionized water and bring volume to 1.0L. Mix thoroughly. Distribute into tubes or flasks. Autoclave for 15 min at 15 psi pressure–121°C.

Use: For the cultivation and maintenance of *Bacillus cereus*.

GAM Agar

Composition per liter:

GAM agar 74.0g

Source: GAM agar is available from Nissui.

Preparation of Medium: Add components to distilled/deionized water and bring volume to 1.0L. Mix thoroughly. Gently heat and bring to boiling. Distribute into tubes or flasks. Autoclave for 15 min at 15 psi pressure–121°C. Pour into sterile Petri dishes or leave in tubes.

Use: For the cultivation and maintenance of *Fusobacterium necrophorum, Pediococcus* species, and *Peptostreptococcus* spp.

GAM Semisolid

Composition per liter:

GAM broth 74.0g
Agar 2.0g

Source: GAM broth is available from Nissui.

Preparation of Medium: Add components to distilled/deionized water and bring volume to 1.0L. Mix thoroughly. Gently heat and bring to boiling. Distribute into tubes or flasks. Autoclave for 15 min at 15 psi pressure–121°C. Pour into sterile Petri dishes or leave in tubes.

Use: For the cultivation and maintenance of actinomycetes.

Gardnerella vaginalis Selective Medium

Composition per liter:

Columbia blood agar base 940.0mL
Rabbit or horse serum 50.0mL
Antibiotic inhibitor solution 10.0mL

pH 7.2 ± 0.2 at 25°C

Source: This medium is available as a premixed powder from Oxoid Unipath.

Columbia Blood Agar Base:
Composition per liter:

Special peptone 23.0g
Agar 10.0g
NaCl 5.0g
Starch 1.0g

Source: Columbia blood agar base is available as a premixed powder from Oxoid Unipath.

Preparation of Columbia Blood Agar Base: Add components to distilled/deionized water and bring volume to 1.0L. Mix thoroughly. Gently heat until boiling. Autoclave for 15 min at 15 psi pressure–121°C. Cool to 45°–50°C.

Antibiotic Inhibitor Solution:
Composition per 10.0mL:

Nalidixic acid 0.035g
Gentamicin sulfate 4.0mg
Amphotericin B 2.0mg
Ethanol 4.0mL

Preparation of Antibiotic Inhibitor Solution: Add components to distilled/deionized water and bring volume to 10.0mL. Mix thoroughly. Filter sterilize.

Preparation of Medium: To 940.0mL of cooled, sterile Columbia blood agar base, aseptically add 50.0mL of rabbit or horse blood serum and 10.0mL of sterile antibiotic inhibitor solution. Pour into sterile Petri dishes or distribute into sterile tubes.

Use: For the selective isolation, cultivation, and differentiation of *Gardnerella vaginalis* from clinical specimens, such as the vaginal discharge of patients with vaginitis. *Gardnerella vaginalis* exhibits β-hemolysis on this medium.

GBNA Medium (Gum Base Nalidixic Acid Medium)

Composition per liter:

Gellan gum 8.0g
Pancreatic digest of casein 5.7g
NaCl 1.7g
Papaic digest of soybean meal 1.0g
Glucose 0.83g
K_2HPO_4 .... 0.83g
$MgCl_2 \cdot 6H_2O$ 0.33g
Nalidixic acid 0.05g

pH 7.2 ± 0.2 at 25°C

Source: This medium is available as a premixed powder from BD Diagnostic Systems.

Preparation of Medium: Add components to tap water and bring volume to 1.0L. Mix thoroughly. Gently heat and bring to boiling. Distribute into tubes or flasks. Autoclave for 15 min at 15 psi pressure–121°C. Pour into sterile Petri dishes or leave in tubes.

Use: For the isolation and cultivation of *Listeria monocytogenes* from clinical and nonclinical specimens.

GBS Agar Base, Islam (Group B Streptococci Agar) (Islam GBS Agar)

Composition per liter:

Proteose peptone	23.0g
Agar	10.0g
Na_2HPO_4	5.75g
Soluble starch	5.0g
NaH_2PO_4	1.5g
Horse serum, heat inactivated	50.0mL

pH 7.5 ± 0.1 at 25°C

Source: This medium is available as a premixed powder from Oxoid Unipath.

Preparation of Medium: Add components, except horse serum, to distilled/deionized water and bring volume to 950.0mL. Mix thoroughly. Gently heat and bring to boiling. Autoclave for 15 min at 15 psi pressure–121°C. Cool to 45°–50°C. Aseptically add 50.0mL of sterile inactivated horse serum. Mix thoroughly. Pour into sterile Petri dishes or distribute into sterile tubes.

Use: For the isolation and detection of group B streptococci from clinical specimens. Group B streptococci produce orange pigmented colonies when incubated under anaerobic conditions.

GBS Medium, Rapid (Group B Streptococci Medium)

Starch	80.0g
Proteose peptone	23.0g
Na_2HPO_4	5.75g
NaH_2PO_4	1.5g
Horse serum, inactivated	50.0mL
Antibiotic inhibitor solution	10.0 mL

pH 7.5 ± 0.2 at 25°C

Source: This medium is available as a premixed powder from Oxoid Unipath.

Antibiotic Inhibitor Solution:

Composition per 10.0mL:

Metronidazole	10.0mg
Gentamicin	2.0mg

Preparation of Antibiotic Inhibitor Solution: Add components to distilled/deionized water and bring volume to 10.0mL. Mix thoroughly. Filter sterilize.

Preparation of Medium: Add components, except horse serum and antibiotic inhibitor solution, to distilled/deionized water and bring volume to 940.0mL. Mix thoroughly. Gently heat and bring to boiling. Autoclave for 15 min at 15 psi pressure–121°C. Cool to 45°–50°C. Aseptically add 50.0mL of sterile heat-inactivated horse serum and 10.0mL of sterile antibiotic inhibitor solution. Mix thoroughly. Aseptically distribute into sterile tubes or flasks. Cool to 5°C and hold at that temperature for 12 h prior to use.

Use: For the rapid isolation and cultivation of group B streptococci from clinical specimens.

GC Agar (LMG Medium 236)

Composition 1010.0mL:

Solution A	500.0mL
Solution B	500.0mL
Supplement solution	10.0mL

pH 7.2 ± 0.2 at 25°C

Solution A:

Agar	5.0g
GC agar base, 2X	500.0mL

GC Agar Base, 2X:

Composition per 500.0mL:

Agar	10.0g
Pancreatic digest of casein	7.5g
Peptic digest of animal tissue	7.5g
NaCl	5.0g
K_2HPO_4	4.0g
Cornstarch	1.0g
KH_2PO_4	1.0g

Source: GC agar base is available as a premixed powder from BD Diagnostic Systems. This base may be replaced by GC medium base available from BD Diagnostic Systems.

Preparation of GC Agar Base, 2X: Add components to distilled/deionized water and bring volume to 500.0mL. Mix thoroughly.

Preparation of Solution A: Add 5.0g agar to 500.0mL GC agar base, 2X. Mix thoroughly. Gently heat until boiling. Autoclave for 15 min at 15 psi pressure–121°C. Cool to 45°–50°C.

Solution B:

Composition per 500.0mL:

Bovine hemoglobin	10.0g

Preparation of Solution B: Add bovine hemoglobin to distilled/deionized water and bring volume to 500.0mL. Mix thoroughly. Autoclave for 15 min at 15 psi pressure–121°C. Cool to 45°–50°C.

Supplement Solution:
Composition per liter:

Glucose	100.0g
L-Cysteine·HCl	25.9g
L-Glutamine	10.0g
L-Cystine	1.1g
Adenine	1.0g
Nicotinamide adenine dinucleotide	0.25g
Vitamin B_{12}	0.1g
Thiamine pyrophosphate	0.1g
Guanine·HCl	0.03g
$Fe(NO_3)_3 \cdot 6H_2O$	0.02g
p-Aminobenzoic acid	0.013g
Thiamine·HCl	3.0mg

Source: The supplement solution (IsoVitaleX enrichment) is available from BD Diagnostic Systems. This enrichment may be replaced by supplement VX from BD Diagnostic Systems.

Preparation of Supplement Solution: Add components to distilled/deionized water and bring volume to 1.0L. Mix thoroughly. Filter sterilize.

Preparation of Medium: To 500.0mL of sterile GC agar base, aseptically add 500.0mL of sterile hemoglobin solution at 45°–50°C. Mix thoroughly. Aseptically add 10.0mL of sterile supplement solution. Mix thoroughly. Pour into sterile Petri dishes or distribute into sterile tubes.

Use: For the cultivation and maintenance of *Haemophilus parainfluenzae* and *Neisseria* spp.

GC Agar (ATCC Medium 814)

Composition per 1010.0mL:

GC agar base, 2×	500.0mL
Hemoglobin solution	500.0mL
Supplement solution	10.0mL

pH 7.2 ± 0.2 at 25°C

GC Agar Base, 2X:
Composition per 500.0mL:

Agar	10.0g
Pancreatic digest of casein	7.5g
Peptic digest of animal tissue	7.5g
NaCl	5.0g
K_2HPO_4	4.0g
Cornstarch	1.0g
KH_2PO_4	1.0g

Source: GC agar base is available as a premixed powder from BD Diagnostic Systems. This base may be replaced by GC medium base available from BD Diagnostic Systems.

Preparation of GC Agar Base, 2X: Add components to distilled/deionized water and bring volume to 500.0mL. Mix thoroughly. Gently heat until boiling. Autoclave for 15 min at 15 psi pressure–121°C. Cool to 45°–50°C.

Hemoglobin Solution:
Composition 500.0mL:

Bovine hemoglobin	10.0g

Preparation of Hemoglobin Solution: Add bovine hemoglobin to distilled/deionized water and bring volume to 500.0mL. Mix thoroughly. Autoclave for 15 min at 15 psi pressure–121°C. Cool to 45°–50°C.

Supplement Solution:
Composition per liter:

Glucose	100.0g
L-Cysteine·HCl	25.9g
L-Glutamine	10.0g
L-Cystine	1.1g
Adenine	1.0g
Nicotinamide adenine dinucleotide	0.25g
Vitamin B_{12}	0.1g
Thiamine pyrophosphate	0.1g
Guanine·HCl	0.03g
$Fe(NO_3)_3 \cdot 6H_2O$	0.02g
p-Aminobenzoic acid	0.013g
Thiamine·HCl	3.0mg

Source: The supplement solution (IsoVitaleX enrichment) is available from BD Diagnostic Systems. This enrichment may be replaced by supplement VX from BD Diagnostic Systems.

Preparation of Supplement Solution: Add components to distilled/deionized water and bring volume to 1.0L. Mix thoroughly. Filter sterilize.

Preparation of Medium: To 500.0mL of sterile GC agar base, aseptically add 500.0mL of sterile hemoglobin solution at 45°–50°C. Mix thoroughly. Aseptically add 10.0mL of sterile supplement solution. Mix thoroughly. Pour into sterile Petri dishes or distribute into sterile tubes.

Use: For the isolation and cultivation of fastidious bacteria, especially *Neisseria* and *Haemophilus* species. For the cultivation and maintenance of *Branhamella catarrhalis, Campylocbacter pylori, and Helicobacter pylori.*

GC Agar (GC Medium) (ATCC Medium 1351)

Composition per liter:

GC agar base	950.0mL
Blood, defibrinated	50.0mL

pH 7.2 ± 0.2 at 25°C

GC Agar Base:

Composition per liter:

Agar	10.0g
Pancreatic digest of casein	7.5g
Peptic digest of animal tissue	7.5g
NaCl	5.0g
K_2HPO_4	4.0g
Cornstarch	1.0g
KH_2PO_4	1.0g

Source: GC agar base is available as a premixed powder from BD Diagnostic Systems. This base may be replaced by GC medium base available from BD Diagnostic Systems.

Preparation of GC Agar Base: Add components to distilled/deionized water and bring volume to 1.0L. Mix thoroughly. Gently heat until boiling. Autoclave for 15 min at 15 psi pressure–121°C. Cool to 75°–80°C.

Preparation of Medium: To 950.0mL of sterile GC agar base aseptically add 50.0mL sterile defibrinated blood with thorough mixing and maintain at 75°–80°C for 15–20 min until the medium is chocolatized. Pour into sterile Petri dishes or distribute into sterile tubes.

Use: For the isolation and cultivation of fastidious bacteria, especially *Neisseria* and *Haemophilus* species. For the cultivation and maintenance of *Branhamella catarrhalis, Campylocbacter pylori, Eikenella corrodens, Helicobacter pylori, Moraxella nonliquefaciens, Morococcus cerebrosis, Oligella ureolytica, Oligella urethralis, Pasteurella volantium, Proteus mirabilis,* and *Taylorella equigenitalis.*

GC Agar with Ampicillin

Composition per 1020.0mL:

GC agar base, 2×	500.0mL
Hemoglobin solution	500.0mL
Supplement solution	10.0mL
Ampicillin solution	10.0mL

pH 7.2 ± 0.2 at 25°C

GC Agar Base, 2X:

Composition per 500.0mL:

Agar	10.0g
Pancreatic digest of casein	7.5g
Peptic digest of animal tissue	7.5g
NaCl	5.0g
K_2HPO_4	4.0g
Cornstarch	1.0g
KH_2PO_4	1.0g

Source: GC agar base is available as a premixed powder from BD Diagnostic Systems. This base may be replaced by GC medium base available from BD Diagnostic Systems.

Preparation of GC Agar Base, 2X: Add components to distilled/deionized water and bring volume to 500.0mL. Mix thoroughly. Gently heat until boiling. Autoclave for 15 min at 15 psi pressure–121°C. Cool to 45°–50°C.

Hemoglobin Solution:

Composition per 500.0mL:

Bovine hemoglobin	10.0g

Preparation of Hemoglobin Solution: Add bovine hemoglobin to distilled/deionized water and bring volume to 500.0mL. Mix thoroughly. Autoclave for 15 min at 15 psi pressure–121°C. Cool to 45°–50°C.

Supplement Solution:

Composition per liter:

Glucose	100.0g
L-Cysteine·HCl	25.9g
L-Glutamine	10.0g
L-Cystine	1.1g
Adenine	1.0g
Nicotinamide adenine dinucleotide	0.25g
Vitamin B_{12}	0.1g
Thiamine pyrophosphate	0.1g
Guanine·HCl	0.03g
$Fe(NO_3)_3 \cdot 6H_2O$	0.02g
p-Aminobenzoic acid	0.013g
Thiamine·HCl	3.0mg

Source: The supplement solution (IsoVitaleX enrichment) is available from BD Diagnostic Systems. This enrichment may be replaced by supplement VX from BD Diagnostic Systems.

Preparation of Supplement Solution: Add components to distilled/deionized water and bring volume to 1.0L. Mix thoroughly. Filter sterilize.

Ampicillin Solution:

Composition per 10.0mL:

Ampicillin	0.02g

Preparation of Ampicillin Solution: Add ampicillin to distilled/deionized water and bring volume to 10.0mL. Mix thoroughly. Filter sterilize.

Preparation of Medium: To 500.0mL of sterile GC agar base, aseptically add 500.0mL of sterile hemoglobin solution at 45°–50°C. Mix thoroughly. Aseptically add 10.0mL of sterile supplement solution and 10.0mL of sterile ampicillin solution. Mix thoroughly. Pour into sterile Petri dishes or distribute into sterile tubes.

Use: For the cultivation and maintenance of *Branhamella catarrhalis, Haemophilus influenzae,* and *Haemophilus parainfluenzae.*

GC Agar with Ampicillin and Gentamicin

Composition per 1030.0mL:

GC agar base, 2×	500.0mL
Hemoglobin solution	500.0mL
Supplement solution	10.0mL
Ampicillin solution	10.0mL
Gentamicin solution	10.0mL

pH 7.2 ± 0.2 at 25°C

GC Agar Base, 2X:
Composition per 500.0mL:

Agar	10.0g
Pancreatic digest of casein	7.5g
Peptic digest of animal tissue	7.5g
NaCl	5.0g
K_2HPO_4	4.0g
Cornstarch	1.0g
KH_2PO_4	1.0g

Source: GC agar base is available as a premixed powder from BD Diagnostic Systems. This base may be replaced by GC medium base available from BD Diagnostic Systems.

Preparation of GC Agar Base, 2X: Add components to distilled/deionized water and bring volume to 500.0mL. Mix thoroughly. Gently heat until boiling. Autoclave for 15 min at 15 psi pressure–121°C. Cool to 45°–50°C.

Hemoglobin Solution:
Composition per 500.0mL:

Bovine hemoglobin	10.0g

Preparation of Hemoglobin Solution: Add bovine hemoglobin to distilled/deionized water and bring volume to 500.0mL. Mix thoroughly. Autoclave for 15 min at 15 psi pressure–121°C. Cool to 45°–50°C.

Supplement Solution:
Composition per liter:

Glucose	100.0g
L-Cysteine·HCl	25.9g
L-Glutamine	10.0g
L-Cystine	1.1g
Adenine	1.0g
Nicotinamide adenine dinucleotide	0.25g
Vitamin B_{12}	0.1g
Thiamine pyrophosphate	0.1g
Guanine·HCl	0.03g
$Fe(NO_3)_3 \cdot 6H_2O$	0.02g
p-Aminobenzoic acid	0.013g
Thiamine·HCl	3.0mg

Source: The supplement solution (IsoVitaleX enrichment) is available from BD Diagnostic Systems. This enrichment may be replaced by supplement VX from BD Diagnostic Systems.

Preparation of Supplement Solution: Add components to distilled/deionized water and bring volume to 1.0L. Mix thoroughly. Filter sterilize.

Ampicillin Solution:
Composition per 10.0mL:

Ampicillin	0.01g

Preparation of Ampicillin Solution: Add ampicillin to distilled/deionized water and bring volume to 10.0mL. Mix thoroughly. Filter sterilize.

Gentamicin Solution:
Composition per 10.0mL:

Gentamicin	2.0mg

Preparation of Gentamicin Solution: Add gentamicin to distilled/deionized water and bring volume to 10.0mL. Mix thoroughly. Filter sterilize.

Preparation of Medium: To 500.0mL of sterile GC agar base, aseptically add 500.0mL of sterile hemoglobin solution at 45°–50°C. Mix thoroughly. Aseptically add 10.0mL of sterile supplement solution, 10.0mL of sterile ampicillin solution, and 10.0mL of sterile gentamicin solution. Mix thoroughly. Pour into sterile Petri dishes or distribute into sterile tubes.

Use: For the cultivation and maintenance of *Haemophilus parainfluenzae.*

GC Agar with Ampicillin and Tetracycline

Composition per 1030.0mL:

GC agar base, 2×	500.0mL
Hemoglobin solution	500.0mL
Supplement solution	10.0mL
Ampicillin solution	10.0mL
Tetracycline solution	10.0mL

pH 7.2 ± 0.2 at 25°C

GC Agar Base, 2X:
Composition per 500.0mL:

Agar	10.0g
Pancreatic digest of casein	7.5g
Peptic digest of animal tissue	7.5g
NaCl	5.0g
K_2HPO_4	4.0g
Cornstarch	1.0g
KH_2PO_4	1.0g

Source: GC agar base is available as a premixed powder from BD Diagnostic Systems. This base may be replaced by GC medium base available from BD Diagnostic Systems.

Preparation of GC Agar Base, 2X: Add components to distilled/deionized water and bring volume to 500.0mL. Mix thoroughly. Gently heat until boiling. Autoclave for 15 min at 15 psi pressure–121°C. Cool to 45°–50°C.

Hemoglobin Solution:
Composition per 500.0mL:

Bovine hemoglobin	10.0g

Preparation of Hemoglobin Solution: Add bovine hemoglobin to distilled/deionized water and bring volume to 500.0mL. Mix thoroughly. Autoclave for 15 min at 15 psi pressure–121°C. Cool to 45°–50°C.

Supplement Solution:
Composition per liter:

Glucose	100.0g
L-Cysteine·HCl	25.9g
L-Glutamine	10.0g
L-Cystine	1.1g
Adenine	1.0g
Nicotinamide adenine dinucleotide	0.25g
Vitamin B_{12}	0.1g
Thiamine pyrophosphate	0.1g
Guanine·HCl	0.03g
$Fe(NO_3)_3 \cdot 6H_2O$	0.02g
p-Aminobenzoic acid	0.013g
Thiamine·HCl	3.0mg

Source: The supplement solution (IsoVitaleX enrichment) is available from BD Diagnostic Systems. This enrichment may be replaced by supplement VX from BD Diagnostic Systems.

Preparation of Supplement Solution: Add components to distilled/deionized water and bring volume to 1.0L. Mix thoroughly. Filter sterilize.

Ampicillin Solution:
Composition per 10.0mL:

Ampicillin	0.01g

Preparation of Ampicillin Solution: Add ampicillin to distilled/deionized water and bring volume to 10.0mL. Mix thoroughly. Filter sterilize.

Tetracycline Solution:
Composition per 10.0mL:

Tetracycline	5.0mg

Preparation of Tetracycline Solution: Add tetracycline to distilled/deionized water and bring volume to 10.0mL. Mix thoroughly. Filter sterilize.

Preparation of Medium: To 500.0mL of sterile GC agar base, aseptically add 500.0mL of sterile hemoglobin solution at 45°–50°C. Mix thoroughly. Aseptically add 10.0mL of sterile supplement solution, 10.0mL of sterile ampicillin solution, and 10.0mL of sterile tetracycline solution. Mix thoroughly. Pour into sterile Petri dishes or distribute into sterile tubes.

Use: For the cultivation and maintenance of *Haemophilus parainfluenzae.*

GC Agar with Chloramphenicol, Tetracycline, and Ampicillin

Composition per 1040.0mL:

GC agar base, 2×	500.0mL
Hemoglobin solution	500.0mL
Supplement solution	10.0mL
Ampicillin solution	10.0mL
Tetracycline solution	10.0mL
Chloramphenicol solution	10.0mL

pH 7.2 ± 0.2 at 25°C

GC Agar Base, 2X:
Composition per 500.0mL:

Agar	10.0g
Pancreatic digest of casein	7.5g
Peptic digest of animal tissue	7.5g
NaCl	5.0g
K_2HPO_4	4.0g
Cornstarch	1.0g
KH_2PO_4	1.0g

Source: GC agar base is available as a premixed powder from BD Diagnostic Systems. This base may be replaced by GC medium base available from BD Diagnostic Systems.

Preparation of GC Agar Base, 2X: Add components to distilled/deionized water and bring volume to 500.0mL. Mix thoroughly. Gently heat until boiling. Autoclave for 15 min at 15 psi pressure–121°C. Cool to 45°–50°C.

Hemoglobin Solution:
Composition per 500.0mL:

Bovine hemoglobin	10.0g

Preparation of Hemoglobin Solution: Add bovine hemoglobin to distilled/deionized water and bring volume to 500.0mL. Mix thoroughly. Autoclave for 15 min at 15 psi pressure–121°C. Cool to 45°–50°C.

Supplement Solution:
Composition per liter:

Glucose	100.0g
L-Cysteine·HCl	25.9g
L-Glutamine	10.0g
L-Cystine	1.1g
Adenine	1.0g
Nicotinamide adenine dinucleotide	0.25g
Vitamin B_{12}	0.1g
Thiamine pyrophosphate	0.1g

Guanine·HCl 0.03g
$Fe(NO_3)_3 \cdot 6H_2O$ 0.02g
p-Aminobenzoic acid 0.013g
Thiamine·HCl 3.0mg

Source: The supplement solution (IsoVitaleX enrichment) is available from BD Diagnostic Systems. This enrichment may be replaced by supplement VX from BD Diagnostic Systems.

Preparation of Supplement Solution: Add components to distilled/deionized water and bring volume to 1.0L. Mix thoroughly. Filter sterilize.

Ampicillin Solution:
Composition per 10.0mL:
Ampicillin 0.01g

Preparation of Ampicillin Solution: Add ampicillin to distilled/deionized water and bring volume to 10.0mL. Mix thoroughly. Filter sterilize.

Tetracycline Solution:
Composition per 10.0mL:
Tetracycline 5.0mg

Preparation of Tetracycline Solution: Add tetracycline to distilled/deionized water and bring volume to 10.0mL. Mix thoroughly. Filter sterilize.

Chloramphenicol Solution:
Composition per 10.0mL:
Chloramphenicol 5.0mg

Preparation of Chloramphenicol Solution: Add chloramphenicol to distilled/deionized water and bring volume to 10.0mL. Mix thoroughly. Filter sterilize.

Preparation of Medium: To 500.0mL of sterile GC agar base, aseptically add 500.0mL of sterile hemoglobin solution at 45°–50°C. Mix thoroughly. Aseptically add 10.0mL of sterile supplement solution, 10.0mL of sterile ampicillin solution, 10.0mL of sterile chloramphenicol, and 10.0mL of sterile tetracycline solution. Mix thoroughly. Pour into sterile Petri dishes or distribute into sterile tubes.

Use: For the cultivation and maintenance of *Haemophilus parainfluenzae.*

GC Agar with Defined Supplements

GC agar base 990.0mL
Defined supplements solution 10.0mL
pH 7.2 ± 0.2 at 25°C

GC Agar Base:
Composition per liter:
Agar 10.0g
Pancreatic digest of casein 7.5g
Peptic digest of animal tissue 7.5g
NaCl 5.0g
K_2HPO_4 .... 4.0g
Cornstarch 1.0g
KH_2PO_4 .... 1.0g

Source: GC agar base is available as a premixed powder from BD Diagnostic Systems. This base may be replaced by GC medium base available from BD Diagnostic Systems.

Preparation of GC Agar Base: Add components to distilled/deionized water and bring volume to 1.0L. Mix thoroughly. Gently heat until boiling. Autoclave for 15 min at 15 psi pressure–121°C. Cool to 45°–50°C.

Defined Supplements Solution:
Composition per 100.0mL:
Glucose 40.0g
Glutamine 1.0g
$Fe(NO_3)_3 \cdot 6H_2O$ 0.05g
Cocarboxylase 2.0mg

Preparation of Defined Supplements Solution: Add components to distilled/deionized water and bring volume to 100.0mL. Mix thoroughly. Filter sterilize.

Preparation of Medium: To 990.0mL of sterile GC agar base, aseptically add 10.0mL of sterile defined supplements solution. Mix thoroughly. Pour into sterile Petri dishes or distribute into sterile tubes.

Use: For the cultivation and maintenance of *Neisseria gonorrhoeae.*

GC Agar with Penicillin G

Composition per 1020.0mL:
GC medium base, 2× 500.0mL
Hemoglobin solution 500.0mL
Supplement solution 10.0mL
Penicillin G solution 10.0mL
pH 7.2 ± 0.2 at 25°C

GC Medium Base, 2X:
Composition per 500.0mL:
Proteose peptone no. 3 .... 15.0g
Agar 10.0g
NaCl 5.0g
K_2HPO_4 .... 4.0g
Cornstarch 1.0g
KH_2PO_4 .... 1.0g

Source: GC medium base is available as a premixed powder from BD Diagnostic Systems.

Preparation of GC Medium Base, 2X: Add components to distilled/deionized water and bring volume to 500.0mL. Mix thoroughly. Gently heat until boiling. Autoclave for 15 min at 15 psi pressure–121°C. Cool to 45°–50°C.

Hemoglobin Solution:
Composition per 500.0mL:

Bovine hemoglobin	10.0g

Preparation of Hemoglobin Solution: Add bovine hemoglobin to distilled/deionized water and bring volume to 500.0mL. Mix thoroughly. Autoclave for 15 min at 15 psi pressure–121°C. Cool to 45°–50°C.

Supplement Solution:
Composition per liter:

Glucose	100.0g
L-Cysteine·HCl	25.9g
L-Glutamine	10.0g
L-Cystine	1.1g
Adenine	1.0g
Nicotinamide adenine dinucleotide	0.25g
Vitamin B_{12}	0.1g
Thiamine pyrophosphate	0.1g
Guanine·HCl	0.03g
$Fe(NO_3)_3 \cdot 6H_2O$	0.02g
p-Aminobenzoic acid	0.013g
Thiamine·HCl	3.0mg

Source: The supplement solution (IsoVitaleX enrichment) is available from BD Diagnostic Systems. This enrichment may be replaced by supplement VX from BD Diagnostic Systems.

Preparation of Supplement Solution: Add components to distilled/deionized water and bring volume to 1.0L. Mix thoroughly. Filter sterilize.

Penicillin G Solution:
Composition per 10.0mL:

Penicillin G	0.05g

Preparation of Penicillin G Solution: Add penicillin G to distilled/deionized water and bring volume to 10.0mL. Mix thoroughly. Filter sterilize.

Preparation of Medium: To 500.0mL of sterile GC medium base, aseptically add 500.0mL of sterile hemoglobin solution at 45°–50°C. Mix thoroughly. Aseptically add 10.0mL of sterile supplement solution and 10.0mL of sterile penicillin G solution. Mix thoroughly. Pour into sterile Petri dishes or distribute into sterile tubes.

Use: For the cultivation and maintenance of *Neisseria gonorrhoeae*.

GC Agar with Supplement A

Composition per 1020.0mL:

GC medium base, 2×	500.0mL
Hemoglobin solution	500.0mL
Supplement solution	10.0mL
Supplement A	10.0mL

pH 7.2 ± 0.2 at 25°C

GC Medium Base, 2X:
Composition per 500.0mL:

Proteose peptone no. 3	15.0g
Agar	10.0g
NaCl	5.0g
K_2HPO_4	4.0g
Cornstarch	1.0g
KH_2PO_4	1.0g

Source: GC medium base and Supplement A are available as a premixed powder from BD Diagnostic Systems.

Preparation of GC Medium Base, 2X: Add components to distilled/deionized water and bring volume to 500.0mL. Mix thoroughly. Gently heat until boiling. Autoclave for 15 min at 15 psi pressure–121°C. Cool to 45°–50°C.

Hemoglobin Solution:
Composition per 500.0mL:

Bovine hemoglobin	10.0g

Preparation of Hemoglobin Solution: Add bovine hemoglobin to distilled/deionized water and bring volume to 500.0mL. Mix thoroughly. Autoclave for 15 min at 15 psi pressure–121°C. Cool to 45°–50°C.

Supplement Solution:
Composition per liter:

Glucose	100.0g
L-Cysteine·HCl	25.9g
L-Glutamine	10.0g
L-Cystine	1.1g
Adenine	1.0g
Nicotinamide adenine dinucleotide	0.25g
Vitamin B_{12}	0.1g
Thiamine pyrophosphate	0.1g
Guanine·HCl	0.03g
$Fe(NO_3)_3 \cdot 6H_2O$	0.02g
p-Aminobenzoic acid	0.013g
Thiamine·HCl	3.0mg

Source: The supplement solution (IsoVitaleX enrichment) is available from BD Diagnostic Systems. This enrichment may be replaced by supplement VX from BD Diagnostic Systems.

Preparation of Supplement Solution: Add components to distilled/deionized water and bring volume to 1.0L. Mix thoroughly. Filter sterilize.

Supplement A:
Composition per 10.0mL:
Supplement A contains yeast concentrate with Crystal Violet.

Preparation of Supplement A: Add components to distilled/deionized water and bring volume to 10.0mL. Mix thoroughly. Filter sterilize.

Preparation of Medium: To 500.0mL of sterile GC medium base, aseptically add 500.0mL of sterile hemoglobin solution at 45°–50°C. Mix thoroughly. Aseptically add 10.0mL of sterile supplement solution and 10.0mL of sterile supplement A solution. Mix thoroughly. Pour into sterile Petri dishes or distribute into sterile tubes.

Use: For the cultivation of *Neisseria gonorrhoeae*, other *Neisseria* species, and *Haemophilus* species.

GC Agar with Supplement A and with VCN Inhibitor

Composition per 1030.0mL:

GC medium base, 2×	500.0mL
Hemoglobin solution	500.0mL
Supplement solution	10.0mL
Supplement A	10.0mL
VCN inhibitor	10.0mL

pH 7.2 ± 0.2 at 25°C

GC Medium Base, 2X:
Composition per 500.0mL:

Proteose peptone no. 3	15.0g
Agar	10.0g
NaCl	5.0g
K_2HPO_4	4.0g
Cornstarch	1.0g
KH_2PO_4	1.0g

Source: GC medium base and Supplement A are available as a premixed powder from BD Diagnostic Systems.

Preparation of GC Medium Base, 2X: Add components to distilled/deionized water and bring volume to 500.0mL. Mix thoroughly. Gently heat until boiling. Autoclave for 15 min at 15 psi pressure–121°C. Cool to 45°–50°C.

Hemoglobin Solution:
Composition per 500.0mL:

Bovine hemoglobin	10.0g

Preparation of Hemoglobin Solution: Add bovine hemoglobin to distilled/deionized water and bring volume to 500.0mL. Mix thoroughly. Autoclave for 15 min at 15 psi pressure–121°C. Cool to 45°–50°C.

Supplement Solution:
Composition per liter:

Glucose	100.0g
L-Cysteine·HCl	25.9g
L-Glutamine	10.0g
L-Cystine	1.1g
Adenine	1.0g
Nicotinamide adenine dinucleotide	0.25g
Vitamin B_{12}	0.1g
Thiamine pyrophosphate	0.1g
Guanine·HCl	0.03g
$Fe(NO_3)_3 \cdot 6H_2O$	0.02g
p-Aminobenzoic acid	0.013g
Thiamine·HCl	3.0mg

Source: The supplement solution (IsoVitaleX enrichment) is available from BD Diagnostic Systems. This enrichment may be replaced by supplement VX from BD Diagnostic Systems.

Preparation of Supplement Solution: Add components to distilled/deionized water and bring volume to 1.0L. Mix thoroughly. Filter sterilize.

Supplement A:
Composition per 10.0mL:

Supplement A contains yeast concentrate with Crystal Violet.

Preparation of Supplement A: Add components to distilled/deionized water and bring volume to 10.0mL. Mix thoroughly. Filter sterilize.

VCN Inhibitor:
Composition per 10.0mL:

Colistin	7.5mg
Vancomycin	3.0mg
Nystatin	12,500U

Preparation of VCN Inhibitor: Add components to distilled/deionized water and bring volume to 10.0mL. Mix thoroughly. Filter sterilize.

Preparation of Medium: To 500.0mL of sterile GC medium base, aseptically add 500.0mL of sterile hemoglobin solution at 45°–50°C. Mix thoroughly. Aseptically add 10.0mL of sterile supplement solution, 10.0mL of sterile supplement A solution, and 10.0mL of sterile VCN inhibitor. Mix thoroughly. Pour into sterile Petri dishes or distribute into sterile tubes.

Use: For the cultivation of *Neisseria gonorrhoeae*, other *Neisseria* species, and *Haemophilus* species.

GC Agar with Supplement A and with VCTN Inhibitor

Composition per 1030.0mL:

GC medium base, 2×	500.0mL
Hemoglobin solution	500.0mL
Supplement solution	10.0mL
Supplement A	10.0mL
VCTN inhibitor	10.0mL

pH 7.2 ± 0.2 at 25°C

GC Medium Base, 2X:
Composition per 500.0mL:

Proteose peptone no. 3	15.0g
Agar	10.0g

NaCl	5.0g
K_2HPO_4	4.0g
Cornstarch	1.0g
KH_2PO_4	1.0g

Source: GC medium base and Supplement A are available as a premixed powder from BD Diagnostic Systems.

Preparation of GC Medium Base, 2X: Add components to distilled/deionized water and bring volume to 500.0mL. Mix thoroughly. Gently heat until boiling. Autoclave for 15 min at 15 psi pressure–121°C. Cool to 45°–50°C.

Hemoglobin Solution:
Composition per 500.0mL:

Bovine hemoglobin	10.0g

Preparation of Hemoglobin Solution: Add bovine hemoglobin to distilled/deionized water and bring volume to 500.0mL. Mix thoroughly. Autoclave for 15 min at 15 psi pressure–121°C. Cool to 45°–50°C.

Supplement Solution:
Composition per liter:

Glucose	100.0g
L-Cysteine·HCl	25.9g
L-Glutamine	10.0g
L-Cystine	1.1g
Adenine	1.0g
Nicotinamide adenine dinucleotide	0.25g
Vitamin B_{12}	0.1g
Thiamine pyrophosphate	0.1g
Guanine·HCl	0.03g
$Fe(NO_3)_3·6H_2O$	0.02g
p-Aminobenzoic acid	0.013g
Thiamine·HCl	3.0mg

Source: The supplement solution (IsoVitaleX enrichment) is available from BD Diagnostic Systems. This enrichment may be replaced by supplement VX from BD Diagnostic Systems.

Preparation of Supplement Solution: Add components to distilled/deionized water and bring volume to 1.0L. Mix thoroughly. Filter sterilize.

Supplement A:
Composition per 10.0mL:
Supplement A contains yeast concentrate with Crystal Violet.

Preparation of Supplement A: Add components to distilled/deionized water and bring volume to 10.0mL. Mix thoroughly. Filter sterilize.

VCTN Inhibitor:
Composition per liter:

Colistin	7.5mg
Trimethoprim lactate	5.0mg
Vancomycin	4.0mg
Nystatin	12,500U

Preparation of VCTN Inhibitor: Add components to distilled/deionized water and bring volume to 10.0mL. Mix thoroughly. Filter sterilize.

Preparation of Medium: To 500.0mL of sterile GC medium base, aseptically add 500.0mL of sterile hemoglobin solution at 45°–50°C. Mix thoroughly. Aseptically add 10.0mL of sterile supplement solution, 10.0mL of sterile supplement A, and 10.0mL of sterile VCTN inhibitor. Mix thoroughly. Pour into sterile Petri dishes or distribute into sterile tubes.

Use: For the cultivation of *Neisseria gonorrhoeae*, other *Neisseria* species, and *Haemophilus* species.

GC Agar with Supplement B

GC agar base	990.0mL
Supplement B	10.0mL

pH 7.2 ± 0.2 at 25°C

GC Agar Base:
Composition per liter:

Agar	10.0g
Pancreatic digest of casein	7.5g
Peptic digest of animal tissue	7.5g
NaCl	5.0g
K_2HPO_4	4.0g
Cornstarch	1.0g
KH_2PO_4	1.0g

Source: GC agar base is available as a premixed powder from BD Diagnostic Systems. This base may be replaced by GC medium base available from BD Diagnostic Systems.

Preparation of GC Agar Base: Add components to distilled/deionized water and bring volume to 1.0L. Mix thoroughly. Gently heat until boiling. Autoclave for 15 min at 15 psi pressure–121°C. Cool to 45°–50°C.

Supplement B:
Composition per 10.0mL:
Supplement B contains yeast concentrate, glutamine, coenzyme, cocarboxylase, hematin, and growth factors.

Preparation of Supplement B: Add components to distilled/deionized water and bring volume to 10.0mL. Mix thoroughly. Filter sterilize.

Preparation of Medium: To 990.0mL of sterile GC agar base, aseptically add 10.0mL of sterile defined supplement B. Mix thoroughly. Pour into sterile Petri dishes or distribute into sterile tubes.

Use: For the cultivation and maintenance of *Neisseria gonorrhoeae.*

GC Medium, New York City Formulation

Composition per liter:

GC agar base	850.0mL
Horse blood, lysed	100.0mL
Yeast autolysate supplement	30.0mL
LCAT antibiotic solution	20.0mL

pH 7.3 ± 0.2 at 25°C

GC Agar Base:

Composition per 850.0mL:

Special peptone	15.0g
Agar	10.0g
NaCl	5.0g
K_2HPO_4	4.0g
Cornstarch	1.0g
KH_2PO_4	1.0g

pH 7.2 ± 0.2 at 25°C

Preparation of GC Agar Base: Add components of GC medium base and the hemoglobin to distilled/deionized water and bring volume to 850.0mL. Mix thoroughly. Gently heat until boiling. Autoclave for 15 min at 15 psi pressure–121°C. Cool to 45°–50°C.

Horse Blood, Lysed:

Composition per 100.0mL:

Saponin	0.5g
Horse blood, defibrinated	100.0mL

Preparation of Horse Blood, Lysed: Add saponin to defibrinated horse blood. Mix thoroughly. Allow blood to lyse.

Yeast Autolysate Supplement:

Composition per 30.0mL:

Yeast autolysate	10.0g
Glucose	1.0g
$NaHCO_3$	0.15g

Preparation of Yeast Autolysate Supplement: Add components to distilled/deionized water and bring volume to 30.0mL. Mix thoroughly. Filter sterilize.

LCAT Antibiotic Solution:

Composition per 20.0mL:

Colistin	6.0mg
Trimethoprim lactate	5.0mg
Lincomycin	1.0mg
Amphotericin B	1.0mg

Preparation of LCAT Antibiotic Solution: Add components to distilled/deionized water and bring volume to 20.0mL. Mix thoroughly. Filter sterilize.

Preparation of Medium: To 850.0mL of cooled sterile GC agar base, aseptically add 100.0mL of sterile lysed horse blood, 30.0mL of sterile yeast autolysate supplement, and 20.0mL of LCAT antibiotic solution. Mix thoroughly. Pour into sterile Petri dishes or distribute into sterile tubes.

Use: For the selective isolation and cultivation of fastidious microorganisms, especially *Neisseria* species.

GCII Agar

Composition per liter:

GCII agar base, 2×	490.0mL
Hemoglobin solution	490.0mL
Supplement solution	10.0mL

pH 7.2 ± 0.2 at 25°C

GCII Agar Base, 2X:

Composition per liter:

Agar	10.0g
Pancreatic digest of casein	7.5g
Selected meat peptone	7.5g
NaCl	5.0g
K_2HPO_4	4.0g
Cornstarch	1.0g
KH_2PO_4	1.0g

Source: GCII agar base is available as a premixed powder from BD Diagnostic Systems.

Preparation of GCII Agar Base, 2X: Add components to distilled/deionized water and bring volume to 500.0mL. Mix thoroughly. Gently heat until boiling. Autoclave for 15 min at 15 psi pressure–121°C. Cool to 45°–50°C.

Hemoglobin Solution:

Composition per 500.0mL:

Hemoglobin	10.0g

Preparation of Hemoglobin Solution: Add hemoglobin to distilled/deionized water and bring volume to 500.0mL. Mix thoroughly. Autoclave for 15 min at 15 psi pressure–121°C. Cool to 45°–50°C.

Supplement Solution:

Composition per liter:

Glucose	100.0g
L-Cysteine·HCl	25.9g
L-Glutamine	10.0g
L-Cystine	1.1g
Adenine	1.0g
Nicotinamide adenine dinucleotide	0.25g
Vitamin B_{12}	0.1g
Thiamine pyrophosphate	0.1g
Guanine·HCl	0.03g
$Fe(NO_3)_3 \cdot 6H_2O$	0.02g

p-Aminobenzoic acid....................................0.013g
Thiamine·HCl 3.0mg

Preparation of Supplement Solution: Add components to distilled/deionized water and bring volume to 1.0L. Mix thoroughly. Filter sterilize.

Preparation of Medium: To 490.0mL of sterile GC II agar base, aseptically add 490.0mL of sterile hemoglobin solution at 45°–50°C. Mix thoroughly. Aseptically add 10.0mL of sterile supplement solution. Mix thoroughly. Pour into sterile Petri dishes or distribute into sterile tubes.

Use: For the isolation and cultivation of fastidious microorganisms, especially *Neisseria* and *Haemophilus* species, from clinical specimens.

GCII Agar

Composition per liter:

GCII agar base 950.0mL
Blood, defibrinated 50.0mL

pH 7.2 ± 0.2 at 25°C

GCII Agar Base with Extra Agar:
Composition per liter:

Agar 12.0g
Pancreatic digest of casein................................7.5g
Selected meat peptone 7.5g
NaCl 5.0g
K_2HPO_4..4.0g
Cornstarch...1.0g
KH_2PO_4..1.0g

Source: GCII agar base is available as a premixed powder from BD Diagnostic Systems.

Preparation of GCII Agar Base Extra Agar: Add components to distilled/deionized water and bring volume to 1.0L. Mix thoroughly. Gently heat until boiling. Autoclave for 15 min at 15 psi pressure–121°C. Cool to 45°–50°C.

Preparation of Medium: To 950.0mL of sterile GC agar base, aseptically add 50.0mL of sterile defibrinated blood with thorough mixing and maintain at 75°–80°C for 15–20 min until the medium is chocolatized. Pour into sterile Petri dishes or distribute into sterile tubes.

Use: For the isolation and cultivation of fastidious microorganisms, especially *Neisseria* and *Haemophilus* species, from clinical specimens.

GC-Lect™ Agar

Composition per liter:

GCII agar base, 2× 500.0mL
Hemoglobin solution................................. 500.0mL
Supplement solution 10.0mL
Selective agent solution 10.0mL

pH 7.2 ± 0.2 at 25°C

Source: This medium is available as a prepared medium from BD Diagnostic Systems.

GCII Agar Base, 2X with Extra Agar:
Composition per liter:

Agar 12.0g
Pancreatic digest of casein................................7.5g
Selected meat peptone 7.5g
NaCl 5.0g
K_2HPO_4..4.0g
Cornstarch...1.0g
KH_2PO_4..1.0g

Source: GCII agar base is available as a premixed powder from BD Diagnostic Systems.

Preparation of GCII Agar Base, 2X with Extra Agar: Add components to distilled/deionized water and bring volume to 500.0mL. Mix thoroughly. Gently heat until boiling. Autoclave for 15 min at 15 psi pressure–121°C. Cool to 45°–50°C.

Hemoglobin Solution:
Composition per 500.0mL:

Hemoglobin 10.0g

Preparation of Hemoglobin Solution: Add hemoglobin to distilled/deionized water and bring volume to 500.0mL. Mix thoroughly. Autoclave for 15 min at 15 psi pressure–121°C. Cool to 45°–50°C.

Supplement Solution:
Composition per liter:

Glucose 100.0g
L-Cysteine·HCl..25.9g
L-Glutamine 10.0g
L-Cystine 1.1g
Adenine..1.0g
Nicotinamide adenine dinucleotide 0.25g
Vitamin B_{12}..0.1g
Thiamine pyrophosphate 0.1g
Guanine·HCl..0.03g
$Fe(NO_3)_3·6H_2O$..0.02g
p-Aminobenzoic acid....................................0.013g
Thiamine·HCl 3.0mg

Source: The supplement solution (IsoVitaleX enrichment) is available from BD Diagnostic Systems. This enrichment may be replaced by supplement VX from BD Diagnostic Systems.

Preparation of Supplement Solution: Add components to distilled/deionized water and bring volume to 1.0L. Mix thoroughly. Filter sterilize.

Selective Agents:

Composition per 10.0mL:

Selective agents 0.017g

Preparation of Selective Agents: Add components to distilled/deionized water and bring volume to 10.0mL. Mix thoroughly. Filter sterilize.

Preparation of Medium: To 500.0mL of sterile GC II agar base, aseptically add 500.0mL of sterile hemoglobin solution at 45°–50°C. Mix thoroughly. Aseptically add 10.0mL of sterile supplement solution and 10.0mL of selective agents solution. Mix thoroughly. Pour into sterile Petri dishes or distribute into sterile tubes.

Use: For the isolation and cultivation of *Neisseria gonorrhoeae* from clinical specimens.

GC Medium with Chloramphenicol

Composition per 1020.0mL:

GC agar base, 2× 500.0mL
Hemoglobin solution 500.0mL
Supplement solution 10.0mL
Chloramphenicol solution 10.0mL

pH 7.2 ± 0.2 at 25°C

GC Agar Base, 2×:

Composition per 500.0mL:

Agar 10.0g
Pancreatic digest of casein 7.5g
Peptic digest of animal tissue 7.5g
NaCl 5.0g
K_2HPO_4 .... 4.0g
Cornstarch 1.0g
KH_2PO_4 .... 1.0g

Source: GC agar base is available as a premixed powder from BD Diagnostic Systems. This base may be replaced by GC medium base available from BD Diagnostic Systems.

Preparation of GC Agar Base, 2×: Add components to distilled/deionized water and bring volume to 500.0mL. Mix thoroughly. Gently heat until boiling. Autoclave for 15 min at 15 psi pressure–121°C. Cool to 45°–50°C.

Hemoglobin Solution:

Composition per 500.0mL:

Bovine hemoglobin 10.0g

Preparation of Hemoglobin Solution: Add bovine hemoglobin to distilled/deionized water and bring volume to 500.0mL. Mix thoroughly. Autoclave for 15 min at 15 psi pressure–121°C. Cool to 45°–50°C.

Supplement Solution:

Composition per liter:

Glucose 100.0g
L-Cysteine·HCl 25.9g
L-Glutamine 10.0g
L-Cystine 1.1g
Adenine 1.0g
Nicotinamide adenine dinucleotide 0.25g
Vitamin B_{12} 0.1g
Thiamine pyrophosphate 0.1g
Guanine·HCl 0.03g
$Fe(NO_3)_3·6H_2O$ 0.02g
p-Aminobenzoic acid 0.013g
Thiamine·HCl 3.0mg

Source: The supplement solution (IsoVitaleX enrichment) is available from BD Diagnostic Systems. This enrichment may be replaced by supplement VX from BD Diagnostic Systems.

Preparation of Supplement Solution: Add components to distilled/deionized water and bring volume to 1.0L. Mix thoroughly. Filter sterilize.

Chloramphenicol Solution:

Composition per 10.0mL:

Chloramphenicol 8.0mg

Preparation of Chloramphenicol Solution: Add chloramphenicol to distilled/deionized water and bring volume to 10.0mL. Mix thoroughly. Filter sterilize.

Preparation of Medium: To 500.0mL of sterile GC agar base, aseptically add 500.0mL of sterile hemoglobin solution at 45°–50°C. Mix thoroughly. Aseptically add 10.0mL of sterile supplement solution and 10.0mL of sterile chloramphenicol solution. Mix thoroughly. Pour into sterile Petri dishes or distribute into sterile tubes.

Use: For the cultivation of *Neisseria gonorrhoeae.*

GCA Agar with Thiamine

Composition per liter:

Glucose 25.0g
Agar 14.0g
Papaic digest of soybean meal 10.0g
NaCl 5.0g
Pancreatic digest of heart muscle 3.0g
Cysteine·HCl·H_2O 1.0g
Thiamine 0.05mg
Rabbit blood, defibrinated 50.0mL

pH 6.8 ± 0.2 at 25°C

Preparation of Medium: Add components, except rabbit blood, to distilled/deionized water and bring volume to 950.0mL. Mix thoroughly. Gently heat and bring to boiling. Autoclave for 15 min at 15 psi pressure–121°C. Cool to 45°–50°C. Aseptically add sterile rabbit blood. Mix thoroughly. Pour into sterile Petri dishes or distribute into sterile tubes.

Use: For the isolation and cultivation of *Francisella tularensis.*

Gelatin Agar

Composition per liter:

Gelatin	30.0g
Agar	15.0g
Pancreatic digest of casein	10.0g
NaCl	10.0g

pH 7.2 ± 0.2 at 25°C

Preparation of Medium: Add components to distilled/deionized water and bring volume to 1.0L. Mix thoroughly. Gently heat and bring to boiling. Distribute into tubes or flasks. Autoclave for 15 min at 15 psi pressure–121°C.

Use: For the cultivation of many bacteria and their differentiation based on proteolytic activity.

Gelatin Agar

Composition per liter:

Agar	15.0g
Gelatin	15.0g
Peptone	4.0g
Yeast extract	1.0g

pH 7.2 ± 0.2 at 25°C

Preparation of Medium: Add components to distilled/deionized water and bring volume to 1.0L. Mix thoroughly. Gently heat and bring to boiling. Distribute into tubes or flasks. Autoclave for 15 min at 15 psi pressure–121°C. Pour into sterile Petri dishes or leave in tubes.

Use: For the cultivation of a variety of heterotrophic bacteria based upon their utilization of gelatin.

Gelatin Agar (GA Medium)

Composition per liter:

Solution 1	950.0mL
Solution 2	50.0mL

pH 7.2 ± 0.2 at 25°C

Solution 1:

Composition per 950.0mL:

Gelatin	30.0g
Agar	15.0g
Pancreatic digest of casein	10.0g
NaCl	2.0g
D-Mannitol	1.0g
Glucose	1.0g
KNO_3	1.0g
Sodium acetate	1.0g
Sodium formate	1.0g
Sodium succinate	1.0g
Yeast extract	1.0g
Sodium lactate (60% solution)	5.0mL

Preparation of Solution 1: Add components to distilled/deionized water and bring volume to 950.0mL. Mix thoroughly. Gently heat and bring to boiling. Autoclave for 15 min at 15 psi pressure–121°C.

Solution 2:

Composition per 50.0mL:

Na_2HPO_4	1.0g
L-Cysteine·HCl·H_2O	0.5g
Na_2CO_3·H_2O	0.5g
Sucrose	0.5g
Dithiothreitol	0.1g
Menadione solution	2.0mL

Preparation of Solution 2: Add components to distilled/deionized water and bring volume to 50.0mL. Mix thoroughly. Filter sterilize.

Menadione Solution:

Composition per 100.0mL:

Menadione (vitamin K_3)	0.05g
Ethanol	99.0mL

Preparation of Menadione Solution: Add menadione to 99.0mL of absolute ethanol. Mix thoroughly.

Preparation of Medium: Aseptically combine sterile solution 1 with sterile solution 2. Mix thoroughly. Pour into sterile Petri dishes.

Use: For the cultivation and differentiation of microorganisms from dental plaque based on their ability to produce gelatinase. For the differentiation of aerobic, anaerobic, and facultative microorganisms of clinical significance.

Gelatin Infusion Broth

Composition per liter:

Beef heart, solids from infusion	500.0g
Gelatin	40.0g
Tryptose	10.0g
NaCl	5.0g

pH 7.4 ± 0.2 at 25°C

Preparation of Medium: Add components to distilled/deionized water and bring volume to 1.0L. Mix thoroughly. Gently heat and bring to boiling. Distribute into tubes or flasks. Autoclave for 15 min at 15 psi pressure–121°C. Pour into sterile Petri dishes or leave in tubes.

Use: For the cultivation and differentiation of a variety of heterotrophic bacteria based upon their production of gelatinase. The gelatinase liquifies the medium.

Gelatin Medium

Composition per liter:

Gelatin	4.0g

pH 7.0 ± 0.2 at 25°C

Preparation of Medium: Add gelatin to distilled/deionized water and bring volume to 1.0L. Mix thoroughly. Gently heat and bring to boiling. Distribute into tubes. Autoclave for 15 min at 15 psi pressure–121°C.

Use: For the cultivation and differentiation of *Nocardia* and *Streptomyces* species based on utilization of gelatin. *Nocardia asteroides* usually exhibits no growth. *Nocardia brasiliensis* shows good growth and round, compact colonies. *Streptomyces* species show varying degrees of growth.

Gelatin Medium

Composition per liter:

Gelatin	120.0g
Pancreatic digest of casein	13.0g
Sodium chloride	5.0g
Yeast extract	5.0g
Heart muscle, solids from infusion	2.0g
Sodium thioglycolate	0.5g

pH 7.0 ± 0.2 at 25°C

Preparation of Medium: Add components to distilled/deionized water and bring volume to 1.0L. Mix thoroughly. Gently heat and bring to boiling. Distribute into tubes. Autoclave for 15 min at 15 psi pressure–121°C. Pour into sterile Petri dishes or leave in tubes.

Use: For the cultivation of gelatin-utilizing *Clostridium* species.

Gelatin Metronidazole Cadmium Medium (GMC Medium)

Composition per liter:

Solution 1	950.0mL
Solution 2	50.0mL

pH 7.2 ± 0.2 at 25°C

Solution 1:

Composition per 950.0mL:

Gelatin	30.0g
Agar	15.0g
Pancreatic digest of casein	10.0g
NaCl	2.0g
D-Mannitol	1.0g
Glucose	1.0g
KNO_3	1.0g
Sodium acetate	1.0g
Sodium formate	1.0g
Sodium succinate	1.0g
Yeast extract	1.0g
$CdSO_4 \cdot 8H_2O$	0.02g
Metronidazole	0.01g
Sodium lactate (60% solution)	5.0mL

Preparation of Solution 1: Add components to distilled/deionized water and bring volume to 950.0mL. Mix thoroughly. Gently heat and bring to boiling. Autoclave for 15 min at 15 psi pressure–121°C.

Solution 2:

Composition per 50.0mL:

Na_2HPO_4	1.0g
L-Cysteine·HCl·H_2O	0.5g
$Na_2CO_3 \cdot H_2O$	0.5g
Sucrose	0.5g
Dithiothreitol	0.1g
Menadione solution	2.0mL

Preparation of Solution 2: Add components to distilled/deionized water and bring volume to 50.0mL. Mix thoroughly. Filter sterilize.

Menadione Solution:

Composition per 100.0mL:

Menadione (vitamin K_3)	0.05g
Ethanol	99.0mL

Preparation of Menadione Solution: Add menadione to 99.0mL of absolute ethanol. Mix thoroughly.

Preparation of Medium: Aseptically combine sterile solution 1 with sterile solution 2. Mix thoroughly. Pour into sterile Petri dishes.

Use: For the cultivation and differentiation of microorganisms from dental plaque based on their ability to produce gelatinase. For the differentiation of aerobic, anaerobic, and facultative microorganisms of clinical significance.

Gelatin Phosphate Salt Agar (GPS Agar)

Composition per liter:

Agar	15.0g
Gelatin	10.0g
NaCl	10.0g
K_2HPO_4	5.0g

pH 7.2 ± 0.2 at 25°C

Preparation of Medium: Add components to distilled/deionized water and bring volume to 1.0L.

Mix thoroughly. Gently heat and bring to boiling. Distribute into tubes or flasks. Autoclave for 15 min at 15 psi pressure–121°C. Pour into sterile Petri dishes.

Use: For the cultivation and differentiation of *Vibrio* species.

Gelatin Phosphate Salt Broth (GPS Broth)

Composition per liter:

Gelatin	10.0g
NaCl	10.0g
K_2HPO_4	5.0g

pH 7.2 ± 0.2 at 25°C

Preparation of Medium: Add components to distilled/deionized water and bring volume to 1.0L. Mix thoroughly. Gently heat and bring to boiling. Distribute into tubes or flasks. Autoclave for 15 min at 15 psi pressure–121°C.

Use: For the cultivation of *Vibrio* species.

Gelatin Salt Agar

Composition per liter:

NaCl	30.0g
Agar	15.0g
Gelatin	15.0g
Peptone	4.0g
Yeast extract	1.0g

pH 7.2 ± 0.2 at 25°C

Preparation of Medium: Add components to distilled/deionized water and bring volume to 1.0L. Mix thoroughly. Gently heat and bring to boiling. Distribute into tubes or flasks. Autoclave for 15 min at 15 psi pressure–121°C. Pour into sterile Petri dishes or leave in tubes.

Use: For the cultivation and differentiation of *Vibrio* species.

Gelatinase Test Medium

Composition per liter:

Gelatin	3.0g
ACES buffer	1.0g
Yeast extract	1.0g
Charcoal, activated	0.15g
α-Ketoglutarate monopotassium salt	0.1g
L-Cysteine·HCl·H_2O (4% solution)	1.0mL
KOH (85% solution)	1.0mL
$Fe_4(P_2O_7)_3$ solution	1.0mL

pH 6.9 ± 0.2 at 25°C

L-Cysteine·HCl·H_2O Solution:

Composition per 10.0mL:

L-Cysteine·HCl·H_2O	0.4g

Preparation of L-Cysteine·HCl·H_2O Solution: Add L-cysteine·HCl·H_2O to distilled/deionized water and bring volume to 10.0mL. Mix thoroughly. Filter sterilize.

$Fe_4(P_2O_7)_3$ Solution:

Composition per 10.0mL:

$Fe_4(P_2O_7)_3$	0.15g

Preparation of $Fe_4(P_2O_7)_3$ Solution: Add $Fe_4(P_2O_7)_3$ to distilled/deionized water and bring volume to 10.0mL. Mix thoroughly. Filter sterilize.

Preparation of Medium: Add ACES buffer to distilled/deionized water and bring volume to 899.0mL. Mix thoroughly. Gently heat to 50°C. Add 1.0mL of KOH solution. Mix thoroughly. Add charcoal, yeast extract and α-ketoglutarate. Add 80.0mL of distilled/deionized water to wash sides of flask. Mix thoroughly. Autoclave for 15 min at 15 psi pressure–121°C. Cool to 50°C. Aseptically add 10.0mL of sterile cysteine solution and 10.0mL of sterile $Fe_4(P_2O_7)_3$ solution. Mix thoroughly. Adjust pH to 6.9. Aseptically distribute into sterile screw-capped tubes.

Use: For the cultivation and differentiation of gelatinase-producing bacteria.

Gluconate Peptone Broth

Composition per liter:

Potassium gluconate	40.0g
Casein peptone	1.5g
K_2HPO_4	1.0g
Yeast extract	1.0g

pH 7.0 ± 0.2 at 25°C

Preparation of Medium: Add components to distilled/deionized water and bring volume to 1.0L. Mix thoroughly. Distribute into tubes or flasks. Autoclave for 15 min at 15 psi pressure–121°C.

Use: For the cultivation and differentiation of Gram-negative bacteria based on their ability to oxidize gluconate to 2-ketogluconate. For the differentiation of fluorescent *Pseudomonas* species. After inoculation with bacteria and 48 hr of growth in this medium, Benedict's reagent is added. Bacteria that produce the reducing sugar 2-ketogluconate turn the reagent yellow-orange to orange-red.

Glucose Nutrient Agar

Composition per liter:

Agar	15.0g
Pancreatic digest of casein	10.0g
Glucose	5.0g
K_2HPO_4	5.0g

NaCl 5.0g
Yeast extract 5.0g

Preparation of Medium: Add components to distilled/deionized water and bring volume to 1.0L. Mix thoroughly. Gently heat and bring to boiling. Distribute into tubes or flasks. Autoclave for 15 min at 15 psi pressure–121°C. Pour into sterile Petri dishes or leave in tubes.

Use: For the isolation and cultivation of *Brochothrix thermosphacta*.

Glucose Phosphate Broth

Composition per liter:

Peptone 10.0g
K_2HPO_4 .... 5.0g
Glucose 5.0g
pH 7.5 ± 0.2 at 25°C

Preparation of Medium: Add components, except glucose, to distilled/deionized water and bring volume to 1.0L. Mix thoroughly. Gently heat and bring to boiling. Filter while hot through Whatman filter paper. Cool to 25°C. Adjust pH to 7.5. Add 5.0g of glucose. Mix thoroughly. Distribute into sterile tubes or flasks. Autoclave for 10 min at 10 psi pressure–115°C.

Use: For the cultivation of a variety of nonfastidious heterotrophic microorganisms.

Glucose Salt Teepol Broth (GSTB)

Composition per liter:

NaCl 30.0g
Peptone 10.0g
Glucose 5.0g
Beef extract 3.0g
Methyl Violet 2.0mg
Sodium lauryl sulfate
(Teepol—0.1% solution) 4.0mL
pH 8.8 ± 0.2 at 25°C

Preparation of Medium: Add components to distilled/deionized water and bring volume to 1.0L. Mix thoroughly. Adjust pH to 8.8. Distribute into tubes or flasks. Autoclave for 15 min at 15 psi pressure–121°C.

Use: For the cultivation of *Vibrio* species.

Glucose Salts Medium

Composition per 1000.5mL:

Glucose 5.0g
$(NH_4)_2SO_4$ 1.0g
$MgSO_4 \cdot 7H_2O$ 0.5g
NaCl 0.5g
$NaH_2PO_4 \cdot 12H_2O$ 0.7mg
$NaH_2PO_4 \cdot 2H_2O$ 0.3mg
Trace elements solution 0.5mL
pH 6.9 ± 0.2 at 25°C

Trace Elements Solution:
Composition per liter:

H_3BO_3 .... 0.3g
$CoCl_2 \cdot 6H_2O$ 0.2g
$ZnSO_4 \cdot 7H_2O$ 0.1g
$MnCl_2 \cdot 4H_2O$ 0.03g
$Na_2MoO_4 \cdot 2H_2O$ 0.03g
$NiCl_2 \cdot 6H_2O$ 0.02g
$CuCl_2 \cdot 2H_2O$ 0.01g

Preparation of Trace Elements Solution: Add components to distilled/deionized water and bring volume to 1.0L. Mix thoroughly. Filter sterilize.

Preparation of Medium: Add components, except trace elements solution, to distilled/deionized water and bring volume to 1.0L. Mix thoroughly. Autoclave for 15 min at 15 psi pressure–121°C. Aseptically add 0.5mL of sterile trace elements solution. Mix thoroughly. Aseptically distribute into sterile tubes or flasks.

Use: For the cultivation of a wide variety of bacteria and fungi.

Glutarate Medium

Composition per liter:

Sodium glutarate 2.6g
NaCl 1.0g
KCl 0.5g
$MgCl_2 \cdot 6H_2O$ 0.4g
NH_4Cl 0.25g
KH_2PO_4 .... 0.2g
$CaCl_2 \cdot 2H_2O$ 0.15g
Resazurin 1.0mg
Rumen fluid 20.0mL
$NaHCO_3$ solution 20.0mL
$Na_2S \cdot 9H_2O$ solution 10.0mL
Trace elements solution SL-10 .... 1.0mL
pH 7.2 ± 0.2 at 25°C

Trace Elements Solution SL-10:
Composition per liter:

$FeCl_2 \cdot 4H_2O$ 1.5g
$CoCl_2 \cdot 6H_2O$ 190.0mg
$MnCl_2 \cdot 4H_2O$ 100.0mg
$ZnCl_2$ 70.0mg
$Na_2MoO_4 \cdot 2H_2O$ 36.0mg
$NiCl_2 \cdot 6H_2O$ 24.0mg
H_3BO_3 .... 6.0mg
$CuCl_2 \cdot 2H_2O$ 2.0mg
HCl (25% solution) 10.0mL

Preparation of Trace Elements Solution SL-10: Add $FeCl_2 \cdot 4H_2O$ to 10.0mL of HCl solution. Mix thoroughly. Add distilled/deionized water and bring volume to 1.0L. Add remaining components. Mix thoroughly. Gas under 80% N_2 + 20% CO_2. Autoclave for 15 min at 15 psi pressure–121°C.

$NaHCO_3$ Solution:

Composition per 20.0mL:

$NaHCO_3$	2.5g

Preparation of $NaHCO_3$ Solution: Add $NaHCO_3$ to distilled/deionized water and bring volume to 20.0mL. Mix thoroughly. Filter sterilize. Gas under 80% N_2 + 20% CO_2.

$Na_2S{\cdot}9H_2O$ Solution:

Composition per 10.0mL:

$Na_2S{\cdot}9H_2O$	0.36g

Preparation of $Na_2S{\cdot}9H_2O$ Solution: Add 0.36g $Na_2S{\cdot}9H_2O$ to distilled/deionized water and bring volume to 10.0mL. Mix thoroughly. Gas under 100% N_2. Autoclave for 15 min at 15 psi pressure–121°C.

Preparation of Medium: Prepare and dispense medium under 80% N_2 + 20% CO_2. Add components, except $NaHCO_3$ solution and $Na_2S{\cdot}9H_2O$ solution, to distilled/deionized water and bring volume to 970.0mL. Mix thoroughly. Sparge with 80% N_2 + 20% CO_2. Autoclave for 15 min at 15 psi pressure–121°C. Aseptically and anaerobically add 20.0mL of sterile $NaHCO_3$ solution and 10.0mL of sterile $Na_2S{\cdot}9H_2O$ solution. Mix thoroughly. Aseptically and anaerobically distribute into sterile tubes or flasks.

Use: For the cultivation of a wide variety of bacteria. that can utilize glutarate as a carbon source.

Glycerol Agar

Composition per liter:

Beef heart, infusion from	300.0g
Glycerol	60.0g
Agar	17.5g
Tryptose	7.0g
NaCl	3.0g
Peptone	2.5g
NaCl	2.5g
Yeast extract	1.0g
Beef extract	0.5g

pH 7.3 ± 0.2 at 25°C

Preparation of Medium: Add components to distilled/deionized water and bring volume to 1.0L. Mix thoroughly. Adjust pH to 7.3. Gently heat and bring to boiling. Distribute into tubes or flasks. Autoclave for 15 min at 15 psi pressure–121°C. Pour into sterile Petri dishes or leave in tubes.

Use: For the cultivation of fastidious bacteria.

Glycerol Asparagine Meat Agar

Composition per liter:

Agar	20.0g
Glycerol	10.0g
Beef extract	10.0g
L-Asparagine	1.0g
K_2HPO_4	1.0g
Trace salts solution	1.0mL

pH 7.4 ± 0.2 at 25°C

Trace Salts Solution:

Composition per 100.0mL:

$FeSO_4{\cdot}7H_2O$	0.1g
$MnCl_2{\cdot}4H_2O$	0.1g
$ZnSO_4{\cdot}7H_2O$	0.1g

Preparation of Trace Salts Solution: Add components to distilled/deionized water and bring volume to 100.0mL. Mix thoroughly. Filter sterilize.

Preparation of Medium: Add components, except trace salts solution, to distilled/deionized water and bring volume to 999.0mL. Mix thoroughly. Gently heat and bring to boiling. Autoclave for 15 min at 15 psi pressure–121°C. Cool to 45°–50°C. Aseptically add 1.0mL of sterile trace salts solution. Mix thoroughly. Pour into sterile Petri dishes or distribute into sterile tubes.

Use: For the cultivation of *Actinomadura* species and *Streptomyces* species.

GN Broth, Hajna

Composition per liter:

Pancreatic digest of casein	10.0g
Peptic digest of animal tissue	10.0g
NaCl	5.0g
Sodium citrate	5.0g
K_2HPO_4	4.0g
D-Mannitol	2.0g
KH_2PO_4	1.5g
Glucose	1.0g
Sodium deoxycholate	0.5g

pH 7.0 ± 0.2 at 25°C

Source: This medium is available as a premixed powder from BD Diagnostic Systems.

Preparation of Medium: Add components to distilled/deionized water and bring volume to 1.0L. Mix thoroughly. Gently heat and bring to boiling. Distribute into tubes or flasks. Autoclave for 15 min at 13 psi pressure–118°C. Pour into sterile Petri dishes or leave in tubes.

Use: For the selective cultivation of *Salmonella* and *Shigella* species.

Gonococcus Medium

Composition per 623.0mL:

Part II	500.0mL
Part I	123.0mL

pH 7.4 ± 0.2 at 25°C

Part I:

Composition per 123.0mL:

K_2HPO_4	10.5g
Glucose	7.5g
NaCl	5.25g
KH_2PO_4	4.5g
Sodium acetate	1.5g
L-Cysteine·HCl·H_2O	1.2g
Sodium citrate	1.13g
$NaHCO_3$	1.0g
K_2SO_4	0.9g
Na_2SO_4	0.75g
$MgCl_2·6H_2O$	0.45g
KCl	0.3g
NH_4Cl	0.3g
L-Arginine·HCl	0.25g
L-Proline	0.25g
Oxaloacetate	0.25g
L-Glutamic acid	0.19g
L-Methionine	0.19g
L-Asparagine·H_2O	0.13g
L-Isoleucine	0.13g
L-Serine	0.13g
L-Cystine	0.05g
Calcium pantothenate	0.02g
Thiamine·HCl	0.02g
Thiamine pyrophosphate chloride	0.02g
Nicotinamide adenine dinucleotide	0.01g
$CaCl_2·2H_2O$	5.0mg
$Fe(NO_3)_3·9H_2O$	5.0mg
Uracil	5.0mg
Biotin	4.0mg
Hypoxanthine	2.5mg
Sodium thioglycolate	0.025mg

Preparation of Part I: Add components to distilled/deionized water and bring volume to 123.0mL. Mix thoroughly. Adjust pH to 7.2 with 6*N* NaOH. Warm to 50°C for 45 min. Filter sterilize.

Part II:

Composition per 500.0mL:

Agar	10.0g
Soluble starch	7.5g

Preparation of Part II: Add components to distilled/deionized water and bring volume to 500.0mL. Mix thoroughly. Gently heat and bring to boiling. Autoclave for 15 min at 15 psi pressure–121°C. Cool to 45°–50°C.

Preparation of Medium: Aseptically combine 123.0mL of sterile part I and 500.0mL of cooled, sterile part II. Mix thoroughly. Pour into sterile Petri dishes.

Use: For the cultivation of *Neisseria gonorrhoeae.*

GP Agar (Glycerol Peptone Agar)

Composition per liter:

Agar	15.0g
Peptone	10.0g
Glycerol	10.0g

Preparation of Medium: Add components to distilled/deionized water and bring volume to 1.0L. Mix thoroughly. Gently heat and bring to boiling. Distribute into tubes or flasks. Autoclave for 15 min at 15 psi pressure–121°C. Pour into sterile Petri dishes or leave in tubes.

Use: For the cultivation of actinomycetes and *Mycobacterium spp.*

GPHF Agar

Composition per liter:

Agar	12.0g
Glucose	10.0g
Beef extract	5.0g
Pancreatic digest of casein	5.0g
Yeast extract	5.0g
$CaCl_2·2H_2O$	0.74g

pH 7.2 ± 0.2 at 25°C

Preparation of Medium: Add components to distilled/deionized water and bring volume to 1.0L. Mix thoroughly. Gently heat and bring to boiling. Distribute into tubes or flasks. Autoclave for 15 min at 15 psi pressure–121°C. Pour into sterile Petri dishes or leave in tubes.

Use: For the cultivation and maintenance of *Actinoplanes caeruleus, Micromonospora* species, *Microtetraspora fusca,* and *Streptomyces yerevanensis.*

Granada Medium

Composition per liter:

Starch, soluble	150.0g
Proteose peptone no. 3	38.0g
NaCl	3.0g
Trimethoprim lactate	0.015g
Sodium phosphate buffer (0.06*M*, pH 7.4)	900.0mL
Horse serum, coagulated	100.0mL

pH 7.4 ± 0.2 at 25°C

Preparation of Medium: Add proteose peptone no. 3 and NaCl to 200.0mL of sodium phosphate buffer and bring to boiling. Add 400.0mL of cold sodium phosphate buffer, starch, and trimethoprim lactate. Mix thoroughly. Bring volume to 900.0mL with sodium phosphate buffer. Gently heat while stirring in a boiling water bath for exactly 20 min. Do not autoclave. Cool to 90°–95°C. Add horse serum. Mix thoroughly. Cool to 60°–65°C while stirring. Pour into sterile Petri dishes. Medium will solidify in 2–3h.

Use: For the early selective isolation and cultivation of Group B streptococci from clinical specimens.

Group A Selective Strep Agar with Sheep Blood

Composition per liter:

Pancreatic digest of casein	14.5g
Agar	14.0g
NaCl	5.0g
Papaic digest of soybean meal	5.0g
Sheep blood	50.0mL
Growth factor solution	10.0mL
Selective agents solution	10.0mL

pH 7.4 ± 0.2 at 25°C

Growth Factor Solution:
Composition per 10.0mL:

Growth factors, BD Diagnostic Systems	1.5g

Preparation of Growth Factor Solution: Add growth factors to distilled/deionized water and bring volume to 10.0mL. Mix thoroughly. Filter sterilize.

Selective Agent Solution:
Composition per 10.0mL:

Selective agents	0.042g

Preparation of Selective Agent Solution: Add selective agents to distilled/deionized water and bring volume to 10.0mL. Mix thoroughly. Filter sterilize.

Preparation of Medium: Add components, except sheep blood, growth factor solution, and selective agent solution, to distilled/deionized water and bring volume to 930.0mL. Mix thoroughly. Gently heat and bring to boiling. Autoclave for 15 min at 15 psi pressure–121°C. Cool to 45°–50°C. Aseptically add 50.0mL of sheep blood, 10.0mL of sterile growth factor solution, and 10.0mL of sterile selective agent components. Mix thoroughly. Pour into sterile Petri dishes or distribute into sterile tubes.

Use: For the selective cultivation and primary isolation of group A streptococci, especially *Streptococcus pyogenes* from clinical specimens.

H Agar

Composition per liter:

Agar	15.0g
Pancreatic digest of casein	10.0g
NaCl	8.0g

Preparation of Medium: Add components to distilled/deionized water and bring volume to 1.0L. Mix thoroughly. Gently heat and bring to boiling. Autoclave for 15 min at 15 psi pressure–121°C. Pour into sterile Petri dishes.

Preparation of Medium: Add components to distilled/deionized water and bring volume to 1.0L. Mix thoroughly. Distribute into tubes or flasks. Autoclave for 15 min at 15 psi pressure–121°C.

Use: For the cultivation of *Escherichia coli* and a variety of other bacteria.

H Agar (Hominis Agar)

Composition per 98.0mL:

Base agar	65.0mL
Horse serum	20.0mL
Yeast dialysate	10.0mL
Penicillin solution	2.0mL
Thallium acetate solution	1.0mL

pH 7.3 ± 0.2 at 25°C

Base Agar:
Composition per liter:

Papaic digest of soybean meal	20.0g
Agarose	10.0g
NaCl	5.0g
Phenol Red (2% solution)	1.0mL

Preparation of Base Agar: Add components to distilled/deionized water and bring volume to 1.0L. Mix thoroughly. Gently heat and bring to boiling. Adjust pH to 7.3. Autoclave for 15 min at 15 psi pressure–121°C. Cool to 45°–50°C.

Yeast Dialysate:
Composition per 10.0mL:

Active, dried yeast	450.0g

Preparation of Yeast Dialysate: Add active, dried yeast to distilled/deionized water and bring volume to 1250.0mL. Gently heat and bring to 40°C. Autoclave for 15 min at 15 psi pressure–121°C. Put into dialysis tubing. Dialyze against 1.0L of distilled/deionized water for 2 days at 4°C. Discard tubing and its contents. Autoclave dialysate for 15 min at 15 psi pressure–121°C. Store at –20°C.

Penicillin Solution:
Composition per 10.0mL:

Penicillin	100,000U

Preparation of Penicillin Solution: Add penicillin to distilled/deionized water and bring volume to 10.0mL. Mix thoroughly. Filter sterilize.

Thallium Acetate Solution:
Composition per 10.0mL:

Thallium acetate	0.33g

Caution: Thallium salts are toxic.

Preparation of Thallium Acetate Solution: Add thallium acetate to distilled/deionized water and bring volume to 10.0mL. Mix thoroughly. Filter sterilize.

Preparation of Medium: To 65.0mL of cooled, sterile base agar, aseptically add 10.0mL of sterile yeast dialysate, 20.0mL of horse serum, 2.0mL of sterile penicillin solution, and 1.0mL of sterile thallium acetate solution. Mix thoroughly. Pour into 10mm × 35mm Petri dishes in 5.0mL volumes. Allow plates to stand overnight at 25°C to remove excess surface moisture.

Use: For the isolation of *Mycoplasma pneumoniae* and *Mycoplasma hominis*.

H Broth

Composition per liter:

NaCl	5.0g
Pancreatic digest of casein	5.0g
Peptone	5.0g
Beef extract	3.0g
K_2HPO_4	2.5g
Glucose	1.0g

pH 7.2 ± 0.2 at 25°C

Preparation of Medium: Add components to distilled/deionized water and bring volume to 1.0L. Mix thoroughly. Distribute into tubes in 4.0mL volumes. Autoclave for 15 min at 10 psi pressure–115°C.

Use: For the preparation of the H agglutination antigen used in the differentiation and identification of *Salmonella* species types and subtypes.

H Broth (Hominis Broth)

Composition per 99.0mL:

Base broth	65.0mL
Horse serum	20.0mL
Yeast dialysate	10.0mL
Penicillin solution	2.0mL
Glucose solution	1.0mL
Thallium acetate solution	1.0mL

pH 7.3 ± 0.2 at 25°C

Base Broth:

Composition per liter:

Papaic digest of soybean meal	20.0g
NaCl	5.0g
Phenol Red (2% solution)	1.0mL

Preparation of Base Broth: Add components to distilled/deionized water and bring volume to 1.0L. Mix thoroughly. Gently heat and bring to boiling. Adjust pH to 7.3. Autoclave for 15 min at 15 psi pressure–121°C. Cool to 25°C.

Yeast Dialysate:

Composition per 10.0mL:

Dried yeast, active	450.0g

Preparation of Yeast Dialysate: Add active, dried yeast to distilled/deionized water and bring volume to 1250.0mL. Gently heat and bring to 40°C. Autoclave for 15 min at 15 psi pressure–121°C. Put into dialysis tubing. Dialyze against 1.0L of distilled/deionized water for 2 days at 4°C. Discard tubing and its contents. Autoclave dialysate for 15 min at 15 psi pressure–121°C. Store at –20°C.

Penicillin Solution:

Composition per 10.0mL:

Penicillin	100,000U

Preparation of Penicillin Solution: Add penicillin to distilled/deionized water and bring volume to 10.0mL. Mix thoroughly. Filter sterilize.

Glucose Solution:

Composition per 10.0mL:

D-Glucose	1.8g

Preparation of Glucose Solution: Add D-glucose to distilled/deionized water and bring volume to 10.0mL. Mix thoroughly. Filter sterilize.

Caution: Thallium salts are toxic.

Thallium Acetate Solution:

Composition per 10.0mL:

Thallium acetate	0.33g

Preparation of Thallium Acetate Solution: Add thallium acetate to distilled/deionized water and bring volume to 10.0mL. Mix thoroughly. Filter sterilize.

Preparation of Medium: To 65.0mL of cooled, sterile base broth, aseptically add 10.0mL of sterile yeast dialysate, 20.0mL of horse serum, 2.0mL of sterile penicillin solution, 1.0mL of sterile glucose solution, and 1.0mL of sterile thallium acetate solution. Mix thoroughly. Aseptically distribute into sterile screw-capped tubes in 5.0mL volumes. Screw caps down tightly.

Use: For the isolation and cultivation of *Mycoplasma pneumoniae*.

H Broth (Hominis Broth)

Composition per 100.0mL:

Base broth	65.0mL
Horse serum	20.0mL
Yeast dialysate	10.0mL
Penicillin solution	2.0mL

Arginine solution 2.0mL
Thallium acetate solution.............................. 1.0mL
pH 7.3 ± 0.2 at 25°C

Base Broth:
Composition per liter:
Papaic digest of soybean meal 20.0g
NaCl 5.0g
Phenol Red (2% solution) 1.0mL

Preparation of Base Broth: Add components to distilled/deionized water and bring volume to 1.0L. Mix thoroughly. Gently heat and bring to boiling. Adjust pH to 7.3. Autoclave for 15 min at 15 psi pressure–121°C. Cool to 25°C.

Yeast Dialysate:
Composition per 10.0mL:
Dried yeast, active..450.0g

Preparation of Yeast Dialysate: Add active, dried yeast to distilled/deionized water and bring volume to 1250.0mL. Gently heat and bring to 40°C. Autoclave for 15 min at 15 psi pressure–121°C. Put into dialysis tubing. Dialyze against 1.0L of distilled/deionized water for 2 days at 4°C. Discard tubing and its contents. Autoclave dialysate for 15 min at 15 psi pressure–121°C. Store at –20°C.

Penicillin Solution:
Composition per 10.0mL:
Penicillin 100,000U

Preparation of Penicillin Solution: Add penicillin to distilled/deionized water and bring volume to 10.0mL. Mix thoroughly. Filter sterilize.

Arginine Solution:
Composition per 10.0mL:
L-Arginine 1.74g

Preparation of Arginine Solution: Add L-arginine to distilled/deionized water and bring volume to 10.0mL. Mix thoroughly. Filter sterilize.

Thallium Acetate Solution:
Composition per 10.0mL:
Thallium acetate..0.33g

Caution: Thallium salts are toxic.

Preparation of Thallium Acetate Solution: Add thallium acetate to distilled/deionized water and bring volume to 10.0mL. Mix thoroughly. Filter sterilize.

Preparation of Medium: To 65.0mL of cooled, sterile base broth, aseptically add 10.0mL of sterile yeast dialysate, 20.0mL of horse serum, 2.0mL of sterile penicillin solution, 1.0mL of sterile glucose solution, and 1.0mL of sterile thallium acetate solution. Mix thoroughly. Aseptically distribute into sterile screw-capped tubes in 5.0mL volumes. Screw caps down tightly.

Use: For the isolation and cultivation of *Mycoplasma hominis.*

H Diphasic Medium

Composition per 197.0mL:
Base agar... 65.0mL
Base broth 65.0mL
Horse serum 40.0mL
Yeast dialysate 20.0mL
Penicillin solution 4.0mL
Thallium acetate solution............................... 2.0mL
Glucose solution 1.0mL
pH 7.3 ± 0.2 at 25°C

Base Agar:
Composition per liter:
Papaic digest of soybean meal 20.0g
Agarose 10.0g
NaCl 5.0g
Phenol Red (2% solution) 1.0mL

Preparation of Base Agar: Add components to distilled/deionized water and bring volume to 1.0L. Mix thoroughly. Gently heat and bring to boiling. Adjust pH to 7.3. Autoclave for 15 min at 15 psi pressure–121°C. Cool to 45°–50°C.

Base Broth:
Composition per liter:
Papaic digest of soybean meal 20.0g
NaCl 5.0g
Phenol Red (2% solution) 1.0mL

Preparation of Base Broth: Add components to distilled/deionized water and bring volume to 1.0L. Mix thoroughly. Gently heat and bring to boiling. Adjust pH to 7.3. Autoclave for 15 min at 15 psi pressure–121°C. Cool to 25°C.

Yeast Dialysate:
Composition per 10.0mL:
Dried yeast, active 450.0g

Preparation of Yeast Dialysate: Add active, dried yeast to distilled/deionized water and bring volume to 1250.0mL. Gently heat and bring to 40°C. Autoclave for 15 min at 15 psi pressure–121°C. Put into dialysis tubing. Dialyze against 1.0L of distilled/deionized water for 2 days at 4°C. Discard tubing and its contents. Autoclave dialysate for 15 min at 15 psi pressure–121°C. Store at –20°C.

Penicillin Solution:
Composition per 10.0mL:
Penicillin 100,000U

Preparation of Penicillin Solution: Add penicillin to distilled/deionized water and bring volume to 10.0mL. Mix thoroughly. Filter sterilize.

Glucose Solution:

Composition per 10.0mL:

D-Glucose	1.8g

Preparation of Glucose Solution: Add D-glucose to distilled/deionized water and bring volume to 10.0mL. Mix thoroughly. Filter sterilize.

Thallium Acetate Solution:

Composition per 10.0mL:

Thallium acetate	0.33g

Caution: Thallium salts are toxic.

Preparation of Thallium Acetate Solution: Add thallium acetate to distilled/deionized water and bring volume to 10.0mL. Mix thoroughly. Filter sterilize.

Preparation of Medium: To 65.0mL of cooled, sterile base agar, aseptically add 10.0mL of sterile yeast dialysate, 20.0mL of horse serum, 2.0mL of sterile penicillin solution, and 1.0mL of sterile thallium acetate solution. Mix thoroughly. Aseptically distribute into screw-capped tubes in 3.0mL volumes. Allow agar to solidify. To 65.0mL of cooled, sterile base broth, aseptically add 10.0mL of sterile yeast dialysate, 20.0mL of horse serum, 2.0mL of sterile penicillin solution, 1.0mL of sterile glucose solution, and 1.0mL of sterile thallium acetate solution. Mix thoroughly. Aseptically distribute 3.0mL of broth solution on top of the 3.0mL of solidified base agar in each tube. Screw caps down tightly.

Use: For the isolation and cultivation of *Mycoplasma pneumoniae*.

Haemophilus ducreyi Medium

Composition per liter:

Columbia blood agar base	675.0mL
Rabbit blood	300.0mL
Fresh yeast extract solution	25.0mL

pH 6.5–7.0 at 25°C

Columbia Blood Agar Base:

Composition per 675.0mL:

Agar	15.0g
Pantone	10.0g
Bitone	10.0g
NaCl	5.0g
Tryptic digest of beef heart	3.0g
Cornstarch	1.0g

Preparation of Columbia Blood Agar Base: Add components to distilled/deionized water and bring volume to 675.0mL. Mix thoroughly. Gently heat until boiling. Autoclave for 15 min at 15 psi pressure–121°C. Cool to 45°–50°C.

Fresh Yeast Extract Solution:

Composition per 100.0mL:

Baker's yeast, live, pressed, starch-free	25.0g

Preparation of Fresh Yeast Extract Solution: Add the live Baker's yeast to 100.0mL of distilled/deionized water. Autoclave for 90 min at 15 psi pressure–121°C. Allow to stand. Remove supernatant solution. Adjust pH to 6.6–6.8. Filter sterilize.

Preparation of Medium: To 675.0mL of cooled, sterile Columbia blood agar base, aseptically add rabbit blood and sterile fresh yeast extract solution. Aseptically adjust pH to 6.5–7.0.

Use: For the cultivation and maintenance of *Haemophilus ducreyi.*

Haemophilus ducreyi Medium, Revised (Ducreyi Medium, Revised)

Composition per 1010.0mL:

Solution B	500.0mL
Solution A	400.0mL
Solution C	110.0mL

pH 7.4 ± 0.2 at 25°C

Solution A:

Composition per 400.0mL:

Beef heart, infusion from	500.0g
Agar	15.0g
Tryptose	10.0g
NaCl	5.0g

Preparation of Solution A: Add components to distilled/deionized water and bring volume to 400.0L. Mix thoroughly. Gently heat and bring to boiling. Autoclave for 15 min at 15 psi pressure–121°C. Cool to 45°–50°C.

Solution B:

Composition per 500.0mL:

Hemoglobin	10.0g

Preparation of Solution B: Add components to distilled/deionized water and bring volume to 500.0L. Mix thoroughly. Gently heat and bring to boiling. Autoclave for 15 min at 15 psi pressure–121°C. Cool to 45°–50°C.

Solution C:

Composition per 110.0mL:

Fetal bovine serum	100.0mL
Supplement solution	10.0mL

Supplement Solution:

Composition per liter:

Glucose	100.0g
L-Cysteine·HCl	25.9g

L-Glutamine 10.0g
L-Cystine 1.1g
Adenine 1.0g
Nicotinamide adenine dinucleotide 0.25g
Vitamin B_{12} 0.1g
Thiamine pyrophosphate 0.1g
Guanine·HCl 0.03g
$Fe(NO_3)_3 \cdot 6H_2O$ 0.02g
p-Aminobenzoic acid 0.013g
Thiamine·HCl 3.0mg

Source: The supplement solution (IsoVitaleX enrichment) is available from BD Diagnostic Systems. This enrichment may be replaced by supplement VX from BD Diagnostic Systems.

Preparation of Supplement Solution: Add components to distilled/deionized water and bring volume to 1.0L. Mix thoroughly. Filter sterilize.

Preparation of Solution C: Combine components. Mix thoroughly. Filter sterilize. Warm to 45°–50°C.

Preparation of Medium: Aseptically combine solution A, solution B, and solution C. Mix thoroughly. Pour into sterile Petri dishes or distribute into sterile tubes.

Use: For the cultivation and maintenance of *Haemophilus ducreyi.*

Haemophilus influenzae Defined Medium MI

Composition per liter:

NaCl 5.8g
K_2HPO_4 .... 3.5g
Glycerol 3.0g
KH_2PO_4 .... 2.7g
Inosine 2.0g
L-Glutamic acid 1.3g
K_2SO_4 .... 1.0g
Sodium lactate 0.8g
L-Aspartic acid 0.5g
Nitrilotriethanol 0.4g
L-Arginine 0.3g
L-Leucine 0.3g
L-Cystine 0.2g
$MgCl_2$ 0.2g
L-Tyrosine 0.2g
L-Methionine 0.1g
L-Serine 0.1g
Uracil 0.1g
L-Lysine 0.05g
Glycine 0.03g
$CaCl_2$ 0.022g
Hypoxanthine 0.02g
Polyvinyl alcohol 0.02g
Tween 80 .... 0.02g
Hemin 0.01g
L-Histidine 0.01g
Calcium pantothenate 4.0mg
Ethylenediaminetetraacetate 4.0mg
Nicotinamide adenine dinucleotide 4.0mg
Thiamine 4.0mg

Preparation of Medium: Add components to distilled/deionized water and bring volume to 1.0L. Mix thoroughly. Filter sterilize.

Use: For the cultivation of *Haemophilus influenzae* in a chemically defined medium.

Haemophilus influenzae Defined Medium MI-Cit

Composition per liter:

NaCl 5.8g
K_2HPO_4 .... 3.5g
Glycerol 3.0g
KH_2PO_4 .... 2.7g
Inosine 2.0g
L-Glutamic acid 1.3g
K_2SO_4 .... 1.0g
Sodium lactate 0.8g
L-Aspartic acid 0.5g
Nitrilotriethanol 0.4g
L-Leucine 0.3g
L-Cystine 0.2g
$MgCl_2$ 0.2g
L-Tyrosine 0.2g
Citrulline 0.15g
L-Methionine 0.1g
L-Serine 0.1g
L-Lysine 0.05g
Glycine 0.03g
$CaCl_2$ 0.022g
Hypoxanthine 0.02g
Polyvinyl alcohol 0.02g
Tween 80 .... 0.02g
Hemin 0.01g
L-Histidine 0.01g
Calcium pantothenate 4.0mg
Ethylenediaminetetraacetate 4.0mg
Nicotinamide adenine dinucleotide 4.0mg
Thiamine 4.0mg

Preparation of Medium: Add components to distilled/deionized water and bring volume to 1.0L. Mix thoroughly. Filter sterilize.

Use: For the cultivation of *Haemophilus influenzae* in a chemically defined medium.

Haemophilus Medium (DSMZ Medium 804)

Composition per liter:

Acid hydrolysate of casein	31.5g
Beef extract	5.4g
Yeast extract	5.0g
Starch	2.7g
NAD solution	1.0mL
Hemin solution	1.0mL

pH 7.3 ± 0.1 at 25°C

NAD Solution:

Composition per 10.0mL:

NAD	150.0mg

Preparation of NAD Solution: Add NAD to distilled/deionized water and bring volume to 10.0mL. Mix thoroughly. Filter sterilize.

Hemin Solution:

Composition per 10.0mL:

Hemin	150.0mg
NaOH (1*N* solution)	2.0mL

Preparation of Hemin Solution: Add hemin to 2.0mL of 1*N* NaOH solution. Mix thoroughly. Bring volume to 10.0mL with distilled/deionized water. Filter sterilize.

Preparation of Medium: Add components, except hemin solution and NAD solution, to distilled/deionized water and bring to 1.0L. Mix thoroughly. Gently heat and bring to boiling. Autoclave for 10 min at 10 psi pressure–115°C. Cool to 25°C. Aseptically add 1.0mL sterile NAD solution and 1.0mL sterile hemin solution. Mix thoroughly. Aseptically distribute into sterile tubes or flasks.

Use: For the cultivation of *Haemophilus influenzae.*

Haemophilus Medium

Composition per 1050.0mL:

Beef heart, infusion from	25.0g
Agar	14.0g
Peptone	5.0g
NaCl	2.5g
Glucose	1.0g
Yeast extract solution	100.0mL
Horse serum	50.0mL

Yeast Extract Solution:

Composition per liter:

Baker's yeast	250.0g

Preparation of Yeast Extract Solution: Add baker's yeast to distilled/deionized water and bring volume to 1.0L. Mix thoroughly. Gently heat and bring to boiling. Filter through a paper filter. Adjust pH to 8.0. Filter through a Seitz filter. Store at -20°C. Check sterility before using.

Preparation of Medium: Add components, except yeast extract solution and horse serum, to distilled/deionized water and bring volume to 850.0mL. Mix thoroughly. Gently heat and bring to boiling. Autoclave for 15 min at 15 psi pressure–121°C. Cool to 50°–55°C. Aseptically add 100.0mL of sterile yeast extract solution and 50.0mL of sterile horse serum. Mix thoroughly. Pour into sterile Petri dishes or distribute into sterile tubes.

Use: For the cultivation and maintenance of *Actinobacillus pleuropneumoniae* and *Haemophilus* spp.

Haemophilus Test Medium (HTM)

Composition per liter:

Beef infusion	300.0g
Acid hydrolysate of casein	17.5g
Agar	17.0g
Yeast extract	5.0g
Starch	1.5g
HTM supplement	10.0mL

pH 7.4 ± 0.2 at 25°C

Source: This medium is available as a premixed powder from Oxoid Unipath.

HTM Supplement:

Composition per 10.0mL:

Nicotinamide adenine dinucleotide	0.03g
Hematin	0.03g

Preparation of HTM Supplement: Add components to distilled/deionized water and bring volume to 10.0mL. Mix thoroughly. Filter sterilize.

Preparation of Medium: Add components, except HTM supplement, to distilled/deionized water and bring volume to 990.0mL. Mix thoroughly. Gently heat and bring to boiling. Autoclave for 15 min at 15 psi pressure–121°C. Cool to 45°–50°C. Aseptically add 10.0mL of sterile HTM supplement. Mix thoroughly. Pour into sterile Petri dishes or distribute into sterile tubes.

Use: For antimicrobial susceptibility testing of *Haemophilus influenzae.*

Haemophilus Test Medium (HTM Base)

Composition per liter:

Beef infusion from	300.0g
Acid hydrolysate of casein	17.5g
Agar	17.0g
Yeast extract, selected for low antagonist levels	5.0g

Starch 1.5g
Supplement solution 10.0mL

pH 7.4 ± 0.2 at 25°C

Source: This medium is available as a premixed powder from Oxoid Unipath.

Supplement Solution:

Composition per 10.0mL:

NAD 15.0mg
Hematin 15.0mg

Preparation of Supplement Solution: Add components to distilled/deionized water and bring volume to 10.0mL. Mix thoroughly. Filter sterilize.

Preparation of Medium: Add components, except supplement solution, to distilled/deionized water and bring volume to 990.0mL. Mix thoroughly. Gently heat while stirring and bring to boiling. Autoclave for 15 min at 15 psi pressure–121°C. Cool to 50°C. Aseptially add 10.0mL supplement solution. Mix thouroughly. Pour into sterile Petri dishes.

Use: For the susceptibility testing of *Haemophilus influenzae*. The medium forms part of the recommended methods of the United States National Committee for Clinical Laboratory Standards (NCCLS). *Haemophilus influenzae* require complex media for growth. These complex media have aggravated the routine susceptibility testing of *Haemophilus influenzae* because of antagonism between some essential nutrients and certain antimicrobial agents. This medium overcomes those limitations. The transparency of the medium allows zones of inhibition to be read easily through the bottom of the Petri dish. HTM contains low levels of antimicrobial antagonists, which allows testing of trimethoprim/sulphamethoxazole to be carried out.

Ham's F-10 Medium

Composition per liter:

NaCl 7.4g
$NaHCO_3$ 1.2g
Glucose 1.1g
$NaH_2PO_4 \cdot H_2O$ 0.29g
KCl 0.28g
L-Arginine·HCl 0.21g
L-Glutamine 0.15g
$MgSO_4 \cdot 7H_2O$ 0.15g
Sodium pyruvate 0.11g
KH_2PO_4 .... 0.08g
$CaCl_2 \cdot 2H_2O$ 0.04g
L-Cystine·2HCl 0.04g
L-Histidine·HCl·H_2O 0.02g
L-Lysine·HCl 0.02g
L-Asparagine-H_2O 0.01g
L-Aspartic acid 0.01g
L-Glutamic acid 0.01g
L-Leucine 0.01g
L-Proline 0.01g
L-Serine 0.01g
L-Alanine 8.9mg
Glycine 7.5mg
D-Phenylalanine 5.0mg
L-Methionine 4.5mg
Hypoxanthine 4.1mg
L-Threonine 3.6mg
L-Valine 3.5mg
L-Isoleucine 2.6mg
L-Tyrosine 1.8mg
Vitamin B_{12} 1.4mg
Folic acid 1.3mg
Phenol Red 1.2mg
Thiamine·HCl 1.0mg
$FeSO_4 \cdot 7H_2O$ 0.8mg
Choline chloride 0.7mg
D-Calcium pantothenate 0.7mg
Thymidine 0.7mg
Niacinamide 0.6mg
L-Tryptophan 0.6mg
Isoinositol 0.5mg
Riboflavin 0.4mg
Lipoic acid 0.2mg
Pyridoxine·HCl 0.2mg
$ZnSO_4 \cdot 7H_2O$ 0.03mg
Biotin 0.02mg
$CuSO_4 \cdot 5H_2O$ 3.0μg

pH 7.0 ± 0.2 at 25°C

Preparation of Medium: Add components to distilled/deionized water and bring volume to 1.0L. Mix thoroughly. Filter sterilize.

Use: For the growth of Y-1 cell cultures used in the mouse adrenal assay for heat-labile toxin of enterotoxigenic *Escherichia coli* and *Vibrio* species.

HBT Bilayer Medium (Human Blood Tween Bilayer Medium)

Composition per 1062.5mL:

Agar 13.5g
Pancreatic digest of casein 12.0g
Casein/meat peptone 10.0g
NaCl 5.0g
Peptic digest of animal tissue 5.0g
Beef extract 3.0g
Yeast extract 3.0g
Cornstarch 1.0g
Human blood, anticoagulated 25.0mL
Colistin solution 10.0mL
Nalidixic acid solution 10.0mL

Amphotericin B solution 10.0mL
Polysorbate 80 (Tween 80) solution 7.5mL
pH 7.3 ± 0.2 at 25°C

Source: This medium is available as a premixed powder from BD Diagnostic Systems.

Colistin Solution:
Composition per liter:
Colistin 0.01g

Preparation of Colistin: Add colistin to distilled/deionized water and bring volume to 10.0mL. Mix thoroughly. Filter sterilize.

Nalidixic Acid Solution:
Composition per liter:
Nalidixic acid 0.02g

Preparation of Nalidixic Acid Solution: Add nalidixic acid to distilled/deionized water and bring volume to 10.0mL. Mix thoroughly. Filter sterilize.

Amphotericin B Solution:
Composition per liter:
Amphotericin B 3.0mg

Preparation of Amphotericin B Solution: Add Amphotericin B to distilled/deionized water and bring volume to 10.0mL. Mix thoroughly. Filter sterilize.

Tween 80 Solution:
Composition per 100.0mL:
Tween 80 .... 1.0mL

Preparation of Tween 80 Solution: Add Tween 80 to distilled/deionized water and bring volume to 100.0mL. Mix thoroughly. Adjust pH to 7.3. Filter sterilize.

Preparation of Medium: Add components, except amphotericin B, Tween 80, and human blood, to distilled/deionized water and bring volume to 1.0L. Mix thoroughly. Gently heat and bring to boiling. Divide the medium into two 500.0mL fractions. Autoclave both flasks of media for 15 min at 15 psi pressure–121°C. Cool to 45°–50°C. To one flask, aseptically add 5.0mL of sterile colistin solution, 5.0mL of sterile nalidixic acid solution, 5.0mL of sterile amphotericin B solution, and 3.75mL of Tween 80 solution. Mix thoroughly. Pour into sterile Petri dishes in 7.0mL volumes. Allow agar to harden. To remaining flask aseptically add 5.0mL of sterile colistin solution, 5.0mL of sterile nalidixic acid solution, 5.0mL of sterile amphotericin B solution, 3.75mL of sterile Tween 80 solution, and 25.0mL of sterile human blood. Mix thoroughly. Pour into the same Petri dishes that each contain 7.0mL of the agar medium without blood. The top layer should be approximately 14.0mL per plate.

Use: For the selective isolation, cultivation, and differentiation of *Gardnerella vaginalis* from clinical specimens.

Heart Infusion Agar

Composition per liter:
Beef heart, infusion from 500.0g
Agar 15.0g
Tryptose 10.0g
NaCl 5.0g
pH 7.4 ± 0.2 at 25°C

Source: This medium is available as a premixed powder from BD Diagnostic Systems.

Preparation of Medium: Add components to distilled/deionized water and bring volume to 1.0L. Mix thoroughly. Gently heat and bring to boiling. Distribute into tubes or flasks. Autoclave for 15 min at 15 psi pressure–121°C. Pour into sterile Petri dishes or leave in tubes.

Use: For the isolation and cultivation of a wide variety of fastidious microorganisms. It can also be used as a base for the preparation of blood agar in determining hemolytic reactions. For the cultivation and maintenance of *Bacillus anthracis, Bacillus cereus, Bacillus mycoides, Serratia rubidaea, Staphylococcus aureus, Tsatumella ptyseos*, and *Vibrio vulnificus*.

Heart Infusion Agar

Composition per liter:
Beef heart, infusion from 500.0g
Agar 15.0g
Tryptose 10.0g
NaCl 5.0g
Horse blood, sterile defibrinated 50.0mL
pH 7.4 ± 0.2 at 25°C

Source: This medium without agar is available as a premixed powder from BD Diagnostic Systems.

Preparation of Medium: Add components, except horse blood, to distilled/deionized water and bring volume to 950.0mL. Mix thoroughly. Gently heat and bring to boiling. Autoclave for 15 min at 15 psi pressure–121°C. Cool to 50°C. Aseptially add 50.0mL sterile defibrinated horse blood. Mix thouroughly. Pour into sterile Petri dishes.

Use: For the isolation and cultivation of a wide variety of fastidious microorganisms. For the cultivation and maintenance of *Bacillus cereus, Staphylococcus aureus*, and *Vibrio vulnificus*, especially for determining hemolytic reactions.

Heart Infusion Agar with Glucose

Composition per liter:

Beef heart, infusion from	500.0g
Agar	15.0g
Tryptose	10.0g
NaCl	5.0g
Glucose	1.0g

pH 7.4 ± 0.2 at 25°C

Preparation of Medium: Add components to distilled/deionized water and bring volume to 1.0L. Mix thoroughly. Gently heat and bring to boiling. Distribute into tubes or flasks. Autoclave for 15 min at 15 psi pressure–121°C. Pour into sterile Petri dishes or leave in tubes.

Use: For the cultivation and maintenance of *Bacillus* species and *Pseudomonas* species.

Heart Infusion Agar with 0.1% Glucose

Composition per liter:

Beef heart, infusion from	500.0g
Agar	15.0g
Tryptose	10.0g
NaCl	5.0g
Glucose	1.0g

pH 7.4 ± 0.2 at 25°C

Source: This medium without glucose is available as a premixed powder from BD Diagnostic Systems.

Preparation of Medium: Add components to distilled/deionized water and bring volume to 1.0L. Mix thoroughly. Gently heat and bring to boiling. Distribute into tubes or flasks. Autoclave for 15 min at 15 psi pressure–121°C. Pour into sterile Petri dishes or leave in tubes.

Use: For the isolation and cultivation of *Bacillus circulans* and *Pseudomonas* spp.

Heart Infusion Agar (pH 7.6) with Inactivated Horse Serum (ATCC Medium 493)

Composition per liter:

Beef heart, infusion from	500.0g
Agar	15.0g
Tryptose	10.0g
NaCl	5.0g
Horse serum, inactivated	100.0mL

pH 7.6 ± 0.2 at 25°C

Preparation of Medium: Add components, except horse serum, to distilled/deionized water and bring volume to 900.0mL. Mix thoroughly. Gently heat and bring to boiling. Autoclave for 15 min at 15 psi pressure–121°C. Cool to 45°–50°C. Aseptically add sterile horse serum. Mix thoroughly. Pour into sterile Petri dishes or distribute into sterile tubes.

Use: For the cultivation and maintenance of *Corynebacterium* species.

Heart Infusion Agar with Inactivated Horse Serum

Composition per liter:

Beef heart, infusion from	500.0g
Agar	15.0g
Tryptose	10.0g
NaCl	5.0g
Horse serum, inactivated	100.0mL

pH 7.4 ± 0.2 at 25°C

Preparation of Medium: Add components, except horse serum, to distilled/deionized water and bring volume to 900.0mL. Mix thoroughly. Gently heat and bring to boiling. Autoclave for 15 min at 15 psi pressure–121°C. Cool to 45°–50°C. Aseptically add sterile horse serum. Mix thoroughly. Pour into sterile Petri dishes or distribute into sterile tubes.

Use: For the cultivation and maintenance of *Corynebacterium* species.

Heart Infusion Agar with Inactivated Horse Serum, NaCl, and Penicillin

Composition per liter:

Beef heart, infusion from	500.0g
NaCl	35.0g
Agar	15.0g
Tryptose	10.0g
Horse serum, inactivated	100.0mL
Penicillin solution	10.0mL

pH 7.4 ± 0.2 at 25°C

Penicillin Solution:

Composition per 10.0mL:

Penicillin	1,000,000U

Preparation of Penicillin Solution: Add penicillin to distilled/deionized water and bring volume to 10.0mL. Mix thoroughly. Filter sterilize.

Preparation of Medium: Add components, except penicillin solution and horse serum, to distilled/deionized water and bring volume to 890.0mL. Mix thoroughly. Gently heat and bring to boiling. Autoclave for 15 min at 15 psi pressure–121°C. Cool to 45°–50°C. Aseptically add 10.0mL of sterile penicillin solution and 100.0mL of sterile horse serum. Mix thoroughly. Pour into sterile Petri dishes or distribute into sterile tubes.

Use: For the cultivation and maintenance of *Corynebacterium* species.

Heart Infusion Agar with Rabbit Blood

Composition per liter:

Beef heart, infusion from 500.0g
Agar 15.0g
Tryptose 10.0g
NaCl 5.0g
Rabbit blood 50.0mL
pH 7.4 ± 0.2 at 25°C

Preparation of Medium: Add components, except rabbit blood, to distilled/deionized water and bring volume to 950.0mL. Mix thoroughly. Gently heat and bring to boiling. Autoclave for 15 min at 15 psi pressure–121°C. Cool to 45°–50°C. Aseptically add sterile rabbit blood. Mix thoroughly. Pour into sterile Petri dishes or distribute into sterile tubes.

Use: For the cultivation and maintenance of *Neisseria lactamica, Bartonella quintana, Bartonella elizabethae,* and *Bartonella henselae.*

Heart Infusion Broth

Composition per liter:

Beef heart, infusion from 500.0g
Tryptose 10.0g
NaCl 5.0g
pH 7.4 ± 0.2 at 25°C

Source: This medium is available as a premixed powder from BD Diagnostic Systems.

Preparation of Medium: Add components to distilled/deionized water and bring volume to 1.0L. Mix thoroughly. Distribute into tubes or flasks. Autoclave for 15 min at 15 psi pressure–121°C.

Use: For the isolation and cultivation of a wide variety of fastidious microorganisms.

Heart Infusion Broth (pH 7.5) with Inactivated Human Serum and Yeast Extract (ATCC Medium 245)

Composition per liter:

Heart infusion broth with yeast extract 800.0mL
Human serum, inactivated 200.0mL
pH 7.5 ± 0.2 at 25°C

Heart Infusion Broth with Yeast Extract:
Composition per liter:

Beef heart, infusion from 500.0g
Tryptose 10.0g
Yeast extract (Oxoid) 6.3g
NaCl 5.0g

Preparation of Heart Infusion Broth with Yeast Extract: Add components to distilled/deionized water and bring volume to 1.0L. Mix thoroughly. Adjust pH to 7.5. Autoclave for 15 min at 15 psi pressure–121°C. Cool to 25°.

Source: Heart infusion broth without yeast extract is available as a premixed powder from BD Diagnostic Systems.

Preparation of Medium: To 800.0mL of sterile cooled heart infusion broth with yeast extract, aseptically add 200.0mL of heat inactivated human serum. Mix thoroughly. Aseptically distribute into sterile tubes or flasks.

Use: For the cultivation and maintenance of *Corynebacterium pseudotuberculosis* and *Streptobacillus moniliformis.*

Heart Infusion Broth with Additives for *Staphylococcus*

Composition per liter:

Beef heart, infusion from 500.0g
NaCl .... 30.0 g
Tryptose 10.0g
Horse serum, inactivated 100.0mL
Penicillin solution 10.0mL
Fresh yeast extract solution 5.0mL
pH 7.4 ± 0.2 at 25°C

Penicillin Solution:
Composition per 10.0mL:

Penicillin 1,000,000U

Preparation of Penicillin Solution: Add penicillin to distilled/deionized water and bring volume to 10.0mL. Mix thoroughly. Filter sterilize.

Fresh Yeast Extract Solution:
Composition per 100.0mL:

Baker's yeast, live, pressed, starch-free 10.0g

Preparation of Fresh Yeast Extract Solution: Add the live Baker's yeast to 100.0mL of distilled/deionized water. Autoclave for 90 min at 15 psi pressure–121°C. Allow to stand. Remove supernatant solution. Adjust pH to 6.6–6.8.

Preparation of Medium: Add components—except horse serum, fresh yeast extract solution, and penicillin solution—to distilled/deionized water and bring volume to 800.0mL. Mix thoroughly. Gently heat and bring to boiling. Autoclave for 15 min at 15 psi pressure–121°C. Cool to 45°–50°C. Aseptically add sterile horse serum. Mix thoroughly. Aseptically distribute into sterile tubes or flasks.

Use: For the cultivation of *Staphylococcus* species.

Heart Infusion Broth with Additives for *Streptobacillus*

Composition per liter:

Beef heart, infusion from 500.0g

Tryptose 10.0g
Peptone...10.0g
NaCl 5.0g
Glucose 0.5g
Horse serum, inactivated........................... 200.0mL

pH 7.5 ± 0.2 at 25°C

Preparation of Medium: Add components, except horse serum, to distilled/deionized water and bring volume to 800.0mL. Mix thoroughly. Gently heat and bring to boiling. Autoclave for 15 min at 15 psi pressure–121°C. Cool to 45°–50°C. Aseptically add sterile horse serum. Mix thoroughly. Aseptically distribute into sterile tubes or flasks.

Use: For the cultivation and maintenance of *Streptobacillus moniliformis.*

Heart Infusion Broth with Inactivated Horse Serum and Fresh Yeast Extract

Composition per liter:

Beef heart, infusion from 500.0g
Tryptose 10.0g
NaCl 5.0g
Horse serum, inactivated........................... 200.0mL
Fresh yeast extract solution........................ 100.0mL

pH 7.5 ± 0.2 at 25°C

Source: This medium is available as a premixed powder from BD Diagnostic Systems.

Fresh Yeast Extract Solution:

Composition per 100.0mL:

Baker's yeast, live, pressed, starch-free............25.0g

Preparation of Fresh Yeast Extract Solution: Add the live Baker's yeast to 100.0mL of distilled/deionized water. Autoclave for 90 min at 15 psi pressure–121°C. Allow to stand. Remove supernatant solution. Adjust pH to 6.6–6.8.

Preparation of Medium: Add components, except horse serum and fresh yeast extract solution, to distilled/deionized water and bring volume to 700.0mL. Mix thoroughly. Gently heat and bring to boiling. Autoclave for 15 min at 15 psi pressure–121°C. Cool to 45°–50°C. Aseptically add sterile horse serum and fresh yeast extract solution. Mix thoroughly. Aseptically distribute into sterile tubes or flasks.

Use: For the cultivation and maintenance of *Acholeplasma* species, *Mycoplasma* species, and *Streptobacillus* species.

Heart Infusion Broth with Inactivated Horse Serum, Fresh Yeast Extract, and Sucrose

Composition per liter:

Beef heart, infusion from 500.0g
Sucrose..40.0g
Tryptose 10.0g
NaCl 5.0g
Horse serum, inactivated 200.0mL
Fresh yeast extract solution 100.0mL

pH 7.5 ± 0.2 at 25°C

Source: This medium is available as a premixed powder from BD Diagnostic Systems.

Fresh Yeast Extract Solution:

Composition per 100.0mL:

Baker's yeast, live, pressed, starch-free............25.0g

Preparation of Fresh Yeast Extract Solution: Add the live Baker's yeast to 100.0mL of distilled/deionized water. Autoclave for 90 min at 15 psi pressure–121°C. Allow to stand. Remove supernatant solution. Adjust pH to 6.6–6.8.

Preparation of Medium: Add components, except horse serum and fresh yeast extract solution, to distilled/deionized water and bring volume to 700.0mL. Mix thoroughly. Gently heat and bring to boiling. Autoclave for 15 min at 15 psi pressure–121°C. Cool to 45°–50°C. Aseptically add sterile horse serum and fresh yeast extract solution. Mix thoroughly. Aseptically distribute into sterile tubes or flasks.

Use: For the cultivation and maintenance of *Acholeplasma* species, *Mycoplasma* species, and *Streptobacillus* species.

Heart Infusion Medium with Fetal Bovine Serum

Composition per liter:

Beef heart, infusion from 500.0g
Agar 15.0g
Tryptose 10.0g
NaCl 5.0g
Fetal bovine serum.................................... 100.0mL

pH 7.4 ± 0.2 at 25°C

Preparation of Medium: Add components, except fetal bovine serum, to distilled/deionized water and bring volume to 900.0mL. Mix thoroughly. Gently heat and bring to boiling. Autoclave for 15 min at 15 psi pressure–121°C. Cool to 45°–50°C. Aseptically add sterile fetal bovine serum. Mix thoroughly. Aseptically distribute into sterile tubes or flasks.

Use: For the cultivation of *Haemophilus ducreyi.*

Heart Infusion Tyrosine Agar

Composition per liter:

Beef heart, infusion from 500.0g
Agar 15.0g
Tryptose 10.0g

NaCl 5.0g
L-Tyrosine 1.0g

pH 7.4 ± 0.2 at 25°C

Preparation of Medium: Add components to distilled/deionized water and bring volume to 1.0L. Mix thoroughly. Gently heat and bring to boiling. Distribute into tubes. Autoclave for 15 min at 15 psi pressure–121°C. Allow tubes to cool in a slanted position.

Use: For the cultivation and differentiation of *Bordetella parapertussis* based on browning of blood-free medium.

Hektoen Enteric Agar

Composition per liter:

Agar 13.5g
Lactose 12.0g
Peptic digest of animal tissue 12.0g
Sucrose 12.0g
Bile salts 9.0g
NaCl 5.0g
$Na_2S_2O_3$ 5.0g
Yeast extract 3.0g
Salicin 2.0g
Ferric ammonium citrate 1.5g
Acid Fuchsin 0.1g
Bromthymol Blue 0.064g

pH 7.6 ± 0.2 at 25°C

Source: This medium is available as a premixed powder from BD Diagnostic Systems and Oxoid Unipath.

Caution: Acid Fuchsin is a potential carcinogen and care must be taken to avoid inhalation of the powdered dye and contact with the skin.

Preparation of Medium: Add components to distilled/deionized water and bring volume to 1.0L. Mix thoroughly. Gently heat while stirring until components are dissolved. Do not autoclave. Pour into sterile Petri dishes. Allow agar to solidify with the Petri dish covers partially off.

Use: For the isolation and cultivation of Gram-negative enteric microorganisms from a variety of clinical and nonclinical specimens based on lactose or sucrose fermentation and H_2S production. For the isolation and differentiation of *Salmonella* and *Shigella*. Bacteria that ferment lactose or sucrose appear as yellow to orange colonies. Bacteria that produce H_2S appear as colonies with black centers.

Helicobacter pylori Isolation Agar

Composition per liter:

Agar 15.0g
Bitone 10.0g
Pancreatic digest of casein 5.0g
NaCl 5.0g
Peptic digest of animal tissue 5.0g
Tryptic digest of beef heart 3.0g
Cornstarch 1.0g
Horse blood, laked 35.0mL
Antibiotic inhibitor solution 10.0mL

pH 7.3 ± 0.2 at 25°C

Antibiotic Inhibitor Solution:
Composition per 10.0mL:

Vancomycin 0.01g
Amphotericin B 5.0mg
Cefsulodin 5.0mg
Trimethoprim lactate 5.0mg

Preparation of Antibiotic Inhibitor Solution: Add components to distilled/deionized water and bring volume to 10.0mL. Mix thoroughly. Filter sterilize.

Preparation of Medium: Add components, except horse blood and antibiotic inhibitor solution, to distilled/deionized water and bring volume to 955.0mL. Mix thoroughly. Gently heat and bring to boiling. Autoclave for 15 min at 15 psi pressure–121°C. Cool to 45°–50°C. Aseptically add sterile horse blood and sterile antibiotic inhibitor solution. Mix thoroughly. Pour into sterile Petri dishes or distribute into sterile tubes.

Use: For the isolation and cultivation of *Helicobacter pylori* from clinical specimens.

Helicobacter pylori Selective Medium

Composition per 1080mL:

Special peptone 23.0g
Agar 10.0g
NaCl 5.0g
Starch 1.0g
Horse blood, laked 70.0mL
Selective supplement solution 10.0mL

pH 7.3 ± 0.2 at 25°C

Source: This medium is available as a premixed powder from Oxoid Unipath.

Horse Blood, Laked:
Composition per 100.0mL:

Horse blood, fresh 100.0mL

Preparation of Horse Blood, Laked: Add blood to a sterile polypropylene bottle. Freeze overnight at –20°C. Thaw at 8°C. Refreeze at –20°C. Thaw again at 8°C.

Selective Supplement Solution:
Composition per 10.0mL:

Vancomycin 10.0mg

Trimethoprim	5.0mg
Cefsulodin	5.0mg
Amphotericin B	5.0mg

Preparation of Selective Supplement Solution: Add components to distilled/deionized water and bring volume to 10.0mL. Mix thoroughly. Filter sterilize.

Preparation of Medium: Add components, except selective supplement solution and laked horse blood, to distilled/deionized water and bring volume to 1.0L. Mix thoroughly. Gently heat while stirring and bring to boiling. Autoclave for 15 min at 15 psi pressure–121°C. Cool to 50°C. Aseptially add 10.0mL selective supplement solution and 70.0mL sterile laked horse blood. Mix thouroughly. Pour into sterile Petri dishes.

Use: For the isolation of *Helicobacter pylori* from clinical specimens. *H. pylori* forms discrete, translucent and non-coalescent colonies.

Hemo ID Quad Plate with Growth Factors (*Hemophilus* Identification Quadrant Plate with Growth Factors)

Composition per plate:

Quadrant I	5.0mL
Quadrant II	5.0mL
Quadrant III	5.0mL
Quadrant IV	5.0mL

Quadrant I:

Composition per 5.0mL:

Hemin	0.1mg
Brain heart infusion agar	5.0mL

Quadrant II:

Composition per 5.0mL:

Brain heart infusion agar	5.0mL
Supplement solution	0.05mL

Quadrant III:

Composition per 5.0mL:

Hemin	0.1mg
Brain heart infusion agar	5.0mL
Supplement solution	0.05mL

Quadrant IV:

Composition per 5.0mL:

Hemin	0.1mg
Brain heart infusion agar	5.0mL
Horse blood	0.25mL
Supplement solution	0.05mL

Source: The supplement solution (IsoVitaleX enrichment) is available from BD Diagnostic Systems. This enrichment may be replaced by supplement VX from BD Diagnostic Systems.

Preparation of Quadrant Media: Sterilize Brain Heart Infusion Agar by autocalving for 15 min at 15 psi pressure–121°C. Cool to 45°–50°C. Add additional components as filter sterilized solutions. Mix and distribute as 5.0mL aliquots into quadrants.

Use: For the differentiation and presumptive identification of *Haemophilus* species. The Hemo ID Quad Plate is a four-sectored plate, each with a different medium.

Hemorrhagic Coli Agar (HC Agar)

Composition per liter:

Sorbitol	20.0g
Pancreatic digest of casein	20.0g
Agar	15.0g
NaCl	5.0g
Bile salts no. 3	1.12g
Bromcresol Purple	0.015g

pH 7.2 ± 0.2 at 25°C

Preparation of Medium: Add components to distilled/deionized water and bring volume to 1.0L. Mix thoroughly. Gently heat and bring to boiling. Distribute into tubes or flasks. Autoclave for 15 min at 15 psi pressure–121°C. Pour into sterile Petri dishes.

Use: For the isolation and cultivation of enterohemorraghic *Escherichia coli.*

Herellea Agar

Composition per liter:

Agar	16.0g
Pancreatic digest of casein	15.0g
Lactose	10.0g
Maltose	10.0g
Enzymatic digest of soybean meal	5.0g
NaCl	5.0g
Bile salts	1.25g
Bromcresol Purple	0.02g

pH 6.8 ± 0.2 at 25°C

Source: This medium is available as a premixed powder from BD Diagnostic Systems.

Preparation of Medium: Add components to distilled/deionized water and bring volume to 1.0L. Mix thoroughly. Gently heat and bring to boiling. Distribute into tubes or flasks. Autoclave for 15 min at 15 psi pressure–121°C. Pour into sterile Petri dishes or leave in tubes.

Use: For the isolation, cultivation, and differentiation of Gram-negative nonfermentative and fermentative bacteria. It is especially recommended for the differentiation of *Acinetobacter (Herellea)* species from *Neisseria gonorrhoeae* in urethral or vaginal specimens. Fermentative bacteria appear as yellow colonies surrounded by yellow zones. Nonfermentative bacteria, such as *Acinetobacter* species, appear as pale lavender colonies.

Hershey's Tris-Buffered Salts Medium

Composition per liter:

Tris(hydroxymethyl)amino-
methane buffer (0.1*M* solution) 12.1g
NaCl 5.4g
KCl 3.0g
NH_4Cl 1.1g
$MgCl_2$ 0.095g
KH_2PO_4 .... 0.087g
Na_2SO_4 .... 0.023g
$CaCl_2$ 0.011g
$FeCl_3$ 0.16mg
Glucose solution 100.0mL
pH 7.4 ± 0.2 at 25°C

Glucose Solution:
Composition per 100.0mL:
Glucose 2.0g

Preparation of Glucose Solution: Add glucose to distilled/deionized water and bring volume to 100.0mL. Mix thoroughly. Autoclave for 15 min at 15 psi pressure–121°C. Cool to 25°C.

Preparation of Medium: Add components, except glucose solution, to distilled/deionized water and bring volume to 900.0mL. Mix thoroughly. Gently heat and bring to boiling. Autoclave for 15 min at 15 psi pressure–121°C. Cool to 25°C. Aseptically add sterile glucose solution. Mix thoroughly. Aseptically distribute into sterile tubes or flasks.

Use: For the cultivation of a variety of heterotrophic microorganisms.

Hexamita Medium

Composition per liter:
TYGM-9 medium 250.0mL
Sonneborn's *Paramecium* medium 750.0mL

TYGM-9 Medium:
Composition per liter:
NaCl 7.5g
K_2HPO_4 .... 2.8g
Casein digest 2.0g
Gastric mucin 2.0g
Yeast extract 1.0g
KH_2PO_4 .... 0.4g
Bovine serum, heat inactivated 30.0mL
Rice starch solution 30.0mL
Tween solution 0.5mL
pH 7.4 ± 0.2 at 25°C

Tween Solution:
Composition per 100.0mL:
Tween 80 .... 10.0mL

Preparation of Tween Solution: Add Tween 80 to absolute ethanol and bring volume to 100.0mL. Mix thoroughly. Filter sterilize.

Rice Starch Solution:
Composition per 100.0mL:
Rice starch 5.0g
Phosphate buffered saline solution 100.0mL

Phosphate Buffered Saline Solution:
Composition per liter:
NaCl 9.0g
$Na_2HPO_4 \cdot 7H_2O$ 0.795g
KH_2PO_4 .... 0.114g

Preparation of Phosphate Buffered Saline Solution: Add components to distilled/deionized water and bring volume to 1.0L. Mix thoroughly. Adjust pH to 7.4. Autoclave for 15 min at 15 psi pressure–121°C. Cool to 25°C.

Preparation of Rice Starch Solution: Heat sterilize rice starch at 150°C for 2h. Aseptically add 100.0mL of sterile phosphate-buffered saline solution. Mix thoroughly. Use immediately.

Preparation of TYGM-9 Medium: Add components, except rice starch solution, Tween solution, and bovine serum, to distilled/deionized water and bring volume to 939.5mL. Mix thoroughly. Autoclave for 15 min at 15 psi pressure–121°C. Cool to 25°C. Aseptically add 30.0mL of sterile bovine serum, 30.0mL of sterile rice starch solution, and 0.5mL of sterile Tween solution. Mix thoroughly. Aseptically distribute into sterile, screw-capped tubes or flasks.

Sonneborn's *Paramecium* Medium:
Composition per liter:
Solution 1 .... 1.0L
Klebsiella pneumoniae
cultured on solution 2 .... variable

Solution 1:
Composition per liter:
Rye grass cerophyll 2.5g
Na_2HPO_4 .... 0.5g

Source: Cerophyll can be obtained from Ward's Natural Science Establishment, Inc. Dairy Goat Nutrition distributes Grass Media Culture, which is equivalent. Cereal Leaf Product from Sigma Chemical is similar to cerophyll.

Preparation of Solution 1: Add cerophyll to distilled/deionized water and bring volume to 1.0L. Mix thoroughly. Gently heat and bring to boil. Boil for 5 min. Filter through Whatman #1 filter paper. Add 0.5g of Na_2HPO_4. Bring volume to 1.0L with distilled/deionized water. Mix thoroughly. Distribute 10.0mL volumes into tubes. Autoclave for 15 min at 15 psi pressure–121°C.

Solution 2:

Composition per liter:

Agar	20.0g
Yeast extract	4.0g
Glucose	0.16g

Preparation of Solution 2: Add components to distilled/deionized water and bring volume to 1.0L. Mix thoroughly. Gently heat and bring to boiling. Distribute 5.0mL into tubes. Autoclave for 15 min at 15 psi pressure–121°C. Allow tubes to cool in a slanted position.

Preparation of Sonneborn's *Paramecium* Medium: Inoculate the surface of agar slants of solution 2 with a culture of *Klebsiella pneumoniae*. Incubate at 37°C for 24–48h. Scrape cells from the surface of the agar slants and add to 10.0mL of solution 1. Incubate at 30°C for 24h.

Preparation of Medium: Aseptically combine 3.0mL of sterile TYGM-9 medium with 9.0mL of Sonneborn's *Paramecium* medium in 16 × 125mm screw-capped test tubes.

Use: For the cultivation of *Hexamita inflata, Hexamita pusilla,* and *Trepomonas agilis*.

HiCrome™ Aureus Agar Base (Aureus Agar Base) (HiCrome *Staphylococcus aureus* Agar) (*Staphylococcus aureus* Agar, HiCrome)

Composition per liter:

Agar	20.0g
Casein enzymic hydrolysate	12.0g
Sodium pyruvate	10.0g
Beef extract	6.0g
LiCl	5.0g
Yeast extract	5.0g
Pancreatic digest of of gelatin	3.0g
Chromogenic mixture	2.1g
Egg tellurite emulsion	50.0mL

pH 7.4 ± 0.2 at 25°C

Caution: Lithium chloride is harmful. Avoid bodily contact and inhalation of vapors. On contact with skin, wash with plenty of water immediately.

Egg Yolk Tellurite Emulsion:

Composition per 100.0mL:

Sterile saline	64.0mL
Egg Yolk	30.0mL
Sterile potassium tellurite solution, 3.5%	6.0mL

Source: This medium is available from Fluka, Sigma-Aldrich.

Preparation of Medium: Add components, except egg yolk tellurite emulsion, to distilled/deionized water and bring volume to 950.0mL. Mix thoroughly. Gently heat and bring to boiling. Autoclave for 15 min at 15 psi pressure–121°C. Cool to 50°C. Aeptically add 50.0mL sterile egg yok tellurite emulsion. Mix thoroughly. Pour into sterile Petri dishes or distribute into sterile tubes.

Use: For the isolation and enumeration of coagulase positive *Staphylococcus aureus*. Coagulase positive *S. aureus* gives brown black colonies where as *S. epidermidis* gives yellow, slightly brownish, colonies.

HiCrome Coliform Agar (Coliform Agar, HiCrome)

Composition per liter:

Agar	12.0g
Sodium chloride	5.0g
Peptone, special	3.0g
K_2HPO_4	3.0g
KH_2PO_4	1.7g
Sodium pyruvate	1.0g
Tryptophan	1.0g
Chromogenic mixture	0.2g
Sodium lauryl sulphate	0.1g

pH 6.8 ± 0.2 at 25°C

Source: This medium is available from Fluka, Sigma-Aldrich.

Preparation of Medium: Add components to distilled/deionized water and bring volume to 1.0L. Mix thoroughly. Gently heat while stirring and bring to boiling. Autoclave for 15 min at 15 psi pressure–121°C. Pour into sterile Petri dishes.

Use: For the simultaneous detection of *Escherichia coli* and total coliforms. Sodium lauryl sulphate inhibits gram-positive organisms. The chromogenic mixture contains two chromogenic substrates as Salmon-GAL and X-glucuronide. The enzyme β-D-galactosidase produced by coliforms cleaves Salmon-GAL, resulting in the salmon to red coloration of coliform colonies. The enzyme β-D-glucuronidase produced by *E. coli*, cleaves Xglucuronide. *Escherichia coli* forms dark blue to violet colored colonies due to cleavage of both Salmon-GAL and X-glucuronide. The addition of tryptophan improves the indole reaction, thereby increasing detection reliability in combination with the two chromogens.

HiCrome Coliform Agar with Novobiocin (Coliform Agar with Novobiocin, HiCrome)

Composition per liter:

Agar	12.0g
Sodium chloride	5.0g
Peptone, special	3.0g
K_2HPO_4	3.0g
KH_2PO_4	1.7g
Sodium pyruvate	1.0g
Tryptophan	1.0g
Chromogenic mixture	0.2g
Sodium lauryl sulphate	0.1g
Novobiocin	5.0mg

pH 6.8 ± 0.2 at 25°C

Source: This medium is available from Fluka, Sigma-Aldrich.

Preparation of Medium: Add components to distilled/deionized water and bring volume to 1.0L. Mix thoroughly. Gently heat while stirring and bring to boiling. Autoclave for 15 min at 15 psi pressure–121°C. Pour into sterile Petri dishes.

Use: For the simultaneous detection of *Escherichia coli* and total coliforms when high numbers of Gram-positive bacteria may be present. Novobiocin and sodium lauryl sulphate inhibit Gram-positive organisms. The chromogenic mixture contains two chromogenic substrates as Salmon-GAL and X-glucuronide. The enzyme β-D-galactosidase produced by coliforms cleaves Salmon-GAL, resulting in the salmon to red coloration of coliform colonies. The enzyme β-D-glucuronidase produced by *E. coli*, cleaves Xglucuronide. *Escherichia coli* forms dark blue to violet colored colonies due to cleavage of both Salmon-GAL and X-glucuronide. The addition of tryptophan improves the indole reaction, thereby increasing detection reliability in combination with the two chromogens.

HiCrome *E. coli* Agar A (*E. coli* Agar A, HiCrome)

Composition per liter:

Casein enzymic hydrolysate	14.0g
Agar	12.0g
Peptone, special	5.0g
NaCl	2.4g
Bile salts mixture	1.5g
Na_2HPO_4	1.0g
NaH_2PO_4	0.6g
X-Glucuronide	0.075g

pH 7.2 ± 0.2 at 25°C

Source: This medium is available from Fluka, Sigma-Aldrich.

Preparation of Medium: Add components to distilled/deionized water and bring volume to 1.0L. Mix thoroughly. Gently heat while stirring and bring to boiling. Autoclave for 15 min at 15 psi pressure–121°C. Cool to 50°C. Pour into sterile Petri dishes.

Use: For the detection and enumeration of *Escherichia coli* without further confirmation on membrane filter or by indole reagent. The chromogenic agent X-glucuronide used in this medium helps to detect glucuronidase activity. *E. coli* cells absorb X-glucuronide and the intracellular glucuronidase splits the bond between the chromophore and the glucuronide. The released chromophore gives coloration to the colonies. Bile salts mixture inhibits gram-positive organisms.

HiCrome *E. coli* Agar B (*E. coli* Agar B, HiCrome)

Composition per liter:

Casein enzymic hydrolysate	20.0g
Agar	15.0g
Bile salts mixture	1.5g
X-Glucuronide	0.075g

pH 7.2 ± 0.2 at 25°C

Source: This medium is available from Fluka, Sigma-Aldrich.

Preparation of Medium: Add components to distilled/deionized water and bring volume to 1.0L. Mix thoroughly. Gently heat while stirring and bring to boiling. Autoclave for 15 min at 15 psi pressure–121°C. Pour into sterile Petri dishes.

Use: For the detection and enumeration of *Escherichia coli* without further confirmation on membrane filter or by indole reagent. The chromogenic agent X-glucuronide used in this medium helps to detect glucuronidase activity. *E. coli* cells absorb X-glucuronide and the intracellular glucuronidase splits the bond between the chromophore and the glucuronide. The released chromophore gives coloration to the colonies. Bile salts mixture inhibits gram-positive organisms.

HiCrome ECC Agar (ECC Agar, HiCrome)

Composition per liter:

Chromogenic mixture	20.3g
Agar	15.0g
Peptone, special	5.0g
NaCl	5.0g
Na_2HPO_4	3.5g

Yeast extract	3.0g
Lactose	2.5g
KH_2PO_4	1.5g
Neutral Red	0.03g

pH 6.8 ± 0.2 at 25°C

Source: This medium is available from Fluka, Sigma-Aldrich.

Preparation of Medium: Add components to distilled/deionized water and bring volume to 1.0L. Mix thoroughly. Gently heat while stirring and bring to boiling. Autoclave for 15 min at 15 psi pressure–121°C. Pour into sterile Petri dishes.

Use: A differential medium for presumptive identification of *E. coli* and other coliforms. The chromogenic mixture contains two chromogens as X-glucuronide and Salmon-GAL. X-glucuronide is cleaved by the the enzyme β-glucuronidase produced by *E. coli*. Salmon-GAL is cleaved by the enzyme galactosidase produced by the majority of coliforms, including *E. coli*.

HiCrome ECC Selective Agar (ECC Selective Agar, HiCrome)

Composition per liter:

Agar	10.0g
Peptone, special	6.0g
Casein enzymic hydrolysate	3.3g
NaCl	2.0g
Sodium pyruvate	1.0g
Tryptophane	1.0g
Sorbitol	1.0g
Na_2HPO_4	1.0g
NaH_2PO_4	0.6g
Tergitol 7	0.15g
Chromogenic mixture	0.43g

pH 6.8 ± 0.2 at 25°C

Source: This medium is available from Fluka, Sigma-Aldrich.

Preparation of Medium: Add components to distilled/deionized water and bring volume to 1.0L. Mix thoroughly and heat with frequent agitation until components are completely dissolved (approximately 35 min.). Do not autoclave. Cool to 45°–50°C. Pour into sterile Petri dishes. Medium may show haziness.

Use: For detection of *Escherichia coli* and coliforms. Tergitol inhibits Ggram-positive as well as some Gram-negative bacteria other than coliforms. The chromogenic mixture contains two chromogenic substrates as Salmon-GAL and X-glucuronide. The enzyme β-D-galactosidase produced by coliforms cleaves Salmon-GAL, resulting in the salmon to red coloration of coliform colonies. The enzyme β-D-glucuronidase produced by *E. coli*, cleaves X-glucuronide. *E. coli* forms dark blue to violet colored colonies due to cleavage of both Salmon-GAL and X- glucuronide. The addition of tryptophan improves the indole reaction.

HiCrome ECC Selective Agar (ECC Selective Agar, HiCrome)

Composition per liter:

Agar	10.0g
Peptone, special	6.0g
Casein enzymic hydrolysate	3.3g
NaCl	2.0g
Sodium pyruvate	1.0g
Tryptophane	1.0g
Sorbitol	1.0g
Na_2HPO_4	1.0g
NaH_2PO_4	0.6g
Tergitol 7	0.15g
Chromogenic mixture	0.43g
Cefulodin	5.0mg

pH 6.8 ± 0.2 at 25°C

Source: This medium is available from Fluka, Sigma-Aldrich.

Preparation of Medium: Add components to distilled/deionized water and bring volume to 1.0L. Mix thoroughly and heat with frequent agitation until components are completely dissolved (approximately 35 min.). Do not autoclave. Cool to 45°–50°C. Pour into sterile Petri dishes. Medium may show haziness.

Use: For detection of *Escherichia coli* and coliforms using pour plate or streak plate methods. Tergitol inhibits Gram-positive as well as some Gram-negative bacteria other than coliforms. The cefsulodin inhibits *Pseudomonas* and *Aeromonas* species. The chromogenic mixture contains two chromogenic substrates as Salmon-GAL and X- glucuronide. The enzyme β-D-galactosidase produced by coliforms cleaves Salmon-GAL, resulting in the salmon to red coloration of coliform colonies. The enzyme β-D-glucuronidase produced by *E. coli*, cleaves X-glucuronide. *E. coli* forms dark blue to violet colored colonies due to cleavage of both Salmon-GAL and X- glucuronide. The addition of Tryptophan improves the indole reaction.

HiCrome ECC Selective Agar (ECC Selective Agar, HiCrome)

Composition per liter:

Agar	10.0g

Peptone, special 6.0g
Casein enzymic hydrolysate 3.3g
NaCl 2.0g
Sodium pyruvate 1.0g
Tryptophane 1.0g
Sorbitol 1.0g
Na_2HPO_4 ....1.0g
NaH_2PO_4 ....0.6g
Chromogenic mixture 0.43g
Tergitol 7 ....0.15g
Cefulodin 10.0mg
pH 6.8 ± 0.2 at 25°C

Source: This medium is available from Fluka, Sigma-Aldrich.

Preparation of Medium: Add components to distilled/deionized water and bring volume to 1.0L. Mix thoroughly and heat with frequent agitation until components are completely dissolved (approximately 35 min.). Do not autoclave. Cool to 45°–50°C. Pour into sterile Petri dishes. Medium may show haziness.

Use: For detection of *Escherichia coli* and coliforms using the membrane filter technique. Tergitol inhibits Gram-positive as well as some Gram-negative bacteria other than coliforms. The cefsulodin inhibits *Pseudomonas* and *Aeromonas* species. The chromogenic mixture contains two chromogenic substrates as Salmon-GAL and X-glucuronide. The enzyme β-D-galactosidase produced by coliforms cleaves Salmon-GAL, resulting in the salmon to red coloration of coliform colonies. The enzyme β-D-glucuronidase produced by *E. coli*, cleaves X-glucuronide. *E. coli* forms dark blue to violet colored colonies due to cleavage of both Salmon-GAL and X-glucuronide. The addition of tryptophan improves the indole reaction.

HiCrome Enterococci Broth (Enterococci HiCrome Broth)

Composition per liter:

Peptone, special 10.0g
NaCl 5.0g
Polysorbate 80 ....2.0g
NaH_2PO_4 ....1.25g
NaN_3 ....0.3g
Chromogenic mixture 0.04g
pH 7.5 ± 0.2 at 25°C

Source: This medium is available from Fluka, Sigma-Aldrich.

Caution: Sodium azide is toxic. It also has a tendency to form explosive metal azides with plumbing materials. It is advisable to use enough water to flush off the disposables.

Preparation of Medium: Add components to distilled/deionized water and bring volume to 1.0L. Mix thoroughly. Distribute into tubes or flasks. Autoclave for 15 min at 15 psi pressure–121°C.

Use: For the rapid and easy identification and differentiation of enterococci. It contains a chromogenic substrate which aids in the detection of enterococci.

HiCrome *Listeria* Agar Base, Modified (*Listeria* HiCrome Agar Base, Modified)

Composition per liter:

Peptone, special 23.0g
Agar 13.0g
Rhamnose 10.0g
Chromogenic mixture 5.13g
LiCl 5.0g
Meat extract 5.0g
NaCl 5.0g
Yeast extract 1.0g
Phenol Red 0.12g
pH 7.3 ± 0.2 at 25°C

Caution: Lithium chloride is harmful. Avoid bodily contact and inhalation of vapors. On contact with skin, wash with plenty of water immediately.

Source: This medium is available from Fluka, Sigma-Aldrich.

Preparation of Medium: Add components to distilled/deionized water and bring volume to 1.0L. Mix thoroughly. Gently heat while stirring and bring to boiling. Autoclave for 15 min at 15 psi pressure–121°C. Pour into sterile Petri dishes.

Use: A selective and differential agar medium recommended for rapid and direct identification of *Listeria* species, specifically *Listeria monocytogenes*.

HiCrome MacConkey-Sorbitol Agar MacConkey-Sorbitol Agar, HiCrome

Composition per liter:

Casein enzymic hydrolysate 17.0g
Agar 13.5g
Sorbitol 10.0g
NaCl 5.0g
Proteose peptone 3.0g
Bile salts mixture 1.5g
5-Bromo-4-chloro-3-indolyl-β-D-glucuronide sodium salt 0.1g
Neutral Red 0.03g
Crystal Violet 0.001g
pH 7.1 ± 0.3 at 25°C

Source: This medium is available from Fluka, Sigma-Aldrich.

Preparation of Medium: Add components to distilled/deionized water and bring volume to 1.0L. Mix thoroughly. Gently heat while stirring and bring to boiling. Mix to completely dissolve components. Do not autoclave. Cool to 50°C. Pour into sterile Petri dishes.

Use: For the direct isolation and differentiation of *E. coli 0157:H7*-strains. The medium contains sorbitol instead of lactose. Enteropathogenic strains of *Escherichia coli* 0157:H7 ferment lactose but does not ferment sorbitol and hence produce colorless colonies. Sorbitol fermenting strains of *Escherichia coli* produce pink-red colonies. The red color is due to production of acid from sorbitol, absorption of neutral red and a subsequent color change of the dye when pH of the medium falls below 6.8. The chromogenic indicator is added to detect the presence of an enzyme β-D-glucuronidase. Strains of *Escherichia coli* possessing β-D-glucuronidase appear as blue colored colonies on the medium. Enteropathogenic strains of *Escherichia coli* 0157 do not possess β-D-glucuronidase activity and thus produce colorless colonies. *Escherichia coli* fermenting sorbitol and possessing β-D-glucuronidase activity produce purple colored colonies. Most of the gram-positive organisms are inhibited by crystal violet and bile salts.

HiCrome MacConkey-Sorbitol Agar with Tellurite-Cefixime Supplement (MacConkey-Sorbitol Agar with Tellurite-Cefixime Supplement)

Composition per 1004mL:

Casein enzymic hydrolysate	17.0g
Agar	13.5g
Sorbitol	10.0g
NaCl	5.0g
Proteose peptone	3.0g
Bile salts mixture	1.5g
5-Bromo-4-chloro-3-indolyl-β-D-glucuronide sodium salt	0.1g
Neutral Red	0.03g
Crystal Violet	0.001g
Tellurite-cefixime supplement	4.0mL

pH 7.1 ± 0.3 at 25°C

Tellurite-Cefixime Supplement:

Composition per 4.0mL:

K_2TeO_3	5.0mg
Cefiximine	0.1mg

Preparation of Tellurite-Cefixime Supplement: Add components to 4.0mL of distilled/deionized water. Mix thoroughly. Filter sterilize.

Caution: Potassium tellurite is toxic.

Source: This medium is available from Fluka, Sigma-Aldrich. Tellurite-cefixime supplement is available from Oxoid.

Preparation of Medium: Add components, except tellurite-cefixime supplement, to distilled/deionized water and bring volume to 1.0L. Mix thoroughly. Gently heat while stirring and bring to boiling. Mix to completely dissolve components. Do not autoclave. Cool to 50°C. Add 4.0mL sterile tellurite-cefixime supplement. Mix thoroughly. Pour into sterile Petri dishes.

Use: For the direct isolation and differentiation of *E. coli 0157:H7*-strains. The medium contains sorbitol instead of lactose. Enteropathogenic strains of *Escherichia coli* 0157:H7 ferment lactose but does not ferment sorbitol and hence produce colorless colonies. Sorbitol fermenting strains of *Escherichia coli* produce pink-red colonies. The red color is due to production of acid from sorbitol, absorption of neutral red and a subsequent color change of the dye when pH of the medium falls below 6.8. the chromogenic indicator is added to detect the presence of an enzyme β-D-glucuronidase. Strains of *Escherichia coli* possessing β-D-glucuronidase appear as blue colored colonies on the medium. Enteropathogenic strains of *Escherichia coli* 0157 do not possess β-D-glucuronidase activity and thus produce colorless colonies. *Escherichia coli* fermenting sorbitol and possessing β-D-glucuronidase activity produce purple colored colonies. Most of the gram-positive organisms are inhibited by crystal violet and bile salts. Addition of tellurite-cefixime supplement makes the medium selective. Potassium tellurite selects the serogroups 0157 from other *E. coli* serogroups and inhibits *Aeromonas* species and *Providencia* species. Cefixime inhibits *Proteus* species.

HiCrome M-CP Agar Base (M-CP HiCrome Agar Base) (Membrane *Clostridium perfringens* HiCrome Agar Base)

Composition per liter:

Tryptose	30.0g
Yeast extract	20.0g
Agar	15.0g
Sucrose	5.0g
L-Cysteine·HCl·H_2O	1.0g
$MgSO_4 \cdot 7H_2O$	0.1g
$FeCl_3 \cdot 6H_2O$	0.09g
Indoxyl-β-D-glucoside	0.06g
Bromocresol Purple	0.04g

pH 7.6 ± 0.2 at 25°C

Source: This medium is available from Fluka, Sigma-Aldrich.

Preparation of Medium: Add components to distilled/deionized water and bring volume to 1.0L. Mix thoroughly. Gently heat while stirring and bring to boiling. Autoclave for 15 min at 15 psi pressure–121°C. Pour into sterile Petri dishes.

Use: For the detection of *Clostridium perfringens.*

HiCrome MS.O157 Agar (MS.O157 Agar HiCrome)

Composition per liter:

Agar	12.0g
Peptone, special	10.0g
Sorbitol	4.0g
Bile salt mixture	1.0g
Chromogenic mixture	0.731g

pH 6.8 ± 0.2 at 25°C

Source: This medium is available from Fluka, Sigma-Aldrich.

Preparation of Medium: Add components to distilled/deionized water and bring volume to 1.0L. Mix thoroughly. Gently heat while stirring and bring to boiling. Do not autoclave. Cool to 50°C. Pour into sterile Petri dishes.

Use: For the simultaneous detection of *Escherichia coli, Escherichia coli 0157:H7* and coliforms. *Escherichia coli 0157:H7* gives colorless colonies because of non-fermentation of sorbitol and absence of β-glucuronidase activity, whereas other strains of *Escherichia coli* having β-glucuronidase activity and fermenting sorbitol appear as steel blue colored colonies. Some non-*Escherichia coli* 0157:H7 may have some colony color.

HiCrome MS.O157 Agar with Tellurite (MS.O157 Agar with Tellurite, HiCrome)

Composition 1000.25mL:

Agar	12.0g
Peptone, special	10.0g
Sorbitol	4.0g
Bile salt mixture	1.0g
Chromogenic mixture	0.731g
Tellurite solution	0.25mL

pH 6.8 ± 0.2 at 25°C

Tellurite Solution:

Composition per 10.0mL:

K_2TeO_3	0.1g

Preparation of Tellurite Solution: Add components to 10.0mL of distilled/deionized water. Mix thoroughly. Filter sterilize.

Caution: Potassium tellurite is toxic.

Source: This medium is available from Fluka, Sigma-Aldrich.

Preparation of Medium: Add components, exept tellurite solution, to distilled/deionized water and bring volume to 1.0L. Mix thoroughly. Gently heat while stirring and bring to boiling. Do not autoclave. Cool to 45°C. Aseptically add 0.25mL sterile tellurite solution. Mix thoroughly. Pour into sterile Petri dishes.

Use: For the simultaneous detection of *Escherichia coli, Escherichia coli 0157:H7,* and coliforms. *Escherichia coli 0157:H7* gives colorless colonies because of non-fermentation of sorbitol and absence of β-glucuronidase activity, whereas other strains of *Escherichia coli* having β-glucuronidase activity and fermenting sorbitol appear as steel blue colored colonies. Addition of tellurite makes the medium much more specific and selective.

HiCrome Rapid Coliform Broth (Coliform Rapid HiCrome Broth) (Rapid Coliform HiCrome Broth)

Composition per liter:

Peptone, special	5.0g
NaCl	5.0g
Na_2HPO_4	2.7g
KH_2PO_4	2.0g
Sorbitol	1.0g
Sodium lauryl sulfate	0.1g
IPTG	0.1g
Chromogenic substrate	0.08g
Fluorogenic substrate	0.05g

pH 6.8 ± 0.3 at 25°C

Source: This medium is available from Fluka, Sigma-Aldrich.

Preparation of Medium: Add components to distilled/deionized water and bring volume to 1.0L. Mix thoroughly. Gently heat while stirring and bring to boiling. Autoclave for 15 min at 15 psi pressure–121°C. Pour into sterile Petri dishes.

Use: For the detection and conformation of *Escherichia coli* and coliforms on the basics of enzyme substrate reaction from water samples, using a combination of chromogenic and fluorogenic substrate.

HiCrome Rapid Enterococci Agar (Enterococci Rapid HiCrome Agar) (Rapid Enterococci HiCrome Agar)

Composition per liter:

Agar	15.0g
Peptone special	10.0g
NaCl	5.0g
Polysorbate 80	2.0g
Na_2HPO_4	1.25g
NaN_3	0.3g
Chromogenic mixture	0.06g

pH 7.5 ± 0.2 at 25°C

Caution: Sodium azide is toxic. It also has a tendency to form explosive metal azides with plumbing materials. It is advisable to use enough water to flush off the disposables.

Source: This medium is available from Fluka, Sigma-Aldrich.

Preparation of Medium: Add components to distilled/deionized water and bring volume to 1.0L. Mix thoroughly. Gently heat while stirring and bring to boiling. Autoclave for 15 min at 15 psi pressure–121°C. Pour into sterile Petri dishes.

Use: For the rapid and easy identification and differentiation of enterococci. It contains a chromogenic substrate, which aids in the detection of enterococci, especially from water samples.

HiCrome *Salmonella* Agar (*Salmonella* Agar, HiCrome)

Composition per liter:

Agar	13.0g
Peptic digest of animal tissue	6.0g
Chromogenic mixture	5.4g
Yeast extract	2.5g
Chromogenic mix	1.5g
Bile salt mixture	1.0g

pH 7.7 ± 0.2 at 25°C

Source: This medium is available from Fluka, Sigma-Aldrich.

Preparation of Medium: Add components to distilled/deionized water and bring volume to 1.0L. Mix thoroughly. Gently heat while stirring and bring to boiling. Mix to completely dissolve components. Do not autoclave. Cool to 50°C. Pour into sterile Petri dishes.

Use: A selective chromgenic medium used for the isolation and differentiation of *Salmonella* species from coliforms. *E. coli* and *Salmonella* are easily distinguishable due to the colony characteristics. *Salmonella* give light purple colonies with a halo. *E. coli* has a characteristic blue color. Other organisms give colorless colonies. The characteristic light purple and blue color is due to the chromogenic mixture. Chromogenic medium for detecting and identifying Enterobacteria, proteus species, and other Gram-positive organisms.

HiCrome *Salmonella* Chromogen Agar (*Salmonella* Chromogen Agar, HiCrome) (Rambach Equivalent Agar)

Composition per liter:

Agar	15.0g
Peptic digest of animal tissue	5.0g
NaCl	5.0g
Yeast extract	2.0g
Meat extract	1.0g
Na-deoxycholate	1.0g
Chromogenic mixture	1 vial

pH 7.3 ± 0.2 at 25°C

Source: This medium is available from Fluka, Sigma-Aldrich.

Preparation of Medium: Add components to distilled/deionized water and bring volume to 1.0L. Mix thoroughly. Gently heat while stirring and bring to boiling. Mix to completely dissolve components. Do not autoclave. Mix and boil in 5 minute sequences for 35-40 min. Cool to 50°C. Shake gently for 30-35 min. Pour into sterile Petri dishes.

Use: A selective medium used for the detectgion of *Salmonella* species. This medium exploits a novel phenotypic characteristic of *Salmonella* spp.: the formation of acid from propylene glycol. This characteristic may be used in combination with a chromogenic indicator of β-galactosidase to differentiate *Salmonella* spp. from *Proteus* spp. and the other members of the Enterobacteriaceae. Desoxycholate is included in the plate medium as an inhibitor of gram-positive organisms. Non-typhi *Salmonella* spp. yield distinct, bright red colonies on this medium, allowing facilitated identification and unambiguous differentiation from *Proteus* spp. Coliforms produce blue-green to blue-violet colonies. Other Enterobacteriaceae and Gram-negative bacteria such as *Proteus, Shigella, Pseudomonas, Salmonella typhi,* and *S. paratyphi* A form colorless or yellow colonies.

HiCrome UTI Agar, Modified (UTI Agar, Modified HiCrome)

Composition per liter:

Peptic digest of animal tissue	8.0g
Agar	15.0g
Chromogenic mixture	12.44g

Beef extract 4.0g
Casein enzymatic hydrolysate 4.0g
pH: 7.2 ± 0.2 at 25°C

Source: This medium is available from Fluka, Sigma-Aldrich.

Preparation of Medium: Add components to distilled/deionized water and bring volume to 1.0L. Mix thoroughly. Gently heat while stirring and bring to boiling. Mix to completely dissolve components. Do not autoclave. Cool to 50°C. Pour into sterile Petri dishes.

Use: A chromgenic medium used for detecting and identifying Enterobacteria, *Proteus* species, and other bacteria involved in urinary tract infections.

Histoplasma capsulatum Agar

Composition per liter:
Agar 12.5g
Glucose 10.0g
Citric acid 10.0g
Potato starch 2.0g
α-Ketoglutaric acid 1.0g
L-Cystine·HCl·H_2O 1.0g
Glutathione, reduced 0.5g
L-Asparagine 0.1g
L-Tryptophan 0.02g
Solution 1 .... 250.0mL
Solution 3 .... 40.0mL
Solution 2 .... 10.0mL
Solution 4 .... 10.0mL
Solution 8 .... 10.0mL
Solution 5 .... 1.0mL
Solution 6 .... 0.1mL
Solution 7 .... 0.1mL
pH 6.5 ± 0.2 at 25°C

Solution 1:
Composition per liter:
KH_2PO_4 .... 8.0g
$(NH_4)_2SO_4$ 8.0g
$MgSO_4·7H_2O$ 0.86g
$CaCl_2$, anhydrous 0.08g
$ZnSO_4·7H_2O$ 0.05g

Preparation of Solution 1: Add components to distilled/deionized water and bring volume to 500.0mL. Mix thoroughly. Bring volume to 1.0L with distilled/deionized water. Store at 5°C.

Solution 2:
Composition per liter:
$FeSO_4·7H_2O$ 5.7g
$MnCl_2·6H_2O$ 0.8g
$NaMoO_4·2H_2O$ 0.15g
HCl, concentrated 1.0mL

Preparation of Solution 2: Add 1.0mL of concentrated HCl to 100.0mL of distilled water in a 1.0L volumetric flask. Dissolve each component completely in the sequence given. Bring volume to 1.0L with distilled/deionized water. Store at 5°C. Discard if red color or red precipitate appears.

Solution 3:
Composition per 100.0mL:
Casein, acid-hydrolyzed, vitamin-free 10.0g

Preparation of Solution 3: Add casein to distilled/deionized water and bring volume to 100.0mL.

Solution 4:
Composition per liter:
Calcium pantothenate 0.2g
Inositol 0.2g
Riboflavin 0.2g
Thiamine·HCl 0.2g
Nicotinamide 0.1g
Biotin 0.01g

Preparation of Solution 4: Add components to distilled/deionized water and bring volume to 1.0L. Mix thoroughly. Store at –20°C.

Solution 5:
Composition per 100.0mL:
Hemin 0.2g
NH_4OH, concentrated 0.3mL

Preparation of Solution 5: Add hemin to approximately 30.0mL of distilled/deionized water. Add NH_4OH. Mix thoroughly until dissolved. Bring volume to 100.0mL with distilled/deionized water. Store at 5° C.

Solution 6:
Composition per 10.0mL:
DL-Thioctic acid 0.01g
Ethanol (95% solution) 10.0mL

Preparation of Solution 6: Add DL-thioctic acid to 10.0mL of ethanol. Mix thoroughly. Store at –20°C.

Solution 7:
Composition per 10.0mL:
Coenzyme A 0.01g
$Na_2S·5H_2O$ (0.05% solution) 0.2mL

Preparation of Solution 7: Prepare $Na_2S·5H_2O$ solution in freshly boiled distilled/deionized water. Add coenzyme A to 9.8mL of distilled/deionized water. Mix thoroughly. Add freshly prepared $Na_2S·5H_2O$ solution. Mix thoroughly. Store the solution at –20°C.

Solution 8:
Composition per 100.0mL:
Oleic acid 0.1g

Preparation of Solution 8: Add oleic acid to 50.0mL of distilled/deionized water. Adjust pH to 9.0 with NaOH. Gently heat until dissolved. Bring volume to 100.0mL with distilled/deionized water. Store at 5°C.

Preparation of Medium: Add components—except agar, potato starch, and solution 8—to distilled/deionized water and bring volume to 400.0mL. Mix thoroughly. Adjust pH to 6.5 with 20% KOH solution. Filter sterilize. In a separate flask, add potato starch to 50.0mL of distilled/deionized water. Add the starch solution to 450.0mL of boiling distilled/deionized water. Add 10.0mL of solution 8 and the agar. Mix thoroughly. Autoclave for 15 min at 15 psi pressure–121°C. Cool to 70°C. Aseptically combine the two sterile solutions. Pour into sterile Petri dishes or distribute into sterile tubes.

Use: For the cultivation and maintenance of *Histoplasma capsulatum* in the yeast phase. For the cultivation of *Histoplasma duboisii, Blastomyces dermatitidis,* and *Sprotrichum schenckii.*

Histoplasma capsulatum Agar

Composition per liter:

Agar	15.0g
Glucose	10.0g
Potato starch	2.0g
α-Ketoglutaric acid	1.0g
L-Cystine·HCl·H_2O	1.0g
Glutathione, reduced	0.5g
L-Asparagine	0.1g
L-Tryptophan	0.02g
Solution 1	250.0mL
Solution 3	40.0mL
Solution 2	10.0mL
Solution 4	10.0mL
Solution 8	10.0mL
Solution 5	1.0mL
Solution 6	0.1mL
Solution 7	0.1mL

pH 6.5 ± 0.2 at 25°C

Solution 1:

Composition per liter:

KH_2PO_4	8.0g
$(NH_4)_2SO_4$	8.0g
$MgSO_4·7H_2O$	0.86g
$CaCl_2$, anhydrous	0.08g

Preparation of Solution 1: Add components to distilled/deionized water and bring volume to 500.0mL. Mix thoroughly. Bring volume to 1.0L with distilled/deionized water. Store at 5°C.

Solution 2:

Composition per liter:

$FeSO_4·7H_2O$	5.7g
$MnCl_2·6H_2O$	0.8g
$NaMoO_4·2H_2O$	0.15g
HCl, concentrated	1.0mL

Preparation of Solution 2: Add the 1.0mL of concentrated HCl to 100.0mL of distilled water in a 1.0L volumetric flask. Dissolve each component completely in the sequence given. Bring volume to 1.0L with distilled/deionized water. Store at 5°C. Discard if red color or red precipitate appears.

Solution 3:

Composition per 100.0mL:

Casein, acid-hydrolyzed, vitamin-free	10.0g

Preparation of Solution 3: Add casein to distilled/deionized water and bring volume to 100.0mL. Do not use enzymatically digested casein.

Solution 4:

Composition per liter:

Calcium pantothenate	0.2g
Inositol	0.2g
Riboflavin	0.2g
Thiamine·HCl	0.2g
Nicotinamide	0.1g
Biotin	0.01g

Preparation of Solution 4: Add components to distilled/deionized water and bring volume to 1.0L. Mix thoroughly. Store at –20°C.

Solution 5:

Composition per 100.0mL:

Hemin	0.2g
NH_4OH, concentrated	0.3mL

Preparation of Solution 5: Add hemin to approximately 30.0mL of distilled/deionized water. Add NH_4OH. Mix thoroughly until dissolved. Bring volume to 100.0mL with distilled/deionized water. Store at 5°C.

Solution 6:

Composition per 10.0mL:

DL-Thioctic acid	0.01g
Ethanol (95% solution)	10.0mL

Preparation of Solution 6: Add DL-thioctic acid to 10.0mL of ethanol. Mix thoroughly. Store solution at –20°C.

Solution 7:

Composition per 10.0mL:

Coenzyme A	0.01g
$Na_2S·5H_2O$ (0.05% solution)	0.2mL

Preparation of Solution 7: Prepare $Na_2S·5H_2O$ solution in freshly boiled distilled/deionized water.

Add coenzyme A to 9.8mL of distilled/deionized water. Mix thoroughly. Add freshly prepared $Na_2S{\cdot}5H_2O$ solution. Mix thoroughly. Store the solution at –20°C.

Solution 8:

Composition per 100.0mL:

Oleic acid	0.1g

Preparation of Solution 8: Add oleic acid to 50.0mL of distilled/deionized water. Adjust pH to 9.0 with NaOH. Gently heat until dissolved. Bring volume to 100.0mL with distilled/deionized water. Store at 5°C.

Preparation of Medium: Add components—except agar, potato starch, and solution 8—to distilled/deionized water and bring volume to 400.0mL. Mix thoroughly. Adjust pH to 6.5 with 20% KOH solution. Filter sterilize. In a separate flask, add potato starch to 50.0mL of distilled/deionized water. Add the starch solution to 450.0mL of boiling distilled/deionized water. Add 10.0mL of solution 8 and the agar. Mix thoroughly. Autoclave for 15 min at 15 psi pressure–121°C. Cool to 70°C. Aseptically combine the two sterile solutions. Pour into sterile Petri dishes or distribute into sterile tubes.

Use: For the cultivation and maintenance of *Histoplasma capsulatum* in the mycelial phase.

Histoplasma capsulatum Broth

Composition per liter:

Glucose	10.0g
Citric acid	10.0g
α-Ketoglutaric acid	1.0g
L-Cystine·HCl·H_2O	1.0g
Potato starch	0.5g
Glutathione, reduced	0.5g
L-Asparagine	0.1g
L-Tryptophan	0.02g
Solution 1	250.0mL
Solution 3	40.0mL
Solution 2	10.0mL
Solution 4	10.0mL
Solution 5	1.0mL
Solution 8	1.0mL
Solution 6	0.1mL
Solution 7	0.1mL

pH 6.5 ± 0.2 at 25°C

Solution 1:

Composition per liter:

KH_2PO_4	8.0g
$(NH_4)_2SO_4$	8.0g
$MgSO_4{\cdot}7H_2O$	0.86g
$CaCl_2$, anhydrous	0.08g
$ZnSO_4{\cdot}7H_2O$	0.05g

Preparation of Solution 1: Add components to distilled/deionized water and bring volume to 500.0mL. Mix thoroughly. Bring volume to 1.0L with distilled/deionized water. Store at 5°C.

Solution 2:

Composition per liter:

$FeSO_4{\cdot}7H_2O$	5.7g
$MnCl_2{\cdot}6H_2O$	0.8g
$NaMoO_4{\cdot}2H_2O$	0.15g
HCl, concentrated	1.0mL

Preparation of Solution 2: Add 1.0mL of concentrated HCl to 100.0mL of distilled water in a 1.0L volumetric flask. Dissolve each component completely in the sequence given. Bring volume to 1.0L with distilled/deionized water. Store at 5°C. Discard if red color or red precipitate appears.

Solution 3:

Composition per 100.0mL:

Casein, acid-hydrolyzed, vitamin-free	10.0g

Preparation of Solution 3: Add casein to distilled/deionized water and bring volume to 100.0mL. Do not use enzymatically digested casein.

Solution 4:

Composition per liter:

Calcium pantothenate	0.2g
Inositol	0.2g
Riboflavin	0.2g
Thiamine·HCl	0.2g
Nicotinamide	0.1g
Biotin	0.01g

Preparation of Solution 4: Add components to distilled/deionized water and bring volume to 1.0L. Mix thoroughly. Store at –20°C.

Solution 5:

Composition per 100.0mL:

Hemin	0.2g
NH_4OH, concentrated	0.3mL

Preparation of Solution 5: Add hemin to approximately 30.0mL of distilled/deionized water. Add NH_4OH. Mix thoroughly until dissolved. Bring volume to 100.0mL with distilled/deionized water. Store at 5°C.

Solution 6:

Composition per 10.0mL:

DL-Thioctic acid	0.01g
Ethanol (95% solution)	10.0mL

Preparation of Solution 6: Add DL-thioctic acid to 10.0mL of ethanol. Mix thoroughly. Store solution at –20°C.

Solution 7:

Composition per 10.0mL:

Coenzyme A	0.01g
$Na_2S \cdot 5H_2O$ (0.05% solution)	0.2mL

Preparation of Solution 7: Prepare $Na_2S \cdot 5H_2O$ solution in freshly boiled distilled/deionized water. Add coenzyme A to 9.8mL of distilled/deionized water. Mix thoroughly. Add freshly prepared $Na_2S \cdot 5H_2O$ solution. Mix thoroughly. Store the solution at −20°C.

Solution 8:

Composition per 100.0mL:

Oleic acid	0.1g

Preparation of Solution 8: Add oleic acid to 50.0mL of distilled/deionized water. Adjust pH to 9.0 with NaOH. Gently heat until dissolved. Bring volume to 100.0mL with distilled/deionized water. Store at 5°C.

Preparation of Medium: Add components—except potato starch and solution 8—to distilled/deionized water and bring volume to 400.0mL. Mix thoroughly. Adjust pH to 6.5 with 20% KOH solution. Filter sterilize. In a separate flask, add potato starch to 50.0mL of distilled/deionized water. Add the starch solution to 450.0mL of boiling distilled/deionized water. Add 1.0mL of solution 8. Mix thoroughly. Autoclave for 15 min at 15 psi pressure–121°C. Cool to 70°C. Aseptically combine the two sterile solutions. Pour into sterile Petri dishes or distribute into sterile tubes.

Use: For the cultivation of *Histoplasma capsulatum* in the yeast phase. For the cultivation of *Histoplasma duboisii, Blastomyces dermatitidis,* and *Sprotrichum schenckii.*

HL Agar

Composition per plate:

Columbia agar base	10.0mL
Columbia blood top agar	5.0mL

pH 7.3 ± 0.2 at 25°C

Columbia Agar Base:

Composition per liter:

Agar	13.5g
Pancreatic digest of casein	12.0g
NaCl	5.0g
Peptic digest of animal tissue	5.0g
Beef extract	3.0g
Yeast extract	3.0g
Cornstarch	1.0g

Preparation of Columbia Agar Base: Add components to distilled/deionized water and bring volume to 1.0L. Mix thoroughly. Gently heat until boiling. Autoclave for 15 min at 15 psi pressure–121°C. Cool to 45°–50°C.

Columbia Blood Top Agar:

Composition per liter:

Agar	13.5g
Pancreatic digest of casein	12.0g
NaCl	5.0g
Peptic digest of animal tissue	5.0g
Beef extract	3.0g
Yeast extract	3.0g
Cornstarch	1.0g
Horse blood, defibrinated	50.0mL

Preparation of Columbia Blood Top Agar: Add components, except horse blood, to distilled/deionized water and bring volume to 950.0mL. Mix thoroughly. Gently heat until boiling. Autoclave for 15 min at 15 psi pressure–121°C. Cool to 45°–50°C. Aseptically add sterile horse blood. Mix thoroughly.

Preparation of Medium: Pour cooled, sterile Columbia agar base into sterile Petri dishes in 10.0mL volumes. Allow agar to solidify. Pour 5.0mL of cooled, sterile Columbia blood top agar over Columbia agar base that has solidified but is still warm.

Use: For the cultivation of *Listeria monocytogenes.*

HO-LE Trace Elements Solution

Composition per liter:

H_3BO_3	2.85g
$MnCl_2 \cdot 4H_2O$	1.8g
Sodium tartrate	1.77g
$FeSO_4$	1.36g
$CoCl_2 \cdot 6H_2O$	0.04g
$CuCl_2 \cdot 2H_2O$	0.026g
$Na_2MoO_4 \cdot 2H_2O$	0.025g
$ZnCl_2$	0.021g

Preparation of Medium: Add components to distilled/deionized water and bring volume to 1.0L. Mix thoroughly. Distribute into tubes or flasks. Autoclave for 15 min at 15 psi pressure–121°C.

Use: For use as an enrichment to other media that require trace minerals.

Horse Blood Agar

Composition per liter:

Beef heart, infusion from	500.0g
Agar	15.0g
Tryptose	10.0g
NaCl	5.0g
Horse blood, defibrinated	50.0mL

pH 6.8 ± 0.2 at 25°C

Preparation of Medium: Add components, except horse blood, to distilled/deionized water and bring volume to 950.0mL. Mix thoroughly. Gently heat and bring to boiling. Autoclave for 15 min at 15 psi pressure–121°C. Cool to 45°–50°C. Aseptically add sterile horse blood. Mix thoroughly. Pour into sterile Petri dishes or distribute into sterile tubes.

Use: For the cultivation and maintenance of *Yersinia pseudotuberculosis*.

Horse Serum Agar

Composition per liter:

Agar	15.0g
Pancreatic digest of gelatin	5.0g
Beef extract	3.0g
Horse serum	200.0mL

pH 6.8 ± 0.2 at 25°C

Preparation of Medium: Add components, except horse serum, to distilled/deionized water and bring volume to 800.0mL. Mix thoroughly. Gently heat and bring to boiling. Autoclave for 15 min at 15 psi pressure–121°C. Cool to 45°–50°C. Aseptically add sterile horse serum. Mix thoroughly. Pour into sterile Petri dishes or distribute into sterile tubes.

Use: For the cultivation and maintenance of *Pseudomonas aeruginosa* and *Streptobacillus moniliformis*.

Horse Serum Broth

Composition per liter:

Pancreatic digest of gelatin	5.0g
Beef extract	3.0g
Horse serum	200.0mL

pH 6.8 ± 0.2 at 25°C

Source: This medium is available as a premixed powder from BD Diagnostic Systems.

Preparation of Medium: Add components, except horse serum, to distilled/deionized water and bring volume to 800.0mL. Mix thoroughly. Gently heat and bring to boiling. Autoclave for 15 min at 15 psi pressure–121°C. Cool to 45°–50°C. Aseptically add sterile horse serum. Mix thoroughly. Aseptically distribute into sterile tubes or flasks.

Use: For the cultivation and maintenance of *Pseudomonas aeruginosa* and *Streptobacillus moniliformis*.

Hoyle Medium

Composition per 1060mL:

Agar	15.0g
Lab-Lemco powder	10.0g
Peptone	10.0g
NaCl	5.0g
Horse blood, laked	50.0mL
Tellurite solution	10.0mL

pH 7.8 ± 0.2 at 25°C

Source: This medium is available as a premixed powder from Oxoid Unipath.

Horse Blood, Laked:
Composition per 50.0mL:

Horse blood, fresh	50.0mL

Preparation of Horse Blood, Laked: Add blood to a sterile polypropylene bottle. Freeze overnight at −20°C. Thaw at 8°C. Refreeze at −20°C. Thaw again at 8°C.

Tellurite Solution:
Composition per 100.0mL:

K_2TeO_3	3.5g

Preparation of Tellurite Solution: Add K_2TeO_3 to distilled/deionized water and bring volume to 100.0mL. Mix thoroughly. Filter sterilize.

Caution: Potassium tellurite is toxic.

Preparation of Medium: Add components, except laked horse blood and tellurite solution, to distilled/deionized water and bring volume to 1.0L Mix thoroughly. Gently heat and bring to boiling. Autoclave for 15 min at 15 psi pressure–121°C. Cool to 50°C. Aseptically add 50.0mL sterile laked horse blood and 10.0mL sterile tellurite solution. Mix thoroughly. Pour into sterile Petri dishes or distribute into sterile tubes.

Use: For the isolation and differentiation of *Corynebacterium diphtheriae* strains. This meidum permits very rapid growth of all types of *Corynebacterium diphtheriae*, so that diagnosis is possible after 18h incubation.

Hoyle Medium Base

Composition per liter:

Agar	15.0g
Beef extract	10.0g
Peptone	10.0g
NaCl	5.0g
Blood, laked	50.0mL
Tellurite solution	10.0mL

pH 7.8 ± 0.2 at 25°C

Source: This medium is available as a premixed powder from Oxoid Unipath.

Tellurite Solution:
Composition per 100.0mL:

K_2TeO_3	3.5g

Caution: Potassium tellurite is toxic.

Preparation of Tellurite Solution: Add K_2TeO_3 to distilled/deionized water and bring volume to 100.0mL. Mix thoroughly. Filter sterilize.

Preparation of Medium: Add components, except laked blood, to distilled/deionized water and bring volume to 940.0mL. Mix thoroughly. Gently heat and bring to boiling. Autoclave for 15 min at 15 psi pressure–121°C. Cool to 45°–50°C. Aseptically add sterile components. Mix thoroughly. Pour into sterile Petri dishes or distribute into sterile tubes.

Use: For the isolation and differentiation of *Corynebacterium diphtheriae.*

HR Antifungal Assay Medium Buffered with MOPS

Composition per liter:

MOPS (3-*N*-Morpholino-propanesulfonic acid) buffer	34.53g
Glucose	10.0g
$(NH_4)_2SO_4$	2.5g
KH_2PO_4	1.0g
$NaHCO_3$	1.0g
Glutamine	0.58g
$MgSO_4{\cdot}7H_2O$	0.5g
$CaCl_2{\cdot}2H_2O$	0.1g
NaCl	0.1g
L-Lysine	0.07g
L-Isoleucine	0.05g
L-Leucine	0.05g
L-Threonine	0.05g
L-Valine	0.05g
L-Arginine	0.04g
L-Histidine	0.02g
L-Methionine	0.01g
L-Tryptophan	8.2mg
DL-Methionine	2.0mg
DL-Tryptophan	2.0mg
Inositol	2.0mg
L-Histidine·HCl	1.0mg
H_3BO_3	0.5mg
Calciun pantothenate	0.4mg
$MnSO_4{\cdot}H_2O$	0.4mg
Niacin	0.4mg
Pyridoxine	0.4mg
Thiamine·HCl	0.4mg
$ZnSO_4{\cdot}7H_2O$	0.4mg
p-Aminobenzoic acid	0.2mg
$FeCl_3$	0.2mg
Riboflavin	0.2mg
Na_2MoO_3	0.2mg
KI	0.1mg
$CuSO_4{\cdot}5H_2O$	0.04mg
Biotin	2.0μg
Folic acid	2.0μg

pH 7.0 ± 0.2 at 25°C

Preparation of Medium: Add components, except $NaHCO_3$ and MOPS buffer, to distilled/deionized water and bring volume to 900.0mL. Mix thoroughly. Add $NaHCO_3$ and MOPS buffer. Mix thoroughly. Adjust pH to 7.0. Bring volume to 1.0L with distilled/deionized water. Filter sterilize.

Use: For testing the effectiveness of antifungal agents against clinical fungal isolates using the broth dilution susceptibility testing method.

Hugh-Leifson's Glucose Broth

Composition per liter:

NaCl	30.0g
Glucose	10.0g
Agar	3.0g
Peptone	2.0g
Yeast extract	0.5g
Bromcresol Purple	0.015g

pH 7.4 ± 0.2 at 25°C

Preparation of Medium: Add components to distilled/deionized water and bring volume to 1.0L. Mix thoroughly. Gently heat while stirring and bring to boiling. Adjust pH to 7.4. Distibute into tubes or flasks. Autoclave for 15 min at 15 psi pressure–121°C.

Use: For the cultivation and differentiation of bacteria based on their ability to ferment glucose. Bacteria that ferment glucose turn the medium yellow.

Hydroxybenzoate Agar

Composition per 1001.0mL:

Solution A	490.0mL
Solution D	500.0mL
Solution B	10.0mL
Solution C	1.0mL

pH 7.0 ± 0.2 at 25°C

Solution A:

Composition per 490.0mL:

4-Hydroxybenzoic acid	3.0g
$(NH_4)_2SO_4$	3.0g
NaCl	2.5g
K_2HPO_4	1.6g
Yeast extract	0.5g

Preparation of Solution A: Add components to distilled/deionized water and bring volume to 490.0mL. Mix thoroughly. Gently heat and bring to boiling. Autoclave for 15 min at 15 psi pressure–121°C. Cool to 45°–50°C.

Solution B:
Composition per 10.0mL:
$MgSO_4 \cdot 7H_2O$ 0.27g

Preparation of Solution B: Add $MgSO_4 \cdot 7H_2O$ to distilled/deionized water and bring volume to 10.0mL. Mix thoroughly. Autoclave for 15 min at 15 psi pressure–121°C. Cool to 45°–50°C.

Solution C:
Composition per 1.0mL:
$Fe(NH_4)_2(SO_4)_2 \cdot 6H_2O$ 0.05g

Preparation of Solution C: Add component to distilled/deionized water and bring volume to 1.0mL. Mix thoroughly. Filter sterilize. Prepare solution immediately before adding to solutions A and B.

Solution D:
Composition per 500.0mL:
Agar 14.0g

Preparation of Solution D: Add agar to distilled/deionized water and bring volume to 500.0mL. Mix thoroughly. Autoclave for 15 min at 15 psi pressure–121°C. Cool to 45°–50°C.

Preparation of Medium: Aseptically combine cooled sterile solution A, cooled sterile solution B, and cooled sterile solution D. Immediately add 1.0mL of freshly prepared sterile solution C. Adjust pH to 7.0 with 6*N* NaOH. Mix thoroughly. Pour into sterile Petri dishes or distribute into sterile tubes.

Use: For the cultivation and maintenance of *Comamonas testosteroni.*

Hydroxybenzoate Broth

Composition per 1001.0mL:
Solution A 990.0mL
Solution B 10.0mL
Solution C 1.0mL
pH 7.0 ± 0.2 at 25°C

Solution A:
Composition per 990.0mL:
4-Hydroxybenzoic acid 3.0g
$(NH_4)_2SO_4$ 3.0g
NaCl 2.5g
K_2HPO_4 .... 1.6g
Yeast extract 0.5g

Preparation of Solution A: Add components to distilled/deionized water and bring volume to 990.0mL. Mix thoroughly. Gently heat and bring to boiling. Autoclave for 15 min at 15 psi pressure–121°C. Cool to 45°–50°C.

Solution B:
Composition per 10.0mL:
$MgSO_4 \cdot 7H_2O$ 0.27g

Preparation of Solution B: Add $MgSO_4 \cdot 7H_2O$ to distilled/deionized water and bring volume to 10.0mL. Mix thoroughly. Autoclave for 15 min at 15 psi pressure–121°C. Cool to 45°–50°C.

Solution C:
Composition per 1.0mL:
$Fe(NH_4)_2(SO_4)_2 \cdot 6H_2O$ 0.05g

Preparation of Solution C: Add component to distilled/deionized water and bring volume to 1.0mL. Mix thoroughly. Filter sterilize. Prepare solution immediately before adding to solutions A and B.

Preparation of Medium: Aseptically combine cooled sterile solution A and cooled sterile solution B. Immediately add 1.0mL of freshly prepared sterile solution C. Adjust pH to 7.0 with 6*N* NaOH. Mix thoroughly. Aseptically distribute into sterile tubes or flasks.

Use: For the cultivation and maintenance of *Comamonas testosteroni.*

Imidazole Utilization Medium

Composition per liter:
Imidazole 5.0g
KH_2PO_4 .... 0.5g
$MgSO_4 \cdot 7H_2O$ 0.5g
$CaCl_2$ 3.0mg
$FeSO_4 \cdot 7H_2O$ 3.0mg
Molybdenum solution 1.0mL
Trace elements solution 1.0mL
pH 6.0 ± 0.2 at 25°C

Molybdenum Solution:
Composition per 18.0mL:
$Na_2MoO_4 \cdot 2H_2O$ 0.5mg

Preparation of Molybdenum Solution: Add components to distilled/deionized water and bring volume to 18.0mL. Mix thoroughly. Filter sterilize.

Trace Elements Solution:
Composition per 18.0mL:
H_3BO_3 .... 11.0mg
$MnCl_2 \cdot 4H_2O$ 7.0mg
$Al_2(SO_4)_3 \cdot 18\ H_2O$ 1.94mg
$Co(NO_3)_2 \cdot 6H_2O$ 1.0mg
$CuSO_4 \cdot 5H_2O$ 1.0mg
$NiSO_4 \cdot 6H_2O$ 1.0mg
$ZnSO_4 \cdot H_2O$ 0.62mg
KBr 0.5mg
KI 0.5mg
LiCl 0.5mg
$SnCl_2 \cdot 2H_2O$ 0.5mg

Preparation of Trace Elements Solution: Add components to distilled/deionized water and bring volume to 18.0mL. Mix thoroughly. Filter sterilize.

Preparation of Medium: Add components, except molybdenum solution and trace elements solution, to distilled/deionized water and bring volume to 998.0mL. Mix thoroughly. Distribute into tubes or flasks. Autoclave for 15 min at 15 psi pressure–121°C. Cool to 25°C. Aseptically add 1.0mL of molybdenum solution and 1.0mL of trace elements solution. Mix thoroughly. Adjust pH to 6.0 with phosphoric acid. Mix thoroughly. Aseptically distribute into sterile tubes or flasks.

Use: For the cultivation and maintenance of *Pseudomonas* species.

Indole Medium

Composition per 200.0mL:

K_2HPO_4	3.13g
L-Tryptophan	1.0g
NaCl	1.0g
KH_2PO_4	0.27g

pH 7.2 ± 0.2 at 25°C

Preparation of Medium: Add components to distilled/deionized water and bring volume to 200.0mL. Mix thoroughly. Distribute into tubes or flasks. Autoclave for 15 min at 15 psi pressure–121°C.

Use: For the differentiation of microorganisms by means of indole production from the tryptophan test.

Indole Medium

Composition per liter:

Pancreatic digest of casein	20.0g

pH 7.3 ± 0.2 at 25°C

Preparation of Medium: Add pancreatic digest of casein to distilled/deionized water and bring volume to 1.0L. Mix thoroughly. Distribute into tubes or flasks. Autoclave for 15 min at 15 psi pressure–121°C.

Use: For the differentiation of microorganisms by means of the indole test.

Indole Nitrite Medium (Trypticase Nitrate Broth)

Composition per liter:

Pancreatic digest of casein	20.0g
Na_2HPO_4	2.0g
Agar	1.0g
Glucose	1.0g
KNO_3	1.0g

pH 7.2 ± 0.2 at 25°C

Source: This medium is available as a premixed powder from BD Diagnostic Systems.

Preparation of Medium: Add components to distilled/deionized water and bring volume to 1.0L. Mix thoroughly. Gently heat and bring to boiling with frequent agitation. Distribute into tubes or flasks. Autoclave for 15 min at 15 psi pressure–121°C.

Use: For the identification of microorganisms by means of the nitrate reduction and indole tests.

Infection Medium (IM)

Composition per 100.0mL:

Pancreatic digest of gelatin	0.05g
Bile salts no. 3	0.05g
Brain heart, solids from infusion	0.02g
Peptic digest of animal tissue	0.02g
NaCl	0.017g
Glucose	0.01g
Na_2HPO_4	8.0mg
Earle's balanced salts solution	80.0mL
Fetal bovine serum, heat inactivated (2 hr at 55°C)	20.0mL

pH 7.4 ± 0.2 at 25°C

Earle's Balanced Salts Solution:

Composition per liter:

NaCl	6.8g
$NaHCO_3$	2.2g
Glucose	1.0g
KCl	0.4g
$CaCl_2 \cdot 2H_2O$	0.265g
$MgSO_4 \cdot 7H_2O$	0.2g
$NaH_2PO_4 \cdot H_2O$	0.14g

Preparation of Earle's Balanced Salts Solution: Add components to distilled/deionized water and bring volume to 1.0L. Mix thoroughly. Filter sterilize.

Preparation of Medium: Combine components. Mix thoroughly. Filter sterilize. Store at 4°–10°C.

Use: For the screening of *Escherichia coli* for pathogenicity using the HeLa cell test for invasiveness.

Infusion Broth

Composition per liter:

Pancreatic digest of casein	13.0g
NaCl	5.0g
Yeast extract	5.0g
Heart muscle, solids from infusion	2.0g

pH 7.4 ± 0.2 at 25°C

Source: This medium is available as a premixed powder from BD Diagnostic Systems.

Preparation of Medium: Add components to distilled/deionized water and bring volume to 1.0L.

Mix thoroughly. Distribute into tubes or flasks. Autoclave for 15 min at 15 psi pressure–121°C.

Use: For the cultivation of a wide variety of microorganisms.

Inhibitory Mold Agar

Composition per liter:

Agar	15.0g
Glucose	5.0g
Yeast extract	5.0g
Pancreatic digest of casein	3.0g
Na_2HPO_4	2.0g
Peptic digest of animal tissue	2.0g
Starch	2.0g
Dextrin	1.0g
$MgSO_4 \cdot 7H_2O$	0.8g
$MnSO_4$	0.16g
Chloramphenicol	0.125g
$FeSO_4$	0.04g
NaCl	0.04g

pH 6.7 ± 0.2 at 25°C

Source: This medium is available as a premixed powder from BD Diagnostic Systems.

Preparation of Medium: Add components to distilled/deionized water and bring volume to 1.0L. Mix thoroughly. Gently heat and bring to boiling with frequent agitation. Distribute into tubes or flasks. Autoclave for 15 min at 15 psi pressure–121°C. Pour into sterile Petri dishes or leave in tubes.

Use: For the isolation of pathogenic fungi.

Inositol Brilliant Green Bile Salts Agar (IBB Agar) (*Pleisomonas* Differential Agar)

Composition per liter:

Agar	15.0g
meso-Inositol	10.0g
Proteose peptone	10.0g
Bile salts no. 3	8.5g
Meat extract	5.0g
NaCL	5.0g
Neutral Red (2% solution)	1.25mL
Brilliant Green (0.1% solution)	0.33mL

pH 7.2 ± 0.1 at 25°C

Preparation of Medium: Add components to distilled/deionized water and bring volume to 1.0L. Mix thoroughly. Gently heat and bring to boiling. Distribute into tubes or flasks. Autoclave for 15 min at 15 psi pressure–121°C. Pour into sterile Petri dishes or leave in tubes.

Use: For the isolation of *Aeromonas* and *Plesiomonas* species.

Inositol Urea Caffeic Acid Medium

Composition per liter:

Agar solution	900.0mL
Base solution	100.0mL

Agar Solution:

Composition per 900.0mL:

Agar	15.0g

Preparation of Agar Solution: Add agar to distilled/deionized water and bring volume to 900.0mL. Mix thoroughly. Gently heat and bring to boiling. Autoclave for 15 min at 15 psi pressure–121°C. Cool to 45°–50°C.

Base Solution:

Composition per 100.0mL:

Inositol	10.0g
Urea	5.0g
KH_2PO_4	1.0g
$MgSO_4 \cdot 7H_2O$	0.5g
Caffeic acid	0.2g
NaCl	0.1g
$CaCl_2 \cdot 2H_2O$	0.1g
Gentamicin sulfate	0.04g
H_3BO_3	0.5mg
$ZnSO_4 \cdot 7H_2O$	0.4mg
$MnSO_4 \cdot 4H_2O$	0.4mg
Thiamine·HCl	0.4mg
Pyroxidine·HCl	0.4mg
Niacin	0.4mg
Calcium pantothenate	0.4mg
p-Aminobenzoic acid	0.2mg
Riboflavin	0.2mg
$FeCl_3$	0.2mg
$Na_2MoO_4 \cdot 4H_2O$	0.2mg
KI	0.1mg
$CuSO_4 \cdot 5H_2O$	0.04mg
Folic acid	2.0μg
Biotin	2.0μg
Ferric citrate solution (1% solution)	1.0mL

Preparation of Base Solution: Add components, except urea, to distilled/deionized water and bring volume to 100.0mL. Mix thoroughly. Gently heat just until components are dissolved. Cool to 75°–80°C. Add urea. Mix thoroughly. Do not heat after addition of urea. Do not autoclave. Filter sterilize.

Preparation of Medium: Aseptically combine the cooled, sterile agar solution with the sterile base solution. Mix thoroughly. Pour into sterile Petri dishes.

Use: For the selective isolation and differentiation of *Cryptococcus* species based on inositol and urea utilization and pigment production from caffeic acid. On this medium, only *Cryptococcus* species utilize inositol as sole carbon source and urea as sole nitrogen source. *Cryptococcus neoformans* appears as dark brown colonies. Other *Cryptococcus* species are unpigmented.

Intracellular Growth Phase Medium (IGP Medium)

Composition per 100.0mL:

Gentamicin sulfate	500.0mg
Lysozyme	30.0mg
Eagle MEM	72.0mL
Dulbecco's phosphate-buffered saline	20.0mL
Fetal bovine serum	8.0mL

pH 7.2–7.4 at 25°C

Eagle MEM:

Composition per liter:

NaCl	8.0g
Glucose	1.0g
KCl	0.4g
$CaCl_2 \cdot 2H_2O$	0.14g
$MgSO_4 \cdot 7H_2O$	0.1g
KH_2PO_4	0.06g
Na_2HPO_4	0.05g
L-Isoleucine	0.026g
L-Leucine	0.026g
L-Lysine	0.026g
L-Threonine	0.024g
L-Valine	0.0235g
L-Tyrosine	0.018g
L-Arginine	0.0174g
L-Phenylalanine	0.0165g
L-Cystine	0.012g
L-Histidine	8.0mg
L-Methionine	7.5mg
Phenol Red	5.0mg
L-Tryptophan	4.0mg
Inositol	1.8mg
Biotin	1.0mg
Folic acid	1.0mg
Calcium pantothenate	1.0mg
Choline chloride	1.0mg
Nicotinamide	1.0mg
Pyridoxal·HCl	1.0mg
Thiamine·HCl	1.0mg
Riboflavin	0.1mg

Preparation of Eagle MEM: Add components to distilled/deionized water and bring volume to 1.0L. Mix thoroughly.

Dulbecco's Phosphate-Buffered Saline:

Composition per liter:

NaCl	8.0g
$Na_2HPO_4 \cdot 7H_2O$	2.16g
KCl	0.2g
KH_2PO_4	0.2g
$CaCl_2$	0.1g
$MnCl_2 \cdot 6H_2O$	0.1g

Preparation of Dulbecco's Phosphate-Buffered Saline: Add components to distilled/deionized water and bring volume to 1.0L. Mix thoroughly.

Preparation of Medium: Combine components. Mix thoroughly. Filter sterilize. Aseptically distribute into sterile tubes or flasks.

Use: For the screening of *Escherichia coli* for pathogenicity using the HeLa cell test for invasiveness.

Ion Agar for *Ureaplasma*

Composition per 101.45mL:

HEPES (*N*-[2-Hydroxyethyl] piperazine-*N'*-[2-ethanesulfonic acid]) buffer	1.19g
Ionagar no. 2	0.75g
Pancreatic digest of casein	0.7g
NaCl	0.5g
Beef extract	0.3g
Yeast extract	0.3g
Beef heart, solids from infusion	0.2g
Yeast extract	0.1g
Horse serum, normal sterile	10.0mL
Ampicillin solution	1.0mL
Urea solution	0.25mL
Nystatin solution	0.1mL
Tripeptide solution	0.1mL

pH 7.2 ± 0.2 at 25°C

Ampicillin Solution:

Composition per 10.0mL:

Ampicillin	1.0g

Preparation of Ampicillin Solution: Add ampicillin to distilled/deionized water and bring volume to 10.0mL. Mix thoroughly. Filter sterilize.

Urea Solution:

Composition per 100.0mL:

Urea	10.0g

Preparation of Urea Solution: Add urea to distilled/deionized water and bring volume to 100.0mL. Filter sterilize. Store at –20°C.

Nystatin Solution:

Composition per 1.0mL:

Nystatin	50,000U

Preparation of Nystatin Solution: Add nystatin to distilled/deionized water and bring volume to 1.0mL. Filter sterilize.

Tripeptide Solution:

Composition per 10.0mL:

Glycyl-L-histidyl-L-lysine acetate	0.2mg

Preparation of Tripeptide Solution: Add glycyl-L-histidyl-L-lysine acetate to distilled/deionized water and bring volume to 10.0mL. Mix thoroughly. Filter sterilize. Store at –20°C.

Preparation of Medium: Add components—except horse serum, ampicillin solution, urea solution, nystatin solution, and tripeptide solution—to distilled/deionized water and bring volume to 90.0mL. Mix thoroughly. Gently heat and bring to boiling. Autoclave for 15 min at 15 psi pressure–121°C. Cool to 45°–50°C. Aseptically add 10.0mL of sterile horse serum, 1.0mL of sterile ampicillin solution, 0.25mL of sterile urea solution, 0.1mL of sterile nystatin solution, and 0.1mL of sterile tripeptide solution. Mix thoroughly. Pour into sterile Petri dishes.

Use: For the cultivation of *Ureaplasma* species from clinical specimens.

Irgasan Ticarcillin Chlorate Broth (ITC Broth)

Composition per liter:

$MgCl_2 \cdot 6H_20$	60.0g
Pancreatic digest of casein	10.0g
NaCl	5.0g
$KClO_4$	1.0g
Yeast extract	1.0g
Malachite Green (0.2% solution)	5.0mL
Irgasan solution	1.0mL
Ticarcillin solution	1.0mL

pH 7.6 ± 0.2 at 25°C

Irgasan Solution:

Composition per 10.0mL:

Irgasan (triclosan)	1.0mg

Preparation of Irgasan Solution: Add Irgasan to distilled/deionized water and bring volume to 10.0mL. Mix thoroughly. Filter sterilize.

Ticarcillin Solution:

Composition per 10.0mL:

Ticarcillin	1.0 mg

Preparation of Ticarcillin Solution: Add ticarcillin to distilled/deionized water and bring volume to 10.0mL. Mix thoroughly. Filter sterilize.

Preparation of Medium: Add components, except Irgasan solution and ticarcillin solution, to distilled/deionized water and bring volume to 998.0mL. Mix thoroughly. Autoclave for 15 min at 15 psi pressure–121°C. Cool to 45°–50°C. Adjust to pH 7.6. Aseptically add 1.0mL of Irgasan solution and 1.0mL of ticarcillin solution. Mix thoroughly. Aseptically distribute into sterile tubes or flasks.

Use: For the selective isolation and cultivation of *Yersinia* species.

Iron Agar, Lyngby (Lyngby Iron Agar)

Composition per liter:

Peptone	20.0g
Agar	12.0g
NaCl	5.0g
Beef extract	3.0g
Yeast extract	3.0g
L-Cysteine	0.6g
Ferric citrate	0.3g
$Na_2S_2O_3$	0.3g

pH 7.4 ± 0.2 at 25°C

Source: This medium is available as a premixed powder from Oxoid Unipath.

Preparation of Medium: Add components to distilled/deionized water and bring volume to 1.0L. Mix thoroughly. Gently heat and bring to boiling. Distribute into tubes or flasks. Autoclave for 15 min at 15 psi pressure–121°C. Pour into sterile Petri dishes or leave in tubes.

Use: For the cultivation and enumeration of H_2S-producing bacteria.

Iron Milk Medium

Composition per liter:

Iron filings	1.0g
Whole milk	1.0L

pH 6.8 ± 0.2 at 25°C

Preparation of Medium: Add iron filings, which may be small balls of steel wool, to whole milk and bring volume to 1.0L. Mix thoroughly. Distribute into tubes or flasks. Autoclave for 15 min at 15 psi pressure–121°C.

Use: For the cultivation of lactic acid bacteria. For the cultivation and differentiation of *Clostridium* species. The medium turns black if H_2S is produced. The medium turns red if acid is produced from milk carbohydrate fermentation. Acid and gas production is characteristic of *Clostridium perfringens* and *Clostridium butyricum*.

Iso-Sensitest Agar

Composition per liter:

Casein, hydrolyzed	11.0g
Agar	8.0g
Peptones	3.0g
NaCl	3.0g
Na_2HPO_4	2.0g
Glucose	2.0g
Sodium acetate	1.0g
Soluble starch	1.0g
Magnesium glycerophosphate	0.2g

Calcium gluconate	0.1g
L-Cysteine·HCl	0.02g
L-Tryptophan	0.02g
Adenine	0.01g
Guanine	0.01g
Xanthine	0.01g
Uracil	0.01g
Nicotinamide	3.0mg
Pantothenate	3.0mg
Pyridoxine	3.0mg
$MnCl_2 \cdot 4H_2O$	2.0mg
$CoSO_4$	1.0mg
$CuSO_4 \cdot 5H_2O$	1.0mg
$FeSO_4 \cdot 7H_2O$	1.0mg
Menadione	1.0mg
Cyanocobalamin	1.0mg
$ZnSO_4 \cdot 7H_2O$	1.0mg
Biotin	0.3mg
Thiamine	0.04mg

pH 7.4 ± 0.2 at 25°C

Preparation of Medium: Add components to distilled/deionized water and bring volume to 1.0L. Mix thoroughly. Gently heat and bring to boiling. Distribute into tubes or flasks. Autoclave for 15 min at 15 psi pressure–121°C. Pour into sterile Petri dishes or leave in tubes.

Use: For antimicrobial susceptibility testing.

Iso-Sensitest Broth

Composition per liter:

Casein, hydrolyzed	11.0g
Peptones	3.0g
NaCl	3.0g
Glucose	2.0g
Na_2HPO_4	2.0g
Sodium acetate	1.0g
Soluble starch	1.0g
Magnesium glycerophosphate	0.2g
Calcium gluconate	0.1g
L-cysteine·HCl	0.02g
L-Tryptophan	0.02g
Adenine	0.01g
Guanine	0.01g
Xanthine	0.01g
Uracil	0.01g
Nicotinamide	3.0mg
Pantothenate	3.0mg
Pyridoxine	3.0mg
$MnCl_2 \cdot 4H_2O$	2.0mg
$CoSO_4$	1.0mg
$CuSO_4 \cdot 5H_2O$	1.0mg
$FeSO_4 \cdot 7H_2O$	1.0mg
Menadione	1.0mg
Cyanocobalamin	1.0mg
$ZnSO_4 \cdot 7H_2O$	1.0mg
Biotin	0.3mg
Thiamine	0.04mg

pH 7.4 ± 0.2 at 25°C

Preparation of Medium: Add components to distilled/deionized water and bring volume to 1.0L. Mix thoroughly. Gently heat and bring to boiling. Distribute into tubes or flasks. Autoclave for 15 min at 15 psi pressure–121°C. Pour into sterile Petri dishes or leave in tubes.

Use: For antimicrobial susceptibility testing.

Isoleucine Hydroxamate Medium

Composition per liter:

Agar	15.0g
K_2HPO_4	7.0g
Glucose	5.0g
KH_2PO_4	3.0g
L-Isoleucine hydroxamate	1.0g
$(NH_4)_2SO_4$	1.0g

pH 7.0 ± 0.2 at 25°C

Preparation of Medium: Add components to distilled/deionized water and bring volume to 1.0L. Mix thoroughly. Gently heat and bring to boiling. Distribute into tubes or flasks. Autoclave for 15 min at 15 psi pressure–121°C. Pour into sterile Petri dishes or leave in tubes.

Use: For the cultivation and maintenance of *Serratia marcescens*.

Jones-Kendrick Pertussis Transport Medium

Composition per liter:

Beef heart, solids from infusion	500.0g
Agar	20.0g
Soluble starch	10.0g
Tryptose	10.0g
NaCl	5.0g
Charcoal powder, activated	4.0g
Yeast extract	3.5g
Penicillin solution	10.0mL

pH 7.4 ± 0.2 at 25°C

Penicillin Solution:

Composition per 10.0mL:

Penicillin	300U

Preparation of Penicillin Solution: Add penicillin to distilled/deionized water and bring volume to 10.0mL. Mix thoroughly. Filter sterilize.

Preparation of Medium: Add components, except penicillin solution, starch, yeast extract, heart infusion, and agar, to water. Boil to dissolve. Add charcoal, mix well, and autoclave. Cool to 50°C, add penicillin, and dispense into small bottles as slants. Cool and seal tightly. Store at 5°C. Stable for 2 to 3 months.

Use: For the cultivation and transport of *Bordetella pertussis* between clinical isolation and laboratory cultivation.

Jordan's Tartrate Agar

Composition per liter:

Agar	15.0g
Pancreatic digest of casein	10.0g
Sodium potassium tartrate	10.0g
NaCl	5.0g
Phenol Red	0.024g

pH 7.7 ± 0.3 at 25°C

Source: This medium is available as a prepared medium in tubes from BD Diagnostic Systems.

Preparation of Medium: Add components to distilled/deionized water and bring volume to 1.0L. Mix thoroughly. Gently heat and bring to boiling. Adjust pH to 7.7. Distribute into tubes. Autoclave for 15 min at 15 psi pressure–121°C.

Use: For the differentiation and identification of members of the Enterobacteriaceae, especially *Salmonella* species, based upon the ability to utilize tartrate. Utilization of tartrate turns the medium yellow. *Salmonella enteritidis* utilizes tartrate. *Salmonella paratyphi* A does not utilize tartrate.

Kanamycin Vancomycin Blood Agar (KVBA)

Composition per liter:

Agar	17.5g
Pancreatic digest of casein	15.0g
Papaic digest of soybean meal	5.0g
NaCl	5.0g
Kanamycin	0.1g
Sheep blood, defibrinated	50.0mL
Vancomycin solution	10.0mL
Vitamin K_1 solution	1.0mL

Vancomycin Solution:

Composition per 10.0mL:

Vancomycin	7.5mg

Preparation of Vancomycin Solution: Add vancomycin to distilled/deionized water and bring volume to 10.0mL. Mix thoroughly. Filter sterilize.

Vitamin K_1 Solution:

Composition per 100.0mL:

Vitamin K_1	1.0g

Preparation of Vitamin K_1 Solution: Add vitamin K_1 to 99.0mL of absolute ethanol. Mix thoroughly. Filter sterilize.

Preparation of Medium: Add components, except sheep blood, vancomycin, and vitamin K_1 solution, to distilled/deionized water and bring volume to 939.0mL. Mix thoroughly. Gently heat and bring to boiling. Autoclave for 15 min at 15 psi pressure–121°C. Cool to 45°–50°C. Aseptically add sheep blood, vancomycin solution, and 1.0mL vitamin K_1 solution. Mix thoroughly. Pour into sterile Petri dishes or distribute into sterile tubes.

Use: For the selective isolation of anaerobes, particularly *Bacteroides*, from clinical specimens.

Kanamycin Vancomycin Laked Blood Agar

Composition per liter:

Agar	17.5g
Pancreatic digest of casein	15.0g
Papaic digest of soybean meal	5.0g
NaCl	5.0g
Kanamycin	0.075g
Sheep blood, laked	50.0mL
Vancomycin solution	10.0mL
Vitamin K_1 solution	1.0mL

Vancomycin Solution:

Composition per 10.0mL:

Vancomycin	7.5mg

Preparation of Vancomycin Solution: Add vancomycin to distilled/deionized water and bring volume to 10.0mL. Mix thoroughly. Filter sterilize.

Vitamin K_1 Solution:

Composition per 100.0mL:

Vitamin K_1	1.0g

Preparation of Vitamin K_1 Solution: Add vitamin K_1 to 99.0mL of absolute ethanol. Mix thoroughly. Filter sterilize.

Preparation of Medium: The blood is laked (hemolyzed) by freezing whole blood overnight and then thawing. Add components, except sheep blood, vancomycin, and vitamin K_1 solution, to distilled/deionized water and bring volume to 939.0mL. Mix thoroughly. Gently heat and bring to boiling. Autoclave for 15 min at 15 psi pressure–121°C. Cool to 45°–50°C. Aseptically add sheep blood, vancomycin solution, and 1.0mL of vitamin K_1 solution. Mix thoroughly. Pour into sterile Petri dishes or distribute into sterile tubes.

Use: For isolation of the *Bacteroides melaninogenicus* group.

Kasai Medium

Composition per liter:

Pancreatic digest of casein	20.0g
Soluble starch	20.0g
L-Cysteine·HCl·H_2O	5.0g
K_2HPO_4	5.0g
NaCl	5.0g
Yeast extract	2.0g

Preparation of Medium: Add components to distilled/deionized water and bring volume to 1.0L. Mix thoroughly. Gently heat and bring to boiling. Distribute into tubes or flasks. Autoclave for 15 min at 15 psi pressure–121°C.

Use: For the isolation and cultivation of *Leptotrichia buccalis* from saliva and plaque.

KCN Broth

Composition per liter:

Na_2HPO_4	5.64g
NaCl	5.0g
Peptone	3.0g
KH_2PO_4	0.225g
KCN (0.5% solution)	15.0mL

pH 7.6 ± 0.2 at 25°C

Caution: Cyanide is toxic.

Preparation of Medium: Add components, except KCN solution, to distilled/deionized water and bring volume to 985.0mL. Mix thoroughly. Autoclave for 15 min at 15 psi pressure–121°C. Cool to 25°C. Aseptically add KCN solution. Mix thoroughly. Aseptically distribute into sterile tubes. Stopper immediately.

Use: For the differentiation of Enterobacteriaceae based upon growth in the presence of potassium cyanide.

Keister's Modified TYI-S-33 Medium

Composition per liter:

Pancreatic digest of casein	20.0g
Glucose	10.0g
Yeast extract	10.0g
L-Cysteine·HCl	2.0g
NaCl	2.0g
K_2HPO_4	1.0g
Bovine bile	0.75g
KH_2PO_4	0.6g
Ascorbic acid	0.2g
Ferric ammonium citrate	22.8mg
Bovine serum, heat inactivated	100.0mL

Preparation of Medium: Add components, except bovine serum, to distilled/deionized water and bring volume to 900.0mL. Mix thoroughly. Autoclave for 15 min at 15 psi pressure–121°C. Aseptically add 1.0L of sterile, heat-inactivated bovine serum. Mix thoroughly. Aseptically distribute into sterile, screw-capped tubes or flasks.

Use: For the cultivation of *Giardia cati, Giardia intestinalis,* and *Hexamita* species.

Kelly Medium, Nonselective Modified

Composition per 1430.0mL:

HEPES buffer (*N*-2-Hydroxyethylpiperazine-*N*-2-ethanesulfonic acid)	6.0g
Proteose peptone no. 2	5.0g
D-Glucose	3.0g
$NaHCO_3$	2.2g
Pancreatic digest of casein	1.0g
Yeast, autolyzed	1.0g
Sodium pyruvate	0.8g
Sodium citrate	0.7g
N-Acetylglucosamine	0.4g
$MgCl_2·6H_2O$	0.3g
Gelatin solution	200.0mL
Bovine serum albumin	143.0mL
CMRL-1066 medium with glutamine, 10X	100.0mL
Rabbit serum, heat inactivated	86.0mL
Hemin solution	1.0mL

pH 7.2 ± 0.2 at 25°C

CMRL-1066 Medium with Glutamine, 10X:
Composition per liter:

NaCl	6.8g
$NaHCO_3$	2.2g
D-Glucose	1.0g
KCl	0.4g
L-Cysteine·HCl·H_2O	0.26g
$CaCl_2$, anhydrous	0.2g
$MgSO_4·7H_2O$	0.2g
$NaH_2PO_4·H_2O$	0.14g
L-Glutamine	0.1g
Sodium acetate·$3H_2O$	0.083g
L-Glutamic acid	0.075g
L-Arginine·HCl	0.07g
L-Lysine·HCl	0.07g
L-Leucine	0.06g
Glycine	0.05g
Ascorbic acid	0.05g
L-Proline	0.04g
L-Tyrosine	0.04g
L-Aspartic acid	0.03g
L-Threonine	0.03g
L-Alanine	0.025g
L-Phenylalanine	0.025g
L-Serine	0.025g
L-Valine	0.025g
L-Cystine	0.02g
L-Histidine·HCl·H_2O	0.02g

L-Isoleucine	0.02g
Phenol Red	0.02g
L-Methionine	0.015g
Deoxyadenosine	0.01g
Deoxycytidine	0.01g
Deoxyguanosine	0.01g
Glutathione, reduced	0.01g
Thymidine	0.01g
Hydroxy-L-proline	0.01g
L-Tryptophan	0.01g
Nicotinamide adenine dinucleotide	7.0mg
Tween 80	5.0mg
Sodium glucuronate·H_2O	4.2mg
Coenzyme A	2.5mg
Cocarboxylase	1.0mg
Flavin adenine dinucleotide	1.0mg
Nicotinamide adenine dinucleotide phosphate	1.0mg
Uridine triphosphate	1.0mg
Choline chloride	0.5mg
Cholesterol	0.2mg
5-Methyldeoxycytidine	0.1mg
Inositol	0.05mg
p-Aminobenzoic acid	0.05mg
Niacin	0.025mg
Niacinamide	0.025mg
Pyridoxine	0.025mg
Pyridoxal·HCl	0.025mg
Biotin	0.01mg
D-Calcium pantothenate	0.01mg
Folic acid	0.01mg
Riboflavin	0.01mg
Thiamine·HCl	0.01mg

pH 7.2 ± 0.2 at 25°C

Source: This solution is available as a premixed powder from BD Diagnostics.

Preparation of CMRL-1066 Medium with Glutamine, 10X: Add components to distilled/deionized water and bring volume to 1.0L. Mix thoroughly. Adjust pH to 7.2. Filter sterilize.

Gelatin Solution:
Composition per 200.0mL:

Gelatin	14.0g

Preparation of Gelatin Solution: Add gelatin to distilled/deionized water and bring volume to 1.0L. Mix thoroughly. Gently heat and bring to boiling. Autoclave for 15 min at 15 psi pressure–121°C. Cool to 50°C.

Hemin Solution:
Composition per 100.0mL:

Hemin	1.0g
NaOH (1*N* solution)	20.0mL

Preparation of Hemin Solution: Add hemin to 20.0mL of 1*N* NaOH solution. Mix thoroughly. Bring volume to 100.0mL with distilled/deionized water.

Bovine Serum Albumin Solution:
Composition per 200.0mL:

Bovine serum albumin	70.0g

Preparation of Bovine Serum Albumin Solution: Add bovine serum albumin to distilled/deionized water and bring volume to 200.0mL. Filter sterilize.

Preparation of Medium: Add components, except gelatin solution, bovine serum albumin solution, and rabbit serum, to distilled/deionized water and bring volume to 1001.0mL. Mix thoroughly. Bring pH to 7.6 with 5*N* NaOH. Filter sterilize. Aseptically add 200.0mL of sterile gelatin solution, 143.0mL of sterile bovine serum albumin, and 86.0mL of sterile heat-inactivated rabbit serum. Mix thoroughly. Aseptically dispense into sterile tubes or flasks.

Use: For the isolation of *Borrelia burgdorferi* and other spirochetes.

Kelly Medium, Selective Modified

Composition per 1270mL:

Bovine serum albumin fraction V	50.0g
HEPES buffer (*N*-2-Hydroxyethylpiperazine-*N*-2-ethanesulfonic acid)	6.0g
Glucose	5.0g
Neopeptone	5.0g
$NaHCO_3$	2.2g
Sodium pyruvate	0.8g
Sodium citrate	0.7g
N-Acetylglucosamine	0.4g
Kanamycin	8.0mg
5-Fluorouracil	2.3mg
Gelatin solution	200.0mL
CMRL-1066 medium with glutamine, 10X	100.0mL
Rabbit serum, partially hemolyzed	70.0mL

pH 7.7 ± 0.2 at 25°C

Gelatin Solution:
Composition per 200.0mL:

Gelatin	14.0g

Preparation of Gelatin Solution: Add gelatin to distilled/deionized water and bring volume to 1.0L. Mix thoroughly. Gently heat and bring to boiling. Autoclave for 15 min at 15 psi pressure–121°C. Cool to 50°C.

CMRL-1066 Medium with Glutamine, 10X:
Composition per liter:

NaCl	6.8g
$NaHCO_3$	2.2g

D-Glucose	1.0g
KCl	0.4g
L-Cysteine·HCl·H_2O	0.26g
$CaCl_2$, anhydrous	0.2g
$MgSO_4 \cdot 7H_2O$	0.2g
$NaH_2PO_4 \cdot H_2O$	0.14g
L-Glutamine	0.1g
Sodium acetate·$3H_2O$	0.083g
L-Glutamic acid	0.075g
L-Arginine·HCl	0.07g
L-Lysine·HCl	0.07g
L-Leucine	0.06g
Glycine	0.05g
Ascorbic acid	0.05g
L-Proline	0.04g
L-Tyrosine	0.04g
L-Aspartic acid	0.03g
L-Threonine	0.03g
L-Alanine	0.025g
L-Phenylalanine	0.025g
L-Serine	0.025g
L-Valine	0.025g
L-Cystine	0.02g
L-Histidine·HCl·H_2O	0.02g
L-Isoleucine	0.02g
Phenol Red	0.02g
L-Methionine	0.015g
Deoxyadenosine	0.01g
Deoxycytidine	0.01g
Deoxyguanosine	0.01g
Glutathione, reduced	0.01g
Thymidine	0.01g
Hydroxy-L-proline	0.01g
L-Tryptophan	0.01g
Nicotinamide adenine dinucleotide	7.0mg
Tween 80	5.0mg
Sodium glucuronate·H_2O	4.2mg
Coenzyme A	2.5mg
Cocarboxylase	1.0mg
Flavin adenine dinucleotide	1.0mg
Nicotinamide adenine dinucleotide phosphate	1.0mg
Uridine triphosphate	1.0mg
Choline chloride	0.5mg
Cholesterol	0.2mg
5-Methyldeoxycytidine	0.1mg
Inositol	0.05mg
p-Aminobenzoic acid	0.05mg
Niacin	0.025mg
Niacinamide	0.025mg
Pyridoxine	0.025mg
Pyridoxal·HCl	0.025mg
Biotin	0.01mg
D-Calcium pantothenate	0.01mg
Folic acid	0.01mg
Riboflavin	0.01mg
Thiamine·HCl	0.01mg

pH 7.2 ± 0.2 at 25°C

Source: This solution is available as a premixed powder from BD Diagnostics.

Preparation of CMRL-1066 Medium with Glutamine, 10X: Add components to distilled/deionized water and bring volume to 1.0L. Mix thoroughly. Adjust pH to 7.2. Filter sterilize.

Preparation of Medium: Add components, except gelatin solution, partially hemolyzed rabbit serum solution, kanamycin, and 5-fluorouracil, to distilled/deionized water and bring volume to 1.0L. Mix thoroughly. Bring pH to 7.6 with 5*N* NaOH. Filter sterilize. Aseptically add 200.0mL of sterile gelatin solution, 70.0mL of partially hemolyzed rabbit serum, 8.0mg of kanamycin, and 230.0mg of 5-fluorouracil. Mix thoroughly. Aseptically distribute into sterile tubes or flasks.

Use: For the isolation of *Borrelia burgdorferi.*

KF *Streptococcus* Agar

Composition per liter:

Agar	20.0g
Maltose	20.0g
Proteose peptone	10.0g
Sodium glycerophosphate	10.0g
Yeast extract	10.0g
NaCl	5.5g
Lactose	1.0g
NaN_3	0.4g
Bromcresol Purple	0.015g
2,3,5-Triphenyltetrazolium chloride solution	10.0mL

pH 7.2 ± 0.2 at 25°C

Caution: Sodium azide is toxic. Azides also react with metals and disposal must be highly diluted.

Source: This medium is available as a premixed powder from BD Diagnostic Systems and Oxoid Unipath.

2,3,5-Triphenyltetrazolium Chloride Solution:

Composition per 10.0mL:

2,3,5-Triphenyltetrazolium chloride	0.1g

Preparation of 2,3,5-Triphenyltetrazolium Chloride Solution: Add 2,3,5-triphenyltetrazolium chloride to distilled/deionized water and bring volume to 10.0mL. Mix thoroughly. Filter sterilize.

Preparation of Medium: Add components, except 2,3,5-triphenyltetrazolium chloride solution, to distilled/deionized water and bring volume to

990.0mL. Mix thoroughly. Gently heat and bring to boiling. Autoclave for 15 min at 15 psi pressure–121°C. Cool to 45°–50°C. Aseptically add 2,3,5-triphenyltetrazolium chloride solution. Mix thoroughly. Pour into sterile Petri dishes or distribute into sterile tubes.

Use: For the isolation and enumeration of enterococci.

KF *Streptococcus* Agar

Composition per liter:

Agar	20.0g
Maltose	20.0g
Sodium glycerophosphate	10.0g
Yeast extract	10.0g
NaCl	5.0g
Pancreatic digest of casein	5.0g
Peptic digest of animal tissue	5.0g
Lactose	1.0g
NaN_3	0.4g
2,3,5-Triphenyltetrazolium chloride solution	10.0mL

pH 7.2 ± 0.2 at 25°C

Caution: Sodium azide is toxic. Azides also react with metals and disposal must be highly diluted.

Source: This medium is available as a premixed powder from BD Diagnostic Systems.

2,3,5-Triphenyltetrazolium Chloride Solution:

Composition per 10.0mL:

2,3,5-Triphenyltetrazolium chloride	0.1g

Preparation of 2,3,5-Triphenyltetrazolium Chloride Solution: Add 2,3,5-triphenyltetrazolium chloride to distilled/deionized water and bring volume to 10.0mL. Mix thoroughly. Filter sterilize.

Preparation of Medium: Add components, except 2,3,5-triphenyltetrazolium chloride solution, to distilled/deionized water and bring volume to 990.0mL. Mix thoroughly. Gently heat and bring to boiling. Autoclave for 15 min at 15 psi pressure–121°C. Cool to 45°–50°C. Aseptically add 2,3,5-triphenyltetrazolium chloride solution. Mix thoroughly. Pour into sterile Petri dishes or distribute into sterile tubes.

Use: For the selective cultivation and enumeration of fecal streptococci.

KF *Streptococcus* Broth

Composition per liter:

Maltose	20.0g
Sodium glycerophosphate	10.0g
Yeast extract	10.0g
NaCl	5.0g
Pancreatic digest of casein	5.0g
Peptic digest of animal tissue	5.0g
Lactose	1.0g
Na_2CO_3	0.636g
NaN_3	0.4g
Phenol Red	0.018g
2,3,5-Triphenyltetrazolium chloride solution	10.0mL

pH 7.2 ± 0.2 at 25°C

Caution: Sodium azide is toxic. Azides also react with metals and disposal must be highly diluted.

Source: This medium is available as a premixed powder from BD Diagnostic Systems.

Preparation of Medium: Add components, except 2,3,5-triphenyltetrazolium chloride solution, to distilled/deionized water and bring volume to 990.0mL. Mix thoroughly. Gently heat and bring to boiling. Autoclave for 15 min at 15 psi pressure–121°C. Cool to 45°–50°C. Aseptically add 2,3,5-triphenyltetrazolium chloride solution. Mix thoroughly. Aseptically distribute into sterile tubes or flasks.

Use: For the selective cultivation of fecal streptococci.

Kirchner's Enrichment Medium

Composition per liter:

$Na_2HPO_4·12H_2O$	19.0g
Asparagine	5.0g
KH_2PO_4	2.5g
Sodium citrate	2.5g
$MgSO_4$	0.6g
Serum	100.0mL
Glycerol	20.0mL
Penicillin solution	10.0mL
Phenol Red (0.4% solution)	3.0mL

pH 7.4–7.6 at 25°C

Penicillin Solution:

Composition per 10.0mL:

Penicillin	100,000U

Preparation of Penicillin Solution: Add penicillin to distilled/deionized water and bring volume to 10.0mL. Mix thoroughly. Filter sterilize.

Preparation of Medium: Add components, except serum and penicillin solution, to distilled/deionized water and bring volume to 890.0mL. Mix thoroughly. Gently heat and bring to boiling. Autoclave for 15 min at 15 psi pressure–121°C. Cool to 45°–50°C. Aseptically add sterile serum and penicillin solution. Mix thoroughly. Aseptically distribute into sterile tubes or flasks.

Use: For the cultivation and enrichment of *Mycobacterium* species.

K-L Virulence Agar (Klebs-Loeffler Virulence Agar) (Elek Agar) (*Corynebacterium diphtheriae* Virulence Test Medium)

Composition per 1300.0mL:

K-L agar base	1.0L
Rabbit serum	200.0mL
K_2TeO_3 solution	100.0mL
K-L filter strips	100

pH 7.8 ± 0.2 at 25°C

Source: This medium is available as a premixed powder from BD Diagnostic Systems.

K-L Agar Base:

Composition per liter:

Meat peptone	20.0g
Agar	15.0g
NaCl	2.5g

Preparation of K-L Agar Base: Add components to distilled/deionized water and bring volume to 1.0L. Mix thoroughly. Gently heat and bring to boiling. Autoclave for 15 min at 15 psi pressure–121°C. Cool to 50°C.

K_2TeO_3 Solution:

Composition per 100.0mL:

K_2TeO_3	0.3g

Preparation of K_2TeO_3 Solution: Add K_2TeO_3 to distilled/deionized water and bring volume to 100.0mL. Mix thoroughly. Filter sterilize.

Caution: Potassium tellurite is toxic.

K-L Filter Strips:

Composition:

Whatman no. 3 filter paper	as needed
Diphtheria toxin solution	10.0mL

Preparation of K-L Strips: Cut Whatman no. 3 filter paper into 1.5cm × 7cm strips. Autoclave for 15 min at 15 psi pressure–121°C. Aseptically dip each strip into a sterile solution containing 1000U of purified diphtheria toxin/mL. Drain off excess liquid.

Preparation of Medium: Filter sterilize rabbit serum. To 1.0L of cooled, sterile K-L agar base, aseptically add sterile rabbit serum and sterile K_2TeO_3 solution. Mix thoroughly. Pour into sterile Petri dishes in 13.0mL volumes. Before the agar solidifies, aseptically add one K-L filter strip across the diameter of the plate. Allow the filter strip to sink to the bottom of the plate or press it down with sterile forceps. Allow the agar to solidify. Dry the surface of the plates by incubating at 35°C with lid of plate ajar for 2h.

Use: For *in vitro* toxigenicity testing of *Corynebacterium diphtheriae* by the agar diffusion technique. *Corynebacterium diphtheriae* that produce toxin form white precipitin lines at approximately 45° angles from the culture streak line.

K-L Virulence Agar (Klebs-Loeffler Virulence Agar)

Composition per 1250.0mL:

K-L agar base	1.0L
K-L enrichment	200.0mL
K_2TeO_3 solution	50.0mL
K-L filter strips	100

pH 7.8 ± 0.2 at 25°C

Source: This medium is available as a premixed powder from BD Diagnostic Systems.

K-L Agar Base:

Composition per liter:

Meat peptone	20.0g
Agar	15.0g
NaCl	2.5g

Preparation of K-L Agar Base: Add components to distilled/deionized water and bring volume to 1.0L. Mix thoroughly. Gently heat and bring to boiling. Autoclave for 15 min at 15 psi pressure–121°C. Cool to 50°C.

K-L Enrichment:

Composition per 200.0mL:

Casamino acids	4.0g
Glycerol	100.0mL
Tween 80	100.0mL

Preparation of K-L Enrichment: Combine components. Mix thoroughly. Filter sterilize.

K_2TeO_3 Solution:

Composition per 100.0mL:

K_2TeO_3	1.0g

Preparation of K_2TeO_3 Solution: Add K_2TeO_3 to distilled/deionized water and bring volume to 100.0mL. Mix thoroughly. Filter sterilize.

Caution: Potassium tellurite is toxic.

K-L Filter Strips:

Composition:

Diphtheria toxin solution	10.0mL
Whatman no. 3 filter paper	as needed

Preparation of K-L Strips: Cut Whatman no. 3 filter paper into 1.5cm × 7cm strips. Autoclave for 15 min at 15 psi pressure–121°C. Aseptically dip each strip into a sterile solution containing 1000U of purified diphtheria toxin/mL. Drain off excess liquid.

Preparation of Medium: To 1.0L of cooled, sterile K-L agar base, aseptically add sterile K-L enrichment and sterile K_2TeO_3 solution. Mix thoroughly.

Pour into sterile Petri dishes in 13.0mL volumes. Before the agar solidifies, aseptically add one K-L filter strip across the diameter of the plate. Allow the filter strip to sink to the bottom of the plate or press it down with sterile forceps. Allow the agar to solidify. Dry the surface of the plates by incubating at 35°C with lid of plate ajar for 2h.

Use: For *in vitro* toxigenicity testing of *Corynebacterium diphtheriae* by the agar diffusion technique. *Corynebacterium diphtheriae* that produce toxin form white precipitin lines at approximately 45° angles from the culture streak line.

Klebsiella Medium (m-*Klebsiella* Medium)

Composition per 1041.0mL:

Agar	15.0g
Adonitol	4.0g
2× Salt solution	500.0mL
Uric acid solution	200.0mL
Sodium taurocholate solution	30.0mL
Phenol Red solution	10.0mL
Carbenicillin solution	1.0mL

2X Salt Solution:

Composition per liter:

KCl	8.0g
K_2HPO_4	3.0g
NaCl	2.0g
KH_2PO_4	1.0g
$MgSO_4 \cdot 7H_2O$	0.2g

Preparation of 2X Salt Solution: Add components to distilled/deionized water and bring volume to 1.0L. Mix thoroughly.

Uric Acid Solution:

Composition per 200.0mL:

Uric acid	0.3g

Preparation of Uric Acid Solution: Dissolve uric acid in a small volume of 1*N* NaOH. Bring volume to 200.0mL with distilled/deionized water. Adjust pH to 7.1 with 1*N* HCl. Filter sterilize.

Phenol Red Solution:

Composition per 10.0mL:

Phenol Red	0.1g

Preparation of Phenol Red Solution: Add Phenol Red to sterile distilled/deionized water and bring volume to 10.0mL. Mix thoroughly.

Sodium Taurocholate Solution:

Composition per 30.0mL:

Sodium taurocholate	0.4g

Preparation of Sodium Taurocholate Solution: Add sodium taurocholate to sterile distilled/deionized water and bring volume to 30.0mL. Mix thoroughly.

Carbenicillin Solution:

Composition per 1.0mL:

Carbenicillin	5.0mg

Preparation of Carbenicillin Solution: Add carbenicillin to distilled/deionized water and bring volume to 1.0mL. Mix thoroughly. Filter sterilize.

Preparation of Medium: Add adonitol and agar to 500.0mL of 2× salt solution. Bring volume to 800.0mL with distilled/deionized water. Mix thoroughly. Gently heat and bring to boiling. Autoclave for 15 min at 15 psi pressure–121°C. Cool to 45°–50°C. Aseptically add 200.0mL of uric acid solution, 30.0mL of sodium taurocholate solution, 10.0mL of Phenol Red solution, and 1.0mL of carbenicillin solution. Mix thoroughly. Pour into sterile Petri dishes or distribute into sterile tubes.

Use: For the enumeration of *Klebsiella* species by the membrane filter method.

Klebsiella Selective Agar

Composition per liter:

Agar	26.0g
DL–Phenylalanine	10.0g
L-Ornithine·HCl	10.0g
Raffinose	7.0g
Pancreatic digest of casein	2.5g
Yeast extract	2.5g
K_2HPO_4	2.0g
Phenol Red solution	10.0mL
Carbenicillin solution	10.0mL

pH 5.6 ± 0.2 at 25°C

Phenol Red Solution:

Composition per 10.0mL:

Phenol Red	0.5g

Preparation of Phenol Red Solution: Add Phenol Red to 50% ethanol and bring volume to 10.0mL. Mix thoroughly.

Preparation of Medium: Add components, except carbenicillin solution, to distilled/deionized water and bring volume to 990.0mL. Mix thoroughly. Gently heat and bring to boiling. Autoclave for 15 min at 15 psi pressure–121°C. Cool to 45°–50°C. Aseptically add 10.0mL carbenicillin solution. Mix thoroughly. Adjust pH to 5.6–5.7 with sterile 1*N* HCl. Pour into sterile Petri dishes or distribute into sterile tubes.

Use: For the isolation and identification of *Klebsiella pneumoniae* from clinical specimens.

Kligler Iron Agar

Composition per liter:

Peptone	20.0g
Agar	12.0g
Lactose	10.0g
NaCl	5.0g
Beef extract	3.0g
Yeast extract	3.0g
Glucose	1.0g
Ferric citrate	0.3g
$Na_2S_2O_3$	0.3g
Phenol Red	0.05g

pH 7.4 ± 0.2 at 25°C

Source: This medium is available as a premixed powder from BD Diagnostic Systems and Oxoid Unipath.

Preparation of Medium: Add components to distilled/deionized water and bring volume to 1.0L. Mix thoroughly. Gently heat and bring to boiling. Distribute into tubes. Autoclave for 15 min at 15 psi pressure–121°C. Pour into sterile Petri dishes or leave in tubes.

Use: For the differentiation and identification of Enterobacteriaceae based upon sugar fermentation and hydrogen sulfide production. Sugar fermentation is indicated by the medium turning yellow. H_2S production results in the medium turning black.

Kligler Iron Agar

Composition per liter:

Agar	15.0g
Lactose	10.0g
Pancreatic digest of casein	10.0g
Peptic digest of animal tissue	10.0g
NaCl	5.0g
Glucose	1.0g
Ferric ammonium citrate	0.5g
$Na_2S_2O_3$	0.5g
Phenol Red	0.025g

pH 7.4 ± 0.2 at 25°C

Source: This medium is available as a premixed powder from BD Diagnostic Systems.

Preparation of Medium: Add components to distilled/deionized water and bring volume to 1.0L. Mix thoroughly. Gently heat and bring to boiling. Distribute into tubes or flasks. Autoclave for 15 min at 15 psi pressure–121°C. Pour into sterile Petri dishes or leave in tubes.

Use: For the differentiation and identification of Enterobacteriaceae based upon sugar fermentation and hydrogen sulfide production. Sugar fermentation is indicated by the medium turning yellow. H_2S production results in the medium turning black.

Kligler Iron Agar (FDA M71)

Composition per liter:

Lactose	20.0g
Agar	15.0g
Pancreatic digest of casein	10.0g
Peptic digest of animal tissue	10.0g
NaCl	5.0g
Glucose	1.0g
Ferric ammonium citrate	0.5g
$Na_2S_2O_3$	0.5g
Phenol Red	0.025g

pH 7.4 ± 0.2 at 25°C

Preparation of Medium: Add components to distilled/deionized water and bring volume to 1.0L. Mix thoroughly. Gently heat and bring to boiling. Distribute into tubes or flasks. Autoclave for 15 min at 15 psi pressure–121°C. Pour into sterile Petri dishes or leave in tubes.

Use: For the differentiation and identification of Enterobacteriaceae based upon sugar fermentation and hydrogen sulfide production. Sugar fermentation is indicated by the medium turning yellow. H_2S production results in the medium turning black.

Knisely Medium for *Bacillus anthracis*

Composition per liter:

Beef heart, solids from infusion	500.0g
Agar	15.0g
Pancreatic digest of casein	10.0g
NaCl	5.0g
EDTA	200.0mg
Lysozyme	40.0mg
Thallous acetate	40.0mg
Polymyxin	30,000U

Preparation of Medium: Add components, except EDTA, lysozyme, thallous acetate, and polymyxin, to distilled/deionized water and bring volume to 1.0mL. Mix thoroughly. Gently heat and bring to boiling. Adjust pH to 7.3. Autoclave for 15 min at 15 psi pressure–121°C. Cool to 45°–50°C. Aseptically add sterile EDTA, lysozyme, thallous acetate, and polymyxin. Mix thoroughly. Pour into sterile Petri dishes or distribute into sterile tubes.

Use: For the cultivation and maintenance of *Bacillus anthracis*.

Koch's K1 Medium

Composition per liter:

Glucose	1.8g
Peptone	0.6g
Yeast extract	0.4g

Preparation of Medium: Add components to distilled/deionized water and bring volume to 1.0L. Mix thoroughly. Distribute into tubes or flasks. Autoclave for 15 min at 15 psi pressure–121°C.

Use: For the cultivation of a variety of fungi.

Korthof Medium

Composition per 1088.0mL:

NaCl	1.4g
$Na_2HPO_4 \cdot 2H_2O$	0.88g
Peptone	0.8g
KH_2PO_4	0.24g
$CaCl_2$	0.04g
KCl	0.04g
$NaHCO_3$	0.02g
Rabbit serum, inactivated	80.0mL
Rabbit hemoglobin solution	8.0mL

pH 7.2 ± 0.2 at 25°C

Rabbit Hemoglobin Solution:

Composition per 20.0mL:

Rabbit blood clot	10.0mL

Preparation of Rabbit Hemoglobin Solution: Add rabbit blood clot to 10.0mL of distilled/deionized water. Lyse the clot by freezing and thawing.

Preparation of Medium: Add components, except rabbit serum and rabbit hemoglobin solution, to distilled/deionized water and bring volume to 1.0L. Mix thoroughly. Gently heat and bring to boiling. Cool to 25°C. Filter through Whatman no. 1 filter paper. Distribute into flasks in 100.0mL volumes. Autoclave for 15 min at 15 psi pressure–121°C. Cool to 45°–50°C. Aseptically add 8.0mL of rabbit serum and 0.8mL of rabbit hemoglobin solution to each flask. Mix thoroughly.

Use: For the cultivation of *Leptospira* species.

Korthof Medium, Modified

Composition per liter:

NaCl	1.4g
$Na_2HPO_4 \cdot 2H_2O$	0.88g
Peptone	0.8g
KH_2PO_4	0.24g
$CaCl_2$	0.04g
KCl	0.04g
$NaHCO_3$	0.02g
Rabbit serum, heat inactivated at 56°C	100.0mL

pH 7.2–7.6 at 25°C

Preparation of Medium: Add components, except rabbit serum, to distilled/deionized water and bring volume to 900.0L. Mix thoroughly. Gently heat and bring to boiling. Boil for 20 min. Cool overnight at 4°C. Filter through Whatman no. 2 filter paper. Distribute into tubes or flasks. Autoclave for 15 min at 15 psi pressure–121°C. Cool to 50°–56°C. Aseptically add 100.0mL of rabbit serum. Mix thoroughly.

Use: For the cultivation of *Leptospira* species.

Koser Citrate Medium

Composition per liter:

Sodium citrate	3.0g
$NaNH_4HPO_4 \cdot 4H_2O$	1.5g
KH_2PO_4	1.0g
$MgSO_4 \cdot 7H_2O$	0.2g

pH 6.7 ± 0.2 at 25°C

Source: This medium is available as a premixed powder from BD Diagnostic Systems.

Preparation of Medium: Add components to distilled/deionized water and bring volume to 1.0L. Mix thoroughly. Gently heat and bring to boiling. Distribute into tubes or flasks. Autoclave for 15 min at 15 psi pressure–121°C. Pour into sterile Petri dishes or leave in tubes.

Use: For the differentiation of *Escherichia coli* and *Enterobacter aerogenes* based on citrate utilization.

Kosmachev's Medium

Composition per liter:

Agar	15.0g
$CaCO_3$	4.0g
KNO_3	1.0g
$(NH_4)_2SO_4$	1.0g
Na_2HPO_4	1.0g
$MgSO_4 \cdot 7H_2O$	0.5g
$FeSO_4 \cdot 7H_2O$	0.01g
Yeast autolysate (30% solution)	15.0mL

Preparation of Medium: Add components to distilled/deionized water and bring volume to 1.0L. Mix thoroughly. Gently heat and bring to boiling. Distribute into tubes or flasks. Autoclave for 15 min at 15 psi pressure–121°C. Pour into sterile Petri dishes or leave in tubes.

Use: For the isolation and cultivation of *Actinomadura* species, *Actinopolyspora* species, *Excellospora* species, and *Microspora* species.

Kupferberg *Trichomonas* Base

Composition per liter:

Pancreatic digest of casein	20.0g
L-Cysteine·HCl·H_2O	1.5g
Agar	1.0g
Maltose	1.0g
Methylene Blue	3.0mg
Bovine serum	50.0mL

pH 6.0 ± 0.2 at 25°C

Source: This medium is available as a premixed powder from BD Diagnostic Systems.

Preparation of Medium: Add components, except bovine serum, to distilled/deionized water and bring volume to 950.0mL. Mix thoroughly. Gently heat and bring to boiling. Autoclave for 15 min at 15 psi pressure–121°C. Cool to 45°–50°C. Aseptically add 50.0mL of bovine serum. If desired, additional selectivity can be obtained by aseptically adding 250,000U of penicillin and 1.0g of streptomycin or 1.0g of chloramphenicol. Mix thoroughly. Pour into sterile Petri dishes or distribute into sterile tubes.

Use: For the cultivation of the *Trichomonas* species from clinical specimens.

Kupferberg *Trichomonas* Broth

Composition per liter:

Enzymatic digest of protein	20.0g
L-Cysteine·HCl·H_2O	1.5g
Agar	1.0g
Maltose	1.0g
Chloramphenicol	0.1g
Methylene Blue	3.0mg
Bovine serum	50.0mL

pH 6.0 ± 0.2 at 25°C

Source: This medium is available as a premixed powder from BD Diagnostic Systems.

Preparation of Medium: Add components, except bovine serum, to distilled/deionized water and bring volume to 950.0mL. Mix thoroughly. Gently heat and bring to boiling. Autoclave for 15 min at 15 psi pressure–121°C. Cool to 45°–50°C. Aseptically add bovine serum. If desired, additional selectivity can be obtained by aseptically adding 250,000U penicillin and 1.0g streptomycin or 1.0g chloramphenicol. Mix thoroughly. Pour into sterile Petri dishes or distribute into sterile tubes.

Use: For the cultivation of the *Trichomonas* species from clinical specimens.

L Agar (Luria Agar)

Composition per liter:

Agar	15.0g
Pancreatic digest of casein	10.0g
Yeast extract	5.0g
NaCl	0.5g
Glucose solution	20.0mL

pH 7.0 ± 0.2 at 25°C

Glucose Solution:

Composition per 100.0mL:

Glucose	10.0g

Preparation of Glucose Solution: Add glucose to distilled/deionized water and bring volume to 100.0mL. Mix thoroughly. Filter sterilize.

Preparation of Medium: Add components, except glucose solution, to distilled/deionized water and bring volume to 980.0mL. Mix thoroughly. Bring pH to 7.0. Gently heat and bring to boiling. Autoclave for 15 min at 15 psi pressure–121°C. Aseptically add 20.0mL of sterile glucose solution. Mix thoroughly. Pour into sterile Petri dishes or leave in tubes.

Use: For the cultivation of *Escherichia coli.*

L Broth (Luria Broth)

Composition per liter:

Pancreatic digest of casein	10.0g
NaCl	5.0g
Yeast extract	5.0g
Glucose	1.0g

Preparation of Medium: Add components to distilled/deionized water and bring volume to 1.0L. Mix thoroughly. Distribute into tubes or flasks. Autoclave for 15 min at 15 psi pressure–121°C.

Use: For the cultivation of *Escherichia coli.*

L Diphasic Blood Agar Medium (ATCC Medium 947)

Composition per 1150.0mL:

Blood agar, diphasic base medium	700.0mL
Rabbit blood, defibrinated	300.0mL
Locke's solution	150.0mL

pH 7.2–7.4 at 25°C

Blood Agar, Diphasic Base Medium:

Composition per 750.0mL:

Beef	25.0g
Agar	10.0g
Neopeptone	10.0g
NaCl	2.5g

Preparation of Blood Agar, Diphasic Base Medium: Trim beef to remove fat. Add 25.0g of lean beef to 250.0mL of distilled/deionized water. Gently heat and bring to boiling. Boil for 2–3 min. Filter through Whatman no. 2 filter paper. Add agar, neopeptone, and NaCl to filtrate. Bring volume to 750.0mL with distilled/deionized water. Mix thoroughly. Adjust pH to 7.2–7.4. Gently heat and bring to boiling. Autoclave for 15 min at 15 psi pressure–121°C. Cool to 50°–55°C.

Locke's Solution:

Composition per liter:

NaCl	8.0g
Glucose	2.5g
KH_2PO_4	0.3g
$CaCl_2$	0.2g
KCl	0.2g

Preparation of Locke's Solution: Add components to distilled/deionized water and bring volume to 1.0L. Mix thoroughly. Autoclave for 15 min at 15 psi pressure–121°C. Cool to 50°–55°C.

Preparation of Medium: Aseptically combine 700.0mL of sterile blood agar, diphasic base medium, with 300.0mL of sterile defibrinated rabbit blood warmed to 50°–55°C. Mix thoroughly. Aseptically distribute 5.0mL volumes into 16 × 125mm screw-capped test tubes. Allow to cool in a slanted position. Overlay the agar in each tube with 1.0mL of sterile Locke's solution.

Use: For the cultivation and maintenance of *Trypanosoma* species, *Leishmania donovani, Herpetomonas* species, and *Trypanosoma neveulemairei.*

L Diphasic Blood Agar Medium (ATCC Medium 1011)

Composition per 1150.0mL:

Blood agar, diphasic base medium	700.0mL
Rabbit blood, defibrinated	300.0mL
Locke's solution	150.0mL

pH 7.2–7.4 at 25°C

Blood Agar, Diphasic Base Medium:

Composition per 750.0mL:

Beef	25.0g
Agar	10.0g
Neopeptone	10.0g
NaCl	2.5g

Preparation of Blood Agar, Diphasic Base Medium: Trim beef to remove fat. Add 25.0g of lean beef to 250.0mL of distilled/deionized water. Gently heat and bring to boiling. Boil for 2–3 min. Filter through Whatman no. 2 filter paper. Add agar, neopeptone, and NaCl to filtrate. Bring volume to 750.0mL with distilled/deionized water. Mix thoroughly. Adjust pH to 7.2–7.4. Gently heat and bring to boiling. Autoclave for 15 min at 15 psi pressure–121°C. Cool to 50°–55°C.

Locke's Solution:

Composition per liter:

NaCl	8.0g
Glucose	2.5g
KH_2PO_4	0.3g
$CaCl_2$	0.2g
KCl	0.2g

Preparation of Locke's Solution: Add components to distilled/deionized water and bring volume to 1.0L. Mix thoroughly. Autoclave for 15 min at 15 psi pressure–121°C. Cool to 50°–55°C.

Preparation of Medium: Aseptically combine 700.0mL of sterile blood agar, diphasic base medium, with 300.0mL of sterile defibrinated rabbit blood warmed to 50°–55°C. Mix thoroughly. Aseptically distribute 5.0mL volumes into 16 × 125mm screw-capped test tubes. Allow to cool in a slanted position. Overlay the agar in each tube with 3.0mL of sterile Locke's solution.

Use: For the cultivation and maintenance of *Trypanosoma* species.

Lab-Lemco Agar

Composition per liter:

Agar	15.0g
Peptone	5.0g
Lab-lemco meat extract	3.0g

pH 7.4 ± 0.2 at 25°C

Preparation of Medium: Add components to distilled/deionized water and bring volume to 1.0L. Mix thoroughly. Gently heat and bring to boiling. Distribute into tubes or flasks. Autoclave for 15 min at 15 psi pressure–121°C. Pour into sterile Petri dishes or leave in tubes.

Use: For the cultivation and maintenance of a variety of heterotrophic microorganisms.

Lactose Broth

Composition per liter:

Lactose	5.0g
Pancreatic digest of gelatin	5.0g
Beef extract	3.0g

pH 6.9 ± 0.2 at 25°C

Source: This medium is available as a premixed powder from BD Diagnostic Systems and Oxoid Unipath.

Preparation of Medium: Add components to distilled/deionized water and bring volume to 1.0L. Mix thoroughly. Distribute into tubes containing an inverted Durham tube in 10.0mL volumes. Autoclave for 12 min at 15 psi pressure–121°C. Cool broth quickly to 25°C. For testing water samples with 10.0mL volumes, prepare medium double strength.

Use: For the detection of lactose-fermenting, Gram-negative coliforms, as a preenrichment broth for *Salmonella* species, and in the study of lactose fermentation of bacteria in general.

Lactose Egg Yolk Milk Agar

Composition per 1206.0mL:

Lactose	12.0g
Agar	1.0g
Columbia blood agar base	800.0mL
Skim milk	150.0mL
Egg yolk emulsion, 50%	36.0mL
Inhibitor solution	20.0mL
Neutral Red (1% solution)	3.25mL

pH 7.0 ± 0.2 at 25°C

Columbia Blood Agar Base:

Composition per 800.0mL:

Agar	15.0g
Pantone	10.0g
Bitone	10.0g
NaCl	5.0g
Tryptic digest of beef heart	3.0g
Cornstarch	1.0g

Preparation of Columbia Blood Agar Base: Add components to distilled/deionized water and bring volume to 1.0L. Mix thoroughly. Gently heat until boiling.

Egg Yolk Emulsion, 50%:

Composition per 100.0mL:

Chicken egg yolks	11
Whole chicken egg	1
NaCl (0.9% solution)	50.0mL

Preparation of Egg Yolk Emulsion, 50%: Soak eggs with 1:100 dilution of saturated mercuric chloride solution for 1 min. Crack eggs and separate yolks from whites. Mix egg yolks with 1 chicken egg. Beat to form emulsion. Measure 50.0mL of egg yolk emulsion and add to 50.0mL of 0.9% NaCl solution. Mix thoroughly. Filter sterilize. Warm to 45°–50°C.

Inhibitor Solution:

Composition per 20.0mL:

Neomycin sulfate	0.18g
NaN_3	0.24g

Caution: Sodium azide is toxic. Azides also react with metals and disposal must be highly diluted.

Preparation of Inhibitor Solution: Add neomycin sulfate and NaN_3 to distilled/deionized water and bring volume to 20.0mL. Mix thoroughly. Filter sterilize.

Preparation of Medium: Combine Columbia blood agar base, lactose, agar, and Neutral Red and bring volume to 1.0L. Adjust pH to 7.0. Autoclave for 15 min at 15 psi pressure–121°C. Cool to 55°C. Filter sterilize skim milk. To 1.0L of cooled, sterile agar mixture, aseptically add 150.0mL of sterile skim milk, 36.0mL of sterile egg yolk emulsion, 50%, and 20.0mL of sterile inhibitor solution. Mix thoroughly. Pour into sterile Petri dishes or distribute into sterile tubes.

Use: For the cultivation and maintenance of *Clostridium* species.

Lactose Gelatin Medium

Composition per liter:

Gelatin	120.0g
Tryptose	15.0g
Lactose	10.0g
Yeast extract	10.0g
Phenol Red (0.5% solution)	10.0mL

pH 7.5 ± 0.2 at 25°C

Preparation of Medium: Add tryptose, yeast extract, and lactose to distilled/deionized water and bring volume to 400.0mL. Mix thoroughly. Add gelatin to distilled/deionized water and bring volume to 590.0mL. Gently heat gelatin solution while stirring and bring to 50°–60°C. Add Phenol Red. Mix the two solutions together. Distribute into tubes in 10.0mL volumes. Autoclave for 10 min at 15 psi pressure–121°C. If medium is not used in 8 hr, deoxygenate by heating to 50°–70°C for 2–3 hr prior to inoculation.

Use: For the cultivation of *Clostridium perfringens.*

Lash Serum Medium

Composition per liter:

Casamino acids	14.0g
NaCl	6.0g
Glucose	2.0g
Maltose	1.5g
Sodium lactate (60% solution)	0.5g
KCl	0.1g
$CaCl_2 \cdot 2H_2O$	0.1g
Serum solution	500.0mL

pH 5.8 ± 0.2 at 25°C

Serum Solution:

Composition per 500.0mL:

$NaHCO_3$	0.1g
Bovine serum	200.0mL

Preparation of Serum Solution: Add components to distilled/deionized water and bring volume to 500.0mL. Mix thoroughly. Filter sterilize.

Preparation of Medium: Add components, except serum solution, to distilled/deionized water and bring volume to 500.0mL. Mix thoroughly. Distribute into tubes in 5.0mL volumes. Autoclave for 15 min at 15 psi pressure–121°C. Cool to 25°C. Aseptically add 5.0mL of sterile serum solution to each tube. Mix thoroughly.

Use: For the cultivation of *Trichomonas vaginalis* from clinical specimens.

Lauryl Sulfate Broth (m-Lauryl Sulfate Broth)

Composition per liter:

Peptone	39.0g
Lactose	30.0g
Yeast extract	6.0g
Sodium lauryl sulfate	1.0g
Phenol Red	0.2g

pH 7.4 ± 0.2 at 25°C

Source: This medium is available as a premixed powder from Oxoid Unipath.

Preparation of Medium: Add components to distilled/deionized water and bring volume to 1.0L. Mix thoroughly. Distribute into bottles or flasks. Autoclave for 15 min at 15 psi pressure–121°C.

Use: For the cultivation and enumeration of coliform bacteria, especially *Escherichia coli*, by the membrane filter method.

Lauryl Sulfate Broth (Lauryl Tryptose Broth)

Composition per liter:

Pancreatic digest of casein	20.0g
Lactose	5.0g
NaCl	5.0g
K_2HPO_4	2.75g
KH_2PO_4	2.75g
Sodium lauryl sulfate	0.1g

pH 6.8 ± 0.2 at 25°C

Source: This medium is available as a premixed powder from BD Diagnostic Systems.

Preparation of Medium: Add components to distilled/deionized water and bring volume to 1.0L. Mix thoroughly. Distribute into tubes containing an inverted Durham tube in 10.0mL volumes. Autoclave for 12 min at 15 psi pressure–121°C. Cool broth quickly to 25°C. For testing water samples with 10.0mL volumes, prepare medium double strength.

Use: For the detection of coliform bacteria in a variety of specimens. Also, for the enumeration of coliform bacteria by the multiple-tube fermentation technique.

Lauryl Sulfate Broth with MUG

Composition per liter:

Pancreatic digest of casein	20.0g
Lactose	5.0g
NaCl	5.0g
K_2HPO_4	2.75g
KH_2PO_4	2.75g
Sodium lauryl sulfate	0.1g
4-Methylumbellferyl-β-D-glucuronide (MUG)	0.05g

pH 6.8 ± 0.2 at 25°C

Source: This medium is available as a premixed powder from BD Diagnostic Systems.

Preparation of Medium: Add components to distilled/deionized water and bring volume to 1.0L. Mix thoroughly. Distribute into tubes containing an inverted Durham tube in 10.0mL volumes. Autoclave for 12 min at 15 psi pressure–121°C. Cool broth quickly to 25°C. For testing water samples with 10.0mL volumes, prepare medium double strength.

Use: For the detection of *Escherichia coli* by a fluorogenic procedure.

LB Agar

Composition per liter:

Agar	15.0g
Pancreatic digest of casein	10.0g
NaCl	5.0g
Yeast extract	5.0g
1*N* NaOH	1.0mL

pH 7.0 ± 0.2 at 25°C

Preparation of Medium: Add components to distilled/deionized water and bring volume to 1.0L. Mix thoroughly. Gently heat and bring to boiling. Adjust pH to 7.0. Distribute into tubes or flasks. Autoclave for 25 min at 15 psi pressure–121°C. Pour into sterile Petri dishes in 35–40.0mL volumes.

Use: For the cultivation of *Escherichia coli.*

Lead Acetate Agar

Composition per liter:

Agar	15.0g
Peptone	15.0g
Proteose peptone	5.0g
Glucose	1.0g
Lead acetate	0.2g
$Na_2S_2O_3$	0.08g

pH 6.6 ± 0.2 at 25°C

Preparation of Medium: Add components to distilled/deionized water and bring volume to 1.0L. Mix thoroughly. Gently heat and bring to boiling. Distribute into tubes or flasks. Autoclave for 15 min at 15 psi pressure–121°C. Pour into sterile Petri dishes or leave in tubes. Allow tubes to cool in a slanted position.

Use: For the cultivation and differentiation of Gram-negative coliform bacteria based on H_2S production. Bacteria that produce H_2S turn the medium brown.

Lecithin Lactose Agar

Composition per liter:

Agar	15.0g
Pancreatic digest of casein	12.7g
Lactose	10.0g
NaCl	5.5g
Peptic digest of animal tissue	5.5g
Yeast extract	3.9g
Pancreatic digest of heart muscle	3.3g
Cornstarch	1.1g
Egg lecithin	0.66g
L-Cysteine·HCl·H_2O	0.5g
NaN_3	0.2g
Neomycin sulfate	0.15g
$CaCl_2$	0.05g
Bromcresol Purple	0.02g

pH 6.8 ± 0.2 at 25°C

Caution: Sodium azide is toxic. Azides also react with metals and disposal must be highly diluted.

Source: This medium is available as a prepared medium from BD Diagnostic Systems.

Preparation of Medium: Add components to distilled/deionized water and bring volume to 1.0L. Mix thoroughly. Gently heat and bring to boiling. Distribute into tubes or flasks. Autoclave for 15 min at 15 psi pressure–121°C. Pour into sterile Petri dishes.

Use: For the isolation, cultivation, and differentiation of histolytic clostridia from clinical specimens based on lecithinase production and lactose fermentation. It is especially useful for the differentiation of *Clostridium perfringens, Clostridium sordelli, Clostridium novyi, Clostridium septicum,* and *Clostridium histolyticum.* Bacteria that produce lecithinase appear as colonies surrounded by an opalescent zone. Bacteria that ferment lactose appear as colonies surrounded by a yellow zone.

Lecithin Lipase Anaerobic Agar

Composition per liter:

Pancreatic digest of casein	40.0g
Agar	25.0g
Yeast extract	5.0g
$Na_2HPO_4·12H_2O$	5.0g
Glucose	2.0g
NaCl	2.0g
KH_2PO_4	1.0g
$MgSO_4·7H_2O$	0.1g
Egg yolk emulsion	100.0mL

pH 7.6 ± 0.2 at 25°C

Egg Yolk Emulsion:

Composition:

Chicken egg yolks	11
Whole chicken egg	1

Preparation of Egg Yolk Emulsion: Soak eggs with 1:100 dilution of saturated mercuric chloride solution for 1 min. Crack eggs and separate yolks from whites. Mix egg yolks with 1 chicken egg. Filter sterilize.

Preparation of Medium: Add components, except egg yolk emulsion, to distilled/deionized water and bring volume to 900.0mL. Mix thoroughly. Gently heat and bring to boiling. Autoclave for 15 min at 15 psi pressure–121°C. Cool to 45°–50°C. Aseptically add sterile egg yolk emulsion. Mix thoroughly. Pour into sterile Petri dishes or distribute into sterile tubes.

Use: For the isolation, cultivation, and differentiation of *Clostridium* species based on lecithinase production and lipase production. Bacteria that produce lecithinase appear as colonies surrounded by a zone of insoluble precipitate. Bacteria that produce lipase appear as colonies with a pearly iridescent sheen.

Lecithin Tween Medium (LT Medium)

Composition per liter:

Tween 80	30.0g
Agar	15.0g
Pancreatic digest of casein	10.0g
Peptic digest of animal tissue	10.0g
NaCl	5.0g
Lecithin	5.0g
$Na_2S_2O_3·5H_2O$	5.0g
Glycerol	3.0g
Histidine, free base	1.0g
Glucose	1.0g

pH 7.5 ± 0.2 at 25°C

Antibiotic Solution:

Composition per 10.0mL:

5–Fluorocytosine	0.2g
Fosfomicin	0.1g
Ticarcillin	0.1g

Preparation of Antibiotic Solution: Add components to distilled/deionized water and bring volume to 10.0mL. Mix thoroughly. Filter sterilize.

Preparation of Medium: Add components, except antibiotic solution, to distilled/deionized water and bring volume to 990.0mL. Mix thoroughly. Gently heat and bring to boiling. Autoclave for 15 min at 15 psi pressure–121°C. Cool to 45°–50°C. Aseptically add sterile antibiotic solution. Mix thoroughly. Pour into sterile Petri dishes in 20.0mL volumes.

Use: For the isolation and cultivation of multiresistant lipophilic *Corynebacterium* species, especially *Corynebacterium* group JK found primarily in infections in immuno-compromised hosts and patients with prosthetic valve endocarditis.

Legionella Agar Base (*Legionella* Medium) (BCYEα Agar, Modified)

Composition per liter:

Agar	17.0g
Yeast extract	10.0g
ACES buffer (*N*-2-acetamido-2-aminoethane sulfonic acid)	6.0g
Charcoal, activated	1.5g
KOH	1.5g
α-Ketoglutarate	1.0g
Legionella agar enrichment	10.0mL

pH 6.85–7.0 at 25°C

Source: This medium is available as a prepared medium from BD Diagnostic Systems.

***Legionella* Agar Enrichment:**
Composition per 10.0mL:

L-Cysteine·HCl·H_2O	0.4g
$Fe_4(P_2O_7)_3$	0.25g

Preparation of *Legionella* Agar Enrichment: Add components to distilled/deionized water and bring volume to 10.0mL. Mix thoroughly. Filter sterilize.

Preparation of Medium: Add components, except *Legionella* agar enrichment, to distilled/deionized water and bring volume to 990.0mL. Mix thoroughly. Gently heat to boiling. Autoclave for 15 min at 15 psi pressure–121°C. Cool to 50° C. Add 10.0mL of sterile *Legionella* agar enrichment. Adjust pH to 6.9 at 50°C by adding 4.0–4.5mL of 1.0*N* KOH. This is a critical step. Mix thoroughly. Pour into sterile Petri dishes in 20.0mL volumes. Swirl medium while pouring to keep charcoal in suspension.

Use: For the preparation of *Legionella* agars. For the isolation and cultivation of *Legionella* species from clinical and nonclinical materials.

Legionella pneumophila Medium (Charcoal Yeast Extract Diphasic Blood Culture Medium) (Diphasic Blood Culture Buffered Charcoal Yeast Extract Medium) (CYE-DBCM)

Composition per liter:

Agar phase	500.0mL
Broth phase	500.0mL

pH 6.9–7.0 at 25°C

Agar Phase:
Composition per 500.0mL:

Agar	17.0g
Charcoal, activated	2.0g

Preparation of Agar Phase: Add components to distilled/deionized water and bring volume to 500.0mL. Mix thoroughly. Gently heat and bring to boiling. Distribute in 20.0mL volumes into 125.0mL serum bottles with aluminum crimp seals and rubber stoppers. Autoclave for 20 min at 15 psi pressure–121°C. Cool to 50°C. Swirl medium to put charcoal in suspension. Allow agar to solidify so that a slant with a 6.0cm height is formed.

Broth Phase:
Composition per 500.0mL:

Yeast extract	20.0g
L-Cysteine·HCl·H_2O solution	0.4g
$Fe(NO_3)_3 \cdot 9H_2O$ solution	0.1g

Preparation of Broth Phase: Add yeast extract to distilled/deionized water and bring volume to 480.0mL. Mix thoroughly. Autoclave for 15 min at 15 psi pressure–121°C. Cool to 25°C. Aseptically add sterile L-cysteine·HCl·H_2O solution and $Fe(NO_3)_3 \cdot 9H_2O$ solution. Mix thoroughly. Adjust pH to 6.9 with 6.0mL of sterile 1*N* KOH.

L-Cysteine·HCl·H_2O Solution:
Composition per 10.0mL:

L-Cysteine·HCl·H_2O	0.04g

Preparation of L-Cysteine·HCl·H_2O Solution: Add L-cysteine·HCl·H_2O to distilled/deionized water and bring volume to 10.0mL. Mix thoroughly. Filter sterilize.

$Fe(NO_3)_3 \cdot 9H_2O$ Solution:
Composition per 10.0mL:

$Fe(NO_3)_3 \cdot 9H_2O$	0.04g

Preparation of $Fe(NO_3)_3 \cdot 9H_2O$ Solution: Add $Fe(NO_3)_3$ to distilled/deionized water and bring volume to 10.0mL. Mix thoroughly. Filter sterilize.

Preparation of Medium: Add 20.0mL of sterile broth phase to 125.0mL serum bottles containing 20.0mL of solidified agar phase. Seal bottles by crimping metal caps over rubber stoppers.

Use: For the isolation and cultivation of *Legionella pneumophila* from blood cultures.

Legionella Selective Agar

Composition per liter:

Agar	15.0g
ACES (2-[(2-Amino-2-oxoethyl)-amino]ethane sulfonic acid) buffer	10.0g
Yeast extract	10.0g

Charcoal, activated 2.0g
α-Ketoglutarate 1.0g
L-Cysteine·HCl·H_2O solution 10.0mL
$Fe_4(P_2O_7)_3$ solution 10.0mL
Antibiotic solution 10.0mL
pH 6.85–7.0 at 25°C

Source: This medium is available as a prepared medium from BD Diagnostic Systems.

L-Cysteine·HCl·H_2O Solution:
Composition per 10.0mL:
L-Cysteine·HCl·H_2O 0.4g

Preparation of L-Cysteine·HCl·H_2O Solution: Add L-cysteine·HCl·H_2O to distilled/deionized water and bring volume to 10.0mL. Mix thoroughly. Filter sterilize.

$Fe_4(P_2O_7)_3$ Solution:
Composition per 10.0mL:
$Fe_4(P_2O_7)_3$ 0.25g

Preparation of $Fe_4(P_2O_7)_3$ Solution: Add $Fe_4(P_2O_7)_3$ to distilled/deionized water and bring volume to 10.0mL. Mix thoroughly. Filter sterilize.

Antibiotic Solution:
Composition per 10.0mL:
Anisomycin 10.0mg
Colistin 3.75mg
Vancomycin 2.0mg

Preparation of Antibiotic Solution: Add components to distilled/deionized water and bring volume to 10.0mL. Mix thoroughly. Filter sterilize.

Preparation of Medium: Add components—except L-cysteine·HCl·H_2O, $Fe_4(P_2O_7)_3$ and antibiotic solutions—to distilled/deionized water and bring volume to 970.0mL. Mix thoroughly. Gently heat and bring to boiling. Autoclave for 15 min at 15 psi pressure–121°C. Cool to 45°–50°C. Aseptically add sterile L-cysteine·HCl·H_2O, $Fe_4(P_2O_7)_3$ and antibiotic solutions. Mix thoroughly. Pour into sterile Petri dishes. Swirl medium while pouring to keep charcoal in suspension.

Use: *Legionella* selective agar is used in qualitative procedures for the isolation of *Legionella* species from clinical and nonclinical specimens.

Leishmania Medium

Composition per 100.0mL:
Sodium citrate 1.2g
NaCl 1.0g
Rabbit blood solution 10.0mL

Rabbit Blood Solution:
Composition per 10.0mL:
Rabbit blood, defibrinated 5.0mL

Preparation of Rabbit Blood Solution: Add 5.0mL of whole rabbit blood to 5.0mL of sterile distilled/deionized water. Freeze and thaw twice to lyse blood cells.

Preparation of Medium: Add components, except rabbit blood solution, to distilled/deionized water and bring volume to 90.0mL. Mix thoroughly. Autoclave for 15 min at 15 psi pressure–121°C. Aseptically add 10.0mL of sterile rabbit blood solution. Mix thoroughly. Aseptically distribute into sterile, screw-capped tubes or flasks.

Use: For the cultivation of *Leishmania donovani*, *Leishmania hertigi*, and *Leishmania tropica*.

Leptospira Medium

Composition per liter:
$(NH_4)_2Fe(SO_4)_2·6H_2O$ 6.0g
NaH_2PO_4 .... 0.53g
L-Asparagine 0.5g
Glycerol 0.2g
Tween 60 .... 0.2g
$MgSO_4·7H_2O$ 0.15g
KH_2PO_4 .... 0.069g
Tween 80 .... 0.05g
EDTA 0.01g
$CaCO_3$ 4.0mg
Thiamine·HCl 1.0mg
Vitamin B_{12} 1.0μg
pH 7.4–7.6 at 25°C

Preparation of Medium: Add components, except thiamine·HCl, to distilled/deionized water and bring volume to 990.0mL. Mix thoroughly. Gently heat and bring to boiling. Autoclave for 15 min at 15 psi pressure–121°C. Aseptically add 1.0mg of thiamine·HCl. Aseptically distribute into sterile tubes or flasks.

Use: For the cultivation of *Leptospira* species.

Leptospira Medium, EMJH (*Leptospira* Medium, Ellinghausen-McCullough/ Johnson-Harris)

Composition per liter:
Na_2HPO_4 .... 1.0g
NaCl 1.0g
KH_2PO_4 .... 0.3g
NH_4Cl 0.25g
Thiamine 5.0mg
Rabbit serum 100.0mL
pH 7.5 ± 0.2 at 25°C

Source: This medium is available as a premixed powder from BD Diagnostic Systems.

Preparation of Medium: Add components, except rabbit serum, to distilled/deionized water and bring volume to 900.0mL. Mix thoroughly. Gently heat and bring to boiling. Autoclave for 15 min at 15 psi pressure–121°C. Cool to 25°C. Aseptically add sterile rabbit serum. Mix thoroughly. Aseptically distribute into sterile tubes or flasks.

Use: For the cultivation and maintenance of *Leptospira* species.

Leptospira Medium, Modified

Composition per liter:

Agar 1.5g
NaCl 0.5g
Peptone 0.3g
Beef extract 0.2g
Hemin solution 2.5mL
Sterile rabbit serum 100.0mL

pH 7.3 ± 0.1 at 25°C

Hemin Solution:

Composition per 100.0mL:

Hemin 0.05g
NaOH (1*N* solution) 1.0mL

Preparation of Hemin Solution: Add hemin to 1.0mL of 1*N* NaOH solution. Mix thoroughly. Bring volume to 100.0mL with distilled/deionized water. Autoclave for 15 min at 15 psi pressure–121°C. Cool to 45°–50°C.

Preparation of Medium: Add components, except hemin solution and rabbit serum, to distilled/deionized water and bring volume to 897.5mL. Mix thoroughly. Gently heat and bring to boiling. Adjust pH to 7.4. Autoclave for 15 min at 15 psi pressure–121°C. Cool to 45°–50°C. Aseptically add 2.5mL of sterile hemin solution and 100.0mL of sterile rabbit serum. Mix thoroughly. The pH of the medium should be 7.3. Store at 4°C for 24h. Inactivate medium at 56°C for 60 min. Aseptically distribute into sterile tubes or flasks.

Use: For the cultivation and maintenance of *Leptospira* species.

Leptospira Protein-Free Medium (*Leptospira* PF Medium)

Composition per liter:

TES (*N*-Tris[hydroxymethyl]methyl-2-aminoethane sulfonic acid) buffer 1.2g
NaCl 0.9g
Sodium pyruvate 0.2g
CT-Tween 60 .... 12.0mL
CT-Tween 40 .... 3.0mL
$MgCl_2$-$CaCl_2$ solution 1.0mL
Cyanocobalamin (0.02% solution) 1.0mL
Glycerol (10% solution) 1.0mL
KH_2PO_4 (1% solution) 1.0mL
$MnSO_4 \cdot H_2O$ (0.1% solution) 1.0mL
$ZnSO_4$ (0.4% solution) 0.1mL

pH 7.6 ± 0.2 at 25°C

CT-Tween 60:

Composition per 200.0mL:

Charcoal, Norit A 40.0g
Tween 60 .... 20.0g

Preparation of CT-Tween 60: Add Tween 60 to 200.0mL of distilled/deionized water. Mix thoroughly. While stirring, add charcoal. Stir mixture for 18 hr at 25°C. Allow charcoal to settle out of suspension for 18 hr at 4°C. Carefully decant the Tween solution off the sediment. Centrifuge the Tween solution at 10,000 × g for 1h. Decant supernatant solution. Pass Tween solution through a thin-channel ultrafiltration XM 100 membrane. Store stock solution at –20°C.

CT-Tween 40:

Composition per 200.0mL:

Charcoal, Norit A 40.0g
Tween 40 .... 20.0g

Preparation of CT-Tween 40: Add Tween 40 to 200.0mL of distilled/deionized water. Mix thoroughly. While stirring, add charcoal. Stir mixture for 18 hr at 25°C. Allow charcoal to settle out of suspension for 18 hr at 4°C. Carefully decant the Tween solution off the sediment. Centrifuge the Tween solution at 10,000 × g for 1h. Decant supernatant solution. Pass Tween solution through a thin-channel ultrafiltration XM 100 membrane. Store stock solution at –20°C.

$MgCl_2$–$CaCl_2$ Solution:

Composition per 100.0mL:

$CaCl_2 \cdot 2H_2O$ 1.5g
$MgCl_2 \cdot 6H_2O$ 1.5g

Preparation of $MgCl_2$-$CaCl_2$ Solution: Add components to distilled/deionized water and bring volume to 100.0mL. Mix thoroughly.

Preparation of Medium: Add components to distilled/deionized water and bring volume to 1.0L. Mix thoroughly. Filter sterilize. Aseptically distribute into sterile tubes or flasks.

Use: For the cultivation of *Leptospira* species.

Leptotrichia buccalis Medium

Composition per liter:

Agar 15.0g
Nutrient broth 8.0g
Yeast extract 2.0g
Glucose solution 10.0mL

L-Cysteine·HCl·H_2O solution........................ 10.0mL
Hemin solution.. 4.0mL

pH 7.2–7.6 at 25°C

Glucose Solution:

Composition per 10.0mL:

D-Glucose...2.0g

Preparation of Glucose Solution: Add glucose to distilled/deionized water and bring volume to 10.0mL. Mix thoroughly. Filter sterilize.

L-Cysteine·HCl·H_2O Solution:

Composition per 10.0mL:

L-Cysteine·HCl·H_2O..1.0g

Preparation of L-Cysteine·HCl·H_2O Solution: Add L-cysteine·HCl·H_2O to distilled/deionized water and bring volume to 10.0mL. Mix thoroughly. Filter sterilize.

Hemin Solution:

Composition per 10.0mL:

Hemin...2.5mg
Triethanolamine (50% solution).................. 10.0mL

Preparation of Hemin Solution: Add hemin to 10.0mL of triethanolamine solution. Mix thoroughly.

Preparation of Medium: Add components, except glucose solution, to distilled/deionized water and bring volume to 990.0mL. Mix thoroughly. Gently heat and bring to boiling. Autoclave for 15 min at 15 psi pressure–121°C. Cool to 45°–50°C. Aseptically add sterile glucose solution. Mix thoroughly. Pour into sterile Petri dishes or distribute into sterile tubes.

Use: For the cultivation and maintenance of *Leptotrichia buccalis*.

Leptotrichia Medium

Composition per liter:

Pancreatic digest of casein.............................. 10.0g
NaCl..5.0g
Peptone..5.0g
Yeast extract...3.0g
Na_2HPO_4...2.5g
L-Cysteine·HCl·H_2O..0.5g
Horse serum.. 100.0mL
Tomato decoction.. 50.0mL

pH 7.2–7.4 at 25°C

Tomato Decoction:

Composition per 100.0mL:

Tomatoes... 50.0mL

Preparation of Tomato Decoction: Mince fresh tomatoes and measure 50.0mL. Add 50.0mL of tap water. Mix thoroughly. Gently heat and bring to boiling. Continue boiling for 10 min. Filter through Whatman #1 filter paper. Autoclave filtrate for 15 min at 15 psi pressure–121°C.

Preparation of Medium: Add components, except horse serum and tomato decoction, to distilled/deionized water and bring volume to 850.0mL. Mix thoroughly. Gently heat and bring to boiling. Adjust pH to 7.2–7.4. Autoclave for 15 min at 15 psi pressure–121°C. Cool to 25°C. Aseptically add sterile horse serum and tomato decoction. Mix thoroughly. Aseptically distribute into sterile tubes or flasks.

Use: For the cultivation and maintenance of *Leptotrichia buccalis*.

Letheen Agar

Composition per liter:

Agar... 15.0g
Tween 80...7.0g
Pancreatic digest of casein................................5.0g
Beef extract...3.0g
Glucose...1.0g
Lecithin...1.0g

pH 7.0 ± 0.2 at 25°C

Source: This medium is available as a premixed powder from BD Diagnostic Systems.

Preparation of Medium: Add components to distilled/deionized water and bring volume to 1.0L. Mix thoroughly. Gently heat and bring to boiling. Distribute into tubes or flasks. Autoclave for 15 min at 15 psi pressure–121°C. Pour into sterile Petri dishes or leave in tubes.

Use: For determination of the antimicrobial activity of quaternary ammonium compounds.

Letheen Agar, Modified

Composition per liter:

Agar... 20.0g
Thiotone.. 10.0g
Pancreatic digest of casein.............................. 10.0g
Tween 80...7.0g
NaCl..5.0g
Beef extract...3.0g
Yeast extract...2.0g
Glucose...1.0g
Lecithin...1.0g
$NaHSO_3$...0.1g

pH 7.2 ± 0.2 at 25°C

Preparation of Medium: Add components to distilled/deionized water and bring volume to 1.0L. Mix thoroughly. Gently heat and bring to boiling. Distribute into tubes or flasks. Autoclave for 15 min at 15 psi pressure–121°C. Pour into sterile Petri dishes.

Use: For determination of the antimicrobial activity of quaternary ammonium compounds.

Letheen Broth

Composition per liter:

Peptic digest of animal tissue	10.0g
Beef extract	5.0g
NaCl	5.0g
Tween 80	5.0g
Lecithin	0.7g

pH 7.0 ± 0.2 at 25°C

Source: This medium is available as a premixed powder from BD Diagnostic Systems.

Preparation of Medium: Add components to distilled/deionized water and bring volume to 1.0L. Mix thoroughly. Distribute into tubes or flasks. Autoclave for 15 min at 15 psi pressure–121°C.

Use: For determination of the antimicrobial activity of quaternary ammonium compounds.

Letheen Broth, Modified

Composition per liter:

Peptic digest of animal tissue	10.0g
Thiotone peptone	10.0g
Beef extract	5.0g
NaCl	5.0g
Tween 80	5.0g
Pancreatic digest of casein	5.0g
Yeast extract	2.0g
Lecithin	0.7g
$NaHSO_3$	0.1g

pH 7.2 ± 0.2 at 25°C

Preparation of Medium: Add components to distilled/deionized water and bring volume to 1.0L. Mix thoroughly. Distribute into screw-capped bottles in 90.0mL volumes. Autoclave for 15 min at 15 psi pressure–121°C.

Use: For determination of the antimicrobial activity of quaternary ammonium compounds.

Levine EMB Agar (Levine Eosin Methylene Blue Agar) (Eosin Methylene Blue Agar, Levine) (LEMB Agar)

Composition per liter:

Agar	15.0g
Lactose	10.0g
Peptone	10.0g
K_2HPO_4	2.0g
Eosin Y	0.4g
Methylene Blue	0.065mg

pH 7.1 ± 0.2 at 25°C

Source: This medium is available as a premixed powder from BD Diagnostic Systems.

Preparation of Medium: Add components to distilled/deionized water and bring volume to 1.0L. Mix thoroughly. Gently heat and bring to boiling. Distribute into tubes or flasks. Autoclave for 15 min at 15 psi pressure–121°C. Pour into sterile Petri dishes or leave in tubes.

Use: For the isolation, cultivation, and differentiation of Gram-negative enteric bacteria based on lactose fermentation. Bacteria that ferment lactose, especially the coliform bacterium *Escherichia coli*, appear as colonies with a green metallic sheen or blue-black to brown color. Bacteria that do not ferment lactose appear as colorless or transparent light purple colonies.

Levinthal's Agar

Composition per 105.0mL:

Nutrient agar, sterile	100.0mL
Rabbit blood or human blood, sterile	5.0mL

pH 6.8 ± 0.2 at 25°C

Nutrient Agar:

Composition per liter:

Agar	15.0g
Pancreatic digest of gelatin	5.0g
Beef extract	3.0g

Source: Nutrient agar is available as a premixed powder from BD Diagnostic Systems.

Preparation of Nutrient Agar: Add components to distilled/deionized water and bring volume to 1.0L. Mix thoroughly. Gently heat while stirring and bring to boiling. Distribute into tubes or flasks. Autoclave for 15 min at 15 psi pressure–121°C. Cool to 45°–50°C.

Preparation of Medium: To 100.0mL of cooled, sterile nutrient agar, aseptically add 5.0mL of human blood or rabbit blood. Mix thoroughly. Heat in a boiling water bath for 5 min. Allow the deposit to settle out of suspension. Pour the clear supernatant solution into sterile Petri dishes or distribute into sterile tubes.

Use: For the cultivation of *Haemophilus* species.

Listeria Enrichment Broth

Composition per liter:

Pancreatic digest of casein	17.0g
Yeast extract	6.0g
NaCl	5.0g
Papaic digest of soybean meal	3.0g

Glucose	2.5g
K_2HPO_4	2.5g
Cycloheximide	0.05g
Nalidixic acid	0.04g
Acriflavine·HCl	0.015g

pH 7.3 ± 0.2 at 25°C

Preparation of Medium: Add components to distilled/deionized water and bring volume to 1.0L. Mix thoroughly. Gently heat and bring to boiling. Distribute into tubes or flasks. Autoclave for 15 min at 15 psi pressure–121°C.

Use: For the isolation and cultivation of *Listeria monocytogenes* according to the FDA formula.

Listeria Fermentation Broth

Composition per liter:

Proteose peptone no. 3	10.0g
NaCl	5.0g
Beef extract	1.0g
Bromcresol Purple	0.1g
Carbohydrate solution	10.0mL

Carbohydrate Solution:
Composition per 10.0mL:

Carbohydrate	5.0g

Preparation of Carbohydrate Solution: Add carbohydrate to distilled/deionized water and bring volume to 10.0mL. Mix thoroughly. Filter sterilize. Use glucose, salicin, rhamnose, dulcitol, or raffinose.

Preparation of Medium: Add components, except carbohydrate solution, to distilled/deionized water and bring volume to 990.0mL. Mix thoroughly. Gently heat and bring to boiling. Autoclave for 15 min at 15 psi pressure–121°C. Cool to 45°–50°C. Aseptically add sterile carbohydrate solution. Mix thoroughly. Aseptically distribute into sterile tubes or flasks.

Use: For the cultivation and differentiation of *Listeria* species based on the fermentation of glucose, salicin, rhamnose, dulcitol, and raffinose.

Listeria Transport Enrichment Medium

Composition per liter:

Sodium glycerophosphate	10.0g
Agar	2.0g
Sodium thioglycolate	1.0g
$CaCl_2$	0.1g
Nalidixic acid	0.04g
Acridine solution	2.0mL

pH 7.4 ± 0.2 at 25°C

Acridine Solution:
Composition per 10.0mL:

Acridine	0.04g

Preparation of Acridine Solution: Add acridine to distilled/deionized water and bring volume to 10.0mL. Mix thoroughly. Autoclave for 15 min at 15 psi pressure–121°C. Cool to 45°–50°C.

Preparation of Medium: Add components, except acridine solution, to distilled/deionized water and bring volume to 998.0mL. Mix thoroughly. Gently heat and bring to boiling. Autoclave for 15 min at 15 psi pressure–121°C. Cool to 45°–50°C. Aseptically add acridine solution. Mix thoroughly. Aseptically distribute into sterile tubes in 10.0mL volumes or fill bottles 4/5s full.

Use: For the maintenance—as a transport medium—and enrichment of *Listeria* species.

LIT Medium

Composition per liter:

Beef liver, infusion from	453.0g
Na_2HPO_4	8.0g
Pancreatic digest of casein	5.0g
Tryptose	5.0g
Glucose	1.0g
NaCl	1.0g
K_2HPO_4	0.5g
KCl	0.4g
Hemin	10.0mg
Fetal bovine serum, heat inactivated	100.0mL

pH 7.2 ± 0.2 at 25°C

Preparation of Medium: Add components to distilled/deionized water and bring volume to 1.0L. Mix thoroughly. Adjust pH to 7.2. Filter sterilize. Aseptically distribute into sterile tubes or flasks.

Use: For the cultivation of *Herpetomonas mariadeanei*, *Trypanoplasma borreli,* and *Trypanosoma cruzi.*

Litmus Milk

Composition per liter:

Skim milk	100.0g
Azolitmin	0.5g
Na_2SO_3	0.5g

pH 6.5 ± 0.2 at 25°C

Source: This medium is available as a premixed powder from BD Diagnostic Systems and Oxoid Unipath.

Preparation of Medium: Add components to distilled/deionized water and bring volume to 1.0L. Mix thoroughly. Gently heat and bring to boiling. Distribute into tubes or flasks. Autoclave for 20 min at 10 psi pressure–115°C.

Use: For the maintenance of lactic acid bacteria and for the differentiation of several bacteria, especially *Clostridium* species, based on their action on milk.

Bacteria that do not ferment carbohydrates, such as *Proteus vulgaris* or *Moraxella lacunata*, show no change in the azolitimin litmus indicator. Bacteria that ferment lactose or glucose with the production of gas, such as *Clostridium perfringens*, turn the medium pink and frothy. Bacteria that proteolytically degrade milk lactalbumin turn the medium blue. Bacteria that coagulate milk casein form a curd or clot. Bacteria that peptonize casein, such as *Pseudomonas aeruginosa*, show a dissolution of the clot.

Littman Oxgall Agar

Composition per liter:

Agar	20.0g
Oxgall	15.0g
Glucose	10.0g
Peptone	10.0g
Crystal Violet	0.01g
Streptomycin solution	10.0mL

pH 6.5 ± 0.2 at 25°C

Source: This medium is available as a premixed powder from BD Diagnostic Systems.

Streptomycin Solution:

Composition per 10.0mL:

Streptomycin	0.03g

Preparation of Streptomycin Solution: Add streptomycin to distilled/deionized water and bring volume to 10.0mL. Mix thoroughly. Filter sterilize.

Preparation of Medium: Add components, except streptomycin solution, to distilled/deionized water and bring volume to 990.0mL. Mix thoroughly. Gently heat and bring to boiling. Autoclave for 15 min at 15 psi pressure–121°C. Cool to 45°–50°C. Aseptically add sterile streptomycin solution. Mix thoroughly. Pour into sterile Petri dishes or distribute into sterile tubes. Allow tubes to cool in a slanted position.

Use: For the selective isolation and cultivation of fungi, especially dermatophytes.

Liver Broth

Composition per liter:

Beef liver, fresh	453.0g
Pancreatic digest of casein	10.0g
K_2HPO_4	1.0g
Soluble starch	1.0g

pH 7.6 ± 0.2 at 25°C

Preparation of Medium: Remove the fat from fresh beef liver. Grind the liver. Add 1.0L of distilled/deionized water. Gently heat and bring to boiling. Continue boiling for 60 min. Adjust pH to 7.6. Filter through cheesecloth. Reserve meat. To filtrate, add pancreatic digest of casein, K_2HPO_4, and soluble starch. Bring volume to 1.0L with distilled/deionized water. Refilter solution. Add meat particles to test tubes to a depth of approximately 2cm. Distribute broth into tubes with meat particles in 15.0mL volumes. Autoclave for 20 min at 15 psi pressure–121°C.

Use: For the isolation and cultivation of anaerobic microorganisms, especially *Clostridium botulinum*.

Liver Infusion Agar

Composition per liter:

Beef liver, infusion from	500.0g
Agar	20.0g
Proteose peptone	10.0g
NaCl	5.0g

pH 6.9 ± 0.2 at 25°C

Source: This medium is available as a premixed powder from BD Diagnostic Systems.

Preparation of Medium: Add components to distilled/deionized water and bring volume to 1.0L. Mix thoroughly. Gently heat and bring to boiling. Distribute into tubes or flasks. Autoclave for 15 min at 15 psi pressure–121°C. Pour into sterile Petri dishes or leave in tubes.

Use: For the cultivation of *Brucella* species and other fastidious pathogenic bacteria.

Liver Infusion Broth

Composition per liter:

Beef liver, infusion from	500.0g
Proteose peptone	10.0g
NaCl	5.0g

pH 6.9 ± 0.2 at 25°C

Source: This medium is available as a premixed powder from BD Diagnostic Systems.

Preparation of Medium: Add components to distilled/deionized water and bring volume to 1.0L. Mix thoroughly. Gently heat and bring to boiling. Distribute into tubes or flasks. Autoclave for 15 min at 15 psi pressure–121°C.

Use: For the cultivation of *Brucella* species and other fastidious pathogenic bacteria.

Liver Veal Agar

Composition per liter:

Veal, infusion from	500.0g
Beef liver, infusion from	50.0g
Gelatin	20.0g
Proteose peptone	20.0g
Agar	15.0g
Soluble starch	10.0g
Glucose	5.0g

NaCl	5.0g
Casein	2.0g
$NaNO_3$	2.0g
Enzymatic digest of protein	1.3g
Pancreatic digest of casein	1.3g

pH 7.3 ± 0.2 at 25°C

Source: This medium is available as a premixed powder from BD Diagnostic Systems.

Preparation of Medium: Add components to distilled/deionized water and bring volume to 1.0L. Mix thoroughly. Gently heat and bring to boiling. Distribute into tubes or flasks. Make sure that some liver and veal particles are transferred to each tube. Autoclave for 15 min at 15 psi pressure–121°C. Pour into sterile Petri dishes or leave in tubes.

Use: For the cultivation of a variety of anaerobic organisms.

Liver Veal Egg Yolk Agar

Composition per 1080.0mL:

Liver veal agar	1.0L
Egg yolk emulsion, 50%	80.0mL

pH 7.3 ± 0.2 at 25°C

Liver Veal Agar:

Composition per liter:

Veal, infusion from	500.0g
Beef liver, infusion from	50.0g
Gelatin	20.0g
Proteose peptone	20.0g
Agar	15.0g
Soluble starch	10.0g
Glucose	5.0g
NaCl	5.0g
Casein	2.0g
$NaNO_3$	2.0g
Enzymatic digest of protein	1.3g
Pancreatic digest of casein	1.3g

Source: This medium is available as a premixed powder from BD Diagnostic Systems.

Preparation of Liver Veal Agar: Add components to distilled/deionized water and bring volume to 1.0L. Mix thoroughly. Gently heat and bring to boiling. Distribute into tubes or flasks. Make sure that some liver and veal particles are transferred to each tube. Autoclave for 15 min at 15 psi pressure–121°C. Cool to 50°C.

Egg Yolk Emulsion, 50%:

Composition per 100.0mL:

Chicken egg yolks	11
Whole chicken egg	1
NaCl (0.9% solution)	50.0mL

Preparation of Egg Yolk Emulsion, 50%: Soak eggs with 1:100 dilution of saturated mercuric chloride solution for 1 min. Crack eggs and separate yolks from whites. Mix egg yolks with 1 chicken egg. Beat to form emulsion. Measure 50.0mL of egg yolk emulsion and add to 50.0mL of 0.9% NaCl solution. Mix thoroughly. Filter sterilize. Warm to 45°–50°C.

Preparation of Medium: To 1.0L of cooled sterile liver veal agar, aseptically add 80.0mL of sterile egg yolk emulsion, 50%. Mix thoroughly. Pour into sterile Petri dishes. Dry plates at 35°C for 24h.

Use: For the cultivation of a variety of anaerobic organisms.

LMX Broth Modified, Fluorocult (Fluorocult LMX Broth, Modified)

Composition per liter:

Tryptose	5.0g
NaCl	5.0g
Sorbitol	1.0g
Tryptophan	1.0g
K_2HPO_4	2.7g
KH_2PO_4	2.0g
Lauryl sulfate sodium salt	0.1g
5-Bromo-4-chloro-3-indolyl-β-D-galactopyran_oside (X-GAL)	0.08
4-Methylumbelliferyl-β-D-glucuronide	0.05g
1-Isopropyl-β-D-1-thio-galactopyranoside IPTG)	0.1

pH: 6.8 ± 0.2 at 25°C

Source: This medium is available from Merck.

Preparation of Medium: Add components to distilled/deionized water and bring volume to 1.0L. Mix thoroughly. Distribute into into test tubes Autoclave for 15 min at 15 psi pressure–121°C. The broth is clear and yellowish brown.

Use: For the simultaneous detection of total coliforms and *E. coli* by the fluorogenic procedure. A color change of the broth from yellow to blue-green indicates the presence of coliforms. A blue fluorescence under long-wave UV light permits the rapid detection of *E.coli*. As tryptophan is added to the broth, the indole reaction is easily done by adding KOVACS reagent. The formation of a red ring additionally confirms the presence of *E.coli*.

Loeffler Blood Serum Medium

Composition per liter:

Beef blood serum	750.0mL
Dextrose broth	250.0mL

pH 7.1 ± 0.2 at 25°C

Source: This medium is available as a premixed powder from BD Diagnostic Systems.

Dextrose Broth:

Composition per liter:

Tryptose 10.0g
Glucose 5.0g
Sodium chloride 5.0g
Beef extract 3.0g

Preparation of Dextrose Broth: Add components to distilled/deionized water and bring volume to 1.0L. Mix thoroughly.

Preparation of Medium: Combine 750.0mL of beef blood serum with 250.0mL of dextrose broth. Mix thoroughly. Distribute into screw-capped tubes. Slant tubes in the autoclave. Close the autoclave door loosely. Autoclave for 10 min at 0 psi pressure–100°C. Close the autoclave door tightly. Autoclave for 15 min at 15 psi pressure–121°C.

Use: For the cultivation of *Corynebacterium diphtheriae*. For demonstration of pigment production and proteolysis by *Corynebacterium diphtheriae*.

Loeffler Blood Serum Medium

Composition per liter:

Beef blood serum 750.0mL
Dextrose broth 250.0mL

pH 7.1 ± 0.2 at 25°C

Dextrose Broth:

Composition per liter:

Enzymatic digest of protein 2.5g
Glucose 1.25g
NaCl 1.25g
Beef extract 0.75g

Preparation of Dextrose Broth: Add components to distilled/deionized water and bring volume to 1.0L. Mix thoroughly.

Preparation of Medium: Combine 750.0mL of beef blood serum with 250.0mL of dextrose broth. Mix thoroughly. Distribute into screw-capped tubes. Slant tubes in the autoclave. Close the autoclave door loosely. Autoclave for 10 min at 0 psi pressure–100°C. Close the autoclave door tightly. Autoclave for 15 min at 15 psi pressure–121°C.

Use: For the cultivation of *Corynebacterium diphtheriae*. For demonstration of pigment production and proteolysis by *Corynebacterium diphtheriae*.

Loeffler Medium

Composition per liter:

Beef serum 70.0g
Egg, dried 7.5g
Heart muscle, solids from infusion 0.72g
Glucose 0.71g
Peptic digest of animal tissue 0.71g
NaCl 0.36g

pH 7.6 ± 0.2 at 25°C

Source: This medium is available as a premixed powder from BD Diagnostic Systems.

Preparation of Medium: Add components to distilled/deionized water and bring volume to 1.0L. Mix thoroughly. Distribute into screw-capped tubes. Slant tubes in the autoclave. Close the autoclave door loosely. Autoclave for 10 min at 0 psi pressure–100°C. Close the autoclave door tightly. Autoclave for 15 min at 15 psi pressure–121°C.

Use: For the cultivation of *Corynebacterium diphtheriae*. For demonstration of pigment production and proteolysis by *Corynebacterium diphtheriae*. For the cultivation and maintenance of *Moraxella lacunata*.

Loeffler Slant

Composition per liter:

Tryptose 5.0g
Glucose 1.0g
Beef serum 750.0mL

Preparation of Medium: Add components to distilled/deionized water and bring volume to 1.0L. Mix thoroughly. Distribute into screw-capped tubes. Slant tubes in the autoclave. Close the autoclave door loosely. Autoclave for 10 min at 0 psi pressure–100°C. Close the autoclave door tightly. Autoclave for 15 min at 15 psi pressure–121°C.

Use: For the cultivation of *Corynebacterium diphtheriae*. For demonstration of pigment production and proteolysis by *Corynebacterium diphtheriae*. For the cultivation and maintenance of *Moraxella lacunata*.

Loeffler Slant, Modified

Composition per liter:

Glucose 1.0g
Peptone 0.5g
Beef serum 300.0mL

pH 7.6 ± 0.2 at 25°C

Preparation of Medium: Add peptone and glucose to distilled/deionized water and bring volume to 100.0mL. Mix thoroughly. Add beef serum. Mix thoroughly. Adjust pH to 7.6. Distribute into screw-capped tubes in 3.0mL volumes. Slant tubes in the autoclave. Autoclave for 30 min at 0 psi pressure–100°C.

Use: For the cultivation of *Corynebacterium diphtheriae*. For demonstration of pigment production and proteolysis by *Corynebacterium diphtheriae*. For the cultivation and maintenance of *Moraxella lacunata*.

Lombard–Dowell Agar (LD Agar)

Agar	20.0g
Pancreatic digest of casein	5.0g
Yeast extract	5.0g
NaCl	2.5g
L-Cystine	0.4g
L-Tryptophan	0.2g
Na_2SO_3	0.1g
Hemin	10.0mg
NaOH (1*N*)	5.0mL
Vitamin K_1 solution	1.0mL

pH 7.5 ± 0.2 at 25°C

Vitamin K_1 Solution:

Composition per 100.0mL:

Vitamin K_1	1.0g
Ethanol	99.0mL

Preparation of Vitamin K_1 Solution: Add vitamin K_1 to 99.0mL of absolute ethanol. Mix thoroughly.

Preparation of Medium: Add hemin and L-cystine to 5.0mL of NaOH. Mix thoroughly. Add remaining components. Bring volume to 1.0L with distilled/deionized water. Mix thoroughly. Gently heat and bring to boiling. Distribute into tubes or flasks. Autoclave for 15 min at 15 psi pressure–121°C. Pour into sterile Petri dishes.

Use: For the cultivation and identification of a variety of obligate anaerobic bacteria. For the cultivation of *Bacteroides species, Fusobacterium species, Clostridium* species, and nonspore-forming Gram-positive anaerobes.

Lombard-Dowell Bile Agar (LD Bile Agar)

Composition per liter:

Agar	20.0g
Oxgall	20.0g
Pancreatic digest of casein	5.0g
Yeast extract	5.0g
NaCl	2.5g
D-Glucose	1.0g
L-Cystine	0.4g
L-Tryptophan	0.2g
Na_2SO_3	0.1g
Hemin	10.0mg
NaOH (1*N*)	5.0mL
Vitamin K_1 solution	1.0mL

pH 7.5 ± 0.2 at 25°C

Vitamin K_1 Solution:

Composition per 100.0mL:

Vitamin K_1	1.0g
Ethanol	99.0mL

Preparation of Vitamin K_1 Solution: Add vitamin K_1 to 99.0mL of absolute ethanol. Mix thoroughly.

Preparation of Medium: Add hemin and L-cystine to 5.0mL of NaOH. Mix thoroughly. Add remaining components. Bring volume to 1.0L with distilled/deionized water. Mix thoroughly. Gently heat and bring to boiling. Distribute into tubes or flasks. Autoclave for 15 min at 15 psi pressure–121°C. Pour into sterile Petri dishes.

Use: For the cultivation and identification of a variety of obligate anaerobic bacteria in the presence of 20% bile.

Lombard-Dowell Broth (LD Broth)

Composition per liter:

Pancreatic digest of casein	5.0g
Yeast extract	5.0g
NaCl	2.5g
Agar	0.7g
L-Tryptophan	0.2g
Na_2SO_3	0.1g
NaOH (1*N*)	5.0mL
Hemin solution	1.0mL
Vitamin K_1 solution	1.0mL

pH 7.5 ± 0.2 at 25°C

Hemin Solution:

Composition per 100.0mL:

Hemin	1.0g
NaOH (1*N* solution)	20.0mL

Preparation of Hemin Solution: Add hemin to 20.0mL of 1*N* NaOH solution. Mix thoroughly. Bring volume to 100.0mL with distilled/deionized water.

Vitamin K_1 Solution:

Composition per 100.0mL:

Vitamin K_1	1.0g
Ethanol	99.0mL

Preparation of Vitamin K_1 Solution: Add vitamin K_1 to 99.0mL of absolute ethanol. Mix thoroughly.

Preparation of Medium: Add tryptophan to 5.0mL of NaOH. Mix thoroughly. Add remaining components. Bring volume to 1.0L with distilled/deionized water. Mix thoroughly. Gently heat and bring to boiling. Adjust pH to 7.5. Distribute into screw-capped tubes in 7.0mL volumes. Autoclave for 15 min at 15 psi pressure–121°C. Cool tubes, with caps loose, under 85% N_2 + 10% H_2 + 5% CO_2. Tighten caps.

Use: For the cultivation of a wide variety of anaerobic bacteria.

Lombard-Dowell Egg Yolk Agar (LD Egg Yolk Agar) (Egg Yolk Agar, Lombard-Dowell)

Composition per 9100.0mL:

$Na_2HPO_4 \cdot 12H_2O$	5.0g
Glucose	2.0g
LD Agar	9000.0mL
Egg yolk emulsion	100.0mL
$MgSO_4 \cdot 7H_2O$ (5% solution)	0.2mL

pH 7.5 ± 0.2 at 25°C

LD Agar:

Composition per liter:

Agar	20.0g
Pancreatic digest of casein	5.0g
Yeast extract	5.0g
NaCl	2.5g
L-Cystine	0.4g
L-Tryptophan	0.2g
Na_2SO_3	0.1g
Hemin	10.0mg
NaOH (1*N* NaOH)	5.0mL
Vitamin K_1 solution	1.0mL

Preparation of LD Agar: Add hemin and L-cystine to 5.0mL of NaOH. Mix thoroughly. Add remaining components. Mix thoroughly. Gently heat and bring to boiling.

Vitamin K_1 Solution:

Composition per 100.0mL:

Vitamin K_1	1.0g
Ethanol	99.0mL

Preparation of Vitamin K_1 Solution: Add vitamin K_1 to 99.0mL of absolute ethanol. Mix thoroughly.

Egg Yolk Emulsion:

Composition:

Chicken egg yolks	11
Whole chicken egg	1

Preparation of Egg Yolk Emulsion: Soak eggs with 1:100 dilution of saturated mercuric chloride solution for 1 min. Crack eggs and separate yolks from whites. Mix egg yolks with 1 chicken egg.

Preparation of Medium: Combine components, except egg yolk emulsion. Mix thoroughly. Autoclave for 15 min at 15 psi pressure–121°C. Cool to 45°–50°C. Aseptically add 100.0mL of egg yolk emulsion. Mix thoroughly. Pour into sterile Petri dishes.

Use: For the cultivation of a wide variety of anaerobic bacteria. For the differentiation of anaerobic bacteria based on lecithinase production, lipase production, and proteolytic ability. Bacteria that produce lecithinase appear as colonies surrounded by a zone of insoluble precipitate. Bacteria that produce lipase appear as colonies with a pearly iridescent sheen. Bacteria that produce proteolytic activity appear as colonies surrounded by a clear zone.

Lombard-Dowell Esculin Agar (LD Esculin Agar) (Esculin Agar, Lombard-Dowell)

Composition per liter:

Agar	20.0g
Pancreatic digest of casein	5.0g
Yeast extract	5.0g
NaCl	2.5g
Esculin	1.0g
Ferric citrate	0.5g
L-Cystine	0.4g
L-Tryptophan	0.2g
Hemin	10.0mg
NaOH (1*N*)	5.0mL
Vitamin K_1 solution	1.0mL

pH 7.5 ± 0.2 at 25°C

Vitamin K_1 Solution:

Composition per 100.0mL:

Vitamin K_1	1.0g
Ethanol	99.0mL

Preparation of Vitamin K_1 Solution: Add vitamin K_1 to 99.0mL of absolute ethanol. Mix thoroughly.

Preparation of Medium: Add hemin and L-cystine to 5.0mL of NaOH. Mix thoroughly. Add remaining components. Bring volume to 1.0L with distilled/deionized water. Mix thoroughly. Gently heat and bring to boiling. Distribute into tubes or flasks. Autoclave for 15 min at 15 psi pressure–121°C. Pour into sterile Petri dishes.

Use: For the cultivation of a wide variety of anaerobic bacteria. For the differentiation of anaerobic bacteria based on esculin hydrolysis, H_2S production, and catalase production. Bacteria that hydrolyze esculin appear as colonies surrounded by a red-brown to dark brown zone. Bacteria that produce H_2S appear as black colonies.

Lombard-Dowell Gelatin Agar (LD Gelatin Agar)

Composition per liter:

Agar	20.0g

Pancreatic digest of casein	5.0g
Yeast extract	5.0g
Gelatin	4.0g
NaCl	2.5g
Glucose	1.0g
L-Cystine	0.4g
L-Tryptophan	0.2g
Na_2SO_3	0.1g
Hemin	10.0mg
NaOH (1*N*)	5.0mL
Vitamin K_1 solution	1.0mL

pH 7.5 ± 0.2 at 25°C

Vitamin K_1 Solution:
Composition per 100.0mL:

Vitamin K_1	1.0g
Ethanol	99.0mL

Preparation of Vitamin K_1 Solution: Add vitamin K_1 to 99.0mL of absolute ethanol. Mix thoroughly.

Preparation of Medium: Add hemin and L-cystine to 5.0mL of NaOH. Mix thoroughly. Add remaining components, except agar and gelatin. Bring volume to 750.0mL with distilled/deionized water. Mix thoroughly. Gently heat and bring to boiling. In a separate flask, add gelatin to 100.0mL of cold distilled/deionized water. Gently heat and bring to 70°C. Add gelatin solution to the 750.0mL of basal medium. Mix thoroughly. Add agar. Bring volume to 1.0L with distilled/deionized water. Autoclave for 15 min at 15 psi pressure–121°C. Pour into sterile Petri dishes.

Use: For the cultivation of a wide variety of anaerobic bacteria. For the differentiation of anaerobic bacteria based on gelatinase production. After incubation of plates, gelatinase activity is determined by the addition of Frazier's reagent. Bacteria that hydrolyze gelatin appear as colonies surrounded by a clear zone.

Lombard-Dowell Neomycin Agar (Egg Yolk Agar with Neomycin)

Composition per 9100.0mL:

$Na_2HPO_4 \cdot 12H_2O$	5.0g
Glucose	2.0g
Neomycin sulfate	0.1g
LD Agar	9000.0mL
Egg yolk emulsion	100.0mL
$MgSO_4 \cdot 7H_20$ (5% solution)	0.2mL

pH 7.5 ± 0.2 at 25°C

LD Agar:
Composition per liter:

Agar	20.0g
Pancreatic digest of casein	5.0g
Yeast extract	5.0g
NaCl	2.5g
L-Cystine	0.4g
L-Tryptophan	0.2g
Na_2SO_3	0.1g
Hemin	10.0mg
NaOH (1*N* NaOH)	5.0mL
Vitamin K_1 solution	1.0mL

Preparation of LD Agar: Add hemin and L-cystine to 5.0mL of NaOH. Mix thoroughly. Add remaining components. Mix thoroughly. Gently heat and bring to boiling.

Vitamin K_1 Solution:
Composition per 100.0mL:

Vitamin K_1	1.0g
Ethanol	99.0mL

Preparation of Vitamin K_1 Solution: Add vitamin K_1 to 99.0mL of absolute ethanol. Mix thoroughly.

Egg Yolk Emulsion:
Composition:

Chicken egg yolks	11
Whole chicken egg	1

Preparation of Egg Yolk Emulsion: Soak eggs with 1:100 dilution of saturated mercuric chloride solution for 1 min. Crack eggs and separate yolks from whites. Mix egg yolks with 1 chicken egg.

Preparation of Medium: Combine components, except egg yolk emulsion and neomycin sulfate. Mix thoroughly. Autoclave for 15 min at 15 psi pressure–121°C. Cool to 45°–50°C. Aseptically add 100.0mL of egg yolk emulsion and neomycin sulfate. Mix thoroughly. Pour into sterile Petri dishes.

Use: For the selective cultivation of a wide variety of anaerobic bacteria. For the differentiation of anaerobic bacteria based on lecithinase production, lipase production, and proteolytic ability. Bacteria that produce lecithinase appear as colonies surrounded by a zone of insoluble precipitate. Bacteria that produce lipase appear as colonies with a pearly iridescent sheen. Bacteria that produce proteolytic activity appear as colonies surrounded by a clear zone.

Long-Term Preservation Medium

Composition per liter:

NaCl	30.0g
Peptone	10.0g
Agar	3.0g
Yeast extract	3.0g

Preparation of Medium: Add components to distilled/deionized water and bring volume to 1.0L. Mix thoroughly. Gently heat and bring to boiling. Distribute into screw-capped tubes in 4.0mL volumes. Autoclave for 15 min at 15 psi pressure–121°C.

Use: For the cultivation and maintenance of a wide variety of bacteria.

Low Iron YC Agar

Composition per 1033.0mL:

Solution 1	1.0L
Solution 4	30.0mL
Solution 2	2.0mL
Solution 3	1.0mL

pH 7.4 ± 0.2 at 25°C

Solution 1:

Composition per liter:

Yeast extract	20.0g
Noble agar	10.0g
Casamino acids	10.0g
KH_2PO_4	5.0g
$CaCl_2$	1.0g
Tryptophan	0.05g

Preparation of Solution 1: Add components to distilled/deionized water and bring volume to 1.0L. Mix thoroughly. Adjust pH to 7.4. Gently heat and bring to boiling. Filter through no. 40 ashless filter paper.

Solution 2:

Composition per 100.0mL:

$MgSO_4·7H_2O$	22.5g
$CuSO_4·5H_2O$	0.5g
$ZnSO_4·5H_2O$	0.2g
β-Alanine	0.115g
Nicotinic acid	0.115g
$MnCl_2·4H_2O$	0.075g
Pimelic acid	7.5mg
HCl, concentrated	3.0mL

Preparation of Solution 2: Add components to distilled/deionized water and bring volume to 100.0mL. Mix thoroughly.

Solution 3:

Composition per 100.0mL:

L-Cystine	20.0g
HCl, concentrated	20.0mL

Preparation of Solution 3: Add components to distilled/deionized water and bring volume to 100.0mL. Mix thoroughly.

Solution 4:

Composition per 100.0mL:

Maltose	50.0g
$CaCl_2·2H_2O$	0.5g

Preparation of Solution 4: Add components to distilled/deionized water and bring volume to 100.0mL. Mix thoroughly. Autoclave for 10 min at 11 psi pressure–116°C. Cool to 45°–50°C.

Preparation of Medium: To 1.0L of solution 1, add 2.0mL of solution 2 and 1.0mL of solution 3. Mix thoroughly. Autoclave for 15 min at 15 psi pressure–121°C. Cool to 45°–50°C. Aseptically add 30.0mL of sterile solution 4. Mix thoroughly. Pour into sterile Petri dishes or distribute into sterile tubes.

Use: For the cultivation and maintenance of *Corynebacterium diphtheriae.*

Low Iron YC Broth

Composition per 1033.0mL:

Solution 1	1.0L
Solution 4	30.0mL
Solution 2	2.0mL
Solution 3	1.0mL

pH 7.4 ± 0.2 at 25°C

Solution 1:

Composition per liter:

Yeast extract	20.0g
Casamino acids	10.0g
KH_2PO_4	5.0g
$CaCl_2·2H_2O$	1.0g
Tryptophan	0.05g

Preparation of Solution 1: Add components to distilled/deionized water and bring volume to 1.0L. Mix thoroughly. Adjust pH to 7.4. Gently heat and bring to boiling. Filter through #40 ashless filter paper.

Solution 2:

Composition per 100.0mL:

$MgSO_4·7H_2O$	22.5g
$CuSO_4·5H_2O$	0.5g
$ZnSO_4·5H_2O$	0.2g
β-Alanine	0.115g
Nicotinic acid	0.115g
$MnCl_2·4H_2O$	0.075g
Pimelic acid	7.5mg
HCl, concentrated	3.0mL

Preparation of Solution 2: Add components to distilled/deionized water and bring volume to 100.0mL. Mix thoroughly.

Solution 3:

Composition per 100.0mL:

L-Cystine	20.0g
HCl, concentrated	20.0mL

Preparation of Solution 3: Add components to distilled/deionized water and bring volume to 100.0mL. Mix thoroughly.

Solution 4:

Composition per 100.0mL:

Maltose	50.0g
$CaCl_2 \cdot 2H_2O$	0.5g

Preparation of Solution 4: Add components to distilled/deionized water and bring volume to 100.0mL. Mix thoroughly. Autoclave for 10 min at 11 psi pressure–116°C. Cool to 25°C.

Preparation of Medium: To 1.0L of solution 1, add 2.0mL of solution 2 and 1.0mL of solution 3. Mix thoroughly. Autoclave for 15 min at 15 psi pressure–121°C. Cool to 25°C. Aseptically add 30.0mL of sterile solution 4. Mix thoroughly. Aseptically distribute into sterile tubes or flasks.

Use: For the cultivation and maintenance of *Corynebacterium diphtheriae.*

Lowenstein-Gruft Medium

Composition per 1600.0mL:

Potato starch	30.0g
Asparagine	3.6g
KH_2PO_4	2.4g
Magnesium citrate	0.6g
Malachite Green	0.4g
$MgSO_4 \cdot 7H_2O$	0.24g
Nalidixic acid	0.056g
Ribonucleic acid	0.08mg
Homogenized whole egg	1.0L
Glycerol	12.0mL
Penicillin	80,000U

Homogenized Whole Egg:

Composition per liter:

Whole eggs	18–24

Preparation of Homogenized Whole Egg: Use fresh eggs, less than 1 week old. Scrub the shells with soap. Let stand in a soap solution for 30 min. Rinse in running water. Soak eggs in 70% ethanol for 15 min. Break the eggs into a sterile container. Homogenize by shaking. Filter through four layers of sterile cheesecloth into a sterile graduated cylinder. Measure out 1.0L.

Preparation of Medium: Add glycerol to 600.0mL of distilled/deionized water. Mix thoroughly. Add remaining components, except fresh egg mixture. Mix thoroughly. Gently heat while stirring and bring to boiling. Autoclave for 15 min at 15 psi pressure–121°C. Cool to 50°C. Aseptically add 1.0L of homogenized whole egg. Mix thoroughly. Distribute into sterile screw-capped tubes. Place tubes in a slanted position. Inspissate at 85°C (moist heat) for 45 min.

Use: For the cultivation and differentiation of *Mycobacterium* species. *Mycobacterium tuberculosis* appears as granular, rough, dry colonies. *Mycobacterium kansasii* appears as smooth to rough photochromogenic colonies. *Mycobacterium gordonae* appears as smooth yellow-orange colonies. *Mycobacterium avium* appears as smooth, colorless colonies. *Mycobacterium smegmatis* appears as wrinkled, creamy white colonies.

Lowenstein-Jensen Medium

Composition per 1600.0mL:

Potato starch	30.0g
Asparagine	3.6g
KH_2PO_4	2.4g
Magnesium citrate	0.6g
Malachite Green	0.4g
$MgSO_4 \cdot 7H_2O$	0.24g
Homogenized whole egg	1.0L
Glycerol	12.0mL

Source: This medium is available as a prepared medium from BD Diagnostic Systems and Oxoid Unipath.

Homogenized Whole Egg:

Composition per liter:

Whole eggs	18-24

Preparation of Homogenized Whole Egg: Use fresh eggs, less than 1 week old. Scrub the shells with soap. Let stand in a soap solution for 30 min. Rinse in running water. Soak eggs in 70% ethanol for 15 min. Break the eggs into a sterile container. Homogenize by shaking. Filter through four layers of sterile cheesecloth into a sterile graduated cylinder. Measure out 1.0L.

Preparation of Medium: Add glycerol to 600.0mL of distilled/deionized water. Mix thoroughly. Add remaining components, except fresh egg mixture. Mix thoroughly. Gently heat while stirring and bring to boiling. Autoclave for 15 min at 15 psi pressure–121°C. Cool to 50°C. Aseptically add 1.0L of homogenized whole egg. Mix thoroughly. Distribute into sterile screw-capped tubes. Place tubes in a slanted position. Inspissate at 85°C (moist heat) for 45 min.

Use: For the cultivation and differentiation of *Mycobacterium* species. *Mycobacterium tuberculosis* appears as granular, rough, dry colonies. *Mycobacterium kansasii* appears as smooth to rough photochromogenic colonies. *Mycobacterium gordonae* appears as smooth yellow-orange colonies. *Mycobacterium avium* appears as smooth, colorless colonies. *Mycobacterium smegmatis* appears as wrinkled, creamy white colonies. Also used for the cultivation and maintenance of *Gordona* species, *Nocardia* species, *Rhodococcus* species, and *Tsukamurella paurometabolum.*

Lowenstein-Jensen Medium without Glycerol

Composition per 1600mL:

Potato starch	30.0g
Asparagine	3.6g
KH_2PO_4	2.4g
Magnesium citrate	0.6g
Malachite green	0.4g
$MgSO_4 \cdot 7H_2O$	0.24g
Homogenized whole egg	1.0L

Homogenized Whole Egg:

Composition per liter:

Whole eggs	18–24

Preparation of Homogenized Whole Egg: Use fresh eggs, less than 1 week old. Scrub the shells with soap. Let stand in a soap solution for 30 min. Rinse in running water. Soak eggs in 70% ethanol for 15 min. Break the eggs into a sterile container. Homogenize by shaking. Filter through four layers of sterile cheesecloth into a sterile graduated cylinder. Measure out 1.0L.

Preparation of Medium: Add components, except fresh egg mixture, to 600.0mL of distilled/deionized water. Mix thoroughly. Gently heat while stirring and bring to boiling. Autoclave for 15 min at 15 psi pressure–121°C. Cool to 50°C. Aseptically add 1.0L of homogenized whole egg. Mix thoroughly. Distribute into sterile screw-capped tubes. Place tubes in a slanted position. Inspissate at 85°C (moist heat) for 45 min.

Use: For the cultivation and maintenance of *Mycobacterium* species, especially *Mycobacterium bovis* and other species that are sensitive to glycerol.

Lowenstein-Jensen Medium with Streptomycin

Composition per 1610.0mL:

Potato starch	30.0g
Asparagine	3.6g
KH_2PO_4	2.4g
Magnesium citrate	0.6g
Malachite Green	0.4g
$MgSO_4 \cdot 7H_2O$	0.24g
Homogenized whole egg	1.0L
Glycerol	12.0mL
Streptomycin solution	10.0mL

Homogenized Whole Egg:

Composition per liter:

Whole eggs	18-24

Preparation of Homogenized Whole Egg: Use fresh eggs, less than 1 week old. Scrub the shells with soap. Let stand in a soap solution for 30 min. Rinse in running water. Soak eggs in 70% ethanol for 15 min. Break the eggs into a sterile container. Homogenize by shaking. Filter through four layers of sterile cheesecloth into a sterile graduated cylinder. Measure out 1.0L.

Streptomycin Solution:

Composition per 10.0mL:

Streptomycin	0.1mg

Preparation of Streptomycin Solution: Add streptomycin to distilled/deionized water and bring volume to 10.0mL. Mix thoroughly. Filter sterilize.

Preparation of Medium: Add glycerol to 600.0mL of distilled/deionized water. Mix thoroughly. Add remaining components, except fresh egg mixture and streptomycin solution. Mix thoroughly. Gently heat while stirring and bring to boiling. Autoclave for 15 min at 15 psi pressure–121°C. Cool to 50°C. Aseptically add 1.0L of homogenized whole egg and 10.0mL of sterile streptomycin solution. Mix thoroughly. Distribute into sterile screw-capped tubes. Place tubes in a slanted position. Inspissate at 85°C (moist heat) for 45 min.

Use: For the cultivation and differentiation of *Mycobacterium* species. *Mycobacterium tuberculosis* appears as granular, rough, dry colonies. *Mycobacterium kansasii* appears as smooth to rough photochromogenic colonies. *Mycobacterium gordonae* appears as smooth yellow-orange colonies. *Mycobacterium avium* appears as smooth, colorless colonies. *Mycobacterium smegmatis* appears as wrinkled, creamy white colonies. Also used for the cultivation and maintenance of *Gordona* species, *Nocardia* species, *Rhodococcus* species, and *Tsukamurella paurometabolum.*

LPM Agar (Lithium Chloride Phenylethanol Moxalactam Plating Agar)

Composition per liter:

Agar	15.0g
Glycine anhydride	10.0g
LiCl	5.0g
NaCl	5.0g
Pancreatic digest of casein	5.0g
Peptic digest of animal tissue	5.0g
Beef extract	3.0g
Phenylethyl alcohol	2.5g
Moxalactam solution	2.0mL

pH 7.3 ± 0.2 at 25°C

Source: This medium is available as a premixed powder from BD Diagnostic Systems.

Moxalactam Solution:

Composition per 10.0mL:

Moxalactam	0.1g

Preparation of Moxalactam Solution: Add moxalactam to distilled/deionized water and bring volume to 10.0mL. Mix thoroughly. Filter sterilize.

Preparation of Medium: Add components, except moxalactam solution, to distilled/deionized water and bring volume to 998.0mL. Mix thoroughly. Gently heat while stirring and bring to boiling. Autoclave for 12 min at 15 psi pressure–121°C. Cool to 45°–50°C. Aseptically add 2.0mL of sterile moxalactam solution. Mix thoroughly. Pour into sterile Petri dishes or distribute into sterile tubes.

Use: For the isolation and cultivation of *Listeria monocytogenes.*

LPM Agar with Esculin and Ferric Iron

Composition per liter:

Agar	15.0g
Glycine anhydride	10.0g
LiCl	5.0g
NaCl	5.0g
Pancreatic digest of casein	5.0g
Peptic digest of animal tissue	5.0g
Beef extract	3.0g
Phenylethyl alcohol	2.5g
Esculin	1.0g
Ferric ammonium citrate	0.5g
Moxalactam solution	2.0mL

pH 7.3 ± 0.2 at 25°C

Moxalactam Solution:

Composition per 10.0mL:

Moxalactam	0.1g

Preparation of Moxalactam Solution: Add moxalactam to distilled/deionized water and bring volume to 10.0mL. Mix thoroughly. Filter sterilize.

Preparation of Medium: Add components, except moxalactam solution, to distilled/deionized water and bring volume to 998.0mL. Mix thoroughly. Gently heat while stirring and bring to boiling. Autoclave for 12 min at 15 psi pressure–121°C. Cool to 45°–50°C. Aseptically add 2.0mL of sterile moxalactam solution. Mix thoroughly. Pour into sterile Petri dishes or distribute into sterile tubes.

Use: For the isolation and cultivation of *Listeria monocytogenes.*

LST-MUG Broth

Composition per liter:

Tryptose	20.0g
Lactose	5.0g
NaCl	5.0g
K_2HPO_4	2.75g
KH_2PO_4	2.75g
L-Tryptophan	1.0g
Sodium lauryl sulfate	0.1g
4-Methylumbelliferyl-β-D-glucuronide	0.1g

pH 6.8 ± 0.2 at 37°C

Source: This medium is available from Fluka, Sigma-Aldrich.

Preparation of Medium: Add components to distilled/deionized water and bring volume to 1.0L. Mix thoroughly. Distribute into test tubes that contain an inverted Durham tube in 10.0mL volumes. Autoclave for 15 min at 15 psi pressure–121°C.

Use: For the detection of *E. coli* by the fluorgenic method. The presence of *E. coli* results in fluorescence in the UV. A positive indole test provide confirmation. β-D-glucoronidase, which is produced by *E. coli*, cleaves 4-methylumbelliferyl-β-D-glucuronide to 4-methylumbelliferone and glucuronide. The fluorogen 4-methylumbelliferone can be detected under a long wavelength UV lamp.

Lysine Arginine Iron Agar

Composition per liter:

Agar	15.0g
L-Arginine	10.0g
L-Lysine	10.0g
Peptone	5.0g
Yeast extract	3.0g
Glucose	1.0g
Ferric ammonium citrate	0.5g
Sodium thiosulfate	0.04g
Bromcresol Purple	0.02g

pH 6.8 ± 0.2 at 25°C

Preparation of Medium: Add components to distilled/deionized water and bring volume to 1.0L. Mix thoroughly. Gently heat and bring to boiling. Adjust pH to 6.8. Distribute into screw-capped tubes in 5.0mL volumes. Autoclave for 12 min at 15 psi pressure–121°C. Allow tubes to cool in a slanted position.

Use: For the cultivation and differentiation of bacteria based on their ability to decarboxylate lysine, decarboxylate arginine, and produce H_2S. Bacteria that decarboxylate lysine or arginine turn the medium purple. Bacteria that produce H_2S appear as black colonies.

Lysine Broth Falkow with NaCl

Composition per liter:

L-Lysine	5.0g
Peptone or gelysate	5.0g
Yeast extract	3.0g
Glucose	1.0g
Bromcresol Purple	0.02g

pH 6.5 ± 0.2 at 25°C

Preparation of Medium: Add components to distilled/deionized water and bring volume to 1.0L. Mix thoroughly. Adjust pH so that it will be 6.5 ± 0.2 after sterilization. Distribute into 16 × 150mm screw-capped tubes in 5.0mL volumes. Autoclave medium with loosely capped tubes for 10 min at 15 psi pressure–121°C. Screw the caps on tightly for storage and after inoculation.

Use: For the cultivation and differentiation of *Vibrio* spp. based on their ability to decarboxylate the amino acid lysine. Bacteria that decarboxylate lysine turn the medium turbid purple.

Lysine Decarboxylase Broth, Falkow

Composition per liter:

Peptone	5.0g
L-Lysine	5.0g
Yeast extract	3.0g
Glucose	1.0g
Bromcresol Purple	0.02g

pH 6.5–6.8 at 25°C

Preparation of Medium: Add components to distilled/deionized water and bring volume to 1.0L. Mix thoroughly. Gently heat and bring to boiling. Adjust pH to 6.5–6.8. Distribute into tubes in 5.0mL volumes. Autoclave for 15 min at 15 psi pressure–121°C.

Use: For the cultivation and differentiation of bacteria, especially *Salmonella*, based on their ability to decarboxylate lysine. Bacteria that decarboxylate lysine turn the medium turbid purple.

Lysine Decarboxylase Broth, Taylor Modification

Composition per liter:

L-Lysine	5.0g
Yeast extract	3.0g
Glucose	1.0g
Bromcresol Purple	0.02g

pH 6.1 ± 0.2 at 25°C

Source: This medium is available as a premixed powder from Oxoid Unipath.

Preparation of Medium: Add components to distilled/deionized water and bring volume to 1.0L. Mix thoroughly. Gently heat and bring to boiling. Adjust pH to 6.1. Distribute into tubes in 5.0mL volumes. Autoclave for 15 min at 15 psi pressure–121°C.

Use: For the cultivation and differentiation of bacteria, especially *Salmonella*, based on their ability to decarboxylate lysine. Bacteria that decarboxylate lysine turn the medium turbid purple.

Lysine Decarboxylase Broth, Taylor Modification (Lysine Decarboxylase Broth)

Composition per liter:

L-Lysine	5.0g
Peptone	5.0g
Yeast extract	3.0g
Glucose	1.0g
Bromcresol Purple	0.02g

pH 6.8 ± 0.2 at 25°C

Source: This medium is available as a premixed powder from BD Diagnostic Systems.

Preparation of Medium: Add components to distilled/deionized water and bring volume to 1.0L. Mix thoroughly. Gently heat and bring to boiling. Adjust pH to 6.1. Distribute into tubes in 5.0mL volumes. Autoclave for 15 min at 15 psi pressure–121°C.

Use: For the cultivation and differentiation of bacteria, especially *Salmonella*, based on their ability to decarboxylate lysine. Bacteria that decarboxylate lysine turn the medium turbid purple.

Lysine Decarboxylase Medium

Composition per liter:

Glucose	0.5g
KH_2PO_4	0.5g
L-Lysine·HCl	0.5g

pH 4.6 ± 0.2 at 25°C

Preparation of Medium: Add components to distilled/deionized water and bring volume to 1.0L. Mix thoroughly. Gently heat and bring to boiling. Adjust pH to 4.6. Autoclave for 15 min at 15 psi pressure–121°C. Aseptically distribute into sterile tubes in 1.0mL volumes.

Use: For the cultivation and differentiation of Gram-negative, nonfermentative bacteria based on their ability to decarboxylate lysine. Bacteria that decarboxylate lysine turn the medium turbid purple.

Lysine Iron Agar

Composition per liter:

Agar	13.5g

L-Lysine 10.0g
Pancreatic digest of gelatin 5.0g
Yeast extract 3.0g
Glucose 1.0g
Ferric ammonium citrate 0.5g
$Na_2S_2O_3 \cdot 5H_2O$ 0.04g
Bromcresol Purple 0.02g

pH 6.7 ± 0.2 at 25°C

Source: This medium is available as a premixed powder from BD Diagnostic Systems and Oxoid Unipath.

Preparation of Medium: Add components to distilled/deionized water and bring volume to 1.0L. Mix thoroughly. Gently heat while stirring and bring to boiling. Distribute into tubes in 10.0mL volumes. Autoclave for 12 min at 15 psi pressure–121°C. Allow tubes to cool in a slanted position.

Use: For the cultivation and differentiation of members of the Enterobacteriaceae based on their ability to decarboxylate lysine and to form H_2S. Bacteria that decarboxylate lysine turn the medium purple. Bacteria that produce H_2S appear as black colonies.

Lysine Ornithine Mannitol Agar (LOM Agar)

Composition per liter:

Agar 13.5g
L–Ornithine·HCl 6.5g
D–Mannitol 5.25g
L–Lysine·HCl 5.0g
NaCl 5.0g
Yeast extract 3.0g
Bromthymol Blue 0.3g
Vancomycin solution 10.0mL

pH 6.5 ± 0.2 at 25°C

Vancomycin Solution:

Composition per 10.0mL:

Vancomycin·HCl 0.03g

Preparation of Vancomycin Solution: Add vancomycin to distilled/deionized water and bring volume to 10.0mL. Mix thoroughly. Filter sterilize.

Preparation of Medium: Add components, except vancomycin solution, to distilled/deionized water and bring volume to 990.0mL. Mix thoroughly. Gently heat and bring to boiling. Autoclave for 15 min at 15 psi pressure–121°C. Cool to 45°–50°C. Aseptically add sterile vancomycin solution. Mix thoroughly. Pour into sterile Petri dishes or distribute into sterile tubes.

Use: For the cultivation and differentiation of Gram-negative bacilli based on their ability to decarboxylate lysine or ornithine and mannitol fermentation. Especially useful for the identification of *Enterobacter agglomerans*. Bacteria that ferment mannitol appear as dark yellow colonies. Bacteria that decarboxylate lysine or ornithine appear as green-yellow colonies.

M7 Medium

Composition per liter:

Yeast extract solution 200.0mL
Dialyzed fetal bovine serum 100.0mL
L-Methionine solution 30.0mL
Buffer solution 20.0mL
Glucose solution 20.0mL

Yeast Extract Solution:

Composition per liter:

Yeast extract 25.0g

Preparation of Yeast Extract Solution: Add yeast extract to distilled/deionized water and bring volume to 1.0L. Mix thoroughly. Autoclave for 15 min at 15 psi pressure–121°C. Cool to 25°C.

L-Methionine Solution:

Composition per liter:

L-Methionine 1.5g

Preparation of L-Methionine Solution: Add L-methionine to distilled/deionized water and bring volume to 1.0L. Mix thoroughly. Autoclave for 15 min at 15 psi pressure–121°C. Cool to 25°C.

Glucose Solution:

Composition per liter:

Glucose 270.0g

Preparation of Glucose Solution: Add glucose to distilled/deionized water and bring volume to 1.0L. Mix thoroughly. Autoclave for 15 min at 15 psi pressure–121°C. Cool to 25°C.

Buffer Solution:

Composition per liter:

Na_2HPO_4 .... 25.0g
KH_2PO_4 .... 18.1g

Preparation of Buffer Solution: Add components to distilled/deionized water and bring volume to 1.0L. Mix thoroughly. Autoclave for 15 min at 15 psi pressure–121°C. Cool to 25°C.

Dialyzed Fetal Bovine Serum:

Composition per 100.0mL:

Fetal bovine serum, heat inactivated 100.0mL

Preparation of Dialyzed Fetal Bovine Serum: Dialyze the heat-inactivated serum at 0°–4°C against 10 volumes of water. Clean the dialysis tubing before use by boiling in a solution of 0.37g/L of EDTA and rinsing with water. Change the water four times at

8–16 hr intervals. Centrifuge the dialyzed serum for 30 min at 35,000 x g. Filter sterilize serum fluid.

Preparation of Medium: Aseptically combine the sterile solutions. Mix thoroughly. Bring volume to 1.0L with sterile distilled/deionized water.

Use: For the cultivation of *Naegleria fowleri*, *Naegleria gruberi*, and *Nuclearia* species.

M9 Medium with Casamino Acids

Composition per liter:

Na_2HPO_4	6.0g
Casamino acids	5.0g
KH_2PO_4	3.0g
NH_4Cl	1.0g
NaCl	0.5g
Glucose solution	10.0mL
$MgSO_4 \cdot 7H_2O$ solution	1.0mL
Thiamine·HCl solution	1.0mL
$CaCl_2$ solution	1.0mL

pH 7.0 ± 0.2 at 25°C

Glucose solution:

Composition per 100.0mL:

D-Glucose	20.0g

Preparation of Glucose Solution: Add glucose to distilled/deionized water and bring volume to 1.0L. Mix thoroughly. Autoclave for 15 min at 15 psi pressure–121°C.

$MgSO_4 \cdot 7H_2O$ Solution:

Composition per liter:

$MgSO_4 \cdot 7H_2O$	246.5g

Preparation of $MgSO_4 \cdot 7H_2O$ Solution: Add $MgSO_4 \cdot 7H_2O$ to distilled/deionized water and bring volume to 1.0L. Mix thoroughly. Autoclave for 15 min at 15 psi pressure–121°C.

Thiamine·HCl Solution:

Composition per 10.0mL:

Thiamine·HCl	10.0mg

Preparation of Thiamine·HCl Solution: Add thiamine·HCl to distilled/deionized water and bring volume to 1.0L. Mix thoroughly. Filter sterilize.

$CaCl_2$ Solution:

Composition per liter:

$CaCl_2$ solution	14.7g

Preparation of $CaCl_2$ Solution: Add $CaCl_2$ solution to distilled/deionized water and bring volume to 1.0L. Mix thoroughly. Autoclave for 15 min at 15 psi pressure–121°C.

Preparation of Medium: Add components, except $MgSO_4 \cdot 7H_2O$ solution, glucose solution, thiamine·HCl solution, and $CaCl_2$ solution, to distilled/deionized water and bring volume to 987.0mL. Mix thoroughly. Adjust pH to 7.0. Autoclave for 15 min at 15 psi pressure–121°C. Cool to room temperature. Aseptically add sterile $MgSO_4 \cdot 7H_2O$ solution, sterile glucose solution, sterile thiamine·HCl solution, and sterile $CaCl_2$ solution. Mix thoroughly. Distribute into tubes or flasks.

Use: For the cultivation and maintenance of *Flavobacterium meningosepticum*.

MacConkey Agar

Composition per liter:

Pancreatic digest of gelatin	17.0g
Agar	13.5g
Lactose	10.0g
NaCl	5.0g
Bile salts	1.5g
Pancreatic digest of casein	1.5g
Peptic digest of animal tissue	1.5g
Neutral Red	0.03g
Crystal Violet	1.0mg

pH 7.1 ± 0.2 at 25°C

Source: This medium is available as a premixed powder from BD Diagnostic Systems.

Preparation of Medium: Add components to distilled/deionized water and bring volume to 1.0L. Mix thoroughly. Gently heat while stirring until boiling. Autoclave for 15 min at 15 psi pressure–121°C. Pour into sterile Petri dishes or distribute into sterile tubes.

Use: For the selective isolation, cultivation, and differentiation of coliforms and enteric pathogens based on the ability to ferment lactose. Lactose-fermenting organisms appear as red to pink colonies. Lactose-nonfermenting organisms appear as colorless or transparent colonies.

MacConkey Agar

Composition per liter:

Peptone	20.0g
Agar	12.0g
Lactose	10.0g
Bile salts	5.0g
NaCl	5.0g
Neutral Red	0.075g

pH 7.4 ± 0.2 at 25°C

Source: This medium is available as a premixed powder from Oxoid Unipath.

Preparation of Medium: Add components to distilled/deionized water and bring volume to 1.0L. Mix thoroughly. Gently heat while stirring until boiling. Autoclave for 15 min at 15 psi pressure–121°C. Pour into sterile Petri dishes or distribute into sterile tubes.

Use: For the selective isolation, cultivation, and differentiation of coliforms and enteric pathogens based on the ability to ferment lactose. Lactose-fermenting organisms appear as red to pink colonies. Lactose-nonfermenting organisms appear as colorless or transparent colonies.

MacConkey Agar, CS

Composition per liter:

Peptone	17.0g
Agar	13.5g
Lactose	10.0g
NaCl	5.0g
Proteose peptone	3.0g
Bile salts	1.5g
Neutral Red	0.03g
Crystal Violet	1.0mg

pH 7.1 ± 0.2 at 25°C

Source: This medium is available as a prepared medium from BD Diagnostic Systems.

Preparation of Medium: Add components to distilled/deionized water and bring volume to 1.0L. Mix thoroughly. Gently heat while stirring until boiling. Autoclave for 15 min at 15 psi pressure–121°C. Pour into sterile Petri dishes or distribute into sterile tubes.

Use: For the cultivation and differentiation of lactose-fermenting and nonfermenting Gram-negative bacteria while also controlling the swarming of *Proteus* species, if present. Lactose-fermenting organisms appear as red to pink colonies. Lactose-nonfermenting organisms appear as colorless or transparent colonies.

MacConkey Agar No. 2 (MacConkey II Agar)

Composition per liter:

Peptone	20.0g
Agar	15.0g
Lactose	10.0g
NaCl	5.0g
Bile salts no. 2	1.5g
Neutral Red	0.05g
Crystal Violet	1.0mg

pH 7.2 ± 0.2 at 25°C

Source: This medium is available as a premixed powder from BD Diagnostic Systems and Oxoid Unipath.

Preparation of Medium: Add components to distilled/deionized water and bring volume to 1.0L. Mix thoroughly. Gently heat while stirring until boiling. Autoclave for 15 min at 15 psi pressure–121°C. Pour into sterile Petri dishes or distribute into sterile tubes.

Use: For the selective isolation, cultivation, and differentiation of enteric pathogens, especially enterococci, in clinical specimens and in materials of sanitary importance.

MacConkey Agar No. 3

Composition per liter:

Peptone	20.0g
Agar	15.0g
Lactose	10.0g
NaCl	5.0g
Bile salts no. 3	1.5g
Neutral Red	0.03g
Crystal Violet	0.001g

pH 7.1 ± 0.2 at 25°C

Source: This medium is available as a premixed powder from Oxoid Unipath.

Preparation of Medium: Add components to distilled/deionized water and bring volume to 1.0L. Mix thoroughly. Gently heat while stirring until boiling. Autoclave for 15 min at 15 psi pressure–121°C. Pour into sterile Petri dishes or distribute into sterile tubes.

Use: For the selective isolation, cultivation, and differentiation of enteric pathogens, especially *Salmonella* and *Shigella*, in clinical specimens.

MacConkey Agar without Crystal Violet

Composition per liter:

Agar	12.0g
Lactose	10.0g
Pancreatic digest of casein	10.0g
Peptic digest of animal tissue	10.0g
Bile salts	5.0g
NaCl	5.0g
Neutral Red	0.05g

pH 7.4 ± 0.2 at 25°C

Source: This medium is available as a premixed powder from BD Diagnostic Systems.

Preparation of Medium: Add components to distilled/deionized water and bring volume to 1.0L. Mix thoroughly. Gently heat while stirring until boiling. Autoclave for 15 min at 15 psi pressure–121°C. Pour into sterile Petri dishes or distribute into sterile tubes.

Use: For the detection of members of the Enterobacteriaceae and enterococci as well as some staphylococci.

MacConkey Agar, Fluorocult (Fluorocult MacConkey Agar)

Composition per liter:

Peptone from casein	17.0g
Agar	13.5g
Lactose	10.0g
NaCl	5.0g
Peptone from meat	3.0g
Bile salt mixture	1.5g
4-Methylumbelliferyl-β-D-glucuronide	0.1g
Neutral Red	0.03g
Crystal Violet	0.001g

pH 7.1 ± 0.2 at 25°C

Source: This medium is available from Merck.

Preparation of Medium: Add components to distilled/deionized water and bring volume to 1.0L. Mix thoroughly. Autoclave for 15 min at 15 psi pressure–121°C. Cool to 45°–50°C. Pour into sterile Petri dishes. The plates are clear and red to red-brown.

Use: For the isolation of *Salmonella*, *Shigella,* and coliform bacteria, in particular *E. coli*, from various specimens. The bile salts and crystal violet largely inhibit the growth of Gram-positive microbial flora. Lactose together with the pH indicator neutral red are used to detect lactose-positive colonies and *E. coli* can be seen among these because of fluorescence under UV light.

MacConkey Agar without Salt

Composition per liter:

Peptone	20.0g
Agar	12.0g
Lactose	10.0g
Bile salts	5.0g
Neutral Red	0.075g

pH 7.4 ± 0.2 at 25°C

Source: This medium is available as a premixed powder from BD Diagnostic Systems and Oxoid Unipath.

Preparation of Medium: Add components to distilled/deionized water and bring volume to 1.0L. Mix thoroughly. Gently heat while stirring until boiling. Autoclave for 15 min at 15 psi pressure–121°C. Pour into sterile Petri dishes or distribute into sterile tubes. Dry the surface of plates before inoculation.

Use: For the isolation and detection of coliforms and enteric pathogens from urine. Provides a low electrolyte medium on which most *Proteus* species will not swarm and therefore avoids overgrowth of the plate.

MacConkey MUG Agar

Composition per liter:

Casein peptone	17.0g
Agar	13.5g
Lactose	10.0g
NaCl	5.0g
Meat peptone	3.0g
Bile salt mixture	1.5g
4-Methylumbelliferyl-β-D-glucuronide	0.1g
Neutral Red	0.03g
Crystal Violet	0.001g

pH 7.1 ± 0.2 at 37°C

Source: This medium is available from Fluka, Sigma-Aldrich.

Preparation of Medium: Add components to distilled/deionized water and bring volume to 1.0L. Mix thoroughly. Gently heat while stirring and bring to boiling. Autoclave for 15 min at 15 psi pressure–121°C. Cool to 50°C. Pour into sterile Petri dishes.

Use: For the isolation of *Salmonella, Shigella,* and coliform bacteria, in particular *E. coli*, from diverse specimens. Bile salts and crystal violet extensively inhibit the Gram-positive flora. The presence of lactose and neutral red indicate lactose-positive colonies from which *E. coli* can be identified by fluorescence in the UV.

Magnesium Oxalate Agar (MOX Agar)

Composition per liter:

Pancreatic digest of casein	15.0g
Agar	15.0g
Papaic digest of soybean meal	5.0g
NaCl	5.0g
$MgCl_2 \cdot 6H_2O$	4.1g
Sodium oxalate	2.68g

pH 7.4–7.6 at 25°C

Preparation of Medium: Add components to distilled/deionized water and bring volume to 1.0L. Mix thoroughly. Gently heat and bring to boiling. Distribute into tubes or flasks. Autoclave for 15 min at 15 psi pressure–121°C. Pour into sterile Petri dishes or leave in tubes.

Use: For the cultivation of *Yersinia enterocolitica.*

Maintenance of L Antigen in *Neisseria*

Composition per liter:

Proteose peptone no. 3	20.0g

Agar	15.0g
Na_2HPO_4	5.0g
NaCl	5.0g
Glucose	0.5g
Rabbit blood, defibrinated	100.0mL

pH 7.4–7.6 at 25°C

Preparation of Medium: Add components, except rabbit blood, to distilled/deionized water and bring volume to 900.0mL. Mix thoroughly. Gently heat while stirring until boiling. Autoclave for 20 min at 15 psi pressure–121°C. Cool to 60°C. Aseptically add 100.0mL of sterile, defibrinated rabbit blood. Maintain at 75°C while shaking for 30 min. Pour into sterile Petri dishes or distribute into sterile tubes.

Use: For the cultivation and maintenance of *Neisseria gonorrhoeae.*

Malachite Green Broth

Composition per liter:

Peptone	15.0g
Beef extract	9.0g
Malachite Green	0.01mg

pH 7.3 ± 0.2 at 25°C

Preparation of Medium: Add components to distilled/deionized water and bring volume to 1.0L. Mix thoroughly. Distribute into tubes or flasks. Autoclave for 15 min at 15 psi pressure–121°C.

Use: For the cultivation of *Pseudomonas aeruginosa.*

Malonate Broth

Composition per liter:

Sodium malonate	3.0g
NaCl	2.0g
$(NH_4)_2SO_4$	2.0g
K_2HPO_4	0.6g
KH_2PO_4	0.4g
Bromthymol Blue	0.025g

pH 6.7 ± 0.2 at 25°C

Source: This medium is available as a premixed powder from BD Diagnostic Systems.

Preparation of Medium: Add components to distilled/deionized water and bring volume to 1.0L. Mix thoroughly. Distribute into tubes or flasks. Autoclave for 15 min at 15 psi pressure–121°C. Avoid introduction of carbon and nitrogen from other sources.

Use: For the cultivation and differentiation of coliforms and other enteric organisms, particularly *Enterobacter* and *Escherichia,* based on their ability to utilize malonate as the sole carbon source and ammonium sulfate as the sole nitrogen source. Malonate-utilizing organisms turn the medium blue.

Malonate Broth, Ewing Modified

Composition per liter:

Sodium malonate	3.0g
NaCl	2.0g
$(NH_4)_2SO_4$	2.0g
Yeast extract	1.0g
K_2HPO_4	0.6g
KH_2PO_4	0.4g
Glucose	0.25g
Bromthymol Blue	0.025g

pH 6.7 ± 0.2 at 25°C

Source: This medium is available as a premixed powder from BD Diagnostic Systems.

Preparation of Medium: Add components to distilled/deionized water and bring volume to 1.0L. Mix thoroughly. Distribute into tubes or flasks. Autoclave for 15 min at 15 psi pressure–121°C.

Use: For the cultivation and differentiation of coliforms and other enteric organisms, particularly *Enterobacter* and *Escherichia,* based on their ability to utilize malonate as a carbon source and ammonium sulfate as a nitrogen source. The small amount of yeast extract and glucose encourages the growth of some organisms that may be distressed or fail to respond. Malonate-utilizing organisms turn the medium blue.

Malt Agar

Composition per liter:

Agar	20.0g
Malt extract	12.5g

Preparation of Medium: Add components to distilled/deionized water and bring volume to 1.0L. Mix thoroughly. Gently heat until boiling. Distribute into tubes or flasks. Autoclave for 15 min at 15 psi pressure–121°C. Pour into sterile Petri dishes or leave in tubes.

Use: For the cultivation and maintenance of fungi.

Mannitol Salt Agar

Composition per liter:

NaCl	75.0g
Agar	15.0g
D-Mannitol	10.0g
Pancreatic digest of casein	5.0g
Peptic digest of animal tissue	5.0g
Beef extract	1.0g
Phenol Red	0.025g

pH 7.4 ± 0.2 at 25°C

Source: This medium is available as a premixed powder from BD Diagnostic Systems and Oxoid Unipath.

Preparation of Medium: Add components to distilled/deionized water and bring volume to 1.0L. Mix thoroughly. Gently heat while stirring and bring to boiling. Distribute into tubes or flasks. Autoclave for 15 min at 15 psi pressure–121°C. Pour into sterile Petri dishes or leave in tubes.

Use: For the selective isolation, cultivation, and enumeration of staphylococci from clinical and nonclinical specimens. Mannitol-utilizing organisms turn the medium yellow.

Martin-Lewis Agar

Composition per liter:

Agar	12.0g
Hemoglobin	10.0g
Pancreatic digest of casein	7.5g
Selected meat peptone	7.5g
NaCl	5.0g
K_2HPO_4	4.0g
Cornstarch	1.0g
KH_2PO_4	1.0g
Supplement solution	10.0mL
VCAT inhibitor	10.0mL

pH 7.2 ± 0.22 at 25°C

Source: Martin-Lewis agar is available as a prepared medium from BD Diagnostic Systems.

Supplement Solution:

Composition per liter:

Glucose	100.0g
L-Cysteine·HCl	25.9g
L-Glutamine	10.0g
L-Cystine	1.1g
Adenine	1.0g
Nicotinamide adenine dinucleotide	0.25g
Vitamin B_{12}	0.1g
Thiamine pyrophosphate	0.1g
Guanine·HCl	0.03g
$Fe(NO_3)_3 \cdot 6H_2O$	0.02g
p-Aminobenzoic acid	0.013g
Thiamine·HCl	3.0mg

Source: The supplement solution IsoVitaleX enrichment is available from BD Diagnostic Systems. This enrichment may be replaced by supplement VX from BD Diagnostic Systems.

Preparation of Supplement Solution: Add components to distilled/deionized water and bring volume to 1.0L. Mix thoroughly. Filter sterilize.

VCAT Inhibitor:

Composition per 10.0mL:

Colistin	7.5mg
Trimethoprim lactate	5.0mg
Vancomycin	4.0mg
Anisomycin	0.02g

Preparation of VCAT Inhibitor: Add components to distilled/deionized water and bring volume to 10.0mL. Mix thoroughly. Filter sterilize.

Preparation of Medium: Add components, except supplement solution enrichment and VCAT inhibitor, to distilled/deionized water and bring volume to 980.0mL. Gently heat while stirring and bring to boiling. Autoclave for 15 min at 15 psi pressure–121°C. Cool to 45°–50°C. Aseptically add sterile supplement solution enrichment and sterile VCAT inhibitor. Mix thoroughly. Pour into sterile Petri dishes.

Use: For the isolation and cultivation of pathogenic *Neisseria* from specimens containing mixed flora of bacteria and fungi.

Martin-Lewis Agar, Enriched

Composition per liter:

Agar	12.0g
Pancreatic digest of casein	7.5g
Selected meat peptone	7.5g
NaCl	5.0g
K_2HPO_4	4.0g
Cornstarch	1.0g
KH_2PO_4	1.0g
Sarcina lutea suspension	20.0mL
Horse serum, inactivated	20.0mL
Supplement solution	10.0mL
PCAT inhibitor	10.0mL

pH 7.2 ± 0.22 at 25°C

Source: The supplement solution (IsoVitaleX enrichment) is available from BD Diagnostic Systems. This enrichment may be replaced by supplement VX from BD Diagnostic Systems.

***Sarcina lutea* Suspension:**

Composition per 20.0mL:

Sarcina lutea FDA 1001	10^6–10^7 cells

Preparation of *Sarcina lutea* Suspension: Aseptically wash the growth of 24-h cultures of *Sarcina lutea* FDA 1001 cells from Thayer-Martin plates with sterile soybean casein digest broth. Standardize the suspension by adding additional sterile tryptic soy broth to yield 40% light transmission at 530nm wavelength.

Soybean Casein Digest Broth:

Composition per liter:

Pancreatic digest of casein	17.0g
NaCl	5.0g
Papaic digest of soybean meal	3.0g
K_2HPO_4	2.5g
Glucose	2.5g

pH 7.3 ± 0.2 at 25°C

Preparation of Soybean Casein Digest Broth: Add components to distilled/deionized water and bring volume to 1.0L. Mix thoroughly. Distribute into tubes or flasks. Autoclave for 15 min at 15 psi pressure–121°C.

Supplement Solution:

Composition per liter:

Glucose	100.0g
L-Cysteine·HCl	25.9g
L-Glutamine	10.0g
L-Cystine	1.1g
Adenine	1.0g
Nicotinamide adenine dinucleotide	0.25g
Vitamin B_{12}	0.1g
Thiamine pyrophosphate	0.1g
Guanine·HCl	0.03g
$Fe(NO_3)_3 \cdot 6H_2O$	0.02g
p-Aminobenzoic acid	0.013g
Thiamine·HCl	3.0mg

Preparation of Supplement Solution: Add components to distilled/deionized water and bring volume to 1.0L. Mix thoroughly. Filter sterilize.

PCAT Inhibitor:

Composition per 10.0mL:

Anisomycin	0.02g
Colistin	7.5mg
Trimethoprim lactate	5.0mg
Penicillin G	25,000U

Preparation of PCAT Inhibitor: Add components to distilled/deionized water and bring volume to 10.0mL. Mix thoroughly. Filter sterilize.

Preparation of Medium: Add components—except *Sarcina lutea* suspension, horse serum, supplement solution, and PCAT inhibitor—to distilled/deionized water and bring volume to 940.0mL. Gently heat while stirring and bring to boiling. Autoclave for 15 min at 15 psi pressure–121°C. Cool to 45°–50°C. Aseptically add 20.0mL of sterile *Sarcina lutea* suspension, 20.0mL of sterile horse serum, 10.0mL of supplement solution, and 10.0mL of sterile PCAT inhibitor. Mix thoroughly. Pour into sterile Petri dishes.

Use: For the isolation and cultivation of pathogenic *Neisseria*, especially penicillinase-producing strains, from specimens containing mixed flora of bacteria and fungi.

Maximum Recovery Diluent

Composition per liter:

NaCl	8.5g
Peptone	1.0g

pH 7.0 ± 0.2 at 25°C

Source: This medium is available as a premixed powder from Oxoid Unipath.

Preparation of Medium: Add components to distilled/deionized water and bring volume to 1.0L. Mix thoroughly. Distribute into tubes or flasks. Autoclave for 15 min at 15 psi pressure–121°C.

Use: This diluent is a physiologically isotonic and protective medium for maximal recovery of microorganisms from a variety of sources.

McBride *Listeria* Agar

Composition per liter:

Agar	15.0g
Glycine	10.0g
Pancreatic digest of casein	5.0g
Peptic digest of animal tissue	5.0g
NaCl	5.0g
Beef extract	3.0g
Phenylethyl alcohol	2.5g
LiCl	0.5g

pH 7.3 ± 0.22 at 25°C

Source: This medium is available as a premixed powder from BD Diagnostic Systems.

Preparation of Medium: Add components to distilled/deionized water and bring volume to 1.0L. Mix thoroughly. Gently heat while stirring and bring to boiling. Distribute into tubes or flasks. Autoclave for 15 min at 15 psi pressure–121°C. Pour into sterile Petri dishes or leave in tubes.

Use: For the selective isolation of *Listeria monocytogenes* from clinical and nonclinical specimens containing mixed flora.

McCarthy Agar

Composition per liter:

Cornstarch	10.0g
Naladixic acid	0.015g
Colistin	0.01g
GC agar base	1.0L

pH 7.2 ± 0.2 at 25°C

GC Agar Base:

Composition per liter:

Agar	10.0g
Pancreatic digest of casein	7.5g
Peptic digest of animal tissue	7.5g
NaCl	5.0g
K_2HPO_4	4.0g
Cornstarch	1.0g
KH_2PO_4	1.0g

Preparation of GC Agar Base: Add components to distilled/deionized water and bring volume to 1.0L. Mix thoroughly.

Preparation of Medium: To 1.0L of GC agar base, add the cornstarch. Gently heat while stirring to dissolve. Add the naladixic acid and colistin. Mix thoroughly. Distribute into tubes or flasks. Autoclave for 15 min at 15 psi pressure–121°C. Pour into sterile Petri dishes or leave in tubes.

Use: For the isolation and differentiation of *Gardnerella vaginalis (Haemophilus vaginalis, Corynebacterium vaginale)* from genitourinary specimens. Bacteria that can utilize starch appear as colonies surrounded by a clear zone.

McClung Carbon-Free Broth

Composition per liter:

$NaNO_3$	2.0g
K_2HPO_4	0.8g
$MgSO_4 \cdot 7H_2O$	0.5g
$FeCl_3$	0.01g
$MnCl_2 \cdot 4H_2O$	8.0mg
$ZnSO_4$	2.0mg

pH 7.2 ± 0.2 at 25°C

Preparation of Medium: Add components to distilled/deionized water and bring volume to 1.0L. Mix thoroughly. Gently heat without boiling until salts dissolve. Cool to 25°C. Adjust pH to 7.2. Filter sterilize.

Use: For use as a basal medium in determining the carbon assimilation capabilities of microorganisms.

McClung-Toabe Agar

Composition per liter:

Proteose peptone	40.0g
Agar	25.0g
Na_2HPO_4	5.0g
Glucose	2.0g
NaCl	2.0g
KH_2PO_4	1.0g
$MgSO_4 \cdot 7H_2O$	0.1g
Egg yolk emulsion, 50%	100.0mL

pH 7.3 ± 0.2 at 25°C

Source: This medium is available as a premixed powder from BD Diagnostic Systems.

Egg Yolk Emulsion, 50%:
Composition per 100.0mL:

Chicken egg yolks	11
Whole chicken egg	1
NaCl (0.9% solution)	50.0mL

Preparation of Egg Yolk Emulsion, 50%: Soak eggs with 1:100 dilution of saturated mercuric chloride solution for 1 min. Crack eggs and separate yolks from whites. Mix egg yolks with 1 chicken egg. Beat to form emulsion. Measure 50.0mL of egg yolk emulsion and add to 50.0mL of 0.9% NaCl solution. Mix thoroughly. Filter sterilize. Warm to 45°–50°C.

Preparation of Medium: Add components, except egg yolk emulsion, 50%, to distilled/deionized water and bring volume to 900.0mL. Mix thoroughly. Gently heat while stirring and bring to boiling. Autoclave for 15 min at 15 psi pressure–121°C. Cool to 50°–55°C. Aseptically add 100.0mL of sterile egg yolk emulsion, 50%. Mix thoroughly. Pour into sterile Petri dishes in 15.0mL volumes.

Use: For the isolation and cultivation of *Clostridium perfringens*.

McClung-Toabe Agar, Modified

Composition per liter:

Proteose peptone no. 2	40.0g
Agar	20.0g
Na_2HPO_4	5.0g
Glucose	2.0g
NaCl	2.0g
KH_2PO_4	1.0g
$MgSO_4 \cdot 7H_2O$	0.1g
Egg yolk emulsion, 50%	100.0mL
Hemin solution	1.0mL

pH 7.6 ± 0.2 at 25°C

Egg Yolk Emulsion, 50%:
Composition per 100.0mL:

Chicken egg yolks	11
Whole chicken egg	1
NaCl (0.9% solution)	50.0mL

Preparation of Egg Yolk Emulsion, 50%: Soak eggs with 1:100 dilution of saturated mercuric chloride solution for 1 min. Crack eggs and separate yolks from whites. Mix egg yolks with 1 chicken egg. Beat to form emulsion. Measure 50.0mL of egg yolk emulsion and add to 50.0mL of 0.9% NaCl solution. Mix thoroughly. Filter sterilize. Warm to 45°–50°C.

Hemin Solution:
Composition per 100.0mL:

Hemin	0.5g
NaOH (1*N* solution)	20.0mL

Preparation of Hemin Solution: Add hemin to 20.0mL of 1*N* NaOH solution. Mix thoroughly. Bring volume to 100.0mL with distilled/deionized water.

Preparation of Medium: Add components, except egg yolk emulsion, 50%, to distilled/deionized water and bring volume to 900.0mL. Mix thoroughly. Gently heat while stirring and bring to boiling. Autoclave for 15 min at 15 psi pressure–121°C. Cool to 50°–55°C. Aseptically add 100.0mL of sterile egg yolk emulsion, 50%. Mix thoroughly. Pour into sterile Petri dishes in 20.0mL volumes.

Use: For the cultivation of a wide variety of anaerobic bacteria. For the differentiation of anaerobic bacteria based on lecithinase production and lipase production. Bacteria that produce lecithinase appear as colonies surrounded by a zone of insoluble precipitate. Bacteria that produce lipase appear as colonies with a pearly iridescent sheen.

McClung-Toabe Agar, Modified

Composition per liter:

Proteose peptone no. 2	40.0g
Agar	20.0g
Na_2HPO_4	5.0g
Glucose	2.0g
NaCl	2.0g
KH_2PO_4	1.0g
$MgSO_4 \cdot 7H_2O$	0.1g
Neomycin	0.1g
Egg yolk emulsion, 50%	100.0mL
Hemin solution	1.0mL

pH 7.6 ± 0.2 at 25°C

Egg Yolk Emulsion, 50%:

Composition per 100.0mL:

Chicken egg yolks	11
Whole chicken egg	1
NaCl (0.9% solution)	50.0mL

Preparation of Egg Yolk Emulsion, 50%: Soak eggs with 1:100 dilution of saturated mercuric chloride solution for 1 min. Crack eggs and separate yolks from whites. Mix egg yolks with 1 chicken egg. Beat to form emulsion. Measure 50.0mL of egg yolk emulsion and add to 50.0mL of 0.9% NaCl solution. Mix thoroughly. Filter sterilize. Warm to 45°–50°C.

Hemin Solution:

Composition per 100.0mL:

Hemin	0.5g
NaOH (1*N* solution)	20.0mL

Preparation of Hemin Solution: Add hemin to 20.0mL of 1*N* NaOH solution. Mix thoroughly. Bring volume to 100.0mL with distilled/deionized water.

Preparation of Medium: Add components, except egg yolk emulsion, 50%, to distilled/deionized water and bring volume to 900.0mL. Mix thoroughly. Gently heat while stirring and bring to boiling. Autoclave for 15 min at 15 psi pressure–121°C. Cool to 50°–55°C. Aseptically add 100.0mL of sterile egg yolk emulsion, 50%. Mix thoroughly. Pour into sterile Petri dishes in 20.0mL volumes.

Use: For the cultivation of *Clostridium* species. For the differentiation of *Clostridium* species based on lecithinase production and lipase production. Bacteria that produce lecithinase appear as colonies surrounded by a zone of insoluble precipitate. Bacteria that produce lipase appear as colonies with a pearly iridescent sheen.

McClung-Toabe Agar, Modified

Composition per liter:

Proteose peptone no. 2	20.0g
Agar	20.0g
Yeast extract	5.0g
Pancreatic digest of casein	5.0g
NaCl	5.0g
Sodium thioglycolate	1.0g
Egg yolk emulsion, 50%	80.0mL

pH 7.6 ± 0.2 at 25°C

Egg Yolk Emulsion, 50%:

Composition per 100.0mL:

Chicken egg yolks	11
Whole chicken egg	1
NaCl (0.9% solution)	50.0mL

Preparation of Egg Yolk Emulsion, 50%: Soak eggs with 1:100 dilution of saturated mercuric chloride solution for 1 min. Crack eggs and separate yolks from whites. Mix egg yolks with 1 chicken egg. Beat to form emulsion. Measure 50.0mL of egg yolk emulsion and add to 50.0mL of 0.9% NaCl solution. Mix thoroughly. Filter sterilize. Warm to 45°–50°C.

Preparation of Medium: Add components, except egg yolk emulsion, 50%, to distilled/deionized water and bring volume to 920.0mL. Mix thoroughly. Gently heat while stirring and bring to boiling. Autoclave for 15 min at 15 psi pressure–121°C. Cool to 50°–55°C. Aseptically add 80.0mL of sterile egg yolk emulsion, 50%. Mix thoroughly. Pour into sterile Petri dishes in 20.0mL volumes.

Use: For the cultivation of *Clostridium botulinum.*

McClung-Toabe Egg Yolk Agar, CDC Modified (CDC Modified McClung-Toabe Egg Yolk Agar)

Composition per liter:

Pancreatic digest of casein	40.0g
Agar	25.0g
Na_2HPO_4	5.0g
Yeast extract	5.0g
Glucose	2.0g
NaCl	2.0g
KH_2PO_4	1.0g

Egg yolk emulsion, 50% 100.0mL
$MgSO_4 \cdot 7H_2O$ (5% solution) 0.2mL
pH 7.3 ± 0.2 at 25°C

Egg Yolk Emulsion, 50%:
Composition per 100.0mL:
Chicken egg yolks 11
Whole chicken egg 1
NaCl (0.9% solution) 50.0mL

Preparation of Egg Yolk Emulsion, 50%: Soak eggs with 1:100 dilution of saturated mercuric chloride solution for 1 min. Crack eggs and separate yolks from whites. Mix egg yolks with 1 chicken egg. Beat to form emulsion. Measure 50.0mL of egg yolk emulsion and add to 50.0mL of 0.9% NaCl solution. Mix thoroughly. Filter sterilize. Warm to 45°–50°C.

Preparation of Medium: Add components—except egg yolk emulsion, 50%—to distilled/deionized water and bring volume to 900.0mL. Mix thoroughly. Gently heat while stirring and bring to boiling. Autoclave for 15 min at 15 psi pressure–121°C. Cool to 50°–55°C. Aseptically add 100.0mL of sterile egg yolk emulsion, 50%. Mix thoroughly. Pour into sterile Petri dishes in 15.0mL volumes.

Use: For the cultivation of a wide variety of anaerobic bacteria. For the differentiation of anaerobic bacteria based on lecithinase production, lipase production, and proteolytic ability. Bacteria that produce lecithinase appear as colonies surrounded by a zone of insoluble precipitate. Bacteria that produce lipase appear as colonies with a pearly iridescent sheen. Bacteria that produce proteolytic activity appear as colonies surrounded by a clear zone.

m-CP Medium

Composition per liter:
Tryptose 30.0g
Yeast extract 20.0g
Agar 15.0g
Sucrose 5.0g
L-Cysteine·HCl·H_2O 1.0g
$MgO_4 \cdot 7H_2O$ 0.1g
Bromcresol Purple 0.04g
Phenolphthalein biphosphate tetrazolium salt solution 10.0mL
Selective supplement solution 4.0mL
Indoxyl-β-D-glucoside solution 4.0mL
Ferric chloride solution 1.0mL
pH 7.6 ± 0.2 at 25°C

Selective Supplement Solution:
Composition per 4.0mL:
D-Cycloserine 0.4g
Polymyxin B sulfate 25.0mg

Preparation of Selective Supplement Solution: Add components to 4.0mL of distilled/deionized water. Mix thoroughly. Filter sterilize.

Phenolphthalein Biphosphate Tetrazolium Salt Solution:
Composition per 10.0mL:
Phenolphthalein biphosphate tetrazolium salt 25.0mg

Preparation of Phenolphthalein Biphosphate Tetrazolium Salt Solution: Add phenolphthalein biphosphate tetrazolium salt to 10.0mL of distilled/deionized water. Mix thoroughly. Filter sterilize.

Indoxyl-β-D-glucoside Solution:
Composition per 10.0mL:
Indoxyl-β-D-glucoside 0.45g

Preparation of Indoxyl-β-D-glucoside Solution: Add indoxyl-β-D-glucoside to 10.0mL of distilled/deionized water. Mix thoroughly. Filter sterilize.

Feric Chloride Solution:
Composition per 4.0mL:
$FeCl_3 \cdot 6H_2O$ 30.0mg

Preparation of Feric Chloride Solution: Add $FeCl_3 \cdot 6H_2O$ to 4.0mL of distilled/deionized water. Mix thoroughly. Filter sterilize.

Preparation of Medium: Add components except selective supplement solution, phenolphthalein biphosphate tetrazolium salt solution, indoxyl-β-D-glucoside solution, and ferric chloride solution, to distilled/deionized water and bring volume to 981.0mL. Mix thoroughly. Autoclave for 15 min at 15 psi pressure–121°C. Cool to 45°–50°C. Aseptically add 4.0mL of selective supplement solution. Mix thoroughly. Aseptically add 10.0mL phenolphthalein biphosphate tetrazolium salt solution, 4.0mL indoxyl-β-D-glucoside solution, and 1.0mL ferric chloride solution. Mix thoroughly. Pour into sterile Petri dishes or aseptically distribute into tubes.

Use: A selective, chromogenic medium for the rapid identification and enumeration of *Clostridium perfringens*.

Membrane Lactose Glucuronide Agar (MLGA)

Composition per liter:
Peptone 40.0g
Lactose 30.0g
Agar 10.0g
Yeast extract 6.0g
Sodium lauryl sulfate 1.0g
Sodium pyruvate 0.5g

5-Bromo-4-chloro-3-indoxyl-
  β-D-glucuronic acid 0.2g
Phenol Red 0.2g

pH 7.4 ± 0.2 at 25°C

Source: This medium is available from Oxoid.

Preparation of Medium: Add components to distilled/deionized water and bring volume to 1.0L. Mix thoroughly. Autoclave for 15 min at 15 psi pressure–121°C. Cool to 45°–50°C. Pour into sterile Petri dishes.

Use: For the direct enumeration of *E.coli* and coliforms by the membrane filtration method. The chromogenic substrate 5-Bromo-4-chloro-3-indoxyl-β-D-glucuronic acid (BCIG), is cleaved by the enzyme β-glucuronidase and produces a blue chromophore that builds up within the bacterial cells. In addition, the incorporation of phenol red detects lactose fermentation and results in yellow colonies when acid is produced. Since coliform colonies are lactose positive, they will appear yellow on this medium and as *E. coli* colonies are both lactose and β-glucuronidase positive, they will appear green.

Methylene Blue Milk Medium (MBM Medium)

Composition per liter:
Skim milk, dehydrated 100.0g
Methylene Blue 10.0g

pH 6.4 ± 0.2 at 25°C

Preparation of Medium: Add components to distilled/deionized water and bring volume to 1.0L. Mix thoroughly. Distribute into tubes or flasks. Autoclave for 20 min at 10 psi pressure–115°C.

Use: For the cultivation and differentiation of group D streptococci (enterococci) from other *Streptococcus* species.

MH IH Agar

Composition per liter:
Solution A 490.0mL
Solution B 490.0mL
Supplement solution 20.0mL

pH 6.9 ± 0.2 at 25°C

Solution A:
Composition per 490.0mL:
Beef infusion 300.0g
Acid hydrolysate of casein 17.5g
Agar 17.0g
Starch 1.5g

Preparation of Solution A: Add components to distilled/deionized water and bring volume to 490.0mL. Mix thoroughly. Gently heat and bring to boiling. Autoclave for 15 min at 15 psi pressure–121°C. Cool to 45°–50°C.

Solution B:
Composition per 490.0mL:
Hemoglobin 10.0g

Preparation of Solution B: Add hemoglobin to distilled/deionized water and bring volume to 490.0mL. Mix thoroughly. Gently heat and bring to boiling. Autoclave for 15 min at 15 psi pressure–121°C. Cool to 45°–50°C.

Supplement Solution:
Composition per liter:
Glucose 100.0g
L-Cysteine·HCl 25.9g
L-Glutamine 10.0g
L-Cystine 1.1g
Adenine 1.0g
Nicotinamide adenine dinucleotide 0.25g
Vitamin B_{12} 0.1g
Thiamine pyrophosphate 0.1g
Guanine·HCl 0.03g
$Fe(NO_3)_3·6H_2O$ 0.02g
p-Aminobenzoic acid 0.013g
Thiamine·HCl 3.0mg

Source: The supplement solution IsoVitaleX enrichment is available from BD Diagnostic Systems. This enrichment may be replaced by supplement VX from BD Diagnostic Systems.

Preparation of Supplement Solution: Add components to distilled/deionized water and bring volume to 1.0L. Mix thoroughly. Filter sterilize.

Preparation of Medium: Aseptically combine cooled, sterile solution A and cooled, sterile solution B. Mix thoroughly. Adjust pH to 6.9 with sterile 1*N* HCl or sterile 1*N* KOH. Aseptically add 20.0mL of sterile supplement solution. Pour into sterile Petri dishes or distribute into sterile tubes.

Use: For the cultivation and differentiation of *Legionella* species.

Middlebrook 13A Medium

Composition per 112.5mL:
Casein hydrolysate 0.1g
Tween 80 .... 0.02g
Sodium polyanetholesulfonate 0.025g
Middlebrook 7H9 broth 100.0mL
Middlebrook 13A enrichment 12.5mL
Catalase 36,000U
^{14}C-substrate 125μCi (185kBq)

pH 6.6 ± 0.2 at 25°C

Middlebrook 7H9 Broth:
Composition per liter:

Na_2HPO_4	2.5g
KH_2PO_4	1.0g
Monosodium glutamate	0.5g
$(NH_4)_2SO_4$	0.5g
Sodium citrate	0.1g
$MgSO_4 \cdot 7H_2O$	0.05g
Ferric ammonium citrate	0.04g
$CuSO_4 \cdot 5H_2O$	1.0mg
Pyridoxine	1.0mg
$ZnSO_4 \cdot 7H_2O$	1.0mg
Biotin	0.5mg
$CaCl_2 \cdot 2H_2O$	0.5mg
Glycerol	2.0mL

Preparation of Middlebrook 7H9 Broth: Add components to distilled/deionized water and bring volume to 1.0L. Mix thoroughly.

Middlebrook 13A Enrichment:
Composition per 20.0mL:

Bovine serum albumin	3.0g

Preparation of Middlebrook 13A Enrichment: Add bovine serum albumin to distilled/deionized water and bring volume to 20.0mL. Mix thoroughly. Filter sterilize.

Preparation of Medium: To 100.0mL of Middlebrook 7H9 broth, add remaining components, except Middlebrook 13A enrichment. Mix thoroughly. Filter sterilize. Aseptically distribute into bottles in 4.0mL volumes. Prior to inoculation, aseptically add 0.5mL of Middlebrook 13A enrichment to each bottle. Mix thoroughly.

Use: For the cultivation of *Mycobacterium* species from the blood of patients suspected of having mycobacteremia.

Middlebrook ADC Enrichment (Middlebrook Albumin Dextrose Catalase Enrichment)

Composition per 100.0mL:

Bovine albumin fraction V	5.0g
Glucose	2.0g
Catalase	0.003g

Source: This medium is available as a prepared enrichment from BD Diagnostic Systems.

Preparation of Enrichment: Add components to distilled/deionized water and bring volume to 100.0mL. Mix thoroughly. Filter sterilize.

Use: For use as a supplement to other Middlebrook media for the isolation, cultivation, and maintenance of *Mycobacterium* species. Also used as a supplement to other Middlebrook media for determining the antimicrobial susceptibility of mycobacteria.

Middlebrook 7H9 Broth with Middlebrook ADC Enrichment

Composition per liter:

Na_2HPO_4	2.5g
KH_2PO_4	1.0g
Monosodium glutamate	0.5g
$(NH_4)_2SO_4$	0.5g
Sodium citrate	0.1g
$MgSO_4 \cdot 7H_2O$	0.05g
Ferric ammonium citrate	0.04g
$CuSO_4 \cdot 5H_2O$	1.0mg
Pyridoxine	1.0mg
$ZnSO_4 \cdot 7H_2O$	1.0mg
Biotin	0.5mg
$CaCl_2 \cdot 2H_2O$	0.5mg
Middlebrook ADC enrichment	100.0mL
Glycerol	2.0mL

pH 6.6 ± 0.2 at 25°C

Source: This medium is available as a premixed powder from BD Diagnostic Systems.

Middlebrook ADC Enrichment:
Composition per 100.0mL:

Bovine albumin fraction V	5.0g
Glucose	2.0g
Catalase	3.0mg

Source: This enrichment is available as a prepared enrichment from BD Diagnostic Systems.

Preparation of Middlebrook ADC Enrichment: Add components to distilled/deionized water and bring volume to 100.0mL. Mix thoroughly. Filter sterilize.

Preparation of Medium: Add glycerol to 900.0mL of distilled/deionized water and add remaining components, except Middlebrook ADC enrichment. Mix thoroughly. Gently heat and bring to boiling. Autoclave for 15 min at 15 psi pressure–121°C. Cool to 50°–55°C. Aseptically add 100.0mL of sterile Middlebrook ADC enrichment. Mix thoroughly. Distribute into sterile tubes or flasks.

Use: For the isolation, cultivation, and maintenance of *Mycobacterium* species, including *Mycobacterium tuberculosis*. Also used for determining the antimicrobial susceptibility of mycobacteria.

Middlebrook 7H9 Broth with Middlebrook OADC Enrichment

Composition per liter:

Na_2HPO_4	2.5g
KH_2PO_4	1.0g

Monosodium glutamate 0.5g
$(NH_4)_2SO_4$ 0.5g
Sodium citrate 0.1g
$MgSO_4 \cdot 7H_2O$ 0.05g
Ferric ammonium citrate 0.04g
$CuSO_4 \cdot 5H_2O$ 1.0mg
Pyridoxine 1.0mg
$ZnSO_4 \cdot 7H_2O$ 1.0mg
Biotin 0.5mg
$CaCl_2 \cdot 2H_2O$ 0.5mg
Middlebrook OADC enrichment 100.0mL
Glycerol 2.0mL

pH 6.6 ± 0.2 at 25°C

Source: This medium is available as a premixed powder from BD Diagnostic Systems.

Middlebrook OADC Enrichment:

Composition per 100.0mL:

Bovine albumin fraction V 5.0g
Glucose 2.0g
NaCl 0.85g
Oleic acid 0.05g
Catalase 4.0mg

Source: This enrichment is available as a prepared enrichment from BD Diagnostic Systems.

Preparation of Middlebrook OADC Enrichment: Add components to distilled/deionized water and bring volume to 100.0mL. Mix thoroughly. Filter sterilize.

Preparation of Medium: Add glycerol to 900.0mL of distilled/deionized water and add remaining components, except Middlebrook OADC enrichment. Mix thoroughly. Gently heat and bring to boiling. Autoclave for 15 min at 15 psi pressure–121°C. Cool to 50°–55°C. Aseptically add 100.0mL of sterile Middlebrook OADC enrichment. Mix thoroughly. Distribute into sterile tubes or flasks.

Use: For the isolation, cultivation, and maintenance of *Mycobacterium* species, including *Mycobacterium tuberculosis.* Also used for determining the antimicrobial susceptibility of mycobacteria.

Middlebrook 7H9 Broth with Middlebrook OADC Enrichment and Triton WR 1339

Composition per liter:

Na_2HPO_4 .... 2.5g
KH_2PO_4 .... 1.0g
Monosodium glutamate 0.5g
$(NH_4)_2SO_4$ 0.5g
Sodium citrate 0.1g
$MgSO_4 \cdot 7H_2O$ 0.05g
Ferric ammonium citrate 0.04g
$CuSO_4 \cdot 5H_2O$ 1.0mg
Pyridoxine 1.0mg
$ZnSO_4 \cdot 7H_2O$ 1.0mg
Biotin 0.5mg
$CaCl_2 \cdot 2H_2O$ 0.5mg
Middlebrook OADC enrichment with Triton WR 1339 .... 100.0mL
Glycerol 2.0mL

pH 6.6 ± 0.2 at 25°C

Source: This medium is available as a premixed powder from BD Diagnostic Systems.

Middlebrook OADC Enrichment with Triton WR 1339:

Composition per 100.0mL:

Bovine albumin fraction V 5.0g
Glucose 2.0g
NaCl 0.85g
Triton WR 1339 .... 0.25g
Oleic acid 0.05g
Catalase 4.0mg

Source: This enrichment is available as a prepared enrichment from BD Diagnostic Systems.

Preparation of Middlebrook OADC Enrichment with Triton WR 1339: Add components to distilled/deionized water and bring volume to 100.0mL. Mix thoroughly. Filter sterilize.

Preparation of Medium: Add glycerol to 900.0mL of distilled/deionized water and add remaining components, except Middlebrook OADC enrichment with Triton WR-1339. Mix thoroughly. Gently heat and bring to boiling. Autoclave for 15 min at 15 psi pressure–121°C. Cool to 50°–55°C. Aseptically add 100.0mL of sterile Middlebrook OADC enrichment with Triton WR-1339. Mix thoroughly. Distribute into sterile tubes or flasks.

Use: For the isolation, cultivation, and maintenance of *Mycobacterium* species, including *Mycobacterium tuberculosis.* Also used for determining the antimicrobial susceptibility of mycobacteria.

Middlebrook 7H9 Broth, Supplemented

Composition per liter:

Na_2HPO_4 .... 2.5g
KH_2PO_4 .... 1.0g
Monosodium glutamate 0.5g
$(NH_4)_2SO_4$ 0.5g
Tween 80 .... 0.5g
Sodium citrate 0.1g
$MgSO_4 \cdot 7H_2O$ 0.05g
Ferric ammonium citrate 0.04g
Mycobactin J 2.0mg
$CuSO_4 \cdot 5H_2O$ 1.0mg

Pyridoxine	1.0mg
$ZnSO_4 \cdot 7H_2O$	1.0mg
Biotin	0.5mg
$CaCl_2 \cdot 2H_2O$	0.5mg
Dubos oleic albumin complex	100.0mL
Glycerol	2.0mL

pH 6.6 ± 0.2 at 25°C

Source: Mycobactin J is available from Allied Laboratories, Inc.

Dubos Oleic Albumin Complex:
Composition per 100.0mL:

Bovine serum albumin, fraction V	5.0g
Oleic acid, sodium salt	0.05g
NaCl (0.85% solution)	100.0mL

Preparation of Dubos Oleic Albumin Complex: Add bovine serum albumin and oleic acid to 100.0mL of NaCl solution. Mix thoroughly. Filter sterilize.

Preparation of Medium: Add components, except Dubos oleic albumin complex, to distilled/deionized water and bring volume to 900.0mL. Mix thoroughly. Gently heat and bring to boiling. Autoclave for 15 min at 15 psi pressure–121°C. Cool to 45°–50°C. Aseptically add sterile Dubos oleic albumin complex. Mix thoroughly. Pour into sterile Petri dishes or distribute into sterile tubes.

Use: For the cultivation and maintenance of *Mycobacterium avium*.

Middlebrook 7H10 Agar with Middlebrook ADC Enrichment

Composition per liter:

Agar	15.0g
Na_2HPO_4	1.5g
KH_2PO_4	1.5g
$(NH_4)_2SO_4$	0.5g
L-Glutamic acid	0.5g
Sodium citrate	0.4g
Ferric ammonium citrate	0.04g
$MgSO_4 \cdot 7H_2O$	0.025g
$ZnSO_4 \cdot 7H_2O$	1.0mg
$CuSO_4 \cdot 5H_2O$	1.0mg
Pyridoxine	1.0mg
Biotin	0.5mg
$CaCl_2 \cdot 2H_2O$	0.5mg
Malachite Green	0.25mg
Middlebrook ADC enrichment	100.0mL
Glycerol	5.0mL

pH 6.6 ± 0.2 at 25°C

Middlebrook ADC Enrichment:
Composition per 100.0mL:

Bovine albumin fraction V	5.0g
Glucose	2.0g
Catalase	0.003g

Source: The medium and enrichment are available as a prepared enrichment from BD Diagnostic Systems.

Preparation of Middlebrook ADC Enrichment: Add components to distilled/deionized water and bring volume to 100.0mL. Mix thoroughly. Filter sterilize.

Preparation of Medium: Add glycerol to 900.0mL of distilled/deionized water and add remaining components, except Middlebrook ADC enrichment. Mix thoroughly. Gently heat and bring to boiling. Autoclave for 15 min at 15 psi pressure–121°C. Cool to 50°–55°C. Aseptically add 100.0mL of sterile Middlebrook ADC enrichment. Mix thoroughly. Pour into sterile Petri dishes or distribute into sterile tubes.

Use: For the isolation, cultivation, and maintenance of *Mycobacterium* species, including *Mycobacterium tuberculosis*. Also used for determining the antimicrobial susceptibility of mycobacteria.

Middlebrook 7H10 Agar with Middlebrook OADC Enrichment (Middlebrook and Cohn 7H10 Agar)

Composition per liter:

Agar	15.0g
Na_2HPO_4	1.5g
KH_2PO_4	1.5g
$(NH_4)_2SO_4$	0.5g
L-Glutamic acid	0.5g
Sodium citrate	0.4g
Ferric ammonium citrate	0.04g
$MgSO_4 \cdot 7H_2O$	0.025g
$ZnSO_4 \cdot 7H_2O$	1.0mg
$CuSO_4 \cdot 5H_2O$	1.0mg
Pyridoxine	1.0mg
Biotin	0.5mg
$CaCl_2 \cdot 2H_2O$	0.5mg
Malachite Green	0.25mg
Middlebrook OADC enrichment	100.0mL
Glycerol	5.0mL

pH 6.6 ± 0.2 at 25°C

Source: This medium is available as a premixed powder from BD Diagnostic Systems.

Middlebrook OADC Enrichment:
Composition per 100.0mL:

Bovine albumin fraction V	5.0g
Glucose	2.0g
NaCl	0.85g
Oleic acid	0.05g
Catalase	4.0mg

Source: This enrichment is available as a prepared enrichment from BD Diagnostic Systems.

Preparation of Middlebrook OADC Enrichment: Add components to distilled/deionized water and bring volume to 100.0mL. Mix thoroughly. Filter sterilize.

Preparation of Medium: Add glycerol to 900.0mL of distilled/deionized water and add remaining components, except Middlebrook OADC enrichment. Mix thoroughly. Gently heat and bring to boiling. Autoclave for 15 min at 15 psi pressure–121°C. Cool to 50°–55°C. Aseptically add 100.0mL of sterile Middlebrook OADC enrichment. Mix thoroughly. Pour into sterile Petri dishes or distribute into sterile tubes.

Use: For the isolation, cultivation, and maintenance of *Mycobacterium* species, including *Mycobacterium tuberculosis.* Also used for determining the antimicrobial susceptibility of mycobacteria.

Middlebrook 7H10 Agar with Middlebrook OADC Enrichment and Hemin (Hemin Medium for *Mycobacterium*)

Composition per liter:

Agar	15.0g
Na_2HPO_4	1.5g
KH_2PO_4	1.5g
$(NH_4)_2SO_4$	0.5g
L-Glutamic acid	0.5g
Sodium citrate	0.4g
Ferric ammonium citrate	0.04g
$MgSO_4 \cdot 7H_2O$	0.025g
$ZnSO_4 \cdot 7H_2O$	1.0mg
$CuSO_4 \cdot 5H_2O$	1.0mg
Pyridoxine	1.0mg
Biotin	0.5mg
$CaCl_2 \cdot 2H_2O$	0.5mg
Malachite Green	0.25mg
Middlebrook OADC enrichment	100.0mL
Glycerol	5.0mL
Hemin solution	3.9mL

pH 6.6 ± 0.2 at 25°C

Source: This medium is available as a premixed powder from BD Diagnostic Systems.

Middlebrook OADC Enrichment:

Composition per 100.0mL:

Bovine albumin fraction V	5.0g
Glucose	2.0g
NaCl	0.85g
Oleic acid	0.05g
Catalase	4.0mg

Preparation of Middlebrook OADC Enrichment: Add components to distilled/deionized water and bring volume to 100.0mL. Mix thoroughly. Filter sterilize.

Hemin Solution:

Composition per 100.0mL:

Hemin	1.0g
NaOH (1*N* solution)	20.0mL

Preparation of Hemin Solution: Add hemin to 20.0mL of 1*N* NaOH solution. Mix thoroughly. Bring volume to 100.0mL with distilled/deionized water.

Preparation of Medium: Add glycerol to 891.1mL of distilled/deionized water and add remaining components, except Middlebrook OADC enrichment. Mix thoroughly. Gently heat and bring to boiling. Autoclave for 15 min at 15 psi pressure–121°C. Cool to 50°–55°C. Aseptically add 100.0mL of sterile Middlebrook OADC enrichment. Mix thoroughly. Pour into sterile Petri dishes or distribute into sterile tubes.

Use: For the isolation, cultivation, and maintenance of *Mycobacterium* species, including *Mycobacterium tuberculosis*. For the cultivation and maintenance of *Mycobacterium haemophilum.* Also used for determining the antimicrobial susceptibility of mycobacteria.

Middlebrook 7H10 Agar with Middlebrook OADC Enrichment and Triton WR 1339

Composition per liter:

Agar	15.0g
Na_2HPO_4	1.5g
KH_2PO_4	1.5g
$(NH_4)_2SO_4$	0.5g
L-Glutamic acid	0.5g
Sodium citrate	0.4g
Ferric ammonium citrate	0.04g
$MgSO_4 \cdot 7H_2O$	0.025g
$ZnSO_4 \cdot 7H_2O$	1.0mg
$CuSO_4 \cdot 5H_2O$	1.0mg
Pyridoxine	1.0mg
Biotin	0.5mg
$CaCl_2 \cdot 2H_2O$	0.5mg
Malachite Green	0.25mg
Middlebrook OADC enrichment with Triton WR 1339	100.0mL
Glycerol	5.0mL

pH 6.6 ± 0.2 at 25°C

Source: This medium is available as a premixed powder from BD Diagnostic Systems.

Middlebrook OADC Enrichment with Triton WR 1339:

Composition per 100.0mL:

Bovine albumin fraction V	5.0g
Glucose	2.0g
NaCl	0.85g
Triton WR 1339	0.25g
Oleic acid	0.05g
Catalase	4.0mg

Source: This enrichment is available as a prepared enrichment from BD Diagnostic Systems.

Preparation of Middlebrook OADC Enrichment with Triton WR 1339: Add components to distilled/deionized water and bring volume to 100.0mL. Mix thoroughly. Filter sterilize.

Preparation of Medium: Add glycerol to 900.0mL of distilled/deionized water and add remaining components, except Middlebrook OADC enrichment with Triton WR-1339. Mix thoroughly. Gently heat and bring to boiling. Autoclave for 15 min at 15 psi pressure–121°C. Cool to 50°–55°C. Aseptically add 100.0mL of sterile Middlebrook OADC enrichment with Triton WR-1339. Mix thoroughly. Pour into sterile Petri dishes or distribute into sterile tubes.

Use: For the isolation, cultivation, and maintenance of *Mycobacterium* species, including *Mycobacterium tuberculosis*. Also used for determining the antimicrobial susceptibility of mycobacteria.

Middlebrook 7H10 Agar with Streptomycin

Composition per liter:

Agar	15.0g
Na_2HPO_4	1.5g
KH_2PO_4	1.5g
$(NH_4)_2SO_4$	0.5g
L-Glutamic acid	0.5g
Sodium citrate	0.4g
Ferric ammonium citrate	0.04g
$MgSO_4 \cdot 7H_2O$	0.025g
$ZnSO_4 \cdot 7H_2O$	1.0mg
$CuSO_4 \cdot 5H_2O$	1.0mg
Pyridoxine	1.0mg
Biotin	0.5mg
$CaCl_2 \cdot 2H_2O$	0.5mg
Malachite Green	0.25mg
Glycerol	5.0mL
Streptomycin	100.0mg

pH 6.6 ± 0.2 at 25°C

Source: This medium is available as a premixed powder from BD Diagnostic Systems.

Preparation of Medium: Add glycerol to 1.0L of distilled/deionized water and add remaining components. Mix thoroughly. Gently heat and bring to boiling. Autoclave for 15 min at 15 psi pressure–121°C. Cool to 50°–55°C. Aseptically add streptomycin. Mix thoroughly. Pour into sterile Petri dishes or distribute into sterile tubes.

Use: For the isolation, cultivation, and maintenance of *Mycobacterium kansasii.*

Middlebrook 7H11 Agar, Selective

Composition per liter:

Agar	15.0g
Na_2HPO_4	1.5g
KH_2PO_4	1.5g
Pancreatic digest of casein	1.0g
$(NH_4)_2SO_4$	0.5g
L-Glutamic acid	0.5g
Sodium citrate	0.4g
$MgSO_4 \cdot 7H_2O$	0.05g
Ferric ammonium citrate	0.04g
Pyridoxine	1.0mg
$ZnSO_4 \cdot 7H_2O$	1.0mg
$CuSO_4 \cdot 5H_2O$	1.0mg
$CaCl_2 \cdot 2H_2O$	0.5mg
Malachite Green	0.25mg
D-Biotin	0.5μg
Middlebrook OADC enrichment	100.0mL
Antibiotic solution	10.0mL
Glycerol	5.0mL

pH 6.6 ± 0.2 at 25°C

Middlebrook OADC Enrichment:

Composition per 100.0mL:

Bovine albumin fraction V	5.0g
Glucose	2.0g
NaCl	0.85g
Oleic acid	0.05g
Catalase	4.0mg

Source: This enrichment is available as a prepared enrichment from BD Diagnostic Systems.

Preparation of Middlebrook OADC Enrichment: Add components to distilled/deionized water and bring volume to 100.0mL. Mix thoroughly. Filter sterilize.

Antibiotic Solution:

Composition per 10.0mL:

Carbenicillin	0.05mg
Trimethoprim lactate	0.02mg
Amphotericin B	0.01mg
Polymyxin B	200,000U

Preparation of Antibiotic Solution: Add components to distilled/deionized water and bring volume to 10.0mL. Mix thoroughly. Filter sterilize.

Preparation of Medium: Add glycerol to 890.0mL of distilled/deionized water and add remaining components, except Middlebrook OADC enrichment and antibiotic solution. Mix thoroughly. Gently heat and bring to boiling. Autoclave for 15 min at 15 psi pressure–121°C. Cool to 50°–55°C. Aseptically add 100.0mL of sterile Middlebrook OADC enrichment and 10.0mL of sterile antibiotic solution. Mix thoroughly. Pour into sterile Petri dishes or distribute into sterile tubes.

Use: For the selective isolation and cultivation of pathogenic mycobacteria from specimens potentially contaminated with bacteria and fungi.

Middlebrook 7H11 Agar with Middlebrook ADC Enrichment (Mycobacteria 7H11 Agar with Middlebrook ADC Enrichment)

Composition per liter:

Agar	15.0g
Na_2HPO_4	1.5g
KH_2PO_4	1.5g
Pancreatic digest of casein	1.0g
$(NH_4)_2SO_4$	0.5g
L-Glutamic acid	0.5g
Sodium citrate	0.4g
$MgSO_4 \cdot 7H_2O$	0.05g
Ferric ammonium citrate	0.04g
Pyridoxine	1.0mg
Malachite Green	0.25mg
D-Biotin	0.5μg
Middlebrook ADC enrichment	100.0mL
Glycerol	5.0mL

pH 6.6 ± 0.2 at 25°C

Source: This medium is available as a premixed powder from BD Diagnostic Systems.

Middlebrook ADC Enrichment:

Composition per 100.0mL:

Bovine albumin fraction V	5.0g
Glucose	2.0g
Catalase	0.003g

Source: This enrichment is available as a prepared enrichment from BD Diagnostic Systems.

Preparation of Middlebrook ADC Enrichment: Add components to distilled/deionized water and bring volume to 100.0mL. Mix thoroughly. Filter sterilize.

Preparation of Medium: Add glycerol to 900.0mL of distilled/deionized water and add remaining components, except Middlebrook ADC enrichment. Mix thoroughly. Gently heat and bring to boiling. Autoclave for 15 min at 15 psi pressure–121°C. Cool to 50°–55°C. Aseptically add 100.0mL of sterile Middlebrook ADC enrichment. Mix thoroughly. Pour into sterile Petri dishes or distribute into sterile tubes.

Use: For the cultivation of drug-resistant (isoniazid [INH]) strains of *Mycobacterium tuberculosis.* For the cultivation of particularly fastidious strains of tubercle bacilli that occur following treatment of tuberculosis patients with secondary antitubercular drugs. Generally, these strains fail to grow on 7H10 medium.

Middlebrook 7H11 Agar with Middlebrook OADC Enrichment (Mycobacteria 7H11 Agar with Middlebrook OADC Enrichment)

Composition per liter:

Agar	15.0g
Na_2HPO_4	1.5g
KH_2PO_4	1.5g
Pancreatic digest of casein	1.0g
$(NH_4)_2SO_4$	0.5g
L-Glutamic acid	0.5g
Sodium citrate	0.4g
$MgSO_4 \cdot 7H_2O$	0.05g
Ferric ammonium citrate	0.04g
Pyridoxine	1.0mg
Malachite Green	0.25mg
D-Biotin	0.5μg
Middlebrook OADC enrichment	100.0mL
Glycerol	5.0mL

pH 6.6 ± 0.2 at 25°C

Source: This medium is available as a premixed powder from BD Diagnostic Systems.

Middlebrook OADC Enrichment:

Composition per 100.0mL:

Bovine albumin fraction V	5.0g
Glucose	2.0g
NaCl	0.85g
Oleic acid	0.05g
Catalase	4.0mg

Source: This enrichment is available as a prepared enrichment from BD Diagnostic Systems.

Preparation of Middlebrook OADC Enrichment: Add components to distilled/deionized water and bring volume to 100.0mL. Mix thoroughly. Filter sterilize.

Preparation of Medium: Add glycerol to 900.0mL of distilled/deionized water and add remaining components, except Middlebrook OADC enrichment. Mix thoroughly. Gently heat and bring to boiling. Autoclave for 15 min at 15 psi pressure–121°C. Cool to 50°–55°C. Aseptically add 100.0mL of sterile Middlebrook OADC enrichment. Mix thoroughly. Pour into sterile Petri dishes or distribute into sterile tubes.

Use: For the cultivation of drug-resistant (isoniazid [INH]) strains of *Mycobacterium tuberculosis.* For the cultivation of particularly fastidious strains of tubercle bacilli that occur following treatment of tuberculosis patients with secondary antitubercular drugs. Generally, these strains fail to grow on 7H10 medium.

Middlebrook 7H11 Agar with Middlebrook OADC Enrichment and Triton WR 1339 (Mycobacteria 7H11 Agar with Middlebrook OADC Enrichment and Triton WR 1339)

Composition per liter:

Agar	15.0g
Na_2HPO_4	1.5g
KH_2PO_4	1.5g
Pancreatic digest of casein	1.0g
$(NH_4)_2SO_4$	0.5g
L-Glutamic acid	0.5g
Sodium citrate	0.4g
$MgSO_4 \cdot 7H_2O$	0.05g
Ferric ammonium citrate	0.04g
Pyridoxine	1.0mg
Malachite Green	0.25mg
D-Biotin	0.5μg
Middlebrook OADC enrichment with Triton WR 1339	100.0mL
Glycerol	5.0mL

pH 6.6 ± 0.2 at 25°C

Source: This medium is available as a premixed powder from BD Diagnostic Systems.

Middlebrook OADC Enrichment with Triton WR 1339:

Composition per 100.0mL:

Bovine albumin fraction V	5.0g
Glucose	2.0g
NaCl	0.85g
Triton WR 1339	0.25g
Oleic acid	0.05g
Catalase	4.0mg

Source: This enrichment is available as a prepared enrichment from BD Diagnostic Systems.

Preparation of Middlebrook OADC Enrichment with Triton WR 1339: Add components to distilled/deionized water and bring volume to 100.0mL. Mix thoroughly. Filter sterilize.

Preparation of Medium: Add glycerol to 900.0mL of distilled/deionized water and add remaining components, except Middlebrook OADC enrichment with Triton WR-1339. Mix thoroughly. Gently heat and bring to boiling. Autoclave for 15 min at 15 psi pressure–121°C. Cool to 50°–55°C. Aseptically add 100.0mL of sterile Middlebrook OADC enrichment with Triton WR-1339. Mix thoroughly. Pour into sterile Petri dishes or distribute into sterile tubes.

Use: For the cultivation of drug-resistant (isoniazid [INH]) strains of *Mycobacterium tuberculosis*. For the cultivation of particularly fastidious strains of tubercle bacilli that occur following treatment of tuberculosis patients with secondary antitubercular drugs. Generally, these strains fail to grow on 7H10 medium.

Middlebrook 7H12 Medium

Composition per 102.5mL:

Bovine serum albumin	0.5g
Casein hydrolyslate	0.1g
Catalase	4800U
^{14}C-Palmitic acid	100μCi
Middlebrook 7H9 broth	100.0mL
Antibiotic solution	2.5mL

pH 6.8 ± 0.1 at 25°C

Middlebrook 7H9 Broth:

Composition per liter:

Na_2HPO_4	2.5g
KH_2PO_4	1.0g
Monosodium glutamate	0.5g
$(NH_4)_2SO_4$	0.5g
Sodium citrate	0.1g
$MgSO_4 \cdot 7H_2O$	0.05g
Ferric ammonium citrate	0.04g
$CuSO_4 \cdot 5H_2O$	1.0mg
Pyridoxine	1.0mg
$ZnSO_4 \cdot 7H_2O$	1.0mg
Biotin	0.5mg
$CaCl_2 \cdot 2H_2O$	0.5mg
Glycerol	2.0mL

Preparation of Middlebrook 7H9 Broth: Add components to distilled/deionized water and bring volume to 1.0L. Mix thoroughly.

Antibiotic Solution:

Composition per 5.0mL:

Nalidixic acid	0.2g
Azlocillin	0.1g
Amphotericin B	0.05g

Trimethoprim	0.05g
Polymyxin B	500,000U

Preparation of Antibiotic Solution: Add components to distilled/deionized water and bring volume to 5.0mL. Mix thoroughly. Filter sterilize.

Preparation of Medium: To 100.0mL of Middlebrook 7H9 broth, add remaining components, except antibiotic solution. Mix thoroughly. Filter sterilize. Aseptically distribute into bottles in 4.0mL volumes. Prior to inoculation, aseptically add 0.1mL of antibiotic solution to each bottle. Mix thoroughly.

Use: For the cultivation of *Mycobacterium* species from the blood of patients suspected of having mycobacteremia.

Middlebrook Medium (DSMZ Medium 645)

Composition per liter:

Agar	15.0g
Na_2HPO_4	1.5g
KH_2PO_4	1.5g
$(NH_4)_2SO_4$	0.5g
L-Glutamic acid	0.5g
Sodium citrate	0.4g
Ferric ammonium citrate	0.04g
$MgSO_4 \cdot 7H_2O$	0.025g
$ZnSO_4 \cdot 7H_2O$	1.0mg
$CuSO_4 \cdot 5H_2O$	1.0mg
Pyridoxine	1.0mg
Biotin	0.5mg
$CaCl_2 \cdot 2H_2O$	0.5mg
Malachite Green	0.25mg
Middlebrook OADC enrichment	100.0mL
Glycerol	5.0mL

pH 6.6 ± 0.2 at 25°C

Source: This medium is available as a premixed powder from BD Diagnostic Systems.

Middlebrook OADC Enrichment:

Composition per 100.0mL:

Bovine albumin fraction V	5.0g
Glucose	2.0g
Catalase	0.003g
Distilled water	100.0mL

Source: This enrichment is available as a prepared enrichment from BD Diagnostic Systems.

Preparation of Middlebrook OADC Enrichment: Add components to distilled/deionized water and bring volume to 100.0mL. Mix thoroughly. Filter sterilize.

Preparation of Medium: Add glycerol to 900.0mL of distilled/deionized water and add remaining components, except Middlebrook OADC enrichment. Mix thoroughly. Gently heat and bring to boiling. Autoclave for 15 min at 15 psi pressure–121°C. Cool to 50°–55°C. Aseptically add 100.0mL of sterile Middlebrook OADC enrichment. Mix thoroughly. Pour into sterile Petri dishes or distribute into sterile tubes.

Use: For the isolation, cultivation, and maintenance of *Mycobacterium* species.

Middlebrook Medium with Mycobactin (DSMZ Medium 780)

Composition per liter:

Agar	15.0g
Na_2HPO_4	1.5g
KH_2PO_4	1.5g
$(NH_4)_2SO_4$	0.5g
L-Glutamic acid	0.5g
Sodium citrate	0.4g
Ferric ammonium citrate	0.04g
$MgSO_4 \cdot 7H_2O$	0.025g
Mycobactin J	2.0mg
$ZnSO_4 \cdot 7H_2O$	1.0mg
$CuSO_4 \cdot 5H_2O$	1.0mg
Pyridoxine	1.0mg
Biotin	0.5mg
$CaCl_2 \cdot 2H_2O$	0.5mg
Malachite Green	0.25mg
Middlebrook ADC enrichment	100.0mL
Glycerol	5.0mL

pH 6.6 ± 0.2 at 25°C

Source: Mycobactin J is available from Allied Laboratories, Inc.

Middlebrook ADC Enrichment:

Composition per 100.0mL:

Bovine albumin fraction V	5.0g
Glucose	2.0g
Catalase	0.003g
Distilled water	100.0mL

Source: This medium and enrichment is available from BD Diagnostic Systems.

Preparation of Middlebrook ADC Enrichment: Add components to distilled/deionized water and bring volume to 100.0mL. Mix thoroughly. Filter sterilize.

Preparation of Medium: Add glycerol to 900.0mL of distilled/deionized water and add remaining components, except Middlebrook ADC enrichment and mycobactin. Mix thoroughly. Gently heat and bring to boiling. Autoclave for 15 min at 15 psi pressure–121°C. Cool to 50°–55°C. Aseptically add 100.0mL of sterile Middlebrook ADC enrichment and mycobactin. The mycobactin is dissolved in 2.0mL ethanol. Be sure to add all of the mycobactin; wash with additional 2.0mL ethanol if needed. Mix thoroughly. Pour into sterile Petri dishes or distribute into sterile tubes.

Use: For the cultivation of *Mycobacterium avium* subsp. *paratuberculosis.*

Middlebrook OADC Enrichment (Middlebrook Oleic Albumin Dextrose Catalase Enrichment)

Composition per 100.0mL:

Bovine albumin fraction V	5.0g
Glucose	2.0g
NaCl	0.85g
Oleic acid	0.05g
Catalase	4.0mg

Source: This medium is available as a prepared enrichment from BD Diagnostic Systems.

Preparation of Enrichment: Add components to distilled/deionized water and bring volume to 100.0mL. Mix thoroughly. Filter sterilize.

Use: For use as a supplement to other Middlebrook media for the isolation, cultivation, and maintenance of *Mycobacterium* species. Also used as a supplement to other Middlebrook media for determining the antimicrobial susceptibility of mycobacteria.

Middlebrook OADC Enrichment with Triton WR 1339 (Middlebrook Oleic Albumin Dextrose Catalase Enrichment with Triton WR 1339)

Composition per 100.0mL:

Bovine albumin fraction V	5.0g
Glucose	2.0g
NaCl	0.85g
Triton WR 1339	0.25g
Oleic acid	0.05g
Catalase	4.0mg

Source: This medium is available as a prepared enrichment from BD Diagnostic Systems.

Preparation of Enrichment: Add components to distilled/deionized water and bring volume to 100.0mL. Mix thoroughly. Filter sterilize.

Use: For use as a supplement to other Middlebrook media for the isolation, cultivation, and maintenance of *Mycobacterium* species. Also used as a supplement to other Middlebrook media for determining the antimicrobial susceptibility of mycobacteria.

MIL Medium (Motility Indole Lysine Medium)

Composition per liter:

Peptone	10.0g
Pancreatic digest of casein	10.0g
L-Lysine·HCl	10.0g
Yeast extract	3.0g
Agar	2.0g
Dextrose	1.0g
Ferric ammonium citrate	0.5g
Bromcresol Purple	0.02g

pH 6.6 ± 0.2 at 25°C

Source: This medium is available as a premixed powder and prepared medium from BD Diagnostic Systems.

Preparation of Medium: Add components to distilled/deionized water and bring volume to 1.0L. Mix thoroughly. Gently heat and bring to boiling. Distribute into tubes in 5.0mL volumes. Autoclave for 15 min at 15 psi pressure–121°C.

Use: For the cultivation and differentiation of members of the Enterobacteriaceae on the basis of motility, lysine decarboxylase activity, lysine deaminase activity, and indole production.

Minimum Essential Medium with Bicarbonate, Serum, and Antibiotics

Composition per 1100.0mL:

Minimum essential medium	950.0mL
Fetal bovine serum, heat inactivated	100.0mL
$NaHCO_3$ solution	40.0mL
Penicillin-streptomycin solution	10.0mL

pH 7.2 ± 0.2 at 25°C

Minimum Essential Medium (MEM):

Composition per liter:

Inorganic salt solution	400.0mL
Other component solution	400.0mL
Amino acid solution	100.0mL
Vitamin solution	100.0mL

Inorganic Salt Solution:

Composition per 400.0mL:

NaCl	6.8g
KCl	0.4g
$CaCl_2$, anhydrous	0.2g
$NaH_2PO_4 \cdot H_2O$	0.14g
$MgSO_4$, anhydrous	97.67mg

Preparation of Inorganic Salt Solution: Add components to distilled/deionized water and bring volume to 400.0mL. Mix thoroughly. Autoclave for 15 min at 15 psi pressure–121°C. Cool to 25°C.

Other Component Solution:
Composition per 400.0mL:

D-Glucose	1.0g
Phenol Red	10.0mg

Preparation of Other Component Solution: Add components to distilled/deionized water and bring volume to 400.0mL. Mix thoroughly. Autoclave for 15 min at 15 psi pressure–121°C. Cool to 25°C.

Amino Acid Solution:
Composition per 100.0mL:

L-Glutamine	292.0mg
L-Arginine·HCl	126.1mg
L-Lysine·HCl	72.5mg
L-Isoleucine	52.0mg
L-Leucine	52.0mg
L-Tyrosine, disodium salt	52.0mg
L-Threonine	48.0mg
L-Valine	46.0mg
L-Histidine·HCl·H_2O	42.0mg
L-Phenylalanine	32.0mg
L-Cysteine·2HCl	31.0mg
L-Methionine	15.0mg
L-Glutamic acid	14.7mg
L-Aspartic acid	13.3mg
L-Asparagine·H_2O	13.2mg
L-Proline	11.5mg
L-Serine	10.5mg
L-Tryptophan	10.0mg
L-Alanine	8.9mg
Glycine	7.5mg

Preparation of Amino Acid Solution: Add components to distilled/deionized water and bring volume to 100.0mL. Mix thoroughly. Filter sterilize.

Vitamin Solution:
Composition per 100.0mL:

i-Inositol	2.0mg
D-Ca pantothenate	1.0mg
Choline chloride	1.0mg
Folic acid	1.0mg
Niacinamide	1.0mg
Pyridoxal·HCl	1.0mg
Thiamine·HCl	1.0mg
Riboflavin	0.1mg

Preparation of Vitamin Solution: Add components to distilled/deionized water and bring volume to 100.0mL. Mix thoroughly. Filter sterilize.

Preparation of Minimum Essential Medium (MEM): Aseptically combine 400.0mL of sterile inorganic salt solution, 400.0mL of sterile other component solution, 100.0mL of sterile amino acid solution, and 100.0mL of sterile vitamin solution.

$NaHCO_3$ Solution:
Composition per 40.0mL:

$NaHCO_3$	2.0g

Preparation of $NaHCO_3$ Solution: Add $NaHCO_3$ to distilled/deionized water and bring volume to 40.0mL. Mix thoroughly. Filter sterilize.

Penicillin-Streptomycin Solution:

Composition per 10.0mL:

Penicillin	0.01g
Streptomycin	0.01g

Preparation of Penicillin-Streptomycin Solution: Add components to distilled/deionized water and bring volume to 10.0mL. Mix thoroughly. Filter sterilize.

Preparation of Medium: Aseptically combine 950.0mL of sterile minimum essential medium, 100.0mL of sterile heat inactivated fetal bovine serum, 40.0mL of sterile $NaHCO_3$ solution, and 10.0mL of sterile penicillin-streptomycin solution. Adjust pH to 7.2 with humidified 10% CO_2 in 90% air.

Use: For the cultivation of *Encephalitozoon cuniculi*, *Encephalitozoon hellem*, *Encephalitozoon intestinalis*, *Naegleria fowleri*, and *Nosema corneum*.

Mitis-Salivarius Agar

Composition per liter:

Sucrose	50.0g
Agar	15.0g
Enzymatic digest of protein	10.0g
Proteose peptone	10.0g
K_2HPO_4	4.0g
Dextrose	1.0g
Trypan Blue	0.08g
Crystal Violet	0.8mg
Na_2TeO_3 solution	1.0mL

pH 7.0 ± 0.2 at 25°C

Source: This medium is available as a premixed powder from BD Diagnostic Systems.

Na_2TeO_3 Solution:
Composition per 10.0mL:

Na_2TeO_3	0.1g

Preparation of Na_2TeO_3 Solution: Add Na_2TeO_3 to 10.0mL of distilled/deionized water. Mix thoroughly. Filter sterilize.

Caution: Potassium tellurite is toxic.

Preparation of Medium: Add components to distilled/deionized water and bring volume to 999.0mL. Mix thoroughly. Gently heat and bring to boiling. Autoclave for 15 min at 15 psi pressure–121°C. Cool medium to 50°–55°C. Aseptically add

1.0mL of the sterile Na_2TeO_3 solution to the cooled basal medium. Mix thoroughly. Pour into sterile Petri dishes or distribute into sterile tubes.

Use: For the selective isolation of *Streptococcus mitis, Streptococcus salivarius,* and other viridans streptococci and enterococci.

Møller Decarboxylase Broth

Composition per liter:

Amino acid	10.0g
Peptic digest of animal tissue	5.0g
Beef extract	5.0g
Glucose	0.5g
Bromcresol Purple	0.01g
Cresol Red	5.0mg
Pyridoxal	5.0mg

pH 6.0 ± 0.2 at 25°C

Source: This medium is available as a premixed powder from BD Diagnostic Systems.

Preparation of Medium: Add components to distilled/deionized water and bring volume to 1.0L. Use L-lysine, L-arginine, or L-ornithine. Mix thoroughly. Gently heat until dissolved. Distribute into screw-capped tubes in 5.0mL volumes. Autoclave for 15 min at 15 psi pressure–121°C. A slight precipitate may form in the ornithine broth.

Use: For the differentiation of Gram-negative enteric bacteria based on the production of arginine dihydrolase, lysine decarboxylase, or ornithine decarboxylase.

Molybdate Agar

Composition per101.5mL:

Base	100.0mL
Phosphomolybdic acid solution	1.5mL

pH 5.3 ± 0.2 at 25°C

Base:

Composition per liter:

Sucrose	40.0g
Agar	15.0g
Meat peptone	10.0g

Preparation of Base: Add components to distilled/deionized water and bring volume to 1.0L. Mix thoroughly. Adjust pH to 7.6. Gently heat and bring to boiling. Autoclave for 15 min at 15 psi pressure–121°C. Cool to 45°–50°C.

Phosphomolybdic Acid Solution:

Composition per 100.0mL:

$P_2O_5 \cdot 20MoO_3$	12.5g

Preparation of Phosphomolybdic Acid Solution: Add $P_2O_5 \cdot 20MoO_3$ (phospho-12-molybdic acid, 12–molybdophosphoric acid, or PMA) to sterile distilled/deionized water. Mix thoroughly. Do not adjust pH.

Preparation of Medium: To 100.0mL of cooled sterile base, add 1.5mL of phosphomolybdic acid solution. Mix thoroughly. Pour into sterile Petri dishes or distribute into sterile tubes.

Use: For the isolation and presumptive identification of yeast, especially *Candida* species. *Candida albicans* appears as smooth, medium olive colonies with medium olive bottoms. *Candida stellatoidea* appears as shiny, light gray colonies with light gray bottoms. *Candida tropicalis* appears as smooth, shiny, dark blue/gray colonies with dark blue/gray bottoms. *Candida krusei* appears as smooth, dull white colonies with white bottoms. *Saccharomyces cerevisiae* appears as smooth, shiny light blue/dark blue colonies with dark blue/green bottoms.

Monsur Agar (Taurocholate Tellurite Gelatin Agar)

Composition per liter:

Gelatin	30.0g
Agar	15.0g
Casein peptone	10.0g
NaCl	10.0g
Sodium taurocholate	5.0g
$Na_2CO_3 \cdot H_2O$	1.0g
K_2TeO_3 solution	10.0mL

pH 8.5 ± 0.2 at 25°C

K_2TeO_3 Solution:

Composition per 10.0mL:

K_2TeO_3	0.02g

Preparation of K_2TeO_3 Solution: Add K_2TeO_3 to 10.0mL of distilled/deionized water. Mix thoroughly. Filter sterilize.

Caution: Potassium tellurite is toxic.

Preparation of Medium: Add components, except K_2TeO_3 solution, to distilled/deionized water and bring volume to 990.0mL. Mix thoroughly. Gently heat and bring to boiling. Autoclave for 15 min at 15 psi pressure–121°C. Cool to 45°–50°C. Add 10.0mL of sterile K_2TeO_3 solution. Mix thoroughly. Pour into sterile Petri dishes or distribute into sterile tubes.

Use: For the isolation of *Vibrio cholerae* from fecal specimens.

Motility GI Medium

Composition per liter:

Gelatin	53.4g
Heart infusion broth	25.0g
Agar	3.0g

pH 7.2 ± 0.2 at 25°C

Source: This medium is available as a premixed powder from BD Diagnostic Systems.

Preparation of Medium: Add components to cold distilled/deionized water and bring to 1.0L. Mix thoroughly. Gently heat and bring to boiling. Distribute into tubes or flasks. Autoclave for 15 min at 15 psi pressure–121°C. Pour into sterile Petri dishes in 20.0mL volumes or leave in tubes.

Use: For demonstrating the motility of microorganisms and for separating organisms in their motile phase.

Motility Indole Ornithine Medium (MIO Medium)

Composition per liter:

Pancreatic digest of gelatin	10.0g
Pancreatic digest of casein	9.5g
L-Ornithine·HCl	5.0g
Yeast extract	3.0g
Agar	2.0g
Glucose	1.5g
Bromcresol Purple	0.02g

pH 6.6 ± 0.2 at 25°C

Source: This medium is available as a premixed powder from BD Diagnostic Systems.

Preparation of Medium: Add components to distilled/deionized water and bring to 1.0L. Mix thoroughly. Gently heat and bring to boiling. Distribute into tubes or flasks. Autoclave for 15 min at 15 psi pressure–121°C.

Use: For the differentiation of Gram-negative enteric bacteria based on their motility, indole production, and ornithine decarboxylase activity.

Motility Medium S

Composition per liter:

Beef heart, solids from infusion	500.0g
Gelatin	30.0g
Enzymatic hydrolyzate of protein	10.0g
NaCl	5.0g
K_2HPO_4	2.0g
KNO_3	2.0g
Agar	1.0g
2,3,5-Triphenyltetrazolium chloride solution	10.0mL

pH 7.2 ± 0.2 at 25°C

2,3,5-Triphenyltetrazolium Chloride Solution:

Composition per 10.0mL:

2,3,5-Triphenyltetrazolium chloride	0.1g

Preparation of 2,3,5-Triphenyltetrazolium Chloride Solution: Add 2,3,5-triphenyltetrazolium chloride to distilled/deionized water and bring volume to 10.0mL. Mix thoroughly. Filter sterilize.

Preparation of Medium: Add components, except 2,3,5-triphenyltetrazolium chloride solution, to distilled/deionized water and bring volume to 990.0mL. Mix thoroughly. Gently heat while stirring and bring to boiling. Autoclave for 15 min at 15 psi pressure–121°C. Cool to 60°C. Aseptically add 10.0mL of the sterile 2,3,5-triphenyltetrazolium chloride solution. Mix thoroughly. Aseptically distribute into sterile tubes. Keep at 4°–8°C until used.

Use: For the determination of bacterial motility.

Motility Nitrate Agar

Composition per liter:

Beef heart, solids from infusion	100.0g
Tryptose	12.0g
Agar	3.0g
NaCl	1.0g
KNO_3	1.0g
Glucose	0.5g

pH 7.4 ± 0.2 at 25°C

Preparation of Medium: Add components to distilled/deionized water and bring volume to 1.0L. Mix thoroughly. Gently heat and bring to boiling. Distribute into tubes in 4.0mL volumes. Autoclave for 15 min at 15 psi pressure–121°C.

Use: For the cultivation and observation of motility and nitrate reduction in a variety of Gram-negative bacteria.

Motility Nitrate Medium (FDA M101)

Composition per liter:

Beef heart, solids from infusion	100.0g
Tryptose	12.0g
Agar	3.0g
NaCl	1.0g
KNO_3, nitrite free	1.0g

pH 7.4 ± 0.2 at 25°C

Preparation of Medium: Add components to distilled/deionized water and bring volume to 1.0L. Mix thoroughly. Gently heat and bring to boiling. Distribute into screw-capped tubes in 4.0mL volumes. Autoclave for 15 min at 15 psi pressure–121°C.

Use: For the cultivation and differentiation of Gram-negative nonfermentative bacteria from cosmetics based on their motility and their ability to reduce nitrate to nitrite.

Motility Nitrate Medium

Composition per liter:

Beef heart, solids from infusion....................100.0g
Tryptose....................12.0g
Agar....................3.0g
NaCl....................1.0g
KNO_3....................1.0g

Preparation of Medium: Add components to distilled/deionized water and bring volume to 1.0L. Mix thoroughly. Gently heat and bring to boiling. Distribute into tubes in 4.0mL volumes. Autoclave for 15 min at 15 psi pressure–121°C.

Use: For the cultivation and observation of motility and nitrate reduction in a variety of Gram-negative nonfermentative bacteria isolated from cosmetics.

Motility Nitrate Medium, Buffered

Composition per liter:

Peptone....................5.0g
Galactose....................5.0g
Agar....................3.0g
Beef extract....................3.0g
Na_2HPO_4....................2.5g
KNO_3....................1.0g
Glycerin....................5.0mL

pH 7.3 ± 0.1 at 25°C

Preparation of Medium: Add components, except agar, to distilled/deionized water and bring volume to 1.0L. Mix thoroughly. Gently heat until dissolved. Add agar. Gently heat until boiling. Distribute into tubes in 11.0mL volumes. Autoclave for 15 min at 15 psi pressure–121°C. If not used within 4 hr, heat tubes to 100°C for 10 min.

Use: For the cultivation and observation of the motility of *Clostridium perfringens.*

Motility Sulfide Medium

Composition per liter:

Gelatin....................80.0g
Proteose peptone....................10.0g
NaCl....................5.0g
Agar....................4.0g
Beef extract....................3.0g
Sodium citrate....................2.0g
L-Cystine....................0.2g
Ferrous ammonium citrate....................0.2g

pH 7.3 ± 0.2 at 25°C

Source: This medium is available as a premixed powder from BD Diagnostic Systems.

Preparation of Medium: Add components to distilled/deionized water and bring volume to 1.0L. Mix thoroughly. Gently heat while stirring and bring to boiling. Distribute into tubes in 4–5.0mL volumes. Autoclave for 15 min at 10 psi pressure–116°C.

Use: For the determination of bacterial motility and the ability of bacteria to produce H_2S from L-cystine. For the differentiation of Gram-negative bacteria of the Enterobacteriaceae.

Motility Test and Maintenance Medium

Composition per liter:

Peptone....................10.0g
NaCl....................5.0g
Agar....................4.0g
Beef extract....................3.0g
2,3,5-Triphenyltetrazolium chloride....................0.05g

Preparation of Medium: Add components to distilled/deionized water and bring volume to 1.0L. Mix thoroughly. Distribute into screw-capped tubes in 8.0mL volumes. Autoclave for 15 min at 15 psi pressure–121°C.

Use: For the cultivation, maintenance, and observation of the motility of *Listeria monocytogenes.*

Motility Test and Maintenance Medium

Composition per liter:

Agar....................9.0g
Tryptose....................8.0g
NaCl....................5.0g
Pancreatic digest of gelatin....................2.5g
Beef extract....................1.5g

pH 7.2 ± 0.1 at 25°C

Preparation of Medium: Add components to distilled/deionized water and bring volume to 1.0L. Mix thoroughly. Gently heat and bring to boiling. Distribute into tubes in 7.0mL volumes. Autoclave for 15 min at 15 psi pressure–121°C. Cool to 45°–50°C. Pass the cooled tubes into an anaerobic chamber containing 85% N_2 + 10% H_2 + 5% CO_2.

Use: For the cultivation, maintenance, and observation of the motility in a variety of anaerobic bacteria.

Motility Test and Maintenance Medium

Composition per liter:

Peptone....................10.0g
NaCl....................5.0g
Agar....................4.0g
Beef extract....................3.0g

pH 7.4 ± 0.1 at 25°C

Preparation of Medium: Add components to distilled/deionized water and bring volume to 1.0L. Mix

thoroughly. Distribute into screw-capped tubes in 8.0mL volumes. Autoclave for 15 min at 15 psi pressure–121°C.

Use: For the cultivation, maintenance, and observation of motility in members of the Enterobacteriaceae.

Motility Test and Maintenance Medium, Gilardi

Composition per liter:

Pancreatic digest of casein	10.0g
NaCl	5.0g
Agar	3.0g
Yeast extract	3.0g

pH 7.2 ± 0.1 at 25°C

Preparation of Medium: Add components to distilled/deionized water and bring volume to 1.0L. Mix thoroughly. Distribute into screw-capped tubes in 3.5mL volumes. Autoclave for 15 min at 15 psi pressure–121°C.

Use: For the cultivation, maintenance, and observation of motility in nonfermenting Gram–negative bacteria.

Motility Test and Maintenance Medium, Tatum

Composition per liter:

Tryptose	8.0g
NaCl	5.0g
Agar	4.0g
Pancreatic digest of gelatin	2.5g
Beef extract	1.5g

pH 6.9 ± 0.2 at 25°C

Preparation of Medium: Add components to distilled/deionized water and bring volume to 1.0L. Mix thoroughly. Distribute into screw-capped tubes in 8.0mL volumes. Autoclave for 15 min at 15 psi pressure–121°C.

Use: For the cultivation, maintenance, and observation of motility in nonfermenting Gram–negative bacteria.

Motility Test Medium

Composition per liter:

Pancreatic digest of gelatin	10.0g
NaCl	5.0g
Agar	4.0g
Beef extract	3.0g

pH 7.3 ± 0.2 at 25°C

Source: This medium is available as a premixed powder from BD Diagnostic Systems.

2,3,5-Triphenyltetrazolium Chloride Solution:

Composition per 10.0mL:

2,3,5-Triphenyltetrazolium chloride	0.1g

Preparation of 2,3,5-Triphenyltetrazolium Chloride Solution: Add 2,3,5-triphenyltetrazolium chloride to distilled/deionized water and bring volume to 10.0mL. Mix thoroughly. Filter sterilize.

Preparation of Medium: Add components to distilled/deionized water and bring volume to 995.0mL. Mix thoroughly. Gently heat while stirring and bring to boiling. Autoclave for 15 min at 15 psi pressure–121°C. Cool to 45°–50°C. Aseptically add 5.0mL of sterile 2,3,5-triphenyltetrazolium chloride solution. Mix thoroughly. Aseptically distribute into sterile tubes.

Use: For detection of the motility of Gram-negative enteric bacteria.

Motility Test Medium

Composition per liter:

Tryptose	10.0g
NaCl	5.0g
Agar	5.0g

pH 7.2 ± 0.2 at 25°C

Source: This medium is available as a premixed powder from BD Diagnostic Systems.

Preparation of Medium: Add components to distilled/deionized water and bring volume to 1.0L. Mix thoroughly. Gently heat while stirring and bring to boiling. Distribute into tubes in 4–5.0mL volumes. Autoclave for 15 min at 15 psi pressure–121°C. Cool tubes quickly in an upright position.

Use: For the determination of bacterial motility.

Motility Test Medium, Semisolid

Composition per liter:

Peptone	10.0g
NaCl	5.0g
Agar	4.0g
Beef extract	3.0g

pH 7.4 ± 0.2 at 25°C

Preparation of Medium: Add components to distilled/deionized water and bring volume to 1.0L. Mix thoroughly. Gently heat while stirring and bring to boiling. Distribute into screw-capped tubes in 8.0mL or 20.0mL volumes. Autoclave for 15 min at 15 psi pressure–121°C. Pour into sterile Petri dishes in 20.0mL volumes or leave in tubes.

Use: For the cultivation and observation of motility in a variety of bacteria, especially *Salmonella* species.

MRS Agar
(DeMan, Rogosa, Sharpe Agar)

Composition per liter:

Glucose	18.5g
Agar	13.5g
Pancreatic digest of gelatin	10.0g
Beef extract	8.0g
Yeast extract	4.0g
Sodium acetate	3.0g
K_2HPO_4	2.0g
Ammonium citrate	2.0g
Polysorbate 80	1.0g
$MgSO_4 \cdot 7H_2O$	0.2g
$MnSO_4 \cdot 4H_2O$	0.05g

pH 6.2 ± 0.2 at 25°C

Source: This medium is available as a premixed powder from BD Diagnostic Systems.

Preparation of Medium: Add components to distilled/deionized water and bring volume to 1.0L. Mix thoroughly. Gently heat while stirring and bring to boiling. Distribute into tubes or flasks. Autoclave for 15 min at 15 psi pressure–121°C. Pour into sterile Petri dishes or leave in tubes.

Use: For the isolation and cultivation of *Lactobacillus* species from clinical specimens.

MRS Broth
(DeMan, Rogosa, Sharpe Broth)

Composition per liter:

Glucose	18.5g
Pancreatic digest of gelatin	10.0g
Beef extract	8.0g
Yeast extract	4.0g
Sodium acetate	3.0g
K_2HPO_4	2.0g
Ammonium citrate	2.0g
Polysorbate 80	1.0g
$MgSO_4 \cdot 7H_2O$	0.2g
$MnSO_4 \cdot 4H_2O$	0.05g

pH 6.2 ± 0.2 at 25°C

Source: This medium is available as a premixed powder from BD Diagnostic Systems.

Preparation of Medium: Add components to distilled/deionized water and bring volume to 1.0L. Mix thoroughly. Distribute into tubes or flasks. Autoclave for 15 min at 15 psi pressure–121°C.

Use: For the isolation and cultivation of *Lactobacillus* species from clinical specimens.

MRVP Broth
(Methyl Red- Voges-Proskauer Broth)

Composition per liter:

Glucose	5.0g
KH_2PO_4	5.0g
Pancreatic digest of casein	3.5g
Peptic digest of animal tissue	3.5g

pH 6.9 ± 0.2 at 25°C

Source: Available as a premixed powder from BD Diagnostic Systems and as a prepared medium from BD Diagnostic Systems.

Preparation of Medium: Add components to distilled/deionized water and bring volume to 1.0L. Mix thoroughly. Distribute into tubes or flasks. Autoclave for 15 min at 15 psi pressure–121°C.

Use: For the differentiation of bacteria based on acid production (Methyl Red test) and acetoin production (Voges-Proskauer reaction).

MRVP Medium
(Methyl Red Voges-Proskauer Medium)

Composition per liter:

Glucose	5.0g
Peptone	5.0g
Phosphate buffer	5.0g

pH 7.5 ± 0.2 at 25°C

Source: This medium is available as a premixed powder from Oxoid Unipath.

Preparation of Medium: Add components to distilled/deionized water and bring volume to 1.0L. Mix thoroughly. Distribute into tubes or flasks. Autoclave for 15 min at 15 psi pressure–121°C.

Use: For the differentiation of bacteria based on acid production (Methyl Red test) and acetoin production (Voges-Proskauer reaction).

Mucate Broth

Composition per liter:

Mucic acid	10.0g
Peptone	10.0g
Bromthymol Blue	0.024g

pH 7.4 ± 0.1 at 25°C.

Preparation of Medium: Add components to distilled/deionized water and bring volume to 1.0L. Mix thoroughly. Add 5*N* NaOH while stirring until mucic acid dissolves. Distribute into screw-capped tubes in 5.0mL volumes. Autoclave for 10 min at 15 psi pressure–121°C.

Use: For the isolation and cultivation of enterovirulent *Escherichia coli* and *Shigella* species.

Mucate Control Broth

Composition per liter:

Peptone	10.0g
Bromthymol Blue	0.024g

pH 7.4 ± 0.1 at 25°C

Preparation of Medium: Add components to distilled/deionized water and bring volume to 1.0L. Mix thoroughly. Distribute into screw-capped tubes in 5.0mL volumes. Autoclave for 10 min at 15 psi pressure–121°C.

Use: For the isolation and cultivation of enterovirulent *Escherichia coli* and *Shigella* species.

Mueller-Hinton Agar

Composition per liter:

Beef infusion	300.0g
Acid hydrolysate of casein	17.5g
Agar	17.0g
Starch	1.5g

pH 7.4 ± 0.2 at 25°C

Source: This medium is available as a premixed powder from BD Diagnostic Systems and Oxoid Unipath.

Preparation of Medium: Add components to distilled/deionized water and bring to 1.0L. Mix thoroughly. Gently heat and bring to boiling. Distribute into tubes or flasks. Autoclave for 15 min at 15 psi pressure–121°C. Pour into sterile Petri dishes or leave in tubes.

Use: For the isolation of pathogenic *Neisseria* species. For antimicrobial susceptibility testing of a variety of bacterial species. For the cultivation and maintenance of *Moraxella osloensis* and *Neisseria meningitidis*.

Mueller-Hinton Agar with IsoVitaleX and Hemoglobin

Composition per liter:

Component A	490.0mL
Component B	490.0mL
IsoVitaleX enrichment	20.0mL

pH 6.9 ± 0.2 at 25°C

Component A:

Composition per 490.0mL:

Beef infusion	300.0g
Acid hydrolysate of casein	17.5g
Agar	17.0g
Starch	1.5g

Preparation of Component A: Add components to distilled/deionized water and bring to 490.0mL. Mix thoroughly. Gently heat and bring to boiling. Autoclave for 15 min at 15 psi pressure–121°C. Cool to 45°–50°C.

Component B:

Composition per 490.0mL:

Hemoglobin	10.0g

Preparation of Component B: Add hemoglobin to distilled/deionized water and bring to 490.0mL. Mix thoroughly. Gently heat and bring to boiling. Autoclave for 15 min at 15 psi pressure–121°C. Cool to 45°–50°C.

IsoVitaleX Enrichment:

Composition per liter:

Glucose	100.0g
L-Cysteine·HCl	25.9g
L-Glutamine	10.0g
L-Cystine	1.1g
Adenine	1.0g
Nicotinamide adenine dinucleotide	0.25g
Vitamin B_{12}	0.1g
Thiamine pyrophosphate	0.1g
Guanine·HCl	0.03g
$Fe(NO_3)_3 \cdot 6H_2O$	0.02g
p-Aminobenzoic acid	0.013g
Thiamine·HCl	3.0mg

Source: The supplement solution IsoVitaleX enrichment is available from BD Diagnostic Systems. This enrichment may be replaced by supplement VX from BD Diagnostic Systems.

Preparation of IsoVitaleX Enrichment: Add components to distilled/deionized water and bring volume to 1.0L. Mix thoroughly. Filter sterilize. Warm to 45°–50°C.

Preparation of Medium: Aseptically combine 490.0mL of component A, 490.0mL of component B, and 20.0mL of IsoVitaleX enrichment. Mix thoroughly. Adjust pH to 6.9. Pour into sterile Petri dishes in 20.0mL volumes.

Use: For the isolation and cultivation of *Legionella pneumophila*.

Mueller-Hinton Broth

Composition per liter:

Acid hydrolysate of casein	17.5g
Beef extract	3.0g
Starch	1.5g

pH 7.3 ± 0.1 at 25°C

Source: This medium is available as a premixed powder from BD Diagnostic Systems and Oxoid Unipath.

Preparation of Medium: Add components to distilled/deionized water and bring to 1.0L. Mix thoroughly. Gently heat and bring to boiling. Distribute into tubes or flasks. Autoclave for 10 min at 10 psi pressure–115°C. Do not overheat.

Use: For the cultivation of a wide variety of microorganisms. For antimicrobial susceptibility testing.

Mueller-Hinton Chocolate Agar

Beef infusion.......300.0g
Acid hydrolysate of casein.......17.5g
Agar.......17.0g
Starch.......1.5g
Sheep blood.......50.0mL
pH 7.4 ± 0.2 at 25°C

Preparation of Medium: Add components, except sheep blood, to distilled/deionized water and bring volume to 950.0mL. Mix thoroughly. Gently heat and bring to boiling. Autoclave for 15 min at 15 psi pressure–121°C. Cool to 45°–50°C. Aseptically add sterile sheep blood. Mix thoroughly. Gently heat to 70°C for 10 min. Pour into sterile Petri dishes or distribute into sterile tubes.

Use: For the cultivation and maintenance of *Neisseria gonorrhoeae* and *Neisseria meningitidis*. For antimicrobial susceptibility testing of fastidious microorganisms.

Mueller-Hinton II Agar

Composition per liter:
Acid hydrolysate of casein.......17.5g
Agar.......17.0g
Beef extract.......2.0g
Starch.......1.5g
pH 7.3 ± 0.1 at 25°C

Source: This medium is available as a premixed powder from BD Diagnostic Systems.

Preparation of Medium: Add components to distilled/deionized water and bring to 1.0L. Mix thoroughly. Gently heat and bring to boiling. Distribute into tubes or flasks. Autoclave for 15 min at 15 psi pressure–121°C. Pour into sterile Petri dishes or leave in tubes.

Use: For antimicrobial disc diffusion susceptibility testing by the Bauer-Kirby method of a variety of bacteria. This medium supplemented with 5% sheep blood is recommended for use in antimicrobial susceptibility testing of *Streptococcus pneumoniae* and *Haemophilus influenzae.*

MUG Tryptone Soya Agar

Composition per liter:
Agar.......15.0g
Casein enzymic hydrolysate.......15.0g
Papaic digest of soyabean meal.......5.0g
NaCl.......5.0g
4-Methylumbelliferyl-β-D-glucuronide.......0.1g
pH 7.3 ± 0.2 at 25°C

Source: This medium is available from Fluka, Sigma-Aldrich.

Preparation of Medium: Add components to distilled/deionized water and bring volume to 1.0L. Mix thoroughly. Gently heat while stirring and bring to boiling. Autoclave for 15 min at 15 psi pressure–121°C. Cool to 50°C. Pour into sterile Petri dishes.

Use: For cultivation of fastidious and nonfastidious microorganisms by fluorogenic method.

Muller-Kauffmann Tetrathionate Broth

Composition per 1028.0mL:
$Na_2S_2O_3$.......40.7g
$CaCO_3$.......25.0g
Pancreatic digest of casein.......7.0g
Ox bile.......4.75g
Soya peptone.......2.3g
NaCl.......2.3g
Iodine solution.......19.0mL
Brilliant Green solution.......9.5mL

Iodine Solution:
Composition per 100.0mL:
Iodine.......20.0g
KI.......25.0g

Preparation of Iodine Solution: Add the KI to approximately 5.0mL of distilled/deionized water. Mix thoroughly. Add the iodine. Gently heat to dissolve. Bring volume to 100.0mL with distilled/deionized water. Filter sterilize.

Brilliant Green Solution:
Composition per 100.0mL:
Brilliant Green.......0.1g

Preparation of Brilliant Green Solution: Add the Brilliant Green to distilled/deionized water and bring volume to 100.0mL. Mix thoroughly. Gently heat while stirring and bring to boiling. Continue boiling for 30 min while stirring until dye has dissolved. Filter sterilize. Store protected from light.

Preparation of Medium: Add components, except iodine solution and Brilliant Green solution, to distilled/deionized water and bring volume to 1.0L. Mix thoroughly. Gently heat and bring to boiling. Do not autoclave. Cool to 45°C. Prior to use, add 19.0mL of iodine solution and 9.5mL of Brilliant Green solution. Mix thoroughly. Aseptically distribute into sterile tubes or flasks.

Use: For the isolation and cultivation of *Salmonella* species from specimens with a mixed flora.

MWY Medium (Wadowsky and Yee Medium, Modified)

Composition per liter:

Agar	13.0g
Yeast extract	10.0g
Glycine	3.0g
ACES buffer (2-[(2-Amino-2-oxoethyl)-amino]-ethane sulfonic acid)	2.0g
Charcoal, activated	2.0g
α-Ketoglutarate	0.2g
$Fe_4(P_2O_7)_3 \cdot 9H_2O$	0.05g
Bromcresol Purple	0.01g
Bromcresol Blue	0.01g
Antibiotic inhibitor	10.0mL
L-Cysteine·HCl·H_2O solution	10.0mL

pH 6.9 ± 0.2 at 25°C

Antibiotic Inhibitor:

Composition per 10.0mL:

Anisomycin	0.16g
Cefamandole	4.0mg
Vancomycin	1.0mg
Polymyxin B	130,000U

Preparation of Antibiotic Inhibitor: Add components to distilled/deionized water and bring volume to 10.0mL. Mix thoroughly. Filter sterilize.

L-Cysteine·HCl·H_2O Solution:

Composition per 10.0mL:

L-Cysteine·HCl·H_2O	0.08g

Preparation of L-Cysteine·HCl·H_2O Solution: Add L-cysteine·HCl·H_2O to distilled/deionized water and bring volume to 10.0mL. Mix thoroughly. Filter sterilize.

Preparation of Medium: Add components, except L-cysteine and antibiotic inhibitor, to distilled/deionized water and bring volume to 980.0mL. Mix thoroughly. Adjust medium to pH 6.9 with 1*N* KOH. Heat gently and bring to boiling for 1 min. Autoclave for 15 min at 15 psi pressure–121°C. Cool to 50°–55°C. Add 10.0mL of the sterile L-cysteine·HCl·H_2O solution and 10.0mL of the sterile antibiotic solution. Mix thoroughly. Pour into sterile Petri dishes with constant agitation to keep charcoal in suspension.

Use: For the selective isolation and cultivation of *Legionella pneumophila* and other *Legionella* species.

Mycin Assay Agar

Composition per liter:

Glucose	15.0g
Peptone	5.0g
Beef extract	3.0g

pH 7.9 ± 0.1 at 25°C

Preparation of Medium: Add components to distilled/deionized water and bring volume to 1.0L. Mix thoroughly. Distribute into tubes or flasks. Autoclave for 15 min at 15 psi pressure–121°C.

Use: For the microbiological assay of antibiotics in pharmaceutical products, body fluids, and other sample materials.

Mycobacterium Agar

Composition per liter:

Agar	12.0g
$Na_2HPO_4 \cdot 12H_2O$	2.5g
Typtic digest of casein	2.0g
Proteose peptone no.3	2.0g
Yeast extract	2.0g
Sodium citrate	1.5g
KH_2PO_4	1.0g
$MgSO_4 \cdot 7H_2O$	0.6g
Glycerol	50.0mL

pH 7.0 ± 0.2 at 25°C

Preparation of Medium: Add components to distilled/deionized water and bring volume to 1.0L. Mix thoroughly. Adjust pH to 7.0. Gently heat and bring to boiling. Distribute into tubes or flasks. Autoclave for 15 min at 15 psi pressure–121°C. Pour into sterile Petri dishes or leave in tubes.

Use: For the cultivation and maintenance of *Actinomyces israelii*, *Mycobacterium* species, and *Nocardia farcinica*.

Mycobactin Medium

Serum Agar Medium:

Composition per liter:

Noble agar	15.0g
Casamino acids	2.5g
Na_2HPO_4, anhydrous	2.5g
Sodium citrate	1.5g
KH_2PO_4	1.0g
$MgSO_4 \cdot 7H_2O$	0.6g
Asparagine	0.3g
Crude mycobactin	0.16g
Chloramphenicol	0.05g
Primaricine (myprozine)	0.05g
Penicillin	100,000U
Bovine serum, 56°C-inactivated	200.0mL
Tween 80 (1% solution)	50.0mL
Glycerol	25.0mL

pH 7.2 ± 0.2 at 25°C

Preparation of Crude Mycobactin: Grow *Mycobacterium phlei* in 600.0mL of mycobactin production broth for 2 weeks at 37°C. Autoclave the culture for 15 min at 15 psi pressure–121°C. Filter the cells and wash with distilled/deionized water. Dry cells under $CaCl_2$. Treat 100.0g of dried culture with three successive acetone extractions—500.0mL of acetone for 30 min in a 1.0L flask fitted with a reflux condenser. Evaporate the acetone to dryness. Extract the residue in a Soxhlet apparatus with petroleum ether for 18–20 hr at 40°–60°C. A hard red residue will remain. Dissolve the residue in warm absolute ethanol. Centrifuge for 30 min at 2250 rpm to remove debris. Evaporate the supernatant to dryness. Grind the residue to a powder of crude mycobactin.

Source: Purified mycobactin is available from Allied Labs, Inc., 2520 Hunt St., Ames, IA 50010.

Mycobactin Production Broth:
Composition per 600.0mL:

Solution B	500.0mL
Solution A	100.0mL

Preparation of Mycobactin Production Broth: Aseptically mix the cooled sterile solution A and solution B.

Solution A:
Composition per 100.0mL:

L-Asparagine	5.0g
Na_2HPO_4	2.0g
KH_2PO_4	1.0g
Glycerol	30.0mL

Preparation of Solution A: Add components to distilled/deionized water and bring volume to 100.0mL. Mix thoroughly. Autoclave for 15 min at 15 psi pressure–121°C. Cool to 45°–50°C.

Solution B:
Composition per 500.0mL:

Glucose	10.0g
$MgSO_4 \cdot 7H_2O$	0.2g

Preparation of Solution B: Add components to distilled/deionized water and bring volume to 500.0mL. Mix thoroughly. Autoclave for 15 min at 15 psi pressure–121°C. Cool to 45°–50°C.

Preparation of Medium: Add components, except penicillin, chloramphenicol, primaricine, and bovine serum, to distilled/deionized water and bring volume to 800.0mL. Mix thoroughly. Gently heat with a minimum of heat. Autoclave for 15 min at 10 psi pressure–116°C. Cool to 50°C. Aseptically add penicillin, chloramphenicol, primaricine, and sterile bovine serum. Mix thoroughly. Adjust pH to 7.2. Distribute into sterile tubes or flasks.

Use: For the cultivation and maintenance of *Mycobacterium avium* and *Mycobacterium paratuberculosis*.

Mycobactosel™ Agar

Composition per liter:

Agar	13.5g
Bovine albumin fraction V	5.0g
Glucose	2.0g
Na_2HPO_4	1.5g
KH_2PO_4	1.5g
Pancreatic digest of casein	1.0g
NaCl	0.85g
$(NH_4)_2SO_4$	0.5g
Monosodium glutamate	0.5g
Sodium citrate	0.4g
$MgSO_4 \cdot 7H_2O$	0.05g
Ferric ammonium citrate	0.04g
Pyridoxine	1.0mg
$ZnSO_4 \cdot 7H_2O$	1.0mg
$CuSO_4 \cdot 5H_2O$	1.0mg
Biotin	0.5mg
$CaCl_2 \cdot H_2O$	0.5mg
Malachite Green	0.25mg
Antibiotic solution	10.0mL
Catalase solution	10.0mL
Glycerol	5.0mL
Oleic acid	0.06mL

pH 6.6 ± 0.2 at 25°C

Source: This medium is available as a prepared medium from BD Diagnostic Systems.

Antibiotic Solution:
Composition per 10.0mL:

Cycloheximide	0.36g
Nalidixic acid	0.02g
Lincomycin	2.0mg

Preparation of Antibiotic Solution: Add components to distilled/deionized water and bring volume to 10.0mL. Mix thoroughly. Filter sterilize.

Catalase Solution:
Composition per 10.0mL:

Catalase	3.0mg

Preparation of Catalase Solution: Add catalase to distilled/deionized water and bring volume to 10.0mL. Mix thoroughly. Filter sterilize.

Preparation of Medium: Add components, except antibiotic solution and catalase solution, to distilled/deionized water and bring volume to 980.0mL. Mix thoroughly. Gently heat and bring to boiling. Autoclave for 15 min at 15 psi pressure–121°C. Cool to 45°–50°C. Aseptically add sterile antibiotic solution and sterile catalase solution. Mix thoroughly. Pour into sterile Petri dishes or distribute into sterile tubes.

Use: For the selective isolation of mycobacteria from specimens containing mixed flora.

Mycobactosel L-J Medium

Composition per liter:

Potato flour	30.0g
L-Asparagine	3.6g
KH_2PO_4, anhydrous	2.5g
Sodium citrate	0.6g
$MgSO_4 \cdot 7H_2O$	0.24g
Homogenized whole egg	1.0L
Malachite Green solution	20.0mL
Glycerol	12.0mL
Antibiotic solution	10.0mL

pH 7.0 ± 0.2 at 25°C

Source: This medium is available as a prepared medium from BD Diagnostic Systems.

Homogenized Whole Egg:

Composition per liter:

Whole eggs	18–24

Preparation of Whole Egg: Use fresh eggs, less than 1 week old. Scrub the shells with soap. Let stand in a soap solution for 30 min. Rinse in running water. Soak eggs in 70% ethanol for 15 min. Break the eggs into a sterile container. Homogenize by shaking. Filter through four layers of sterile cheesecloth into a sterile graduated cylinder. Measure out 1.0L.

Malachite Green Solution:

Composition per 20.0mL:

Malachite Green	0.4g

Preparation of Malachite Green Solution: Add Malachite Green to sterile distilled/deionized water and bring volume to 20.0mL in a sterile container. Mix thoroughly.

Antibiotic Solution:

Composition per 10.0mL:

Cycloheximide	0.64g
Nalidixic acid	0.056g
Lincomycin	3.2mg

Preparation of Antibiotic Solution: Add components to distilled/deionized water and bring volume to 10.0mL. Mix thoroughly. Filter sterilize.

Preparation of Medium: Add components—except whole egg, Malachite Green solution, and antibiotic solution—to distilled/deionized water and bring volume to 600.0mL. Mix thoroughly. Autoclave for 30 min at 15 psi pressure–121°C. Cool to room temperature. Add the homogenized whole egg, Malachite Green solution, and antibiotic solution. Distribute into sterile tubes in 8.0mL volumes. Coagulate medium in a slanted position at 85°C (moist heat) for 50 min.

Use: For the isolation and cultivation of *Mycobacterium* species from clinical specimens.

Mycobiotic Agar (Cycloheximide Chloramphenicol Agar)

Composition per liter:

Agar	15.0g
Enzymatic hydrolysate of soybean meal	10.0g
Glucose	10.0g
Cycloheximide	0.5g
Chloramphenicol	0.05g

pH 6.5 ± 0.2 at 25°C

Source: This medium is available as a premixed powder from BD Diagnostic Systems.

Preparation of Medium: Add components to distilled/deionized water and bring volume to 1.0L. Mix thoroughly. Gently heat and bring to boiling. Distribute into tubes or flasks. Autoclave for 15 min at 15 psi pressure–121°C. Cool tubes quickly in a slanted position.

Use: For the selective isolation and cultivation of pathogenic fungi.

Mycological Agar

Composition per liter:

Agar	15.0g
Enzymatic hydrolysate of soybean meal	10.0g
Glucose	10.0g

pH 7.0 ± 0.2 at 25°C

Source: This medium is available as a premixed powder from BD Diagnostic Systems.

Preparation of Medium: Add components to distilled/deionized water and bring volume to 1.0L. Mix thoroughly. Gently heat and bring to boiling. Distribute into tubes or flasks. Autoclave for 15 min at 15 psi pressure–121°C.

Use: For the selective isolation, cultivation, and maintenance of pathogenic fungi.

Mycological Agar with Low pH

Composition per liter:

Agar	15.0g
Enzymatic hydrolysate of soybean meal	10.0g
Glucose	10.0g

pH 4.8 ± 0.2 at 25°C

Source: This medium is available as a premixed powder from BD Diagnostic Systems.

Preparation of Medium: Add components to distilled/deionized water and bring volume to 1.0L. Mix thoroughly. Gently heat and bring to boiling. Distribute into tubes or flasks. Autoclave for 15 min at 15 psi pressure–121°C.

Use: For the selective isolation, cultivation, and maintenance of pathogenic fungi.

Mycological Broth

Composition per liter:

Glucose	40.0g
Enzymatic hydrolysate of soybean meal	10.0g

pH 7.0 ± 0.2 at 25°C

Source: This medium is available as a premixed powder from BD Diagnostic Systems.

Preparation of Medium: Add components to distilled/deionized water and bring volume to 1.0L. Mix thoroughly. Gently heat and bring to boiling. Distribute into tubes or flasks. Autoclave for 15 min at 15 psi pressure–121°C.

Use: For the cultivation of fungi.

Mycophil™ Broth

Composition per liter:

Glucose	40.0g
Pancreatic digest of casein	5.0g
Peptic digest of animal tissue	5.0g

pH 7.0 ± 0.2 at 25°C

Source: This medium is available as a premixed powder from BD Diagnostic Systems.

Preparation of Medium: Add components to distilled/deionized water and bring volume to 1.0L. Mix thoroughly. Gently heat and bring to boiling. Distribute into tubes or flasks. Autoclave for 15 min at 15 psi pressure–118°C. Do not overheat.

Use: For the isolation and cultivation of a wide variety of fungi.

Mycoplasma Agar (ATCC Medium 555)

Composition per 103.0mL:

Noble agar	0.7g
Hartley's digest broth	30.0mL
Pig serum	20.0mL
Enzymatic hydrolysate of lactalbumin	10.0mL
Hanks' balanced salt solution, 10X	4.0mL
Fresh yeast extract solution	2.0mL
Phenol Red (0.25% solution)	1.0mL

pH 7.4 ± 0.2 at 25°C

Hartley's Digest Broth:

Composition per 10.0L:

Ox heart	3,000.0g
Pancreatin	50.0g
Na_2CO_3, anhydrous (0.8% solution)	5.0L
HCl, concentrated	80.0mL

pH 7.5 ± 0.2 at 25°C

Preparation of Hartley's Digest Broth: Finely mince the ox heart. Add the meat to 5.0L of distilled/deionized water. Gently heat and bring to 80°C. Add the 5.0L of Na_2CO_3 solution. Cool to 45°C. Add pancreatin and maintain at 45°C for 4 hr while stirring. Add the HCl and steam at 100°C for 30 min. Cool to room temperature. Adjust pH to 8.0 with 1*N* NaOH. Gently heat and bring to boiling. Continue boiling for 25 min. Filter while hot. Cool to room temperature. Adjust pH to 7.5. Autoclave for 15 min at 15 psi pressure–121°C.

Pig Serum:

Composition per 100.0mL:

Pig serum	100.0mL

Preparation of Pig Serum: Adjust pH of pig serum to 4.4 with sterile 1*N* HCl. Do not let pH go below 4.2. Let serum stand at 4°C for 18-20h. Adjust pH to 7.0 with sterile 1*N* NaOH. Centrifuge at 9000 rpm for 20 min. Discard pellet. Filter supernatant solution through a 0.2μm membrane. Store at –70°C.

Enzymatic Hydrolysate of Lactalbumin:

Composition per 100.0mL:

Enzymatic hydrolysate of lactalbumin	5.0g

Preparation of Enzymatic Hydrolysate of Lactalbumin: Add enzymatic hydrolysate of lactalbumin to 100.0mL of phosphate buffered saline, 1X, pH 7.0.

Phosphate Buffered Saline Solution, 1X:

Composition per liter:

NaCl	8.0g
$Na_2HPO_4 \cdot 7H_2O$	2.16g
KCl	0.2g
KH_2PO_4	0.2g
$MgCl_2 \cdot 6H_2O$	0.1g
$CaCl_2$	0.1g

Hanks' Balanced Salt Solution, 10X:

Composition per liter:

NaCl	80.0g
Glucose	10.0g
KCl	4.0g
$CaCl_2$	1.4g
$MgCl_2 \cdot 6H_2O$	1.0g
$MgSO_4 \cdot 7H_2O$	1.0g
$Na_2HPO_4 \cdot 7H_2O$	0.9g
KH_2PO_4	0.6g

Preparation of Hanks' Balanced Salt Solution, 10X: Add components to distilled/deionized water and bring volume to 1.0L. Mix thoroughly.

Preparation of Medium: Combine components, except agar, in the following order: Hanks' balanced salt solution, 10X, Phenol Red, Hartley's digest broth, pig serum, enzymatic hydrolysate of lactalbumin, and fresh yeast extract solution. Mix thoroughly. Adjust pH to 7.4 with 1*N* NaOH. Filter sterilize through a 0.2μm membrane. Add 0.7g of Noble agar to 36.0mL of distilled/deionized water. Autoclave for 15 min at 15 psi pressure–121°C. Cool to 56°C. Warm basal medium to 56°C. Aseptically combine the two solutions. Pour into sterile Petri dishes or distribute into sterile tubes.

Use: For the cultivation of *Mycoplasma* species.

Mycoplasma Agar Base (PPLO Agar Base)

Composition per 1300.0mL:

Agar	14.0g
Pancreatic digest of casein	7.0g
NaCl	5.0g
Beef extract	3.0g
Yeast extract	3.0g
Beef heart, solids from infusion	2.0g
Horse serum	260.0mL
Fresh yeast extract solution	65.0mL

pH 7.8 ± 0.2 at 25°C

Source: This medium is available as a premixed powder from BD Diagnostic Systems.

Fresh Yeast Extract Solution:
Composition per 100.0mL:

Baker's yeast, live, pressed, starch-free	25.0g

Preparation of Fresh Yeast Extract Solution: Add the live Baker's yeast to 100.0mL of distilled/deionized water. Autoclave for 90 min at 15 psi pressure–121°C. Allow to stand. Remove supernatant solution. Adjust pH to 6.6–6.8.

Preparation of Medium: Add components, except horse serum and fresh yeast extract, to distilled/deionized water and bring volume to 1.0L. Mix thoroughly. Gently heat and bring to boiling. Distribute into tubes or flasks. Autoclave for 15 min at 15 psi pressure–121°C. Cool to 50°C. To each 75.0mL of cooled, sterile basal medium, add 20.0mL of sterile horse serum and 5.0mL of special yeast extract. Mix thoroughly. Pour into sterile Petri dishes or distribute into sterile tubes.

Use: For the preparation of media for the cultivation of *Mycoplasma.*

Mycoplasma Agar with Increased Selectivity

Composition per 1300.0mL:

Agar	14.0g
Pancreatic digest of casein	7.0g
NaCl	5.0g
Beef extract	3.0g
Yeast extract	3.0g
Beef heart, solids from infusion	2.0g
Thallous acetate	0.7g
Penicillin	70,000U
Horse serum	260.0mL
Fresh yeast extract solution	65.0mL

pH 7.8 ± 0.2 at 25°C

Caution: Thallous acetate is a poison.

Fresh Yeast Extract Solution:
Composition per 100.0mL:

Baker's yeast, live, pressed, starch-free	25.0g

Preparation of Fresh Yeast Extract Solution: Add the live Baker's yeast to 100.0mL of distilled/deionized water. Autoclave for 90 min at 15 psi pressure–121°C. Allow to stand. Remove supernatant solution. Adjust pH to 6.6–6.8.

Preparation of Medium: Add components—except horse serum, special yeast extract, thallous acetate, and penicillin—to distilled/deionized water and bring volume to 1.0L. Mix thoroughly. Gently heat and bring to boiling. Distribute into tubes or flasks. Autoclave for 15 min at 15 psi pressure–121°C. Cool to 50°C. To each of 70.0mL of cooled, sterile basal medium, add 20.0mL of sterile horse serum, 10.0mL of special yeast extract, 50.0mg of thallous acetate, and 5000U of penicillin. Mix thoroughly. Pour into sterile Petri dishes or distribute into sterile tubes.

Use: For the preparation of media for the cultivation of *Mycoplasma* species.

Mycoplasma Agar with Supplement G

Composition per liter:

Agar	10.0g
Bacteriological peptone	10.0g
Beef extract	10.0g
NaCl	5.0g
Special mineral supplement	0.5g
Mycoplasma supplement G	250.0mL

pH 7.8 ± 0.2 at 25°C

Source: This medium is available as a premixed powder from Oxoid Unipath.

***Mycoplasma* Supplement G:**
Composition per 20.0mL:

Thallous acetate	25.0mg
Horse serum	20.0mL
Yeast extract (25% solution)	10.0mL
Penicillin	20,000U

Preparation of *Mycoplasma* Supplement G: Add components to distilled/deionized water and bring volume to 20.0mL. Mix thoroughly. Filter sterilize.

Caution: Thallous acetate is a poison.

Preparation of Medium: Add components, except *Mycoplasma* supplement G, to distilled/deionized water and bring volume to 1.0L. Mix thoroughly. Gently heat and bring to boiling. Distribute into flasks in 80.0mL volumes. Autoclave for 15 min at 15 psi pressure–121°C. Cool to 50°C. Aseptically add 20.0mL of sterile *Mycoplasma* supplement G to each 80.0mL of basal medium. Mix thoroughly.

Use: For the growth of *Mycoplasma* species.

Mycoplasma Agar with Supplement P

Composition per liter:

Agar	10.0g
Bacteriological peptone	10.0g
Beef extract	10.0g
NaCl	5.0g
Special mineral supplement	0.5g
Mycoplasma supplement P	250.0mL

pH 7.8 ± 0.2 at 25°C

Source: This medium is available as a premixed powder from Oxoid Unipath.

***Mycoplasma* Supplement P:**
Composition per 20.0mL:

Glucose	0.3g
Mycoplasma broth base	0.145g
Thallous acetate	8.0mg
Phenol Red	1.2mg
Methylene Blue chloride	0.3mg
Penicillin	12,000U
Horse serum	6.0mL
Yeast extract (25% solution)	3.0mL

Preparation of *Mycoplasma* Supplement P: Add components to distilled/deionized water and bring volume to 20.0mL. Mix thoroughly. Filter sterilize.

Caution: Thallous acetate is a poison.

Fresh Yeast Extract Solution:
Composition per 100.0mL:

Baker's yeast, live, pressed, starch-free	25.0g

Preparation of Fresh Yeast Extract Solution: Add the live Baker's yeast to 100.0mL of distilled/deionized water. Autoclave for 90 min at 15 psi pressure–121°C. Allow to stand. Remove supernatant solution. Adjust pH to 6.6–6.8.

Preparation of Medium: Add components, except *Mycoplasma* supplement P, to distilled/deionized water and bring volume to 1.0L. Mix thoroughly. Gently heat and bring to boiling. Distribute into bottles in 1.0mL volumes. Autoclave for 15 min at 15 psi pressure–121°C. Cool to room temperature. Aseptically add 2.0mL of sterile *Mycoplasma* supplement P to each bottle.

Use: For the growth of *Mycoplasma* species.

Mycoplasma Broth (ATCC Medium 555)

Composition per 103.0mL:

Hartley's digest broth	30.0mL
Pig serum	20.0mL
Enzymatic hydrolysate of lactalbumin	10.0mL
Hanks' balanced salt solution, 10X	4.0mL
Fresh yeast extract solution	2.0mL
Phenol Red (0.25% solution)	1.0mL

pH 7.4 ± 0.2 at 25°C

Hartley's Digest Broth:
Composition per 10.0L:

Ox heart	3,000.0g
Pancreatin	50.0g
Na_2CO_3, anhydrous (0.8% solution)	5.0L
HCl, concentrated	80.0mL

pH 7.5 ± 0.2 at 25°C

Preparation of Hartley's Digest Broth: Finely mince the ox heart. Add the meat to 5.0L of distilled/deionized water. Gently heat and bring to 80°C. Add the 5.0L of Na_2CO_3 solution. Cool to 45°C. Add pancreatin and maintain at 45°C for 4 hr while stirring. Add the HCl and steam at 100°C for 30 min. Cool to room temperature. Adjust pH to 8.0 with 1*N* NaOH. Gently heat and bring to boiling. Continue boiling for 25 min. Filter while hot. Cool to room temperature. Adjust pH to 7.5. Autoclave for 15 min at 15 psi pressure–121°C.

Pig Serum:
Composition per 100.0mL:

Pig serum	100.0mL

Preparation of Pig Serum: Adjust pH of pig serum to 4.4 with sterile 1*N* HCl. Do not let pH go below 4.2. Let serum stand at 4°C for 18-20h. Adjust pH to 7.0 with sterile 1*N* NaOH. Centrifuge at 9000 rpm for 20 min. Discard pellet. Filter supernatant solution through a 0.2μm membrane. Store at –70°C.

Enzymatic Hydrolysate of Lactalbumin:
Composition per 100.0mL:

Enzymatic hydrolysate of lactalbumin	5.0g

Preparation of Enzymatic Hydrolysate of Lactalbumin: Add enzymatic hydrolysate of lactalbumin to 100.0mL of phosphate buffered saline, 1X, pH 7.0.

Phosphate Buffered Saline Solution, 1X:
Composition per liter:

NaCl	8.0g
$Na_2HPO_4 \cdot 7H_2O$	2.16g
KCl	0.2g
KH_2PO_4	0.2g

$MgCl_2 \cdot 6H_2O$ 0.1g
$CaCl_2$ 0.1g

Preparation of Buffered Saline Solution, 1X: Add components to distilled/deionized water and bring volume to 1.0L. Mix thoroughly.

Hanks' Balanced Salt Solution, 10X:
Composition per liter:
NaCl 80.0g
Glucose 10.0g
KCl 4.0g
$CaCl_2$ 1.4g
$MgCl_2 \cdot 6H_2O$ 1.0g
$MgSO_4 \cdot 7H_2O$ 1.0g
$Na_2HPO_4 \cdot 7H_2O$ 0.9g
KH_2PO_4 .... 0.6g

Preparation of Hanks' Balanced Salt Solution, 10X: Add components to distilled/deionized water and bring volume to 1.0L. Mix thoroughly.

Preparation of Medium: Combine components in the following order: Hanks' balanced salt solution, 10X, Phenol Red, Hartley's digest broth, pig serum, enzymatic hydrolysate of lactalbumin, and fresh yeast extract solution. Mix thoroughly. Add 36.0mL of distilled/deionized water. Adjust pH to 7.4 with 1*N* NaOH. Filter sterilize through a 0.2µm membrane. Store at 4°C for up to 3 weeks.

Use: For the cultivation of *Mycoplasma* species.

Mycoplasma Broth Base (PPLO Broth Base without Crystal Violet)

Composition per liter:
Pancreatic digest of casein 7.0g
NaCl 5.0g
Beef extract 3.0g
Yeast extract 3.0g
Beef heart, solids from infusion 2.0g
pH 7.8 ± 0.2 at 25°C

Source: This medium is available as a premixed powder from BD Diagnostic Systems.

Preparation of Medium: Add components to distilled/deionized water and bring volume to 1.0L. Gently heat and bring to boiling. Mix thoroughly. Distribute into tubes or flasks. Autoclave for 15 min at 15 psi pressure–121°C.

Use: Used as a basal medium that should be enriched for the isolation and cultivation of *Mycoplasma* species.

Mycoplasma Broth, Supplemented

Composition per liter:
Pancreatic digest of casein 7.0g
NaCl 5.0g
Beef extract 3.0g
Yeast extract 3.0g
Beef heart, solids from infusion 2.0g
Horse serum 260.0mL
Fresh yeast extract solution 65.0mL
pH 7.8 ± 0.2 at 25°C

Fresh Yeast Extract Solution:
Composition per 100.0mL:
Baker's yeast, live, pressed, starch-free 25.0g

Preparation of Fresh Yeast Extract Solution: Add the live Baker's yeast to 100.0mL of distilled/deionized water. Autoclave for 90 min at 15 psi pressure–121°C. Allow to stand. Remove supernatant solution. Adjust pH to 6.6–6.8.

Preparation of Medium: Add components, except horse serum and fresh yeast extract, to distilled/deionized water and bring volume to 1.0L. Mix thoroughly. Gently heat and bring to boiling. Distribute into tubes or flasks. Autoclave for 15 min at 15 psi pressure–121°C. Cool to 50°C. To each 75.0mL of cooled, sterile basal medium, add 20.0mL of sterile horse serum and 5.0mL of fresh yeast extract. Mix thoroughly. Aseptically distribute into sterile tubes.

Use: For the isolation and cultivation of *Mycoplasma* species.

Mycoplasma Broth with Supplement G

Composition per liter:
Bacteriological peptone 10.0g
Beef extract 10.0g
NaCl 5.0g
Special mineral supplement, Oxoid Unipath 0.5g
Mycoplasma supplement G 250.0mL
pH 7.8 ± 0.2 at 25°C

Source: This medium is available as a premixed powder from Oxoid Unipath.

***Mycoplasma* Supplement G:**
Composition per 20.0mL:
Thallous acetate 25.0mg
Horse serum 20.0mL
Yeast extract (25% solution) 10.0mL
Penicillin 20,000U

Preparation of *Mycoplasma* Supplement G: Add components to distilled/deionized water and bring volume to 20.0mL. Mix thoroughly. Filter sterilize.

Caution: Thallous acetate is a poison.

Preparation of Medium: Add components, except *Mycoplasma* supplement G, to distilled/deionized water and bring volume to 1.0L. Mix thoroughly. Gently heat and bring to boiling. Distribute into flasks in 80.0mL volumes. Autoclave for 15 min at 15 psi pressure–121°C. Cool to 50°C. Aseptically add 20.0mL of sterile *Mycoplasma* supplement G to each 80.0mL of basal medium. Mix thoroughly.

Use: For the growth of *Mycoplasma* species.

Mycoplasma Broth with Supplement P

Composition per liter:

Bacteriological peptone	10.0g
Beef extract	10.0g
NaCl	5.0g
Special mineral supplement, Oxoid Unipath	0.5g
Mycoplasma supplement P	250.0mL

pH 7.8 ± 0.2 at 25°C

Source: This medium is available as a premixed powder from Oxoid Unipath.

***Mycoplasma* Supplement P:**

Composition per 20.0mL:

Glucose	0.3g
Mycoplasma broth base	0.145g
Thallous acetate	8.0mg
Phenol Red	1.2mg
Methylene Blue chloride	0.3mg
Penicillin	12,000U
Horse serum	6.0mL
Yeast extract (25% solution)	3.0mL

Preparation of *Mycoplasma* Supplement P: Add components to distilled/deionized water and bring volume to 20.0mL. Mix thoroughly. Filter sterilize.

Caution: Thallous acetate is a poison.

Preparation of Medium: Add components, except *Mycoplasma* supplement P, to distilled/deionized water and bring volume to 1.0L. Mix thoroughly. Gently heat and bring to boiling. Distribute into bottles in 1.0mL volumes. Autoclave for 15 min at 15 psi pressure–121°C. Cool to room temperature. Aseptically add 2.0mL of sterile *Mycoplasma* supplement P to each bottle.

Use: For the cultivation of *Mycoplasma* species.

Mycoplasma Horse Serum Broth (ATCC Medium 1959)

Mycoplasma broth base	660.0mL
Horse serum	200.0mL
Yeast extract solution	100.0mL
Phenol Red (0.1%)	20.0mL
Glucose solution	10.0mL
NaOH (1*N* solution)	6.25mL
Arginine solution	5.0mL

***Mycoplasma* Broth Base:**

Composition per liter:

Pancreatic digest of casein	7.0g
NaCl	5.0g
Beef extract	3.0g
Yeast extract	3.0g
Beef heart, solids from infusion	2.0g

Preparation of *Mycoplasma* Broth Base: Add components to distilled/deionized water and bring volume to 1.0L. Mix thoroughly.

Yeast Extract Solution:

Composition per 100.0mL:

Baker's yeast, live, pressed, starch-free	25.0g

Preparation of Fresh Yeast Extract Solution: Add the live Baker's yeast to 100.0mL of distilled/deionized water. Autoclave for 90 min at 15 psi pressure–121°C. Allow to stand. Remove supernatant solution. Adjust pH to 6.6–6.8.

Glucose Solution:

Composition per 10.0mL:

Glucose	5.0g

Preparation of Glucose Solution: Add glucose to 10.0mL of distilled/deionized water. Mix thoroughly. Filter sterilize.

Arginine Solution:

Composition per 10.0mL:

L-Arginine	4.2g

Preparation of Arginine Solution: Add arginine to 10.0mL of distilled/deionized water. Mix thoroughly. Filter sterilize.

Preparation of Medium: Combine 660.0mL Mycoplasma broth base, 20.0mL Phenol Red, and 6.25mL 1N NaOH. Mix thoroughly. Autoclave for 15 min at 15 psi pressure–121°C. Cool to 25°C. Aseptically add 5.0mL sterile arginine solution, 100.0mL sterile yeast extract solution, 10.0mL sterile glucose solution, and 200.0mL filter sterilized horse serum. Mix thoroughly. Aseptically distribute into sterile tubes or flasks.

Use: For the preparation of media for the cultivation of *Mycoplasma* spp.

Mycoplasma pneumoniae Isolation Medium

Composition per 1200.0mL:

Beef heart for infusion	50.0g
Peptone	10.0g
NaCl	5.0g
Water	900.0mL

Yeast extract solution 100.0mL
α-Gamma horse serum, unheated 200.0mL
pH 7.6–7.8 at 25°C

Yeast Extract Solution:
Composition per 10.0mL:
Yeast, active dry Baker's 250.0g

Preparation of Yeast Extract Solution: Add yeast to 1.0L of distilled/deionized water. Mix thoroughly. Gently heat and bring to boiling. Filter through Whatman no. 2 filter paper. Adjust the pH of the filtrate to 8.0 with NaOH. Distribute into tubes in 10.0mL volumes. Autoclave for 15 min at 15 psi pressure–121°C. Store at –20°C.

Preparation of Medium: Add components, except yeast extract solution and α-gamma horse serum, to distilled/deionized water and bring volume to 990.0mL. Mix thoroughly. Gently heat and bring to boiling. Autoclave for 15 min at 15 psi pressure–121°C. Cool to 45°–50°C. Aseptically add sterile yeast extract solution and α-gamma horse serum. Mix thoroughly. Aseptically distribute into sterile tubes.

Use: For the isolation and cultivation of *Mycoplasma pneumoniae.*

Mycoplasmal Agar

Composition per liter:
Papaic digest of soy meal 20.0g
Agarose 10.0g
NaCl 5.0g
Phenol Red (2% solution) 1.0mL
pH 7.3 ± 0.2 at 25°C

Preparation of Medium: Add components, except agarose, to distilled/deionized water and bring volume to 1.0L. Mix thoroughly. Adjust pH to 7.3 with 1*N* NaOH. Add agarose. Mix thoroughly. Gently heat and bring to boiling. Distribute into tubes or flasks. Autoclave for 15 min at 15 psi pressure–121°C. Pour into sterile Petri dishes or leave in tubes.

Use: For the isolation and cultivation of human mycoplasmas and ureaplasmas.

Neisseria Medium

Composition per liter:
Biosate 10.0g
Polypeptone 10.0g
NaCl 5.0g
Myosate 3.0g
Agar 1.5g
Phenol Red 0.017g
Carbohydrate solution 50.0mL
pH 7.4–7.6 at 25°C

Carbohydrate Solution:
Composition per 50.0mL:
Carbohydrate 10.0g

Preparation of Carbohydrate Solution: Add glucose, sucrose, or maltose to distilled/deionized water and bring volume to 50.0mL. Mix thoroughly. Filter sterilize.

Preparation of Medium: Add components, except carbohydrate solution, to distilled/deionized water and bring volume to 950.0mL. Mix thoroughly. Gently heat and bring to boiling. Autoclave for 15 min at 15 psi pressure–121°C. Cool to 45°–50°C. Aseptically add sterile carbohydrate solution. Mix thoroughly. Pour into sterile Petri dishes or distribute into sterile tubes.

Use: For the cultivation of *Neisseria* species.

Neisseria meningitidis Medium

Composition per liter:
Beef infusion 300.0g
Acid hydrolysate of casein 17.5g
Agar 17.0g
Starch 1.5g
Antibiotic solution 10.0mL
pH 7.4 ± 0.2 at 25°C

Antibiotic Solution:
Composition per 10.0mL:
Vancomycin 3.0mg
Colistin 7.5mg
Nystatin 12,500U

Preparation of Antibiotic Solution: Add components to distilled/deionized water and bring volume to 10.0mL. Mix thoroughly. Filter sterilize.

Preparation of Medium: Add components, except antibiotic solution, to distilled/deionized water and bring volume to 990.0mL. Mix thoroughly. Gently heat and bring to boiling. Autoclave for 15 min at 15 psi pressure–121°C. Cool to 45°–50°C. Aseptically add sterile antibiotic solution. Mix thoroughly. Pour into sterile Petri dishes or distribute into sterile tubes.

Use: For the selective isolation and cultivation of *Neisseria meningitidis.*

Nelson Culture Medium for *Naegleria*

Composition per 100.0mL:
Glucose 0.17g
Panmede 0.17g
Na_2HPO_4 .... 14.2mg
KH_2PO_4 .... 13.6mg
NaCl 12.0mg
$CaCl_2 \cdot 2H_2O$ 0.4mg
$MgSO_4 \cdot 7H_2O$ 0.4mg
Bovine serum, heat-inactivated fetal 10.0mL

Source: Panmede is available from Paines and Byrne Ltd., Greenford, England, and Harrisons and Crosfield, Inc., Bronxville, NY.

Preparation of Medium: Add components, except bovine serum, to distilled/deionized water and bring volume to 90.0mL. Mix thoroughly. Autoclave for 15 min at 15 psi pressure–121°C. Cool to 25°C. Aseptically add 10.0mL of sterile, heat-inactivated fetal bovine serum. Mix thoroughly. Aseptically distribute into sterile tubes or flasks. Use immediately.

Use: For the cultivation of *Naegleria fowleri* and *Paratetramitus jugosus*.

Nelson Culture Medium for *Naegleria*

Composition per 100.0mL:

Glucose	0.17g
Liver infusion	0.17g
Na_2HPO_4	14.2mg
KH_2PO_4	13.6mg
NaCl	12.0mg
$CaCl_2 \cdot 2H_2O$	0.4mg
$MgSO_4 \cdot 7H_2O$	0.4mg
Bovine serum, heat-inactivated fetal	10.0mL

Preparation of Medium: Add components, except bovine serum, to distilled/deionized water and bring volume to 90.0mL. Mix thoroughly. Autoclave for 15 min at 15 psi pressure–121°C. Cool to 25°C. Aseptically add 10.0mL of sterile, heat-inactivated fetal bovine serum. Mix thoroughly. Aseptically distribute into sterile tubes or flasks. Use immediately.

Use: For the cultivation of *Naegleria fowleri* and *Paratetramitus jugosus*.

Nelson Medium for *Naegleria fowleri*

Composition per liter:

Glucose	1.0g
Ox liver digest	1.0g
Page's amoeba saline	1.0L
Fetal calf serum, inactivated	20.0mL

Page's Amoeba Saline:
Composition per liter:

Na_2HPO_4	0.142g
KH_2PO_4	0.136g
NaCl	0.12g
$MgSO_4 \cdot 7H_2O$	4.0mg
$CaCl_2 \cdot 2H_2O$	4.0mg

Preparation of Page's Amoeba Saline: Add components to distilled/deionized water and bring volume to 10.0mL. Mix thoroughly.

Preparation of Medium: Add the glucose and ox liver digest to 1.0L of Page's amoeba saline. Mix thoroughly. Distribute into screw-capped tubes in 10.0mL volumes. Autoclave for 15 min at 15 psi pressure–121°C. Cool to 25°C. Aseptically add 0.2mL of sterile fetal calf serum to each tube. Mix thoroughly.

Use: For the cultivation of *Naegleria fowleri*.

Neomycin Agar, Modified

Composition per liter:

Agar	15.0g
Peptone	6.0g
Pancreatic digest of casein	4.0g
Yeast extract	3.0g
Beef extract	1.5g
Glucose	1.0g
Methanol	20.0mL
Neomycin solution	10.0mL

pH 7.0 ± 0.2 at 25°C

Neomycin Solution:
Composition per 10.0mL:

Neomycin sulfate	1.0g

Preparation of Neomycin Solution: Add neomycin sulfate to distilled/deionized water and bring volume to 10.0mL. Mix thoroughly. Filter sterilize.

Preparation of Medium: Add components, except methanol and neomycin solution, to distilled/deionized water and bring volume to 970.0mL. Mix thoroughly. Gently heat and bring to boiling. Autoclave for 15 min at 15 psi pressure–121°C. Cool to 45°–50°C. Filter sterilize methanol. To cooled, sterile basal medium, aseptically add sterile methanol and sterile neomycin solution. Mix thoroughly. Pour into sterile Petri dishes or distribute into sterile tubes.

Use: For the cultivation and maintenance of *Bordetella bronchiseptica*.

Neomycin Blood Agar

Composition per liter:

Pancreatic digest of casein	14.5g
Agar	14.0g
Papaic digest of soybean meal	5.0g
NaCl	5.0g
Growth factors	1.5g
Sheep blood, defibrinated	50.0mL
Neomycin solution	10.0mL

pH 7.3 ± 0.2 at 25°C

Source: This medium is available as a premixed powder from BD Diagnostic Systems.

Neomycin Solution:
Composition per 10.0mL:

Neomycin sulfate	0.03g

Preparation of Neomycin Solution: Add neomycin sulfate to distilled/deionized water and bring volume to 10.0mL. Mix thoroughly. Filter sterilize.

Preparation of Medium: Add components, except sheep blood and neomycin solution, to distilled/deionized water and bring volume to 940.0mL. Mix thoroughly. Gently heat and bring to boiling. Autoclave for 15 min at 15 psi pressure–121°C. Cool to 45°–50°C. Aseptically add sterile sheep blood and sterile neomycin solution. Mix thoroughly. Pour into sterile Petri dishes or distribute into sterile tubes.

Use: For the isolation and cultivation of group A streptococci (*Streptococcus pyogenes)* and group B streptococci (*Streptococcus agalactiae*) from throat cultures and other clinical specimens.

Neomycin Luria Agar

Composition per liter:

Agar	15.0g
Pancreatic digest of casein	10.0g
Yeast extract	5.0g
NaCl	0.5g
Glucose solution	20.0mL
Neomycin	10.0mL

pH 7.0 ± 0.1 at 25°C

Glucose Solution:

Composition per 100.0mL:

Glucose	10.0g

Preparation of Glucose Solution: Add glucose to distilled/deionized water and bring volume to 100.0mL. Mix thoroughly. Filter sterilize.

Neomycin Solution:

Composition per 10.0mL:

Neomycin sulfate	12.0mg

Preparation of Neomycin Solution: Add neomycin sulfate to distilled/deionized water and bring volume to 10.0mL. Mix thoroughly. Filter sterilize.

Preparation of Medium: Add components, except glucose solution and neomycin solution, to distilled/deionized water and bring volume to 970.0mL. Mix thoroughly. Gently heat and bring to boiling. Autoclave for 15 min at 15 psi pressure–121°C. Aseptically add 20.0mL of sterile glucose solution and 10.0mL of sterile neomycin solution. Mix thoroughly. Pour into sterile Petri dishes or distribute into sterile tubes.

Use: For the cultivation of *Escherichia coli.*

Neomycin Medium No. 1

Composition per liter:

Agar	15.0g
Peptone	5.0g
NaCl	5.0g
Yeast extract	2.0g
Beef extract	1.0g
Sucrose solution	20.0mL
Neomycin	10.0mL

pH 7.0 ± 0.1 at 25°C

Sucrose Solution:

Composition per 100.0mL:

Sucrose	2.5g

Preparation of Sucrose Solution: Add sucrose to distilled/deionized water and bring volume to 100.0mL. Mix thoroughly. Filter sterilize.

Neomycin Solution:

Composition per 10.0mL:

Neomycin sulfate	500.0mg

Preparation of Neomycin Solution: Add neomycin sulfate to distilled/deionized water and bring volume to 10.0mL. Mix thoroughly. Filter sterilize.

Preparation of Medium: Add components, except sucrose solution and neomycin solution, to distilled/deionized water and bring volume to 970.0mL. Mix thoroughly. Gently heat and bring to boiling. Autoclave for 15 min at 15 psi pressure–121°C. Aseptically add 20.0mL of sterile sucrose solution and 10.0mL of sterile neomycin solution. Mix thoroughly. Pour into sterile Petri dishes or distribute into sterile tubes.

Use: For the cultivation of *Pseudomonas aeruginosa.*

Neopeptone Glucose Agar

Composition per liter:

Agar	20.0g
Glucose	10.0g
Neopeptone	5.0g

pH 6.5 ± 0.2 at 25°C

Preparation of Medium: Add components to distilled/deionized water and bring volume to 1.0L. Mix thoroughly. Gently heat and bring to boiling. Distribute into tubes or flasks. Autoclave for 15 min at 15 psi pressure–121°C. Adjust pH to 6.5. Pour into sterile Petri dishes or leave in tubes.

Use: For the maintenance of stock cultures of a variety of microorganisms.

Neopeptone Glucose Rose Bengal Aureomycin® Agar

Composition per liter:

Agar	20.0g
Neopeptone	5.0g
Glucose	1.0g

Tetracycline solution 5.0mL
Rose Bengal solution 3.5mL
pH 6.5 ± 0.2 at 25°C

Tetracycline Solution:
Composition per 150.0mL:
Tetracycline 1.0g

Preparation of Tetracycline Solution: Add tetracycline to distilled/deionized water and bring volume to 150.0mL. Mix thoroughly. Filter sterilize.

Rose Bengal Solution:
Composition per 100.0mL:
Rose Bengal 1.0g

Preparation of Rose Bengal Solution: Add Rose Bengal to distilled/deionized water and bring volume to 100.0mL. Mix thoroughly. Filter sterilize.

Preparation of Medium: Add components, except tetracycline solution, to distilled/deionized water and bring volume to 995.0mL. Mix thoroughly. Gently heat and bring to boiling. Autoclave for 15 min at 15 psi pressure–121°C. Cool to 45°–50°C. Aseptically add 5.0mL of sterile tetracycline solution. Mix thoroughly. Pour into sterile Petri dishes or distribute into sterile tubes.

Use: For the isolation and cultivation of a wide variety of fungal species.

Neopeptone Infusion Agar

Composition per liter:
Beef heart, infusion from 500.0g
Neopeptone 20.0g
Agar 20.0g
NaCl 5.0g
Sheep blood, defibrinated 50.0mL
pH 7.4 ± 0.2 at 25°C

Preparation of Medium: Add components, except sheep blood, to distilled/deionized water and bring volume to 950.0mL. Mix thoroughly. Gently heat and bring to boiling. Autoclave for 15 min at 15 psi pressure–121°C. Cool to 45°–50°C. Aseptically add sterile sheep blood. Mix thoroughly. Pour into sterile Petri dishes or distribute into sterile tubes.

Use: For the cultivation of a wide variety of fastidious microorganisms.

New York City Medium

Composition per liter:
NYC basal medium 640.0mL
Horse blood cells 200.0mL
Horse plasma, citrated 120.0mL
Yeast dialysate 25.0mL
Glucose solution 10.0mL
Antibiotic VCNT solution 5.0mL
pH 7.4 ± 0.2 at 25°C

NYC Basal Medium:
Composition per 640.0mL:
Solution 1 .... 400.0mL
Solution 3 .... 200.0mL
Solution 2 .... 40.0mL

Preparation of NYC Basal Medium: Combine solution 1, solution 2, and solution 3. Mix thoroughly. Autoclave for 15 min at 15 psi pressure–121°C. Cool to 45°–50°C.

Solution 1:
Composition per 400.0mL:
Agar 20.0g

Preparation of Solution 1: Add agar to distilled/deionized water and bring volume to 400.0mL. Mix thoroughly. Melt agar in autoclave for 10 min at 0 psi pressure–100°C. Cool to 45°–50°C.

Solution 2:
Composition per 40.0mL:
Cornstarch 1.0g

Preparation of Solution 2: Add cornstarch to distilled/deionized water and bring volume to 40.0mL. Mix thoroughly. Warm to 45°–50°C.

Solution 3:
Composition per 200.0mL:
Proteose peptone no. 3 .... 15.0g
NaCl 5.0g
K_2HPO_4 .... 4.0g
KH_2PO_4 .... 1.0g

Preparation of Solution 3: Add components to distilled/deionized water and bring volume to 1.0L. Mix thoroughly. Gently heat and bring to boiling. Cool to 45°–50°C.

Horse Blood Cells:
Composition per 200.0mL:
Horse blood cells, sedimented 6.0mL

Preparation of Horse Blood Cells: Cow blood may be used instead of horse blood but do not use sheep blood. Use cells freshly packed by sedimentation. Do not pack by centrifugation. Aseptically add 6.0mL of sedimented blood cells to 200.0mL of sterile distilled/deionized water. Mix thoroughly.

Horse Plasma, Citrated:
Composition per 6.0L:
Horse blood 5400.0mL
Citrate solution 600.0mL

Preparation of Horse Plasma, Citrated: Place 600.0mL of sterile citrate solution into a receiving bottle. Draw horse blood to the 6.0L mark. Allow cells to sediment out. Aseptically remove plasma.

Citrate Solution:
Composition per liter:

Sodium citrate	150.0g
NaCl	81.13g

Preparation of Citrate Solution: Add components to distilled/deionized water and bring volume to 1.0L. Mix thoroughly. Filter sterilize.

Glucose Solution:
Composition per 10.0mL:

D-Glucose	5.0g

Preparation of Glucose Solution: Add D-glucose to distilled/deionized water and bring volume to 10.0mL. Mix thoroughly. Autoclave for 10 min at 10 psi pressure–115°C. Cool to 45°–50°C.

Yeast Dialysate:
Composition per 2500.0mL:

Baker's yeast, fresh	908.0g

Preparation of Yeast Dialysate: Add fresh baker's yeast to 2500.0mL of distilled/deionized water. Mix thoroughly. Autoclave for 10 min at 15 psi pressure–121°C. Cool to 25°C. Put into dialysis tubing. Dialyze against 2.0L of distilled/deionized water for 48h at 4°C.

Antibiotic VCNT Solution:
Composition per 5.0mL:

Colistin	7.5mg
Trimethorprim lactate	3.0mg
Vancomycin·HCl	2.0mg
Nystatin	12.5U

Preparation of Antibiotic Solution: Add components to distilled/deionized water and bring volume to 5.0mL. Mix thoroughly. Filter sterilize.

Preparation of Medium: Have all solutions prepared and at 45°–50°C. Aseptically combine components. Mix thoroughly. Pour into sterile Petri dishes.

Use: For the isolation and cultivation of pathogenic *Neisseria* species. Used as a transport medium for urogenital and other clinical specimens. For the isolation and presumptive identification of Mycoplasmatales, including large-colony species (*Mycoplasma pneumoniae*) and T–mycoplasmas from urogenital specimens.

New York City Medium, Modified

Composition per liter:

NYC basal medium	840.0mL
α-Gamma horse serum (Flow Labs)	120.0mL
Yeast dialysate	25.0mL
Glucose solution	10.0mL
Antibiotic LCNT solution	5.0mL

pH 7.4 ± 0.2 at 25°C

NYC Basal Medium:
Composition per 840.0mL:

Horse blood	5400.0mL
Solution 1	600.0mL
Solution 3	200.0mL
Solution 2	40.0mL

Preparation of NYC Basal Medium: Combine solution 1, solution 2, and solution 3. Mix thoroughly. Autoclave for 15 min at 15 psi pressure–121°C. Cool to 45°–50°C.

Solution 1:
Composition per 600.0mL:

Agar	20.0g

Preparation of Solution 1: Add agar to distilled/deionized water and bring volume to 600.0mL. Mix thoroughly. Melt agar in autoclave for 10 min at 0 psi pressure–100°C. Cool to 45°–50°C.

Solution 2:
Composition per 40.0mL:

Cornstarch	1.0g

Preparation of Solution 2: Add cornstarch to distilled/deionized water and bring volume to 40.0mL. Mix thoroughly. Warm to 45°–50°C.

Solution 3:
Composition per 200.0mL:

Proteose peptone no. 3	15.0g
NaCl	5.0g
K_2HPO_4	4.0g
KH_2PO_4	1.0g

Preparation of Solution 3: Add components to distilled/deionized water and bring volume to 200.0mL. Mix thoroughly. Gently heat and bring to boiling. Cool to 45°–50°C.

Glucose Solution:
Composition per 10.0mL:

D-Glucose	5.0g

Preparation of Glucose Solution: Add glucose to distilled/deionized water and bring volume to 10.0mL. Mix thoroughly. Autoclave for 10 min at 10 psi pressure–115°C. Cool to 45°–50°C.

Yeast Dialysate:
Composition per 2500.0mL:

Baker's yeast, fresh	908.0g

Preparation of Yeast Dialysate: Add fresh Baker's yeast to 2500.0mL of distilled/deionized water. Mix thoroughly. Autoclave for 10 min at 15 psi pressure–121°C. Cool to 25°C. Put into dialysis tubing.

Dialyze against 2.0L of distilled/deionized water for 48h at 4°C.

Antibiotic LCNT Solution:

Composition per 5.0mL:

Colistin 7.5mg
Lincomycin·HCl 4.0mg
Trimethorprim lactate 3.0mg
Nystatin 12.5U

Preparation of Antibiotic LCNT Solution: Add the components to distilled/deionized water and bring volume to 5.0mL. Mix thoroughly. Filter sterilize the solution.

Preparation of Medium: Have all solutions prepared and at 45°–50°C. Aseptically combine components. Mix thoroughly. Pour into sterile Petri dishes.

Use: For the isolation and cultivation of pathogenic *Neisseria* species. Used as a transport medium for urogenital and other clinical specimens. For the isolation and presumptive identification of Mycoplasmatales, including large-colony species (*Mycoplasma pneumoniae*) and T–mycoplasmas from urogenital specimens.

NIH Agar

Composition per liter:

Pancreatic digest of casein 15.0g
Agar 15.0g
Glucose 5.5g
Yeast extract 5.0g
NaCl 2.5g
L-Cystine 0.05g
pH 7.1 ± 0.2 at 25°C

Source: This medium is available as a premixed powder from BD Diagnostic Systems.

Preparation of Medium: Add components to distilled/deionized water and bring volume to 1.0L. Mix thoroughly. Gently heat while stirring and bring to boiling. Distribute into tubes or flasks. Autoclave for 15 min at 15 psi pressure–121°C. Pour into sterile Petri dishes or leave in tubes.

Use: For the cultivation and maintenance of microorganisms isolated from sterility testing of biological products. Also used as a solid medium for sterility testing.

NIH Thioglycolate Broth

Composition per liter:

Pancreatic digest of casein 15.0g
Glucose 5.5g
Yeast extract 5.0g
NaCl 2.5g
L-Cystine 0.5g
Sodium thioglycolate 0.5g
pH 7.1 ± 0.2 at 25°C

Source: This medium is available as a premixed powder from BD Diagnostic Systems.

Preparation of Medium: Add components to distilled/deionized water and bring volume to 1.0L. Mix thoroughly. Gently heat while stirring and bring to boiling. Distribute into tubes or flasks. Autoclave for 15 min at 15 psi pressure–121°C.

Use: For sterility testing of biological products that are turbid or otherwise cannot be cultivated in fluid thioglycolate broth because of its viscosity.

Nitrate Agar

Composition per liter:

Agar 12.0g
Peptone 5.0g
Beef extract 3.0g
KNO_3 .... 1.0g
pH 6.8 ± 0.2 at 25°C

Preparation of Medium: Add components to distilled/deionized water and bring volume to 1.0L. Mix thoroughly. Gently heat and bring to boiling. Distribute into tubes. Autoclave for 15 min at 15 psi pressure–121°C. Allow tubes to cool in a slanted position.

Use: For the differentiation of aerobic and facultative Gram-negative microorganisms based on their ability to reduce nitrate. Test for nitrates with sulfanilic acid and α-naphthylamine reagents. Bacteria that reduce nitrate to nitrite turn the reagents red or pink.

Nitrate Broth (International Streptomyces Project Medium 8) (ISP Medium 8) (ATCC Medium 872)

Composition per liter:

Peptone 5.0g
Beef extract 3.0g
KNO_3 .... 1.0g
pH 7.0 ± 0.2 at 25°C

Source: This medium is available as a premixed powder from BD Diagnostic Systems.

Preparation of Medium: Add components to distilled/deionized water and bring volume to 1.0L. Mix thoroughly. Distribute into tubes or flasks. Autoclave for 15 min at 15 psi pressure–121°C.

Use: For the differentiation of aerobic and facultative Gram-negative microorganisms based on their ability to reduce nitrate. Test for nitrates with sulfanilic acid and α-naphthylamine reagents. Bacteria that reduce nitrate to nitrite turn the reagents red or pink.

Nitrate Broth

Composition per liter:

Pancreatic digest of gelatin 20.0g
KNO_3 .... 2.0g

pH 7.2 ± 0.2 at 25°C

Source: This medium is available as a premixed powder from BD Diagnostic Systems.

Preparation of Medium: Add components to distilled/deionized water and bring volume to 1.0L. Mix thoroughly. Distribute into tubes or flasks. Autoclave for 15 min at 15 psi pressure–121°C.

Use: For the differentiation of aerobic and facultative Gram-negative microorganisms based on their ability to reduce nitrate. Test for nitrates with sulfanilic acid and α-naphthylamine reagents. Bacteria that reduce nitrate to nitrite turn the reagents red or pink.

Nitrate Broth, *Campylobacter*

Composition per liter:

Beef heart, solids from infusion 500.0g
Tryptose 10.0g
NaCl 5.0g
KNO_3 .... 2.0g

pH 7.0 ± 0.2 at 25°C

Preparation of Medium: Add components to distilled/deionized water and bring volume to 1.0L. Mix thoroughly. Adjust pH to 7.0. Distribute 4.0mL volumes into test tubes that contain inverted Durham tubes. Autoclave for 15 min at 15 psi pressure–121°C.

Use: For the differentiation of *Campylobacter* species based on their ability to reduce nitrate.

Nitrate Reduction Broth

Composition per liter:

Pancreatic digest of gelatin 5.0g
Beef extract 3.0g
KNO_3 .... 1.0g

pH 6.9 ± 0.2 at 25°C

Preparation of Medium: Add components to distilled/deionized water and bring volume to 1.0L. Mix thoroughly. Distribute into test tubes that contain an inverted Durham tube. Autoclave for 15 min at 15 psi pressure–121°C.

Use: For the differentiation of members of the Pseudomonadaceae based on their ability to reduce nitrate to nitrite or form N_2 gas. Test for nitrates with sulfanilic acid and α-naphthylamine reagents. Bacteria that reduce nitrate to nitrite turn the reagents red or pink.

Nitrate Reduction Broth

Composition per liter:

Pancreatic digest of casein 13.0g
NaCl 5.0g
Yeast extract 5.0g
Heart muscle, solids from infusion 2.0g
KNO_3 or $NaNO_3$ 2.0g

pH 7.4 ± 0.2 at 25°C

Preparation of Medium: Add components to distilled/deionized water and bring volume to 1.0L. Mix thoroughly. Distribute into test tubes that contain an inverted Durham tube. Autoclave for 15 min at 15 psi pressure–121°C.

Use: For the differentiation of a variety of Gram-negative bacteria based on their ability to reduce nitrate to nitrite or form N_2 gas. Test for nitrates with sulfanilic acid and α-naphthylamine reagents. Bacteria that reduce nitrate to nitrite turn the reagents red or pink.

Nitrate Reduction Broth

Composition per liter:

Pancreatic digest of casein 13.0g
NaCl 5.0g
Yeast extract 5.0g
Heart muscle, solids from infusion 2.0g
KNO_3 or $NaNO_3$ 2.0g

pH 7.4 ± 0.2 at 25°C

Preparation of Medium: Dispense in 3.0mL amounts into screw-capped tubes, add inverted vials, and autoclave at 121°C for 15 min.

Use: For the differentiation of a variety of nonfermenting Gram-negative bacteria based on their ability to reduce nitrate to nitrite or form N_2 gas. Test for nitrates with sulfanilic acid and α-naphthylamine reagents. Bacteria that reduce nitrate to nitrite turn the reagents red or pink.

Nitrate Reduction Broth, Clark

Composition per liter:

Peptone 20.0g
KNO_3 or $NaNO_3$ 2.0g

Preparation of Medium: Add components to distilled/deionized water and bring volume to 1.0L. Mix thoroughly. Distribute into test tubes that contain an inverted Durham tube. Autoclave for 15 min at 15 psi pressure–121°C.

Use: For the differentiation of a variety of Gram-negative bacteria based on their ability to reduce nitrate to nitrite or form N_2 gas. Test for nitrates with sulfanilic acid and α-naphthylamine reagents. Bacteria that reduce nitrate to nitrite turn the reagents red or pink.

Nitrate Reduction Broth for *Pseudomonas* and Related Genera

Composition per liter:

Peptone	5.0g
NaCl	5.0g
Yeast extract	2.0g
Beef extract	1.0g
$NaNO_3$	0.1g

pH 7.4 ± 0.2 at 25°C

Preparation of Medium: Add components to distilled/deionized water and bring volume to 1.0L. Mix thoroughly. Distribute into test tubes that contain an inverted Durham tube. Autoclave for 15 min at 15 psi pressure–121°C.

Use: For the differentiation of members of the Pseudomonadaceae based on their ability to reduce nitrate to nitrite or form N_2 gas. Test for nitrates with sulfanilic acid and α-naphthylamine reagents. Bacteria that reduce nitrate to nitrite turn the reagents red or pink.

NNN Medium (Novy, MacNeal, and Nicole Medium)

Composition per liter:

Agar	7.0g
NaCl	3.0g
Rabbit blood, defibrinated	150.0mL

Preparation of Medium: Add components, except rabbit blood, to distilled/deionized water and bring volume to 850.0mL. Mix thoroughly. Gently heat and bring to boiling. Autoclave for 15 min at 15 psi pressure–121°C. Cool to 50°C. Aseptically add sterile rabbit blood. Mix thoroughly. Aseptically distribute into sterile tubes in 5.0mL volumes. Allow tubes to cool in a slanted position at 4°C.

Use: For the cultivation and maintenance of *Leishmania* species and *Trypanosoma cruzi.*

Nonfat Dry Milk, Reconstituted

Composition per liter:

Milk, nonfat dry	100.0g

pH 6.8 ± 0.2 at 25°C

Preparation of Medium: Add 100.0g of nonfat dry milk to distilled/deionized water and bring volume to 1.0L. Mix thoroughly. Distribute into tubes or flasks. Autoclave for 15 min at 15 psi pressure–121°C.

Use: For the cultivation of *Salmonella* species and monkey kidney cells in tissue culture.

Nonnutrient Agar Plates

Composition per liter:

Agar	15.0g
Page's amoeba saline	1.0L

Page's Amoeba Saline:

Composition per liter:

Na_2HPO_4	0.142g
KH_2PO_4	0.136g
NaCl	0.12g
$MgSO_4 \cdot 7H_2O$	4.0mg
$CaCl_2 \cdot 2H_2O$	4.0mg

Preparation of Page's Amoeba Saline: Add components to distilled/deionized water and bring volume to 10.0mL. Mix thoroughly.

Preparation of Medium: Add agar to 1.0L of Page's amoeba saline. Mix thoroughly. Gently heat and bring to boiling. Autoclave for 15 min at 15 psi pressure–121°C. Cool to 60°C. Pour into sterile Petri dishes in 20.0mL volumes. Store at 4°C for up to 3 months.

Use: For the isolation and cultivation of pathogenic free-living amoebae.

NOS Medium, Modified

Composition per 100.67mL:

Basal medium	94.0mL
$NaHCO_3$ solution	2.67mL
TPP/VFA mixture	2.0mL
Rabbit serum, heat inactivated	2.0mL

pH 7.4 ± 0.2 at 25°C

Basal Medium:

Composition per 94.0mL:

Pancreatic digest of casein	1.0g
Pancreatic digest of gelatin	0.48g
Yeast extract	0.25g
Brain heart, solids from infusion	0.2g
Peptic digest of animal tissue	0.2g
D-Glucose	0.2g
NaCl	0.17g
Glucose	0.1g
L-Cysteine·HCl·H_2O	0.1g
Na_2HPO_4	0.085g
Sodium thioglycolate	0.05g
L-Asparagine	0.025g
Resazurin (0.1% w/v solution)	0.1mL

Preparation of Basal Medium: Add components to distilled/deionized water and bring volume to 94.0mL. Mix thoroughly. Gently heat and bring to boiling. Gas under O_2-free 85% N_2 + 10% CO_2 + 5% H_2. Stopper and wire flask closed. Autoclave for 20 min at 15 psi pressure–121°C. Cool to 45°–50°C.

$NaHCO_3$ Solution:

Composition per 10.0mL:

$NaHCO_3$	0.75g

Preparation of $NaHCO_3$ Solution: Add the $NaHCO_3$ to distilled/deionized water and bring volume to 10.0mL. Mix thoroughly. Filter sterilize.

TPP/VFA Mixture:

Composition per 10.9mL:

Thiamine pyrophosphate (0.2% solution)	1.5mL
VFA solution	1.0mL

Preparation of TPP/VFA Mixture: Add components to distilled/deionized water and bring volume to 10.9mL. Mix thoroughly. Filter sterilize. Store at –20°C.

VFA Solution:

Composition per 100.0mL:

NaOH (0.1*N* solution)	98.0mL
Isobutyric acid	0.5mL
2-Methylbutyric acid	0.5mL
Isovaleric acid	0.5mL
Valeric acid	0.5mL

Preparation of VFA Solution: Add volatile fatty acids to 98.0mL of NaOH solution. Mix thoroughly. Filter sterilize. Store at 4°C.

Preparation of Medium: Open the flask containing 94.0mL of cooled sterile basal medium while flushing with O_2-free 85% N_2 + 10% CO_2 + 5% H_2. Aseptically add sterile $NaHCO_3$ solution, sterile TPP/VFA mixture, and filter-sterilized rabbit serum. Mix thoroughly.

Use: For the cultivation and maintenance of *Treponema vincentii* and other *Treponema* species.

NOS Spirochete Medium

Composition per 1045.0mL:

Basal medium	1.0L
$NaHCO_3$ (10% solution)	20.0mL
Rabbit serum, heat inactivated	20.0mL
Thiamine pyrophosphate (0.2% solution)	3.0mL
VFA solution	2.0mL

pH 7.4 ± 0.2 at 25°C

Basal Medium:

Composition per liter:

Pancreatic digest of casein	10.0g
Pancreatic digest of gelatin	4.85g
Noble agar	3.0g
Yeast extract	2.5g
Brain heart, solids from infusion	2.0g
Peptic digest of animal tissue	2.0g
Glucose	2.0g
NaCl	1.65g
Glucose	1.0g
L-Cysteine·HCl·H_2O	1.0g
Na_2HPO_4	0.85g
Sodium thioglycolate	0.5g
L-Asparagine	0.25g

Preparation of Basal Medium: Add components to distilled/deionized water and bring volume to 1.0L. Mix thoroughly. Gently heat and bring to boiling. Gas under O_2-free 80% N_2 + 10% CO_2 + 10% H_2. Stopper and wire flask closed. Autoclave for 20 min at 15 psi pressure–121°C. Cool to 45°–50°C.

VFA Solution:

Composition per 100.0mL:

KOH (0.1*N* solution)	98.0mL
Isobutyric acid	0.5mL
2-Methylbutyric acid	0.5mL
Isovaleric acid	0.5mL
Valeric acid	0.5mL

Preparation of VFA Solution: Add volatile fatty acids to 98.0mL of KOH solution. Mix thoroughly. Filter sterilize. Store at 4°C.

Preparation of Medium: Combine 20.0mL of $NaHCO_3$ solution, 20.0mL of rabbit serum, 3.0mL of thiamine pyrophosphate solution, and 2.0mL of VFA solution. Mix thoroughly. Filter sterilize. Open the flask containing 1.0L of cooled, sterile basal medium while flushing with O_2-free 85% N_2 + 10% CO_2 + 5% H_2. Aseptically add the filter-sterilized mixture. Mix thoroughly. Aseptically and anaerobically distribute into sterile tubes or flasks.

Use: For the cultivation and maintenance of *Treponema denticola* and *Treponema socranskii*.

Novobiocin Agar

Composition per liter:

Agar	15.0g
Peptone	5.0g
NaCl	5.0g
Yeast extract	2.0g
Beef extract	1.0g
Novobiocin solution	10.0mL

Novobiocin Solution:

Composition per 10.0mL:

Novobiocin	10.0mg

Preparation of Novobiocin Solution: Add novobiocin to distilled/deionized water and bring volume to 10.0mL. Mix thoroughly. Filter sterilize.

Preparation of Medium: Add components, except novobiocin solution, to distilled/deionized water and bring volume to 990.0mL. Mix thoroughly. Gently heat and bring to boiling. Autoclave for 15 min at 15 psi pressure–121°C. Aseptically add 10.0mL of sterile novobiocin solution. Mix thoroughly. Pour into sterile Petri dishes or distribute into sterile tubes.

Use: For the cultivation of *Staphylococcus aureus.*

Nutrient Agar (LMG Medium 160)

Composition per liter:

Agar	3.0g
Lab-Lemco beef extract	1.0g
Peptone	1.0g
NaCl	0.5g

pH 7.3 ± 0.2 at 25°C

Preparation of Medium: Add components to distilled/deionized water and bring volume to 1.0L. Mix thoroughly. Distribute into tubes or flasks. Autoclave for 15 min at 15 psi pressure–121°C.

Use: For the cultivation of heterotrophic bacteria.

Nutrient Agar

Composition per liter:

Agar	15.0g
Peptone	5.0g
NaCl	5.0g
Yeast extract	2.0g
Beef extract	1.0g

pH 7.4 ± 0.2 at 25°C

Source: This medium is available as a premixed powder from Oxoid Unipath.

Preparation of Medium: Add components to distilled/deionized water and bring volume to 1.0L. Mix thoroughly. Gently heat and bring to boiling. Distribute into tubes or flasks. Autoclave for 15 min at 15 psi pressure–121°C. Pour into sterile Petri dishes or leave in tubes.

Use: For the cultivation and maintenance of a wide variety of microorganisms.

Nutrient Agar (ATCC Medium 3)

Composition per liter:

Agar	15.0g
Pancreatic digest of gelatin	5.0g
Beef extract	3.0g

pH 6.8 ± 0.2 at 25°C

Source: This medium is available as a premixed powder from BD Diagnostic Systems.

Preparation of Medium: Add components to distilled/deionized water and bring volume to 1.0L. Mix thoroughly. Gently heat while stirring and bring to boiling. Distribute into tubes or flasks. Autoclave for 15 min at 15 psi pressure–121°C. Pour into sterile Petri dishes or leave in tubes.

Use: For the cultivation of a wide variety of bacteria.

Nutrient Agar with Formate, Fumarate, and Horse Blood (LMG Medium 250)

Composition per liter:

Agar	15.0g
Peptone	5.0g
NaCl	5.0g
Fumaric acid	3.0g
Sodium formate	2.0g
Yeast extract	2.0g
Lab-Lemco beef extract	1.0g
Horse blood, sterile defibrinated	50.0mL

pH 7.2 ± 0.2 at 25°C

Preparation of Medium: Add components, except horse blood, to distilled/deionized water and bring volume to 950.0mL. Mix thoroughly. Adjust pH to 7.2. Gently heat and bring to boiling. Autoclave for 15 min at 15 psi pressure–121°C. Cool to 45°C. Aseptically add 50.0mL sterile defibrinated horse blood. Mix thoroughly. Pour into sterile Petri dishes or distribute into sterile tubes.

Use: For the cultivation and maintenance of *Campylobacter rectus* and *Campylobacter gracilis.*

Nutrient Agar No. 2

Composition per liter:

Agar	20.0g
Meat extract	10.0g
Peptone	10.0g
NaCl	5.0g

pH 7.0 ± 0.2 at 25°C

Preparation of Medium: Add components to distilled/deionized water and bring volume to 1.0L. Mix thoroughly. Gently heat and bring to boiling. Distribute into tubes or flasks. Autoclave for 15 min at 15 psi pressure–121°C. Pour into sterile Petri dishes or leave in tubes.

Use: For the cultivation and maintenance of a wide variety of bacteria.

Nutrient Agar with Dihydrostreptomycin

Composition per liter:

Agar	15.0g
Pancreatic digest of gelatin	5.0g

Beef extract 3.0g
Dihydrostreptomycin solution 10.0mL
pH 6.8 ± 0.2 at 25°C

Source: Nutrient agar is available as a premixed powder from BD Diagnostic Systems.

Dihydrostreptomycin Solution:
Composition per 10.0mL:
Dihydrostreptomycin 0.625g

Preparation of Dihydrostreptomycin Solution: Add dihydrostreptomycin to distilled/deionized water and bring volume to 10.0mL. Mix thoroughly. Filter sterilize.

Preparation of Medium: Add components, except dihydrostreptomycin solution, to distilled/deionized water and bring volume to 990.0mL. Mix thoroughly. Gently heat and bring to boiling. Autoclave for 15 min at 15 psi pressure–121°C. Cool to 45°–50°C. Aseptically add sterile dihydrostreptomycin solution. Mix thoroughly. Pour into sterile Petri dishes or distribute into sterile tubes.

Use: For the cultivation and maintenance of *Escherichia coli, Micrococcus luteus, Shigella* species, and *Vibrio cholerae.*

Nutrient Agar with Horse Serum (LMG Medium 38)

Composition per liter:
Agar 15.0g
Peptone.. 5.0g
NaCl...5.0g
Yeast extract...2.0g
Lab-Lemco beef extract 1.0g
Horse serum 100.0mL
pH 7.4 ± 0.2 at 25°C

Preparation of Medium: Add components, except horse serum, to distilled/deionized water and bring volume to 1.0L. Mix thoroughly. Gently heat and bring to boiling. Autoclave for 15 min at 15 psi pressure–121°C. Cool medium to 45-50°C. Aseptically add 100.0mL of sterile horse serum. Mix thoroughly. Pour into sterile Petri dishes or distribute into sterile tubes.

Use: For the cultivation of fastidious bacteria.

Nutrient Agar with 0.5% NaCl

Composition per liter:
Agar 15.0g
NaCl...5.0g
Pancreatic digest of gelatin 5.0g
Beef extract...3.0g
pH 6.8 ± 0.2 at 25°C

Source: Nutrient agar is available as a premixed powder from BD Diagnostic Systems.

Preparation of Medium: Add components to distilled/deionized water and bring volume to 1.0L. Mix thoroughly. Gently heat and bring to boiling. Distribute into tubes or flasks. Autoclave for 15 min at 15 psi pressure–121°C. Pour into sterile Petri dishes or leave in tubes.

Use: For the cultivation and maintenance of *Pseudomonas aeruginosa, Shigella dysenteriae, Shigella flexneri, Vibrio* species, and *Yersinia* species.

Nutrient Agar with 2% NaCl

Composition per liter:
NaCl...20.0g
Agar 15.0g
Pancreatic digest of gelatin 5.0g
Beef extract 3.0g
pH 6.8 ± 0.2 at 25°C

Preparation of Medium: Add components to distilled/deionized water and bring volume to 1.0L. Mix thoroughly. Gently heat and bring to boiling. Distribute into tubes or flasks. Autoclave for 15 min at 15 psi pressure–121°C. Pour into sterile Petri dishes or leave in tubes.

Use: For the cultivation and maintenance of *Vibrio parahaemolyticus* and *Vibrio vulnificus.*

Nutrient Agar with Yeast Extract

Composition per liter:
Yeast extract...20.0g
Agar 15.0g
Pancreatic digest of gelatin 5.0g
Beef extract 3.0g
pH 6.8 ± 0.2 at 25°C

Source: Nutrient agar is available as a premixed powder from BD Diagnostic Systems.

Preparation of Medium: Add components to distilled/deionized water and bring volume to 1.0L. Mix thoroughly. Gently heat and bring to boiling. Distribute into tubes or flasks. Autoclave for 15 min at 15 psi pressure–121°C. Pour into sterile Petri dishes or leave in tubes.

Use: For the cultivation and maintenance of *Bacillus anthracis* and *Comamonas testosteroni.*

Nutrient Broth

Composition per liter:
Peptone 5.0g
NaCl...5.0g

Yeast extract..2.0g
Beef extract..1.0g

pH 7.4 ± 0.2 at 25°C

Source: This medium is available as a premixed powder from Oxoid Unipath.

Preparation of Medium: Add components to distilled/deionized water and bring volume to 1.0L. Mix thoroughly. Distribute into tubes or flasks. Autoclave for 15 min at 15 psi pressure–121°C.

Use: For the cultivation of a wide variety of nonfastidious microorganisms.

Nutrient Broth

Composition per liter:

Pancreatic digest of gelatin...............................5.0g
Beef extract..3.0g

pH 6.9 ± 0.2 at 25°C

Source: This medium is available as a premixed powder from BD Diagnostic Systems.

Preparation of Medium: Add components to distilled/deionized water and bring volume to 1.0L. Mix thoroughly. Distribute into tubes or flasks. Autoclave for 15 min at 15 psi pressure–121°C.

Use: For the cultivation of a wide variety of nonfastidious microorganisms.

Nutrient Broth No. 2

Composition per liter:

Beef extract..10.0g
Peptone..10.0g
NaCl...5.0g

pH 7.5 ± 0.2 at 25°C

Source: This medium is available as a premixed powder from Oxoid Unipath.

Preparation of Medium: Add components to distilled/deionized water and bring volume to 1.0L. Mix thoroughly. Distribute into tubes or flasks. Autoclave for 15 min at 15 psi pressure–121°C.

Use: For the cultivation of a variety of fastidious and nonfastidious microorganisms.

Nutrient Broth, Standard II

Composition per liter:

Special peptone...8.6g
NaCl...6.4g

pH 7.5 ± 0.1 at 37°C

Preparation of Medium: Add components to distilled/deionized water and bring volume to 1.0L. Mix thoroughly. Distribute into tubes or flasks. Autoclave for 15 min at 15 psi pressure–121°C.

Use: For the cultivation of a variety of fastidious and nonfastidious microorganisms.

Nutrient Gelatin

Composition per liter:

Gelatin..120.0g
Pancreatic digest of gelatin...............................5.0g
Beef extract..3.0g

pH 6.8 ± 0.2 at 25°C

Source: This medium is available as a premixed powder from BD Diagnostic Systems and Oxoid Unipath.

Preparation of Medium: Add components to distilled/deionized water and bring volume to 1.0L. Mix thoroughly. Gently heat while stirring to 50°C. Distribute into tubes. Autoclave for 15 min at 15 psi pressure–121°C.

Use: For the cultivation and differentiation of bacteria based on their ability to liquefy gelatin.

N-Z Amine A™ Glycerol Agar

Composition per liter:

Agar...15.0g
N-Z-Amine A..5.0g
Beef extract..1.0g
Glycerol...70.0mL

pH 6.5–7.0 at 25°C

Preparation of Medium: Add components to distilled/deionized water and bring volume to 1.0L. Mix thoroughly. Gently heat and bring to boiling. Distribute into tubes or flasks. Autoclave for 15 min at 15 psi pressure–121°C. Pour into sterile Petri dishes or leave in tubes.

Use: For the isolation and cultivation of *Actinomadura* species, *Actinopolyspora* species, *Excellospora* species, and *Microspora* species.

N-Z Amine A Medium with Soluble Starch and Glucose

Composition per liter:

Soluble starch..20.0g
Agar...15.0g
Glucose..10.0g
Yeast extract..5.0g
N-Z-Amine A..5.0g
$CaCO_3$...1.0g

Preparation of Medium: Add components to distilled/deionized water and bring volume to 1.0L. Mix thoroughly. Gently heat and bring to boiling. Distribute into tubes or flasks. Autoclave for 15 min at 15 psi pressure–121°C. Pour into sterile Petri dishes or leave in tubes.

Use: For the cultivation and maintenance of *Actinomadura* species.

NZY Agar

Composition per liter:

Agar	20.0g
N-Z- Amine A	10.0g
NaCl	5.0g
Yeast extract	5.0g
$MgCl_2 \cdot 6H_2O$	2.0g

Preparation of Medium: Add components to distilled/deionized water and bring volume to 1.0L. Mix thoroughly. Gently heat and bring to boiling. Distribute into tubes. Autoclave for 15 min at 15 psi pressure–121°C. Allow tubes to cool in a slanted position.

Use: For the cultivation and maintenance of a variety of microorganisms.

NZY Agar

Composition per liter:

Agar	11.0g
N-Z-Amine A	10.0g
NaCl	5.0g
Yeast extract	5.0g
$MgCl_2 \cdot 6H_2O$	2.0g

Preparation of Medium: Add components to distilled/deionized water and bring volume to 1.0L. Mix thoroughly. Gently heat and bring to boiling. Distribute into tubes or flasks. Autoclave for 15 min at 15 psi pressure–121°C. Pour into sterile Petri dishes.

Use: For the cultivation and enumeration of a variety of microorganisms.

NZY Broth

Composition per liter:

N-Z-Amine A	10.0g
NaCl	5.0g
Yeast extract	5.0g
$MgCl_2 \cdot 6H_2O$	2.0g

Preparation of Medium: Add components to distilled/deionized water and bring volume to 1.0L. Mix thoroughly. Gently heat and bring to boiling. Distribute into tubes. Autoclave for 15 min at 15 psi pressure–121°C.

Use: For the cultivation of a variety of microorganisms.

O157:H7(+) Plating Medium

Composition per liter:

Agar	15.0g
Sorbitol	12.0g
Salicin	10.0g
Inositol	10.0g
Peptone	10.0g
Adonitol	8.0g
NaCl	5.0g
Tryptone	5.0g
Proteose peptone	3.0g
Bile salts no. 3	1.25g
Indoxyl-β-D-galactopyranoside	0.12g
5-Bromo-4-chloro-3-indoxyl-β-D-galactopyranoside	0.12g
Phenol Red	0.1g
Isopropyl-β-D-thiogalactopyranoside	0.1g
Novobiocin solution	1.0mL
Tellurite solution	1.0mL

pH 6.8 ± 0.2 at 25°C

Source: This medium is available as a premixed powder from BIOSYNTH International, Inc.

Novobiocin Solution:

Composition per 10.0mL:

Novobiocin	0.1g

Preparation of Novobiocin Solution: Add novobiocin to distilled/deionized water and bring volume to 10.0mL. Mix thoroughly. Filter sterilize.

Tellurite Solution:

Composition per 10.0mL:

K_2TeO_3	0.01g

Preparation of Tellurite Solution: Add K_2TeO_3 to distilled/deionized water and bring volume to 100.0mL. Mix thoroughly. Filter sterilize.

Caution: Potassium tellurite is toxic.

Preparation of Medium: Add components, except novobiocin solution and tellurite solution, to distilled/deionized water and bring volume to 998.0mL. Mix thoroughly. Gently heat while stirring and bring to boiling. Autoclave for 15 min at 15 psi pressure–121°C. Cool to 50°C. Aseptially add 1.0mL sterile novobiocin solution and 1.0 sterile tellurite solution. Mix thoroughly. Pour into sterile Petri dishes.

Use: For the detection of *Escherichia coli* O157:H7. *E.coli* O157:H7 grow with blue-black colonies and *E.coli* non-O157 with green-yellow colonies.

O157:H7 ID Agar

Composition per liter:

Proprietary

Source: This medium is available from bioMérieux.

Use: A new chromogenic medium for the detection of *Escherichia coli* O157:H7.

Oatmeal Agar

Composition per 750.0mL:

Oatmeal, instant for babies 40.0g
Agar 5.0g

Preparation of Medium: Add agar to distilled/deionized water and bring volume to 500.0mL. Mix thoroughly. Gently heat and bring to boiling. Add instant oatmeal for babies to distilled/deionized water and bring volume to 250.0mL. Mix thoroughly. Add the oatmeal solution to the melted agar solution. Mix thoroughly. Autoclave for 15 min at 15 psi pressure–121°C. Pour into sterile Petri dishes or aseptically distribute into sterile tubes.

Use: For the cultivation of fungi.

Oatmeal Agar (ATCC 551)

Composition per liter:

Oatmeal 60.0g
Agar 12.5g

pH 6.0 ± 0.2 at 25°C

Source: This medium is available as a premixed powder from BD Diagnostic Systems.

Preparation of Medium: Add components to distilled/deionized water and bring volume to 1.0L. Mix thoroughly. Gently heat and bring to boiling. Distribute into tubes or flasks. Autoclave for 15 min at 15 psi pressure–121°C. Pour into sterile Petri dishes or leave in tubes.

Use: For cultivation of fungi and actinomycetes, particularly for macrospore formation.

Oatmeal Agar (OA)

Composition per liter:

Agar 12.0g
Oatmeal, rolled oats 10.0g
Glycerin 5.0mL
Lactic acid 0.2mL

Preparation of Medium: Add components to distilled/deionized water and bring volume to 1.0L. Mix thoroughly. Gently heat and bring to boiling. Distribute into tubes or flasks. Autoclave for 15 min at 15 psi pressure–121°C. Pour into sterile Petri dishes or leave in tubes.

Use: For the cultivation of many filamentous fungi.

Ogawa TB Medium

Composition per 300.0mL:

KH_2PO_4 .... 1.0g
Homogenized whole egg 200.0mL
Glycerol 6.0mL
Malachite Green (2% solution) 6.0mL

pH 6.5 ± 0.2 at 25°C

Homogenized Whole Egg:

Composition per liter:

Whole eggs 18–24

Preparation of Homogenized Whole Egg: Use fresh eggs, less than 1 week old. Scrub the shells with soap. Let stand in a soap solution for 30 min. Rinse in running water. Soak eggs in 70% ethanol for 15 min. Break the eggs into a sterile container. Homogenize by shaking. Filter through four layers of sterile cheesecloth into a sterile graduated cylinder. Measure out 1.0L.

Preparation of Medium: Add components, except homogenized whole egg, to distilled/deionized water and bring volume to 100.0mL. Mix thoroughly. Autoclave for 15 min at 15 psi pressure–121°C. Cool to 45°–50°C. Aseptically add 200.0mL of sterile homogenized whole egg. Mix thoroughly. Aseptically distribute into sterile screw-capped tubes in 7.0mL volumes. Inspissate at 85°–90°C (moist heat) for 60 min.

Use: For the isolation and cultivation of *Mycobacterium* species, except for *Mycobacterium leprae*.

Oleic Albumin Complex

Composition per liter:

Bovine albumin fraction V 50.0g
NaCl 8.5g
Oleic acid 0.6mL

Preparation of Medium: Add components to distilled/deionized water and bring volume to 1.0L. Mix thoroughly. Filter sterilize.

Use: For use in media employed for the cultivation of mycobacteria.

Önöz *Salmonella* Agar

Composition per liter:

Agar 15.0g
Sucrose 13.0g
Lactose 11.5g
Trisodium citrate–5,5–hydrate 9.3g
Meat peptone 6.8g
Beef extract 6.0g
L-Phenylalanine 5.0g
$Na_2S_2O_3 \cdot 5H_2O$ 4.25g
Bile salt mixture 3.825g
Yeast extract 3.0g
$Na_2HPO_4 \cdot 2H_2O$ 1.0g
Ferric citrate 0.5g
Metachrome Yellow 0.47g

$MgSO_4 \cdot 7H_2O$ 0.4g
Aniline Blue 0.25g
Neutral Red 0.022g
Brilliant Green 0.00166g
pH 7.1 ± 0.2 at 25°C

Preparation of Medium: Add components to distilled/deionized water and bring volume to 1.0L. Mix thoroughly. Gently heat and bring to boiling. Distribute into tubes or flasks. Autoclave for 15 min at 15 psi pressure–121°C. Pour into sterile Petri dishes or leave in tubes.

Use: For the isolation and cultivation of *Salmonella* from feces.

ONPG Broth

Composition per liter:
Peptone water 750.0mL
ONPG solution 250.0mL
pH 7.2–7.4 at 25°C

ONPG Solution:
Composition per 250.0mL:
ONPG (*o*-Nitrophenyl-β-D-galactopyranoside) 1.5g
Sodium phosphate buffer (0.01*M*, pH 7.5) 250.0mL

Preparation of ONPG Solution: Add ONPG to 250.0mL of sodium phosphate buffer. Mix thoroughly. Filter sterilize.

Peptone Water:
Composition per 750.0mL:
Peptone 7.5g
NaCl 3.75g

Preparation of Peptone Water: Add components to distilled/deionized water and bring volume to 750.0mL. Mix thoroughly. Gently heat and bring to boiling. Adjust pH to 8.0–8.4. Continue boiling for 10 min. Filter through Whatman no. 1 filter paper. Readjust pH of filtrate to 7.2–7.4. Autoclave for 20 min at 10 psi pressure–115°C. Cool to 25°C.

Preparation of Medium: Aseptically combine the sterile ONPG solution with the cooled, sterile peptone water. Mix thoroughly. Aseptically distribute into tubes in 2.5–3.0mL volumes. Store at 4°C for up to 1 month.

Use: For the differentiation of a variety of Gram-negative bacteria based on production of β-galactosidase. For the differentiation of lactose-delayed bacteria from lactose-negative bacteria. For the differentiation of *Pseudomonas cepacia* (positive) and *Pseudomonas maltophila* (positive) from other *Pseudomonas* species (negative). Bacteria that produce β-galactosidase turn the medium yellow.

OR Indicator Agar (Oxidation-Reduction Indicator Agar)

Composition per liter:
Agar 15.0g
Sodium glycerol phosphate 10.0g
Sodium thioglycolate 1.7g
$CaCl_2 \cdot 2H_2O$ 0.1g
Methylene Blue 6.0mg

Preparation of Medium: Add components to distilled/deionized water and bring volume to 1.0L. Mix thoroughly. Gently heat and bring to boiling. Distribute into tubes or flasks. Autoclave for 15 min at 15 psi pressure–121°C. Pour into sterile Petri dishes or leave in tubes.

Use: For use as an indicator of oxygen-free conditions in anaerobic culture chambers.

Oral *Fusobacterium* Medium

Composition per liter:
Agar 15.0g
Proteose peptone 10.0g
Na_2HPO_4 .... 5.0g
Glucose 5.0g
Beef extract 3.0g
Soluble starch 2.0g
$NaNO_3$ 1.0g
Yeast extract 1.0g
L-Cysteine·HCl·H_2O 0.5g
Ethyl Violet solution 10.0mL
Bacitracin solution 10.0mL
pH 7.6 ± 0.2 at 25°C

Ethyl Violet Solution:
Composition per 10.0mL:
Ethyl Violet 0.04g

Preparation of Ethyl Violet Solution: Add Ethyl Violet to distilled/deionized water and bring volume to 10.0mL. Mix thoroughly. Filter sterilize.

Bacitracin Solution:
Composition per 10.0mL:
Bacitracin 1.0mg

Preparation of Bacitracin Solution: Add bacitracin to distilled/deionized water and bring volume to 10.0mL. Mix thoroughly. Filter sterilize.

Preparation of Medium: Add components, except Ethyl Violet solution and bacitracin solution, to distilled/deionized water and bring volume to 980.0mL. Mix thoroughly. Gently heat and bring to boiling. Autoclave for 15 min at 15 psi pressure–121°C. Cool to 45°–50°C. Aseptically add sterile Ethyl Violet solution and bacitracin solution. Mix thoroughly. Pour into sterile Petri dishes or distribute into sterile tubes.

Use: For the selective isolation and cultivation of oral *Fusobacterium* species, especially *Fusobacterium nucleatum.*

Oral *Treponema* Medium

Composition per 1250.0mL:

Veal heart, fresh ground	1.0Kg
Thiopeptone	20.0g
NaCl	10.0g
Ionagar no. 2	2.0g
Glutathione (1% solution)	100.0mL
Rabbit serum or ascitic fluid	100.0mL
Eggs, whole fresh	2

pH 6.8–7.0 at 25°C

Preparation of Medium: Add agar to 50.0mL of distilled/deionized water. Gently heat and bring to boiling. Autoclave for 15 min at 15 psi pressure–121°C. Cool to 45°–50°C. In a separate flask, add finely ground veal heart to 1.0L of distilled/deionized water. Add remaining components, except glutathione, rabbit serum, and agar. Gently heat and bring to 70°C. Adjust pH to 7.4. Gently heat and bring to 100°C. Continue heating at 100°C for 2h. Skim off fat. Filter through glass wool. Adjust pH to 7.6. Gently heat and bring to 100°C. Maintain at 100°C for 30 min. Store at 4°C for 18h. Sterilize in an Arnold sterilizer for 30 min at 100°C on two consecutive days. Cool to 45°–50°C. Aseptically add rabbit serum, glutathione solution, and sterile cooled agar solution. Mix thoroughly. Aseptically distribute into sterile tubes or flasks.

Use: For the isolation and cultivation of *Treponema denticola* and *Treponema oralis.*

Organic Acid Medium KP (Organic Acid Medium, Kauffmann and Petersen)

Composition per liter:

Gelatin	10.0g
Bromthymol Blue	0.024g
Organic acid solution	100.0mL

pH 7.4 ± 0.2 at 25°C

Organic Acid Solution:

Composition per 100.0mL:

Organic acid	10.0g

Preparation of Organic Acid Solution: Add organic acid to distilled/deionized water and bring volume to 100.0mL. Sodium potassium D-tartrate, sodium citrate, or mucic acid may be used. Mix thoroughly.

Preparation of Medium: Add components, except organic acid solution, to distilled/deionized water and bring volume to 900.0mL. Mix thoroughly. Gently heat and bring to boiling. Autoclave for 15 min at 15 psi pressure–121°C. Cool to 45°–50°C. Aseptically add sterile organic acid solution. Mix thoroughly. Aseptically distribute into sterile tubes or flasks.

Use: For the cultivation and differentiation of members of the Enterobacteriaceae based on their ability to utilize different organic acids as carbon source. Bacteria that utilize tartrate, citrate, or mucate turn the medium yellow.

Ornithine Broth

Composition per liter:

L-Ornithine	5.0g
Peptone or gelysate	5.0g
Yeast extract	3.0g
Glucose	1.0g
Bromcresol Purple	0.02g

pH 6.5 ± 0.2 at 25°C

Preparation of Medium: Add components to distilled/deionized water and bring volume to 1.0L. Mix thoroughly. Adjust pH so that is will be 6.5 ± 0.2 after sterilization. Distribute into 16 × 150mm screw-capped tubes in 5.0mL volumes. Autoclave medium with loosely capped tubes for 10 min at 15 psi pressure–121°C. Screw the caps on tightly for storage and after inoculation.

Use: For the cultivation and differentiation of bacteria based on their ability to decarboxylate the amino acid ornithine. Bacteria that decarboxylate ornithine turn the medium turbid purple.

Ornithine Broth with NaCl

Composition per liter:

L-Ornithine	5.0g
Peptone or gelysate	5.0g
Yeast extract	3.0g
Glucose	1.0g
Bromcresol Purple	0.02g

pH 6.5 ± 0.2 at 25°C

Preparation of Medium: Add components to distilled/deionized water and bring volume to 1.0L. Mix thoroughly. Adjust pH so that is will be 6.5 ± 0.2 after sterilization. Distribute into 16 × 150mm screw-capped tubes in 5.0mL volumes. Autoclave medium with loosely capped tubes for 10 min at 15 psi pressure–121°C. Screw the caps on tightly for storage and after inoculation.

Use: For the cultivation and differentiation of *Vibrio* spp. based on their ability to decarboxylate the amino acid ornithine. Bacteria that decarboxylate ornithine turn the medium turbid purple.

Oxford Agar
(*Listeria* Selective Agar, Oxford)

Composition per liter:

Special peptone	23.0g
LiCl	15.0g
Agar	10.0g
NaCl	5.0g
Cornstarch	1.0g
Esculin	1.0g
Ferric ammonium citrate	0.5g
Antibiotic inhibitor	10.0mL

pH 7.0 ± 0.2 at 25°C

Source: This medium is available as a premixed powder from Oxoid Unipath.

Antibiotic Inhibitor:

Composition per 10.0mL:

Cycloheximide	0.4g
Colistin sulfate	0.02g
Fosfomycin	0.01g
Acriflavine	5.0mg
Cefotetan	2.0mg
Ethanol (50% solution)	10.0mL

Preparation of Antibiotic Inhibitor: Add antibiotics to 10.0mL of ethanol. Mix thoroughly. Filter sterilize.

Preparation of Medium: Add components, except antibiotic inhibitor, to distilled/deionized water and bring volume to 990.0mL. Mix thoroughly. Gently heat and bring to boiling. Autoclave for 15 min at 15 psi pressure–121°C. Cool to 45°–50°C. Aseptically add 10.0mL of sterile antibiotic inhibitor. Mix thoroughly. Pour into sterile Petri dishes or distribute into sterile tubes.

Use: For the isolation and cultivation of *Listeria monocytogenes* from specimens containing a mixed bacterial flora.

Oxford Agar, Modified
(*Listeria* Selective Agar, Modified Oxford)
(MOX Agar)

Composition per liter:

Special peptone	23.0g
LiCl	15.0g
Agar	12.0g
NaCl	5.0g
Cornstarch	1.0g
Esculin	1.0g
Ferric ammonium citrate	0.5g
Antibiotic inhibitor	10.0mL

pH 7.0 ± 0.2 at 25°C

Antibiotic Inhibitor:

Composition per 10.0mL:

Moxalactam	0.015g
Colistin sulfate	0.01g

Preparation of Antibiotic Inhibitor: Add components to distilled /deionized water and bring volume to 10.0mL. Mix thoroughly. Filter sterilize.

Preparation of Medium: Add components, except antibiotic inhibitor, to distilled/deionized water and bring volume to 990.0mL. Mix thoroughly. Gently heat and bring to boiling. Autoclave for 10 min at 15 psi pressure–121°C. Cool to 45°–50°C. Aseptically add 10.0mL of sterile antibiotic inhibitor. Mix thoroughly. Pour into sterile Petri dishes or distribute into sterile tubes.

Use: For the isolation and cultivation of *Listeria monocytogenes* from specimens containing a mixed bacterial flora.

Oxidation-Fermentation Medium
(OF Medium)

Composition per liter:

NaCl	5.0g
Agar	2.5g
Pancreatic digest of casein	2.0g
K_2HPO_4	0.3g
Bromthymol Blue	0.03g
Carbohydrate solution	100.0mL

pH 6.8 ± 0.1 at 25°C

Source: This medium is available as a premixed powder from BD Diagnostic Systems.

Carbohydrate Solution:

Composition per 100.0mL:

Carbohydrate	10.0g

Preparation of Carbohydrate Solution: Add carbohydrate to distilled/deionized water and bring volume to 100.0mL. Mix thoroughly. Filter sterilize.

Preparation of Medium: Add components, except carbohydrate solution, to distilled/deionized water and bring volume to 900.0mL. Mix thoroughly. Gently heat and bring to boiling. Autoclave for 15 min at 15 psi pressure–121°C. Cool to 45°–50°C. Aseptically add 100.0mL of sterile carbohydrate solution. Mix thoroughly. Pour into sterile Petri dishes or distribute into sterile tubes.

Use: For differentiating Gram-negative bacteria based upon determining the oxidative and fermentative metabolism of carbohydrates.

Oxidation-Fermentation Medium, Hugh-Leifson's (Hugh-Leifson's Oxidation Fermentation Medium)

Composition per liter:

NaCl	5.0g
Agar	3.0g
Peptone	2.0g
K_2HPO_4	0.3g
Carbohydrate solution	100.0mL
Bromthymol Blue solution (0.2%)	15.0mL

pH 7.1 ± 0.2 at 25°C

Carbohydrate Solution:

Composition per 100.0mL:

Carbohydrate	10.0g

Preparation of Carbohydrate Solution: Add carbohydrate to distilled/deionized water and bring volume to 100.0mL. Mix thoroughly. Filter sterilize.

Preparation of Medium: Add components, except carbohydrate solution, to distilled/deionized water and bring volume to 900.0mL. Mix thoroughly. Gently heat and bring to boiling. Autoclave for 15 min at 15 psi pressure–121°C. Cool to 45°–50°C. Aseptically add 100.0mL of sterile carbohydrate solution. Mix thoroughly. Pour into sterile Petri dishes or distribute into sterile tubes.

Use: For differentiating Gram-negative bacteria, such as *Vibrio* species, based upon determining the oxidative and fermentative metabolism of carbohydrates. Bacteria that ferment the carbohydrate turn the medium yellow.

Oxidation-Fermentation Medium, King's (King's OF Medium)

Composition per liter:

Base solution	900.0mL
Carbohydrate solution	100.0mL

pH to 7.3 ± 0.2

Base Solution:

Composition per liter:

Agar	3.0g
Pancreatic digest of casein	2.0g
Carbohydrate solution	100.0mL
Phenol Red (1.5% solution)	2.0mL

Preparation of Base Solution: Add carbohydrate to distilled/deionized water and bring volume to 1.0L. Mix thoroughly.

Carbohydrate Solution:

Composition per 100.0mL:

Carbohydrate	10.0g

Preparation of Carbohydrate Solution: Add carbohydrate to distilled/deionized water and bring volume to 100.0mL. Mix thoroughly. Filter sterilize.

Preparation of Medium: Add components, except carbohydrate solution, to distilled/deionized water and bring volume to 900.0mL. Mix thoroughly. Gently heat and bring to boiling. Autoclave for 15 min at 15 psi pressure–121°C. Cool to 45°–50°C. Aseptically add 100.0mL of sterile carbohydrate solution. Mix thoroughly. Pour into sterile Petri dishes or distribute into sterile tubes.

Use: For differentiating bacteria based upon determining the oxidative and fermentative metabolism of carbohydrates. Bacteria that ferment the carbohydrate turn the medium yellow.

Oxidative-Fermentative Medium (OF Medium)

Composition per liter:

NaCl	5.0g
Agar	2.0g
Pancreatic digest of casein	2.0g
K_2HPO_4	0.3g
Bromthymol Blue	0.08g
Carbohydrate solution	100.0mL

pH 6.8 ± 0.1 at 25°C

Source: This medium is available as a premixed powder from BD Diagnostic Systems and Oxoid Unipath.

Carbohydrate Solution:

Composition per 100.0mL:

Carbohydrate	10.0g

Preparation of Carbohydrate Solution: Add carbohydrate to distilled/deionized water and bring volume to 100.0mL. Mix thoroughly. Filter sterilize.

Preparation of Medium: Add components, except carbohydrate solution, to distilled/deionized water and bring volume to 900.0mL. Mix thoroughly. Gently heat and bring to boiling. Autoclave for 15 min at 15 psi pressure–121°C. Cool to 45°–50°C. Aseptically add 100.0mL of sterile carbohydrate solution. Mix thoroughly. Pour into sterile Petri dishes or distribute into sterile tubes.

Use: For differentiating bacteria based upon determining the oxidative and fermentative metabolism of carbohydrates. Bacteria that ferment the carbohydrate turn the medium yellow.

Oxidative-Fermentative Glucose Medium, Semisolid (OF Glucose Medium, Semisolid)

Composition per liter:

Glucose	10.0g
NaCl	5.0g
Agar	2.0g
Pancreatic digest of casein	2.0g
K_2HPO_4	0.3g
Bromthymol Blue dye	0.08g

pH 6.8 ± 0.2 at 25°C

Preparation of Medium: Add components to distilled/deionized water and bring volume to 1.0L. Mix thoroughly. Gently heat and bring to boiling. Distribute into tubes or flasks. Autoclave for 15 min at 15 psi pressure–121°C. Pour into sterile Petri dishes or leave in tubes.

Use: For differentiating Gram-negative bacteria based upon determining the oxidative and fermentative metabolism of glucose. Bacteria that ferment glucose turn the medium yellow.

Oxidative-Fermentative Glucose Medium, Semisolid, with NaCl (OF Glucose Medium, Semisolid with NaCl)

Composition per liter:

NaCl	20.0g
Glucose	10.0g
Agar	2.0g
Pancreatic digest of casein	2.0g
K_2HPO_4	0.3g
Bromthymol Blue dye	0.08g

pH 6.8 ± 0.2 at 25°C

Preparation of Medium: Add components to distilled/deionized water and bring volume to 1.0L. Mix thoroughly. Gently heat and bring to boiling. Distribute into tubes or flasks. Autoclave for 15 min at 15 psi pressure–121°C. Pour into sterile Petri dishes or leave in tubes.

Use: For differentiating *Vibrio* species based upon determining the oxidative and fermentative metabolism of glucose. Bacteria that ferment glucose turn the medium yellow.

Oxidative-Fermentative Test Medium (OF Test Medium)

Composition per liter:

NaCl	5.0g
Agar	3.0g
Peptone	2.0g
K_2HPO_4	0.3g
Bromthymol Blue	0.03g
Carbohydrate solution	100.0mL

Carbohydrate Solution:

Composition per 100.0mL:

Carbohydrate	10.0g

Preparation of Carbohydrate Solution: Add carbohydrate to distilled/deionized water and bring volume to 100.0mL. Mix thoroughly. Filter sterilize.

Preparation of Medium: Add components, except carbohydrate solution, to distilled/deionized water and bring volume to 1.0L. Mix thoroughly. Gently heat and bring to boiling. Distribute into tubes in 3.0mL volumes. Autoclave for 15 min at 15 psi pressure–121°C. Cool to 45°–50°C. Aseptically add 0.3mL of sterile carbohydrate solution to each tube. Mix thoroughly.

Use: For the cultivation and differentiation of a variety of microorganisms based on their ability to ferment a specific carbohydrate. Bacteria that ferment the specific carbohydrate turn the medium yellow.

P Agar

Composition per liter:

Agar	15.0g
Peptone	10.0g
NaCl	5.0g
Yeast extract	5.0g
Glucose	1.0g

pH 7.5 ± 0.2 at 25°C

Preparation of Medium: Add components to distilled/deionized water and bring volume to 1.0L. Mix thoroughly. Gently heat and bring to boiling. Distribute into tubes or flasks. Autoclave for 15 min at 15 psi pressure–121°C. Pour into sterile Petri dishes or leave in tubes.

Use: For the cultivation of *Staphylococcus* species.

Pagano Levin Agar

Composition per liter:

Glucose	40.0g
Agar	15.0g
Peptone	10.0g
Yeast extract	1.0g
Neomycin	0.5g
2,3,5-Triphenyltetrazolium chloride	0.1g

pH 6.0 ± 0.1 at 25°C

Source: This medium is available as a premixed powder from BD Diagnostic Systems.

Preparation of Medium: Add components, except neomycin and 2,3,5-triphenyltetrazolium chloride, to distilled/deionized water and bring volume to

1.0L. Mix thoroughly. Gently heat and bring to boiling. Autoclave for 15 min at 15 psi pressure–121°C. Cool to 45°–50°C. Aseptically add neomycin and 2,3,5-triphenyltetrazolium chloride. Mix thoroughly. Pour into sterile Petri dishes or distribute into sterile tubes. Allow tubes to cool in a slanted position.

Use: For the isolation, cultivation, and differentiation of *Candida* species. *Candida albicans* appears as smooth, shiny, cream-light pink colonies.

Pai Medium

Composition per liter:

Homogenized whole egg	666.0mL
NaCl (0.85% solution)	334.0mL

pH 6.75 ± 0.2 at 25°C

Homogenized Whole Egg:

Composition per liter:

Whole eggs	18-24

Preparation of Homogenized Whole Egg: Use fresh eggs, less than 1 week old. Scrub the shells with soap. Let stand in a soap solution for 30 min. Rinse in running water. Soak eggs in 70% ethanol for 15 min. Break the eggs into a sterile container. Homogenize by shaking. Filter through four layers of sterile cheesecloth into a sterile graduated cylinder. Measure out 1.0L.

Preparation of Medium: Combine components. Mix thoroughly. Aseptically distribute into sterile tubes. Inspissate tubes in a slanted position at 80°–90°C (moist heat) for 30 min.

Use: For the maintenance of stock cultures of *Salmonella typhi* and other *Salmonella* species.

Pai Medium

Composition per 1620.0mL:

Glucose	5.0g
Homogenized whole egg	1.0L
Glycerol	120.0mL

pH 6.75 ± 0.2 at 25°C

Homogenized Whole Egg:

Composition per liter:

Whole eggs	18–24

Preparation of Homogenized Whole Egg: Use fresh eggs, less than 1 week old. Scrub the shells with soap. Let stand in a soap solution for 30 min. Rinse in running water. Soak eggs in 70% ethanol for 15 min. Break the eggs into a sterile container. Homogenize by shaking. Filter through four layers of sterile cheesecloth into a sterile graduated cylinder. Measure out 1.0L.

Preparation of Medium: Combine components. Mix thoroughly. Aseptically distribute into sterile tubes. Inspissate tubes in a slanted position at 80°–90°C (moist heat) for 30 min.

Use: For the isolation and cultivation of *Corynebacterium* spp.

PALCAM Agar (Polymyxin Acriflavine LiCl Ceftazidime Esculin Mannitol Agar)

Composition per liter:

Peptone	23.0g
LiCl	15.0g
Agar	10.0g
Mannitol	10.0g
NaCl	5.0g
Yeast extract	3.0g
Starch	1.0g
Esculin	0.8g
Ferric ammonium citrate	0.5g
Glucose	0.5g
Phenol Red	0.08g
PALCAM selective supplement	10.0mL

pH 7.2 ± 0.2 at 25°C

Source: This medium is available as a premixed powder from Oxoid Unipath.

PALCAM Selective Supplement:

Composition per 10.0mL:

Ceftazidime	20.0mg
Polymyxin B	10.0mg
Acriflavine·HCl	5.0mg

Preparation of PALCAM Selective Supplement: Add components to distilled/deionized water and bring volume to 10.0mL. Mix thoroughly. Filter sterilize.

Egg Yolk Emulsion:

Composition:

Chicken egg yolks	11
Whole chicken egg	1

Preparation of Egg Yolk Emulsion: Soak eggs with 1:100 dilution of saturated mercuric chloride solution for 1 min. Crack eggs and separate yolks from whites. Mix egg yolks with 1 chicken egg.

Preparation of Medium: Add components, except PALCAM selective supplement, to distilled/deionized water and bring volume to 990.0mL. Mix thoroughly. Gently heat and bring to boiling. Autoclave for 15 min at 15 psi pressure–121°C. Cool to 45°–50°C. Aseptically add sterile PALCAM selective supplement. Mix thoroughly. The addition of 25.0mL of egg yolk emulsion may aid in the recovery of damaged *Listeria*. Pour into sterile Petri dishes or distribute into sterile tubes.

Use: For the selective isolation, cultivation, and differentiation of *Listeria monocytogenes* and other *Listeria* species.

Pasteurella haemolytica Selective Medium

Composition per 1010.0mL:

Tryptose agar with peptic digest of blood 1.0L
Antibiotic solution 10.0mL
pH 7.2 ± 0.2 at 25°C

Tryptose Agar With Peptic Digest of Blood:

Composition per liter:

Agar 15.0g
Pancreatic digest of casein 10.0g
Peptic digest of animal tissue 10.0g
NaCl 5.0g
Glucose 1.0g
Peptic digest of blood 50.0mL

Preparation of Tryptose Agar With Peptic Digest of Blood: Add components to distilled/deionized water and bring volume to 950.0mL. Mix thoroughly. Gently heat and bring to boiling. Autoclave for 15 min at 15 psi pressure–121°C. Cool to 45°–50°C. Aseptically add peptic digest of blood. Mix thoroughly.

Antibiotic Solution:

Composition per 10.0mL:

Actidione (cycloheximide) 0.1g
Novobiocin 2.0mg
Neomycin 1.5mg

Preparation of Antibiotic Solution: Add components to distilled/deionized water and bring volume to 10.0mL. Mix thoroughly. Filter sterilize.

Preparation of Medium: To 1.0L of cooled, sterile tryptose agar with peptic digest of blood, aseptically add 10.0mL of sterile antibiotic solution. Mix thoroughly. Pour into sterile Petri dishes or distribute into sterile tubes.

Use: For the selective cultivation of *Pasteurella haemolytica.*

Pasteurella multocida Selective Medium

Composition per 1020.0mL:

Tryptose agar with peptic digest of blood .. 1.0L
Antibiotic solution 10.0mL
K_2TeO_3 solution 10.0mL
pH 7.2 ± 0.2 at 25°C

Tryptose Agar With Peptic Digest of Blood:

Composition per liter:

Agar 15.0g
Pancreatic digest of casein 10.0g
Peptic digest of animal tissue 10.0g
NaCl 5.0g
Glucose 1.0g
Peptic digest of blood 50.0mL

Preparation of Tryptose Agar with Peptic Digest of Blood: Add components to distilled/deionized water and bring volume to 950.0mL. Mix thoroughly. Gently heat and bring to boiling. Autoclave for 15 min at 15 psi pressure–121°C. Cool to 45°–50°C. Aseptically add peptic digest of blood. Mix thoroughly.

Antibiotic Solution:

Composition per 10.0mL:

Actidione (cycloheximide) 0.1g
Novobiocin 0.01g
Erythrocin 5.0mg

Preparation of Antibiotic Solution: Add components to distilled/deionized water and bring volume to 10.0mL. Mix thoroughly. Filter sterilize.

K_2TeO_3 Solution:

Composition per 10.0mL:

K_2TeO_3 .... 5.0mg

Preparation of K_2TeO_3 Solution: Add K_2TeO_3 to distilled/deionized water and bring volume to 10.0mL. Mix thoroughly. Filter sterilize.

Caution: Potassium tellurite is toxic.

Preparation of Medium: To 1.0L of cooled, sterile tryptose agar with peptic digest of blood, aseptically add 10.0mL of sterile antibiotic solution and 10.0mL of sterile K_2TeO_3 solution. Mix thoroughly. Pour into sterile Petri dishes or distribute into sterile tubes.

Use: For the selective cultivation of *Pasteurella multocida.*

PDM-114 Medium

Composition per liter:

Base solution 770.0mL
Bovine serum, heat inactivated 100.0mL
Solution A 50.0mL
Solution B 50.0mL
Special 107 vitamin mix 30.0mL
pH 6.8 ± 0.2 at 25°C

Base Solution:

Composition per 770.0mL:

Pancreatic digest of casein 20.0g
NaCl 2.0g
L-Cysteine·HCl 1.0g
Ascorbic acid 0.2g
Ammonium chloride 0.16g
Adenine 73.0mg
Guanosine 59.0mg

Adenosine 5′-monophosphate 52.0mg
Uracil 37.0mg
Cytosine 30.0mg
Ferric ammonium citrate...............................22.8mg
Adenosine 5′-diphosphate 16.6mg
Adenosine 5′-triphosphate..............................7.6mg

Preparation of Base Solution: Add components to distilled/deionized water and bring volume to 770.0mL. Mix thoroughly. Adjust pH to 6.8 with NaOH. Autoclave for 15 min at 15 psi pressure–121°C. Cool to room temperature.

Solution A:

Composition per 50.0.0mL:

Glucose 10.0g

Preparation of Solution A: Add glucose to distilled/deionized water and bring volume to 50.0mL. Mix thoroughly. Filter sterilize.

Solution B:

Composition per 50.0.0mL:

K_2HPO_4...1.0g
KH_2PO_4..0.6g

Preparation of Solution B: Add components to distilled/deionized water and bring volume to 50.0mL. Mix thoroughly. Autoclave for 15 min at 15 psi pressure–121°C. Cool to room temperature.

Special 107 Vitamin Mix:

Composition per 120.0.0mL:

Solution 4 vitamins 100.0mL
Solution 2 .... 1.2mL
Solution 1 .... 0.4mL
Solution 3 .... 0.4mL

Solution 1:

Composition per 100.0.0mL:

Absolute ethanol 100.0mL
DL-6,8-Thioctic acid, oxidized 100.0mg

Preparation of Solution 1: Add DL-6,8-thioctic acid to absolute ethanol and bring volume to 100.0mL. Mix thoroughly. Filter sterilize.

Solution 2:

Vitamin B_{12} 40.0mg

Preparation of Solution 2: Add vitamin B_{12} to distilled/deionized water and bring volume to 100.0mL. Mix thoroughly. Filter sterilize.

Solution 3:

Tween 80 ....50.0g
Absolute ethanol 100.0mL

Preparation of Solution 3: Add Tween 80 to absolute ethanol and bring volume to 100.0mL. Mix thoroughly. Filter sterilize.

Solution 4 Vitamins:

Composition per liter:

Vitamin B_{12} 10.0mg
Calcium D-(+)-pantothenate 2.3mg
Choline chloride...1.25mg
Riboflavin 0.7mg
Vitamin A, crystallized alcohol 0.25mg
Calciferol (vitamin D_2) 0.25mg
i-Inositol...0.125mg
p-Aminobenzoic acid.................................0.125mg
Niacin..0.0625mg
Niacinamide...0.0625mg
Pyridoxal·HCl 0.0625mg
Pyridoxine·HCl..0.0625mg
α-Tocopherol phosphate, disodium salt 0.025mg
Biotin 0.025mg
Folic acid 0.025mg
Menadione (vitamin K_3) 0.025mg
Thiamine·HCl 0.025mg

Preparation of Solution 4 Vitamins: Add components to distilled/deionized water and bring volume to 1.0L. Mix thoroughly. Filter sterilize.

Preparation of Special 107 Vitamin Mix: Aseptically combine 0.4mL of sterile solution 1, 1.2mL of sterile solution 2, 0.4mL of sterile solution 3, and 100.0mL of sterile solution 4 vitamins. Bring volume to 120.0mL with sterile distilled/deionized water.

Preparation of Medium: Aseptically combine 770.0mL of sterile base solution, 100.0mL of heat-inactivated bovine serum, 50.0mL of sterile solution A, 50.0mL of sterile solution B, and 30.0mL of sterile special 107 vitamin mix. Aseptically distribute into sterile tubes or flasks.

Use: For the cultivation of *Entamoeba histolytica.*

Peptone Iron Agar

Composition per liter:

Agar 15.0g
Peptone 15.0g
Proteose peptone 5.0g
Sodium glycerophosphate.................................1.0g
Ferric ammonium citrate...................................0.5g
$Na_2S_2O_3$ 0.08g

pH 6.7 ± 0.2 at 25°C

Source: This medium is available as a premixed powder from BD Diagnostic Systems.

Preparation of Medium: Add components to distilled/deionized water and bring volume to 1.0L. Mix thoroughly. Gently heat and bring to boiling. Distribute into tubes. Autoclave for 15 min at 15 psi pressure–121°C. Allow tubes to cool in an upright position.

Use: For the cultivation and differentiation of microorganisms based on their ability to produce H_2S. Microorganisms that produce H_2S turn the medium black.

Peptone Meat Extract Soil Extract Agar (PFE Agar)

Composition per liter:

Agar	12.0g
Proteose peptone no. 3	5.0g
Meat extract	3.0g
Tap water	850.0mL
Soil extract	150.0mL
Glycerol	20.0mL

pH 7.0 ± 0.2 at 25°C

Soil Extract:

Composition per liter:

Garden soil, air dried	400.0g

Preparation of Soil Extract: Pass 400.0g of air-dried garden soil through a coarse sieve. Add soil to 960.0mL of tap water. Mix thoroughly. Autoclave for 60 min at 15 psi pressure–121°C. Cool to room temperature. Allow residue to settle. Decant supernatant solution. Filter through Whatman filter paper. Distribute into bottles in 200.0mL volumes. Autoclave for 15 min at 15 psi pressure–121°C. Store at room temperature until clear.

Preparation of Medium: Add components to tap water and bring volume to 1.0L. Mix thoroughly. Gently heat and bring to boiling. Distribute into tubes or flasks. Autoclave for 15 min at 15 psi pressure–121°C. Pour into sterile Petri dishes or leave in tubes.

Use: For the cultivation and maintenance of *Mycobacterium avium, Mycobacterium gastri, Mycobacterium kansasii, Mycobacterium marinum, Mycobacterium scrofulaceum,* and *Mycobacterium terrae.*

Peptone Starch Dextrose Agar (PSD Agar) (Dunkelberg Agar)

Composition per liter:

Proteose peptone no. 3	20.0g
Agar	15.0g
Soluble starch	10.0g
Glucose	2.0g
Na_2HPO_4	1.0g
NaH_2PO_4	1.0g

pH 6.8 ± 0.2 at 25°C

Preparation of Medium: Add starch to approximately 100.0mL of cold distilled/deionized water. Mix thoroughly. Add starch solution to 400.0mL of boiling distilled/deionized water. Add remaining components. Mix thoroughly. Bring volume to 1.0L with distilled/deionized water. Autoclave for 12 min at 8 psi pressure–112°C. Pour into sterile Petri dishes or distribute into screw-capped tubes.

Use: For the selective isolation and cultivation of *Gardnerella vaginalis.*

Peptone Water

Composition per liter:

Peptone	10.0g
NaCl	5.0g

pH 7.2 ± 0.2 at 25°C

Source: This medium is available as a premixed powder from BD Diagnostic Systems and Oxoid Unipath.

Preparation of Medium: Add components to distilled/deionized water and bring volume to 1.0L. Mix thoroughly. Distribute into tubes or flasks. Autoclave for 15 min at 15 psi pressure–121°C.

Use: For the cultivation of nonfastidious microorganisms, for carbohydrate fermentation tests, and for performing the indole test.

Peptone Water with Andrade's Indicator

Composition per liter:

Peptone	10.0g
NaCl	5.0g
Andrade's indicator	100.0mL
Carbohydrate solution	20.0mL

pH 7.4 ± 0.2 at 25°C

Source: This medium is available as a premixed powder from Oxoid Unipath.

Andrade's Indicator:

Composition per 100.0mL:

NaOH (1*N* solution)	16.0mL
Acid Fuchsin	0.1 g

Caution: Acid Fuchsin is a potential carcinogen and care must be taken to avoid inhalation of the powdered dye and contact with the skin.

Carbohydrate Solution:

Composition per 20.0mL:

Carbohydrate	5.0–10.0g

Preparation of Carbohydrate Solution: Add carbohydrate to distilled/deionized water and bring volume to 20.0mL. Mix thoroughly. Filter sterilize.

Preparation of Medium: Add components, except carbohydrate solution, to distilled/deionized water and bring volume to 980.0mL. Mix thoroughly. Adjust pH to 7.4 if necessary. Distribute into tubes containing an inverted Durham tube. Fill each tube with 9.8mL of medium. Autoclave for 15 min at 15 psi pressure–121°C. Aseptically add 0.2mL of sterile carbohydrate solution to each tube.

Use: For use in carbohydrate fermentation tests. Fermentation is determined by the production of acid—broth turns pink—and formation of gas—bubble trapped in Durham tube.

Peptone Yeast Trypticase Agar (ATCC Medium 118)

Composition per liter:

Agar	15.0g
Peptone	6.0g
Trypticase (pancreatic digest of casein)	4.0g
Yeast extract	3.0g
Beef extract	1.5g
Glucose	1.0g

pH 7.0 ± 0.2 at 25°C

Preparation of Medium: Add components to distilled/deionized water and bring volume to 1.0L. Mix thoroughly. Gently heat and bring to boiling. Distribute into tubes or flasks. Autoclave for 15 min at 15 psi pressure–121°C. Pour into sterile Petri dishes or leave in tubes.

Use: For the cultivation of a variety of heterotrophic bacteria.

Petragnani Medium

Composition per 2398.0mL:

Skim milk	100.0g
Potato flour	36.4g
L-Asparagine	5.1g
Pancreatic digest of casein	5.1g
Malachite Green	1.2g
Whole egg	1277.0mL
Egg yolk	121.0mL
Glycerol	60.0mL

pH 7.0 ± 0.2 at 25°C

Source: This medium is available as a prepared medium from BD Diagnostic Systems.

Preparation of Medium: Add components—except whole egg, egg yolk, and glycerol—to distilled/deionized water and bring volume to 940.0mL. Mix thoroughly. Add glycerol. Gently heat while stirring and bring to boiling. Autoclave for 15 min at 15 psi pressure–121°C. Cool to 45°–50°C. Scrub the eggshells with soap. Let stand in a soap solution for 30 min. Rinse in running water. Soak eggs in 70% ethanol for 15 min. Break the eggs into a sterile container. Homogenize by shaking. Filter through four layers of sterile cheesecloth into a sterile graduated cylinder. Measure out 1277.0mL. Add separated egg yolks to another sterile container. Measure out 121.0mL. Aseptically add homogenized whole egg and egg yolk to cooled sterile basal medium. Mix thoroughly. Aseptically distribute into sterile tubes. Inspissate at 85°–90°C (moist heat) for 45 min.

Use: For the isolation and cultivation of *Mycobacterium* species from clinical specimens.

Petragnani Medium

Composition per 2285.0mL:

Potato	500.0g
Potato flour	36.0g
Malachite Green	1.2g
Whole egg	1200.0mL
Whole milk	900.0mL
Egg yolk	115.0mL
Glycerol	70.0mL

pH 7.2 ± 0.2 at 25°C

Source: This medium is available as a prepared medium from BD Diagnostic Systems.

Preparation of Medium: Peel and dice potato. Add potato to 500.0mL of distilled/deionized water. Gently heat and bring to boiling. Continue boiling for 30 min. Filter solids through two layers of cheesecloth. Combine potato solids with remaining components, except whole egg, egg yolk, and glycerol. Mix thoroughly. Add glycerol. Gently heat while stirring and bring to boiling. Autoclave for 15 min at 15 psi pressure–121°C. Cool to 45°–50°C. Scrub the eggshells with soap. Let stand in a soap solution for 30 min. Rinse in running water. Soak eggs in 70% ethanol for 15 min. Break the eggs into a sterile container. Homogenize by shaking. Filter through four layers of sterile cheesecloth into a sterile graduated cylinder. Measure out 1200.0mL. Add separated egg yolks to another sterile container. Measure out 115.0mL. Aseptically add homogenized whole egg and egg yolk to cooled sterile basal medium. Mix thoroughly. Aseptically distribute into sterile tubes. Inspissate at 85°–90°C (moist heat) for 45 min.

Use: For the isolation and cultivation of *Mycobacterium* species from clinical specimens.

Pfizer Selective *Enterococcus* Agar (PSE Agar)

Peptone C	17.0g
Agar	15.0g
Bile	10.0g
NaCl	5.0g
Yeast extract	5.0g
Peptone B	3.0g
Esculin	1.0g
Sodium citrate	1.0g

Ferric ammonium citrate 0.5g
NaN_3 .... 0.25g

pH 7.1 ± 0.2 at 25°C

Caution: Sodium azide is toxic. Azides also react with metals and disposal must be highly diluted.

Preparation of Medium: Add components to distilled/deionized water and bring volume to 1.0L. Mix thoroughly. Gently heat and bring to boiling. Distribute into tubes or flasks. Autoclave for 15 min at 15 psi pressure–121°C. Pour into sterile Petri dishes or leave in tubes.

Use: For the selective isolation, cultivation, and enumeration of *Enterococcus* species by the multiple tube technique.

PGA (Potato Glucose Agar)

Composition per liter:

Potatoes 500.0g
Glucose 20.0g
Agar 15.0g
Yeast extract 5.0g

Preparation of Medium: Slice potatoes with skin. Place 500.0g of potatoes in 1.0L of distilled/deionized water. Gently heat and bring to boiling. Allow to boil for 20 min. Filter through Whatman filter paper. Add agar and other components to filtrate. Bring volume to 1.0L with distilled/deionized water. Mix thoroughly. Gently heat and bring to boiling. Distribute into tubes or flasks. Autoclave for 15 min at 15 psi pressure–121°C. Pour into sterile Petri dishes or leave in tubes.

Use: For the cultivation of numerous filamentous fungi.

PGP Broth (Peptone Glycerol Phosphate Broth)

Composition per liter:

Peptone 5.0g
K_2HPO_4 .... 2.0g
Glycerol 10.0mL

Preparation of Medium: Add components to distilled/deionized water and bring volume to 1.0L. Mix thoroughly. Distribute into tubes or flasks. Autoclave for 15 min at 15 psi pressure–121°C.

Use: For the cultivation and maintenance of *Serratia marcescens*.

PGT Medium

Composition per liter:

Casamino acids 30.0g
L-Glutamic acid 0.5g
$MgSO_4 \cdot 7H_2O$ 0.45g
Maltose 0.2g
L-Cystine 0.2g
DL-Tryptophan 0.1g
Solution 3 .... 100.0mL
Solution 2 .... 2.0mL
Calcium pantothenate (0.1% solution) 0.5mL

pH 6.8 ± 0.2 at 25°C

Solution 3:

Composition per 500.0mL:

Maltose 200.0g
$CaCl_2$ 1.5g
Calcium pantothenate (0.1% solution) 3.0mL
$FeSO_4$ (1% in 1*N* HCl) 0.2mL

Preparation of Solution 3: Add components to distilled/deionized water and bring volume to 500.0mL. Mix thoroughly. Autoclave for 15 min at 7 psi pressure–111°C. Cool to 45°–50°C.

Solution 2:

Composition per 100.0mL:

β-Alanine 0.115g
Nicotinic acid 0.115g
$CuSO_4 \cdot 5H_2O$ 0.05g
$ZnSO_4 \cdot 7H_2O$ 0.045g
$MnCl_2 \cdot 4H_2O$ 0.015g
Pimelic acid 7.5mg
HCl, concentrated 3.0mL

Preparation of Solution 2: Add components to distilled/deionized water and bring volume to 100.0mL. Mix thoroughly.

Preparation of Medium: Add components, except solution 3, to distilled/deionized water and bring volume to 900.0mL. Mix thoroughly. Adjust pH to 6.8 with 50% KOH. Gently heat and bring to boiling. Autoclave for 15 min at 15 psi pressure–121°C. Cool to 45°–50°C. Aseptically add sterile solution 3. Mix thoroughly. Aseptically distribute into sterile tubes or flasks.

Use: For the cultivation and maintenance of *Corynebacterium diphtheriae*.

Phenol Red Agar

Composition per liter:

Agar 15.0g
Pancreatic digest of casein 10.0g
NaCl 5.0g
Phenol Red 0.018g
Carbohydrate solution 20.0mL

pH 7.4 ± 0.2 at 25°C

Source: This medium is available as a premixed powder from BD Diagnostic Systems.

Carbohydrate Solution:

Composition per 20.0mL:

Carbohydrate..5.0–10.0g

Preparation of Carbohydrate Solution: Add carbohydrate to distilled/deionized water and bring volume to 20.0mL. Mix thoroughly. Filter sterilize.

Preparation of Medium: Add components, except carbohydrate solution, to distilled/deionized water and bring volume to 980.0mL. Mix thoroughly. Adjust pH to 7.4 if necessary. Autoclave for 15 min at 15 psi pressure–121°C. Cool to 45°–50°C. Aseptically add 20.0mL of sterile carbohydrate solution. Pour into sterile Petri dishes or distribute into sterile tubes. Allow tubes to cool in a slanted position.

Use: For the determination of fermentation reactions. Bacteria that can ferment the added carbohydrate turn the medium yellow.

Phenol Red Agar

Composition per liter:

Agar..15.0g
Proteose peptone no. 3....................................10.0g
NaCl..5.0g
Beef extract...1.0g
Phenol Red...0.025g
Carbohydrate solution.................................20.0mL
pH 7.4 ± 0.2 at 25°C

Source: This medium is available as a premixed powder from BD Diagnostic Systems.

Carbohydrate Solution:

Composition per 20.0mL:

Carbohydrate..5.0–10.0g

Preparation of Carbohydrate Solution: Add carbohydrate to distilled/deionized water and bring volume to 20.0mL. Mix thoroughly. Filter sterilize.

Preparation of Medium: Add components, except carbohydrate solution, to distilled/deionized water and bring volume to 980.0mL. Mix thoroughly. Adjust pH to 7.4 if necessary. Autoclave for 15 min at 15 psi pressure–121°C. Cool to 45°–50°C. Aseptically add 20.0mL of sterile carbohydrate solution. Pour into sterile Petri dishes or distribute into sterile tubes. Allow tubes to cool in a slanted position.

Use: For the determination of fermentation reactions. Bacteria that can ferment the added carbohydrate turn the medium yellow.

Phenol Red Broth

Composition per liter:

Pancreatic digest of casein..............................10.0g
NaCl..5.0g
Phenol Red...0.018g
Carbohydrate solution.................................20.0mL
pH 7.4 ± 0.2 at 25°C

Source: This medium is available as a premixed powder from BD Diagnostic Systems.

Carbohydrate Solution:

Composition per 20.0mL:

Carbohydrate..5.0–10.0g

Preparation of Carbohydrate Solution: Add carbohydrate to distilled/deionized water and bring volume to 20.0mL. Mix thoroughly. Filter sterilize.

Preparation of Medium: Add components, except carbohydrate solution, to distilled/deionized water and bring volume to 980.0mL. Mix thoroughly. Adjust pH to 7.4 if necessary. Distribute into tubes containing an inverted Durham tube. Fill each tube with 9.8mL of medium. Autoclave for 15 min at 13 psi pressure–118°C. Cool to 45°–50°C. Aseptically add 0.2mL of sterile carbohydrate solution to each tube.

Use: For the determination of fermentation reactions in the differentiation of microorganisms. Fermentation is determined by the production of acid—broth turns yellow—and formation of gas—bubble trapped in Durham tube.

Phenol Red Glucose Broth

Composition per liter:

Pancreatic digest of casein..............................10.0g
Glucose...5.0g
NaCl..5.0g
Phenol Red...0.018g
pH 7.3 ± 0.2 at 25°C

Source: This medium is available as a premixed powder from BD Diagnostic Systems.

Preparation of Medium: Add components to distilled/deionized water and bring volume to 1.0L. Mix thoroughly. Adjust pH to 7.3 if necessary. Distribute into tubes containing an inverted Durham tube. Fill each tube with 10.0mL of medium. Autoclave for 15 min at 13 psi pressure–118°C.

Use: For determination of the ability of a microorganism to ferment glucose. Fermentation is determined by the production of acid—broth turns yellow—and formation of gas—bubble trapped in Durham tube.

Phenol Red Lactose Agar

Composition per liter:

Agar..15.0g
Lactose..10.0g
Proteose peptone no. 3....................................10.0g

NaCl 5.0g
Beef extract 1.0g
Phenol Red 25.0mg

pH 7.4 ± 0.2 at 25°C

Source: This medium is available as a premixed powder from BD Diagnostic Systems.

Preparation of Medium: Add components to distilled/deionized water and bring volume to 1.0L. Mix thoroughly. Gently heat and bring to boiling. Distribute into tubes or flasks. Autoclave for 15 min at 13 psi pressure–118°C. Pour into sterile Petri dishes or leave in tubes. Allow tubes to cool in a slanted position.

Use: For determination of the ability of a microorganism to ferment lactose. Fermentation is determined by the production of acid—medium turns yellow.

Phenol Red Lactose Broth

Composition per liter:

Pancreatic digest of casein 10.0g
Lactose 5.0g
NaCl 5.0g
Phenol Red 0.018g

pH 7.3 ± 0.2 at 25°C

Source: This medium is available as a premixed powder from BD Diagnostic Systems.

Preparation of Medium: Add components to distilled/deionized water and bring volume to 1.0L. Mix thoroughly. Adjust pH to 7.3 if necessary. Distribute into tubes containing an inverted Durham tube. Fill each tube with 10.0mL of medium. Autoclave for 15 min at 13 psi pressure–118°C.

Use: For determination of the ability of a microorganism to ferment lactose. Fermentation is determined by the production of acid—broth turns yellow—and formation of gas—bubble trapped in Durham tube.

Phenol Red Mannitol Agar

Composition per liter:

Agar 15.0g
Mannitol 10.0g
Proteose peptone no. 3 .... 10.0g
NaCl 5.0g
Beef extract 1.0g
Phenol Red 25.0mg

pH 7.4 ± 0.2 at 25°C

Source: This medium is available as a premixed powder from BD Diagnostic Systems.

Preparation of Medium: Add components to distilled/deionized water and bring volume to 1.0L. Mix thoroughly. Gently heat and bring to boiling. Distribute into tubes or flasks. Autoclave for 15 min at 13 psi pressure–118°C. Pour into sterile Petri dishes or leave in tubes. Allow tubes to cool in a slanted position.

Use: For determination of the ability of a microorganism to ferment mannitol. Fermentation is determined by the production of acid—medium turns yellow.

Phenol Red Mannitol Broth

Composition per liter:

Pancreatic digest of casein 10.0g
D-Mannitol 5.0g
NaCl 5.0g
Phenol Red 0.018g

pH 7.3 ± 0.2 at 25°C

Source: This medium is available as a premixed powder from BD Diagnostic Systems.

Preparation of Medium: Add components to distilled/deionized water and bring volume to 1.0L. Mix thoroughly. Adjust pH to 7.3 if necessary. Distribute into tubes containing an inverted Durham tube. Fill each tube with 10.0mL of medium. Autoclave for 15 min at 13 psi pressure–118°C.

Use: For determination of the ability of a microorganism to ferment mannitol. Fermentation is determined by the production of acid—broth turns yellow—and formation of gas—bubble trapped in Durham tube.

Phenol Red Sucrose Broth

Composition per liter:

Pancreatic digest of casein 10.0g
NaCl 5.0g
Sucrose 5.0g
Phenol Red 0.018g

pH 7.3 ± 0.2 at 25°C

Source: This medium is available as a premixed powder from BD Diagnostic Systems.

Preparation of Medium: Add components to distilled/deionized water and bring volume to 1.0L. Mix thoroughly. Adjust pH to 7.3 if necessary. Distribute into tubes containing an inverted Durham tube. Fill each tube with 10.0mL of medium. Autoclave for 15 min at 13 psi pressure–118°C.

Use: For determination of the ability of a microorganism to ferment sucrose. Fermentation is determined by the production of acid—broth turns yellow—and formation of gas—bubble trapped in Durham tube.

Phenol Red Tartrate Agar

Composition per liter:

Agar	15.0g
Peptone	10.0g
Potassium tartrate	10.0g
NaCl	5.0g
Phenol Red	0.024g

pH 7.6 ± 0.2 at 25°C

Source: This medium is available as a premixed powder from BD Diagnostic Systems.

Preparation of Medium: Add components to cold distilled/deionized water and bring volume to 1.0L. Mix thoroughly. Gently heat and bring to boiling. Distribute into tubes or flasks. Autoclave for 15 min at 13 psi pressure–118°C. Pour into sterile Petri dishes or leave in tubes. Allow tubes to cool in an upright position.

Use: For the differentiation of Gram-negative bacteria of the intestinal groups, particularly members of the *Salmonella* (paratyphoid) group based on their ability to ferment tartrate.

Phenol Red Tartrate Broth

Composition per liter:

Pancreatic digest of casein	10.0g
Potassium tartrate	10.0g
Agar	5.0g
NaCl	5.0g
Phenol Red	0.024g

pH 7.6 ± 0.2 at 25°C

Preparation of Medium: Add components to distilled/deionized water and bring volume to 1.0L. Mix thoroughly. Distribute into tubes or flasks. Autoclave for 15 min at 15 psi pressure–121°C.

Use: For the differentiation of Gram-negative bacteria of the intestinal groups, particularly members of the *Salmonella* (paratyphoid) group based on their ability to ferment tartrate.

Phenylalanine Agar (Phenylalanine Deaminase Medium)

Composition per liter:

Agar	12.0g
NaCl	5.0g
Yeast extract	3.0g
DL-Phenylalanine	2.0g
Na_2HPO_4	1.0g

pH 7.3 ± 0.2 at 25°C

Source: This medium is available as a premixed powder from BD Diagnostic Systems.

Preparation of Medium: Add components to distilled/deionized water and bring volume to 1.0L. Mix thoroughly. Gently heat while stirring and bring to boiling. Distribute into tubes or flasks. Autoclave for 10 min at 15 psi pressure–121°C. Pour into sterile Petri dishes or leave in tubes.

Use: For the differentiation of enteric Gram-negative bacilli on the basis of their ability to produce phenylpyruvic acid from phenylalanine. After appropriate incubation of bacteria, ferric chloride reagent is added on the agar. Formation of a green color in 1–5 min indicates the production of phenylpyruvic acid.

Phenylalanine Malonate Broth

Composition per liter:

Sodium malonate	3.0g
DL-Phenylalanine	2.0g
NaCl	2.0g
$(NH_4)_2SO_4$	2.0g
Yeast extract	1.0g
K_2HPO_4	0.6g
KH_2PO_4	0.4g
Bromthymol Blue	0.025g

pH 7.3 ± 0.2 at 25°C

Source: This medium is available as a premixed powder from BD Diagnostic Systems.

Preparation of Medium: Add components to distilled/deionized water and bring volume to 1.0L. Mix thoroughly. Distribute into tubes or flasks. Autoclave for 10 min at 10 psi pressure–115°C.

Use: For the differentiation of Gram-negative enteric bacilli on the basis of malonate utilization and formation of pyruvic acid from phenylalanine.

Phenylethanol Agar

Composition per liter:

Agar	15.0g
Tryptose	10.0g
NaCl	5.0g
Beef extract	3.0g
Phenylethanol	2.5g

pH 7.3 ± 0.2 at 25°C

Source: This medium is available as a premixed powder from BD Diagnostic Systems.

Preparation of Medium: Add components to distilled/deionized water and bring volume to 1.0L. Mix thoroughly. Gently heat and bring to boiling. Distribute into tubes or flasks. Autoclave for 15 min at 15 psi pressure–121°C. Pour into sterile Petri dishes or leave in tubes.

Use: For the isolation of staphylococci and streptococci from specimens containing a mixed flora.

Phenylethanol Blood Agar

Composition per liter:

Agar	15.0g
Tryptose	10.0g
NaCl	5.0g
Beef extract	3.0g
Phenylethanol	2.5g
Blood, defibrinated	50.0mL

pH 7.3 ± 0.2 at 25°C

Preparation of Medium: Add components, except blood, to distilled/deionized water and bring volume to 950.0mL. Mix thoroughly. Gently heat and bring to boiling. Autoclave for 15 min at 13 psi pressure–118°C. Cool to 45°–50°C. Aseptically add sterile defibrinated blood. Mix thoroughly. Pour into sterile Petri dishes or distribute into sterile tubes.

Use: For the isolation of staphylococci and streptococci from specimens containing a mixed flora.

PHYG Medium

Composition per 110.1mL:

Beef heart, solids from infusion	10.0g
Polypeptone	2.0g
Gelatin	1.0g
Glucose	1.0g
Yeast extract	1.0g
$NaHCO_3$	0.5g
Tryptose	0.2g
Agar	0.16g
NaCl	0.1g
L-Cysteine·HCl·H_2O	0.09g
$(NH_4)_2SO_4$	0.05g
Resazurin	0.16mg
Salts solution	50.0mL
Rabbit serum, inactivated	10.0mL
Thiamine pyrophosphate solution	0.1mL

pH 7.2–7.5 at 25°C

Salts Solution:

Composition per 400.0mL:

K_2HPO_4	0.9g
NaCl	0.8g
KH_2PO_4	0.4g
$MnCl_2 \cdot 4H_2O$	0.16g
$MgSO_4$	0.08g

Preparation of Salts Solution: Add components to distilled/deionized water and bring volume to 400.0mL. Mix thoroughly.

Thiamine Pyrophosphate Solution:

Composition per 10.0mL:

Thiamine pyrophosphate	0.05g

Preparation of Thiamine Pyrophosphate Solution: Add thiamine pyrophosphate to distilled/deionized water and bring volume to 10.0mL. Mix thoroughly. Filter sterilize.

Preparation of Medium: Add components, except rabbit serum and thiamine pyrophosphate solution, to distilled/deionized water and bring volume to 100.0mL. Mix thoroughly. Gently heat and bring to boiling. Autoclave for 15 min at 15 psi pressure–121°C. Cool to 45°–50°C. Aseptically add 10.0mL of sterile rabbit serum and 0.1mL of sterile thiamine pyrophosphate solution. Mix thoroughly. Pour into sterile Petri dishes or distribute into sterile tubes.

Use: For the cultivation of treponemes.

Pike Streptococcal Broth

Composition per liter:

Pancreatic digest of casein	10.0g
Tryptose	10.0g
Yeast extract	10.0g
Glucose	0.2g
NaN_3	0.065g
Crystal Violet	2.0mg
Rabbit blood, defibrinated	50.0mL

pH 7.4 ± 0.2 at 25°C

Caution: Sodium azide is toxic. Azides also react with metals and disposal must be highly diluted.

Preparation of Medium: Add components, except rabbit blood, to distilled/deionized water and bring volume to 950.0mL. Mix thoroughly. Gently heat and bring to boiling. Distribute into flasks in 100.0mL volumes. Autoclave for 15 min at 15 psi pressure–121°C. Cool to 45°–50°C. Aseptically add 5.0mL of sterile rabbit blood to each flask. Mix thoroughly.

Use: For the isolation and enrichment of hemolytic streptococci from throat swabs and other clinical specimens. After incubation of bacteria for 18–24h in this medium, they may be isolated by streaking the culture onto blood agar plates.

Pine and Drouhet's *Histoplasma* Yeast Phase Medium

Composition per liter:

Casein hydrolysate-vitamin-glucose base	500.0mL
Agar starch base	500.0mL

Casein Hydrolysate-Vitamin-Glucose Base:
Composition per 500.0mL:

Glucose	10.0g
Citric acid	10.0g
L-Cysteine·HCl	1.0g
L-Asparagine	1.0g
Glutathione, reduced	0.5g
L-Tryptophan	0.02g
Solution 1	250.0mL
Solution 2	10.0mL
Solution 3	40.0mL
Solution 4	10.0mL
Solution 5	1.0mL
Solution 6	0.1mL
Solution 7	0.1mL

Solution 1:
Composition per liter:

KH_2PO_4	8.0g
$(NH_4)_2SO_4$	8.0g
$MgSO_4 \cdot 7H_2O$	0.86g
$CaCl_2$	0.08g

Preparation of Solution 1: Add components to distilled/deionized water and bring volume to 1.0L. Mix thoroughly. Store at 5°C.

Solution 2:
Composition per liter:

$FeSO_4 \cdot 7H_2O$	5.7g
$MnCl_2 \cdot 6H_2O$	0.8g
$Na_2MoO_4 \cdot 2H_2O$	0.15g
HCl, concentrated	1.0mL

Preparation of Solution 2: Add the concentrated HCl to 100.0mL of distilled/deionized water. Add $FeSO_4 \cdot 7H_2O$, $MnCl_2 \cdot 6H_2O$, and $Na_2MoO_4 \cdot 2H_2O$, in that order. Make sure each salt is completely dissolved before adding the next. Mix thoroughly. Bring volume to 1.0L with distilled/deionized water. Mix thoroughly. Store at 5°C. Solution is good until a red precipitate forms.

Solution 3
Composition per 100.0mL:

Acid hydrolysate of casein, vitamin free	10.0g

Preparation of Solution 3: Add acid hydrolysate of casein to distilled/deionized water and bring volume to 100.0mL. Mix thoroughly. Gently heat until dissolved.

Solution 4:
Composition per liter:

Inositol	0.2g
Thiamine·HCl	0.2g
Calcium pantothenate	0.2g
Riboflavin	0.2g
Nicotinamide	0.1g
Biotin	10.0mg

Preparation of Solution 4: Add components to distilled/deionized water and bring volume to 1.0L. Mix thoroughly. Store at 5°C. Suspension is good for 6 months to 1 year.

Solution 5:
Composition per 100.0mL:

Hemin	0.2g
NH_4OH, concentrated solution	2–3 drops

Preparation of Solution 5: Add hemin to 10.0mL of distilled/deionized water. Dissolve by adding a few drops of concentrated NH_4OH. Mix thoroughly. Bring volume to 100.0mL with distilled/deionized water. Store at 5°C.

Solution 6
Composition per 100.0mL:

DL-Thioctic acid	10.0mg
Ethanol (95% solution)	10.0mL

Preparation of Solution 6: Add thioctic acid to 10.0mL of 95% ethanol. Mix thoroughly. Store at −20°C.

Solution 7:
Composition per 10.0mL:

Coenzyme A	10.0mg
$Na_2S \cdot 5H_2O$ (0.05% solution in freshly boiled, distilled/deionized H_2O)	2 drops

Preparation of Solution 7: Add coenzyme A to distilled/deionized water and bring volume to 10.0mL. Mix thoroughly. Add 2 drops of 0.05% $Na_2S \cdot 5H_2O$ solution. Store at −20°C.

Preparation of Casein Hydrolysate-Vitamin-Glucose Base: Add components to distilled/deionized water and bring volume to 450.0mL. Mix thoroughly. Adjust pH to 6.5 with 20% KOH solution. Bring volume to 500.0mL with distilled/deionized water. Mix thoroughly. Filter sterilize.

Agar Starch Base:
Composition per 500.0mL:

Agar	12.5g
Potato starch, insoluble	2.0g
Solution 8	10.0mL

Solution 8:
Composition per 100.0mL:

Oleic acid	0.1g

Preparation of Solution 8: Add oleic acid to distilled/deionized water and bring volume to 50.0mL. Mix thoroughly. Adjust pH to 7.0 with NaOH. Bring volume to 100.0mL with distilled/deionized water.

Preparation of Agar Starch Base: Add potato starch to distilled/deionized water and bring volume to 50.0mL. Mix thoroughly. Pour into 440.0mL of boiling, distilled/deionized water. Add 10.0mL of solution 8 and the agar. Mix thoroughly. Autoclave for 20 min at 15 psi pressure–121°C.

Preparation of Medium: Aseptically combine 500.0mL of filter-sterilized casein hydrolysate-vitamin-glucose base with 500.0mL of hot agar starch base. Mix thoroughly. Heat nearly to boiling while stirring constantly. Aseptically distribute into sterile tubes. Allow tubes to cool in a slanted position.

Use: For the cultivation and maintenance of *Blastomyces dermatitidis* and *Histoplasma capsulatum* in the yeast phase.

Pisu Medium

Composition per 1512.0mL:

Agar base	960.0mL
Horse serum, sterile	360.0mL
L-Cystine solution	180.0mL
Lead acetate solution	12.0mL

pH 6.8 ± 0.2 at 25°C

Agar Base:

Composition per liter:

Proteose peptone no. 3	20.0g
Agar	7.0g
NaCl	5.0g
Meat extract	4.0g

Preparation for Agar Base: Add components to distilled/deionized water and bring volume to 1.0L. Mix thoroughly. Dissolve in steam for 15 min at 0 psi pressure–100°C. Cool to 45°–50°C. Adjust pH to 7.5. Filter sterilize. Distribute into 200.0mL Erlenmeyer flasks in 80.0mL volumes. Autoclave for 60 min at 0 psi pressure–100°C. Cool to 60°C.

L-Cystine Solution:

Composition per 200.0mL:

L-Cystine	2.0g

Preparation of L-Cystine Solution: Add L-cystine to distilled/deionized water and bring volume to 200.0mL. Mix thoroughly. Filter sterilize.

Lead Acetate Solution:

Composition per 100.0mL:

Lead acetate	10.0g

Preparation of Lead Acetate Solution: Add lead acetate to distilled/deionized water and bring volume to 100.0mL. Mix thoroughly. Filter sterilize.

Preparation for Medium: To each flask containing 80.0mL of cooled, sterile agar base, aseptically add 30.0mL of sterile horse serum, 15.0mL of sterile L-cystine solution, and 1.0mL of sterile lead acetate solution. Mix thoroughly. Aseptically distribute into small sterile tubes in 2.0–3.0mL volumes.

Use: For the cultivation and differentiation of bacteria based on their ability to produce cystinase. Cystinase-producing bacteria turn the medium black.

PKU Test Agar (Phenylketonuria Test Agar)

Composition per liter:

Agar	15.0g
K_2HPO_4	15.0g
Glucose	10.0g
KH_2PO_4	5.0g
$(NH_4)Cl$	2.5g
$(NH_4)NO_3$	0.5g
Asparagine	0.5g
DL-Alanine	0.5g
L-Glutamic acid	0.5g
Na_2SO_4	0.5g
$MgSO_4 \cdot 7H_2O$	0.05g
$FeCl_3$	5.0mg
$MnCl_2 \cdot 4H_2O$	5.0mg
β-2-Thienylalanine	3.3mg
$CaCl_2 \cdot 2H_2O$	2.5mg
Bacillus subtilis spore suspension	10.0mL

pH 7.0 ± 0.2 at 25°C

Source: This medium is available as a premixed powder from BD Diagnostic Systems.

Preparation of Medium: Add components to distilled/deionized water and bring volume to 1.0L. Mix thoroughly. Gently heat and bring to boiling. Continue boiling for 5 min. Do not autoclave. Cool to 50°C. Add 10.0mL of a suspension of *Bacillus subtilis* ATCC 6633 spores. Mix thoroughly. Pour into sterile Petri dishes or other containers.

Use: For the determination of phenylalanine concentrations in serum or urine. Used in the Guthrie-modified bacterial-inhibition assay procedure for screening newborn infants for phenylketonuria (PKU).

PKU Test Agar (Phenylketonuria Test Agar)

Composition per liter:

K_2HPO_4	15.0g
Agar	13.5g
Glucose	5.0g
KH_2PO_4	5.0g
$(NH_4)Cl$	2.5g
L-Asparagine	0.5g
L-Glutamic acid	0.5g

Na_2SO_4 .... 0.5g
$(NH_4)NO_3$ 0.5g
L-Alanine 0.25g
$MgSO_4 \cdot 7H_2O$ 0.05g
$FeCl_3$ 0.005g
$MnSO_4$ 0.005g
$CaCl_2 \cdot 2H_2O$ 2.5mg
β-2-Thienylalanine solution 1.0mL
pH 6.9 ± 0.2 at 25°C

Source: This medium is available as a premixed powder from BD Diagnostic Systems.

β-2-Thienylalanine Solution:

Composition per 100.0mL:

β-2-Thienylalanine 0.33g

Preparation of β-2-Thienylalanine Solution: Add β-2-thienylalanine to distilled/deionized water and bring volume to 100.0mL. Mix thoroughly. Filter sterilize.

Preparation of Medium: Add components, except β-2-thienylalanine solution, to distilled/deionized water and bring volume to 1.0L. Mix thoroughly. Gently heat and bring to boiling. Continue boiling for 5 min. Do not autoclave. Cool to 50°C. Add 10.0mL of a suspension of *Bacillus subtilis* ATCC 6633 spores and 1.0mL of sterile β-2-thienylalanine solution. Mix thoroughly. Pour into sterile Petri dishes or other containers.

Use: For the determination of phenylalanine concentrations in serum or urine. Used in the Guthrie-modified bacterial-inhibition assay procedure for screening newborn infants for phenylketonuria (PKU).

Plate Count MUG Agar

Composition per liter:

Agar 14.0g
Casein peptone 5.0g
Yeast extract 2.5g
D(+)-Glucose 1.0g
Tryptophan 1.0g
4-Methylumbelliferyl-β-D-glucuronide 0.07g
pH 7.0 ± 0.2 at 37°C

Source: This medium is available from Fluka, Sigma-Aldrich.

Preparation of Medium: Add components to distilled/deionized water and bring volume to 1.0L. Mix thoroughly. Gently heat while stirring and bring to boiling. Autoclave for 15 min at 15 psi pressure–121°C. Cool to 50°C. Pour into sterile Petri dishes.

Use: For the determination of bacterial counts in milk, dairy products, water, and other material. *E. coli* can be identified by fluorescence in the UV and verified by means of a positive indole test.

Plate Count MUG Agar

Composition per liter:

Agar 14.0g
Casein peptone 5.0g
Yeast extract 2.5g
D(+)-Glucose 1.0g
Tryptophan 1.0g
4-Methylumbelliferyl-β-D-glucuronide 0.07g
pH 7.0 ± 0.2 at 37°C

Source: This medium is available from Fluka, Sigma-Aldrich.

Preparation of Medium: Add components to distilled/deionized water and bring volume to 1.0L. Mix thoroughly. Gently heat while stirring and bring to boiling. Autoclave for 15 min at 15 psi pressure–121°C. Cool to 50°C. Pour into sterile Petri dishes.

Use: For the determination of bacterial counts in milk, dairy products, water, and other material. *E. coli* can be identified by fluorescence in the UV and verified by means of a positive indole test.

PLET Agar

Composition per liter:

Beef heart, solids from infusion 500.0g
Agar 15.0g
Tryptose 10.0g
NaCl 5.0g
Ethylenediamine tetracetic acid (EDTA) 0.3g
Thallous acetate 0.04g
Antibiotic inhibitor 10.0mL

Antibiotic Inhibitor:

Composition per 10.0mL:

Lysozyme 300,000U
Polymyxin 30,000U

Preparation of Antibiotic Inhibitor: Add components to distilled/deionized water and bring volume to 10.0mL. Mix thoroughly. Filter sterilize.

Preparation of Medium: Add components, except antibiotic inhibitor, to distilled/deionized water and bring volume to 990.0mL. Mix thoroughly. Gently heat and bring to boiling. Autoclave for 15 min at 15 psi pressure–121°C. Cool to 50°C. Aseptically add sterile antibiotic inhibitor. Mix thoroughly. Pour into sterile Petri dishes or distribute into sterile tubes.

Use: For the selective isolation and cultivation of *Bacillus anthracis*.

Poly-β-Hydroxybutyrate Medium (PHB Medium)

Composition per liter:

Part A	900.0mL
Part B	100.0mL

pH 7.2 ± 0.2 at 25°C

Part A:

Composition per 900.0mL:

$K_2HPO_4 \cdot 3H_2O$	0.6g
KH_2PO_4	0.2g
$MgSO_4 \cdot 7H_2O$	0.2g
$(NH_4)_2SO_4$	0.2g

Preparation of Medium: Add components to distilled/deionized water and bring volume to 900.0mL. Mix thoroughly. Adjust pH to 7.2. Autoclave for 15 min at 15 psi pressure–121°C. Cool to 25°C.

Part B:

Composition per 100.0mL:

Glucose	10.0g

Preparation of Medium: Add glucose to distilled/deionized water and bring volume to 100.0mL. Mix thoroughly. Autoclave for 15 min at 15 psi pressure–121°C. Cool to 25°C.

Preparation of Medium: Aseptically combine 900.0mL of cooled, sterile part A and 100.0mL of cooled, sterile part B. Mix thoroughly. Aseptically distribute into sterile tubes or flasks.

Use: For the cultivation and differentiation of *Pseudomonas* species based on their ability to produce intracellular poly-β-hydroxybutyrate. Production of poly-β-hydroxybutyrate is determined by staining cells with Sudan Black B.

Polymyxin *Staphylococcus* Medium

Composition per liter:

Agar	15.0g
Pancreatic digest of gelatin	5.0g
Beef extract	3.0g
Lecithin	0.7g
Polymyxin	0.075g
Tween 80	10.2mL

pH 6.8 ± 0.2 at 25°C

Preparation of Medium: Add components to distilled/deionized water and bring volume to 1.0L. Mix thoroughly. Gently heat and bring to boiling. Distribute into tubes or flasks. Autoclave for 15 min at 15 psi pressure–121°C. Pour into sterile Petri dishes.

Use: For the selective isolation and cultivation of pathogenic, coagulase-positive *Staphylococcus aureus. Proteus* species will grow on this medium but appear as translucent colonies.

Porcine Heart Agar

Composition per liter:

Porcine heart, infusion from	375.0g
Agar	15.0g
Papaic digest of soybean meal	6.5g
Glucose	5.0g
NaCl	5.0g
Proteose peptone no. 3	5.0g
Yeast extract	3.5g
Sheep blood, defibrinated	50.0mL

pH 7.2 ± 0.2 at 25°C

Source: This medium is available as a premixed powder from BD Diagnostic Systems.

Preparation of Medium: Add components, except sheep blood, to distilled/deionized water and bring volume to 950.0mL. Mix thoroughly. Gently heat and bring to boiling. Autoclave for 15 min at 15 psi pressure–121°C. Cool to 45°–50°C. Aseptically add sterile sheep blood. Mix thoroughly. Pour into sterile Petri dishes.

Use: For determination of the sensitivity of microorganisms to antimicrobics using the disc plate technique.

Potassium Cyanide Broth

Composition per liter:

Na_2HPO_4	5.64g
NaCl	5.0g
Proteose peptone no. 3	3.0g
KH_2PO_4	0.225g
KCN solution	15.0mL

pH 7.6 ± 0.2 at 25°C

KCN Solution:

Composition per 100.0mL:

KCN	0.5g

Preparation of KCN Solution: Add KCN to distilled/deionized water and bring volume to 100.0mL. Mix thoroughly.

Caution: Cyanide is toxic.

Preparation of Medium: Add components, except KCN solution, to distilled/deionized water and bring volume to 985.0mL. Mix thoroughly. Gently heat and bring to boiling. Autoclave for 15 min at 15 psi pressure–121°C. Cool to 25°C. Aseptically add 15.0mL of KCN solution. Mix thoroughly. Distribute into sterile screw-capped tubes or flasks in 1.0–1.5mL volumes. Close caps tightly.

Use: For the cultivation and differentiation of urease-negative, Gram-negative enteric bacteria. *Salmonella* species and *Shigella* species are nonmotile in this medium. *Proteus* species are motile in this medium.

Potassium Tellurite Agar

Composition per liter:

Beef heart, solids from infusion....................500.0g
Agar 15.0g
Tryptose 10.0g
NaCl....................5.0g
Blood, defibrinated 50.0mL
K_2TeO_3 solution....................20.0mL

pH 6.0 ± 0.2 at 25°C

K_2TeO_3 Solution:

Composition per 20.0mL:

K_2TeO_3....................0.5g

Preparation of K_2TeO_3 Solution: Add K_2TeO_3 to distilled/deionized water and bring volume to 10.0mL. Mix thoroughly. Filter sterilize.

Caution: Potassium tellurite is toxic.

Preparation of Medium: Add components, except K_2TeO_3 solution, to distilled/deionized water and bring volume to 930.0mL. Mix thoroughly. Gently heat and bring to boiling. Autoclave for 15 min at 15 psi pressure–121°C. Cool to 45°–50°C. Aseptically add sterile K_2TeO_3 solution and 50.0mL of blood. Rabbit or sheep blood may be used. Mix thoroughly. Pour into sterile Petri dishes or distribute into sterile tubes. Allow tubes to cool in a slanted position.

Use: For the cultivation and differentiation of *Enterococcus faecalis*. *Enterococcus faecalis* appears as black colonies.

Potato Dextrose Yeast Agar (PDY Agar)

Composition per liter:

Glucose 20.0g
Agar 15.0g
Yeast extract....................5.0g
Potato infusion 500.0mL

pH 5.6 ± 0.2 at 25°C

Potato Infusion:

Composition per 500.0mL:

Potatoes....................300.0g

Preparation of Potato Infusion: Peel and dice potatoes. Add 500.0mL of distilled/deionized water. Gently heat and bring to boiling. Continue boiling for 30 min. Filter through cheesecloth. Reserve filtrate.

Preparation of Medium: Add components to distilled/deionized water and bring volume to 1.0L. Mix thoroughly. Gently heat and bring to boiling. Distribute into tubes or flasks. Autoclave for 15 min at 15 psi pressure–121°C. Pour into sterile Petri dishes or leave in tubes.

Use: For the cultivation and maintenance of *Bacillus* species and fungi.

Potato Extract Agar

Composition per liter:

Agar 15.0g
Peptone 5.0g
NaCl....................5.0g
Yeast extract....................2.0g
Beef extract powder....................1.0g
Potato extract 20.0mL

pH 7.4 ± 0.2 at 25°C

Potato Extract:

Composition per liter:

Potatoes....................300.0g

Preparation of Potato Extract: Peel and dice potatoes. Add 500.0mL of distilled/deionized water. Gently heat and bring to boiling. Continue boiling for 30 min. Filter through cheesecloth.

Use: For the cultivation of a wide variety of yeasts and molds.

PPLO Agar

Composition per liter:

Beef heart, infusion from....................50.0g
Agar 14.0g
Peptone 10.0g
NaCl....................5.0g
Bovine serum....................100.0mL

pH 7.8 ± 0.2 at 25°C

Source: This medium is available as a premixed powder from BD Diagnostic Systems.

Preparation of Medium: Add components, except bovine serum, to distilled/deionized water and bring volume to 900.0mL. Mix thoroughly. Gently heat and bring to boiling. Autoclave for 15 min at 15 psi pressure–121°C. Cool to 45°–50°C. Aseptically add sterile bovine serum. Mix thoroughly. Pour into sterile Petri dishes or distribute into sterile tubes.

Use: For the isolation and cultivation of *Mycoplasma* species (pleuro-pneumonia-like organisms).

PPLO Agar

Composition per liter:

Beef heart, infusion from....................50.0g
Agar 14.0g
Peptone 10.0g

NaCl 5.0g
Ascitic fluid 250.0mL
pH 7.8 ± 0.2 at 25°C

Source: This medium is available as a premixed powder from BD Diagnostic Systems.

Preparation of Medium: Add components, except ascitic fluid, to distilled/deionized water and bring volume to 750.0mL. Mix thoroughly. Gently heat and bring to boiling. Autoclave for 15 min at 15 psi pressure–121°C. Cool to 45°–50°C. Aseptically add sterile ascitic fluid. Mix thoroughly. Pour into sterile Petri dishes or distribute into sterile tubes.

Use: For the isolation and cultivation of *Mycoplasma* species (pleuro-pneumonia-like organisms).

PPLO Broth with Crystal Violet

Composition per liter:
Beef heart, infusion from 50.0g
Peptone 10.0g
NaCl 5.0g
Crystal Violet 0.01g
Ascitic fluid 250.0mL
Chapman tellurite solution 2.85mL
pH 7.8 ± 0.2 at 25°C

Source: This medium is available as a premixed powder from BD Diagnostic Systems.

Chapman Tellurite Solution:
Composition per 100.0mL:
K_2TeO_3 .... 1.0g

Preparation of Chapman Tellurite Solution: Add K_2TeO_3 to distilled/deionized water and bring volume to 100.0mL. Mix thoroughly. Filter sterilize.

Caution: Potassium tellurite is toxic.

Preparation of Medium: Add components, except ascitic fluid and Chapman tellurite solution, to distilled/deionized water and bring volume to 747.15mL. Mix thoroughly. Autoclave for 15 min at 15 psi pressure–121°C. Cool to less than 37°C. Aseptically add sterile ascitic fluid and 2.85mL of Chapman tellurite solution. Mix thoroughly. Aseptically distribute into sterile tubes or flasks.

Use: For the isolation of *Mycoplasma* species from clinical specimens.

PPLO Broth without Crystal Violet

Composition per liter:
Beef heart, infusion from 50.0g
Peptone 10.0g
NaCl 5.0g
Thallium acetate (optional) 0.5g
Penicillin (optional) 100,000U
Ascitic fluid 250.0mL
pH 7.8 ± 0.2 at 25°C

Source: This medium is available as a premixed powder from BD Diagnostic Systems.

Preparation of Medium: Add components, except ascitic fluid, to distilled/deionized water and bring volume to 750.0mL. Mix thoroughly. Autoclave for 15 min at 15 psi pressure–121°C. Cool to less than 37°C. Aseptically add sterile ascitic fluid. If desired, 0.5g of thallium acetate or 100,000U of penicillin may be added for a more selective medium. Mix thoroughly. Aseptically distribute into sterile tubes or flasks.

Use: For the enrichment of pleuro-pneumonia-like organisms (PPLOs) and *Mycoplasma* species from clinical specimens.

PPLO Broth without Crystal Violet with Calf Serum, Fresh Yeast Extract, and Sodium Acetate

Composition per liter:
Beef heart, infusion from 50.0g
Peptone 10.0g
Sodium acetate 9.0g
NaCl 5.0g
Yeast extract solution, fresh 250.0mL
Calf serum 100.0mL
pH 7.8 ± 0.2 at 25°C

Yeast Extract Solution:
Composition per 300.0mL:
Baker's yeast, live, pressed, starch-free 75.0g

Preparation of Yeast Extract Solution: Add the live Baker's yeast to 300.0mL of distilled/deionized water. Autoclave for 90 min at 15 psi pressure–121°C. Allow to stand. Remove supernatant solution. Adjust pH to 6.6–6.8. Filter sterilize.

Preparation of Medium: Add components, except fresh yeast extract solution and calf serum, to distilled/deionized water and bring volume to 550.0mL. Mix thoroughly. Autoclave for 15 min at 15 psi pressure–121°C. Cool to 45°–50°C. Aseptically add sterile fresh yeast extract solution and calf serum. Mix thoroughly. Aseptically distribute into sterile tubes or flasks.

Use: For the cultivation and maintenance of *Mycoplasma* species.

Presumpto Media

Composition per plate:
Quadrant 1 .... 5.0mL
Quadrant 2 .... 5.0mL

Quadrant 3 .... 5.0mL
Quadrant 4 .... 5.0mL

Quadrant 1:
Composition per 5.0mL:
Lombard-Dowell agar 5.0mL

Quadrant 2:
Composition per 5.0mL:
Lombard-Dowell bile agar 5.0mL

Quadrant 3:
Composition per 5.0mL:
Lombard-Dowell egg yolk agar 5.0mL

Quadrant 4:
Composition per 5.0mL:
Lombard-Dowell esculin agar 5.0mL

Preparation of Quadrant Media: Sterilize Lombard-Dowell Agar by autocalving for 15 min at 15 psi pressure–121°C. Cool to 45°-50°C. Add additional components as filter sterilized solutions. Mix and distribute as 5.0mL aliquots into quadrants.

Use: For the differentiation and presumptive identification of anaerobic bacteria. The Presumpto media is a four-sectored plate each containing a different medium.

Proteose No. 3 Agar

Composition per 1010.0mL:

Proteose peptone no. 3	20.0g
Agar	15.0g
Na_2HPO_4	5.0g
NaCl	5.0g
Glucose	0.5g
Hemoglobin solution	500.0mL
Supplement A	10.0mL

pH 7.3 ± 0.2 at 25°C

Source: This medium is available as a premixed powder from BD Diagnostic Systems.

Hemoglobin Solution:
Composition per 500.0mL:
Hemoglobin 10.0g

Preparation of Hemoglobin Solution: Add hemoglobin to distilled/deionized water and bring volume to 500.0mL. Mix thoroughly. Autoclave for 15 min at 15 psi pressure–121°C. Cool to 45°–50°C.

Supplement A:
Composition per 100.mL:
Supplement A contains yeast concentrate with Crystal Violet.

Preparation of Supplement A: Add components to distilled/deionized water and bring volume to 10.0mL. Mix thoroughly. Filter sterilize.

Preparation of Medium: Add components, except hemoglobin solution and supplement A, to distilled/deionized water and bring volume to 500.0mL. Mix thoroughly. Gently heat and bring to boiling. Autoclave for 15 min at 15 psi pressure–121°C. Cool to 50°–60°C. Aseptically add 500.0mL of sterile hemoglobin solution and 10.0mL of sterile supplement A. Mix thoroughly. Pour into sterile Petri dishes or distribute into sterile tubes.

Use: For the isolation and cultivation of *Neisseria* species, *Hemophilus* species, and other fastidious bacteria. For the cultivation and maintenance of *Escherichia coli.*

Proteose No. 3 Agar

Composition per 1010.0mL:

Proteose peptone no. 3	20.0g
Agar	15.0g
Na_2HPO_4	5.0g
NaCl	5.0g
Glucose	0.5g
Hemoglobin solution	500.0mL
Supplement B	10.0mL

pH 7.3 ± 0.2 at 25°C

Source: This medium is available as a premixed powder from BD Diagnostic Systems.

Hemoglobin Solution:
Composition per 500.0mL:
Hemoglobin 10.0g

Preparation of Hemoglobin Solution: Add hemoglobin to distilled/deionized water and bring volume to 500.0mL. Mix thoroughly. Autoclave for 15 min at 15 psi pressure–121°C. Cool to 45°–50°C.

Supplement B:
Composition per 10.0mL:
Supplement B contains yeast concentrate, glutamine, coenzyme, cocarboxylase, hematin, and growth factors.

Preparation of Supplement B: Add components to distilled/deionized water and bring volume to 10.0mL. Mix thoroughly. Filter sterilize.

Preparation of Medium: Add components, except hemoglobin solution and supplement B, to distilled/deionized water and bring volume to 500.0mL. Mix thoroughly. Gently heat and bring to boiling. Autoclave for 15 min at 15 psi pressure–121°C. Cool to 50°–60°C. Aseptically add 500.0mL of sterile hemoglobin solution and 10.0mL of sterile supplement B. Mix thoroughly. Pour into sterile Petri dishes or distribute into sterile tubes.

Use: For the isolation and cultivation of *Neisseria* species, *Hemophilus* species, and other fastidious bacteria. For the cultivation and maintenance of *Escherichia coli.*

Proteose No. 3 Agar

Composition per 1010.0mL:

Proteose peptone no. 3	20.0g
Agar	15.0g
Na_2HPO_4	5.0g
NaCl	5.0g
Glucose	0.5g
Hemoglobin solution	500.0mL
Supplement VX	10.0mL

pH 7.3 ± 0.2 at 25°C

Source: This medium is available as a premixed powder from BD Diagnostic Systems.

Hemoglobin Solution:

Composition per 500.0mL:

Hemoglobin	10.0g

Preparation of Hemoglobin Solution: Add hemoglobin to distilled/deionized water and bring volume to 500.0mL. Mix thoroughly. Autoclave for 15 min at 15 psi pressure–121°C. Cool to 45°–50°C.

Supplement VX:

Composition per 10.0mL:

Supplement B contains essential growth factors.

Preparation of Supplement VX: Add components to distilled/deionized water and bring volume to 10.0mL. Mix thoroughly. Filter sterilize.

Preparation of Medium: Add components, except hemoglobin solution and supplement VX, to distilled/deionized water and bring volume to 500.0mL. Mix thoroughly. Gently heat and bring to boiling. Autoclave for 15 min at 15 psi pressure–121°C. Cool to 50°–60°C. Aseptically add 500.0mL of sterile hemoglobin solution and 10.0mL of sterile supplement VX. Mix thoroughly. Pour into sterile Petri dishes or distribute into sterile tubes.

Use: For the isolation and cultivation of *Neisseria* species, *Hemophilus* species, and other fastidious bacteria. For the cultivation and maintenance of *Escherichia coli.*

Proteose Yeast Extract Medium

Composition per liter:

Proteose peptone	20.0g
Glucose	10.0g
Yeast extract	5.0g

Preparation of Medium: Add components to distilled/deionized water and bring volume to 1.0L. Mix thoroughly. Distribute into tubes or flasks. Autoclave for 15 min at 15 psi pressure–121°C.

Use: For the cultivation of a variety of bacteria.

Pseudomonas Agar F

Composition per liter:

Proteose peptone no. 3	20.0g
Agar	15.0g
Glycerol	10.0g
Pancreatic digest of casein	10.0g
K_2HPO_4	1.5g
$MgSO_4 \cdot 7H_2O$	0.73g

pH 7.0 ± 0.2 at 25°C

Preparation of Medium: Add components to distilled/deionized water and bring volume to 1.0L. Mix thoroughly. Gently heat and bring to boiling. Distribute into tubes or flasks. Autoclave for 15 min at 15 psi pressure–121°C. Pour into sterile Petri dishes or leave in tubes.

Use: For the cultivation and observation of fluorescein production in *Pseudomonas* species.

Pseudomonas Agar F

Composition per liter:

Agar	15.0g
Glycerol	10.0g
Proteose peptone no. 3	10.0g
Pancreatic digest of casein	10.0g
K_2HPO_4	1.5g
$MgSO_4 \cdot 7H_2O$	1.5g

pH 7.0 ± 0.2 at 25°C

Source: This medium is available as a premixed powder from BD Diagnostic Systems.

Preparation of Medium: Add components to distilled/deionized water and bring volume to 1.0L. Mix thoroughly. Gently heat and bring to boiling. Distribute into tubes or flasks. Autoclave for 15 min at 15 psi pressure–121°C. Pour into sterile Petri dishes or leave in tubes.

Use: For the isolation, cultivation, and differentiation of *Pseudomonas aeruginosa* on the basis of pigment production.

Pseudomonas Agar P

Composition per liter:

Proteose peptone no. 3	20.0g
Agar	15.0g
Glycerol	10.0g
K_2HPO_4	10.0g
$MgCl_2 \cdot 6H_2O$	1.4g

pH 7.0 ± 0.2 at 25°C

Source: This medium is available as a premixed powder from BD Diagnostic Systems.

Preparation of Medium: Add components to distilled/deionized water and bring volume to 1.0L. Mix thoroughly. Gently heat and bring to boiling. Distribute into tubes or flasks. Autoclave for 15 min at 15 psi pressure–121°C. Pour into sterile Petri dishes or leave in tubes.

Use: For the isolation, cultivation, and differentiation of *Pseudomonas aeruginosa* on the basis of pigment production.

Pseudomonas Basal Mineral Medium

Composition per liter:

K_2HPO_4	12.5g
KH_2PO_4	3.8g
$(NH_4)_2SO_4$	1.0g
$MgSO_4 \cdot 7H_2O$	0.1g
Carbon source (0.8*M* solution)	100.0mL
Trace elements solution	5.0mL

pH 7.2 ± 0.2 at 25°C

Trace ElementsSolution:

Composition per liter:

H_3BO_3	0.232g
$ZnSO_4 \cdot 7H_2O$	0.174g
$FeSO_4(NH_4)_2SO_4 \cdot 6H_2O$	0.116g
$CoSO_4 \cdot 7H_2O$	0.096g
$(NH_4)_6Mo_7O_{24} \cdot 4H_2O$	0.022g
$CuSO_4 \cdot 5H_2O$	8.0mg
$MnSO_4 \cdot 4H_2O$	8.0mg

Preparation of Trace Elements Solution: Add components to distilled/deionized water and bring volume to 1.0L. Mix thoroughly.

Carbon Source:

Composition per 100.0mL:

Glucose	14.4g

Preparation of Carbon Source: Add glucose to distilled/deionized water and bring volume to 100.0mL. Mix thoroughly. Filter sterilize. Other carbon sources may replace glucose. Prepare 0.8*M* carbon source solution.

Preparation of Medium: Add components, except carbon source, to distilled/deionized water and bring volume to 900.0mL. Mix thoroughly. Gently heat and bring to boiling. Autoclave for 15 min at 15 psi pressure–121°C. Cool to 45°–50°C. Aseptically add 100.0mL of sterile carbon source. Mix thoroughly. Aseptically distribute into sterile tubes or flasks.

Use: For the cultivation and differentiation of *Pseudomonas* species based on their ability to grow on different carbon sources.

Pseudomonas CFC Agar

Composition per liter:

Pancreatic digest of gelatin	16.0g
Agar	11.0g
Pancreatic digest of casein	10.0g
K_2SO_4	10.0g
$MgCl_2 \cdot 6H_2O$	1.4g
CFC selective supplement	10.0mL
Glycerol	10.0mL

pH 7.1 ± 0.2 at 25°C

Source: This medium is available as a premixed powder from Oxoid Unipath.

CFC Selective Supplement:

Composition per 10.0mL:

Cephaloridine	0.05g
Fucidin	0.01g
Cetrimide	0.01g

Preparation of CFC Selective Supplement: Add components to distilled/deionized water and bring volume to 10.0mL. Mix thoroughly. Filter sterilize.

Preparation of Medium: Add components, except CFC selective supplement, to distilled/deionized water and bring volume to 990.0mL. Mix thoroughly. Gently heat and bring to boiling. Autoclave for 15 min at 15 psi pressure–121°C. Cool to 45°–50°C. Aseptically add sterile CFC selective supplement. Mix thoroughly. Pour into sterile Petri dishes or distribute into sterile tubes.

Use: For the selective isolation and cultivation of *Pseudomonas* species.

Pseudomonas CN Agar

Composition per liter:

Pancreatic digest of gelatin	16.0g
Agar	11.0g
Pancreatic digest of casein	10.0g
K_2SO_4	10.0g
$MgCl_2 \cdot 6H_2O$	1.4g
CN selective supplement	10.0mL
Glycerol	10.0mL

CN Selective Supplement:

Cetrimide	0.1g
Sodium nalidixate	7.5mg

pH 7.1 ± 0.2 at 25°C

Source: This medium is available as a premixed powder from Oxoid Unipath.

Preparation of Medium: Add components, except CN selective supplement, to distilled/deionized water and bring volume to 990.0mL. Mix thoroughly. Gently heat and bring to boiling. Autoclave for 15 min at 15 psi pressure–121°C. Cool to 45°–50°C. Aseptically add sterile CN selective supplement. Mix thoroughly. Pour into sterile Petri dishes or distribute into sterile tubes.

Use: For the selective isolation and cultivation of *Pseudomonas* species.

Purple Agar

Composition per liter:

Agar	15.0g
Proteose peptone no. 3	10.0g
NaCl	5.0g
Beef extract	1.0g
Bromcresol Purple	0.02g
Carbohydrate solution	20.0mL

pH 6.8 ± 0.2 at 25°C

Source: This medium is available as a premixed powder from BD Diagnostic Systems.

Carbohydrate Solution:

Composition per 20.0mL:

Carbohydrate	10.0g

Preparation of Carbohydrate Solution: Add carbohydrate to distilled/deionized water and bring volume to 20.0mL. For expensive carbohydrates, 5.0g may be used instead of 10.0g. Mix thoroughly. Filter sterilize.

Preparation of Medium: Add components, except carbohydrate solution, to distilled/deionized water and bring volume to 980.0mL. Mix thoroughly. Gently heat and bring to boiling. Distribute into tubes in 9.8mL volumes. Autoclave for 15 min at 15 psi pressure–121°C. Cool to 45°–50°C. Aseptically add 0.2mL of sterile carbohydrate solution to each tube. Mix thoroughly. Allow tubes to cool in a slanted position.

Use: For the preparation of carbohydrate media used in fermentation studies for the identification of bacteria, especially members of the Enterobacteriaceae. Bacteria that can ferment the carbohydrate turn the medium yellow.

Purple Broth (Purple Carbohydrate Broth)

Composition per liter:

Proteose peptone no. 3	10.0g
NaCl	5.0g
Beef extract	1.0g
Bromcresol Purple	0.015g
Carbohydrate solution	20.0mL

pH 6.8 ± 0.2 at 25°C

Source: This medium is available as a premixed powder from BD Diagnostic Systems.

Carbohydrate Solution:

Composition per 20.0mL:

Carbohydrate	10.0g

Preparation of Carbohydrate Solution: Add carbohydrate to distilled/deionized water and bring volume to 20.0mL. For expensive carbohydrates, 5.0g may be used instead of 10.0g. Mix thoroughly. Filter sterilize.

Preparation of Medium: Add components, except carbohydrate solution, to distilled/deionized water and bring volume to 980.0mL. Mix thoroughly. Gently heat and bring to boiling. Distribute into tubes in 9.8mL volumes. Autoclave for 15 min at 15 psi pressure–121°C. Cool to 25°C. Aseptically add 0.2mL of sterile carbohydrate solution to each tube. Mix thoroughly.

Use: For the preparation of carbohydrate media used in fermentation studies for the identification of bacteria, especially members of the Enterobacteriaceae. Bacteria that can ferment the carbohydrate turn the medium yellow.

Purple Broth

Composition per liter:

Pancreatic digest of gelatin	10.0g
NaCl	5.0g
Bromcresol Purple	0.02g
Carbohydrate solution	20.0mL

pH 6.8 ± 0.2 at 25°C

Source: This medium is available as a premixed powder from BD Diagnostic Systems.

Carbohydrate Solution:

Composition per 20.0mL:

Carbohydrate	10.0g

Preparation of Carbohydrate Solution: Add carbohydrate to distilled/deionized water and bring volume to 20.0mL. For expensive carbohydrates, 5.0g may be used instead of 10.0g. Mix thoroughly. Filter sterilize.

Preparation of Medium: Add components, except carbohydrate solution, to distilled/deionized water and bring volume to 980.0mL. Mix thoroughly. Adjust pH to 7.4 if necessary. Distribute into tubes containing an inverted Durham tube. Fill each tube with 9.8mL of medium. Autoclave for 15 min at 15 psi pressure–121°C. Aseptically add 0.2mL of sterile carbohydrate solution to each tube.

Use: For the preparation of liquid fermentation media. Bacteria that can ferment the carbohydrate turn the medium yellow.

Purple Lactose Agar

Composition per liter:

Agar	10.0g
Lactose	10.0g
Peptone	5.0g
Beef extract	3.0g
Bromcresol Purple	0.025g

pH 6.8 ± 0.1 at 25°C

Source: This medium is available as a premixed powder from BD Diagnostic Systems.

Preparation of Medium: Add components to distilled/deionized water and bring volume to 1.0L. Mix thoroughly. Gently heat and bring to boiling. Distribute into tubes or flasks. Autoclave for 15 min at 15 psi pressure–121°C. Pour into sterile Petri dishes or leave in tubes. Allow tubes to cool in a slanted position.

Use: For the detection and differentiation of members of the Enterobacteriaceae. Bacteria that can ferment lactose turn the medium yellow.

Purple Serum Agar Base

Composition per liter:

Agar	20.0g
Lactose	20.0g
Peptone	20.0g
NaCl	5.0g
Bromcresol Purple	0.03g
Phenol Red	0.024g

pH 7.6 ± 0.2 at 25°C

Preparation of Medium: Add components to distilled/deionized water and bring volume to 1.0L. Mix thoroughly. Gently heat and bring to boiling. Distribute into tubes or flasks. Autoclave for 15 min at 15 psi pressure–121°C. Pour into sterile Petri dishes or leave in tubes.

Use: For the cultivation and differentiation of Gram-negative bacteria isolated from the urinary tract. Bacteria that can ferment lactose turn the medium yellow.

PV Blood Agar (Paromomycin Vancomycin Blood Agar)

Composition per liter:

Agar	20.0g
Pancreatic digest of casein	15.0g
NaCl	5.0g
Papaic digest of soybean meal	5.0g
Yeast extract	5.0g
L-Cystine	0.4g
Paromomycin	0.1g
Vancomycin	7.5mg
Hemin	5.0mg
Sheep blood, defibrinated	50.0mL
Vitamin K_1 solution	10.0mL

pH 7.5 ± 0.2 at 25°C

Vitamin K_1 Solution:

Composition per 10.0mL:

Vitamin K_1	0.01g
Ethanol	10.0mL

Preparation of Vitamin K_1 Solution: Add vitamin K_1 to 10.0mL of absolute ethanol. Mix thoroughly. Filter sterilize.

Preparation of Medium: Add components—except vitamin K_1 solution, sheep blood, paromomycin, and vancomycin—to distilled/deionized water and bring volume to 940.0mL. Mix thoroughly. Gently heat and bring to boiling for 1 min. Autoclave for 15 min at 15 psi pressure–121°C. Cool to 50°–55°C. Aseptically add the sterile vitamin K_1 solution, sheep blood, vancomycin, and paromomycin. Mix thoroughly. Pour into sterile Petri dishes.

Use: For the selective cultivation of fastidious anaerobic bacteria.

PY Basal Medium

Composition per 104.0mL:

Yeast extract	1.0g
L-Cysteine·HCl·H_2O	0.5g
Peptone	0.5g
Pancreatic digest of casein	0.5g
Resazurin	0.16mg
Salts solution	4.0mL

Salts Solution:

Composition per liter:

$NaHCO_3$	10.0g
NaCl	2.0g
K_2HPO_4	1.0g
KH_2PO_4	1.0g
$CaCl_2$	0.2g
$MgSO_4$	0.2g

Preparation of Salts Solution: Add components to distilled/deionized water and bring volume to 1.0L. Mix thoroughly.

Preparation of Medium: Add components to distilled/deionized water and bring volume to

104.0mL. Mix thoroughly. Distribute into tubes or flasks. Autoclave for 15 min at 15 psi pressure–121°C.

Use: For the identification of treponemes.

Pyrazinamidase Agar (Pyrazinamide Medium)

Composition per liter:

Pancreatic digest of casein	15.0g
Agar	15.0g
Papaic digest of soybean meal	5.0g
NaCl	5.0g
Yeast extract	3.0g
Pyrazinecarboxamide	1.0g
Tris(hydroxymethyl)amino-methane maleate buffer (0.2*M*, pH 6.0)	1.0L

pH 6.0 ± 0.2 at 25°C

Preparation of Medium: Combine components. Mix thoroughly. Gently heat and bring to boiling. Distribute into tubes in 5.0mL volumes. Autoclave for 15 min at 15 psi pressure–121°C. Allow tubes to cool in a slanted position.

Use: For the cultivation, differentiation, and maintenance of pathogenic *Yersinia* species. Bacteria that produce pyrazinamidase turn the medium pink.

Pyrazinamide Medium

Composition per liter:

Agar	15.0g
Na_2HPO_4	2.5g
Sodium pyruvate	2.0g
L-Asparagine	2.0g
KH_2PO_4	1.0g
Pancreatic digest of casein	0.5g
Tween 80	0.2g
Pyrazinamide	0.1g
$CaCl_2 \cdot 2H_2O$	0.5mg
$CuSO_4 \cdot 5H_2O$	0.1mg
$ZnSO_4 \cdot 7H_2O$	0.1mg
Ferric ammonium citrate	0.05g
$MgSO_4 \cdot 7H_2O$	0.01g

pH 6.6 ± 0.2 at 25°C

Preparation of Medium: Combine components. Mix thoroughly. Gently heat and bring to boiling. Distribute into tubes in 5.0mL volumes. Autoclave for 15 min at 15 psi pressure–121°C. Allow tubes to cool in a slanted position.

Use: For the cultivation and differentiation of *Corynebacterium* species and related organisms. Bacteria that produce pyrazinamidase turn the medium pink.

Pyruvate Utilization Medium

Composition per liter:

Sodium pyruvate	10.0g
Pancreatic digest of casein	10.0g
K_2HPO_4	5.0g
NaCl	5.0g
Yeast extract	5.0g
Bromthymol Blue	0.1g

pH 7.1–7.4 at 25°C

Preparation of Medium: Add components to distilled/deionized water and bring volume to 1.0L. Mix thoroughly. Gently heat and bring to boiling. Adjust pH to 7.1–7.4. Distribute into tubes in 5.0mL volumes. Autoclave for 15 min at 15 psi pressure–121°C.

Use: For the cultivation of bacteria that can metabolize pyruvate. Bacteria that can utilize pyruvate turn the medium yellow.

Pyruvic Acid Egg Medium

Composition per 1640.0mL:

KH_2PO_4	11.4g
D-Glucose	10.0g
Na_2HPO_4	6.0g
Pyruvic acid	3.0g
$MgSO_4 \cdot 7H_2O$	0.3g
Malachite Green	0.125g
Egg, homogenized whole	1.0L
Penicillin solution	10.0mL

Source: This medium is available as a prepared medium from Oxoid Unipath.

Penicillin Solution:

Composition per 10.0mL:

Penicillin G	100,000U

Preparation of Penicillin Solution: Add penicillin G to distilled/deionized water and bring volume to 10.0mL. Mix thoroughly. Filter sterilize.

Homogenized Whole Egg:

Composition per liter:

Whole eggs	18-24

Preparation of Homogenized Whole Egg: Use fresh eggs, less than 1 week old. Scrub the shells with soap. Let stand in a soap solution for 30 min. Rinse in running water. Soak eggs in 70% ethanol for 15 min. Break the eggs into a sterile container. Homogenize by shaking. Filter through four layers of sterile cheesecloth into a sterile graduated cylinder.

Preparation of Medium: Add components, except homogenized whole egg and penicillin solution, to distilled/deionized water and bring volume to 630.0mL. Mix thoroughly. Autoclave for 15 min at 15 psi pressure–121°C. Cool to 45°–50°C. Aseptically add 1.0L homogenized whole egg and penicillin solution to cooled sterile basal medium. Mix thoroughly. Aseptically distribute into sterile tubes. Inspissate at 85°–90°C (moist heat) for 45 min.

Use: For the isolation and cultivation of *Mycobacterium* species, especially ones that are drug resistant and difficult to grow.

R Medium

Composition per liter:

Agar	30.0g
$NaHCO_3$	8.0g
K_2HPO_4	3.0g
Glucose	2.5g
Glutamine	0.61g
Serine	0.24g
Leucine	0.23g
Lysine	0.23g
Asparagine	0.18g
Valine	0.17g
Isoleucine	0.17g
Tyrosine	0.14g
Arginine·HCl	0.125g
Phenylalanine	0.125g
Threonine	0.12g
Methionine	0.073g
Glycine	0.065g
Histidine·HCl	0.055g
Proline	0.043g
Tryptophan	0.035g
L-Cystine	0.025g
$MgSO_4 \cdot H_2O$	9.9mg
$CaCl_2 \cdot 2H_2O$	7.4mg
Adenine sulfate	2.1mg
Uracil	1.4mg
Thiamine·HCl	1.0mg
$MnSO_4 \cdot H_2O$	0.9mg

pH 8.0 ± 0.2 at 25°C

Preparation of Medium: Add components, except agar, to distilled/deionized water and bring volume to 500.0mL. Mix thoroughly. Filter sterilize. Warm to 45°–50°C. Add agar to distilled/deionized water and bring volume to 500.0mL. Mix thoroughly. Autoclave for 15 min at 15 psi pressure–121°C. Cool to 45°–50°C. Aseptically combine both solutions. Mix thoroughly. Pour into sterile Petri dishes or distribute into sterile tubes.

Use: For the cultivation of *Bacillus anthracis*.

Rabbit Blood Agar

Composition per 1250.0mL:

Pancreatic digest of casein	16.0g
Agar	13.5g
Brain heart, solids from infusion	8.0g
Peptic digest of animal tissue	5.0g
NaCl	5.0g
Na_2HPO_4	2.5g
Glucose	2.0g
Rabbit blood, defibrinated	250.0mL

pH 7.4 ± 0.2 at 25°C

Preparation of Medium: Add components, except rabbit blood, to distilled/deionized water and bring volume to 1.0L. Mix thoroughly. Autoclave for 15 min at 15 psi–121°C. Aseptically add sterile rabbit blood. Pour into sterile Petri dishes or aseptically distribute into sterile tubes or flasks while shaking.

Use: For the cultivation and maintenance of *Corynebacterium diphtheriae*, *Haemophilus ducreyi,* and *Actinobacillus lignieresii.*

Rabbit Heart Infusion Agar

Composition per liter:

Beef heart, infusion from	500.0g
Agar	15.0g
Tryptose	10.0g
NaCl	5.0g
Rabbit blood, defibrinated	50.0mL

pH 7.4 ± 0.2 at 25°C

Preparation of Medium: Add components, except rabbit blood, to distilled/deionized water and bring volume to 950.0mL. Mix thoroughly. Gently heat and bring to boiling. Autoclave for 15 min at 15 psi pressure–121°C. Cool to 50°–55°C. Aseptically add 50.0mL of sterile rabbit blood. Mix thoroughly. Pour into sterile Petri dishes or distribute into sterile tubes.

Use: For the culture of *Bartonella quintana.*

Rabbit Serum Medium (Rabbit Serum Bovine Serum Albumin Tween 80 Medium) (Rabbit Serum BSA Tween 80 Medium)

Composition per liter:

Basal medium	900.0mL
Rabbit serum with supplements	100.0mL

pH 7.4 ± 0.2 at 25°C

Basal Medium:

Composition per 900.0mL:

Na_2HPO_4	1.0g
NaCl	1.0g
KH_2PO_4	0.3g

Glycerol (10% solution)	1.0mL
NH_4Cl (25% solution)	1.0mL
Sodium pyruvate (10% solution)	1.0mL
Thiamine (0.5% solution)	1.0mL

Preparation of Basal Medium: Add components to distilled/deionized water and bring volume to 900.0mL. Mix thoroughly. Gently heat and bring to boiling. Autoclave for 20 min at 15 psi pressure–121°C. Cool to 25°C.

Rabbit Serum with Supplements:
Composition per 106.0mL:

Rabbit serum	100.0mL
L-Asparagine (3% solution)	5.0mL
$MgCl_2$-$CaCl_2$ solution	1.0mL

Preparation of Rabbit Serum with Supplements: Combine the three solutions. Mix thoroughly. Filter sterilize.

$MgCl_2$-$CaCl_2$ Solution:
Composition per 100.0mL:

$CaCl_2 \cdot 2H_2O$	1.5g
$MgCl_2 \cdot 6H_2O$	1.5g

Preparation of $MgCl_2$-$CaCl_2$ Solution: Add components to distilled/deionized water and bring volume to 100.0mL. Mix thoroughly.

Preparation of Medium: Aseptically combine 900.0mL of cooled sterile basal medium and 100.0mL of sterile rabbit serum with supplements. Mix thoroughly. Aseptically distribute into sterile tubes or flasks.

Use: For the cultivation of *Leptospira* species.

Rainbow Agar O157

Composition per liter:
Proprietary.

Source: This medium is available as a premixed powder from Biolog Inc.

Preparation: Suspend 60.0g of the proprietary mixture in distilled/deionized water and bring volume to 1.0L Mix thoroughly. Gently heat and bring to boiling. Autoclave for 15 min at 15 psi pressure–121°C. Cool to 45-50°C. Mix thoroughly. Pour into sterile Petri dishes or distribute into sterile tubes. The final medium should be clear and virtually colorless. No pH adjustment is needed. The final pH should be pH 7.9-8.3. To increase the selectivity of the medium, a sterile solution containing 0.8 mg potassium tellurite and 10 mg novobiocin can be added. Caution must be used because tellurite is toxic.

Use: For the detection, isolation, and presumptive identification of verotoxin-producing strains of *Escherichia coli*, particularly serotype O157:H7. The medium contains chromogenic substrates that are specific for two *E. coli*-associated enzymes: β-galactosidase (a blue-black chromogenic substrate) and β-glucuronidase (a red chromogenic substrate). The distinctive black or gray coloration of *E. coli* O157:H7 colonies is easily viewed by laying the Petri plate against a white background. When O157 is surrounded by pink or magenta non-toxigenic colonies, it may have a bluish hue. The addition of selective agents improves performance. *E. coli* O157:H7 colony coloration will be slightly bluer with these selective agents added.Tellurite is highly selective for *E. coli* O157:H7 and can reduce background flora considerably. Novobiocin inhibits *Proteus* swarming and the growth of tellurite-reducing bacteria. Rare strains of O157:H7 are tellurite sensitives.

RAMBACH® Agar

Composition per liter:

NaC	15.0g
Agar	15.0g
Proplylene glycol	10.5g
Peptone	8.0g
Chromogenic mix	1.5g
Sodium deoxycholate	1.0g

pH 7.3 ± 0.2 at 25°C

Source: This medium is available from Merck.

Preparation of Medium: Add components to distilled/deionized water and bring volume to 1.0L. Mix thoroughly. Heat in a boiling water-batch or in a current of steam, while shaking from time to time. The medium is totally suspended, if no visual particles stick to the glass-wall. The medium should not be heat-treated further. Complete dissolution with shaking in 5-min. sequences is approximately 35-40 min. Do not autoclave. Do not overheat. Cool as fast as possible to 45°–50°C while gently shaking from time to time. Pour into sterile Petri dishes. To prevent any precipitate or clotting of the chromogenic-mix in the plates, place Petri dishes during pouring procedure on a cool (max. 25°C) surface. The plates are opaque and pink.

Use: For the detection of enteric bacteria, including coliforms and *Salmonella* spp. Sodium desoxycholate inhibits the accompanying Gram-positive flora. This medium enables *Salmonella* spp. to be differentiated unambiguously from other bacteria. *Salmonella* spp. form a characteristic red color. In order to differentiate coliforms from Salmonellae, the medium contains a chromogene indicating the presence of β-galactosidase splitting, a characteristic for coliforms. Coliform microorganisms grow as blue-green or blue-violet colonies. Other Enterobacteriaceae and Gram-negative bacteria, such as *Proteus, Pseudomo-*

nas, Shigella, S. typhi and S. parathyphi A grow as colorless-yellow colonies.

Rapid Fermentation Medium

Composition per liter:

Pancreatic digest of casein	20.0g
NaCl	5.0g
Agar	3.5g
L-Cystine	0.5g
Na_2SO_3	0.5g
Phenol Red	0.017g

pH 7.3 ± 0.2 at 25°C

Source: This medium is available as a premixed powder from BD Diagnostic Systems.

Preparation of Medium: Add components to distilled/deionized water and bring volume to 1.0L. Mix thoroughly. Distribute into tubes or flasks. Autoclave for 15 min at 15 psi pressure–121°C.

Use: For the differentiation of *Neisseria* species isolated from clinical specimens.

RAPID′*E. coli* 2 Agar

Composition per liter:

Proprietary

Source: This medium is available from Biorad.

Use: For the direct enumeration of *E.coli* and coliforms. Selectivity and electivity are based on the detection of glucuronidase and galactosidase activities. Hydrolysis of chromogenic substrate results in purple to pink *E.coli* colonies (gluc+/gal+) and blue-green coliform colonies (gluc-/gal+). RAPID′*E.coli* 2 agar is AFNOR validated according to ISO 16140 protocol to enumerate *E.coli* and coliforms on the same plate at 37°C, without any further confirmation of characteristic colonies.

RAPID′*Enterococcus* Agar

Composition per liter:

Proprietary

Source: This medium is available from Biorad.

Use: For the direct enumeration, without confirmation, of enterococci. The cleavage of the chromogenic substrate by glucosidase activity of Enterococci leads to specific blue colonies. RAPID′*Enterococcus* totally inhibits growth of Gram-negative flora and that of practically all Gram-positive bacteria other than Enterococci, due to the combined action of temperature and selective media.

RAPID′*L. mono* Medium

Composition per liter:

Peptones	30.0g
Agar B, proprietary	13.0g
D-Xylose	10.0g
LiCl	9.0g
Meat extract	5.0g
Yeast extract	1.0g
Phenol Red	0.12g
Selective supplement, proprietary	20.0g
Chromogenic substrate, proprietary	1.0mL

pH 7.3 ± 0.1 at 25°C

Source: This medium is available from Biorad.

Use: For the detection and differentiation of Listeria spp., including *L. ivanovii* and *L. monocytogenes*.

Rappaport Broth, Modified (Rap Broth, Modified)

Composition per 250.2mL:

Solution A	155.0mL
Solution C	53.0mL
Solution B	40.0mL
Solution D	1.6mL
Solution E	0.6mL

Solution A:

Composition per liter:

Pancreatic digest of casein	10.0g

Preparation of Solution A: Add pancreatic digest of casein to distilled/deionized water and bring volume to 1.0L. Mix thoroughly.

Solution B:

Composition per liter:

Na_2HPO_4	9.5g

Preparation of Solution B: Add Na_2HPO_4 to distilled/deionized water and bring volume to 1.0L. Mix thoroughly.

Solution C:

Composition per 100.0mL:

$MgCl_2{\cdot}6H_2O$	40.0g

Preparation of Solution C: Add $MgCl_2{\cdot}6H_2O$ to distilled/deionized water and bring volume to 100.0mL. Mix thoroughly. Autoclave for 15 min at 15 psi pressure–121°C. Cool to 25°C.

Solution D:

Composition per 100.0mL:

Malachite Green	0.2g

Preparation of Solution D: Add Malachite Green to sterile distilled/deionized water and bring volume to 100.0mL. Mix thoroughly. Do not sterilize.

Solution E:

Composition per 10.0mL:

Carbenicillin	0.01g

Preparation of Solution E: Add carbenicillin to distilled/deionized water and bring volume to 10.0mL. Mix thoroughly. Filter sterilize.

Preparation of Medium: Combine 155.0mL of solution A and 40.0mL of solution B. Mix thoroughly. Autoclave for 15 min at 15 psi pressure–121°C. Cool to 45°–50°C. Aseptically add 53.0mL of sterile solution C, 1.6mL of solution D, and 0.6mL of sterile solution E. Mix thoroughly. Aseptically distribute into sterile tubes or flasks.

Use: For the isolation and cultivation of *Yersinia enterocolitica*.

Rappaport-Vassiliadis Enrichment Broth (RV Enrichment Broth)

Composition per 1110.0mL:

NaCl	8.0g
Papaic digest of soybean meal	5.0g
KH_2PO_4	1.6g
Magnesium chloride solution	100.0mL
Malachite Green solution	10.0mL

pH 5.2 ± 0.2 at 25°C

Source: This medium is available as a premixed powder from Oxoid Unipath.

Magnesium Chloride Solution:

Composition per 100.0mL:

$MgCl_2 \cdot 6H_2O$	40.0g

Preparation of Magnesium Chloride Solution: Add $MgCl_2 \cdot 6H_2O$ to distilled/deionized water and bring volume to 100.0mL. Mix thoroughly. Autoclave for 15 min at 15 psi pressure–121°C. Cool to 45°–50°C.

Malachite Green Solution:

Composition per 10.0mL:

Malachite Green oxalate	0.04g

Preparation of Malachite Green Solution: Add Malachite Green to distilled/deionized water and bring volume to 10.0mL. Mix thoroughly. Autoclave for 15 min at 15 psi pressure–121°C. Cool to 45°–50°C.

Preparation of Medium: Add components to distilled/deionized water and bring volume to 1.0L. Mix thoroughly. Distribute into tubes in 10.0mL volumes. Autoclave for 15 min at 10 psi pressure–115°C.

Use: For the isolation and cultivation of *Salmonella* species.

Rappaport-Vassiliadis R10 Broth

Composition per liter:

$MgCl_2$, anhydrous	13.4g
NaCl	7.2g
Papaic digest of soybean meal	4.54g
KH_2PO_4	1.45g
Malachite Green oxalate	0.036g

pH 5.1 ± 0.2 at 25°C

Preparation of Medium: Add components to distilled/deionized water and bring volume to 1.0L. Mix thoroughly. Distribute into screw-capped tubes in 10.0mL volumes. Autoclave for 15 min at 10 psi pressure–116°C.

Use: For the isolation and cultivation of *Salmonella* species.

Rappaport-Vassiliadis Soya Peptone Broth (RVS Broth)

Composition per liter:

$MgCl_2$, anhydrous	13.58g
NaCl	7.2g
Papaic digest of soybean meal	4.5g
KH_2PO_4	1.26g
K_2HPO_4	0.18g
Malachite Green	0.036g

pH 5.2 ± 0.2 at 25°C

Source: This medium is available as a premixed powder from Oxoid Unipath.

Preparation of Medium: Add components to distilled/deionized water and bring volume to 1.0L. Mix thoroughly. Distribute into screw-capped tubes in 10.0mL volumes. Autoclave for 15 min at 10 psi pressure–115°C.

Use: For the isolation and cultivation of *Salmonella* species.

Reactivation with Tryptone Soya Broth (DSMZ Medium 220a)

Composition per liter:

Peptone from casein	17.0g
Peptone from soymeal	3.0g
NaCl	5.0g
D(+)-Glucose	2.5g
K_2HPO_4	2.5g

pH 7.3 ± 0.2 at 25°C

Preparation of Medium: Add components to distilled/deionized water and bring volume to 1.0L. Mix thoroughly. Gently heat and bring to boiling. Distribute into tubes or flasks. Autoclave for 15 min at 15 psi pressure–121°C.

Use: For the reactivation and cultivation of *Aeromonas* spp. *and Burkholderia* spp.

Reduced Transport Fluid

Composition per liter:

$(NH_4)_2SO_4$	9.0g
NaCl	9.0g
K_2HPO_4	4.5g
KH_2PO_4	4.5g
Na_2CO_3	4.0g
EDTA (ethylenediamine tetraacetic acid)	3.8g
Dithiothreitol	2.0g
$MgSO_4 \cdot 7H_2O$	1.8g

pH 8.0 ± 0.2 at 25°C

Preparation of Medium: Add components to distilled/deionized water and bring volume to 1.0L. Mix thoroughly. Filter sterilize. Aseptically distribute into sterile tubes with rubber stoppers.

Use: For the transport and isolation of bacteria from dental plaque, especially *Streptococcus mutans* and *Streptococcus sanguis*. Also used for the cultivation of a variety of Gram-positive bacteria from the oral cavity, especially streptococci, actinomycetes, lactobacilli, clostridia, *Bacteroides* species, *Fusobacterium* species, and *Veillonella* species.

Reduced Transport Fluid

Composition per liter:

Stock mineral salt solution no. 1	75.0mL
Stock mineral salt solution no. 2	75.0mL
Dithiothreitol (1% solution)	20.0mL
Ethylenediamine tetraacetic acid (1*M* solution)	10.0mL
Na_2CO_3 (8% solution)	5.0mL
Resazurin (0.1% solution)	1.0mL

pH 8.0 ± 0.2 at 25°C

Stock Mineral Salt Solution No. 1:
Composition per 100.0mL:

K_2HPO_4	0.6g

Preparation of Stock Mineral Salt Solution No. 1: Add K_2HPO_4 to distilled/deionized water and bring volume to 100.0mL. Mix thoroughly.

Stock Mineral Salt Solution No. 2:
Composition per 100.0mL:

NaCl	1.2g
$(NH_4)_2SO_4$	1.2g
K_2HPO_4	0.6g
$MgSO_4 \cdot 7H_2O$	0.25g

Preparation of Stock Mineral Salt Solution No. 2: Add components to distilled/deionized water and bring volume to 100.0mL. Mix thoroughly.

Preparation of Medium: Add components to distilled/deionized water and bring volume to 1.0L. Mix thoroughly. Filter sterilize. Aseptically distribute into sterile tubes with rubber stoppers.

Use: For the transport and isolation of bacteria from dental plaque, especially *Streptococcus mutans* and *Streptococcus sanguis*. Also used for the cultivation of a variety of Gram-positive bacteria from the oral cavity, especially streptococci, actinomycetes, lactobacilli, clostrida, *Bacteroides*, Fusobacteria, and *Veillonela*.

Regan-Lowe Charcoal Agar (Regan-Lowe Medium)

Composition per liter:

Agar	12.0g
Beef extract	10.0g
Pancreatic digest of gelatin	10.0g
Soluble starch	10.0g
NaCl	5.0g
Charcoal	4.0g
Niacin	0.01g
Horse blood, defibrinated	100.0mL
Cephalexin solution	10.0mL

pH 7.4 ± 0.2 at 25°C

Source: This medium is available as a premixed powder from BD Diagnostic Systems.

Cephalexin Solution:
Composition per 10.0mL:

Cephalexin	0.04g

Preparation of Cephalexin Solution: Add cephalexin to distilled/deionized water and bring volume to 10.0mL. Mix thoroughly. Filter sterilize.

Preparation of Medium: Add components, except horse blood and cephalexin solution, to distilled/deionized water and bring volume to 890.0mL. Mix thoroughly. Gently heat and bring to boiling. Autoclave for 15 min at 15 psi pressure–121°C. Cool to 45°–50°C. Aseptically add sterile horse blood and sterile cephalexin solution. Mix thoroughly. Pour into sterile Petri dishes or distribute into sterile tubes. Swirl medium while dispensing to keep charcoal in suspension.

Use: For the selective isolation and cultivation of *Bordetella pertussis* and *Bordetella parapertussis* from clinical specimens.

Regan-Lowe Semisolid Transport Medium

Composition per liter:

Agar	6.0g
Beef extract	5.0g
Pancreatic digest of gelatin	5.0g
Soluble starch	5.0g
NaCl	2.5g
Charcoal	2.0g

Niacin	0.01g
Horse blood, defibrinated	100.0mL
Cephalexin solution	10.0mL

pH 7.4 ± 0.2 at 25°C

Cephalexin Solution:
Composition per 10.0mL:

Cephalexin	0.04g

Preparation of Cephalexin Solution: Add cephalexin to distilled/deionized water and bring volume to 10.0mL. Mix thoroughly. Filter sterilize.

Preparation of Medium: Add components, except horse blood and cephalexin solution, to distilled/deionized water and bring volume to 890.0mL. Mix thoroughly. Gently heat and bring to boiling. Autoclave for 15 min at 15 psi pressure–121°C. Cool to 45°–50°C. Aseptically add sterile horse blood and sterile cephalexin solution. Mix thoroughly. Aseptically distribute into small, sterile, screw-capped tubes. Fill tubes half-full. Swirl medium while dispensing to keep charcoal in suspension.

Use: For the transport of *Bordetella pertussis* and *Bordetella parapertussis* isolated from clinical specimens.

Reinforced Clostridial Agar

Composition per liter:

Agar	13.5g
Beef extract	10.0g
Pancreatic digest of casein	10.0g
NaCl	5.0g
Glucose	5.0g
Yeast extract	3.0g
Sodium acetate	3.0g
Soluble starch	1.0g
L-Cysteine·HCl·H_2O	0.5g

pH 6.8 ± 0.2 at 25°C

Source: This medium is available as a premixed powder from BD Diagnostic Systems and Oxoid Unipath.

Preparation of Medium: Add components to distilled/deionized water and bring volume to 1.0L. Mix thoroughly. Gently heat and bring to boiling. Distribute into tubes or flasks. Autoclave for 15 min at 10 psi pressure–115°C. Pour into sterile Petri dishes or leave in tubes.

Use: For the cultivation and enumeration of *Clostridium* species, *Bifidobacterium* species, other anaerobes (e.g., lactobacilli), and facultative organisms from clinical specimens.

Reinforced Clostridial Medium

Composition per liter:

Tryptose	10.0g
Beef extract	10.0g
Glucose	5.0g
NaCl	5.0g
Yeast extract	3.0g
Sodium acetate	3.0g
Soluble starch	1.0g
L-Cysteine·HCl·H_2O	0.5g
Agar	0.5g

pH 6.8 ± 0.2 at 25°C

Source: This medium is available as a premixed powder from BD Diagnostic Systems and Oxoid Unipath.

Preparation of Medium: Add components to distilled/deionized water and bring volume to 1.0L. Mix thoroughly. Gently heat and bring to boiling. Distribute into tubes or flasks. Autoclave for 15 min at 10 psi pressure–115°C. Pour into sterile Petri dishes or leave in tubes.

Use: For the nonselective cultivation and enumeration of *Clostridium* species, other anaerobes such as lactobacilli, and facultative organisms from clinical specimens.

Reinforced Clostridial Medium with Sodium Lactate

Composition per liter:

Tryptose	10.0g
Beef extract	10.0g
Glucose	5.0g
NaCl	5.0g
Yeast extract	3.0g
Sodium acetate	3.0g
Soluble starch	1.0g
L-Cysteine·HCl·H_2O	0.5g
Agar	0.5g
Sodium lactate (60% solution)	15.0mL

pH 6.8 ± 0.2 at 25°C

Preparation of Medium: Add components to distilled/deionized water and bring volume to 1.0L. Mix thoroughly. Gently heat and bring to boiling. Distribute into tubes or flasks. Autoclave for 15 min at 10 psi pressure–115°C. Pour into sterile Petri dishes or leave in tubes.

Use: For the nonselective cultivation and enumeration of *Clostridium* species, other anaerobes such as lactobacilli, from clinical specimens.

Rice Extract Agar

Composition per liter:

Agar	20.0g
White rice, solids from extract	5.0g
Polysorbate 80	10.0mL

pH 6.6 ± 0.2 at 25°C

Source: This medium is available as a premixed powder from BD Diagnostic Systems.

Preparation of Medium: Add components, except polysorbate 80, to distilled/deionized water and bring volume to 990.0mL. Mix thoroughly. Gently heat and bring to boiling. Add polysorbate 80. Mix thoroughly. Distribute into tubes or flasks. Autoclave for 15 min at 15 psi pressure–121°C. Pour into sterile Petri dishes.

Use: For the cultivation and differentiation of *Candida albicans* and *Candida stellatoidea* from other *Candida* species based on chlamydospore formation.

Rice Extract Agar

Composition per liter:

Agar	20.0g
White rice, solids from extract	20.0g

pH 7.1 ± 0.2 at 25°C

Source: This medium is available as a premixed powder from BD Diagnostic Systems.

Preparation of Medium: Add components, except polysorbate 80, to distilled/deionized water and bring volume to 990.0mL. Mix thoroughly. Gently heat and bring to boiling. Add polysorbate 80. Mix thoroughly. Distribute into tubes or flasks. Autoclave for 15 min at 15 psi pressure–121°C. Pour into sterile Petri dishes.

Use: For the cultivation and differentiation of *Candida albicans* and *Candida stellatoidea* from other *Candida* species based on chlamydospore formation.

Rimler-Shotts Medium (RS Medium)

Composition per liter:

Agar	13.5g
$Na_2S_2O_3 \cdot 5H_2O$	6.8g
L-Ornithine·HCl	6.5g
NaCl	5.0g
L-Lysine·HCl	5.0g
Maltose	3.5g
Yeast extract	3.0g
Sodium deoxycholate	1.0g
Ferric ammonium citrate	0.8g
L-Cysteine·HCl	0.3g
Bromthymol Blue	0.03g
Novobiocin solution	10.0mL

pH 7.0 ± 0.2 at 25°C

Novobiocin Solution:

Composition per 10.0mL:

Novobiocin	5.0mg

Preparation of Novobiocin Solution: Add novobiocin to distilled/deionized water and bring volume to 10.0mL. Mix thoroughly. Filter sterilize.

Preparation of Medium: Add components, except novobiocin solution, to distilled/deionized water and bring volume to 990.0mL. Mix thoroughly. Gently heat and bring to boiling. Autoclave for 15 min at 15 psi pressure–121°C. Cool to 45°–50°C. Aseptically add sterile components. Mix thoroughly. Pour into sterile Petri dishes or distribute into sterile tubes.

Use: For the selective isolation, cultivation, and presumptive identification of *Aeromonas hydrophila* and other Gram-negative bacteria based on their ability to decarboxylate lysine and ornithine, ferment maltose, and produce H_2S. Maltose-fermenting bacteria appear as yellow colonies. Bacteria that produce lysine or ornithine decarboxylase turn the medium greenish-yellow to yellow. Bacteria that produce H_2S appear as colonies with black centers.

RIOT Agar (Rice Infusion Oxgall Tween 80 Agar)

Composition per 1010.0mL:

Agar	10.0g
Oxgall	10.0g
Rice extract	1.0L
Tween 80	10.0mL

pH 7.3 ± 0.2 at 25°C

Rice Extract:

Composition per liter:

Cream of rice cereal	10.0g

Preparation of Rice Extract: Add cream of rice cereal to 1.0L of boiling tap water. Mix thoroughly. Filter quickly through cheesecloth. Bring volume of filtrate to 1.0L with tap water.

Preparation of Medium: Combine components. Mix thoroughly. Gently heat and bring to boiling. Distribute into tubes or flasks. Autoclave for 15 min at 15 psi pressure–121°C. Pour into sterile Petri dishes or leave in tubes.

Use: For the cultivation and differentiation of *Candida albicans* and *Candida stellatoidea* from other *Candida* species based on chlamydospore formation.

Rogosa Agar

Composition per liter:

Sodium acetate	25.0g
Agar	20.0g
Glucose	20.0g
Pancreatic digest of casein	10.0g
KH_2PO_4	6.0g
Yeast extract	5.0g

Ammonium citrate 2.0g
Sorbitan monooleate 1.0g
$MgSO_4 \cdot 7H_2O$ 0.575g
$MnSO_4 \cdot H_2O$ 0.12g
$FeSO_4 \cdot 7H_2O$ 0.4mg
Acetic acid, glacial 1.32mL

pH 5.4 ± 0.2 at 25°C

Source: This medium is available as a premixed powder from Oxoid Unipath.

Preparation of Medium: Add components, except acetic acid, to distilled/deionized water and bring volume to 998.7mL. Mix thoroughly. Gently heat and bring to boiling. Add glacial acetic acid. Mix thoroughly. Gently heat while stirring and bring to 90°–100°C for 2–3 min. Do not autoclave. Pour into sterile Petri dishes or distribute into sterile tubes.

Use: For the isolation, cultivation, and enumeration of lactobacilli, especially from feces, saliva, and vaginal specimens.

Rogosa SL Agar (Rogosa Selective *Lactobacillus* Agar)

Composition per liter:

Agar 15.0g
Sodium acetate 15.0g
Glucose 10.0g
Pancreatic digest of casein 10.0g
K_2HPO_4 .... 6.0g
Yeast extract 5.0g
Arabinose 5.0g
Sucrose 5.0g
Ammonium citrate 2.0g
Sorbitan monooleate 1.0g
$MgSO_4 \cdot 7H_2O$ 0.57g
$MnSO_4 \cdot 7H_2O$ 0.12g
$FeSO_4 \cdot H_2O$ 0.03g
Acetic acid, glacial 1.32mL

pH 5.4 ± 0.2 at 25°C

Source: This medium is available as a premixed powder from BD Diagnostic Systems.

Preparation of Medium: Add components, except glacial acetic acid, to distilled/deionized water and bring volume to 998.7mL. Mix thoroughly. Gently heat and bring to boiling. Add glacial acetic acid. Mix thoroughly. Gently heat while stirring and bring to 90°–100°C for 2–3 min. Do not autoclave. Pour into sterile Petri dishes or distribute into sterile tubes.

Use: For the isolation, cultivation, and enumeration of lactobacilli, especially from feces, saliva, and vaginal specimens.

Rogosa SL Broth (Rogosa Selective *Lactobacillus* Broth)

Composition per liter:

Sodium acetate 15.0g
Glucose 10.0g
Pancreatic digest of casein 10.0g
K_2HPO_4 .... 6.0g
Yeast extract 5.0g
Arabinose 5.0g
Sucrose 5.0g
Ammonium citrate 2.0g
Sorbitan monooleate 1.0g
$MgSO_4 \cdot 7H_2O$ 0.57g
$MnSO_4 \cdot 7H_2O$ 0.12g
$FeSO_4 \cdot H_2O$ 0.03g
Acetic acid, glacial 1.32mL

pH 5.4 ± 0.2 at 25°C

Source: This medium is available as a premixed powder from BD Diagnostic Systems.

Preparation of Medium: Add components, except glacial acetic acid, to distilled/deionized water and bring volume to 998.7mL. Mix thoroughly. Gently heat and bring to boiling. Add glacial acetic acid. Mix thoroughly. Gently heat while stirring and bring to 90°–100°C for 2–3 min. Do not autoclave. Aseptically distribute into sterile tubes.

Use: For the isolation, cultivation, and enumeration of lactobacilli, especially from feces, saliva, and vaginal specimens.

RPMI 1640 Medium with L-Glutamine

Composition per liter:

NaCl 6.0g
$NaHCO_3$ 2.0g
D-Glucose 2.0g
$Na_2HPO_4 \cdot 7H_2O$ 1.5g
KCl 0.4g
L-Glutamine 0.3g
L-Arginine 0.2g
$Ca(NO_3)_2 \cdot 4H_2O$ 0.1g
$MgSO_4 \cdot 7H_2O$ 0.1g
L-Asparagine 0.05g
L-Cystine 0.05g
L-Isoleucine, allo free 0.05g
L-Leucine, methionine free 0.05g
L-Lysine·HCl 0.04g
i-Inositol 0.035g
L-Serine 0.03g
L-Aspartic acid 0.02g
L-Glutamic acid 0.02g
L-Hydroxyproline 0.02g
L-Proline, hydroxy-L-proline free 0.02g
L-Threonine, allo free 0.02g

L-Tyrosine	0.02g
L-Valine	0.02g
L-Histidine, free base	0.015g
L-Methionine	0.015g
L-Phenylalanine	0.015g
Glycine	0.01g
L-Tryptophan	5.0mg
Phenol Red	5.0mg
Choline chloride	3.0mg
Glutathione, reduced	1.0mg
p-Aminobenzoic acid	1.0mg
Folic acid	1.0mg
Nicotinamide	1.0mg
Pyridoxine·HCl	1.0mg
Thiamine·HCl	1.0mg
D-Calcium pantothenate	0.25mg
Biotin	0.2mg
Riboflavin	0.2mg
Vitamin B_{12}	5.0μg

pH 7.3 ± 0.2 at 25°C

Preparation of Medium: Add components to distilled/deionized water and bring volume to 1.0L. Adjust pH to 7.3 with 1*N* HCl or 1*N* NaOH. Filter sterilize. Aseptically distribute into sterile tubes or flasks.

Use: For the cultivation of mammalian cells in tissue culture. Culture media for human immunodeficiency viruses.

Russell Double-Sugar Agar

Composition per liter:

Agar	15.0g
Proteose peptone no. 3	12.0g
Lactose	10.0g
NaCl	5.0g
Beef extract	1.0g
Glucose	1.0g
Phenol Red	0.025g

pH 7.5 ± 0.2 at 25°C

Preparation of Medium: Add components to distilled/deionized water and bring volume to 1.0L. Mix thoroughly. Gently heat and bring to boiling. Distribute into tubes. Autoclave for 15 min at 15 psi pressure–121°C. Allow tubes to cool in a slanted position.

Use: For the identification of Gram-negative enteric bacilli based on their fermentation of glucose and lactose. Bacteria that ferment both glucose and lactose produce a yellow slant and yellow butt. Bacteria that ferment glucose but do not ferment lactose produce a red slant and a yellow butt. Bacteria that ferment neither glucose nor lactose produce an unchanged pink-orange color.

SABHI Agar (Sabouraud Glucose and Brain Heart Infusion Agar)

Composition per liter:

Glucose	21.0g
Agar	15.0g
Pancreatic digest of casein	10.5g
Peptic digest of animal tissue	5.0g
Brain heart, solids from infusion	4.0g
NaCl	2.5g
Na_2HPO_4	1.25g

pH 6.8 ± 0.2 at 25°C

Source: This medium is available as a premixed powder from BD Diagnostic Systems.

Preparation of Medium: Add components to distilled/deionized water and bring volume to 1.0L. Mix thoroughly. Gently heat and bring to boiling. Distribute into tubes or flasks. Autoclave for 15 min at 15 psi pressure–121°C. Pour into sterile Petri dishes in 20.0mL volumes or leave in tubes.

Use: For the cultivation of dermatophytes and other pathogenic and nonpathogenic fungi from clinical and nonclinical specimens.

SABHI Agar

Composition per liter:

Beef heart, infusion from	125.0g
Calf brains, infusion from	100.0g
Glucose	21.0g
Agar	15.0g
Neopeptone	5.0g
Proteose peptone	5.0g
NaCl	2.5g
Na_2HPO_4	1.25g
Chloromycetin solution	1.0mL

pH 7.0 ± 0.2 at 25°C

Source: This medium is available as a premixed powder from BD Diagnostic Systems.

Chloromycetin Solution:

Composition per 10.0mL:

Chloromycetin	1.0g

Preparation of Chloromycetin Solution: Add chloromycetin to distilled/deionized water and bring volume to 10.0mL. Mix thoroughly. Filter sterilize.

Preparation of Medium: Add components, except chloromycetin solution, to distilled/deionized water and bring volume to 999.0mL. Mix thoroughly. Gently heat and bring to boiling. Autoclave for 15 min at 15 psi pressure–121°C. Cool to 45°–50°C. Aseptically add 1.0mL of sterile chloromycetin solution. Mix thoroughly. Aseptically distribute into sterile tubes in 5.0mL volumes.

Use: For the cultivation of dermatophytes and other pathogenic and nonpathogenic fungi from clinical and nonclinical specimens.

SABHI Agar, Modified

Composition per liter:

Beef heart, infusion from	62.5g
Calf brain, infusion from	50.0g
Glucose	20.5g
Brain heart infusion broth	18.6g
Agar	7.5g
Neopeptone	5.0g
Pancreatic digest of gelatin	2.5g
NaCl	1.25g
Na_2HPO_4	0.625g

pH 6.8 ± 0.2 at 25°C

Preparation of Medium: Dissolve, then autoclave at 15 psi pressure–121°C for 15 min. Cool to 50°C and add 1.0mL of sterile chloramphenicol solution (100.0mg/mL). Mix well and dispense into sterile tubes. Slant and allow to harden. Refrigerate until needed.

Use: For the cultivation of dermatophytes and other pathogenic and nonpathogenic fungi from clinical and nonclinical specimens.

SABHI Blood Agar

Composition per liter:

Beef heart, infusion from	125.0g
Calf brains, infusion from	100.0g
Glucose	21.0g
Agar	15.0g
Neopeptone	5.0g
Proteose peptone	5.0g
NaCl	2.5g
Na_2HPO_4	1.25g
Blood	100.0mL
Chloromycetin solution	1.0mL

pH 7.0 ± 0.2 at 25°C

Source: This medium is available as a premixed powder from BD Diagnostic Systems.

Chloromycetin Solution:
Composition per 10.0mL:

Chloromycetin	1.0g

Preparation of Chloromycetin Solution: Add chloromycetin to distilled/deionized water and bring volume to 10.0mL. Mix thoroughly. Filter sterilize.

Preparation of Medium: Add components, except blood and chloromycetin solution, to distilled/deionized water and bring volume to 899.0mL. Mix thoroughly. Gently heat and bring to boiling. Autoclave for 15 min at 15 psi pressure–121°C. Cool to 45°–50°C. Aseptically add 100.0mL of sterile blood and 1.0mL of sterile chloromycetin solution. Sheep blood or human blood may be used. Mix thoroughly. Aseptically distribute into sterile tubes in 5.0mL volumes.

Use: For the cultivation of dermatophytes and other pathogenic and nonpathogenic fungi from clinical and nonclinical specimens. Blood enhances the recovery of *Blastomyces dermatitidis* and *Histoplasma capsulatum* and their conversion to the yeast phase.

Sabouraud Agar

Composition per liter:

Neopeptone	30.0g
Agar	20.0g

Preparation of Medium: Add components to tap water and bring volume to 1.0L. Mix thoroughly. Gently heat and bring to boiling. Distribute into tubes or flasks. Autoclave for 15 min at 15 psi pressure–121°C. Pour into sterile Petri dishes or leave in tubes.

Use: For the cultivation of yeasts and molds.

Sabouraud Agar with CCG and 3% NaCl

Composition per 3031.5mL:

Glucose	120.0g
NaCl	90.0g
Agar	45.0g
Peptone	30.0g
Chloramphenicol solution	15.0mL
Cycloheximide solution	15.0mL
Gentamicin solution	1.5mL

Chloramphenicol Solution:
Composition per 15.0mL:

Chloramphenicol	0.15g

Preparation of Chloramphenicol Solution: Add chloramphenicol to distilled/deionized water and bring volume to 15.0mL. Mix thoroughly. Filter sterilize.

Cycloheximide Solution:
Composition per 15.0mL:

Cycloheximide	0.3g

Preparation of Cycloheximide Solution: Add cycloheximide to distilled/deionized water and bring volume to 15.0mL. Mix thoroughly. Filter sterilize.

Gentamicin Solution:
Composition per 10.0mL:

Gentamicin	0.4g

Preparation of Gentamicin Solution: Add gentamicin to distilled/deionized water and bring volume to 10.0mL. Mix thoroughly. Filter sterilize.

Preparation of Medium: Add components—except chloramphenicol solution, cycloheximide solution, and gentamicin solution—to distilled/deionized water and bring volume to 3.0L. Mix thoroughly. Gently heat and bring to boiling. Autoclave for 15 min at 15 psi pressure–121°C. Cool to 45°–50°C. Aseptically add 15.0mL of sterile chloramphenicol solution, 15.0mL of sterile cycloheximide solution, and 1.5mL of sterile gentamicin solution. Mix thoroughly. Aseptically distribute into sterile tubes. Allow tubes to cool in a slanted position.

Use: For the selective isolation and cultivation of fungi from specimens with a mixed flora.

Sabouraud Agar with CCG and 5% NaCl

Composition per 3031.5mL:

NaCl	150.0g
Glucose	120.0g
Agar	45.0g
Peptone	30.0g
Chloramphenicol solution	15.0mL
Cycloheximide solution	15.0mL
Gentamicin solution	1.5mL

Chloramphenicol Solution:
Composition per 15.0mL:

Chloramphenicol	0.15g

Preparation of Chloramphenicol Solution: Add chloramphenicol to distilled/deionized water and bring volume to 15.0mL. Mix thoroughly. Filter sterilize.

Cycloheximide Solution:
Composition per 15.0mL:

Cycloheximide	0.3g

Preparation of Cycloheximide Solution: Add cycloheximide to distilled/deionized water and bring volume to 15.0mL. Mix thoroughly. Filter sterilize.

Gentamicin Solution:
Composition per 10.0mL:

Gentamicin	0.4g

Preparation of Gentamicin Solution: Add gentamicin to distilled/deionized water and bring volume to 10.0mL. Mix thoroughly. Filter sterilize.

Preparation of Medium: Add components—except chloramphenicol solution, cycloheximide solution, and gentamicin solution—to distilled/deionized water and bring volume to 3.0L. Mix thoroughly. Gently heat and bring to boiling. Autoclave for 15 min at 15 psi pressure–121°C. Cool to 45°–50°C. Aseptically add 15.0mL of sterile chloramphenicol solution, 15.0mL of sterile cycloheximide solution, and 1.5mL of sterile gentamicin solution. Mix thoroughly. Aseptically distribute into sterile tubes. Allow tubes to cool in a slanted position.

Use: For the selective isolation and cultivation of fungi from specimens with a mixed flora.

Sabouraud Glucose Agar, Emmons

Composition per liter:

Glucose	20.0g
Agar	17.0g
Pancreatic digest of casein	5.0g
Peptic digest of animal tissue	5.0g

pH 6.9 ± 0.2 at 25°C

Source: This medium is available as a premixed powder from BD Diagnostic Systems.

Preparation of Medium: Add components to distilled/deionized water and bring volume to 1.0L. Mix thoroughly. Gently heat and bring to boiling. Distribute into tubes or flasks. Autoclave for 15 min at 13 psi pressure–118°C. Pour into sterile Petri dishes or leave in tubes.

Use: For the cultivation of dermatophytes and other pathogenic and nonpathogenic fungi from clinical and nonclinical specimens. For the cultivation of yeast and filamentous fungi.

Sabouraud Glucose Broth

Composition per liter:

Glucose	20.0g
Neopeptone	10.0g

pH 5.6 ± 0.2 at 25°C

Source: This medium is available as a premixed powder from BD Diagnostic Systems.

Preparation of Medium: Add components to distilled/deionized water and bring volume to 1.0L. Mix thoroughly. Distribute into tubes or flasks. Autoclave for 15 min at 15 psi pressure–121°C. Avoid overheating.

Use: For the cultivation of pathogenic and nonpathogenic fungi, especially dermatophytes. The medium may be made more selective for fungi by the addition of chloramphenicol.

Sabouraud Maltose Agar

Composition per liter:

Maltose	40.0g
Agar	15.0g
Pancreatic digest of casein	5.0g
Peptic digest of animal tissue	5.0g

pH 5.6 ± 0.2 at 25°C

Source: This medium is available as a premixed powder from BD Diagnostic Systems and Oxoid Unipath.

Preparation of Medium: Add components to distilled/deionized water and bring volume to 1.0L. Mix thoroughly. Gently heat and bring to boiling. Distribute into tubes or flasks. Autoclave for 15 min at 15 psi pressure–121°C. Avoid overheating. Pour into sterile Petri dishes or leave in tubes.

Use: For the cultivation and maintenance of a variety of fungi.

Sabouraud Maltose Broth

Composition per liter:

Maltose	40.0g
Neopeptone	10.0g

pH 5.6 ± 0.2 at 25°C

Source: This medium is available as a premixed powder from BD Diagnostic Systems.

Preparation of Medium: Add components to distilled/deionized water and bring volume to 1.0L. Mix thoroughly. Distribute into tubes or flasks. Autoclave for 15 min at 15 psi pressure–121°C. Avoid overheating.

Use: For the cultivation of a variety of fungi.

Sabouraud Medium, Fluid

Composition per liter:

Glucose	20.0g
Pancreatic digest of casein	5.0g
Peptamin	5.0g

pH 5.7 ± 0.2at 25°C

Source: This medium is available as a premixed powder from BD Diagnostic Systems and Oxoid Unipath.

Preparation of Medium: Add components to distilled/deionized water and bring volume to 1.0L. Mix thoroughly. Distribute into tubes or flasks. Autoclave for 15 min at 15 psi pressure–121°C. Avoid overheating.

Use: For the isolation and cultivation of yeasts and molds.

Salmonella Chromogen Agar (Rambach Equivalent Agar)

Composition per liter:

Agar	15.0g
Peptone	5.0g
NaCl	5.0g
Yeast extract	2.0g
Meat extract	1.0g
Sodium deoxycholate	1.0g

pH 7.3 ± 0.2 at 25°C

Source: This medium is available from Fluka, Sigma-Aldrich.

Preparation of Medium: Add components to distilled/deionized water and bring volume to 1.0L. Mix thoroughly. Gently heat while stirring and bring to boiling. Autoclave for 15 min at 15 psi pressure–121°C. Pour into sterile Petri dishes.

Use: For the detection of *Salmonella* spp.

Salmonella Chromogenic Agar

Composition per liter:

Chromogenic mix	28.0g
Agar	12.0g
Special peptone	10.0g
Selective supplement solution	10.0mL

pH 7.2 ± 0.2 at 25°C

Source: This medium is available as a premixed powder from Oxoid Unipath.

Selective Supplement Solution:

Composition per 10.0mL:

Cefsulodin	12.0mg
Novobiocin	5.0mg

Preparation of Selective Supplement Solution: Add components to distilled/deionized water and bring volume to 10.0mL. Mix thoroughly. Filter sterilize.

Preparation of Medium: Add components to distilled/deionized water and bring volume to 1.0mL Mix thoroughly. Gently heat while stirring and bring to boiling. Do not autoclave. Cool quickly to 50°C. Mix thoroughly. Pour into sterile Petri dishes.

Use: For the identification of *Salmonella* species and differentiation of *Salmonella* spp. other organisms in the family Enterobacteriaceae. This medium combines two chromogens for the detection of *Salmonella* spp., 5-bromo-6-chloro-3-indolyl caprylate (Magenta-caprylate) and 5-bromo-4-chloro-3-Indolyl β-D galactopyranoside (X-gal). X-gal is a substrate for the enzyme β-D-galactosidase. Hydrolysis of the chromogen, Mag-caprylate, by lactose negative *Salmonella* species results in magenta colonies. The medium contains bile salts to inhibit the growth of Gram-positive organisms and the addition of the selective supplement solution increases the selectivity of the medium. Novobiocin inhibits *Proteus* growth and cefsulodin inhibits growth of pseudomonads.

Salmonella Rapid Test Elective Medium

Composition per liter:

Tryptone	10.0g
Na_2HPO_4	9.0g
Sodium chloride	5.0g
Casein	5.0g
KH_2PO_4	1.5g
Malachite Green	0.0025g

pH 6.5 ± 0.2 at 25°C

Preparation of Medium: Add components to distilled/deionized water and bring volume to 1.0L. Mix thoroughly. Autoclave for 15 min at 15 psi pressure–121°C.

Use: For the Oxoid *Salmonella* Rapid Test which is for the presumptive detection of motile *Salmonella.*

Salmonella Rapid Test Elective Medium, 2X

Composition per liter:

Tryptone	20.0g
Na_2HPO_4	18.0g
Sodium chloride	10.0g
Casein	10.0g
KH_2PO_4	3.0g
Malachite Green	0.005g

pH 6.5 ± 0.2 at 25°C

Preparation of Medium: Add components to distilled/deionized water and bring volume to 1.0L. Mix thoroughly. Autoclave for 15 min at 15 psi pressure–121°C.

Use: For the presumptive detection of motile *Salmonella.*

Salmonella Shigella Agar (SS Agar)

Composition per liter:

Agar	13.5g
Lactose	10.0g
Bile salts	8.5g
$Na_2S_2O_3$	8.5g
Sodium citrate	8.5g
Beef extract	5.0g
Pancreatic digest of casein	2.5g
Peptic digest of animal tissue	2.5g
Ferric citrate	1.0g
Neutral Red	0.025g
Brilliant Green	0.33mg

pH 7.0 ± 0.2 at 25°C

Source: This medium is available as a premixed powder from BD Diagnostic Systems and Oxoid Unipath.

Preparation of Medium: Add components to distilled/deionized water and bring volume to 1.0L. Mix thoroughly. Gently heat while stirring and bring to boiling. Do not autoclave. Cool to 45°–50°C. Pour into sterile Petri dishes in 20.0mL volumes. Allow the surface of the plates to dry before inoculation.

Use: For the selective isolation and differentiation of pathogenic enteric bacilli, especially those belonging to the genus *Salmonella.* This medium is not recommended for the primary isolation of *Shigella* species. Lactose-fermenting bacteria such as *Escherichia coli* or *Klebsiella pneumoniae* appear as small pink or red colonies. Lactose-nonfermenting bacteria—such as *Salmonella* species, *Proteus* species, and *Shigella* species—appear as colorless colonies. Production of H_2S by *Salmonella* species turns the center of the colonies black.

Salmonella Shigella Agar, Modified (SS Agar, Modified)

Composition per liter:

Agar	12.0g
Lactose	10.0g
Sodium citrate	10.0g
$Na_2S_2O_3$	8.5g
Bile salts	5.5g
Beef extract	5.0g
Peptone	5.0g
Ferric citrate	1.0g
Neutral Red	0.025g
Brilliant Green	0.33mg

pH 7.3 ± 0.2 at 25°C

Source: This medium is available as a premixed powder from Oxoid Unipath.

Preparation of Medium: Add components to distilled/deionized water and bring volume to 1.0L. Mix thoroughly. Gently heat while stirring and bring to boiling. Do not autoclave. Cool to 45°–50°C. Pour into sterile Petri dishes in 20.0mL volumes. Allow the surface of the plates to dry before inoculation.

Use: For the selective isolation and differentiation of pathogenic enteric bacilli, especially those belonging to the genus *Salmonella.* This medium provides better growth of *Shigella* species. Lactose-fermenting bacteria such as *Escherichia coli* or *Klebsiella pneumoniae* appear as small pink or red colonies. Lactose-nonfermenting bacteria—such as *Salmonella* species, *Proteus* species, and *Shigella* species—appear as colorless colonies. Production of H_2S by *Salmonella* species turns the center of the colonies black.

Salt Meat Broth

Composition per liter:

NaCl	100.0g

Neutral ox-heart tissue......30.0g
Beef extract......10.0g
Peptone......10.0g

pH 7.6 ± 0.2 at 25°C

Source: This medium is available as tablets from Oxoid Unipath.

Preparation of Medium: Add components to distilled/deionized water and bring volume to 1.0L. Mix thoroughly. Distribute into tubes or flasks. Autoclave for 15 min at 15 psi pressure–121°C.

Use: For the isolation and cultivation of staphylococci from specimens with a mixed flora such as fecal specimens, especially during the investigation of staphylococcal food poisoning.

Salt Tolerance Medium

Composition per liter:

Beef heart, infusion from......500.0g
NaCl......65.0g
Tryptose......10.0g
Glucose......1.0g
Indicator solution......1.0mL

pH 7.4 ± 0.2 at 25°C

Indicator Solution:

Composition per 100.0mL:

Bromcresol Purple......1.6g
Ethanol (95% solution)......100.0mL

Preparation of Indicator Solution: Add Bromcresol Purple to ethanol. Mix thoroughly.

Preparation of Medium: Add components to distilled/deionized water and bring volume to 1.0L. Mix thoroughly. Distribute into tubes or flasks. Autoclave for 15 min at 15 psi pressure–121°C.

Use: For the cultivation of salt-tolerant *Streptococcus* species and other salt-tolerant Gram-positive cocci. For the differentiation of Gram-positive cocci based on salt tolerance.

Salt Tolerance Medium

Composition per liter:

NaCl......60.0g
Peptone......5.0g
Yeast extract......2.0g
Beef extract......1.0g

pH 7.4 ± 0.2 at 25°C

Preparation of Medium: Add components to distilled/deionized water and bring volume to 1.0L. Mix thoroughly. Distribute into tubes or flasks. Autoclave for 15 min at 15 psi pressure–121°C.

Use: For the cultivation and differentiation of *Aeromonas* and *Plesiomonas* species based on salt tolerance.

Salt Tolerance Medium

Composition per liter:

Beef heart, solids from infusion......500.0g
NaCl......65.0g
Tryptose......10.0g

pH 7.4 ± 0.2 at 25°C

Preparation of Medium: Add components to distilled/deionized water and bring volume to 1.0L. Mix thoroughly. Distribute into tubes or flasks. Autoclave for 15 min at 15 psi pressure–121°C.

Use: For testing the salt tolerance of a variety of microorganisms.

Salt Tolerance Medium, Gilardi

Composition per liter:

NaCl......65.0g
Pancreatic digest of casein......15.0g
Agar......15.0g
Papaic digest of soybean meal......5.0g

pH 7.3 ± 0.2 at 25°C

Preparation of Medium: Add components to distilled/deionized water and bring volume to 1.0L. Mix thoroughly. Gently heat and bring to boiling. Distribute into tubes or flasks. Autoclave for 15 min at 15 psi pressure–121°C. Do not overheat. Pour into sterile Petri dishes or leave in tubes.

Use: For the cultivation and maintenance of salt-tolerant, nonfermenting Gram-negative bacteria. For the differentiation of nonfermenting Gram-negative bacteria based on salt tolerance.

Salt Tolerance Medium, Tatum

Composition per liter:

NaCl......65.0g
Peptone......5.0g
Yeast extract......2.0g
Beef extract......1.0g

pH 7.4 ± 0.2 at 25°C

Preparation of Medium: Add components to distilled/deionized water and bring volume to 1.0L. Mix thoroughly. Distribute into tubes or flasks. Autoclave for 15 min at 15 psi pressure–121°C.

Use: For the cultivation of salt-tolerant, nonfermenting Gram-negative bacteria. For the differentiation of nonfermenting Gram-negative bacteria based on salt tolerance.

Sauton's Medium

Composition per liter:

L-Asparagine	4.0g
Citric acid	2.0g
K_2HPO_4	0.5g
$MgSO_4$	0.5g
Triton WR 1339	0.25g
Ferric ammonium citrate	0.05g
Glycerol	40.0mL

Preparation of Medium: Add components to distilled/deionized water and bring volume to 1.0L. Mix thoroughly. Distribute into tubes or flasks. Autoclave for 15 min at 15 psi pressure–121°C.

Use: For the cultivation of *Mycobacterium tuberculosis* Bacille Calmette-Guèrin (BCG) for vaccine production.

SBG Enrichment Broth (Selenite Brilliant Green Enrichment Broth)

Composition per liter:

D-Mannitol	5.0g
Peptone	5.0g
Yeast extract	5.0g
$Na_2SeO_3 \cdot 5H_2O$	4.0g
K_2HPO_4	2.65g
KH_2PO_4	1.02g
Sodium taurocholate	1.0g
Brilliant Green	5.0mg

pH 7.2 ± 0.2 at 25°C

Source: This medium is available as a premixed powder from BD Diagnostic Systems.

Preparation of Medium: Add components to distilled/deionized water and bring volume to 1.0L. Mix thoroughly. Gently heat and bring to boiling. Continue boiling for 5–10 min. Do not autoclave. Distribute into sterile tubes or flasks.

Use: For the selective isolation of *Salmonella* species.

SBG Sulfa Enrichment

Composition per liter:

D-Mannitol	5.0g
Peptone	5.0g
Yeast extract	5.0g
$Na_2SeO_3 \cdot 5H_2O$	4.0g
K_2HPO_4	2.65g
KH_2PO_4	1.02g
Sodium taurocholate	1.0g
Sodium sulfapyridine	0.5g
Brilliant Green	5.0mg

pH 7.2 ± 0.2 at 25°C

Source: This medium is available as a premixed powder from BD Diagnostic Systems.

Preparation of Medium: Add components to distilled/deionized water and bring volume to 1.0L. Mix thoroughly. Gently heat and bring to boiling. Continue boiling for 5–10 min. Do not autoclave. Distribute into sterile tubes or flasks.

Use: For the selective isolation of *Salmonella* species.

SCGYEM Medium

Composition per liter:

Casein, isoelectric	10.0g
Yeast extract	5.0g
Glucose	2.5g
Na_2HPO_4	1.325g
KH_2PO_4	0.8g
Calf serum, heat inactivated	100.0mL

Preparation of Medium: Add components, except calf serum, to distilled/deionized water and bring volume to 900.0mL. Mix thoroughly. Autoclave for 30 min at 15 psi pressure–121°C. Cool to 25°C. Aseptically add 100.0mL of sterile, heat-inactivated calf serum. Mix thoroughly. Aseptically distribute into sterile, screw-capped tubes or flasks.

Use: For the cultivation of *Naegleria australiensis* and *Naegleria fowleri*.

Schaedler Agar (Schaedler Anaerobic Agar)

Composition per liter:

Agar	13.5g
Glucose	5.83g
Pancreatic digest of casein	5.7g
Proteose peptone no. 3	5.0g
Yeast extract	5.0g
Tris(hydroxymethyl)aminomethane buffer	3.0g
NaCl	1.65g
Papaic digest of soybean meal	1.0g
K_2HPO_4	0.83g
L-Cystine	0.4g
Hemin	0.01g

pH 7.6 ± 0.2 at 25°C

Source: This medium is available as a premixed powder from BD Diagnostic Systems and Oxoid Unipath.

Preparation of Medium: Add components to distilled/deionized water and bring volume to 1.0L. Mix thoroughly. Gently heat and bring to boiling. Distribute into tubes or flasks. Autoclave for 15 min at 15 psi pressure–121°C. Pour into sterile Petri dishes or leave in tubes.

Use: For the isolation, cultivation, and enumeration of anaerobic and aerobic microorganisms.

Schaedler Agar

Composition per liter:

Agar	13.5g
Pancreatic digest of casein	8.2g
Glucose	5.8g
Yeast extract	5.0g
Tris(hydroxymethyl)aminomethane buffer	3.0g
Peptic digest of animal tissue	2.5g
NaCl	1.7g
Papaic digest of soybean meal	1.0g
K_2HPO_4	0.8g
L-Cystine	0.4g
Hemin	0.01g

pH 7.6 ± 0.2 at 25°C

Source: This medium is available as a premixed powder from BD Diagnostic Systems.

Preparation of Medium: Add components to distilled/deionized water and bring volume to 1.0L. Mix thoroughly. Gently heat and bring to boiling. Distribute into tubes or flasks. Autoclave for 15 min at 15 psi pressure–121°C. Pour into sterile Petri dishes or leave in tubes.

Use: For the isolation, cultivation, and enumeration of anaerobic and aerobic microorganisms.

Schaedler Agar with Vitamin K_1 and Sheep Blood

Composition per liter:

Agar	13.5g
Pancreatic digest of casein	8.2g
Glucose	5.8g
Yeast extract	5.0g
Tris(hydroxymethyl)aminomethane buffer	3.0g
Peptic digest of animal tissue	2.5g
Papaic digest of soybean meal	1.0g
NaCl	1.7g
K_2HPO_4	0.8g
L-Cystine	0.4g
Hemin	0.01g
Sheep blood, defibrinated	50.0mL
Vitamin K_1 solution	1.0mL

pH 7.6 ± 0.2 at 25°C

Vitamin K_1 Solution:
Composition per 10.0mL:

Vitamin K_1	5.0g
Ethanol, absolute	10.0mL

Preparation of Vitamin K_1 Solution: Add vitamin K_1 to ethanol. Mix thoroughly.

Preparation of Medium: Add components, except sheep blood, to distilled/deionized water and bring volume to 950.0mL. Mix thoroughly. Gently heat and bring to boiling. Autoclave for 15 min at 15 psi pressure–121°C. Cool to 45°–50°C. Aseptically add sterile sheep blood. Mix thoroughly. Pour into sterile Petri dishes or distribute into sterile tubes.

Use: For the recovery of fastidious anaerobic bacteria such as *Bacteroides* species.

Schaedler Broth (Schaedler Anaerobic Broth)

Composition per liter:

Pancreatic digest of casein	8.2g
Glucose	5.8g
Yeast extract	5.0g
Tris(hydroxymethyl)aminomethane buffer	3.0g
Peptic digest of animal tissue	2.5g
NaCl	1.7g
Papaic digest of soybean meal	1.0g
K_2HPO_4	0.8g
L-Cystine	0.4g
Hemin	0.01g

pH 7.6 ± 0.2 at 25°C

Source: This medium is available as a premixed powder from BD Diagnostic Systems and Oxoid Unipath.

Preparation of Medium: Add components to distilled/deionized water and bring volume to 1.0L. Mix thoroughly. Distribute into tubes or flasks. Autoclave for 15 min at 15 psi pressure–121°C.

Use: For the cultivation and maintenance of *Eubacterium combesii, Eubacterium contortum*, and a variety of other anaerobic bacteria.

Schaedler CNA Agar with Vitamin K_1 and Sheep Blood

Composition per liter:

Agar	13.5g
Pancreatic digest of casein	8.2g
Glucose	5.8g
Yeast extract	5.0g
Tris(hydroxymethyl)aminomethane buffer	3.0g
Peptic digest of animal tissue	2.5g
Papaic digest of soybean meal	1.0g
NaCl	1.7g
K_2HPO_4	0.8g
L-Cystine	0.4g
Hemin	0.01g
Colistin	0.01g
Nalidixic acid	0.01g

Sheep blood, defibrinated	50.0mL
Vitamin K_1 solution	1.0mL

pH 7.6 ± 0.2 at 25°C

Vitamin K_1 Solution:
Composition per 10.0mL:

Vitamin K_1	5.0g
Ethanol, absolute	10.0mL

Preparation of Vitamin K_1 Solution: Add vitamin K_1 to ethanol. Mix thoroughly.

Preparation of Medium: Add components, except sheep blood, to distilled/deionized water and bring volume to 950.0mL. Mix thoroughly. Gently heat and bring to boiling. Autoclave for 15 min at 15 psi pressure–121°C. Cool to 45°–50°C. Aseptically add sterile sheep blood. Mix thoroughly. Pour into sterile Petri dishes or distribute into sterile tubes.

Use: For the selective isolation of anaerobic, Gram-positive cocci, especially *Peptococcus* species and *Peptostreptococcus* species.

Schaedler KV Agar with Vitamin K_1 and Sheep Blood

Composition per liter:

Agar	13.5g
Pancreatic digest of casein	8.2g
Glucose	5.8g
Yeast extract	5.0g
Tris(hydroxymethyl)aminomethane buffer	3.0g
Peptic digest of animal tissue	2.5g
NaCl	1.7g
Papaic digest of soybean meal	1.0g
K_2HPO_4	0.8g
L-Cystine	0.4g
Hemin	0.01g
Kanamycin	0.01g
Vancomycin	7.5mg
Sheep blood, defibrinated	50.0 mL
Vitamin K_1 solution	1.0mL

pH 7.6 ± 0.2 at 25°C

Vitamin K_1 Solution:
Composition per 10.0mL:

Vitamin K_1	5.0g
Ethanol, absolute	10.0mL

Preparation of Vitamin K_1 Solution: Add vitamin K_1 to ethanol. Mix thoroughly.

Preparation of Medium: Add components, except sheep blood, to distilled/deionized water and bring volume to 950.0mL. Mix thoroughly. Gently heat and bring to boiling. Autoclave for 15 min at 15 psi pressure–121°C. Cool to 45°–50°C. Aseptically add sterile sheep blood. Mix thoroughly. Pour into sterile Petri dishes or distribute into sterile tubes.

Use: For the selective isolation of Gram-negative anaerobic bacteria.

Schleifer-Krämer Agar (SK Agar)

Composition per liter:

Agar	13.0g
Glycerol	10.0g
Sodium pyruvate	10.0g
Pancreatic digest of casein	10.0g
Beef extract	5.0g
Yeast extract	3.0g
Potassium isothiocyanate	2.25g
LiCl	2.0g
$Na_2HPO_4 \cdot 2H_2O$	0.9g
$NaH_2PO_4 \cdot H_2O$	0.6g
Glycine	0.5g
NaN_3 solution	10.0mL

pH 7.2 ± 0.2 at 25°C

NaN_3 Solution:
Composition per 10.0mL:

NaN_3	0.045g

Preparation of NaN_3 Solution: Add NaN_3 to distilled/deionized water and bring volume to 10.0mL. Mix thoroughly. Filter sterilize.

Preparation of Medium: Add components, except NaN_3 solution, to distilled/deionized water and bring volume to 990.0mL. Mix thoroughly. Adjust pH to 7.2. Gently heat and bring to boiling. Autoclave for 15 min at 15 psi pressure–121°C. Cool to 45°–50°C. Aseptically add sterile NaN_3 solution. Mix thoroughly. Pour into sterile Petri dishes or distribute into sterile tubes.

Use: For the isolation and cultivation of *Staphylococcus* species.

Selenite Broth (Selenite Broth, Lactose) (Selenite F Enrichment Medium) (Sodium Biselenite Medium) (Sodium Hydrogen Selenite Medium)

Composition per liter:

Na_2HPO_4	10.0g
Pancreatic digest of casein	5.0g
Lactose	4.0g
$NaHSeO_3 \cdot 5H_2O$	4.0g

pH 7.0 ± 0.2 at 25°C

Source: This medium is available as a premixed powder from BD Diagnostic Systems and a prepared medium from Oxoid Unipath.

Caution: Sodium biselenite is toxic and a potential teratogen and care must be taken to avoid inhalation of the powdered dye, contact with the skin, or ingestion, especially in pregnant laboratory workers.

Preparation of Medium: Add components to distilled/deionized water and bring volume to 1.0L. Mix thoroughly. Gently heat and bring to boiling. Do not autoclave. Distribute into sterile tubes in 10.0mL volumes.

Use: For the isolation and enrichment of *Salmonella* species from clinical specimens.

Selenite Broth Base, Mannitol

Composition per liter:

Na_2HPO_4	10.0g
Peptone	5.0g
Mannitol	4.0g
$NaHSeO_3 \cdot 5H_2O$	4.0g

pH 7.1 ± 0.2 at 25°C

Source: This medium is available as a premixed powder from Oxoid Unipath.

Caution: Sodium selenite is toxic and a potential teratogen and care must be taken to avoid inhalation of the powdered dye, contact with the skin, or ingestion, especially in pregnant laboratory workers.

Preparation of Medium: Add components to distilled/deionized water and bring volume to 1.0L. Mix thoroughly. Gently heat. Do not autoclave. Distribute into sterile tubes in 10.0mL volumes. Sterilize for 10 min at 0 psi pressure–100°C.

Use: For the isolation and cultivation of *Salmonella typhi* and *Salmonella paratyphi*.

Selenite Cystine Broth

Composition per liter:

Na_2HPO_4	10.0g
Pancreatic digest of casein	5.0g
Lactose	4.0g
$Na_2SeO_3 \cdot 5H_2O$	4.0g
L-Cystine	0.02g

pH 7.0 ± 0.2 at 25°C

Source: This medium is available as a premixed powder from BD Diagnostic Systems and Oxoid Unipath.

Caution: Sodium selenite is toxic and a potential teratogen and care must be taken to avoid inhalation of the powdered dye, contact with the skin, or ingestion, especially in pregnant laboratory workers.

Preparation of Medium: Add components to distilled/deionized water and bring volume to 1.0L. Mix thoroughly. Gently heat. Do not autoclave. Distribute into sterile tubes in 10.0mL volumes. Sterilize for 15 min at 0 psi pressure–100°C.

Use: For the isolation and cultivation of *Salmonella* species from feces and other specimens.

Selenite F Broth

Composition per liter:

KH_2PO_4	7.0g
Pancreatic digest of casein	5.0g
Lactose	4.0g
$Na_2SeO_3 \cdot 5H_2O$	4.0g
Na_2HPO_4	3.0g

pH 7.0 ± 0.2 at 25°C

Source: This medium is available as a premixed powder from BD Diagnostic Systems.

Caution: Sodium selenite is toxic and a potential teratogen and care must be taken to avoid inhalation of the powdered dye, contact with the skin, or ingestion, especially in pregnant laboratory workers.

Preparation of Medium: Add components to distilled/deionized water and bring volume to 1.0L. Mix thoroughly. Gently heat. Do not autoclave. Distribute into sterile tubes in 10.0mL volumes. Sterilize for 30 min at 0 psi pressure–100°C.

Use: For the isolation and cultivation of *Salmonella* species from feces and other specimens.

Semisolid *Brucella* Broth

Composition per liter:

Peptone	10.0g
Pancreatic digest of casein	10.0g
NaCl	5.0g
Yeast extract	2.0g
Agar	1.6g
D-Glucose	1.0g

Preparation of Medium: Add components to distilled/deionized water and bring volume to 1.0L. Mix thoroughly. Distribute into tubes or flasks. Autoclave for 15 min at 15 psi pressure–121°C.

Use: For the cultivation of *Campylobacter* species.

Semisolid Medium for Motility

Composition per liter:

Biosate	5.0g
Polypeptone	5.0g
NaCl	5.0g
Agar	4.0g
Myosate	1.5g
Triphenyltetrazolium chloride solution	2.5mL

pH 6.9–7.1at 25°C

Triphenyltetrazolium Chloride Solution:
Composition per 10.0mL:
Triphenyltetrazolium chloride 0.1g
Ethanol (95% solution) 10.0mL

Preparation of Triphenyltetrazolium Chloride Solution: Add triphenyltetrazolium chloride to 10.0mL of ethanol. Mix thoroughly.

Preparation of Medium: Add components, except triphenyltetrazolium chloride solution, to distilled/deionized water and bring volume to 997.5mL. Mix thoroughly. Gently heat and bring to boiling. Add 2.5mL of triphenyltetrazolium chloride solution. Mix thoroughly. Distribute into tubes in 10.0mL volumes. Autoclave for 15 min at 15 psi pressure–121°C.

Use: For the differentiation of bacteria based on motility.

Sensitest Agar

Composition per liter:
Pancreatic digest of casein 11.0g
Agar 8.0g
Buffer salts 3.3g
Peptone..3.0g
NaCl 3.0g
Glucose 2.0g
Starch 1.0g
Nucleoside bases 0.02g
Thiamine 0.02mg
pH 7.4 ± 0.2 at 25°C

Source: This medium is available as a premixed powder from Oxoid Unipath.

Preparation of Medium: Add components to distilled/deionized water and bring volume to 1.0L. Mix thoroughly. Gently heat and bring to boiling. Distribute into tubes or flasks. Autoclave for 15 min at 15 psi pressure–121°C. Pour into sterile Petri dishes.

Use: For the performance of antibiotic sensitivity assays.

Serratia Differential Medium (SD Medium)

Composition per 102.0mL:
Solution A 92.0mL
Solution B 10.0mL
pH 6.7 ± 0.2 at 25°C

Solution A:
Composition per 92.0mL:
Yeast extract 1.0g
L-Ornithine 1.0g
NaCl 0.5g
Agar 0.4g
Irgasan inhibitor... 1.0mL
Indicator solution 1.0mL

Preparation of Solution A: Add components to distilled/deionized water and bring volume to 92.0mL. Mix thoroughly. Adjust pH to 6.7 with 1*N* NaOH.

Irgasan Inhibitor:
Composition per 100.0mL:
Irgasan-DP-300 (4,2′, 4′-trichloro-
2-hydroxydiphenylether)...............................0.1g
NaOH (1*N* solution).................................... 10.0mL

Preparation of Irgasan Inhibitor: Add irgasan to 10.0mL of NaOH solution. Mix thoroughly. Gently heat to dissolve. Bring volume to 100.0mL with distilled/deionized water.

Indicator Solution:
Composition per 100.0mL:
Bromthymol Blue 0.2g
Phenol Red..0.1g

Preparation of Indicator Solution: Add components to 50.0mL of distilled/deionized water. Mix thoroughly for 1h. Bring volume to 100.0mL with distilled/deionized water.

Solution B:
Composition per 10.0mL:
L-Arabinose 1.0g

Preparation of Solution B: Add L-arabinose to distilled/deionized water and bring volume to 10.0mL. Mix thoroughly.

Preparation of Medium: Combine 92.0mL of solution A with 10.0mL of solution B. Mix thoroughly. Distribute into tubes. Autoclave for 15 min at 15 psi pressure–121°C. Allow tubes to cool in an upright position.

Use: For the cultivation and differentiation of *Serratia* species based on the fermentation of arabinose and production of ornithine decarboxylase. *Serratia marcescens* changes the medium to purple throughout the tube. *Serratia liquefaciens* changes the medium to a band of purple at the top of the tube with a green/yellow butt. *Serratia rubidaea* changes the medium to yellow throughout the tube.

Serratia Hd-MHr

Composition per liter:
Agar 15.0g
K_2HPO_4...7.0g
Glucose 5.0g
KH_2PO_4...3.0g
2-Methyl-DL-histidine·2HCl 1.0g

$(NH_4)_2SO_4$ 1.0g
$MgSO_4 \cdot 7H_2O$ 0.5g

Preparation of Medium: Add components to distilled/deionized water and bring volume to 1.0L. Mix thoroughly. Gently heat and bring to boiling. Distribute into tubes or flasks. Autoclave for 15 min at 15 psi pressure–121°C. Pour into sterile Petri dishes or leave in tubes.

Use: For the cultivation and maintenance of *Serratia marcescens*.

Serratia Medium (ATCC Medium 181)

Composition per liter:

Agar 20.0g
Pancreatic digest of casein 5.0g
Yeast extract 5.0g
Glucose 1.0g
K_2HPO_4 .... 1.0g
pH 7.0 ± 0.2 at 25°C

Preparation of Medium: Add components to distilled/deionized water and bring volume to 1.0L. Mix thoroughly. Gently heat and bring to boiling. Distribute into tubes or flasks. Autoclave for 15 min at 15 psi pressure–121°C. Pour into sterile Petri dishes or leave in tubes.

Use: For the cultivation and maintenance of *Serratia marcescens*.

Serratia Medium (ATCC Medium 1399)

Composition per liter:

Agar 15.0g
K_2HPO_4 .... 7.0g
Glucose 5.0g
KH_2PO_4 .... 3.0g
Casein hydrolysate 1.0g
$(NH_4)_2SO_4$ 1.0g
Yeast extract 1.0g
$MgSO_4 \cdot 7H_2O$ 0.1g
pH 7.0 ± 0.2 at 25°C

Preparation of Medium: Add components to distilled/deionized water and bring volume to 1.0L. Mix thoroughly. Gently heat and bring to boiling. Distribute into tubes or flasks. Autoclave for 15 min at 15 psi pressure–121°C. Pour into sterile Petri dishes or leave in tubes.

Use: For the cultivation and maintenance of *Serratia marcescens*.

Serum Glucose Agar (Serum Dextrose Agar) (ATCC Medium 287)

Composition per 1060.0mL:

Agar 15.0g
Peptone 10.0g
Beef extract 5.0g
NaCl 5.0g
Serum-glucose solution 60.0mL
pH 7.3 ± 0.2 at 25°C

Serum-Glucose Solution:
Composition per 60.0mL:

D-Glucose 10.0g
Serum (inactivated at 56°C, 30 min) 50.0mL

Preparation of Serum-Glucose Solution: Add glucose to 50.0mL of heat-inactivated serum. Horse serum or ox serum may be used. Mix thoroughly. Filter sterilize.

Preparation of Medium: Add components, except serum-glucose solution, to distilled/deionized water and bring volume to 1.0L. Mix thoroughly. Gently heat and bring to boiling. Autoclave for 15 min at 10 psi pressure–115°C. Cool to 50°C. Aseptically add 60.0mL of sterile serum-glucose solution. Mix thoroughly. Pour into sterile Petri dishes or distribute into sterile tubes. Allow tubes to cool in a slanted position.

Use: For the cultivation and maintenance of *Brucella* species.

Serum Glucose Agar, Farrell Modified

Composition per 1086.9mL:

Agar 15.0g
Peptone 10.0g
Beef extract 5.0g
NaCl 5.0g
Serum-glucose solution 60.0mL
Bacitracin solution 12.5mL
Cycloheximide solution 10.0mL
Nystatin solution 2.0mL
Polymyxin B solution 1.0mL
Nalidixic acid solution 1.0mL
Vancomycin solution 0.4mL
pH 7.3 ± 0.2 at 25°C

Serum-Glucose Solution:
Composition per 60.0mL:

D-Glucose 10.0g
Serum (inactivated at 56°C, 30 min) 50.0mL

Preparation of Serum-Glucose Solution: Add glucose to 50.0mL of heat-inactivated serum. Horse serum or ox serum may be used. Mix thoroughly. Filter sterilize.

Bacitracin Solution:
Composition per 12.5mL:
Bacitracin 25,000U

Preparation of Bacitracin Solution: Add bacitracin to distilled/deionized water and bring volume to 12.5mL. Mix thoroughly. Filter sterilize.

Cycloheximide Solution:
Composition per 100.0mL:
Cycloheximide 1.0g
Acetone 5.0mL

Preparation of Cycloheximide Solution: Add cycloheximide to 5.0mL of acetone. Mix thoroughly. Bring volume to 100.0mL with distilled/deionized water. Mix thoroughly. Filter sterilize.

Nystatin Solution:
Composition per 5.0mL:
Nystatin 250,000U

Preparation of Nystatin Solution: Add nystatin to distilled/deionized water and bring volume to 5.0mL. Mix thoroughly. Filter sterilize.

Polymyxin B Solution:
Composition per 2.0mL:
Polymyxin B 10,000U

Preparation of Polymyxin B Solution: Add polymyxin B to distilled/deionized water and bring volume to 2.0mL. Mix thoroughly. Filter sterilize.

Nalidixic Acid Solution:
Composition per 2.0mL:
Nalidixic acid 0.1g
NaOH (0.5*N* solution) 2.0mL

Preparation of Nalidixic Acid Solution: Add nalidixic acid to 2.0mL of NaOH solution. Mix thoroughly. Immediately before use, add 1.0mL of this stock solution to 9.0mL of distilled/deionized water. Mix thoroughly. Filter sterilize.

Vancomycin Solution:
Composition per 1.0mL:
Vancomycin 0.05g

Preparation of Vancomycin Solution: Add vancomycin to distilled/deionized water and bring volume to 1.0mL. Mix thoroughly. Filter sterilize.

Preparation of Medium: Add components—except serum-glucose solution, bacitracin solution, cycloheximide solution, nystatin solution, polymyxin B solution, nalidixic acid solution, and vancomycin solution—to distilled/deionized water and bring volume to 1.0L. Mix thoroughly. Gently heat and bring to boiling. Autoclave for 15 min at 10 psi pressure–115°C. Cool to 50°C. Aseptically add 60.0mL of sterile serum-glucose solution, 12.5mL of sterile bacitracin solution, 10.0mL of sterile cycloheximide solution, 2.0mL of sterile nystatin solution, 1.0mL of sterile polymyxin B solution, 1.0mL of sterile nalidixic acid solution, and 0.4mL of sterile vancomycin solution. Mix thoroughly. Pour into sterile Petri dishes or distribute into sterile tubes. Allow tubes to cool in a slanted position.

Use: For the selective isolation and cultivation of *Brucella* species.

Serum Potato Infusion Agar

Composition per 1120.0mL:
Agar 15.0g
Peptone 10.0g
Meat extract 5.0g
NaCl 5.0g
Potato infusion 1.0L
Horse serum, heat inactivated 100.0mL
Glycerol 20.0mL
pH 6.8 ± 0.2 at 25°C

Potato Infusion:
Composition per 10.0mL:
Potatoes 250.0g

Preparation of Potato Infusion: Add peeled, thinly sliced potatoes to 1.0L of distilled/deionized water. Infuse overnight at 60°C. Filter through Whatman no. 1 filter paper. Bring volume to 1.0L with distilled/deionized water.

Preparation of Medium: Combine components, except horse serum. Mix thoroughly. Gently heat and bring to boiling. Autoclave for 15 min at 15 psi pressure–121°C. Cool to 45°–50°C. Aseptically add 100.0mL of sterile horse serum. Mix thoroughly. Pour into sterile Petri dishes or distribute into sterile tubes.

Use: For the cultivation of *Brucella* species.

Serum Tellurite Agar

Composition per liter:
Agar 20.0g
Pancreatic digest of casein 10.0g
Peptic digest of animal tissue 10.0g
NaCl 5.0g
Glucose 2.0g
Lamb serum 50.0mL
Chapman tellurite solution 10.0mL
pH 7.5 ± 0.2 at 25°C

Source: This medium is available as a premixed powder from BD Diagnostic Systems.

Chapman Tellurite Solution:
Composition per 100.0mL:
K_2TeO_3 .... 1.0g

Preparation of Chapman Tellurite Solution: Add K_2TeO_3 to distilled/deionized water and bring volume to 100.0mL. Mix thoroughly. Filter sterilize.

Preparation of Medium: Add components, except lamb serum and Chapman tellurite solution, to distilled/deionized water and bring volume to 940.0mL. Mix thoroughly. Gently heat and bring to boiling. Autoclave for 15 min at 15 psi pressure–121°C. Cool to 45°–50°C. Aseptically add sterile lamb serum and 10.0mL of sterile Chapman tellurite solution. Mix thoroughly. Pour into sterile Petri dishes or distribute into sterile tubes.

Use: For the isolation and cultivation of *Corynebacterium* species, especially in the laboratory diagnosis of diphtheria.

Seven H11 Agar (Selective 7H11 Agar)

Composition per 1010.0mL:

Agar	13.5g
KH_2PO_4	1.5g
Na_2HPO_4	1.5g
Pancreatic digest of casein	1.0g
NaCl	0.85g
Monosodium glutamate	0.5g
$(NH_4)_2SO_4$	0.5g
Sodium citrate	0.4g
$MgSO_4 \cdot 7H_2O$	0.05g
Ferric ammonium citrate	0.04g
$CuSO_4 \cdot 5H_2O$	1.0mg
Pyridoxine	1.0mg
$ZnSO_4 \cdot 7H_2O$	1.0mg
Biotin	0.5mg
$CaCl_2 \cdot 2H_2O$	0.5mg
Malachite Green	0.25mg
Middlebrook OADC enrichment	100.0mL
Antibiotic inhibitor	10.0mL
Glycerol	5.0mL

pH 6.6 ± 0.2 at 25°C

Source: This medium is available as a prepared medium from BD Diagnostic Systems.

Middlebrook OADC Enrichment:

Bovine albumin fraction V	5.0g
Glucose	2.0g
NaCl	0.85g
Catalase	3.0mg
Oleic acid	0.06mL

Preparation of Middlebrook OADC Enrichment: Add components to distilled/deionized water and bring volume to 100.0mL. Mix thoroughly. Filter sterilize.

Antibiotic Inhibitor:
Composition per 10.0mL:

Carbenicillin	0.05g
Trimethoprim lactate	0.02g
Amphotericin B	0.01g
Polymyxin B	200,000U

Preparation of Antibiotic Inhibitor: Add components to distilled/deionized water and bring volume to 10.0mL. Mix thoroughly. Filter sterilize.

Preparation of Medium: Add glycerol to 900.0mL of distilled/deionized water. Mix thoroughly. Add remaining components, except Middlebrook OADC enrichment and antibiotic inhibitor. Mix thoroughly. Gently heat. Do not boil. Autoclave for 10 min at 15 psi pressure–121°C. Cool to 50°–55°C. Aseptically add 100.0mL of sterile Middlebrook OADC enrichment and 10.0mL of sterile antibiotic solution. Mix thoroughly. Pour into sterile Petri dishes or distribute into sterile tubes.

Use: For the isolation and cultivation of *Mycobacterium* species from specimens with a mixed flora.

SF Broth (*Streptococcus faecalis* Broth)

Composition per liter:

Pancreatic digest of casein	20.0g
Glucose	5.0g
NaCl	5.0g
K_2HPO_4	4.0g
KH_2PO_4	1.5g
NaN_3	0.5g
Bromcresol Purple	0.032g

pH 6.9 ± 0.2 at 25°C

Source: This medium is available as a premixed powder from BD Diagnostic Systems.

Preparation of Medium: Add components to distilled/deionized water and bring volume to 1.0L. Mix thoroughly. Distribute into tubes or flasks. Autoclave for 15 min at 15 psi pressure–121°C.

Use: For the cultivation and differentiation of group D enterococci (*Streptococcus faecalis* and *Streptococcus faecium*) from group D nonenterococci and from other *Streptococcus* species. Group D enterococci turn the medium turbid and yellow-brown.

Shigella Broth

Composition per liter:

Pancreatic digest of casein	20.0g
NaCl	5.0g
K_2HPO_4	2.0g
KH_2PO_4	2.0g
Glucose	1.0g

Novobiocin solution 11.1mL
Tween 80 .... 1.5mL

pH 7.0 ± 0.2 at 25°C

Novobiocin Solution:

Composition per liter:

Novobiocin 0.05g

Preparation of Novobiocin Solution: Add novobiocin to distilled/deionized water and bring volume to 1.0L. Mix thoroughly. Filter sterilize.

Preparation of Medium: Add components, except novobiocin solution, to distilled/deionized water and bring volume to 988.9mL. Mix thoroughly. Gently heat and bring to boiling. Autoclave for 15 min at 15 psi pressure–121°C. Cool to 45°–50°C. Aseptically add sterile novobiocin solution. Mix thoroughly. Aseptically distribute into sterile tubes.

Use: For the isolation and cultivation of *Shigella* species.

Sierra Medium

Composition per liter:

Agar 15.0g
Peptone 10.0g
NaCl 5.0g
$CaCl_2 \cdot H_2O$ 0.1g
Tween 80 .... 10.0mL

pH 7.4 ± 0.2 at 25°C

Preparation of Medium: Add components, except Tween 80, to distilled/deionized water and bring volume to 990.0mL. Mix thoroughly. Gently heat and bring to boiling. Autoclave for 15 min at 15 psi pressure–121°C. Cool to 45°–50°C. Separately autoclave Tween 80 for 15 min at 15 psi pressure–121°C. Cool to 45°–50°C. Aseptically add 10.0mL of sterile Tween 80. Mix thoroughly. Pour into sterile Petri dishes.

Use: For the differentiation of bacteria based on lipase activity. Bacteria with lipase activity form colonies surrounded by a white precipitate.

SIM Medium

Composition per liter:

Peptone 30.0g
Agar 3.0g
Beef extract 3.0g
Peptonized iron 0.2g
$Na_2S_2O_3 \cdot 5H_2O$ 0.025g

pH 7.3 ± 0.2 at 25°C

Source: This medium is available as a premixed powder from BD Diagnostic Systems.

Preparation of Medium: Add components to distilled/deionized water and bring volume to 1.0L. Mix thoroughly. Gently heat and bring to boiling. Distribute into tubes in 15.0mL volumes. Autoclave for 15 min at 15 psi pressure–121°C. Allow tubes to cool in an upright position.

Use: For the differentiation of members of the Enterobacteriaceae based on H_2S production, indole production, and motility.

SIM Medium

Composition per liter:

Pancreatic digest of casein 20.0g
Peptic digest of animal tissue 6.1g
Agar 3.5g
$Fe(NH_4)_2(SO_4)_2 \cdot 6H_2O$ 0.2g
$Na_2S_2O_3 \cdot 5H_2O$ 0.2g

pH 7.3 ± 0.2 at 25°C

Source: This medium is available as a premixed powder from BD Diagnostic Systems and Oxoid Unipath.

Preparation of Medium: Add components to distilled/deionized water and bring volume to 1.0L. Mix thoroughly. Gently heat and bring to boiling. Distribute into tubes in 15.0mL volumes. Autoclave for 15 min at 15 psi pressure–121°C. Allow tubes to cool in an upright position.

Use: For the differentiation of members of the Enterobacteriaceae based on H_2S production, indole production, and motility.

Simmons' Citrate Agar (Citrate Agar)

Composition per liter:

Agar 15.0g
NaCl 5.0g
Sodium citrate 2.0g
K_2HPO_4 .... 1.0g
$(NH_4)H_2PO_4$ 1.0g
$MgSO_4 \cdot 7H_2O$ 0.2g
Bromthymol Blue 0.08g

pH 6.9 ± 0.2 at 25°C

Source: This medium is available as a premixed powder from BD Diagnostic Systems and Oxoid Unipath.

Preparation of Medium: Add components to distilled/deionized water and bring volume to 1.0L. Mix thoroughly. Gently heat while stirring and bring to boiling. Distribute into tubes or flasks. Autoclave for 15 min at 15 psi pressure–121°C. Pour into sterile Petri dishes or leave in tubes.

Use: For the differentiation of Gram-negative bacteria on the basis of citrate utilization. Bacteria that can utilize citrate as sole carbon source turn the medium blue.

Skim Milk Agar

Composition per 1100.0mL:

Agar	15.0g
Pancretic digest of casein	5.0g
Yeast extract	2.5g
Glucose	1.0g
Skim milk solution	100.0mL

pH 7.0 ± 0.1 at 25°C

Preparation of Skim Milk Solution: Add skim milk solids to distilled/deionized water and bring volume to 100.0mL. Mix thoroughly. Autoclave for 15 min at 15 psi pressure–121°C. Cool to 45°–50°C.

Preparation of Medium: Add components, except skim milk solution, to distilled/deionized water and bring volume to 1.0L. Mix thoroughly. Gently heat and bring to boiling. Distribute into tubes or flasks. Autoclave for 15 min at 15 psi pressure–121°C. Cool to 45°–50°C. Aseptically add 100.0mL of cooled, sterile skim milk solution. Mix thoroughly. Pour into sterile Petri dishes or aseptically distribute into sterile tubes.

Use: For the cultivation and differentiation of bacteria based on proteolytic activity.

Skirrow *Brucella* Medium

Composition per liter:

Blood agar base no. 2	940.0mL
Horse blood, lysed defibrinated	50.0mL
Antibiotic solution	10.0mL

pH 7.4 ± 0.2 at 25°C

Blood Agar Base No. 2

Composition per 940.0mL:

Proteose peptone	15.0g
Agar	12.0g
NaCl	5.0g
Yeast extract	5.0g
Liver digest	2.5g

pH 7.4 ± 0.2 at 25°C

Preparation of Blood Agar Base No. 2: Add components to distilled/deionized water and bring volume to 940.0mL. Mix thoroughly. Gently heat while stirring and bring to boiling. Autoclave for 15 min at 15 psi pressure–121°C. Cool to 45°–50°C.

Antibiotic Solution:

Composition per 10.0mL:

Vancomycin	0.01g
Trimethoprim	5.0mg
Polymyxin B	2500U

Preparation of Antibiotic Solution: Add components to distilled/deionized water and bring volume to 10.0mL. Mix thoroughly. Filter sterilize.

Preparation of Medium: To 940.0mL of sterile cooled blood agar base no. 2, aseptically add 50.0mL of sterile, lysed defibrinated horse blood and 10.0mL of sterile antibiotic solution. Pour into sterile Petri dishes or distribute into sterile tubes.

Use: For the selective isolation and cultivation of *Campylobacter* species.

SM ID2 Agar

Composition per liter:

Proprietary

Source: This medium is available from bioMérieux.

Use: A new chromogenic medium for the selective isolation and detection of *Salmonella*, *S.* typhi, and *S. paratyphi* and most Lactose(+) *Salmonella* present pale pink to mauve colonies. Other organisms are either inhibited, colorless, or pale blue in appearance.

Snyder Agar

Composition per liter:

Glucose	20.0g
Agar	16.0g
Pancreatic digest of casein	13.5g
Yeast extract	6.5g
NaCl	5.0g
Bromcresol Green	0.02g

pH 4.8 ± 0.2 at 25°C

Source: This medium is available as a premixed powder from BD Diagnostic Systems.

Preparation of Medium: Add components to distilled/deionized water and bring volume to 1.0L. Mix thoroughly. Gently heat and bring to boiling. Distribute into tubes in 10.0mL volumes. Autoclave for 15 min at 13 psi pressure–118°C. Do not overheat. Pour into sterile Petri dishes or leave in tubes.

Use: For the cultivation and enumeration of lactobacilli in saliva and indication of dental caries activity.

Snyder Test Agar

Composition per liter:

Agar	20.0g
Glucose	20.0g
Tryptose	20.0g
NaCl	5.0g
Bromcresol Green	0.02g

pH 4.8 ± 0.2 at 25°C

Source: This medium is available as a premixed powder from BD Diagnostic Systems.

Preparation of Medium: Add components to distilled/deionized water and bring volume to 1.0L. Mix thoroughly. Gently heat and bring to boiling. Distribute into tubes in 10.0mL volumes. Autoclave for 15 min at 13 psi pressure–118°C. Do not overheat. Pour into sterile Petri dishes or leave in tubes.

Use: For the cultivation and enumeration of lactobacilli in saliva and indication of dental caries activity.

Sodium Chloride Broth, 6.5

Composition per liter:

Beef heart, solids from infusion	500.0g
NaCl	65.0g
Tryptose	10.0g

pH 7.4 ± 0.2 at 25°C

Preparation of Medium: Add components to distilled/deionized water and bring volume to 1.0L. Mix thoroughly. Distribute into tubes or flasks. Autoclave for 15 min at 15 psi pressure–121°C.

Use: For the cultivation of enterococci and other salt-tolerant organisms. For the differentiation of microorganisms based on salt tolerance.

Sodium Chloride Sucrose Medium 900 (Sodium Chloride SUC Medium 900)

Composition per liter:

Sucrose	97.3g
Pancreatic digest of gelatin	14.5g
NaCl	14.3g
Agar	13.3g
Brain heart, solids from infusion	6.0g
Peptic digest of animal tissue	6.0g
Yeast extract	5.0g
Glucose	3.0g
Na_2HPO_4	2.5g
$MgSO_4$	0.25g
Horse serum (γ-globulin free, inactivated 30 min at 56°C)	100.0mL
Carbenicillin solution	10.0mL

pH 7.4 ± 0.2 at 25°C

Carbenicillin Solution:

Composition per 10.0mL:

Carbenicillin	5.0g

Preparation of Carbenicillin Solution: Add carbenicillin to distilled/deionized water and bring volume to 10.0mL. Mix thoroughly. Filter sterilize.

Preparation of Medium: Add components, except carbenicillin solution and horse serum, to distilled/deionized water and bring volume to 890.0mL. Mix thoroughly. Gently heat and bring to boiling. Autoclave for 15 min at 15 psi pressure–121°C. Cool to 45°–50°C. Aseptically add carbenicillin solution and horse serum. Mix thoroughly. Pour into sterile Petri dishes or distribute into sterile tubes.

Use: For the cultivation of *Pseudomonas aeruginosa.*

Sodium Chloride Sucrose Medium 900 with Penicillin G

Composition per liter:

Sucrose	97.3g
Pancreatic digest of gelatin	14.5g
NaCl	14.3g
Agar	13.3g
Brain heart, solids from infusion	6.0g
Peptic digest of animal tissue	6.0g
Yeast extract	5.0g
Glucose	3.0g
Na_2HPO_4	2.5g
$MgSO_4$	0.25g
Horse serum (γ-globulin free, inactivated 30 min at 56°C)	100.0mL
Penicillin solution	10.0mL

pH 7.4 ± 0.2 at 25°C

Penicillin Solution:

Composition per 10.0mL:

Penicillin G	500,000U

Preparation of Penicillin Solution: Add penicillin G to distilled/deionized water and bring volume to 10.0mL. Mix thoroughly. Filter sterilize.

Preparation of Medium: Add components, except penicillin solution and horse serum, to distilled/deionized water and bring volume to 900.0mL. Mix thoroughly. Gently heat and bring to boiling. Autoclave for 15 min at 15 psi pressure–121°C. Cool to 45°–50°C. Aseptically add penicillin solution and horse serum. Mix thoroughly. Pour into sterile Petri dishes or distribute into sterile tubes.

Use: For the cultivation of *Pseudomonas aeruginosa.*

Sodium Hippurate Broth (Hippurate Broth)

Composition per liter:

Beef heart, solids from infusion	500.0g
Tryptose	10.0g
Sodium hippurate	10.0g
NaCl	5.0g

pH 7.4 ± 0.2 at 25°C

Source: Heart infusion broth is available as a premixed powder from BD Diagnostic Systems.

Preparation of Medium: Add components to distilled/deionized water and bring volume to 1.0L.

Mix thoroughly. Gently heat and bring to boiling. Distribute into screw-capped tubes or flasks. Autoclave for 15 min at 15 psi pressure–121°C. Tighten caps to prevent drying.

Use: For the identification and differentiation of β-hemolytic streptococci based on hippurate hydrolysis. After inoculation and incubation, tubes are treated with $FeCl_3$ reagent. A heavy precipitate remaining after 10–15 min indicates that hippurate has been hydrolyzed.

Soft Agar Gelatin Overlay

Composition per plate:

Base agar	15.0mL
Soft agar gelatin overlay	2.5mL

pH 7.0 ± 0.2 at 25°C

Base Agar:

Composition per liter:

Agar	15.0g
Peptone	5.0g
NaCl	5.0g
Beef extract	3.0g
$MnSO_4 \cdot H_2O$	0.05g

Preparation of Base Agar: Add components to distilled/deionized water and bring volume to 1.0L. Mix thoroughly. Gently heat and bring to boiling. Autoclave for 15 min at 15 psi pressure–121°C. Cool to 45°–50°C.

Soft Agar Gelatin Overlay:

Composition per liter:

Gelatin	15.0g
Agar	8.0g
Peptone	5.0g
NaCl	5.0g
Beef extract	3.0g
$MnSO_4 \cdot H_2O$	0.05g

Preparation of Soft Agar Gelatin Overlay: Add components to distilled/deionized water and bring volume to 1.0L. Mix thoroughly. Gently heat and bring to boiling. Autoclave for 15 min at 15 psi pressure–121°C. Cool to 45°–50°C.

Preparation of Medium: Aseptically pour cooled, sterile base agar into sterile Petri dishes in 15.0mL volumes. Allow agar to solidify. Inoculate plates with samples. Overlay each plate with 2.5mL of soft agar gelatin overlay.

Use: For the cultivation and differentiation of microorganisms based on proteolytic activity.

Sorbitol MacConkey Agar (MacConkey Agar with Sorbitol)

Composition per liter:

Peptone	20.0g
Agar	15.0g
Sorbitol	10.0g
NaCl	5.0g
Bile salts no. 3	1.5g
Neutral Red	0.03g
Crystal Violet	1.0mg

pH 7.1 ± 0.2 at 25°C

Source: This medium is available as a premixed powder from BD Diagnostic Systems and Oxoid Unipath.

Preparation of Medium: Add components to distilled/deionized water and bring volume to 1.0L. Mix thoroughly. Gently heat and bring to boiling. Distribute into tubes or flasks. Autoclave for 15 min at 15 psi pressure–121°C. Pour into sterile Petri dishes or leave in tubes.

Use: For the isolation and cultivation of pathogenic *Escherichia coli.*

Sorbitol MacConkey Agar with BCIG (SMAC with BCIG)

Composition per liter:

Peptone	20.0g
Agar	15.0g
Sorbitol	10.0g
NaCl	5.0g
Proteose peptone	3.0g
Bile salts mixture	1.5g
5-Bromo-4-chloro-3-indolyl-β-D-glucuronide sodium salt	0.1g
Neutral Red	0.03g

pH 7.1 ± 0.2 at 25°C

Source: This medium is available from Oxoid.

Preparation of Medium: Add components to distilled/deionized water and bring volume to 1.0L. Mix thoroughly. Gently heat while stirring and bring to boiling. Autoclave for 15 min at 15 psi pressure–121°C. Pour into sterile Petri dishes.

Use: A selective and differential medium for the detection of *Escherichia coli* O157 incorporating the chromogen 5-bromo-4-chloro-3-indolyl-β-D-glucuronide (BCIG). The medium combines two different screening mechanisms for the detection of *E. coli* O157, the failure to ferment sorbitol and the absence of β-glucuronidase activity. The nonsorbitol-fermenting and β-glucuronidase-negative *E. coli* O157 will appear as straw colored colonies. Organisms with β-glucuronidase activity will cleave the substrate, leading to a distinct blue-green coloration of the colonies.

Soy Peptone Broth

Composition per liter:

Papaic digest of soybean meal	20.0g
NaCl	5.0g
Phenol Red (2% solution)	1.0mL

pH 7.3 ± 0.2 at 25°C

Preparation of Medium: Add components to distilled/deionized water and bring volume to 1.0L. Mix thoroughly. Adjust pH to 7.3. Distribute into tubes or flasks. Autoclave for 15 min at 15 psi pressure–121°C.

Use: For the isolation and cultivation of *Mycoplasma* species and *Ureaplasma* species.

Specimen Preservative Medium

Composition per liter:

NaCl	5.0g
Sodium citrate·$2H_2O$	5.0g
$(NH_4)_2HPO_4$	4.0g
KH_2PO_4	2.0g
Yeast extract	1.0g
Sodium deoxycholate	0.5g
$MgSO_4 \cdot 7H_2O$	0.4g
Glycerol	300.0mL

pH 7.0 ± 0.2 at 25°C

Preparation of Medium: Add components, except glycerol, to distilled/deionized water and bring volume to 700.0mL. Mix thoroughly. Gently heat and bring to boiling. Add 300.0mL of glycerol. Mix thoroughly. Distribute into tubes or flasks. Autoclave for 10 min at 11 psi pressure–116°C.

Use: For the preservation of viable microorganisms in stool specimens. For the transport of fecal material.

Spray's Fermentation Medium

Composition per 1100.0mL:

Neopeptone	10.0g
Pancreatic digest of casein	10.0g
Agar	2.0g
Sodium thioglycolate	0.025g
Carbohydrate solution	110.0mL

pH 7.4 ± 0.1 at 25°C

Carbohydrate Solution:

Composition per 200.0mL:

Carbohydrate	20.0g

Preparation of Carbohydrate Solution: Add carbohydrate to distilled/deionized water and bring volume to 200.0mL. Glucose or glycerol may be used. Mix thoroughly. Filter sterilize.

Preparation of Medium: Add components, except agar and carbohydrate solution, to distilled/deionized water and bring volume to 990.0mL. Mix thoroughly. Adjust pH to 7.4. Add agar. Gently heat and bring to boiling. Distribute into tubes in 9.0mL volumes. Autoclave for 15 min at 15 psi pressure–121°C. Cool to 25°C. Immediately prior to use heat tubes in a boiling water bath for 10 min. Cool to 45°C. Aseptically add 1.0mL of sterile carbohydrate solution. Mix thoroughly.

Use: For the cultivation and differentiation of *Clostridium perfringens* based on carbohydrate fermentation patterns.

SPS Agar (Sulfite Polymyxin Sulfadiazine Agar)

Composition per liter:

Pancreatic digest of casein	15.0g
Agar	13.9g
Yeast extract	10.0g
Ferric citrate	0.5g
Na_2SO_3	0.5g
Sulfadiazine	0.12g
Polymyxin sulfate	0.01g

pH 7.0 ± 0.2 at 25°C

Source: This medium is available as a premixed powder from BD Diagnostic Systems.

Preparation of Medium: Add components to distilled/deionized water and bring volume to 1.0L. Mix thoroughly. Gently heat while stirring and bring to boiling. Distribute into tubes or flasks. Autoclave for 15 min at 13 psi pressure–118°C. Pour into sterile Petri dishes or leave in tubes.

Use: For the isolation and detection of *Clostridium perfringens* and *Clostridium botulinum.*

SS Deoxycholate Agar (*Salmonella-Shigella* Deoxycholate Agar) (SSDC)

Composition per liter:

Agar	13.5g
Lactose	10.0g
Sodium desoxycholate	10.0g
Bile salts	8.5g
$Na_2S_2O_3$	8.5g
Sodium citrate	8.5g
Beef extract	5.0g
Pancreatic digest of casein	2.5g
Peptic digest of animal tissue	2.5g
$CaCl_2 \cdot 2H_2O$	1.0g
Ferric citrate	1.0g

Neutral Red..0.025g
Brilliant Green 0.33mg
pH 7.0 ± 0.2 at 25°C

Preparation of Medium: Add components to distilled/deionized water and bring volume to 1.0L. Mix thoroughly. Gently heat while stirring and bring to boiling. Do not autoclave. Cool to 45°–50°C. Pour into sterile Petri dishes in 20.0mL volumes. Allow the surface of the plates to dry before inoculation.

Use: For the isolation and cultivation of *Yersinia enterocolitica.*

ST Holding Medium (m-ST Holding Medium)

Composition per liter:
KH_2PO_4...3.0g
Tris(hydroxymethyl)aminomethane buffer.........3.0g
Sulfanilamide...1.5g
$NaH_2PO_4{\cdot}H_2O$..0.1g
Ethanol (95% solution) 10.0mL
pH 8.6 ± 0.2 at 25°C

Preparation of Medium: Dissolve the sulfanilamide in the ethanol. Add all components to distilled/deionized water and bring volume to 1.0L. Mix thoroughly. Autoclave for 15 min at 15 psi pressure–121°C. Distribute in 1.8mL volumes to sterile Petri dishes with tight-fitting lids and an absorbent filter.

Use: For the cultivation and enumeration of coliform bacteria by the delayed-incubation total coliform procedure. For use as a holding or transport medium to keep coliform bacteria viable between sampling and laboratory culture.

Standard Fluid Medium 10B (Shepard's M10 Medium)

Composition per 102.5mL:
Base solution... 70.0mL
Horse serum, unheated................................. 20.0mL
Fresh yeast extract solution.......................... 10.0mL
Penicillin solution 1.0mL
CVA enrichment... 0.5mL
L-Cysteine·HCl·H_2O solution.......................... 0.5mL
Urea solution... 0.4mL
Phenol Red.. 0.1mL
pH 6.0 ± 0.2 at 25°C

Base Solution:
Composition per 70.0mL:
Beef heart, solids from infusion..........................5.0g
Peptone...1.0g
NaCl 0.5g

Preparation of Base Solution: Add components to distilled/deionized water and bring volume to 70.0mL. Mix thoroughly. Adjust pH to 5.5 with 2*N* HCl. Autoclave for 15 min at 15 psi pressure–121°C. Cool to 45°–50°C.

Fresh Yeast Extract Solution:
Composition per 100.0mL:
Baker's yeast, live, pressed, starch-free,...........25.0g

Preparation of Fresh Yeast Extract Solution: Add the live Baker's yeast to 100.0mL of distilled/deionized water. Autoclave for 90 min at 15 psi pressure–121°C. Allow to stand. Remove supernatant solution. Adjust pH to 6.6–6.8.

Penicillin Solution:
Composition per 10.0mL:
Penicillin G 1,000,000U

Preparation of Penicillin Solution: Add penicillin to distilled/deionized water and bring volume to 10.0mL. Mix thoroughly. Filter sterilize.

CVA Enrichment:
Composition per liter:
Glucose 100.0g
L-Cysteine·HCl·H_2O...25.9g
L-Glutamine 10.0g
L-Cystine·2HCl 1.0g
Adenine..1.0g
Nicotinamide adenine dinucleotide 0.25g
Cocarboxylase...0.1g
Guanine·HCl 0.03g
$Fe(NO_3)_3$ 0.02g
p-Aminobenzoic acid....................................0.013g
Vitamin B_{12} 0.01g
Thiamine·HCl 3.0mg

Preparation of CVA Enrichment: Add components to distilled/deionized water and bring volume to 1.0L. Mix thoroughly. Filter sterilize.

L-Cysteine·HCl·H_2O Solution:
Composition per 10.0mL:
L-Cysteine·HCl·H_2O 0.2g

Preparation of L-Cysteine·HCl·H_2O Solution: Add L-cysteine·HCl·H_2O to distilled/deionized water and bring volume to 10.0mL. Mix thoroughly. Filter sterilize.

Urea Solution:
Composition per 10.0mL:
Urea...1.0g

Preparation of Urea Solution: Add urea to distilled/deionized water and bring volume to 10.0mL. Mix thoroughly. Filter sterilize.

Phenol Red Solution:
Composition per 10.0mL:
Phenol Red...0.1g

Preparation of Phenol Red Solution: Add Phenol Red to distilled/deionized water and bring volume to 10.0mL. Mix thoroughly. Autoclave for 15 min at 15 psi pressure–121°C.

Preparation of Medium: To 70.0mL of cooled, sterile base solution, aseptically add 20.0mL of sterile horse serum, 10.0mL of sterile fresh yeast extract solution, 1.0mL of sterile penicillin solution, 0.5mL of sterile CVA enrichment, 0.5mL of sterile L-cysteine·HCl·H_2O solution, 0.4mL of sterile urea solution, and 0.1mL of sterile Phenol Red solution. Mix thoroughly. Aseptically distribute into sterile tubes or flasks.

Use: For the isolation and cultivation of *Ureaplasma urealyticum* from clinical specimens.

Standard II Nutrient Agar

Composition per liter:

Agar	13.0g
Tryptose	7.0g
NaCl	5.0g

pH 7.5 ± 0.2 at 25°C

Source: This medium is available as a premixed powder from BD Diagnostic Systems.

Preparation of Medium: Add components to distilled/deionized water and bring volume to 1.0L. Mix thoroughly. Gently heat and bring to boiling. Distribute into tubes or flasks. Autoclave for 15 min at 15 psi pressure–121°C. Pour into sterile Petri dishes or leave in tubes.

Use: For the cultivation of nonfastidious microorganisms. For the maintenance of cultures of a wide variety of nonfastidious bacteria. May also be used as a base for blood and other enrichments for the cultivation of fastidious microorganisms. May be used to determine indole production.

Staphylococcus Agar No. 110

Composition per liter:

NaCl	75.0g
Gelatin	30.0g
Agar	15.0g
D-Mannitol	10.0g
Pancreatic digest of casein	10.0g
K_2HPO_4	5.0g
Yeast extract	2.5g
Lactose	2.0g

pH 7.0 ± 0.2 at 25°C

Source: This medium is available as a premixed powder from BD Diagnostic Systems and Oxoid Unipath.

Preparation of Medium: Add components to distilled/deionized water and bring volume to 1.0L. Mix thoroughly. Gently heat and bring to boiling. Distribute into tubes or flasks. Autoclave for 15 min at 15 psi pressure–121°C. Pour into sterile Petri dishes or leave in tubes. Swirl flask while pouring plates to disperse precipitate.

Use: For the isolation and enumeration of staphylococci from clinical and nonclinical specimens.

Staphylococcus Broth (m-*Staphylococcus* Broth)

Composition per liter:

NaCl	75.0g
Mannitol	10.0g
Pancreatic digest of casein	10.0g
K_2HPO_4	5.0g
Yeast extract	2.5g
Lactose	2.0g

pH 7.0 ± 0.2 at 25°C

Source: This medium is available as a premixed powder from BD Diagnostic Systems.

Preparation of Medium: Add components to distilled/deionized water and bring volume to 1.0L. Mix thoroughly. Distribute into tubes or flasks. Autoclave for 15 min at 15 psi pressure–121°C.

Use: For the cultivation and enumeration of pathogenic and enterotoxigenic staphylococci by the membrane filter method.

Staphylococcus Medium

Composition per liter:

Agar	15.0g
Peptone	6.0g
Pancreatic digest of casein	4.0g
Yeast extract	3.0g
Beef extract	1.5g
Glucose	1.0g

pH 6.6 ± 0.2 at 25°C

Preparation of Medium: Add components to distilled/deionized water and bring volume to 1.0L. Mix thoroughly. Gently heat and bring to boiling. Distribute into tubes or flasks. Autoclave for 15 min at 15 psi pressure–121°C. Pour into sterile Petri dishes or leave in tubes.

Use: For the cultivation and maintenance of *Staphylococcus aureus*. For the enumeration of pathogenic and enterotoxigenic staphylococci by the membrane filter method.

Staphylococcus Medium No. 110

Composition per liter:

NaCl	75.0g

Gelatin	30.0g
Agar	15.0g
Mannitol	10.0g
Pancreatic digest of casein	10.0g
K_2HPO_4	5.0g
Yeast extract	2.5g
Lactose	2.0g

pH 7.1 ± 0.2 at 25°C

Source: This medium is available as a premixed powder from BD Diagnostic Systems.

Preparation of Medium: Add components to distilled/deionized water and bring volume to 1.0L. Mix thoroughly. Gently heat and bring to boiling. Distribute into tubes or flasks. Autoclave for 15 min at 15 psi pressure–121°C. Mix thoroughly. Pour into sterile Petri dishes or leave in tubes.

Use: For the selective isolation, cultivation, and maintenance of *Staphylococcus* species.

Staphylococcus/Streptococcus Selective Medium

Composition per 1060.0mL:

Columbia blood agar base	1.0L
Horse blood, defibrinated	50.0mL
Antibiotic inhibitor	10.0mL

pH 7.3 ± 0.2 at 25°C

Columbia Blood Agar Base:

Composition per liter:

Special peptone	23.0g
Agar	10.0g
NaCl	5.0g
Starch	1.0g

Source: Columbia blood agar base is available as a premixed powder from Oxoid Unipath.

Preparation of Columbia Blood Agar Base: Add components to distilled/deionized water and bring volume to 1.0L. Mix thoroughly. Gently heat and bring to boiling. Autoclave for 15 min at 15 psi pressure–121°C. Cool to 45°–50°C.

Antibiotic Inhibitor:

Composition per 10.0mL:

Nalidixic acid	0.015g
Colistin sulfate	0.01g
Ethanol (95% solution)	10.0mL

Preparation of Antibiotic Inhibitor: Add components to 10.0mL ethanol. Mix thoroughly. Filter sterilize.

Preparation of Medium: To 1.0L of cooled sterile Columbia blood agar base, aseptically add sterile horse blood and sterile antibiotic inhibitor. Mix thoroughly. Pour into sterile Petri dishes or distribute into sterile tubes.

Use: For the selective isolation of *Staphylococcus aureus* and streptococci from clinical specimens.

Starch Agar

Composition per liter:

Agar	12.0g
Soluble starch	10.0g
Beef extract	3.0g

pH 7.5 ± 0.2 at 25°C

Source: This medium is available as a premixed powder from BD Diagnostic Systems.

Preparation of Medium: Add components to distilled/deionized water and bring volume to 1.0L. Mix thoroughly. Gently heat and bring to boiling. Distribute into tubes or flasks. Autoclave for 15 min at 15 psi pressure–121°C. Pour into sterile Petri dishes.

Use: For the cultivation and differentiation of a variety of microorganisms based on amylase production. After incubation, starch hydrolysis is determined by the addition of Gram's or Lugol's iodine solution. Organisms that produce amylase appear as colonies surrounded by a clear zone.

Starch Agar with Bromcresol Purple

Composition per liter:

Agar	15.0g
Cornstarch	10.0g
Meat peptone	10.0g
Bromcresol Purple solution	1.2mL

pH 6.8 ± 0.2 at 25°C

Bromcresol Purple Solution:

Composition per 10.0mL:

Bromcresol Purple	0.16g
Ethanol (95% solution)	10.0mL

Preparation of Bromcresol Purple Solution: Add Bromcresol Purple to 10.0mL of 95% ethanol. Mix thoroughly.

Preparation of Medium: Add components to distilled/deionized water and bring volume to 1.0L. Mix thoroughly. Gently heat and bring to boiling. Distribute into tubes or flasks. Autoclave for 15 min at 15 psi pressure–121°C. Pour into sterile Petri dishes or leave in tubes.

Use: For the differentiation of *Gardnerella vaginalis (Haemophilus vaginalis, Corynebacterium vaginale)* from other microorganisms found in the genitourinary tract, with the exception of some strains of *Streptococcus* and *Loctobacillus*. Differentiation is based on starch hydrolysis. Bacteria that can hydrolyze starch appear as colonies surrounded by a yellow zone.

Starch Agar with Bromcresol Purple

Composition per liter:

Solution 1	200.0mL
Solution 2	20.0mL

pH 7.8 ± 0.2 at 25°C

Solution 1:

Composition per 200.0mL:

Heart infusion agar	5.0mL
Bromcresol Purple solution	0.2mL

Preparation of Solution 1: Add components to distilled/deionized water and bring volume to 200.0mL. Mix thoroughly. Gently heat while stirring and bring to boiling.

Heart Infusion Agar:

Composition per liter:

Beef heart, solids from infusion	500.0g
Agar	15.0g
Tryptose	10.0g
NaCl	5.0g

Preparation of Heart Infusion Agar: Add components to distilled/deionized water and bring volume to 1.0L. Mix thoroughly. Gently heat and bring to boiling.

Bromcresol Purple Solution:

Composition per 10.0mL:

Bromcresol Purple	0.16g
Ethanol (95% solution)	10.0mL

Preparation of Bromcresol Purple Solution: Add Bromcresol Purple to 10.0mL of ethanol. Mix thoroughly.

Solution 2:

Composition per 20.0mL:

Starch	0.4g

Preparation of Solution 2: Add starch to distilled/deionized water and bring volume to 20.0mL. Mix thoroughly. Gently heat while stirring and bring to boiling.

Preparation of Medium: Combine solution 1 and solution 2. Mix thoroughly. Autoclave for 15 min at 15 psi pressure–121°C. Pour into sterile Petri dishes or distribute into sterile tubes.

Use: For the differentiation of *Gardnerella vaginalis (Haemophilus vaginalis, Corynebacterium vaginale)* from other microorganisms found in the genitourinary tract, with the exception of some strains of *Streptococcus* and *Lactobacillus*. Differentiation is based on starch hydrolysis. Bacteria that can hydrolyze starch appear as colonies surrounded by a yellow zone.

Starch Hydrolysis Agar

Composition per liter:

Beef heart, infusion from	500.0g
Soluble starch	20.0g
Agar	15.0g
Tryptose	10.0g
NaCl	5.0g

pH 7.4 ± 0.2 at 25°C

Preparation of Medium: Add components to distilled/deionized water and bring volume to 1.0L. Mix thoroughly. Gently heat and bring to boiling. Distribute into tubes or flasks. Autoclave for 15 min at 15 psi pressure–121°C. Pour into sterile Petri dishes or leave in tubes.

Use: For the cultivation and differentiation of a variety of microorganisms based on amylase production. After incubation, starch hydrolysis is determined by the addition of Gram's or Lugol's iodine solution. Organisms that produce amylase appear as colonies surrounded by a clear zone.

Sterility Test Broth (USP Alternative Thioglycolate Medium)

Composition per liter:

Pancreatic digest of casein	15.0g
Glucose	5.0g
Yeast extract	5.0g
NaCl	2.5g
L-Cystine	0.5g
Sodium thioglycolate	0.5g

pH 7.1 ± 0.2 at 25°C

Source: This medium is available as a premixed powder from BD Diagnostic Systems.

Preparation of Medium: Add components to distilled/deionized water and bring volume to 1.0L. Mix thoroughly. Gently heat and bring to boiling. Distribute into tubes or flasks. Autoclave for 15 min at 15 psi pressure–121°C. Cool to 25°C. If not used immediately, prior to inoculation heat tubes in a boiling water bath for 5–10 min. Cool to 25°C.

Use: As an alternate medium, instead of fluid thioglycolate broth, for testing the sterility of a variety of specimens.

Stock Culture Agar

Composition per liter:

Beef heart infusion	500.0g
Gelatin	10.0g

Proteose peptone	10.0g
Agar	7.5g
Casein	5.0g
Na_2HPO_4	4.0g
Sodium citrate	3.0g
Glucose	0.5g

pH 7.5 ± 0.2 at 25°C

Source: This medium is available as a premixed powder from BD Diagnostic Systems.

Preparation of Medium: Add components to cold distilled/deionized water and bring volume to 1.0L. Mix thoroughly. Gently heat while stirring and bring to boiling. Distribute into tubes or flasks. Autoclave for 15 min at 15 psi pressure–121°C. Pour into sterile Petri dishes or leave in tubes.

Use: For the maintenance of pathogenic and nonpathogenic bacteria, especially streptococci.

Stock Culture Agar with L-Asparagine

Composition per liter:

Beef heart infusion	500.0g
Gelatin	10.0g
Proteose peptone	10.0g
Agar	7.5g
Casein	5.0g
Na_2HPO_4	4.0g
Sodium citrate	3.0g
L-Asparagine	1.0g
Glucose	0.5g

pH 7.5 ± 0.2 at 25°C

Preparation of Medium: Add components to cold distilled/deionized water and bring volume to 1.0L. Mix thoroughly. Gently heat while stirring and bring to boiling. Distribute into tubes or flasks. Autoclave for 15 min at 15 psi pressure–121°C. Pour into sterile Petri dishes or leave in tubes.

Use: For the maintenance of pathogenic and nonpathogenic bacteria, especiallly streptococci.

Stonebrink's Medium

Composition per 1600.0mL:

Homogenized whole egg	2.0L
Mineral salts solution	1.0L
Malachite Green solution	40.0mL

Mineral Salts Solution:

Na-pyruvate	12.5g
KH_2PO_4	7.0g
$Na_2HPO_4 \cdot 7H_2O$	4.0g

Preparation of Mineral Salts Solution: Add components to distilled/deionized water and bring volume to 1.0L. Mix thoroughly. Autoclave for 15 min at 15 psi pressure–121°C. Cool to 50°C.

Malchite Green Solution:
Composition per 100.0mL:

Malachite green	2.0g

Preparation of Malchite Green Solution: Add malachite to distilled/deionized water and bring volume to 100.0mL. Mix thoroughly. Autoclave for 15 min at 15 psi pressure–121°C. Cool to 50°C.

Homogenized Whole Egg:
Composition per 2.0L:

Whole eggs	36-48

Preparation of Homogenized Whole Egg: Use fresh eggs, less than 1 week old. Scrub the shells with soap. Let stand in a soap solution for 30 min. Rinse in running water. Soak eggs in 70% ethanol for 15 min. Break the eggs into a sterile container. Homogenize by shaking. Filter through four layers of sterile cheesecloth into a sterile graduated cylinder. Measure out 2.0L.

Preparation of Medium: Aseptically add 40.0mL sterile malachite to 1.0L of sterile mineral salts solution. Mix thoroughly. Aseptically add 2.0L of homogenized whole egg. Mix thoroughly. Distribute into sterile screw-capped tubes. Place tubes in a slanted position. Inspissate at 85°C (moist heat) for 45 min.

Use: For the cultivation of *Mycobacterium* species. For the isolation of *Mycobacterium bovis.*

Strep ID Quad Plate

Composition per liter:

Quadrant I	5.0mL
Quadrant II	5.0mL
Quadrant III	5.0mL
Quadrant IV	5.0mL

Source: Available as a prepared medium from BD Diagnostic Systems.

Quadrant I:
Composition per 5.0mL:

Bacitracin	0.5mg
TSA II agar	5.0mL

Quadrant II:
Composition per 5.0mL:

TSA II agar	5.0mL
Sheep blood, defibrinated	0.25mL

Quadrant III:
Composition per 5.0mL:

Bile esculin agar	5.0mL

Quadrant IV:
Composition per 5.0mL:

Blood agar base with 6.5% NaCl	5.0mL

Preparation of Quadrant Media: Sterilize agars by autoclaving for 15 min at 15 psi pressure–121°C. Cool to 45°–50°C. Add additional components as filter sterilized solutions. Mix and distribute as 5.0mL aliquots into quadrants.

Use: For the differentiation and presumptive identification of streptococci. The Strep (*Streptococcus)* ID (Identification) Quad Plate is a four-sectored plate, each containing a different medium.

StrepB Carrot Broth™

Composition per liter:

Proteose peptone no. 3	25.0g
Soluble starch	20.0g
$MgSO_4$	20.0g
Selective agents	12.2g
Morpholinepropanesulfonic acid (MOPS)	11.0g
Na_2HPO_4	8.5g
Glucose	2.5g
Sodium Pyruvate	1.0g
StrepB carrot broth disk growth promoting factors	variable

pH 7.4 ± 0.1 at 25°C

Source: This medium is available from Hardy Diagnostics.

Preparation: This medium is supplied as a prepared broth in tubes. The StrepB carrot broth disk is added to a tube just prior to inoculation with a vaginal swab. The disk must remain submerged in the broth.

Use: For detecting the presence of Group B *Streptococcus* infections in pregnant women. This new screening test is an improvement over conventional methods, by increasing sensitivity, decreasing turn around time, while lowering overall cost. Tubes show an orange to red color change, typical of group B streptococci.

Streptococcal Growth Medium

Composition per liter:

Beef heart, solids from infusion	500.0g
Tryptose	10.0g
NaCl	5.0g
Glucose	1.0g
Bromcresol Purple solution	1.0mL

pH 7.4 ± 0.2 at 25°C

Bromcresol Purple Solution:

Composition per 10.0mL:

Bromcresol Purple	0.16g
Ethanol (95% solution)	10.0mL

Preparation of Bromcresol Purple Solution: Add Bromcresol Purple to 10.0mL of ethanol. Mix thoroughly.

Preparation of Medium: Add components to distilled/deionized water and bring volume to 1.0L. Mix thoroughly. Distribute into tubes in 5.0mL volumes. Autoclave for 15 min at 15 psi pressure–121°C.

Use: For the cultivation of *Streptococcus* species and other Gram-positive cocci. Growth in this medium turns the indicator yellow and the solution turbid.

Streptococcus Agar

Composition per liter:

Glucose	20.0g
Pancreatic digest of casein	20.0g
Agar	15.0g
K_2HPO_4	2.0g
$MgSO_4 \cdot 7H_2O$	0.1g

pH 6.8 ± 0.2 at 25°C

Preparation of Medium: Add components to distilled/deionized water and bring volume to 1.0L. Mix thoroughly. Gently heat and bring to boiling. Distribute into tubes or flasks. Autoclave for 15 min at 15 psi pressure–121°C. Pour into sterile Petri dishes or leave in tubes.

Use: For the cultivation and maintenance of *Streptococcus* species.

Streptococcus Blood Agar, Selective

Composition per liter:

Agar	15.0g
Pancreatic digest of casein	10.0g
Beef extract	6.7g
Nucleic acid	6.0g
NaCl	5.0g
Sheep blood, defibrinated	50.0mL
Maltose solution	10.0mL
Antibiotic inhibitor	10.0mL

pH 7.3 ± 0.2 at 25°C

Maltose Solution:

Composition per 10.0mL:

Maltose	0.25–5.0g

Preparation of Maltose Solution: Add maltose to distilled/deionized water and bring volume to 10.0mL. Mix thoroughly. Filter sterilize.

Antibiotic Inhibitor:

Composition per 10.0mL:

Polymyxin B sulfate	0.02g
Neomycin sulfate	0.01g

Preparation of Antibiotic Inhibitor: Add components to distilled/deionized water and bring volume to 10.0mL. Mix thoroughly. Filter sterilize.

Preparation of Medium: Add components—except sheep blood, maltose solution, and antibiotic inhibitor—to distilled/deionized water and bring volume to 930.0mL. Mix thoroughly. Gently heat and bring to boiling. Autoclave for 15 min at 15 psi pressure–121°C. Cool to 45°–50°C. Aseptically add sterile sheep blood, sterile maltose solution, and sterile antibiotic inhibitor. Mix thoroughly. Pour into sterile Petri dishes or distribute into sterile tubes.

Use: For the isolation and cultivation of group A hemolytic *Streptococcus* species from the human respiratory tract.

Streptococcus Medium

Composition per liter:

Agar	15.0g
Glucose	4.0g
K_2HPO_4	3.8g
Pancreatic digest of casein	2.5g
Yeast extract	2.5g

pH 7.6 ± 0.2 at 25°C

Preparation of Medium: Add components to distilled/deionized water and bring volume to 1.0L. Mix thoroughly. Gently heat and bring to boiling. Distribute into tubes or flasks. Autoclave for 15 min at 15 psi pressure–121°C. Pour into sterile Petri dishes or leave in tubes.

Use: For the cultivation and maintenance of *Enterococcus faecalis.*

Streptococcus pneumoniae Medium

Composition per liter:

Pancreatic digest of casein	17.0g
Glucose	10.0g
NaCl	5.0g
Papaic digest of soybean meal	3.0g
Yeast extract	3.0g
K_2HPO_4	2.5g

pH 7.2 ± 0.2 at 25°C

Preparation of Medium: Add components to distilled/deionized water and bring volume to 1.0L. Mix thoroughly. Adjust pH to 7.2. Distribute into tubes or flasks. Autoclave for 15 min at 15 psi pressure–121°C.

Use: For the cultivation of *Streptococcus pneumoniae.*

Streptococcus Selective Medium

Composition per liter:

Special peptone	23.0g
Agar	10.0g
NaCl	5.0g
Starch	1.0g
Horse blood, defibrinated	50.0mL
Antibiotic inhibitor	10.0mL

pH 7.3± 0.2 at 25°C

Source: This medium is available as a premixed powder from Oxoid Unipath.

Antibiotic Inhibitor:
Composition per 10.0mL:

Colistin sulfate	10.0mg
Oxolinic acid	5.0mg

Preparation of Antibiotic Inhibitor: Add components to distilled/deionized water and bring volume to 10.0mL. Mix thoroughly. Filter sterilize.

Preparation of Medium: Add components, except horse blood and antibiotic inhibitor, to distilled/deionized water and bring volume to 940.0mL. Mix thoroughly. Gently heat and bring to boiling. Autoclave for 15 min at 15 psi pressure–121°C. Cool to 45°–50°C. Aseptically add sterile horse blood and sterile antibiotic inhibitor. Mix thoroughly. Pour into sterile Petri dishes or distribute into sterile tubes.

Use: For the selective isolation of streptococci from clinical specimens.

Streptosel™ Agar

Composition per liter:

Pancreatic digest of casein	15.0g
Agar	12.0g
Glucose	5.0g
Papaic digest of soybean meal	5.0g
NaCl	4.0g
Sodium citrate	1.0g
L-Cystine	0.2g
NaN_3	0.2g
Na_2SO_3	0.2g
Crystal Violet	0.2mg

pH 7.4 ± 0.2 at 25°C

Source: This medium is available as a premixed powder from BD Diagnostic Systems.

Preparation of Medium: Add components to distilled/deionized water and bring volume to 1.0L. Mix thoroughly. Gently heat and bring to boiling. Distribute into tubes or flasks. If medium is used the same day, do not autoclave. Pour into sterile Petri dishes or leave in tubes. If medium is to be stored, autoclave for 15 min at 13 psi pressure–118°C. Pour into sterile Petri dishes or leave in tubes.

Use: For the selective isolation, cultivation, and enumeration of streptococci from specimens containing a mixed flora.

Streptosel Broth

Composition per liter:

Pancreatic digest of casein 15.0g
Glucose 5.0g
Papaic digest of soybean meal 5.0g
NaCl 4.0g
Sodium citrate 1.0g
L-Cystine 0.2g
Na_2SO_3 .... 0.2g
NaN_3 .... 0.2g
Crystal Violet 0.2mg

pH 7.4 ± 0.2 at 25°C

Source: This medium is available as a premixed powder from BD Diagnostic Systems.

Preparation of Medium: Add components to distilled/deionized water and bring volume to 1.0L. Mix thoroughly. Distribute into tubes or flasks. Autoclave for 15 min at 13 psi pressure–118°C.

Use: For the selective isolation and cultivation of streptococci from specimens containing a mixed flora.

Stuart *Leptospira* Broth, Modified

Composition per liter:

NaCl 1.93g
Na_2HPO_4 .... 0.66g
NH_4Cl 0.34g
$MgCl_2·6H_2O$ 0.19g
L-Asparagine 0.13g
KH_2PO_4 .... 0.08g
Rabbit serum,
inactivated at 56°C, 30 min 100.0mL
Glycerol 5.0mL

pH 7.4 ± 0.2 at 25°C

Preparation of Medium: Add each component, except rabbit serum, to distilled/deionized water in separate flasks and bring each volume to 100.0mL. Mix thoroughly. Combine the seven solutions, except the rabbit serum. Mix thoroughly. Gently heat and bring to boiling. Autoclave for 15 min at 15 psi pressure–121°C. Cool to 45°–50°C. Aseptically add sterile rabbit serum. Mix thoroughly. Aseptically distribute into sterile tubes or flasks.

Use: For the cultivation of *Leptospira* species.

Stuart Medium Base

Composition per 1100.0mL:

NaCl 1.8g
Na_2HPO_4 .... 0.67g
$MgCl_2·6H_2O$ 0.41g
NH_4Cl 0.27g
Asparagine 0.13g
KH_2PO_4 .... 0.09g
Phenol Red 0.01g
Leptospira enrichment 100.0mL
Glycerol 5.0mL

pH 7.6 ± 0.2 at 25°C

Source: This medium is available as a premixed powder from BD Diagnostic Systems. *Leptospira* enrichment contains rabbit serum and hemoglobin and is available from BD Diagnostic Systems.

Preparation of Medium: Add components, except glycerol and *Leptospira* enrichment, to distilled/deionized water and bring volume to 995.0mL. Mix thoroughly. Add glycerol. Mix thoroughly. Autoclave for 15 min at 15 psi pressure–121°C. Cool to 45°–50°C. Aseptically add *Leptospira* enrichment. Mix thoroughly. Aseptically distribute into sterile screw-capped tubes in 10.0mL volumes.

Use: For the cultivation of *Leptospira* species.

Stuart Transport Medium

Composition per liter:

Sodium glycerophosphate 10.0g
Sodium thioglycolate 1.0g
$CaCl_2·2H_2O$ 0.1g
Methylene Blue 2.0mg

pH 7.4 ± 0.2 at 25°C

Preparation of Medium: Add components to distilled/deionized water and bring volume to 1.0L. Mix thoroughly. Gently heat and bring to boiling. Distribute into 7.0mL screw-capped tubes. Fill tubes to capacity. Autoclave for 15 min at 15 psi pressure–121°C.

Use: For the preservation of *Neisseria* species and other fastidious organisms during their transport from clinic to laboratory.

Stuart Transport Medium, Modified

Composition per liter:

Sodium glycerophosphate 10.0g
Agar 5.0g
L-Cysteine·HCl·H_2O 0.5g
Sodium thioglycolate 0.5g
$CaCl_2·2H_2O$ 0.1g
Methylene Blue 1.0mg

pH 7.4 ± 0.2 at 25°C

Source: This medium is available as a premixed powder from Oxoid Unipath.

Preparation of Medium: Add components to distilled/deionized water and bring volume to 1.0L. Mix thoroughly. Gently heat and bring to boiling. Distribute into 7.0mL screw-capped tubes. Fill tubes to capacity. Autoclave for 15 min at 15 psi pressure–121°C.

Use: For the preservation of *Neisseria* species and other fastidious organisms during their transport from clinic to laboratory.

Sucrose Phosphate Glutamate Transport Medium

Composition per liter:

Sucrose	75.0g
Na_2HPO_4	1.22g
Glutamic acid	0.72g
K_2HPO_4	0.52g
Bovine serum	50.0mL
Antibiotic inhibitor	10.0mL

pH 7.4–7.6 at 25°C

Antibiotic Inhibitor:

Composition per 10.0mL:

Vancomycin	0.1g
Streptomycin	0.05g
Nystatin	25000U

Preparation of Antibiotic Inhibitor: Add components to distilled/deionized water and bring volume to 10.0mL. Mix thoroughly. Filter sterilize.

Preparation of Medium: Add components, except bovine serum and antibiotic inhibitor, to distilled/deionized water and bring volume to 940.0mL. Mix thoroughly. Gently heat and bring to boiling. Adjust pH to 7.4–7.6. Autoclave for 15 min at 15 psi pressure–121°C. Cool to 45°–50°C. Aseptically add sterile bovine serum and sterile antibiotic inhibitor. Mix thoroughly. Aseptically distribute into sterile tubes or flasks.

Use: For the maintenance of *Chlamydia* species during transport.

Sucrose Phosphate Transport Medium

Composition per liter:

Sucrose	68.5g
K_2HPO_4	2.1g
KH_2PO_4	1.1g
Bovine serum	50.0mL
Antibiotic inhibitor	10.0mL

pH 7.0 ± 0.2 at 25°C

Antibiotic Inhibitor:

Composition per 10.0mL:

Vancomycin	0.1g
Streptomycin	0.05g
Nystatin	25,000U

Preparation of Antibiotic Inhibitor: Add components to distilled/deionized water and bring volume to 10.0mL. Mix thoroughly. Filter sterilize.

Preparation of Medium: Add components, except bovine serum and antibiotic inhibitor, to distilled/deionized water and bring volume to 940.0mL. Mix thoroughly. Gently heat and bring to boiling. Adjust pH to 7.0. Autoclave for 15 min at 15 psi pressure–121°C. Cool to 45°–50°C. Aseptically add sterile bovine serum and sterile antibiotic inhibitor. Mix thoroughly. Aseptically distribute into sterile tubes or flasks.

Use: For the maintenance of *Chlamydia* species during transport.

Sucrose Teepol Tellurite Agar (STT Agar)

Composition per liter:

Agar	20.0g
Beef extract	1.0g
Peptone	1.0g
Sucrose	1.0g
NaCl	0.5g
Bromthymol Blue (0.2% solution)	2.5mL
Tellurite solution	2.5mL
Sodium lauryl sulfate (Teepol, 0.1% solution)	0.2mL

pH 8.0 ± 0.2 at 25°C

Tellurite Solution:

Composition per 100.0mL:

K_2TeO_3	0.05g

Preparation of Tellurite Solution: Add the K_2TeO_3 to distilled/deionized water and bring the volume to 100.0mL. Mix thoroughly. Filter sterilize. Use freshly prepared solution.

Caution: Potassium tellurite is toxic.

Preparation of Medium: Add components to distilled/deionized water and bring volume to 1.0L. Mix thoroughly. Gently heat and bring to boiling. Do not autoclave. Pour into sterile Petri dishes.

Use: For the selective isolation, cultivation, and differentiation of *Vibrio* species based on their ability to ferment sucrose. *Vibrio cholerae* appears as flat yellow colonies. *Vibrio parahaemolyticus* appears as elevated green-yellow mucoid colonies.

Sucrose Yeast Extract Medium

Composition per liter:

Sucrose	20.0g
Yeast extract	4.0g
K_2HPO_4	2.5g
$MgSO_4 \cdot 7H_2O$	1.0g
Trace elements solution	4.0mL

pH 7.0 ± 0.2 at 25°C

Trace Elements Solution:

Composition per liter:

$ZnSO_4 \cdot 7H_2O$	18.0g

$FeSO_4 \cdot 7H_2O$ 9.0g
$MnSO_4 \cdot 4H_2O$ 3.0g
$CoCl_2 \cdot 6H_2O$ 0.9g
$CuSO_4 \cdot 5H_2O$ 0.8g
Conc. H_2SO_4 .... 5.0mL

Preparation of Trace Elements Solution: Add components to distilled/deionized water and bring volume to 1.0L. Mix thoroughly.

Preparation of Medium: Add components to distilled/deionized water and bring volume to 1.0L. Mix thoroughly. Distribute into tubes or flasks. Autoclave for 15 min at 15 psi pressure–121°C.

Use: For the cultivation of a variety of bacteria that can utilize sucrose as carbon source.

Tap Water Agar

Composition per liter:

Agar 15.0g
Tap water 1.0L

Preparation of Medium: Add agar to 1.0L of tap water. Mix thoroughly. Gently heat and bring to boiling. Autoclave for 15 min at 15 psi pressure–121°C. Pour into sterile Petri dishes.

Use: For the cultivation and differentiation of fungi and aerobic actinomycetes based on filament and aerial hyphae morphology.

Tarshis Blood Agar

Composition per 1050.0mL:

Beef heart infusion 500.0g
Agar 15.0g
Meat peptone 10.0g
NaCl 5.0g
Penicillin G, sterile 100,000U
Sheep blood, sterile 300.0mL
Glycerol 10.0mL

pH 6.6 ± 0.2 at 25°C

Preparation of Medium: Add components, except sheep blood and penicillin G, to distilled/deionized water and bring volume to 750.0mL. Mix thoroughly. Gently heat and bring to boiling. Autoclave for 15 min at 15 psi pressure–121°C. Cool to 45°–50°C. Aseptically add sterile sheep blood and sterile penicillin G. Mix thoroughly. Pour into sterile Petri dishes or distribute into sterile tubes.

Use: For the isolation and cultivation of *Mycobacterium tuberculosis.*

TB Nitrate Reduction Broth

Composition per 100.0mL:

$Na_2HPO_4 \cdot 12H_2O$ 0.485g
KH_2PO_4 .... 0.117g
$NaNO_3$ 0.085g

pH 7.0 ± 0.2 at 25°C

Preparation of Medium: Add components to distilled/deionized water and bring volume to 100.0mL. Mix thoroughly. Distribute into tubes or flasks. Autoclave for 15 min at 15 psi pressure–121°C.

Use: For the differentiation of *Mycobacterium* species based on nitrate reduction. After growth of cells in appropriate medium, nitrate reduction is determined by making a suspension of cells in TB nitrate reduction broth and adding hydrochloric acid, sulfanilamide, and *N*-naphylenendiamine. Nitrate reduction turns the medium pink. *Mycobacterium tuberculosis* reduces nitrate and turns the medium deep pink within 1 min. *Mycobacterium bovis* does not reduce nitrate and does not change the medium.

TBX Agar (Tryptone Bile X-glucuronide Agar)

Composition per liter:

Peptone 20.0g
Agar 15.0g
Bile salts 1.5g
X-β-D-glucuronide 0.075g

pH 7.2 ± 0.2 at 25°C

Source: This medium is available from Fluka, Sigma-Aldrich.

Preparation of Medium: Add components to distilled/deionized water and bring volume to 1.0L. Mix thoroughly. Gently heat while stirring and bring to boiling. Autoclave for 15 min at 15 psi pressure–121°C. Pour into sterile Petri dishes.

Use: For the detection and enumeration of *E. coli* without further confirmation. *E. coli* colonies are colored blue-green. The presence of the enzyme β-D-glucuronidase differentiates most *E. coli* ssp. from other coliforms. *E. coli* absorbs the chromogenic substrate 5-bromo-4-chloro-3-indolyl-β-D-glucuronide. The enzyme β-glucuronidase splits the bond between the chromophore 5-bromo-4-chloro-3-indolyle- and the β-D-glucuronide. *E. coli* colonies are colored blue-green. Growth of accompanying Gram-positive flora is largely inhibited by the use of bile salts.

TCBS Agar (Thiosulfate Citrate Bile Salt Sucrose Agar)

Composition per liter:

Sucrose 20.0g
Agar 14.0g

NaCl	10.0g
Sodium citrate	10.0g
$Na_2S_2O_3$	10.0g
Yeast extract	5.0g
Pancreatic digest of casein	5.0g
Peptic digest of animal tissue	5.0g
Oxgall	5.0g
Sodium cholate	3.0g
Ferric citrate	1.0g
Thymol Blue	0.04g
Bromthymol Blue	0.04g

pH 8.6 ± 0.2 at 25°C

Source: This medium is available as a premixed powder from BD Diagnostic Systems.

Preparation of Medium: Add components to distilled/deionized water and bring volume to 1.0L. Mix thoroughly. Gently heat while stirring and bring to boiling. Do not autoclave. Cool to 45°–50°C. Pour into sterile Petri dishes or distribute into sterile tubes.

Use: For the selective isolation of *Vibrio cholerae* and *Vibrio parahaemolyticus* from a variety of clinical and nonclinical specimens.

TCH Medium (Thiophene 2 Carboxylic Acid Hydrazide Medium)

Composition per 1105.0mL:

Thiophene-2-carboxylic acid hydrazide	1.1mg
Middlebrook 7H10 agar base	1.0L
OADC enrichment	100.0mL
Glycerol	5.0mL

pH 6.6 ± 0.2 at 25°C

Middlebrook 7H10 Agar Base:

Composition per liter:

Agar	15.0g
Na_2HPO_4	1.5g
KH_2PO_4	1.5g
$(NH_4)_2SO_4$	0.5g
L-Glutamic acid	0.5g
Sodium citrate	0.4g
Ferric ammonium citrate	0.04g
$MgSO_4 \cdot 7H_2O$	0.025g
$ZnSO_4 \cdot 7H_2O$	1.0mg
$CuSO_4 \cdot 5H_2O$	1.0mg
Pyridoxine	1.0mg
Biotin	0.5mg
$CaCl_2 \cdot 2H_2O$	0.5mg
Malachite Green	0.25mg

Preparation of Middlebrook 7H10 Agar Base: Add glycerol to 900.0mL of distilled/deionized water and add remaining components. Mix thoroughly. Gently heat and bring to boiling.

Middlebrook OADC Enrichment:

Composition per 100.0mL:

Bovine albumin fraction V	5.0g
Glucose	2.0g
NaCl	0.85g
Oleic acid	0.05g
Catalase	4.0mg

Source: This enrichment is available as a prepared enrichment from BD Diagnostic Systems.

Preparation of Middlebrook OADC Enrichment: Add components to distilled/deionized water and bring volume to 100.0mL. Mix thoroughly. Filter sterilize.

Preparation for Medium: Combine components. Mix thoroughly. Distribute into tubes or flasks. Autoclave for 15 min at 15 psi pressure–121°C. Pour into sterile Petri dishes or leave in tubes.

Use: For the differentiation of *Mycobacterium* species based on sensitivity to TCH. *Mycobacterium bovis* is inhibited by TCH. *Mycobacterium tuberculosis* and other mycobacteria are generally resistant to low concentrations of TCH. This distinguishes *Mycobacterium bovis* from other nonchromogenic, slow-growing mycobacteria.

Tellurite Glycine Agar

Composition per liter:

Agar	17.5g
Pancreatic digest of casein	10.0g
Glycine	10.0g
Yeast extract	6.5g
D-Mannitol	5.0g
K_2HPO_4	5.0g
LiCl	5.0g
Enzymatic hydrolysate of soybean meal	3.5g
Chapman tellurite solution	10.0mL

pH 7.2 ± 0.2 at 25°C

Source: This medium is available as a premixed powder from BD Diagnostic Systems.

Chapman Tellurite Solution:

Composition per 100.0mL:

K_2TeO_3	1.0g

Preparation of Chapman Tellurite Solution: Add K_2TeO_3 to distilled/deionized water and bring volume to 100.0mL. Mix thoroughly. Filter sterilize.

Caution: Potassium tellurite is toxic.

Preparation of Medium: Add components, except Chapman tellurite solution, to distilled/deionized water and bring volume to 990.0mL. Mix thoroughly. Gently heat and bring to boiling. Autoclave for 15 min at 15 psi pressure–121°C. Cool to 50°–55°C. Aseptically add 10.0mL of sterile Chapman tellurite solution. Mix thoroughly. Pour into sterile Petri dishes or distribute into sterile tubes. Allow the surface of the plates to dry before inoculating.

Use: For the isolation and cultivation of coagulase-positive staphylococci.

Tellurite Glycine Agar

Composition per liter:

Agar	16.0g
Pancreatic digest of casein	10.0g
Glycine	10.0g
Yeast extract	5.0g
D-Mannitol	5.0g
K_2HPO_4	5.0g
LiCl	5.0g
Chapman tellurite solution	20.0mL

pH 7.2 ± 0.2 at 25°C

Source: This medium is available as a premixed powder from BD Diagnostic Systems.

Chapman Tellurite Solution:

Composition per 100.0mL:

K_2TeO_3	1.0g

Preparation of Chapman Tellurite Solution: Add K_2TeO_3 to distilled/deionized water and bring volume to 100.0mL. Mix thoroughly. Filter sterilize.

Caution: Potassium tellurite is toxic.

Preparation of Medium: Add components, except Chapman tellurite solution, to distilled/deionized water and bring volume to 980.0mL. Mix thoroughly. Gently heat and bring to boiling. Autoclave for 15 min at 15 psi pressure–121°C. Cool to 50°–55°C. Aseptically add 20.0mL of sterile Chapman tellurite solution. Mix thoroughly. Pour into sterile Petri dishes or distribute into sterile tubes. Allow the surface of the plates to dry before inoculating.

Use: For the quantitative detection of coagulase-positive *Staphyloccous* species.

Tergitol 7 Agar

Composition per liter:

Lactose	20.0g
Agar	13.0g
Peptone	10.0g
Yeast extract	6.0g
Meat extract	5.0g
Tergitol-7	0.1g
Bromthymol Blue	0.05g
TTC solution	5.0mL

pH 7.2 ± 0.2 at 25°C

Source: This medium is available as a premixed powder from Oxoid Unipath.

TTC Solution:

Composition per 100.0mL:

Triphenyltetrazolium chloride	0.05g

Preparation of TTC Solution: Add triphenyltetrazolium chloride to distilled/deionized water and bring volume to 100.0mL. Mix thoroughly. Filter sterilize.

Preparation of Medium: Add components to distilled/deionized water and bring volume to 995.0mL. Mix thoroughly. Gently heat and bring to boiling. Autoclave for 15 min at 15 psi pressure–121°C. Cool to 50°C. Aseptically add 5.0mL of sterile TTC solution. Mix thoroughly. Pour into sterile Petri dishes or distribute into sterile tubes.

Use: For the detection and enumeration of coliforms. Lactose-fermenting bacteria appear as yellow colonies. Lactose-nonfermenting bacteria appear as blue colonies.

Tergitol 7 Agar

Composition per liter:

Agar	15.0g
Lactose	10.0g
Yeast extract	3.0g
Pancreatic digest of casein	2.5g
Peptic digest of animal tissue	2.5g
Tergitol 7	0.1g
Bromthymol Blue	25.0mg
TTC solution	3.0mL

pH 6.9 ± 0.2 at 25°C

Source: This medium is available as a premixed powder from BD Diagnostic Systems.

TTC Solution:

Composition per 100.0mL:

Triphenyltetrazolium chloride	1.0g

Preparation of TTC Solution: Add triphenyltetrazolium chloride to distilled/deionized water and bring volume to 100.0mL. Mix thoroughly. Filter sterilize.

Preparation of Medium: Add components to distilled/deionized water and bring volume to 997.0mL. Mix thoroughly. Gently heat and bring to boiling. Autoclave for 15 min at 15 psi pressure–121°C. Cool to 50°C. Aseptically add 3.0mL of sterile TTC solution. Mix thoroughly. Pour into sterile Petri dishes or distribute into sterile tubes.

Use: For the selective isolation and differentiation of coliform bacteria based on lactose fermentation. Lactose-fermenting bacteria appear as yellow colonies. Lactose-nonfermenting bacteria appear as blue colonies.

Tergitol 7 Agar H

Composition per liter:

Agar	15.0g
Lactose	10.0g
Yeast extract	3.0g
Pancreatic digest of casein	2.5g
Peptic digest of animal tissue	2.5g
Ferric ammonium citrate	0.5g
$Na_2S_2O_3$	0.5g
Tergitol 7	0.1g
Bromthymol Blue	0.025g

pH 7.2 ± 0.2

Preparation of Medium: Add components to distilled/deionized water and bring volume to 1.0L. Mix thoroughly. Gently heat and bring to boiling. Distribute into tubes or flasks. Autoclave for 15 min at 15 psi pressure–121°C. Pour into sterile Petri dishes or leave in tubes.

Use: For the selective isolation and differentiation of enteric bacteria from urine.

Tergitol 7 Broth

Composition per liter:

Lactose	10.0g
Yeast extract	3.0g
Pancreatic digest of casein	2.5g
Peptic digest of animal tissue	2.5g
Tergitol 7	0.1g
Bromthymol Blue	25.0mg
TTC solution	3.0mL

pH 6.9 ± 0.2 at 25°C

Source: This medium is available as a premixed powder from BD Diagnostic Systems.

TTC Solution:

Composition per 100.0mL:

Triphenyltetrazolium chloride	1.0g

Preparation of TTC Solution: Add triphenyltetrazolium chloride to distilled/deionized water and bring volume to 100.0mL. Mix thoroughly. Filter sterilize.

Preparation of Medium: Add components to distilled/deionized water and bring volume to 997.0mL. Mix thoroughly. Gently heat while stirring and bring to boiling. Autoclave for 15 min at 15 psi pressure–121°C. Cool to 25°C. Aseptically add 3.0mL of sterile TTC solution. Mix thoroughly.

Use: For the isolation and cultivation of coliforms, *Salmonella,* and other enteric bacteria.

Tetrathionate Broth

Composition per liter:

$Na_2S_2O_3$	40.7g
$CaCO_3$	25.0g
NaCl	4.5g
Peptone	4.5g
Yeast extract	1.8g
Beef extract	0.9g
Iodine solution	20.0mL

Iodine Solution:

Composition per 20.0mL:

Iodine	6.0g
KI	5.0g

Preparation of Iodine Solution: Add iodine and KI to distilled/deionized water and bring volume to 20.0mL. Mix thoroughly.

Preparation of Medium: Add components, except iodine solution, to distilled/deionized water and bring volume to 980.0mL. Mix thoroughly. Gently heat and bring to boiling. Do not autoclave. Cool to 40°C. Add 20.0mL of iodine solution. Mix thoroughly. Distribute into tubes in 10.0mL volumes. Use medium the same day it is prepared.

Use: For the selective isolation and enrichment of *Salmonella typhi* and other salmonellae from fecal specimens.

Tetrathionate Broth (FDA M145)

Composition per 1030.0mL:

Tetrathionate broth base	1.0L
Iodine-potassium iodide solution	20.0mL
Brilliant Green solution	10.0mL

pH 8.4 ± 0.2 at 25°C

Tetrathionate Broth Base:

Composition per liter:

$Na_2S_2O_3 \cdot 5H_2O$	30.0g
$CaCO_3$	10.0g
Polypeptone	5.0g
Bile salts	1.0g

Preparation of Tetrathionate Broth Base: Add components to distilled/deionized water and bring volume to 1.0L. Mix thoroughly. Gently heat and bring to boiling. A slight precipitate will remain. Do not autoclave. Cool to 25°C. Store at 4°C.

Iodine–Potassium Iodide Solution:

Composition per 20.0mL:

Iodine, resublimed	6.0g
KI	5.0g

Preparation of Iodine–Potassium Iodide Solution: Add KI to 5.0mL of sterile distilled/deionized water. Mix thoroughly. Add iodine. Mix thoroughly. Bring volume to 20.0mL with sterile distilled/deionized water.

Brilliant Green Solution:

Composition per 100.0mL:

Brilliant Green 0.1g

Preparation of Brilliant Green Solution: Add Brilliant Green to sterile distilled/deionized water. Mix thoroughly.

Preparation of Medium: Combine 1.0L of tetrathionate broth base, 20.0mL of iodine-potassium iodide solution, and 10.0mL of Brilliant Green solution. Mix thoroughly. Aseptically distribute into tubes in 10.0mL volumes. Do not heat medium after it has been mixed.

Use: For the selective isolation and cultivation of *Salmonella* species.

Tetrathionate Broth (TT Broth)

Composition per liter:

$Na_2S_2O_3$ 30.0g
$CaCO_3$ 10.0g
Proteose peptone 5.0g
Bile salts 1.0g
Iodine solution 20.0mL

pH 8.4± 0.2 at 25°C

Source: This medium is available as a premixed powder from BD Diagnostic Systems.

Iodine Solution:

Composition per 20.0mL:

Iodine 6.0g
KI 5.0g

Preparation of Iodine Solution: Add iodine and KI to distilled/deionized water and bring volume to 20.0mL. Mix thoroughly.

Preparation of Medium: Add components, except iodine solution, to distilled/deionized water and bring volume to 980.0mL. Mix thoroughly. Gently heat and bring to boiling. Do not autoclave. Cool to 40°C. Add 20.0mL of iodine solution. Mix thoroughly. Distribute into tubes in 10.0mL volumes. Use medium the same day it is prepared.

Use: For the selective isolation and enrichment of *Salmonella typhi* and other salmonellae from infectious material.

Tetrathionate Broth (m-Tetrathionate Broth) (m-TT Broth)

Composition per liter:

$Na_2S_2O_3$ 30.0g
$CaCO_3$ 10.0g
Pancreatic digest of casein 2.5g
Peptic digest of animal tissue 2.5g
Iodine-iodide solution 20.0mL

pH 8.0 ± 0.2 at 25°C

Iodine-Iodide Solution:

Composition per 20.0mL:

Iodine 6.0g
KI 5.0g

Preparation of Iodine-Iodide Solution: Add iodine and KI to distilled/deionized water and bring volume to 20.0mL. Mix thoroughly.

Preparation of Medium: Add components, except iodine-iodide solution, to distilled/deionized water and bring volume to 980.0mL. Mix thoroughly. Gently heat and bring to boiling. Do not autoclave. Cool to 40°C. Add 20.0mL of iodine-iodide solution. Mix thoroughly. Distribute into tubes in 10.0mL volumes. Use medium the same day it is prepared.

Use: For the selective isolation in the membrane filter method of *Salmonella* species from feces, urine, and other specimens.

Tetrathionate Broth (m-Tetrathionate Broth)

Composition per liter:

$Na_2S_2O_3$ 30.0g
Proteose peptone 5.0g
Bile salts 1.0g
Iodine solution 20.0mL

pH 8.0 ± 0.2 at 25°C

Source: This medium is available as a premixed powder from BD Diagnostic Systems.

Iodine Solution:

Composition per 20.0mL:

Iodine 6.0g
KI 5.0g

Preparation of Iodine Solution: Add iodine and KI to distilled/deionized water and bring volume to 20.0mL. Mix thoroughly.

Preparation of Medium: Add components, except iodine solution, to distilled/deionized water and bring volume to 980.0mL. Mix thoroughly. Gently heat and bring to boiling. Do not autoclave. Cool to 40°C. Add 20.0mL of iodine solution. Mix thoroughly. Use medium the same day it is prepared.

Use: For the enrichment of *Salmonella* species in the membrane filter method prior to placing the filter on selective media such as Brilliant Green broth.

Tetrathionate Broth, Hajna (TT Broth, Hajna)

Composition per liter:

$Na_2S_2O_3$	38.0g
$CaCO_3$	25.0g
Casein/meat peptone (50/50)	18.0g
NaCl	5.0g
D-Mannitol	2.5g
Yeast extract	2.0g
Glucose	0.5g
Sodium deoxycholate	0.5g
Brilliant Green	0.01g
Iodine solution	40.0mL

pH 7.5–7.8 at 25°C

Source: This medium is available as a premixed powder from BD Diagnostic Systems.

Iodine Solution:
Composition per 40.0mL:

KI	8.0g
Iodine	5.0g

Preparation of Iodine Solution: Add iodine and KI to distilled/deionized water and bring volume to 40.0mL. Mix thoroughly.

Preparation of Medium: Add components, except iodine solution, to distilled/deionized water and bring volume to 960.0mL. Mix thoroughly. Gently heat and bring to boiling. Do not autoclave. Cool to 40°C. Add 40.0mL of iodine solution. Mix thoroughly. Distribute into tubes in 10.0mL volumes. Use medium the same day it is prepared.

Use: For the isolation of *Salmonella* species, except *Salmonella typhi*, and *Arizona* species from fecal specimens, urine, and other specimens.

Tetrathionate Broth with Novobiocin

Composition per liter:

$Na_2S_2O_3$	38.0g
$CaCO_3$	25.0g
Casein/meat peptone (50/50)	18.0g
NaCl	5.0g
Yeast extract	2.0g
D-Mannitol	0.5g
Glucose	0.5g
Sodium deoxycholate	0.5g
Brilliant Green	0.01g
Novobiocin	4.0mg
Iodine solution	40.0mL

pH 7.5–7.8 at 25°C

Iodine Solution:
Composition per 40.0mL:

KI	8.0g
Iodine	5.0g

Preparation of Iodine Solution: Add iodine and KI to distilled/deionized water and bring volume to 20.0mL. Mix thoroughly.

Preparation of Medium: Add components, except iodine solution, to distilled/deionized water and bring volume to 960.0mL. Mix thoroughly. Gently heat and bring to boiling. Do not autoclave. Cool to 40°C. Add 40.0mL of iodine solution. Mix thoroughly. Distribute into tubes in 10.0mL volumes. Use medium the same day it is prepared.

Use: For the isolation of *Salmonella* species, except *Salmonella typhi*, and *Arizona* species from fecal specimens. Novobiocin suppresses the growth of *Proteus* species.

Tetrathionate Broth, USA (TT Broth, USA)

$Na_2S_2O_3$	30.0g
$CaCO_3$	10.0g
Casein peptone	2.5g
Meat peptone	2.5g
Bile salts	1.0g

Source: This medium is available as a premixed powder from Oxoid Unipath.

Iodine-Iodide Solution:
Composition per 20.0mL:

Iodine	6.0g
KI	5.0g

Preparation of Iodine-Iodide Solution: Add iodine and KI to distilled/deionized water and bring volume to 20.0mL. Mix thoroughly.

Preparation of Medium: Add components, except iodine solution, to distilled/deionized water and bring volume to 980.0mL. Mix thoroughly. Gently heat and bring to boiling. Do not autoclave. Cool to 40°C. Add 20.0mL of iodine solution. Mix thoroughly. Distribute into tubes in 10.0mL volumes. Use medium the same day it is prepared.

Use: For the selective enrichment of *Salmonella* species from feces, urine, and other specimens.

Tetrathionate Crystal Violet Enhancement Broth

Composition per liter:

Potassium tetrathionate	20.0g
Casein/meat peptone (50/50)	8.6g

NaCl 6.4g
Crystal Violet 0.005g
pH 6.5 ± 0.2 at 25°C

Preparation of Medium: Add components to distilled/deionized water and bring volume to 1.0L. Mix thoroughly. Distribute into tubes or flasks. Autoclave for 15 min at 15 psi pressure–121°C.

Use: For the isolation of *Salmonella* species, except *Salmonella typhi*, and *Arizona* species from fecal specimens, urine, and other specimens.

Tetrathionate Reductase Test Medium

Composition per 1025.0mL:
$K_2S_4O_6$ 5.0g
Peptone water 1.0L
Bromthymol Blue (0.2% solution) 25.0mL
pH 7.4 ± 0.2 at 25°C

Peptone Water:
Composition per liter:
Peptone 10.0g
NaCl 5.0g

Preparation of Peptone Water: Add components to distilled/deionized water and bring volume to 1.0L. Mix thoroughly.

Preparation of Medium: Combine components. Mix thoroughly. Adjust pH to 7.4. Filter sterilize. Dispense into tubes in 1.0mL volumes or into wells of sterile microculture plates for replica inoculation.

Use: For the cultivation and identification of *Serratia* species based on their ability to reduce tetrathionate. Bacteria that reduce tetrathionate turn the medium yellow.

Tetrazolium Thallium Glucose Agar

Composition per liter:
Agar 14.0g
Beef extract 10.0g
Peptone 10.0g
Glucose solution 100.0mL
2,3,5-Triphenyltetrazolium·HCl solution 10.0mL
Thallous acetate solution 10.0mL

2,3,5-Triphenyltetrazolium·HCl Solution:
Composition per 10.0mL:
2,3,5-Triphenyltetrazolium·HCl 0.1g

Preparation of 2,3,5-Triphenyltetrazolium-HCl Solution: Add 2,3,5-triphenyltetrazolium·HCl to distilled/deionized water and bring volume to 10.0mL. Mix thoroughly. Autoclave for 7 min at 15 psi pressure–121°C.

Glucose Solution:
Composition per 100.0mL:
Glucose 10.0g

Preparation of Glucose Solution: Add glucose to distilled/deionized water and bring volume to 100.0mL. Mix thoroughly. Filter sterilize.

Thallous Acetate Solution:
Composition per 10.0mL:
Thallous acetate 1.0g

Preparation of Thallous Acetate Solution: Add thallous acetate to distilled/deionized water and bring volume to 10.0mL. Mix thoroughly. Filter sterilize.

Preparation of Medium: Add components—except glucose solution, 2,3,5-triphenyltetrazolium·HCl solution, and thallous acetate solution—to distilled/deionized water and bring volume to 880.0mL. Mix thoroughly. Gently heat and bring to boiling. Autoclave for 15 min at 15 psi pressure–121°C. Cool to 45°–50°C. Aseptically add sterile glucose solution, 2,3,5-triphenyltetrazolium·HCl solution, and thallous acetate solution. Mix thoroughly. Pour into sterile Petri dishes or distribute into sterile tubes.

Use: For the cultivation of *Streptococcus* species.

Tetrazolium Tolerance Agar (TTC Agar)

Composition per liter:
Pancreatic digest of casein 15.0g
Agar 15.0g
Triphenyltetrazolium chloride 10.0g
Papaic digest of soybean meal 5.0g
NaCl 5.0g
pH 7.3 ± 0.2 at 25°C

Preparation of Medium: Add components to distilled/deionized water and bring volume to 1.0L. Mix thoroughly. Gently heat and bring to boiling. Distribute into tubes or flasks. Autoclave for 15 min at 15 psi pressure–121°C. Do not overheat. Pour into sterile Petri dishes or leave in tubes.

Use: For the differentiation of bacteria based upon the ability to tolerate and grow in the presence of tetrazolium. *Streptococcus faecalis* (enterococci) rapidly reduces tetrazolium.

Thayer-Martin Agar, Modified (MTM II) (Modified Thayer-Martin Agar)

Composition per liter:
Agar 12.0g
Hemoglobin 10.0g
Pancreatic digest of casein 7.5g
Selected meat peptone 7.5g
NaCl 5.0g
K_2HPO_4 .... 4.0g

Cornstarch 1.0g
KH_2PO_4 .... 1.0g
CNVT inhibitor 10.0mL
Supplement solution 10.0mL
pH 7.2 ± 0.2 at 25°C

CNVT Inhibitor:
Composition per 10.0mL:
Colistin sulfate 7.5mg
Trimethoprim lactate 5.0mg
Vancomycin 3.0mg
Nystatin 12,500U

Preparation of CNVT Inhibitor: Add components to distilled/deionized water and bring volume to 10.0mL. Mix thoroughly. Filter sterilize.

Supplement Solution:
Composition per liter:
Glucose 100.0g
L-Cysteine·HCl 25.9g
L-Glutamine 10.0g
L-Cystine 1.1g
Adenine 1.0g
Nicotinamide adenine dinucleotide 0.25g
Vitamin B_{12} 0.1g
Thiamine pyrophosphate 0.1g
Guanine·HCl 0.03g
$Fe(NO_3)_3 \cdot 6H_2O$ 0.02g
p-Aminobenzoic acid 0.013g
Thiamine·HCl 3.0mg

Source: The supplement solution IsoVitaleX enrichment is available from BD Diagnostic Systems. This enrichment may be replaced by supplement VX from BD Diagnostic Systems.

Preparation of Supplement Solution: Add components to distilled/deionized water and bring volume to 1.0L. Mix thoroughly. Filter sterilize.

Preparation of Medium: Add components, except CNVT inhibitor and supplement solution, to distilled/deionized water and bring volume to 990.0mL. Mix thoroughly. Gently heat and bring to boiling. Distribute into tubes or flasks. Autoclave for 15 min at 15 psi pressure–121°C. Cool to 45°–50°C. Aseptically add 10.0mL of sterile CNVT inhibitor and 10.0mL of sterile supplement solution. Mix thoroughly. Pour into sterile Petri dishes or distribute into sterile tubes.

Use: For the isolation of *Neisseria* species from specimens containing mixed flora of bacteria and fungi.

Thayer-Martin Medium

Composition per liter:
GC agar base 740.0mL
Hemoglobin solution 250.0mL
Vitox supplement 10.0mL
pH 7.3 ± 0.2 at 25°C

GC Agar Base:
Composition per 740.0mL:
Special peptone 15.0g
Agar 10.0g
NaCl 5.0g
K_2HPO_4 .... 4.0g
Cornstarch 1.0g
KH_2PO_4 .... 1.0g
pH 7.2 ± 0.2 at 25°C

Preparation of GC Agar Base: Add components of GC medium base and the hemoglobin to distilled/deionized water and bring volume to 740.0mL. Mix thoroughly. Gently heat until boiling. Autoclave for 15 min at 15 psi pressure–121°C. Cool to 50°C.

Hemoglobin Solution:
Composition per 250.0mL:
Hemoglobin 5.0g

Preparation of Hemoglobin Solution: Add hemoglobin to distilled/deionized water and bring volume to 250.0mL. Mix thoroughly. Autoclave for 15 min at 15 psi pressure–121°C. Cool to 45°–50°C.

Vitox Supplement:
Composition per 10.0mL:
Glucose 2.0g
L-Cysteine·HCl 0.518g
L-Glutamine 0.2g
L-Cystine 0.022g
Adenine sulfate 0.01g
Nicotinamide adenine dinucleotide 5.0mg
Cocarboxylase 2.0mg
Guanine·HCl 0.6mg
$Fe(NO_3)_3 \cdot 6H_2O$ 0.4mg
p-Aminobenzoic acid 0.26mg
Vitamin B_{12} 0.2mg
Thiamine·HCl 0.06mg

Preparation of Vitox Supplement: Add components to distilled/deionized water and bring volume to 10.0mL. Mix thoroughly. Filter sterilize.

Preparation of Medium: To 740.0mL of cooled sterile GC agar base, aseptically add 250.0mL of sterile hemoglobin solution and 10.0mL of sterile Vitox supplement. Mix thoroughly. Pour into sterile Petri dishes or distribute into sterile tubes.

Use: For the isolation and cultivation of fastidious microorganisms, especially *Neisseria* species.

Thayer-Martin Medium

Composition per liter:
Hemoglobin 10.0g

GC medium base..980.0mL
CNVT inhibitor..10.0mL
Supplement B..10.0mL

pH 7.3 ± 0.2 at 25°C

Source: This medium is available as a prepared medium in tubes from BD Diagnostic Systems.

GC Medium Base:
Composition per 980.0mL:

Proteose peptone no. 3....................................15.0g
Agar..10.0g
NaCl...5.0g
K_2HPO_4..4.0g
Cornstarch..1.0g
KH_2PO_4..1.0g

pH 7.2 ± 0.2 at 25°C

Preparation of GC Medium Base: Add components of GC medium base and the hemoglobin to distilled/deionized water and bring volume to 1.0L. Mix thoroughly. Gently heat until boiling. Autoclave for 15 min at 15 psi pressure–121°C. Cool to 45°–50°C.

CNVT Inhibitor:
Composition per 10.0mL:

Colistin sulfate..7.5mg
Trimethoprim lactate..5.0mg
Vancomycin...3.0mg
Nystatin..12,500U

Preparation of CNVT Inhibitor: Add components to distilled/deionized water and bring volume to 10.0mL. Mix thoroughly. Filter sterilize.

Preparation of Medium: To 980.0mL of cooled sterile GC medium base, aseptically add 10.0mL of sterile CNVT inhibitor and 10.0mL of sterile supplement B. Mix thoroughly. Pour into sterile Petri dishes or distribute into sterile tubes.

Use: For the isolation and cultivation of fastidious microorganisms, especially *Neisseria* species.

Thayer-Martin Medium, Modified (Modified Thayer-Martin Agar)

Composition per liter:

GC agar base..720.0mL
Hemoglobin solution....................................250.0mL
GC supplement...30.0mL

pH 7.3 ± 0.2 at 25°C

GC Agar Base:
Composition per 720.0mL:

Special peptone..15.0g
Agar..10.0g
NaCl...5.0g
K_2HPO_4..4.0g
Cornstarch..1.0g
KH_2PO_4..1.0g

pH 7.2 ± 0.2 at 25°C

Preparation of GC Agar Base: Add components of GC medium base and the hemoglobin to distilled/deionized water and bring volume to 720.0mL. Mix thoroughly. Gently heat until boiling. Autoclave for 15 min at 15 psi pressure–121°C. Cool to 50°C.

Hemoglobin Solution:
Composition per 250.0mL:

Hemoglobin..5.0g

Preparation of Hemoglobin Solution: Add hemoglobin to distilled/deionized water and bring volume to 250.0mL. Mix thoroughly. Autoclave for 15 min at 15 psi pressure–121°C. Cool to 45°–50°C.

GC Supplement:
Composition per 30.0mL:

Yeast autolysate..10.0g
Glucose..1.5g
$NaHCO_3$...0.15g
Colistin sulfate..7.5mg
Trimethoprim lactate..5.0mg
Vancomycin...3.0mg
Nystatin..12,500U

Preparation of GC Supplement: Add components to distilled/deionized water and bring volume to 30.0mL. Mix thoroughly. Filter sterilize.

Preparation of Medium: To 720.0mL of cooled sterile GC agar base, aseptically add 250.0mL of sterile hemoglobin solution and 30.0mL of sterile GC supplement. Mix thoroughly. Pour into sterile Petri dishes or distribute into sterile tubes.

Use: For the selective isolation and cultivation of fastidious microorganisms, especially *Neisseria* species.

Thayer-Martin Medium, Modified (Modified Thayer-Martin Agar)

Composition per liter:

GC agar base..730.0mL
Hemoglobin solution....................................250.0mL
Vitox supplement..10.0mL
VCNT antibiotic solution...............................10.0mL

pH 7.3 ± 0.2 at 25°C

GC Agar Base:
Composition per 730.0mL:

Special peptone..15.0g
Agar..10.0g
NaCl...5.0g
K_2HPO_4..4.0g
Cornstarch..1.0g
KH_2PO_4..1.0g

pH 7.2 ± 0.2 at 25°C

Preparation of GC Agar Base: Add components of GC medium base and the hemoglobin to distilled/deionized water and bring volume to 730.0mL. Mix thoroughly. Gently heat until boiling. Autoclave for 15 min at 15 psi pressure–121°C. Cool to 45°–50°C.

Hemoglobin Solution:
Composition per 250.0mL:

Hemoglobin	5.0g

Preparation of Hemoglobin Solution: Add hemoglobin to distilled/deionized water and bring volume to 250.0mL. Mix thoroughly. Autoclave for 15 min at 15 psi pressure–121°C. Cool to 45°–50°C.

Vitox Supplement:
Composition per 10.0mL:

Glucose	2.0g
L-Cysteine·HCl	0.518g
L-Glutamine	0.2g
L-Cystine	0.022g
Adenine sulfate	0.01g
Nicotinamide adenine dinucleotide	5.0mg
Cocarboxylase	2.0mg
Guanine·HCl	0.6mg
$Fe(NO_3)_3 \cdot 6H_2O$	0.4mg
p-Aminobenzoic acid	0.26mg
Vitamin B_{12}	0.2mg
Thiamine·HCl	0.06mg

Preparation of Vitox Supplement: Add components to distilled/deionized water and bring volume to 10.0mL. Mix thoroughly. Filter sterilize.

VCNT Antibiotic Solution:
Composition per 10.0mL:

Colistin methane sulfonate	7.5mg
Trimethoprim lactate	5.0mg
Vancomycin	3.0mg
Nystatin	12500U

Preparation of VCNT Antibiotic Solution: Add components to distilled/deionized water and bring volume to 10.0mL. Mix thoroughly. Filter sterilize.

Preparation of Medium: To 730.0mL of cooled, sterile GC agar base, aseptically add 250.0mL of sterile hemoglobin solution, 10.0mL of sterile Vitox supplement, and 10.0mL of VCNT antibiotic solution. Mix thoroughly. Pour into sterile Petri dishes or distribute into sterile tubes.

Use: For the selective isolation and cultivation of fastidious microorganisms, especially *Neisseria* species.

Thayer-Martin Medium, Selective

Composition per liter:

GC agar base	730.0mL
Hemoglobin solution	250.0mL
Vitox supplement	10.0mL
VCN antibiotic solution	10.0mL

pH 7.3 ± 0.2 at 25°C

GC Agar Base:
Composition per 730.0mL:

Special peptone	15.0g
Agar	10.0g
NaCl	5.0g
K_2HPO_4	4.0g
Cornstarch	1.0g
KH_2PO_4	1.0g

pH 7.2 ± 0.2 at 25°C

Preparation of GC Agar Base: Add components of GC medium base and the hemoglobin to distilled/deionized water and bring volume to 730.0mL. Mix thoroughly. Gently heat until boiling. Autoclave for 15 min at 15 psi pressure–121°C. Cool to 45°–50°C.

Hemoglobin Solution:
Composition per 250.0mL:

Hemoglobin	5.0g

Preparation of Hemoglobin Solution: Add hemoglobin to distilled/deionized water and bring volume to 250.0mL. Mix thoroughly. Autoclave for 15 min at 15 psi pressure–121°C. Cool to 45°–50°C.

Vitox Supplement:
Composition per 10.0mL:

Glucose	2.0g
L-Cysteine·HCl	0.518g
L-Glutamine	0.2g
L-Cystine	0.022g
Adenine sulfate	0.01g
Nicotinamide adenine dinucleotide	5.0mg
Cocarboxylase	2.0mg
Guanine·HCl	0.6mg
$Fe(NO_3)_3 \cdot 6H_2O$	0.4mg
p-Aminobenzoic acid	0.26mg
Vitamin B_{12}	0.2mg
Thiamine·HCl	0.06mg

Preparation of Vitox Supplement: Add components to distilled/deionized water and bring volume to 10.0mL. Mix thoroughly. Filter sterilize.

VCN Antibiotic Solution:
Composition per 10.0mL:

Colistin methane sulfonate	7.5mg
Vancomycin	3.0mg
Nystatin	12,500U

Preparation of VCN Antibiotic Solution: Add components to distilled/deionized water and bring volume to 10.0mL. Mix thoroughly. Filter sterilize.

Preparation of Medium: To 730.0mL of cooled, sterile GC agar base, aseptically add 250.0mL of sterile hemoglobin solution, 10.0mL of sterile Vitox supplement, and 10.0mL of VCN antibiotic solution. Mix thoroughly. Pour into sterile Petri dishes or distribute into sterile tubes.

Use: For the selective isolation and cultivation of fastidious microorganisms, especially *Neisseria* species.

Thayer-Martin Selective Agar

Composition per liter:

Agar	12.0g
Hemoglobin	10.0g
Pancreatic digest of casein	7.5g
Selected meat peptone	7.5g
NaCl	5.0g
K_2HPO_4	4.0g
Cornstarch	1.0g
KH_2PO_4	1.0g
Supplement solution	10.0mL
VCN inhibitor	10.0mL

pH 7.2 ± 0.2 at 25°C

Source: This medium is available as a premixed powder from BD Diagnostic Systems.

Supplement Solution:

Composition per liter:

Glucose	100.0g
L-Cysteine·HCl	25.9g
L-Glutamine	10.0g
L-Cystine	1.1g
Adenine	1.0g
Nicotinamide adenine dinucleotide	0.25g
Vitamin B_{12}	0.1g
Thiamine pyrophosphate	0.1g
Guanine·HCl	0.03g
$Fe(NO_3)_3 \cdot 6H_2O$	0.02g
p-Aminobenzoic acid	0.013g
Thiamine·HCl	3.0mg

Source: The supplement solution IsoVitaleX enrichment is available from BD Diagnostic Systems. This enrichment may be replaced by supplement VX from BD Diagnostic Systems.

Preparation of Supplement Solution: Add components to distilled/deionized water and bring volume to 1.0L. Mix thoroughly. Filter sterilize.

VCN Inhibitor:

Composition per 10.0mL:

Colistin	7.5mg
Vancomycin	3.0mg
Nystatin	12,500U

Preparation of VCN Inhibitor: Add components to distilled/deionized water and bring volume to 10.0mL. Mix thoroughly. Filter sterilize.

Preparation of Medium: Add components, except supplement solution and VCN inhibitor, to distilled/deionized water and bring volume to 980.0mL. Mix thoroughly. Gently heat and bring to boiling. Autoclave for 15 min at 15 psi pressure–121°C. Cool to 45°–50°C. Aseptically add sterile VCN inhibitor and sterile supplement solution. Mix thoroughly. Pour into sterile Petri dishes or distribute into sterile tubes.

Use: For the selective isolation of *Neisseria gonorrhoeae* and *Neisseria meningitidis* from specimens containing mixed flora of bacteria and fungi.

Thiogel® Medium

Composition per liter:

Gelatin	50.0g
Pancreatic digest of casein	17.0g
Glucose	6.0g
Papaic digest of soybean meal	3.0g
NaCl	2.5g
Agar	0.7g
Sodium thioglycolate	0.5g
L-Cystine	0.25g
Na_2SO_3	0.1g

pH 7.0 ± 0.2 at 25°C

Source: This medium is available as a premixed powder from BD Diagnostic Systems.

Preparation of Medium: Add components to distilled/deionized water preheated to 50°C and bring volume to 1.0L. Mix thoroughly. Let stand for 5 min. Gently heat while stirring and bring to boiling. Distribute into tubes, filling them half full. Autoclave for 15 min at 13 psi pressure–118°C. Pour into sterile Petri dishes or leave in tubes.

Use: For the differentiation of microorganisms based on their ability to liquefy gelatin.

Thioglycolate Bile Broth

Composition per 1050.0mL:

Pancreatic digest of casein	15.0g
Glucose	5.5g
Yeast extract	5.0g
NaCl	2.5g
Agar	0.75g
L-Cystine	0.5g
Sodium thioglycolate	0.5g
Bile solution	50.0mL

pH 7.1 ± 0.2 at 25°C

Bile Solution:

Composition per 100.0mL:

Oxgall	40.0g
Sodium deoxycholate	2.0g

Preparation of Bile Solution: Add components to distilled/deionized water and bring volume to 100.0mL. Mix thoroughly. Filter sterilize.

Preparation of Medium: Add components, except bile solution, to distilled/deionized water and bring volume to 1.0L. Mix thoroughly. Gently heat and bring to boiling. Distribute into tubes in 10.0mL volumes. Autoclave for 15 min at 15 psi pressure–121°C. Cool to 45°–50°C. Aseptically add 0.5mL of sterile bile solution to each tube. Mix thoroughly.

Use: For the cultivation of *Bacteroides fragilis* and *Clostridium perfringens* from clinical specimens.

Thioglycolate Broth USP, Alternative

Composition per liter:

Pancreatic digest of casein	15.0g
Glucose	5.5g
Yeast extract	5.0g
NaCl	2.5g
L-Cystine	0.5g
Sodium thioglycolate	0.5g

pH 7.1 ± 0.2 at 25°C

Source: This medium is available as a premixed powder from Oxoid Unipath.

Preparation of Medium: Add components to distilled/deionized water and bring volume to 1.0L. Mix thoroughly. Distribute into tubes or flasks. Autoclave for 15 min at 15 psi pressure–121°C. Prepare freshly or boil and cool the medium just before use.

Use: For the cultivation of both aerobic and anaerobic organisms in the performance of sterility tests of turbid or viscous specimens.

Thioglycolate Gelatin Medium

Composition per liter:

Gelatin	50.0g
Pancreatic digest of casein	15.0g
Yeast extract	5.0g
NaCl	2.5g
Glucose	2.0g
Agar	0.75g
L-Cystine	0.25g
Na_2SO_3	0.1g
Thioglycollic acid	0.3mL

pH 7.0 ± 0.2 at 25°C

Source: This medium is available as a premixed powder from BD Diagnostic Systems.

Preparation of Medium: Add components to distilled/deionized water and bring volume to 1.0L. Mix thoroughly. Gently heat and bring to 50°C. Let stand 5 min. Gently heat and bring to boiling. Distribute into tubes or flasks. Autoclave for 15 min at 15 psi pressure–121°C.

Use: For the determination of gelatin liquefaction by aerobes, microaerophiles, and anaerobes without special incubation.

Thioglycolate Medium (DSMZ Medium 530)

Composition per liter:

Agar	15.0g
Peptone from casein	5.0g
Meat extract	3.0g
Na-thioglycolate solution	100.0mL

pH 5.5 ± 0.2 at 25°C

Thioglycolate Solution :

Composition per 100.0mL:

Na-Thioglycolate	0.5g

Preparation of Thioglycolate Solution: Add thioglycolate to distilled/deionized water and bring volume to 100.0mL. Mix thoroughly. Filter sterilize. Warm to 50°C.

Preparation of Medium: Add components, except thiglycolate solution, to distilled/deionized water and bring volume to 900.0mL. Mix thoroughly. Gently heat and bring to boiling. Autoclave for 15 min at 15 psi pressure–121°C. Cool to 50°C. Aseptically add 100.0mL warm thioglycolate solution. Adjust pH to 5.5. Pour into sterile Petri dishes or distribute into sterile tubes.

Use: For the cultivation and maintenance of various anaerobic bacteria.

Thioglycolate Medium with 20% Bile (THIO + Bile Medium)

Composition per liter:

Oxgall	20.0g
Thioglycolate medium without indicator	1.0L
Hemin solution	0.5mL
Vitamin K_1 solution	0.1mL

pH 7.0 ± 0.2 at 25°C

Thioglycolate Medium without Indicator:

Composition per liter:

Pancreatic digest of casein	17.0g
Glucose	6.0g
Papaic digest of soybean meal	3.0g
NaCl	2.5g
Agar	0.7g
Sodium thioglycolate	0.5g

L-Cystine 0.25g
Na_2SO_3 .... 0.1g

Preparation of Thioglycolate Medium Without Indicator: Add components to distilled/deionized water and bring volume to 1.0L. Mix thoroughly.

Vitamin K_1 Solution:
Composition per 100.0mL:
Vitamin K_1 .... 1.0g

Preparation of Vitamin K_1 Solution: Add vitamin K_1 to 99.0mL of absolute ethanol. Mix thoroughly.

Hemin Solution:
Composition per 100.0mL:
Hemin 1.0g
NaOH (1*N* solution) 20.0mL

Preparation of Hemin Solution: Add hemin to 20.0mL of 1*N* NaOH solution. Mix thoroughly. Bring volume to 100.0mL with distilled/deionized water.

Preparation of Medium: Add 0.5mL of hemin solution, 0.1mL of vitamin K_1 solution, and oxgall to 1.0L of thioglycolate medium without indicator. Mix thoroughly. Distribute into screw-capped tubes or flasks. Autoclave for 15 min at 15 psi pressure–121°C. Cool tubes or flasks under 85% N_2 + 10% H_2 + 5% CO_2. Tighten caps.

Use: For the isolation, cultivation, and identification of a variety of obligate anaerobic bile-tolerant bacteria.

Thioglycolate Medium, Brewer

Composition per liter:
Glucose 5.0g
Peptone 5.0g
NaCl 5.0g
Yeast extract 2.0g
Sodium thioglycolate 1.1g
Agar 1.0g
Beef extract 1.0g
Methylene Blue 2.0mg
pH 7.2 ± 0.2 at 25°C

Source: This medium is available as a premixed powder from Oxoid Unipath.

Preparation of Medium: Add components to distilled/deionized water and bring volume to 1.0L. Mix thoroughly. Gently heat and bring to boiling. Distribute into tubes or flasks. Autoclave for 15 min at 15 psi pressure–121°C.

Use: For determination of the sterility of solutions containing mercurial preservatives.

Thioglycolate Medium, Brewer Modified

Composition per liter:
Pancreatic digest of casein 17.5g
Glucose 10.0g
NaCl 5.0g
Papaic digest of soybean meal 2.5g
K_2HPO_4 .... 2.0g
Sodium thioglycolate 1.0g
Agar 0.5g
Methylene Blue 0.002g
pH 7.2 ± 0.2 at 25°C

Source: This medium is available as a premixed powder from BD Diagnostic Systems.

Preparation of Medium: Add components to distilled/deionized water and bring volume to 1.0L. Mix thoroughly. Gently heat while stirring and bring to boiling. Distribute into tubes or flasks, filling them half full. Autoclave for 15 min at 15 psi pressure–121°C.

Use: For the cultivation of obligate anaerobes, microaerophiles, and facultative organisms.

Thioglycolate Medium, Enriched (THIO Medium) (Thioglycolate Medium with Vitamin K_1 and Hemin)

Composition per liter:
Thioglycolate medium without indicator 1.0L
Hemin solution 0.5mL
Vitamin K_1 solution 0.1mL
pH 7.0 ± 0.2 at 25°C

Thioglycolate Medium without Indicator:
Composition per liter:
Pancreatic digest of casein 17.0g
Glucose 6.0g
Papaic digest of soybean meal 3.0g
NaCl 2.5g
Agar 0.7g
Sodium thioglycolate 0.5g
L-Cystine 0.25g
Na_2SO_3 .... 0.1g

Source: Thioglycolate medium without indicator is available as a premixed powder from Oxoid Unipath and BD Diagnostic Systems.

Preparation of Thioglycolate Medium without Indicator: Add components to distilled/deionized water and bring volume to 1.0L. Mix thoroughly.

Vitamin K_1 Solution:
Composition per 100.0mL:
Vitamin K_1 .... 1.0g

Preparation of Vitamin K_1 Solution: Add vitamin K_1 to 99.0mL of absolute ethanol. Mix thoroughly.

Hemin Solution:
Composition per 100.0mL:

Hemin	1.0g
NaOH (1*N* solution)	20.0mL

Preparation of Hemin Solution: Add hemin to 20.0mL of 1*N* NaOH solution. Mix thoroughly. Bring volume to 100.0mL with distilled/deionized water.

Preparation of Medium: Add 0.5mL of hemin solution and 0.1mL of vitamin K_1 solution to 1.0L of thioglycolate medium without indicator. Mix thoroughly. Distribute into screw-capped tubes or flasks. Autoclave for 15 min at 15 psi pressure–121°C. Cool tubes or flasks under 85% N_2 + 10% H_2 + 5% CO_2. Tighten caps.

Use: For the isolation, cultivation, and identification of a wide variety of obligate anaerobic bacteria.

Thioglycolate Medium without Glucose

Composition per liter:

Pancreatic digest of casein	15.0g
Yeast extract	5.0g
NaCl	2.5g
Agar	0.75g
L-Cystine	0.25g
Methylene Blue	2.0mg
Thioglycollic acid	0.3mL

pH 7.2 ± 0.2 at 25°C

Source: This medium is available as a premixed powder from BD Diagnostic Systems.

Preparation of Medium: Add components to distilled/deionized water and bring volume to 1.0L. Mix thoroughly. Gently heat and bring to boiling. Distribute into tubes or flasks. Autoclave for 15 min at 15 psi pressure–121°C. If medium becomes oxidized before use (Methylene Blue turns blue), heat in a boiling water bath to expel absorbed O_2. Cool to 25°C.

Use: For the cultivation of anaerobic, microaerophilic, and aerobic microorganisms. For use in sterility testing of a variety of specimens.

Thioglycolate Medium without Glucose and Indicator

Composition per liter:

Pancreatic digest of casein	15.0g
Yeast extract	5.0g
NaCl	2.5g
Agar	0.75g
L-Cystine	0.25g
Thioglycollic acid	0.3mL

pH 7.2 ± 0.2 at 25°C

Source: This medium is available as a premixed powder from BD Diagnostic Systems.

Preparation of Medium: Add components to distilled/deionized water and bring volume to 1.0L. Mix thoroughly. Gently heat and bring to boiling. Distribute into tubes or flasks. Autoclave for 15 min at 15 psi pressure–121°C. If medium becomes oxidized before use, heat in a boiling water bath to expel absorbed O_2. Cool to 25°C.

Use: For the cultivation of anaerobic, microaerophilic, and aerobic microorganisms. For use in sterility testing of a variety of specimens.

Thioglycolate Medium without Indicator

Composition per liter:

Pancreatic digest of casein	15.0g
Yeast extract	5.0g
Glucose	5.0g
NaCl	2.5g
Agar	0.75g
Sodium thioglycolate	0.5g
L-Cystine	0.25g

pH 7.2 ± 0.2 at 25°C

Source: This medium is available as a premixed powder from BD Diagnostic Systems.

Preparation of Medium: Add components to distilled/deionized water and bring volume to 1.0L. Mix thoroughly. Gently heat and bring to boiling. Distribute into tubes or flasks. Autoclave for 15 min at 15 psi pressure–121°C. If medium becomes oxidized before use, heat in a boiling water bath to expel absorbed O_2. Cool to 25°C.

Use: For the cultivation of anaerobic, microaerophilic, and aerobic microorganisms. For use in sterility testing of a variety of specimens.

Thioglycolate Medium without Indicator

Composition per liter:

Pancreatic digest of casein	17.0g
Glucose	6.0g
Papaic digest of soybean meal	3.0g
NaCl	2.5g
Agar	0.7g
Sodium thioglycolate	0.5g
L-Cystine	0.25g
Na_2SO_3	0.1g

pH 7.0 ± 0.2 at 25°C

Source: This medium is available as a premixed powder from Oxoid Unipath.

Preparation of Medium: Add components to distilled/deionized water and bring volume to 1.0L. Mix thoroughly. Distribute into tubes or flasks. Autoclave for 15 min at 15 psi pressure–121°C. Prepare freshly or boil and cool the medium just before use.

Use: For the growth of aerobic and anaerobic microorganisms in diagnostic bacteriology.

Thioglycolate Medium without Indicator-135C

Composition per liter:

Pancreatic digest of casein	17.0g
Glucose	6.0g
Papaic digest of soybean meal	3.0g
NaCl	2.5g
Agar	0.7g
Sodium thioglycolate	0.5g
Na_2SO_3	0.1g
L-Cystine	0.25g

pH 7.0 ± 0.2 at 25°C

Source: This medium is available as a premixed powder from BD Diagnostic Systems.

Preparation of Medium: Add components to distilled/deionized water and bring volume to 1.0L. Mix thoroughly. Gently heat while stirring and bring to boiling. Distribute into tubes or flasks, filling them half full. For maintenance of cultures, a small quantity of $CaCO_3$ may be added to tubes before adding medium. Autoclave for 15 min at 13 psi pressure–118°C. Prepare freshly or boil and cool the medium just before use. Store prepared medium at 2°–8°C in the dark.

Use: For the isolation and cultivation of a wide variety of microorganisms, particularly obligate anaerobes, from clinical specimens and other materials.

Thioglycolate Medium without Indicator with Hemin (Thioglycolate Medium, Supplemented)

Composition per liter:

Pancreatic digest of casein	17.0g
$CaCO_3$, chips or powder	10.0g
Glucose	6.0g
Papaic digest of soybean meal	3.0g
NaCl	2.5g
Agar	0.7g
Sodium thioglycolate	0.5g
L-Cystine	0.25g
Na_2SO_3	0.1g
Hemin	5.0mg
Na_2CO_3 solution	10.0mL
Vitamin K_1 solution	10.0mL

pH 7.2 ± 0.2 at 25°C

Na_2CO_3 Solution:

Composition per 10.0mL:

Na_2CO_3	1.0g

Preparation of Na_2CO_3 Solution: Add Na_2CO_3 to distilled/deionized water and bring volume to 10.0mL. Mix thoroughly. Filter sterilize.

Vitamin K_1 Solution:

Composition per 100.0mL:

Vitamin K_1	1.0g
Ethanol, absolute	99.0mL

Preparation of Vitamin K_1 Solution: Add vitamin K_1 to 99.0mL of absolute ethanol. Mix thoroughly.

Preparation of Medium: Add components, except $CaCO_3$, Na_2CO_3 solution, and vitamin K_1 solution, to distilled/deionized water and bring volume to 990.0mL. Mix thoroughly. Gently heat and bring to boiling. Add 0.1g of $CaCO_3$ chips or powder to each of 100 test tubes. Distribute broth into the same tubes in 10.0mL volumes. Autoclave for 15 min at 15 psi pressure–121°C. Cool to 45°–50°C. Aseptically add 0.1mL of sterile Na_2CO_3 solution and 0.1mL of sterile vitamin K_1 solution to each tube. Mix thoroughly.

Use: For the cultivation of a wide variety of obligate anaerobes.

Thioglycolate Medium, USP

Composition per liter:

Pancreatic digest of casein	15.0g
Glucose	5.5g
Yeast extract	5.0g
NaCl	2.5g
Agar	0.5g
L-Cystine	0.5g
Sodium thioglycolate	0.5g
Resazurin	1.0mg

pH 7.1 ± 0.2 at 25°C

Source: This medium is available as a premixed powder from Oxoid Unipath.

Preparation of Medium: Add components to distilled/deionized water and bring volume to 1.0L. Mix thoroughly. Gently heat and bring to boiling. Distribute into tubes or flasks. Autoclave for 15 min at 15 psi pressure–121°C.

Use: For the cultivation of both aerobic and anaerobic organisms in the performance of sterility tests.

Thiol Medium

Composition per liter:

Proteose peptone no. 3	10.0g
Thiol complex	8.0g
Yeast extract	5.0g
NaCl	5.0g
Glucose	1.0g
Agar	1.0g
p-Aminobenzoic acid	0.05g

pH 7.1 ± 0.2 at 25°C

Source: This medium is available as a premixed powder from BD Diagnostic Systems.

Preparation of Medium: Add components to distilled/deionized water and bring volume to 1.0L. Mix thoroughly. Gently heat and bring to boiling. Distribute into tubes or flasks. For neutralization of penicillin, distribute medium into tubes to a depth of 60 mm. For neutralization of streptomycin, distribute medium into tubes in shallow layers. Autoclave for 15 min at 15 psi pressure–121°C.

Use: For the cultivation of bacteria from body fluids and other materials containing penicillin, streptomycin, or sulfonamides. Also used for the cultivation and maintenance of *Bifidobacterium* species.

Tinsdale Agar

Composition per 1100.0mL:

Proteose peptone	20.0g
Agar	15.0g
NaCl	5.0g
Yeast extract	5.0g
L-Cystine	0.24g
Tinsdale supplement	150.0mL

pH 7.4 ± 0.2 at 25°C

Source: This medium is available as a premixed powder from BD Diagnostic Systems and Oxoid Unipath.

Tinsdale Supplement:
Composition per 100.0mL:

$Na_2S_2O_3$	0.43g
K_2TeO_3	0.35g
Serum	100.0mL

Caution: Potassium tellurite is toxic.

Preparation of Tinsdale Supplement: Add $Na_2S_2O_3$ and K_2TeO_3 to serum. Mix thoroughly. Filter sterilize.

Preparation of Medium: Add components, except Tinsdale supplement, to distilled/deionized water and bring volume to 1.0L. Mix thoroughly. Gently heat and bring to boiling. Autoclave for 15 min at 15 psi pressure–121°C. Cool to 50°–55°C. Aseptically add 100.0mL of sterile Tinsdale supplement. Mix thoroughly. Pour into sterile Petri dishes or distribute into sterile tubes.

Use: For the primary isolation and identification of *Corynebacterium diphtheriae*.

Tissue Culture Amino Acids, HeLa 100X (TC Amino Acids, HeLa 100X)

Composition per liter:

L-Lysine	0.029g
L-Isoleucine	0.026g
L-Leucine	0.026g
L-Threonine	0.023g
L-Valine	0.023g
L-Tyrosine	0.018g
L-Arginine	0.017g
L-Phenylalanine	0.016g
L-Cystine	0.012g
L-Histidine	7.8mg
L-Methionine	7.5mg
L-Tryptophan	4.1mg

pH 7.2–7.4 at 25°C

Preparation of Tissue Culture Amino Acids, HeLa 100X: Add components to distilled/deionized water and bring volume to 1.0L. Mix thoroughly. Adjust pH to 7.2–7.4. Filter sterilize.

Use: For the preparation of Eagle HeLa medium for tissue culture procedures and virual culture.

Tissue Culture Amino Acids, Minimal Eagle 50X (TC Amino Acids, Minimal Eagle 50X)

Composition per liter:

L-Arginine	0.1g
L-Lysine	0.058g
L-Isoleucine	0.052g
L-Leucine	0.052g
L-Threonine	0.048g
L-Valine	0.046g
L-Tyrosine	0.036g
L-Phenylalanine	0.032g
L-Histidine	0.031g
L-Cystine	0.024g
L-Methionine	0.015g
L-Tryptophan	0.01g

pH 7.2–7.4 at 25°C

Preparation of Tissue Culture Amino Acids, Minimal Eagle 50X: Add components to distilled/deionized water and bring volume to 1.0L. Mix thoroughly. Adjust pH to 7.2–7.4. Filter sterilize.

Use: For the preparation of TC minimal medium Eagle for tissue culture procedures and virus studies.

Tissue Culture Dulbecco Solution (TC Dulbecco Solution)

Composition per liter:

NaCl	8.0g
Na_2HPO_4	1.15g
KH_2PO_4	0.2g
KCl	0.2g
$CaCl_2 \cdot 2H_2O$	0.1g
$MgCl_2 \cdot 6H_2O$	0.1g

pH 7.2–7.4 at 25°C

Preparation of Tissue Culture Dulbecco Solution: Add components to distilled/deionized water and bring volume to 1.0L. Mix thoroughly. Adjust pH to 7.2–7.4. Filter sterilize.

Use: For use in tissue culture and virus preparations.

Tissue Culture Earle Solution (TC Earle Solution)

Composition per 1002.0mL:

NaCl	6.8g
$NaHCO_3$	2.2g
Glucose	1.0g
KCl	0.4g
$CaCl_2 \cdot 2H_2O$	0.2g
NaH_2PO_4	0.125g
$MgSO_4 \cdot 7H_2O$	0.1g
Phenol Red (1% solution)	2.0mL

pH 7.2–7.4 at 25°C

Preparation of Tissue Culture Earle Solution: Add components, except Phenol Red, to distilled/deionized water and bring volume to 1.0L. Mix thoroughly. Add 2.0mL of Phenol Red solution. Adjust pH to 7.2–7.4. Filter sterilize.

Use: For use in tissue culture and virus preparations.

Tissue Culture Hanks Solution (TC Hanks Solution)

Composition per liter:

NaCl	8.0g
Glucose	1.0g
KCl	0.4g
$NaHCO_3$	0.35g
$CaCl_2 \cdot 2H_2O$	0.14g
$MgCl_2 \cdot 6H_2O$	0.1g
$MgSO_4 \cdot 7H_2O$	0.1g
Na_2HPO_4	0.06g
KH_2PO_4	0.06g
Phenol Red	0.02g

pH 7.2–7.4 at 25°C

Source: This medium is available as a premixed solution from BD Diagnostic Systems.

Preparation of Tissue Culture Hanks Solution: Add components to distilled/deionized water and bring volume to 1.0L. Mix thoroughly. Adjust pH to 7.2–7.4. Filter sterilize.

Use: For use in tissue culture procedures.

Tissue Culture Medium 199 (TC Medium 199)

Composition per 1050.0mL:

NaCl	8.0g
Glucose	1.0g
KCl	0.4g
$NaHCO_3$	0.35g
DL-Glutamic acid	0.15g
$CaCl_2 \cdot 2H_2O$	0.14g
DL-Leucine	0.12g
L-Glutamine	0.1g
$MgSO_4 \cdot 7H_2O$	0.1g
L-Arginine	0.07g
L-Lysine	0.07g
DL-Aspartic acid	0.06g
Na_2HPO_4	0.06g
KH_2PO_4	0.06g
DL-Threonine	0.06g
DL-Alanine	0.05g
Glycine	0.05g
DL-Phenylalanine	0.05g
DL-Serine	0.05g
Sodium acetate	0.05g
DL-Valine	0.05g
DL-Isoleucine	0.04g
L-Proline	0.04g
L-Tyrosine	0.04g
DL-Methionine	0.03g
L-Cystine	0.02g
L-Histidine	0.02g
Phenol Red	0.02g
DL-Tryptophan	0.02g
Adenine	0.01g
L-Hydroxyproline	0.01g
Tween 80	5.0mg
Adenosine triphosphate	1.0mg
Choline	0.5mg
Deoxyribose	0.5mg
Ribose	0.5mg
Guanine	0.3mg
Hypoxanthine	0.3mg
Thymine	0.3mg
Uracil	0.3mg
Xanthine	0.3mg
Adenylic acid	0.2mg
Cholesterol	0.2mg
Calciferol	0.1mg
$Fe(NO_3)_3 \cdot 9H_2O$	0.1mg

L-Cysteine	0.1mg
Vitamin A	0.1mg
p-Aminobenzoic acid	0.05mg
Ascorbic acid	0.05mg
Glutathione	0.05mg
Inositol	0.05mg
Niacin	0.025mg
Niacinamide	0.025mg
Pyridoxine·HCl	0.025mg
Pyridoxal·HCl	0.025mg
α-Tocopherol phosphate	0.01mg
Biotin	0.01mg
Calcium pantothenate	0.01mg
Folic acid	0.01mg
Menadione	0.01mg
Riboflavin	0.01mg
Thiamine·HCl	0.01mg
Serum	50.0–100.0mL

pH 7.2–7.4 at 25°C

Preparation of Medium: Add components, except serum, to distilled/deionized water and bring volume to 1.0L. Mix thoroughly. Adjust pH to 7.2–7.4 with 10% Na_2CO_3 solution. Filter sterilize. Aseptically add 50.0–100.0mL of sterile serum. Human serum, bovine serum, horse serum, or fetal calf serum may be used. Mix thoroughly. If desired, antibacterial inhibitors may be added. Aseptically add 500,000U of penicillin and 0.5g of streptomycin to 1050.0mL of the complete medium to increase selectivity.

Use: For the cultivation of a wide variety of cell lines in tissue culture. It is especially useful for the detection, titering, and identification of viruses in tissue culture cells.

Tissue Culture Medium Eagle with Earle Balanced Salt Solution (TC Medium Eagle with Earle BSS)

Composition per 1056.0mL:

NaCl	6.8g
$NaHCO_3$	2.2g
Glucose	1.0g
KCl	0.4g
$CaCl_2 \cdot 2H_2O$	0.2g
NaH_2PO_4	0.125g
$MgSO_4 \cdot 7H_2O$	0.1g
L-Isoleucine	0.026g
L-Leucine	0.026g
L-Lysine	0.026g
L-Threonine	0.024g
L-Valine	0.0235g
L-Tyrosine	0.018g
L-Arginine	0.0174g
L-Phenylalanine	0.0165g
L-Cystine	0.012g
L-Histidine	8.0mg
L-Methionine	7.5mg
Phenol Red	5.0mg
L-Tryptophan	4.0mg
Inositol	1.8mg
Biotin	1.0mg
Calcium pantothenate	1.0mg
Choline chloride	1.0mg
Folic acid	1.0mg
Nicotinamide	1.0mg
Pyridoxal·HCl	1.0mg
Thiamine·HCl	1.0mg
Riboflavin	0.1mg
Serum	50.0–100.0mL
Glutamine solution	6.0mL

pH 7.2–7.4 at 25°C

Glutamine Solution:

Composition per 100.0mL:

L-Glutamine	5.0g
NaCl (0.85% solution)	100.0mL

Preparation of Glutamine Solution: Add the glutamine to the 0.85% NaCl solution. Mix thoroughly. Filter sterilize.

Preparation of Medium: Add components, except glutamine and serum, to distilled/deionized water and bring volume to 1.0L. Mix thoroughly. Adjust pH to 7.2–7.4. Filter sterilize. Aseptically add 6.0mL of sterile glutamine solution and 50.0–100.0mL of sterile serum. Human serum, bovine serum, horse serum, or fetal calf serum may be used. Mix thoroughly.

Use: For the cultivation of HeLa, KB, and other tissue culture cell lines.

Tissue Culture Medium Eagle with Hanks Balanced Salt Solution (TC Medium Eagle with Hanks BSS)

Composition per 1056.0mL:

NaCl	8.0g
Glucose	1.0g
KCl	0.4g
$CaCl_2 \cdot 2H_2O$	0.14g
$MgSO_4 \cdot 7H_2O$	0.1g
KH_2PO_4	0.06g
Na_2HPO_4	0.05g
L-Isoleucine	0.026g
L-Leucine	0.026g
L-Lysine	0.026g
L-Threonine	0.024g
L-Valine	0.0235g
L-Tyrosine	0.018g
L-Arginine	0.0174g
L-Phenylalanine	0.0165g

L-Cystine	0.012g
L-Histidine	8.0mg
L-Methionine	7.5mg
Phenol Red	5.0mg
L-Tryptophan	4.0mg
Inositol	1.8mg
Biotin	1.0mg
Folic acid	1.0mg
Calcium pantothenate	1.0mg
Choline chloride	1.0mg
Nicotinamide	1.0mg
Pyridoxal·HCl	1.0mg
Thiamine·HCl	1.0mg
Riboflavin	0.1mg
Serum	50.0–100.0mL
Glutamine solution	6.0mL

pH 7.2–7.4 at 25°C

Glutamine Solution:
Composition per 100.0mL:

L-Glutamine	5.0g
NaCl (0.85% solution)	100.0mL

Preparation of Glutamine Solution: Add the glutamine to the 0.85% NaCl solution. Mix thoroughly. Filter sterilize.

Preparation of Medium: Add components, except glutamine and serum, to distilled/deionized water and bring volume to 1.0L. Mix thoroughly. Adjust pH to 7.2–7.4. Filter sterilize. Aseptically add 6.0mL of sterile glutamine solution and 50.0–100.0mL of sterile serum. Human serum, bovine serum, horse serum, or fetal calf serum may be used. Mix thoroughly.

Use: For use as a base in the preparation of liquid media used for the cultivation of tissue culture cell lines.

Tissue Culture Medium Eagle, HeLa (TC Medium Eagle, HeLa)

Composition per 1056.0mL:

NaCl	5.85g
$NaHCO_3$	1.68g
Glucose	0.9g
KCl	0.373g
NaH_2PO_4	0.138g
$MgCl_2 \cdot 6H_2O$	0.12g
$CaCl_2 \cdot 2H_2O$	0.11g
L-Lysine	0.0269g
L-Isoleucine	0.0262g
L-Leucine	0.0262g
L-Threonine	0.0238g
L-Valine	0.0234g
L-Tyrosine	0.0181g
L-Arginine	0.0174g
L-Phenylalanine	0.0165g
L-Cystine	0.012g
L-Histidine	7.8mg
L-Methionine	7.5mg
Phenol Red	5.0mg
L-Tryptophan	4.1mg
Folic acid	0.44mg
Thiamine·HCl	0.34mg
Biotin	0.24mg
Pantothenic acid	0.22mg
Pyridoxal·HCl	0.2mg
Choline chloride	0.14mg
Nicotinamide	0.12mg
Riboflavin	0.04mg
Serum	50.0mL–100.0mL
Glutamine solution	6.0mL

pH 7.2–7.4 at 25°C

Glutamine Solution:
Composition per 100.0mL:

L-Glutamine	5.0g
NaCl (0.85% solution)	100.0mL

Preparation of Glutamine Solution: Add the glutamine to the 0.85% NaCl solution. Mix thoroughly. Filter sterilize.

Preparation of Medium: Add components, except glutamine and serum, to distilled/deionized water and bring volume to 1.0L. Mix thoroughly. Adjust pH to 7.2–7.4. Filter sterilize. Aseptically add 6.0mL of sterile glutamine solution and 50.0–100.0mL of sterile serum. Human serum, bovine serum, horse serum, or fetal calf serum may be used. Mix thoroughly.

Use: For the cultivation and maintenance of HeLa and other cell lines in tissue culture, and for studying the cytopathogenicity of viral agents.

Tissue Culture Medium Ham F10 (TC Medium Ham F10)

Composition per 1050.0mL:

NaCl	7.4g
Glucose	1.1g
Na_2HPO_4	0.29g
KCl	0.285g
L-Arginine	0.211g
$MgSO_4 \cdot 7H_2O$	0.153g
L-Glutamine	0.1462g
Sodium pyruvate	0.11g
KH_2PO_4	0.083g
$CaCl_2 \cdot 2H_2O$	0.044g
L-Cystine	0.0315g
L-Lysine	0.0293g
L-Histidine	0.021g
L-Asparagine	0.015g
L-Glutamic acid	0.0147g
L-Aspartic acid	0.0133g

L-Leucine	0.0131g
L-Proline	0.0115g
L-Serine	0.0105g
L-Alanine	8.91mg
Glycine	7.51mg
L-Phenylalanine	4.96mg
L-Methionine	4.48mg
Hypoxanthine	4.0mg
L-Threonine	3.57mg
L-Valine	3.5mg
L-Isoleucine	2.6mg
L-Tyrosine	1.81mg
Cyanocobalamin	1.3mg
Folic acid	1.3mg
Phenol Red	1.2mg
Thiamine·HCl	1.0mg
$FeSO_4 \cdot 7H_2O$	0.83mg
Calcium pantothenate	0.7mg
Thymidine	0.7mg
Choline chloride	0.69mg
Niacinamide	0.6mg
L-Tryptophan	0.6mg
i-Inositol	0.54mg
Riboflavin	0.37mg
Lipoic acid	0.2mg
Pyridoxine·HCl	0.2mg
$ZnSO_4 \cdot 7H_2O$	0.028mg
Biotin	0.024mg
$CuSO_4 \cdot 5H_2O$	2.5μg
Fetal calf serum	50.0–100.0mL

pH 7.2–7.4 at 25°C

Preparation of Medium: Add components, except fetal calf serum, to distilled/deionized water and bring volume to 1.0L. Mix thoroughly. Adjust pH to 7.2–7.4 with 10% Na_2CO_3 solution. Filter sterilize. Aseptically add 50.0–100.0mL of sterile fetal calf serum. Mix thoroughly.

Use: For the cultivation of a wide variety of cell lines in tissue culture.

Tissue Culture Medium NCTC 109 (TC Medium NCTC 109)

Composition per 1050.0mL:

NaCl	6.8g
$NaHCO_3$	2.2g
Glucose	1.0g
KCl	0.4g
L-Cysteine	0.26g
$CaCl_2 \cdot 2H_2O$	0.2g
NaH_2PO_4	0.14g
L-Glutamine	0.14g
$MgSO_4 \cdot 7H_2O$	0.1g
Sodium acetate	0.05g
Ascorbic acid	0.05g
L-Alanine	0.03g
L-Lysine	0.03g
L-Arginine	0.026g
L-Valine	0.025g
L-Leucine	0.02g
Phenol Red	0.02g
L-Histidine	0.019g
L-Threonine	0.019g
L-Isoleucine	0.018g
L-Tryptophan	0.017g
L-Phenylalanine	0.017g
L-Tyrosine	0.016g
Glycine	0.014g
Tween 80	0.012g
L-Serine	0.011g
L-Cystine	0.01g
Glutathione	0.01g
Cyanocobalamin	0.01g
Deoxycytidine	0.01g
Deoxyguanosine	0.01g
Deoxyadenosine	0.01g
Thymidine	0.01g
L-Aspartic acid	9.91mg
L-Glutamic acid	8.26mg
L-Arginine	8.09mg
L-Ornithine	7.38mg
Nicotinamide adenine dinucleotide	7.0mg
L-Proline	6.13mg
L-α-N-Butyric acid	5.51mg
L-Methionine	4.44mg
L-Taurine	4.18mg
L-Hydroxyproline	4.09mg
D-Glucosamine	3.2mg
Coenzyme A	2.5mg
Glucuronolactone	1.8mg
Sodium glucuronate	1.8mg
Choline chloride	1.25mg
Cocarboxylase	1.0mg
Flavin adenine dinucleotide	1.0mg
Uridine triphosphate	1.0mg
Nicotinamide adenine dinucleotide phosphate	1.0mg
Vitamin A	0.25mg
Calciferol	0.25mg
i-Inositol	0.125mg
p-Aminobenzoic acid	0.125mg
5-Methylcytosine	0.1mg
Pyridoxine·HCl	0.0625mg
Pyridoxal·HCl	0.0625mg
Niacin	0.0625mg
Niacinamide	0.0625mg
Biotin	0.025mg
Folic acid	0.025mg
Menadione	0.025mg
Pantothenate	0.025mg

Riboflavin 0.025mg
Thiamine·HCl 0.025mg
α-Tocopherol phosphate 0.025mg
Serum 50.0–100.0mL

pH 7.2–7.4 at 25°C

Preparation of Medium: Add components, except serum, to distilled/deionized water and bring volume to 1.0L. Mix thoroughly. Adjust pH to 7.2–7.4 with 10% Na_2CO_3 solution. Filter sterilize. Aseptically add 50.0–100.0mL of sterile serum. Human serum, bovine serum, horse serum, or fetal calf serum may be used. Mix thoroughly.

Use: For the cultivation of a wide variety of cell lines in tissue culture.

Tissue Culture Medium RPMI #1640 (TC Medium RPMI #1640)

Composition per liter:

NaCl 6.46g
Glucose 2.0g
$NaHCO_3$ 2.0g
NaH_2PO_4 ......1.512g
KCl 0.4g
L-Glutamine 0.3g
L-Arginine 0.2g
Calcium nitrate 0.1g
$MgSO_4 \cdot 7H_2O$ 0.1g
L-Asparagine 0.05g
L-Cystine 0.05g
L-Isoleucine 0.05g
L-Leucine 0.05g
L-Lysine·HCl 0.04g
Inositol 0.035g
L-Serine 0.03g
Hydroxy-L-Proline 0.02g
L-Aspartic acid 0.02g
L-Glutamic acid 0.02g
L-Proline 0.02g
L-Threonine 0.02g
L-Tyrosine 0.02g
L-Valine 0.02g
L-Histidine 0.015g
L-Methionine 0.015g
L-Phenylalanine 0.015g
Glycine 0.01g
L-Tryptophan 5.0mg
Phenol Red 5.0mg
Vitamin B_{12} 5.0mg
Choline chloride 3.0mg
p-Aminobenzoic acid 1.0mg
Folic acid 1.0mg
Glutathione 1.0mg
Nicotinamide 1.0mg
Pyridoxine·HCl 1.0mg
Thiamine·HCl 1.0mg
Calcium pantothenate 0.25mg
Biotin 0.2mg
Riboflavin 0.2mg
Serum 50.0–100.0mL

pH 7.2–7.4 at 25°C

Source: This medium is available as a premixed powder and solution from BD Diagnostic Systems.

Preparation of Medium: Add components, except serum, to distilled/deionized water and bring volume to 1.0L. Mix thoroughly. Adjust pH to 7.2–7.4 with 10% Na_2CO_3 solution. Filter sterilize. Aseptically add 50.0–100.0mL of sterile serum. Human serum, bovine serum, horse serum, or fetal calf serum may be used. Mix thoroughly.

Use: For the cultivation of a wide variety of cell lines in tissue culture.

Tissue Culture Minimal Medium Eagle

Composition per liter:

Sterile salt solution 944.0mL
TC amino acids, minimal Eagle 50X...... 20.0mL
TC $NaHCO_3$, 10% 20.0mL
TC vitamins, minimal Eagle 100X 10.0mL
TC glutamine, 5% 6.0mL

pH 7.2–7.4 at 25°C

Sterile Salt Solution:

Composition per 944.0mL:

NaCl 6.8g
Glucose 1.0g
KCl 0.4g
$CaCl_2$ 0.2g
$MgCl_2$ 0.2g
NaH_2PO_4 ......0.15g

Preparation of Sterile Salt Solution: Add components to distilled/deionized water and bring volume to 944.0mL. Mix thoroughly. Filter sterilize.

Tissue Culture Amino Acids, Minimal Eagle 50X:

Composition per liter:

L-Arginine 0.1g
L-Lysine 0.06g
L-Isoleucine 0.05g
L-Leucine 0.05g
L-Threonine 0.05g
L-Valine 0.05g
L-Tyrosine 0.04g
L-Phenylalanine 0.03g
L-Histidine 0.03g
L-Cystine 0.02g

L-Methionine..0.02g
L-Tryptophan..0.01g

Preparation of Tissue Culture Amino Acids, Minimal Eagle 50X: Add components to distilled/deionized water and bring volume to 1.0L. Mix thoroughly. Adjust pH to 7.2–7.4. Filter sterilize.

TC $NaHCO_3$, 10%:
Composition per 100.0mL:
$NaHCO_3$..10.0g

Preparation of TC $NaHCO_3$, 10%: Add $NaHCO_3$ to distilled/deionized water and bring volume to 100.0mL. Mix thoroughly. Filter sterilize.

TC Vitamins, Minimal Eagle 100X:
Composition per liter:
Inositol..2.0mg
Calcium pantothenate..1.0mg
Choline chloride..1.0mg
Folic acid..1.0mg
Nicotinamide..1.0mg
Pyridoxal..1.0mg
Thiamine·HCl..1.0mg
Riboflavin..0.1mg

Preparation of TC Vitamins, Minimal Eagle 100X: Add components to distilled/deionized water and bring volume to 1.0L. Mix thoroughly. Filter sterilize.

TC Glutamine, 5%:
Composition per 100.0mL:
L-Glutamine..5.0g
NaCl (0.85% solution)..100.0mL

Preparation of TC Glutamine, 5%: Add the glutamine to the 0.85% NaCl solution. Mix thoroughly. Filter sterilize.

Preparation of Medium: Aseptically combine 944.0mL of sterile salt solution, 20.0mL of sterile TC amino acids, minimal Eagle 50X, 20.0mL of sterile TC $NaHCO_3$, 10%, 10.0mL of sterile TC vitamins, minimal Eagle 100X, and 6.0mL of sterile TC glutamine, 5%. Mix thoroughly. Adjust pH to 7.2–7.4, if necessary.

Use: For the cultivation of mammalian cells in monolayer or suspension for tissue culture procedures and virus preparation.

Tissue Culture Minimal Medium Eagle with Earle Balanced Salts Solution (TC Minimal Medium Eagle with Earle BSS)

Composition per 1056.0mL:
NaCl..6.8g
Glucose..1.0g
KCl..0.4g
$CaCl_2 \cdot 2H_2O$..0.2g
$MgCl_2 \cdot 6H_2O$..0.2g
NaH_2PO_4..0.15g
L-Arginine..0.1g
L-Lysine..0.06g
L-Isoleucine..0.05g
L-Leucine..0.05g
L-Threonine..0.05g
L-Valine..0.05g
L-Tyrosine..0.04g
L-Phenylalanine..0.03g
L-Histidine..0.03g
L-Cystine..0.02g
L-Methionine..0.02g
L-Tryptophan..0.01g
i-Inositol..2.0mg
Calcium pantothenate..1.0mg
Choline chloride..1.0mg
Folic acid..1.0mg
Nicotinamide..1.0mg
Pyridoxal..1.0mg
Thiamine·HCl..1.0mg
Riboflavin..0.1mg
Serum..50.0–100.0mL
Glutamine solution..6.0mL
$CaCl_2 \cdot 2H_2O$ solution (optional)..2.0mL
pH 7.2–7.4 at 25°C

Glutamine Solution:
Composition per 100.0mL:
L-Glutamine..5.0g
NaCl (0.85% solution)..100.0mL

Preparation of Glutamine Solution: Add the glutamine to the 0.85% NaCl solution. Mix thoroughly. Filter sterilize.

Preparation of Medium: Add components, except glutamine and serum, to distilled/deionized water and bring volume to 1.0L. Mix thoroughly. Adjust pH to 7.2–7.4 with 10% Na_2CO_3 solution. Filter sterilize. Aseptically add 6.0mL of sterile glutamine solution and 50.0–100.0mL of sterile serum. Human serum, bovine serum, horse serum, or fetal calf serum may be used. Mix thoroughly. To grow cells in a monolayer, aseptically add 2.0mL of a sterile 10% $CaCl_2 \cdot 2H_2O$ solution. To grow cells in suspension, omit the $CaCl_2 \cdot 2H_2O$ solution.

Use: For preparation of Eagle's minimal medium for the cultivation of cells in monolayer or suspension in tissue culture.

Tissue Culture Minimal Medium Eagle Spinner Modified (TC Minimal Medium Eagle Spinner Modified MEM-S)

Composition per 1056.0mL:

NaCl	6.8g
$NaHCO_3$	2.2g
NaH_2PO_4	1.35g
Glucose	1.0g
KCl	0.4g
$CaCl_2 \cdot 2H_2O$	0.2g
NaH_2PO_4	0.125g
$MgSO_4 \cdot 7H_2O$	0.1g
L-Isoleucine	0.026g
L-Leucine	0.026g
L-Lysine	0.026g
L-Threonine	0.024g
L-Valine	0.0235g
L-Tyrosine	0.018g
L-Arginine	0.0174g
L-Phenylalanine	0.0165g
L-Cystine	0.012g
L-Histidine	8.0mg
L-Methionine	7.5mg
Phenol Red	5.0mg
L-Tryptophan	4.0mg
Inositol	1.8mg
Biotin	1.0mg
Calcium pantothenate	1.0mg
Choline chloride	1.0mg
Folic acid	1.0mg
Nicotinamide	1.0mg
Pyridoxal·HCl	1.0mg
Thiamine·HCl	1.0mg
Riboflavin	0.1mg
Serum	50.0mL–100.0mL
Glutamine solution	6.0mL

pH 7.2–7.4 at 25°C

Glutamine Solution:

Composition per 100.0mL:

L-Glutamine	5.0g
NaCl (0.85% solution)	100.0mL

Preparation of Glutamine Solution: Add the glutamine to the 0.85% NaCl solution. Mix thoroughly. Filter sterilize.

Preparation of Medium: Add components, except glutamine and serum, to distilled/deionized water and bring volume to 1.0L. Mix thoroughly. Adjust pH to 7.2–7.4 with 10% Na_2CO_3 solution. Filter sterilize. Aseptically add 6.0mL of sterile glutamine solution and 50.0–100.0mL of sterile serum. Human serum, bovine serum, horse serum, or fetal calf serum may be used. Mix thoroughly.

Use: For the cultivation of mammalian cells in suspension.

Tissue Culture Tyrode Solution (TC Tyrode Solution)

Composition per 1002.0mL:

NaCl	8.0g
Glucose	1.0g
$NaHCO_3$	1.0g
$CaCl_2 \cdot 2H_2O$	0.2g
KCl	0.2g
$MgCl_2 \cdot 6H_2O$	0.1g
NaH_2PO_4	0.05g
Phenol Red (1% solution)	2.0mL

pH 7.2–7.4 at 25°C

Preparation of Tissue Culture Tyrode Solution: Add components, except Phenol Red, to distilled/deionized water and bring volume to 1.0L. Mix thoroughly. Add 2.0mL of Phenol Red solution. Adjust pH to 7.2–7.4. Filter sterilize.

Use: For use in tissue culture procedures.

Tissue Culture Vitamins Minimal Eagle, 100X (TC Vitamins Minimal Eagle, 100X)

Composition per liter:

Inositol	2.0mg
Calcium pantothenate	1.0mg
Choline chloride	1.0mg
Folic acid	1.0mg
Nicotinamide	1.0mg
Pyridoxal	1.0mg
Thiamine·HCl	1.0mg
Riboflavin	0.1mg

pH 7.2–7.4 at 25°C

Preparation of TC Vitamins, Minimal Eagle 100X: Add components to distilled/deionized water and bring volume to 1.0L. Mix thoroughly. Filter sterilize.

Use: For the preparation of TC minimal medium Eagle used in tissue culture procedures.

T_1N_0 Broth (Tryptone Broth)

Composition per liter:

Pancreatic digest of casein	10.0g

pH 7.1 ± 0.2 at 25°C

Preparation of Medium: Add components to distilled/deionized water and bring volume to 1.0L. Mix thoroughly. Gently heat and bring to boiling.

Distribute into tubes or flasks. Autoclave for 15 min at 15 psi pressure–121°C.

Use: For the cultivation of *Vibrio cholerae* and other *Vibrio* species.

T_1N_1 Agar (Tryptone Salt Agar)

Composition per liter:

Agar	20.0g
NaCl	10.0g
Pancreatic digest of casein	10.0g

pH 7.1 ± 0.2 at 25°C

Preparation of Medium: Add components to distilled/deionized water and bring volume to 1.0L. Mix thoroughly. Gently heat and bring to boiling. Distribute into tubes or flasks. Autoclave for 15 min at 15 psi pressure–121°C. Pour into sterile Petri dishes or leave in tubes. Allow tubes to cool in a slanted position.

Use: For the cultivation of *Vibrio cholerae* and other *Vibrio* species.

T_1N_2 Agar (Tryptone Salt Agar)

Composition per liter:

Agar	20.0g
NaCl	20.0g
Pancreatic digest of casein	10.0g

pH 7.1 ± 0.2 at 25°C

Preparation of Medium: Add components to distilled/deionized water and bring volume to 1.0L. Mix thoroughly. Gently heat and bring to boiling. Distribute into tubes or flasks. Autoclave for 15 min at 15 psi pressure–121°C. Pour into sterile Petri dishes or leave in tubes. Allow tubes to cool in a slanted position.

Use: For the cultivation of *Vibrio cholerae* and other *Vibrio* species.

T_1N_1 Broth (Tryptone Salt Broth)

Composition per liter:

NaCl	10.0g
Pancreatic digest of casein	10.0g

pH 7.1 ± 0.2 at 25°C

Preparation of Medium: Add components to distilled/deionized water and bring volume to 1.0L. Mix thoroughly. Gently heat and bring to boiling. Distribute into tubes or flasks. Autoclave for 15 min at 15 psi pressure–121°C.

Use: For the cultivation of *Vibrio cholerae* and other *Vibrio* species.

T_1N_3 Broth (Tryptone Salt Broth)

Composition per liter:

NaCl	30.0g
Pancreatic digest of casein	10.0g

pH 7.1 ± 0.2 at 25°C

Preparation of Medium: Add components to distilled/deionized water and bring volume to 1.0L. Mix thoroughly. Gently heat and bring to boiling. Distribute into tubes or flasks. Autoclave for 15 min at 15 psi pressure–121°C.

Use: For the cultivation of *Vibrio cholerae* and other *Vibrio* species.

T_1N_6 Broth (Tryptone Salt Broth)

Composition per liter:

NaCl	60.0g
Pancreatic digest of casein	10.0g

pH 7.1 ± 0.2 at 25°C

Preparation of Medium: Add components to distilled/deionized water and bring volume to 1.0L. Mix thoroughly. Gently heat and bring to boiling. Distribute into tubes or flasks. Autoclave for 15 min at 15 psi pressure–121°C.

Use: For the cultivation of *Vibrio cholerae* and other *Vibrio* species.

T_1N_8 Broth (Tryptone Salt Broth)

Composition per liter:

NaCl	80.0g
Pancreatic digest of casein	10.0g

pH 7.1 ± 0.2 at 25°C

Preparation of Medium: Add components to distilled/deionized water and bring volume to 1.0L. Mix thoroughly. Gently heat and bring to boiling. Distribute into tubes or flasks. Autoclave for 15 min at 15 psi pressure–121°C.

Use: For the cultivation of *Vibrio cholerae* and other *Vibrio* species.

T_1N_{10} Broth (Tryptone Salt Broth)

Composition per liter:

NaCl	100.0g
Pancreatic digest of casein	10.0g

pH 7.1 ± 0.2 at 25°C

Preparation of Medium: Add components to distilled/deionized water and bring volume to 1.0L. Mix thoroughly. Gently heat and bring to boiling. Distribute into tubes or flasks. Autoclave for 15 min at 15 psi pressure–121°C.

Use: For the cultivation of *Vibrio cholerae* and other *Vibrio* species.

TOC Agar (Tween 80 Oxgall Caffeic Acid Agar)

Composition per liter:

Agar	20.0g
Oxgall	10.0g
Caffeic acid	0.3g
Tween 80	10.0mL

Source: This medium is available as a prepared medium from BD Diagnostic Systems.

Preparation of Medium: Add components to distilled/deionized water and bring volume to 1.0L. Mix thoroughly. Gently heat and bring to boiling. Autoclave for 15 min at 15 psi pressure–121°C. Pour into sterile Petri dishes.

Use: For the differentiation and identification of *Candida albicans* and *Cryptococcus neoformans. Cryptococcus albicans* produces germ tubes and chlamydospores when grown on this medium. *Cryptococcus neoformans* appears as tan to brown colonies.

Todd-Hewitt Broth

Composition per liter:

Beef heart, infusion from	500.0g
Neopeptone	20.0g
Na_2CO_3	2.5g
Glucose	2.0g
NaCl	2.0g
Na_2HPO_4	0.4g

pH 7.8 ± 0.2 at 25°C

Source: This medium is available as a premixed powder from BD Diagnostic Systems.

Preparation of Medium: Add components to distilled/deionized water and bring volume to 1.0L. Mix thoroughly. Distribute into tubes or flasks. Autoclave for 15 min at 15 psi pressure–121°C.

Use: For the cultivation of group A streptococci used in serological typing, and for the cultivation of a variety of pathogenic microorganisms.

Todd-Hewitt Broth

Composition per liter:

Pancreatic digest of casein	20.0g
Infusion from 450.0g fat-free minced meat	10.0g
Glucose	2.0g
$NaHCO_3$	2.0g
NaCl	2.0g
Na_2HPO_4	0.4g

pH 7.8 ± 0.2 at 25°C

Source: This medium is available as a premixed powder from Oxoid Unipath.

Preparation of Medium: Add components to distilled/deionized water and bring volume to 1.0L. Mix thoroughly. Distribute into tubes or flasks. Autoclave for 10 min at 10 psi pressure–115°C.

Use: For the cultivation of group A streptococci used in serological typing, and for the cultivation of a variety of pathogenic microorganisms.

Todd-Hewitt Broth (ATCC Medium 235)

Composition per liter:

Peptone	20.0g
Beef heart, solids from infusion	3.1g
Na_2CO_3	2.5g
Glucose	2.0g
NaCl	2.0g
Na_2HPO_4	0.4g

pH 7.8 ± 0.2 at 25°C

Source: This medium is available as a premixed powder from BD Diagnostic Systems.

Preparation of Medium: Add components to distilled/deionized water and bring volume to 1.0L. Mix thoroughly. Distribute into tubes or flasks. Autoclave for 15 min at 15 psi pressure–121°C.

Use: For the cultivation of Group A streptococci used in serological typing, and for the cultivation of a variety of pathogenic microorganisms.

Todd-Hewitt Broth, Modified

Composition per liter:

Neopeptone	20.0g
Glucose	2.0g
$NaHCO_3$	2.0g
NaCl	2.0g
Na_2HPO_4	0.4g
Beef heart infusion	1.0L

pH 7.8 ± 0.2 at 25°C

Preparation of Medium: Add components to distilled/deionized water and bring volume to 1.0L. Mix thoroughly. Distribute into tubes or flasks. Autoclave for 10 min at 10 psi pressure–115°C.

Use: For the cultivation of streptococci for serological identification.

Toluidine Blue DNA Agar

Composition per liter:

Agar	10.0g
NaCl	10.0g
Tris(hydroxymethyl)aminomethane buffer	6.1g
Deoxyribonucleic acid	0.3g
Toluidine Blue O	0.083g
$CaCl_2$, anhydrous	1.1mg

pH 9.0 ± 0.2 at 25°C

Preparation of Medium: Add tris(hydroxymethyl)aminomethane buffer to distilled/deionized water and bring volume to 1.0L. Mix thoroughly. Adjust pH to 9.0. Add the remaining components, except Toluidine Blue O. Mix thoroughly. Gently heat and bring to boiling. Add Toluidine Blue O. Mix thoroughly. If used the same day, sterilization is not necessary. Cool to 50°C. Pour into sterile Petri dishes or distribute into sterile tubes.

Use: For the cultivation and differentiation of *Staphylococcus aureus.*

Toluidine Blue DNA Agar

Composition per liter:

Agar	10.0g
NaCl	10.0g
Tris(hydroxymethyl)aminomethane buffer	6.1g
Deoxyribonucleic acid (DNA)	0.3g
Toluidine Blue O	0.083g
$CaCl_2$, anhydrous	1.1mg

pH 7.3 ± 0.2 at 25°C

Preparation of Medium: Add components, except Toluidine Blue O, to distilled/deionized water and bring volume to 1.0L. Mix thoroughly. Gently heat and bring to boiling. Add Toluidine Blue O. Mix thoroughly. Medium does not have to be sterilized if used immediately. Pour into sterile Petri dishes or distribute into sterile tubes. Allow tubes to cool in a slanted position.

Use: For the cultivation and differentiation of bacteria based on their production of deoxyribonuclease (DNase). Bacteria that produce DNase turn the medium pink.

Toxoplasma Medium

Composition per liter:

NaCl	6.8g
$NaHCO_3$	2.2g
Glucose	1.0g
KCl	0.4g
Glutamine	0.292g
$CaCl_2$	0.2g
$NaH_2PO_4 \cdot H_2O$	0.125g
Arginine	0.105g
$MgSO_4$	0.1g
Lysine	0.058g
Isoleucine	0.052g
Leucine	0.052g
Phenol Red	0.050g
Threonine	0.048g
Valine	0.046g
Tyrosine	0.036g
Phenylalanine	0.032g
Histidine	0.031g
L-Cystine	0.024g
Methionine	0.015g
Tryptophan	0.010g
Inositol	2.0mg
Choline	1.0mg
Folic acid	1.0mg
Nicotinamide	1.0mg
Pantothenic acid	1.0mg
Pyridoxal·HCl	1.0mg
Thiamine·HCl	1.0mg
Riboflavin	0.1mg
Fetal bovine serum, heat inactivated	100.0mL

pH 7.2–7.4 at 25°C

Preparation of Medium: Add components, except fetal bovine serum, to distilled/deionized water and bring volume to 905.0mL. Mix thoroughly. Adjust pH to 7.2–7.4. Autoclave for 15 min at 15 psi pressure–121°C. Aseptically add 100.0mL of sterile, heat-inactivated fetal bovine serum. Mix thoroughly. Aseptically distribute into sterile tubes or flasks.

Use: For the cultivation of *Toxoplasma gondii.*

TPEY Agar
(Tellurite Polymyxin Egg Yolk Agar)

Composition per liter:

NaCl	20.0g
Agar	15.5g
Pancreatic digest of casein	10.0g
Yeast extract	5.0g
D-Mannitol	5.0g
LiCl	2.0g
Egg yolk emulsion (30% solution)	100.0mL
Chapman tellurite solution	10.0mL
Polymyxin B solution	0.4mL

pH 7.1 ± 0.2 at 25°C

Source: This medium is available as a premixed powder from BD Diagnostic Systems.

Egg Yolk Emulsion (30% Solution):

Composition per 100.0mL:

NaCl	0.6g
Egg yolk	30.0mL

Preparation of Egg Yolk Emulsion (30% Solution): Add NaCl and egg yolk to distilled/deionized water and bring volume to 100.0mL. Mix thoroughly. Filter sterilize.

Chapman Tellurite Solution:
Composition per 100.0mL:
K_2TeO_3 .... 1.0g

Preparation of Chapman Tellurite Solution: Add K_2TeO_3 to distilled/deionized water and bring volume to 100.0mL. Mix thoroughly. Filter sterilize.

Polymyxin B Solution:
Composition per 100.0mL:
Polymyxin B 1.0g

Preparation of Polymyxin B Solution: Add polymyxin B to distilled/deionized water and bring volume to 100.0mL. Mix thoroughly. Filter sterilize.

Caution: Potassium tellurite is toxic.

Preparation of Medium: Add components—except 30% egg yolk emulsion, Chapman tellurite solution, and polymyxin B solution—to distilled/deionized water and bring volume to 890.0mL. Mix thoroughly. Gently heat and bring to boiling. Autoclave for 15 min at 15 psi pressure–121°C. Cool to 45°–50°C. Aseptically add 100.0mL of sterile 30% egg yolk emulsion, 10.0mL of sterile Chapman tellurite solution, and 0.4mL of sterile polymyxin B solution. Mix thoroughly. Pour into sterile Petri dishes or distribute into sterile tubes.

Use: For the recovery of staphylococci.

TPGY Medium (Thioglycolate Peptone Glucose Yeast Extract Medium)

Composition per liter:
Pancreatic digest of casein 50.0g
Peptone 5.0g
Yeast extract 5.0g
Glucose 1.0g
Sodium thioglycolate 1.0g
pH 7.1 ± 0.2 at 25°C

Preparation of Medium: Add components to distilled/deionized water and bring volume to 1.0L. Mix thoroughly. Distribute into tubes or flasks. Autoclave for 15 min at 15 psi pressure–121°C.

Use: For the cultivation of a variety of anaerobic bacteria.

Trace Elements Solution HO-LE

Composition per liter:
H_3BO_3 .... 2.85g
$MnCl_2 \cdot 4H_2O$ 1.8g
Sodium tartrate 1.77g
$FeSO_4 \cdot 7H_2O$ 1.36g
$CoCl_2 \cdot 6H_2O$ 0.04g
$CuCl_2 . 2H_2O$ 0.027g
$Na_2MoO_4 \cdot 2H_2O$ 0.025g
$ZnCl_2$ 0.02g

Preparation of Trace Elements Solution HO-LE: Add components to distilled/deionized water and bring volume to 1.0L. Mix thoroughly. Filter sterilize.

Use: For the enrichment of other media requiring added trace metals.

Transgrow Medium

Composition per liter:
GC agar base 730.0mL
Hemoglobin solution 250.0mL
Vitox supplement 10.0mL
VCN antibiotic solution 10.0mL
pH 7.3 ± 0.2 at 25°C

GC Agar Base:
Composition per 730.0mL:
Special peptone 15.0g
Agar 20.0g
NaCl 5.0g
K_2HPO_4 .... 4.0g
Cornstarch 1.0g
KH_2PO_4 .... 1.0g
pH 7.2 ± 0.2 at 25°C

Preparation of GC Agar Base: Add components of GC medium base and the hemoglobin to distilled/deionized water and bring volume to 730.0mL. Mix thoroughly. Gently heat until boiling. Autoclave for 15 min at 15 psi pressure–121°C. Cool to 45°–50°C.

Hemoglobin Solution:
Composition per 250.0mL:
Hemoglobin 5.0g

Preparation of Hemoglobin Solution: Add hemoglobin to distilled/deionized water and bring volume to 250.0mL. Mix thoroughly. Autoclave for 15 min at 15 psi pressure–121°C. Cool to 45°–50°C.

Vitox Supplement:
Composition per 10.0mL:
Glucose 2.0g
L-Cysteine·HCl 0.518g
L-Glutamine 0.2g
L-Cystine 0.022g
Adenine sulfate 0.01g
Nicotinamide adenine dinucleotide 5.0mg
Cocarboxylase 2.0mg
Guanine·HCl 0.6mg
$Fe(NO_3)_3 \cdot 6H_2O$ 0.4mg
p-Aminobenzoic acid 0.26mg

Vitamin B_{12}	0.2mg
Thiamine·HCl	0.06mg

Preparation of Vitox Supplement: Add components to distilled/deionized water and bring volume to 10.0mL. Mix thoroughly. Filter sterilize.

VCN Antibiotic Solution:
Composition per 10.0mL:

Colistin methane sulfonate	7.5mg
Vancomycin	3.0mg
Nystatin	12,500U

Preparation of VCN Antibiotic Solution: Add components to distilled/deionized water and bring volume to 10.0mL. Mix thoroughly. Filter sterilize.

Preparation of Medium: To 730.0mL of cooled, sterile GC agar base, aseptically add 250.0mL of sterile hemoglobin solution, 10.0mL of sterile Vitox supplement, and 10.0mL of VCN antibiotic solution. Mix thoroughly. Pour into sterile Petri dishes or distribute into sterile tubes.

Use: For the cultivation and transport of fastidious microorganisms, especially *Neisseria* species.

Transgrow Medium

Composition per liter:

GC medium base	730.0mL
Hemoglobin solution	250.0mL
Supplement B	10.0mL
VCNT antibiotic solution	10.0mL

pH 7.3 ± 0.2 at 25°C

GC Medium Base:
Composition per 730.0mL:

Proteose peptone no. 3	15.0g
Agar	20.0g
NaCl	5.0g
K_2HPO_4	4.0g
Glucose	1.5g
Cornstarch	1.0g
KH_2PO_4	1.0g

pH 7.2 ± 0.2 at 25°C

Preparation of GC Medium Base: Add components to distilled/deionized water and bring volume to 730.0mL. Mix thoroughly. Gently heat until boiling. Autoclave for 15 min at 15 psi pressure–121°C. Cool to 45°–50°C.

Hemoglobin Solution:
Composition per 250.0mL:

Hemoglobin	10.0g

Preparation of Hemoglobin Solution: Add hemoglobin to distilled/deionized water and bring volume to 250.0mL. Mix thoroughly. Autoclave for 15 min at 15 psi pressure–121°C. Cool to 45°–50°C.

Supplement B:
Composition per 10.0mL:

Supplement B contains yeast concentrate, glutamine, coenzyme, cocarboxylase, hematin, and growth factors.

Preparation of Supplement B: Add components to distilled/deionized water and bring volume to 10.0mL. Mix thoroughly. Filter sterilize.

Source: Supplement B is available as a premixed powder from BD Diagnostic Systems.

VCNT Antibiotic Solution:
Composition per 10.0mL:

Colistin methane sulfonate	7.5mg
Trimethoprim lactate	5.0mg
Vancomycin	3.0mg
Nystatin	12,500U

Preparation of VCNT Antibiotic Solution: Add components to distilled/deionized water and bring volume to 10.0mL. Mix thoroughly. Filter sterilize.

Preparation of Medium: To 730.0mL of cooled, sterile GC medium base, aseptically add 250.0mL of sterile hemoglobin solution, 10.0mL of sterile supplement B, and 10.0mL of VCNT antibiotic solution. Mix thoroughly. Pour into sterile Petri dishes or distribute into sterile tubes.

Use: For the cultivation and transport of fastidious microorganisms, especially *Neisseria* species.

Transgrow Medium with Trimethoprim

Composition per liter:

Agar	20.0g
Hemoglobin	10.0g
Pancreatic digest of casein	7.5g
Selected meat peptone	7.5g
NaCl	5.0g
K_2HPO_4	4.0g
Glucose	1.5g
Cornstarch	1.0g
KH_2PO_4	1.0g
Supplement solution	10.0mL
VCNT inhibitor	10.0mL

pH 6.7 ± 0.2 at 25°C

Source: This medium is available as a prepared medium from BD Diagnostic Systems.

Supplement Solution:
Composition per liter:

Glucose	100.0g
L-Cysteine·HCl	25.9g
L-Glutamine	10.0g
L-Cystine	1.1g
Adenine	1.0g

Nicotinamide adenine dinucleotide 0.25g
Vitamin B_{12} 0.1g
Thiamine pyrophosphate.................0.1g
Guanine·HCl 0.03g
$Fe(NO_3)_3 \cdot 6H_2O$ 0.02g
p-Aminobenzoic acid.................0.013g
Thiamine·HCl 3.0mg

Source: The supplement solution (IsoVitaleX enrichment) is available from BD Diagnostic Systems. This enrichment may be replaced by supplement VX from BD Diagnostic Systems.

Preparation of Supplement Solution: Add components to distilled/deionized water and bring volume to 1.0L. Mix thoroughly. Filter sterilize.

VCNT Inhibitor:
Composition per 10.0mL:
Colistin.................7.5mg
Trimethoprim lactate.................5.0mg
Vancomycin 3.0mg
Nystatin.................12,500U

Preparation of VCNT Inhibitor: Add components to distilled/deionized water and bring volume to 10.0mL. Mix thoroughly. Filter sterilize.

Preparation of Medium: Add components, except supplement solution and VCNT inhibitor, to distilled/deionized water and bring volume to 980.0mL. Mix thoroughly. Gently heat and bring to boiling. Autoclave for 15 min at 15 psi pressure–121°C. Cool to 45°–50°C under 5–30% CO_2. Aseptically add 10.0mL of sterile supplement solution and 10.0mL of sterile VCNT inhibitor. Mix thoroughly. Aseptically distribute under 5–30% CO_2 into sterile screw-capped tubes.

Use: For the transportation and recovery of pathogenic *Neisseria* species.

Transgrow Medium without Trimethoprim

Composition per liter:
Agar 20.0g
Hemoglobin 10.0g
Pancreatic digest of casein.................7.5g
Selected meat peptone 7.5g
NaCl.................5.0g
K_2HPO_4.................4.0g
Glucose 1.5g
Cornstarch.................1.0g
KH_2PO_4.................1.0g
Supplement solution 10.0mL
VCN inhibitor 10.0mL
pH 6.7 ± 0.2 at 25°C

Source: This medium is available as a prepared medium from BD Diagnostic Systems.

Supplemement Solution:
Composition per liter:
Glucose 100.0g
L-Cysteine·HCl.................25.9g
L-Glutamine 10.0g
L-Cystine 1.1g
Adenine.................1.0g
Nicotinamide adenine dinucleotide 0.25g
Vitamin B_{12}.................0.1g
Thiamine pyrophosphate 0.1g
Guanine·HCl 0.03g
$Fe(NO_3)_3 \cdot 6H_2O$.................0.02g
p-Aminobenzoic acid.................0.013g
Thiamine·HCl 3.0mg

Source: The supplement solution IsoVitaleX enrichment is available from BD Diagnostic Systems. This enrichment may be replaced by supplement VX from BD Diagnostic Systems.

Preparation of Supplement Solution: Add components to distilled/deionized water and bring volume to 1.0L. Mix thoroughly. Filter sterilize.

VCN Inhibitor:
Composition per 10.0mL:
Colistin.................7.5mg
Vancomycin 3.0mg
Nystatin.................12,500U

Preparation of VCN Inhibitor: Add components to distilled/deionized water and bring volume to 10.0mL. Mix thoroughly. Filter sterilize.

Preparation of Medium: Add components, except supplement solution and VCN inhibitor, to distilled/deionized water and bring volume to 980.0mL. Mix thoroughly. Gently heat and bring to boiling. Autoclave for 15 min at 15 psi pressure–121°C. Cool to 45°–50°C under 5–30% CO_2. Aseptically add 10.0mL of sterile supplement solution and 10.0mL of sterile VCN inhibitor. Mix thoroughly. Aseptically distribute under 5–30% CO_2 into sterile screw-capped tubes.

Use: For the transportation and recovery of pathogenic *Neisseria* species.

Transport Medium

Composition per liter:
Sodium glycerophosphate.................10.0g
Agar 3.0g
Sodium thioglycolate.................1.0g
$CaCl_2 \cdot 2H_2O$ 0.1g
Methylene Blue.................2.0mg
pH 7.3 ± 0.2 at 25°C

Source: This medium is available as a premixed powder from BD Diagnostic Systems.

Preparation of Medium: Add components to distilled/deionized water and bring volume to 1.0L. Mix thoroughly. Gently heat while stirring and bring to boiling. Distribute into screw-capped tubes or vials. Fill tubes nearly to capacity. Leave only enough space so that when a small swab is introduced the tube does not overflow. Autoclave for 10 min at 15 psi pressure–121°C. Tighten caps on tubes.

Use: For the transportation of swab specimens for the recovery of a wide variety of microorganisms, including *Neisseria gonorrhoeae.*

Transport Medium Stuart

Composition per liter:

Sodium glycerophosphate	10.0g
Agar	3.0g
Sodium thioglycolate	0.9g
$CaCl_2 \cdot 2H_2O$	0.1g
Methylene Blue	2.0mg

pH 7.3 ± 0.2 at 25°C

Source: This medium is available as a premixed powder from BD Diagnostic Systems.

Preparation of Medium: Add components to distilled/deionized water and bring volume to 1.0L. Mix thoroughly. Gently heat while stirring and bring to boiling. Distribute into screw-capped tubes or vials. Fill tubes nearly to capacity. Leave only enough space so that when a small swab is introduced the tube does not overflow. Autoclave for 10 min at 15 psi pressure–121°C. Tighten caps on tubes.

Use: For the transportation of swab specimens for the recovery of a wide variety of microorganisms, including *Neisseria gonorrhoeae.*

Treponema Isolation Medium

Composition per liter:

Solution A	450.0mL
Spirolate broth	450.0mL
Rabbit serum, inactivated at 56°C for 30 min	100.0mL

pH 7.4 ± 0.2 at 25°C

Solution A:

Composition per 450.0mL:

Agar	8.0g
Asparagine	0.25g
Sodium thioglycolate	0.25g
Pancreatic digest of casein	0.25g
Brain heart infusion broth	450.0mL

Preparation of Solution A: Combine components. Mix thoroughly. Gently heat and bring to boiling. Autoclave for 15 min at 15 psi pressure–121°C. Cool to 45°–50°C.

Brain Heart Infusion Broth:

Composition per liter:

Pancreatic digest of gelatin	14.5g
Brain heart, solids from infusion	6.0g
Peptic digest of animal tissue	6.0g
NaCl	5.0g
Casein	5.0g
Glucose	3.0g
Na_2HPO_4	2.5g

Preparation of Brain Heart Infusion Broth: Add components to distilled/deionized water and bring volume to 1.0L. Mix thoroughly.

Spirolate Broth:

Composition per liter:

Pancreatic digest of casein	15.0g
Glucose	5.0g
Yeast extract	5.0g
NaCl	2.5g
L-Cysteine·HCl·H_2O	1.0g
Sodium thioglycolate	0.5g
Palmitic acid	0.05g
Stearic acid	0.05g
Oleic acid	0.05g
Linoleic acid	0.05g

Preparation of Spirolate Broth: Add components to distilled/deionized water and bring volume to 1.0L. Mix thoroughly. Autoclave for 15 min at 15 psi pressure–121°C. Cool to 25°C.

Preparation of Medium: Combine 450.0mL of sterile solution A, 450.0mL of sterile spirolate broth, and 100.0mL of rabbit serum. Mix thoroughly. Aseptically distribute into sterile tubes or flasks.

Use: For the isolation and cultivation of oral, genital, and fecal treponemes.

Treponema Isolation Medium

Composition per liter:

Beef heart, solids from infusion	20.0g
Ionagar no. 2	7.2g
K_2HPO_4	2.0g
Arabinose	0.8g
Glucose	0.8g
Maltose	0.8g
Polypeptone	0.8g
Pyruvate	0.8g
Starch, soluble	0.8g
Sucrose	0.8g
Cysteine·HCl	0.68g
$(NH_4)_2SO_4$	0.6g
Serine	0.4g

Tryptose 0.4g
Yeast extract 0.4g
NaCl 0.2g
Rumen fluid 500.0mL
Rabbit serum–cocarboxylase solution 100.0mL
pH 7.2 ± 0.2 at 25°C

Rabbit Serum-Cocarboxylase Solution:
Composition per liter:
Rabbit serum, heat inactivated 100.0mL
Cocarboxylase solution 1.0mL

Preparation of Rabbit Serum-Cocarboxylase Solution: Heat rabbit serum at 56°C for 1h. Add 1.0mL of cocarboxylase solution. Mix thoroughly.

Cocarboxylase Solution:
Composition per 1.0mL:
Cocarboxylase 0.5g

Preparation of Cocarboxylase Solution: Add cocarboxylase to 1.0mL of distilled/deionized water. Mix thoroughly. Filter sterilize.

Preparation of Medium: Add components, except rumen fluid and rabbit serum-cocarboxylase solution, to distilled/deionized water and bring volume to 400.0mL. Mix thoroughly. Gently heat and bring to boiling. Autoclave for 15 min at 15 psi pressure–121°C. Cool to 45°–50°C. Aseptically add 500.0mL of sterile rumen fluid and 100.0mL of sterile rabbit serum–cocarboxylase solution. Mix thoroughly. Pour into sterile Petri dishes or distribute into sterile tubes.

Use: For the isolation of oral treponemes.

Treponema Medium

Composition per liter:
Pancreatic digest of casein 30.0g
Ionagar no. 2 .... 8.0g
Glucose 5.0g
Yeast extract 5.0g
NaCl 2.5g
Cysteine·HCl 0.75g
Horse serum, inactivated 100.0mL
pH 7.4 ± 0.2 at 25°C

Preparation of Medium: Add components, except horse serum, to distilled/deionized water and bring volume to 900.0mL. Mix thoroughly. Gently heat and bring to boiling. Autoclave for 15 min at 15 psi pressure–121°C. Cool to 45°–50°C. Aseptically add 100.0mL of sterile horse serum. Mix thoroughly. Pour into sterile Petri dishes or distribute into sterile tubes.

Use: For the isolation and cultivation of oral treponemes.

Treponema Medium

Composition per liter:
Spirolate agar 900.0mL
Rabbit serum,
inactivated at 56°C for 30 min 100.0mL

Spirolate Agar:
Composition per liter:
Pancreatic digest of casein 15.0g
Agar 14.0g
Glucose 5.0g
Yeast extract 5.0g
NaCl 2.5g
L-Cysteine·HCl·H_2O 1.0g
Sodium thioglycolate 0.5g
Palmitic acid 0.05g
Stearic acid 0.05g
Oleic acid 0.05g
Linoleic acid 0.05g

Preparation of Spirolate Agar: Add components to distilled/deionized water and bring volume to 1.0L. Mix thoroughly. Gently heat and bring to boiling. Autoclave for 15 min at 15 psi pressure–121°C. Cool to 45°–50°C.

Preparation of Medium: To 900.0mL of cooled, sterile spirolate agar, aseptically add 100.0mL of rabbit serum. Mix thoroughly. Aseptically distribute into sterile tubes or flasks.

Use: For the isolation of oral treponemes.

Treponema Medium

Composition per liter:
Spirolate agar 675.0mL
Brain heart infusion broth 225.0mL
Rabbit serum,
inactivated at 56°C for 30 min 100.0mL
pH 7.0–7.2 ± 0.2 at 25°C

Spirolate Agar:
Composition per 675.0mL:
Pancreatic digest of casein 15.0g
Ionagar no. 2 .... 8.0g
Glucose 5.0g
Yeast extract 5.0g
NaCl 2.5g
L-Cysteine·HCl·H_2O 1.0g
Sodium thioglycolate 0.5g
Palmitic acid 0.05g
Stearic acid 0.05g
Oleic acid 0.05g
Linoleic acid 0.05g

Preparation of Spirolate Agar: Add components to distilled/deionized water and bring volume to 675.0mL. Mix thoroughly. Autoclave for 15 min at 15 psi pressure–121°C. Cool to 45°–50°C.

Brain Heart Infusion Broth:

Composition per liter:

Pancreatic digest of gelatin	14.5g
Brain heart, solids from infusion	6.0g
Peptic digest of animal tissue	6.0g
NaCl	5.0g
Casein	5.0g
Glucose	3.0g
Na_2HPO_4	2.5g

Preparation of Brain Heart Infusion Broth: Add components to distilled/deionized water and bring volume to 1.0L. Mix thoroughly.

Preparation of Medium: Aseptically combine 675.0mL of cooled, sterile spirolate agar, 225.0mL of cooled, sterile brain heart infusion broth, and 100.0mL of rabbit serum. Mix thoroughly. Pour into sterile Petri dishes or distribute into sterile tubes.

Use: For the isolation of oral treponemes.

Treponema Medium

Composition per liter:

Solution A	440.0mL
Spirolate broth	440.0mL
Rabbit serum, inactivated at 56°C for 30 min	100.0mL
Mucin solution	20.0mL

pH 7.8 ± 0.2 at 25°C

Solution A:

Composition per 440.0mL:

Ionagar no. 2	8.0g
Brain heart infusion broth	440.0mL

Brain Heart Infusion Broth:

Composition per liter:

Pancreatic digest of gelatin	14.5g
Brain heart, solids from infusion	6.0g
Peptic digest of animal tissue	6.0g
NaCl	5.0g
Casein	5.0g
Glucose	3.0g
Na_2HPO_4	2.5g

Preparation of Brain Heart Infusion Broth: Add components to distilled/deionized water and bring volume to 1.0L. Mix thoroughly.

Preparation of Solution A: Add 8.0g of ionagar to 440.0mL of brain heart infusion broth. Mix thoroughly. Gently heat and bring to boiling. Autoclave for 15 min at 15 psi pressure–121°C. Cool to 45°–50°C.

Spirolate Broth:

Composition per liter:

Pancreatic digest of casein	15.0g
Glucose	5.0g
Yeast extract	5.0g
NaCl	2.5g
L-Cysteine·HCl·H_2O	1.0g
Sodium thioglycolate	0.5g
Palmitic acid	0.05g
Stearic acid	0.05g
Oleic acid	0.05g
Linoleic acid	0.05g

Preparation of Spirolate Broth: Add components to distilled/deionized water and bring volume to 1.0L. Mix thoroughly. Autoclave for 15 min at 15 psi pressure–121°C. Cool to 25°.

Mucin Solution:

Composition per 20.0mL:

Mucin	0.2g

Preparation of Mucin Solution: Add mucin to distilled/deionized water and bring volume to 20.0mL. Mix thoroughly. Filter sterilize.

Preparation of Medium: Aseptically combine 440.0mL of solution A, 440.0mL of spirolate broth, 100.0mL of rabbit serum, and 20.0mL of mucin solution. Mix thoroughly. Aseptically distribute into sterile tubes or flasks.

Use: For the isolation of intestinal treponemes.

Treponema Medium

Composition per liter:

Agar	13.0g
Glucose	1.4g
Cysteine·HCl	0.64g
$(NH_4)_2SO_4$	0.5g
Polypeptone	0.5g
Starch, soluble	0.5g
Yeast extract	0.5g
Resazurin	1.6mg
Salts solution	500.0mL
Bovine rumen fluid	280.0mL

pH 7.2–7.5 at 25°C

Salts Solution:

Composition per liter:

$NaHCO_3$	10.0g
NaCl	2.0g
K_2HPO_4	1.0g
KH_2PO_4	1.0g
$CaCl_2$	0.2g
$MgSO_4$	0.2g
CoCl	3.4mg

$MnSO_4$ 3.4mg
$NaMoO_4$ 3.4mg

Preparation of Salts Solution: Add components to distilled/deionized water and bring volume to 1.0L. Mix thoroughly.

Preparation of Medium: Add components, except bovine rumen fluid, to distilled/deionized water and bring volume to 720.0mL. Mix thoroughly. Gently heat and bring to boiling. Autoclave for 15 min at 15 psi pressure–121°C. Cool to 45°–50°C. Aseptically add bovine rumen fluid. Mix thoroughly. Pour into sterile Petri dishes or distribute into sterile tubes.

Use: For the isolation of intestinal treponemes.

Treponema Medium

Composition per liter:

Cysteine·HCl·H_2O 1.0g
Glucose 1.0g
Nicotinamide 0.4g
Spermidine·4HCl 0.15g
Sodium isobutyrate 0.02g
Thiamine pyrophosphate 5.0mg
PPLO broth 900.0mL
Rabbit serum, inactivated 100.0mL
pH 7.8 ± 0.2 at 25°C

PPLO Broth:

Composition per 900.0mL:

Beef heart, infusion from solids 50.0g
Peptone 10.0g
NaCl 5.0g

Preparation of PPLO Broth: Add components to distilled/deionized water and bring volume to 900.0mL. Mix thoroughly.

Preparation of Medium: Combine components, except rabbit serum. Mix thoroughly. Filter sterilize. Aseptically add sterile rabbit serum. Mix thoroughly. Aseptically distribute into sterile tubes or flasks.

Use: For the cultivation of oral treponemes. For the cultivation of *Treponema denticola, Treponema macrodentium,* and *Treponema oralis.*

Treponema Medium

Composition per liter:

Spirolate broth 675.0mL
Brain heart infusion broth 225.0mL
Rabbit serum 100.0mL

Spirolate Broth:

Composition per liter:

Pancreatic digest of casein 15.0g
Glucose 5.0g
Yeast extract 5.0g
NaCl 2.5g
L-Cysteine·HCl·H_2O 1.0g
Sodium thioglycolate 0.5g
Palmitic acid 0.05g
Stearic acid 0.05g
Oleic acid 0.05g
Linoleic acid 0.05g

Preparation of Spirolate Broth: Add components to distilled/deionized water and bring volume to 1.0L. Mix thoroughly. Autoclave for 15 min at 15 psi pressure–121°C. Cool to 25°C.

Brain Heart Infusion Broth:

Composition per liter:

Pancreatic digest of gelatin 14.5g
Brain heart, solids from infusion 6.0g
Peptic digest of animal tissue 6.0g
NaCl 5.0g
Casein 5.0g
Glucose 3.0g
Na_2HPO_4 .... 2.5g

Preparation of Brain Heart Infusion Broth: Add components to distilled/deionized water and bring volume to 1.0L. Mix thoroughly. Autoclave for 15 min at 15 psi pressure–121°C. Cool to 25°C.

Preparation of Medium: Aseptically combine 675.0mL of cooled, sterile spirolate broth, 225.0mL of cooled, sterile brain heart infusion broth, and 100.0mL of rabbit serum. Mix thoroughly.

Use: For the cultivation of oral treponemes.

Treponema Medium

Composition per liter:

Heart infusion broth, modified 450.0mL
Spirolate broth 450.0mL
Rabbit serum, inactivated 100.0mL
pH 7.4 ± 0.2 at 25°C

Heart Infusion Broth, Modified:

Composition per liter:

Beef heart, solids from infusion 500.0g
Tryptose 10.0g
NaCl 5.0g
Asparagine 2.5g
Sodium thioglycolate 2.5g
Pancreatic digest of casein 2.5g

Preparation of Heart Infusion Broth, Modified: Add components to distilled/deionized water and bring volume to 1.0L. Mix thoroughly. Gently heat and bring to boiling. Autoclave for 15 min at 15 psi pressure–121°C. Cool to 25°C.

Spirolate Broth:

Composition per liter:

Pancreatic digest of casein 15.0g

Glucose	5.0g
Yeast extract	5.0g
NaCl	2.5g
L-Cysteine·HCl·H_2O	1.0g
Sodium thioglycolate	0.5g
Palmitic acid	0.05g
Stearic acid	0.05g
Oleic acid	0.05g
Linoleic acid	0.05g

Preparation of Spirolate Broth: Add components to distilled/deionized water and bring volume to 1.0L. Mix thoroughly. Autoclave for 15 min at 15 psi pressure–121°C. Cool to 25°C.

Preparation of Medium: Aseptically combine 450.0mL of cooled, sterile spirolate broth, 450.0mL of cooled, sterile heart infusion broth, modified, and 100.0mL of rabbit serum. Mix thoroughly. Aseptically distribute into sterile tubes or flasks.

Use: For the cultivation of treponemes.

Treponema Medium

Composition per 500.0mL:

Beef heart, solids from infusion	250.0g
Sucrose	50.0g
Tryptose	5.0g
NaCl	2.5g
Yeast extract	2.5g
Agar	0.5g
Sodium thioglycolate	0.38g
$MgSO_4$	0.05g
Horse serum, inactivated	100.0mL

pH 7.4 ± 0.2 at 25°C

Preparation of Medium: Add components, except horse serum, to distilled/deionized water and bring volume to 400.0mL. Mix thoroughly. Adjust pH to 7.4. Gently heat and bring to boiling. Distribute into tubes in 4.0mL volumes. Autoclave for 15 min at 15 psi pressure–121°C. Cool to 25°C. Prior to inoculation, add 1.0mL sterile horse serum to each tube.

Use: For the cultivation and maintenance of *Treponema pallidum* and other *Treponema* species.

Treponema Medium 1

Composition per liter:

Thioglycolate agar USP, alternate	900.0mL
Normal calf serum	100.0mL

pH 7.1 ± 0.2 at 25°C

Thioglycolate Agar USP, Alternate:

Composition per 900.0mL:

Pancreatic digest of casein	15.0g
Ionagar no. 2	7.0g
Glucose	5.5g
Yeast extract	5.0g
NaCl	2.5g
L-Cystine	0.5g
Sodium thioglycolate	0.5g

Preparation of Thioglycolate Agar USP, Alternate: Add components to distilled/deionized water and bring volume to 1.0L. Mix thoroughly. Autoclave for 15 min at 15 psi pressure–121°C. Cool to 45°–50°C.

Preparation of Medium: Aseptically combine 900.0mL of cooled sterile thioglycolate agar USP, alternate, and 100.0mL of calf serum. Mix thoroughly. Pour into sterile Petri dishes or distribute into sterile tubes.

Use: For the cultivation of treponemes.

Treponema Medium 2

Composition per liter:

Pancreatic digest of casein	30.0g
Ionagar no. 2	7.0g
Glucose	5.0g
Yeast extract	5.0g
NaCl	2.5g
L-Cysteine·HCl·H_2O	2.0g
Rabbit serum	100.0mL

pH 7.2 ± 0.2 at 25°C

Preparation of Medium: Add components, except rabbit serum, to distilled/deionized water and bring volume to 900.0mL. Mix thoroughly. Gently heat and bring to boiling. Autoclave for 15 min at 15 psi pressure–121°C. Cool to 45°–50°C. Aseptically add sterile rabbit serum. Mix thoroughly. Pour into sterile Petri dishes or distribute into sterile tubes.

Use: For the cultivation of treponemes.

Treponema Medium 3

Composition per liter:

Spirolate agar	675.0mL
Brain heart infusion broth	225.0mL
Rabbit serum	100.0mL

Spirolate Agar:

Composition per liter:

Pancreatic digest of casein	15.0g
Ionagar no. 2	7.0g
Glucose	5.0g
Yeast extract	5.0g
NaCl	2.5g
L-Cysteine·HCl·H_2O	1.0g
Sodium thioglycolate	0.5g
Palmitic acid	0.05g

Stearic acid 0.05g
Oleic acid 0.05g
Linoleic acid 0.05g

Preparation of Spirolate Agar: Add components to distilled/deionized water and bring volume to 1.0L. Mix thoroughly. Gently heat and bring to boiling. Autoclave for 15 min at 15 psi pressure–121°C. Cool to 25°C.

Brain Heart Infusion Broth:
Composition per liter:
Pancreatic digest of gelatin 14.5g
Brain heart, solids from infusion 6.0g
Peptic digest of animal tissue 6.0g
NaCl 5.0g
Casein 5.0g
Glucose 3.0g
Na_2HPO_4 .... 2.5g

Preparation of Brain Heart Infusion Broth: Add components to distilled/deionized water and bring volume to 1.0L. Mix thoroughly. Gently heat and bring to boiling. Autoclave for 15 min at 15 psi pressure–121°C. Cool to 25°C.

Preparation of Medium: Aseptically combine 675.0mL of cooled, sterile spirolate broth, 225.0mL of cooled, sterile brain heart infusion broth, and 100.0mL of rabbit serum. Mix thoroughly.

Use: For the cultivation of treponemes.

Treponema Medium, Prereduced

Composition per liter:
Agar 1.6g
Glucose 1.4g
Cysteine·HCl·H_2O 0.64g
$(NH_4)_2SO_4$ 0.5g
Polypeptone 0.5g
Starch, soluble 0.5g
Yeast extract 0.5g
Resazurin 1.6mg
Salts solution 500.0mL
Bovine rumen fluid 280.0mL
pH 7.2–7.5 at 25°C

Salts Solution:
Composition per liter:
$NaHCO_3$ 10.0g
NaCl 2.0g
K_2HPO_4 .... 1.0g
KH_2PO_4 .... 1.0g
$CaCl_2$ 0.2g
$MgSO_4$ 0.2g
CoCl 3.4mg
$MnSO_4$ 3.4mg
$NaMoO_4$ 3.4mg

Preparation of Salts Solution: Add components to distilled/deionized water and bring volume to 1.0L. Mix thoroughly.

Preparation of Medium: Add components, except bovine rumen fluid, to distilled/deionized water and bring volume to 720.0mL. Mix thoroughly. Gently heat and bring to boiling. Autoclave for 15 min at 15 psi pressure–121°C. Cool to 45°–50°C. Aseptically add 280.0mL of sterile bovine rumen fluid. Mix thoroughly. Aseptically and anaerobically distribute into sterile tubes or flasks under 100% N_2.

Use: For the cultivation of fecal and intestinal treponemes.

Trichomonas Medium

Composition per liter:
Liver digest 25.0g
NaCl 6.5g
Glucose 5.0g
Agar 1.0g
Horse serum 80.0mL
pH 6.4 ± 0.2 at 25°C

Source: This medium is available as a premixed powder from Oxoid Unipath.

Horse Serum:
Composition per 80.0mL:
Horse serum 80.0mL

Preparation of Horse Serum: Gently heat sterile horse serum to 56°C for 30 min. Aseptically adjust pH to 6.0 with 0.1*N* HCl. Use immediately.

Preparation of Medium: Add components, except horse serum, to distilled/deionized water and bring volume to 920.0mL. Mix thoroughly. Gently heat and bring to boiling. Autoclave for 15 min at 15 psi pressure–121°C. Cool to 45°–50°C. Aseptically add 80.0mL of freshly prepared sterile horse serum. Mix thoroughly. Aseptically distribute into sterile tubes or flasks.

Use: For the cultivation of *Trichomonas vaginalis*.

Trichomonas Medium No. 2

Composition per liter:
Glucose 22.5g
Liver digest 18.0g
Pancreatic digest of casein 17.0g
NaCl 5.0g
Pancreatic digest of soybean meal 3.0g
K_2HPO_4 .... 2.5g
Chloramphenicol 0.125g
Horse serum 250.0mL
Calcium pantothenate (0.5% solution) 1.0mL
pH 6.2 ± 0.2 at 25°C

Source: This medium is available as a prepared medium from Oxoid Unipath.

Preparation of Medium: Add components, except horse serum, to distilled/deionized water and bring volume to 750.0mL. Mix thoroughly. Autoclave for 15 min at 5 psi pressure–108°C. Cool to 45°–50°C. Aseptically add 250.0mL of sterile horse serum. Mix thoroughly. Aseptically distribute into sterile tubes or flasks.

Use: For the isolation of *Trichomonas vaginalis*.

Trichomonas Selective Medium

Composition per liter:

Liver digest	25.0g
NaCl	6.5g
Glucose	5.0g
Agar	1.0g
Horse serum	80.0mL
Antibiotic inhibitor	10.0mL

pH 6.4 ± 0.2 at 25°C

Source: This medium is available as a premixed powder from Oxoid Unipath.

Horse Serum:

Composition per 80.0mL:

Horse serum	80.0mL

Preparation of Horse Serum: Gently heat sterile horse serum to 56°C for 30 min. Aseptically adjust pH to 6.0 with 0.1*N* HCl. Use immediately.

Antibiotic Inhibitor:

Composition per 10.0mL:

Streptomycin	500.0mg
Penicillin G	1,000,000U

Preparation of Antibiotic Inhibitor: Add components to distilled/deionized water and bring volume to 10.0mL. Mix thoroughly. Filter sterilize.

Preparation of Medium: Add components, except horse serum, to distilled/deionized water and bring volume to 910.0mL. Mix thoroughly. Gently heat and bring to boiling. Autoclave for 15 min at 15 psi pressure–121°C. Cool to 45°–50°C. Aseptically add 80.0mL of freshly prepared sterile horse serum and 10.0mL of sterile antibiotic inhibitor. Mix thoroughly. Aseptically distribute into sterile tubes or flasks.

Use: For the cultivation of *Trichomonas vaginalis* from specimens with a mixed bacterial flora.

Trichomonas Selective Medium

Composition per liter:

Liver digest	25.0g
NaCl	6.5g
Glucose	5.0g
Agar	1.0g
Horse serum	80.0mL
Antibiotic inhibitor	10.0mL

pH 6.4 ± 0.2 at 25°C

Horse Serum:

Composition per 80.0mL:

Horse serum	80.0mL

Preparation of Horse Serum: Gently heat sterile horse serum to 56°C for 30 min. Aseptically adjust pH to 6.0 with 0.1*N* HCl. Use immediately.

Antibiotic Inhibitor:

Composition per 10.0mL:

Chloramphenicol	100.0mg

Preparation of Antibiotic Inhibitor: Add chloramphenicol to distilled/deionized water and bring volume to 10.0mL. Mix thoroughly. Filter sterilize.

Preparation of Medium: Add components, except horse serum, to distilled/deionized water and bring volume to 910.0mL. Mix thoroughly. Gently heat and bring to boiling. Autoclave for 15 min at 15 psi pressure–121°C. Cool to 45°–50°C. Aseptically add 80.0mL of freshly prepared sterile horse serum and 10.0mL of sterile antibiotic inhibitor. Mix thoroughly. Aseptically distribute into sterile tubes or flasks.

Use: For the cultivation of *Trichomonas vaginalis* from specimens with a mixed bacterial flora.

Trichosel™ Broth, Modified

Composition per liter:

Pancreatic digest of casein	12.0g
Yeast extract	5.0g
Liver extract	2.0g
Maltose	2.0g
L-Cysteine·HCl	1.0g
Agar	1.0g
Chloramphenicol	0.1g
Methylene Blue	3.0mg
Horse serum	50.0mL

pH 6.0 ± 0.2 at 25°C

Source: This medium is available as a premixed powder from BD Diagnostic Systems.

Preparation of Medium: Add components to distilled/deionized water and bring volume to 950.0mL. Mix thoroughly. Gently heat while stirring and bring to boiling. Autoclave for 15 min at 13 psi pressure–118°C. Cool to 45°–50°C. Aseptically add 50.0mL of sterile horse serum. Mix thoroughly. Aseptically distribute into sterile tubes or flasks.

Use: For the isolation and cultivation of *Trichomonas* species.

Trimethylamine *N*-Oxide Medium (TMAO Medium)

Composition per liter:

Beef extract	10.0g
Peptone	10.0g
NaCl	5.0g
Agar	2.0g
Trimethylamine N-oxide	1.0g
Yeast extract	1.0g

pH 7.5 ± 0.2 at 25°C

Source: This medium is available as a premixed powder from Oxoid Unipath.

Preparation of Medium: Add components to distilled/deionized water and bring volume to 1.0L. Mix thoroughly. Gently heat and bring to boiling. Distribute into screw-capped tubes in 4.0mL volumes. Autoclave for 15 min at 15 psi pressure–121°C. Allow tubes to cool in an upright position.

Use: For the cultivation and differentiation of *Campylobacter. Campylobacter jejuni* and *Campylobacter coli* will not grow.

Triple Sugar Iron Agar (TSI Agar)

Composition per liter:

Peptone	20.0g
Agar	12.0g
Lactose	10.0g
Sucrose	10.0g
NaCl	5.0g
Beef extract	3.0g
Yeast extract	3.0g
Glucose	1.0g
Ferric citrate	0.3g
$Na_2S_2O_3$	0.3g
Phenol Red	0.025g

pH 7.4 ± 0.2 at 25°C

Source: This medium is available as a premixed powder from BD Diagnostic Systems and Oxoid Unipath.

Preparation of Medium: Add components to distilled/deionized water and bring volume to 1.0L. Mix thoroughly. Gently heat and bring to boiling. Distribute into tubes or flasks. Autoclave for 15 min at 15 psi pressure–121°C. Allow tubes to cool in a slanted position to form a 1.0-inch butt.

Use: For the differentiation of members of the Enterobacteriaceae based on their fermentation of lactose, sucrose, and glucose and the production of H_2S.

Triple Sugar Iron Agar (TSI Agar)

Composition per liter:

Agar	13.0g
Pancreatic digest of casein	10.0g
Peptic digest of animal tissue	10.0g
Lactose	10.0g
Sucrose	10.0g
NaCl	5.0g
Glucose	1.0g
$Fe(NH_4)_2(SO_4)_2 \cdot 6H_2O$	0.2g
$Na_2S_2O_3$	0.2g
Phenol Red	0.025g

pH 7.3 ± 0.2 at 25°C

Source: This medium is available as a premixed powder from BD Diagnostic Systems.

Preparation of Medium: Add components to distilled/deionized water and bring volume to 1.0L. Mix thoroughly. Gently heat and bring to boiling. Distribute into tubes or flasks. Autoclave for 15 min at 15 psi pressure–121°C. Allow tubes to cool in a slanted position to form a 1.0-inch butt.

Use: For the differentiation of members of the Enterobacteriaceae based on their fermentation of lactose, sucrose, and glucose and the production of H_2S.

Trypaflavin Nalidixic Acid Serum Agar (TNSA Agar)

Composition per liter:

Ionagar no. 2	12.0g
Peptone	10.0g
Beef extract	3.0g
H_2O	926.5mL
Bovine serum, heat inactivated	50.0mL
Nalidixic acid solution	20.0mL
Trypaflavin solution	3.5mL

pH 7.2–7.4 at 25°C

Nalidixic Acid Solution:
Composition per 10.0mL:

Nalidixic acid	0.02g

Preparation of Nalidixic Acid Solution: Add nalidixic acid to distilled/deionized water and bring volume to 10.0mL. Mix thoroughly. Filter sterilize.

Trypaflavin Solution:
Composition per 10.0mL:

Trypaflavin	0.1g

Preparation of Trypaflavin Solution: Add trypaflavin to distilled/deionized water and bring volume to 10.0mL. Mix thoroughly. Filter sterilize.

Preparation of Medium: Add components—except bovine serum, nalidixic acid solution, and trypaflavin solution—to distilled/deionized water and bring volume to 926.5mL. Mix thoroughly. Gently heat and bring to boiling. Autoclave for 15 min at 15 psi pressure–121°C. Cool to 45°–50°C. Aseptically add 50.0mL of sterile bovine serum, 20.0mL of sterile nalidixic acid solution, and 3.5mL of trypaflavin solution. Mix thoroughly. Pour into sterile Petri dishes or distribute into sterile tubes.

Use: For the isolation and cultivation of *Listeria* species from preenriched specimens.

Trypanosome Medium

Composition per 1300.0mL:

Solid phase	1.0L
Liquid phase (Locke's solution)	300.0mL

pH 7.2–7.4 at 25°C

Solid Phase:

Composition per liter:

Agar	15.0g
NaCl	8.0g
Peptone	5.0g
Beef extract	3.0g
Rabbit blood, defibrinated	300.0mL

Preparation of Solid Phase: Add components, except rabbit blood, to distilled/deionized water and bring volume to 700.0mL. Mix thoroughly. Adjust pH to 7.2–7.4. Gently heat and bring to boiling. Autoclave for 15 min at 15 psi pressure–121°C. Cool to 50°–55°C. Aseptically add 300.0mL of sterile defibrinated rabbit blood. Mix thoroughly. Distribute 10.0mL aliquots into sterile screw-capped tubes. Allow to solidify in a slanted position.

Liquid Phase (Locke's Solution):

Composition per liter:

NaCl	8.0g
Glucose	2.5g
KH_2PO_4	0.3g
$CaCl_2$	0.2g
KCl	0.2g

Preparation of Liquid Phase (Locke's Solution): Add components to distilled/deionized water and bring volume to 1.0L. Mix thoroughly. Autoclave for 15 min at 15 psi pressure–121°C.

Preparation of Medium: Aseptically overlay agar slants (solid phase) with 3.0mL per tube of sterile liquid phase (Locke's solution).

Use: For the cultivation of *Leishmania donovani*, *Leishmania braziliensis*, *Trypanosoma gambiense*, and *Trypanosoma rhodesiense*.

Tryptic Digest Broth

Composition per liter:

Tryptic digest of beef heart	10.0g
NaCl	5.0g
Glucose	1.0g

pH 7.6 ± 0.2 at 25°C

Source: This medium is available as a premixed powder from BD Diagnostic Systems.

Preparation of Medium: Add components to distilled/deionized water and bring volume to 1.0L. Mix thoroughly. Distribute into tubes or flasks. Autoclave for 15 min at 15 psi pressure–121°C.

Use: For use as a base medium to which enrichments are added. For the cultivation of fastidious microorganisms.

Tryptic Nitrate Medium

Composition per liter:

Tryptose	20.0g
Na_2HPO_4	2.0g
Agar	1.0g
Glucose	1.0g
KNO_3	1.0g

pH 7.6 ± 0.2 at 25°C

Source: This medium is available as a premixed powder from BD Diagnostic Systems.

Preparation of Medium: Add components to distilled/deionized water and bring volume to 1.0L. Mix thoroughly. Gently heat and bring to boiling. Distribute into tubes in 10.0mL volumes. Autoclave for 15 min at 15 psi pressure–121°C.

Use: For the cultivation and differentiation of *Pseudomonas* and related genera. For the differentiation of bacteria based on their reduction of nitrate to nitrite. After incubation of the bacterium in tryptic nitrate medium for 18–24 hr, sulfanillic acid and α-naphthol reagents are added. Nitrate reduction is indicated by the development of a red to violet color.

Tryptic Soy Agar with Sodium Chloride (ATCC Medium 2276)

Composition per liter:

Pancreatic digest of casein	50.0g
NaCl	20.0g
Agar	15.0g
Pancreatic digest of soybean meal	5.0g
$MgSO_4 \cdot 7H_2O$	1.5g

pH 7.3 ± 0.2 at 25°C

Preparation of Medium: Add components to distilled/deionized water and bring volume to 1.0L. Mix thoroughly. Gently heat and bring to boiling. Autoclave for 15 min at 15 psi pressure–121°C. Pour into sterile Petri dishes or leave in tubes.

Use: For the cultivation of *Vibrio* species.

Tryptic Soy Fast Green Agar (TSFA)

Composition per liter:

Pancreatic digest of casein	17.0g
Agar	15.0g
NaCl	5.0g
Papaic digest of soybean meal	3.0g
K_2HPO_4	2.5g
Glucose	2.5g
Fast Green FCF	0.25g

pH 7.3 ± 0.2 at 25°C

Preparation of Medium: Add components to distilled/deionized water and bring volume to 1.0L. Mix thoroughly. Gently heat and bring to boiling. Distribute into tubes or flasks. Autoclave for 15 min at 15 psi pressure–121°C. Cool to 45°–50°C. Aseptically adjust pH to 7.3. Pour into sterile Petri dishes.

Use: For the isolation and cultivation of *Salmonella* species.

Trypticase Agar Base

Composition per liter:

Pancreatic digest of casein	20.0g
Agar	3.5g
Phenol Red	0.02g

pH 7.4 ± 0.2 at 25°C

Source: This medium is available as a premixed powder from BD Diagnostic Systems and Oxoid Unipath.

Preparation of Medium: Add components to distilled/deionized water and bring volume to 1.0L. Mix thoroughly. Gently heat and bring to boiling. Distribute into tubes or flasks. Autoclave for 15 min at 15 psi pressure–121°C. Pour into sterile Petri dishes or leave in tubes.

Use: For the differentiation of microorganisms based on their motility.

Trypticase Agar Base with Carbohydrate

Composition per liter:

Pancreatic digest of casein	20.0g
Carbohydrate	5.0g
Agar	3.5g
Phenol Red	0.02g

pH 7.4 ± 0.2 at 25°C

Preparation of Medium: Add components to distilled/deionized water and bring volume to 1.0L. Mix thoroughly. Gently heat and bring to boiling. Distribute into tubes. Autoclave for 15 min at 13 psi pressure–118°C. Do not overheat. Pour into sterile Petri dishes or leave in tubes.

Use: For differentiation of microorganisms based on their motility and fermentation reactions. Fermentation of carbohydrate changes the medium yellow.

Trypticase Novobiocin Broth (TN Broth)

Composition per liter:

Pancreatic digest of casein	17.0g
NaCl	5.0g
Papaic digest of soybean meal	3.0g
K_2HPO_4	2.5g
Glucose	2.5g
Bile salts no. 3	1.5g
K_2HPO_4	1.5g
Novobiocin solution	10.0mL

pH 7.3 ± 0.2 at 25°C

Novobiocin Solution:

Composition per 10.0mL:

Novobiocin	0.02g

Preparation of Novobiocin Solution: Add novobiocin to distilled/deionized water and bring volume to 10.0mL. Mix thoroughly. Filter sterilize.

Preparation of Medium: Add components, except novobiocin solution, to distilled/deionized water and bring volume to 990.0mL. Mix thoroughly. Gently heat and bring to boiling. Autoclave for 15 min at 15 psi pressure–121°C. Cool to 45°–50°C. Aseptically add sterile novobiocin solution. Mix thoroughly. Pour into sterile Petri dishes or distribute into sterile tubes.

Use: For the cultivation of verotoxin-producing *Escherichia coli.*

Trypticase Peptone Glucose Yeast Extract Broth (TPGY Broth)

Composition per liter:

Pancreatic digest of casein	50.0g
Yeast extract	20.0g
Peptone	5.0g
Glucose	4.0g
Sodium thioglycolate	1.0g

pH 7.0 ± 0.2 at 25°C

Preparation of Medium: Add components to distilled/deionized water and bring volume to 1.0L. Mix thoroughly. Distribute into tubes in 15.0mL volumes. Autoclave for 10 min at 15 psi pressure–121°C.

Use: For the cultivation of *Clostridium botulinum.*

Trypticase Peptone Glucose Yeast Extract Broth, Buffered

Composition per liter:

Pancreatic digest of casein	50.0g
Yeast extract	20.0g
Na_2HPO_4	5.0g
Peptone	5.0g
Glucose	4.0g
Sodium thioglycolate	1.0g

pH 7.3 ± 0.2 at 25°C

Preparation of Medium: Add components to distilled/deionized water and bring volume to 1.0L. Mix thoroughly. Gently heat until dissolved. Adjust pH to 7.3. Distribute into tubes in 15.0mL volumes. Autoclave for 8 min at 15 psi pressure–121°C.

Use: For the isolation and cultivation of *Clostridium perfringens.*

Trypticase Peptone Glucose Yeast Extract Broth with Trypsin (TPGYT Broth)

Composition per 1067.0mL:

Pancreatic digest of casein	50.0g
Yeast extract	20.0g
Peptone	5.0g
Glucose	4.0g
Sodium thioglycolate	1.0g
Trypsin solution	67.0mL

pH 7.0 ± 0.2 at 25°C

Trypsin Solution:
Composition per 100.0mL:

Trypsin	1.5g

Preparation of Trypsin Solution: Add trypsin to distilled/deionized water and bring volume to 100.0mL. Mix thoroughly. Filter sterilize.

Preparation of Medium: Add components, except trypsin solution, to distilled/deionized water and bring volume to 1.0L. Mix thoroughly. Gently heat and bring to boiling. Distribute into tubes in 15.0mL volumes. Autoclave for 10 min at 15 psi pressure–121°C. Immediately prior to use, aseptically add 1.0mL of sterile trypsin solution to each tube. Mix thoroughly.

Use: For the cultivation of *Clostridium botulinum.*

Trypticase Soy Agar (Tryptic Soy Agar) (Soybean Casein Digest Agar) (ATCC Medium 77)

Composition per liter:

Pancreatic digest of casein	15.0g
Agar	15.0g
Papaic digest of soybean meal	5.0g
NaCl	5.0g

pH 7.3 ± 0.2 at 25°C

Source: This medium is available as a premixed powder from BD Diagnostic Systems.

Preparation of Medium: Add components to distilled/deionized water and bring volume to 1.0L. Mix thoroughly. Gently heat and bring to boiling. Distribute into tubes or flasks. Autoclave for 15 min at 15 psi pressure–121°C. Do not overheat. Pour into sterile Petri dishes or leave in tubes.

Use: For the isolation and cultivation of a wide variety of fastidious as well as nonfastidious microorganisms.

Trypticase Soy Agar with Human Blood

Composition per liter:

Pancreatic digest of casein	15.0g
Agar	15.0g
Papaic digest of soybean meal	5.0g
NaCl	5.0g
Human blood, defibrinated	50.0mL

pH 7.3 ± 0.2 at 25°C

Preparation of Medium: Add components, except human blood, to distilled/deionized water and bring volume to 950.0mL. Mix thoroughly. Gently heat and bring to boiling. Autoclave for 15 min at 15 psi pressure–121°C. Cool to 45°–50°C. Aseptically add sterile human blood. Mix thoroughly. Pour into sterile Petri dishes in 17.0mL volumes or distribute into sterile tubes.

Use: For the cultivation of a wide variety of fastidious microorganisms. For the observation of hemolytic reactions of a variety of bacteria. May be used to perform the CAMP test for the presumptive identification of group B streptococci (*Streptococcus agalactiae*).

Trypticase Soy Agar with Lecithin and Polysorbate 80 (Microbial Content Test Agar)

Composition per liter:

Pancreatic digest of casein	15.0g
Agar	15.0g
Papaic digest of soybean meal	5.0g
NaCl	5.0g
Polysorbate 80 (Tween 80)	5.0g
Lecithin	0.7g

pH 7.3 ± 0.2 at 25°C

Source: This medium is available as a premixed powder from BD Diagnostic Systems.

Preparation of Medium: Add components to distilled/deionized water and bring volume to 1.0L. Mix thoroughly. Gently heat and bring to boiling. Distribute into tubes or flasks. Autoclave for 15 min at 13 psi pressure–118°C. Cool to 45°–50°C. Pour into sterile Petri dishes in 17.0mL volumes or leave in tubes.

Use: For the detection and enumeration of microorganisms in replicate plating techniques. Also used for the detection and enumeration of microorganisms present on surfaces of sanitary importance.

Trypticase Soy Agar, Modified (ATCC Medium 1481)

Composition per liter:

Pancreatic digest of casein	17.0g
Agar	15.0g
NaCl	5.0g
Papaic digest of soybean meal	3.0g
K_2HPO_4	2.5g
Glucose	2.5g
L-Glutamine	10.0mL

pH 6.5 ± 0.2 at 25°C

Preparation of Medium: Add components, except glutamine, to distilled/deionized water and bring volume to 990.0mL. Mix thoroughly. Gently heat and bring to boiling. Autoclave for 15 min at 15 psi pressure–121°C. Cool to 45°–50°C. Aseptically add 10.0mL of sterile glutamine. Mix thoroughly. Adjust pH to 6.5. Pour into sterile Petri dishes or distribute into sterile tubes.

Use: Used as a base that is supplemented. For the cultivation of fastidious microorganisms. When supplemented with sheep blood, this medium is useful for the observation of hemolytic reactions of a variety of bacteria.

Trypticase Soy Agar, Modified (TSA II™)

Composition per liter:

Pancreatic digest of casein	14.5g
Agar	14.0g
Papaic digest of soybean meal	5.0g
NaCl	5.0g
Growth factors (BD Diagnostic Systems)	1.5g

pH 7.3 ± 0.2 at 25°C

Source: This medium is available as a premixed powder from BD Diagnostic Systems.

Preparation of Medium: Add components to distilled/deionized water and bring volume to 1.0L. Mix thoroughly. Gently heat while stirring and bring to boiling. Distribute into tubes or flasks. Autoclave for 15 min at 15 psi pressure–121°C. Do not overheat. Pour into sterile Petri dishes or leave in tubes. For blood plates, 50.0–100.0mL of sterile defibrinated sheep blood may be added to sterile medium that has been melted and cooled to 45°–50°C.

Use: Used as a base that is supplemented. For the cultivation of fastidious microorganisms. When supplemented with sheep blood, this medium is useful for the observation of hemolytic reactions of a variety of bacteria. It may be used to perform the CAMP test for the presumptive identification of group B streptococci (*Streptococcus agalactiae*).

Trypticase Soy Agar with Sheep Blood (Tryptic Soy Blood Agar) (TSA Blood Agar)

Composition per liter:

Pancreatic digest of casein	15.0g
Agar	15.0g
Papaic digest of soybean meal	5.0g
NaCl	5.0g
Sheep blood, defibrinated	50.0mL

pH 7.3 ± 0.2 at 25°C

Preparation of Medium: Add components, except sheep blood, to distilled/deionized water and bring volume to 950.0mL. Mix thoroughly. Gently heat and bring to boiling. Autoclave for 15 min at 15 psi pressure–121°C. Cool to 45°–50°C. Aseptically add sterile sheep blood. Mix thoroughly. Pour into sterile Petri dishes in 17.0mL volumes or distribute into sterile tubes.

Use: For the cultivation of a wide variety of fastidious microorganisms. For the observation of hemolytic reactions of a variety of bacteria. May be used to perform the CAMP test for the presumptive identification of group B streptococci (*Streptococcus agalactiae*).

Trypticase Soy Agar with Sheep Blood, Formate, and Fumarate

Composition per liter:

Pancreatic digest of casein	14.5g
Agar	14.0g
Papaic digest of soybean meal	5.0g
NaCl	5.0g
Sucrose	2.0g
Growth factors	1.5g
Sheep blood, defibrinated	50.0mL
Formate-fumarate solution	13.0mL

pH 7.3 ± 0.2 at 25°C

Source: Growth factors are available as a premixed powder from BD Diagnostic Systems.

Formate-Fumarate Solution:
Composition per 100.0mL:

Sodium formate	6.0g
Fumaric acid	6.0g

Preparation of Formate-Fumarate Solution: Add components to distilled/deionized water and bring volume to 100.0mL. Mix thoroughly. Adjust pH to 7.0. Filter sterilize.

Preparation of Medium: Add components, except sheep blood, to distilled/deionized water and bring volume to 950.0mL. Mix thoroughly. Gently heat and bring to boiling. Autoclave for 15 min at 15 psi pressure–121°C. Cool to 45°–50°C. Aseptically add sterile sheep blood. Mix thoroughly. Pour into sterile Petri dishes. Prior to inoculation, aseptically spread 0.2mL of sterile formate-fumarate solution on each plate.

Use: For the isolation of *Streptococcus pneumoniae* from a variety of clinical specimens.

Trypticase Soy Agar with Sheep Blood and Gentamicin (TSA II with Sheep Blood and Gentamicin)

Composition per liter:

Pancreatic digest of casein	14.5g
Agar	14.0g
Papaic digest of soybean meal	5.0g
NaCl	5.0g
Growth factors	1.5g
Sheep blood, defibrinated	50.0mL
Gentamicin solution	10.0mL

pH 7.3 ± 0.2 at 25°C

Source: This medium is available as a premixed powder from BD Diagnostic Systems.

Gentamicin Solution:
Composition per 10.0mL:

Gentamicin	2.5mg

Preparation of Gentamicin Solution: Add gentamicin to distilled/deionized water and bring volume to 10.0mL. Mix thoroughly. Filter sterilize.

Preparation of Medium: Add components, except sheep blood and gentamicin solution, to distilled/deionized water and bring volume to 940.0mL. Mix thoroughly. Gently heat and bring to boiling. Autoclave for 15 min at 15 psi pressure–121°C. Cool to 45°–50°C. Aseptically add sterile sheep blood and sterile gentamicin solution. Mix thoroughly. Pour into sterile Petri dishes or distribute into sterile tubes.

Use: For the isolation of *Streptococcus pneumoniae* from a variety of clinical specimens.

Trypticase Soy Agar with Sheep Blood, Sucrose, and Tetracycline

Composition per liter:

Pancreatic digest of casein	14.5g
Agar	14.0g
Papaic digest of soybean meal	5.0g
NaCl	5.0g
Sucrose	2.0g
Growth factors	1.5g
Sheep blood, defibrinated	50.0mL
Tetracycline solution	10.0mL

pH 7.3 ± 0.2 at 25°C

Source: Growth factors are available as a premixed powder from BD Diagnostic Systems.

Tetracycline Solution:
Composition per 10.0mL:

Tetracycline	0.5mg

Preparation of Tetracycline Solution: Add tetracycline to distilled/deionized water and bring volume to 10.0mL. Mix thoroughly. Filter sterilize.

Preparation of Medium: Add components, except sheep blood and tetracycline solution, to distilled/deionized water and bring volume to 940.0mL. Mix thoroughly. Gently heat and bring to boiling. Autoclave for 15 min at 15 psi pressure–121°C. Cool to 45°–50°C. Aseptically add sterile sheep blood and sterile tetracycline solution. Mix thoroughly. Pour into sterile Petri dishes or distribute into sterile tubes.

Use: For the isolation of *Streptococcus pneumoniae* from a variety of clinical specimens.

Trypticase Soy Agar with Tobramycin

Composition per liter:

Pancreatic digest of casein	17.0g
Agar	15.0g
NaCl	5.0g
Papaic digest of soybean meal	3.0g
K_2HPO_4	2.5g
Glucose	2.5g
Tobramycin solution	10.0mL

pH 7.3 ± 0.2 at 25°C

Tobromycin Solution:
Composition per 10.0mL:

Tobramycin	8.0mg

Preparation of Tobromycin Solution: Add tobramycin to distilled/deionized water and bring volume to 10.0mL. Mix thoroughly. Filter sterilize.

Preparation of Medium: Add components, except tobramycin solution, to distilled/deionized water and bring volume to 990.0mL. Mix thoroughly. Gently heat and bring to boiling. Autoclave for 15 min at 15 psi pressure–121°C. Cool to 45°–50°C. Aseptically add sterile tobramycin solution. Mix thoroughly. Pour into sterile Petri dishes or distribute into sterile tubes.

Use: For the cultivation and maintenance of *Serratia marcescens*.

Trypticase Soy Agar with Sheep Blood and Tween 80 (ATCC Medium 1893)

Composition per liter:

Pancreatic digest of casein	15.0g
Agar	15.0g
Papaic digest of soybean meal	5.0g
NaCl	5.0g
Sheep blood, defibrinated	50.0mL
Tween 80	10.0mL

pH 7.3 ± 0.2 at 25°C

Preparation of Medium: Add components, except sheep blood, to distilled/deionized water and bring volume to 950.0mL. Mix thoroughly. Gently heat and bring to boiling. Autoclave for 15 min at 15 psi pressure–121°C. Cool to 45°–50°C. Aseptically add sterile sheep blood. Mix thoroughly. Pour into sterile Petri dishes in 17.0mL volumes or distribute into sterile tubes.

Use: For the cultivation of a wide variety of fastidious microorganisms.

Trypticase Soy Agar with Sheep Blood and Vacnomycin (ATCC Medium 1976)

Composition per liter:

Pancreatic digest of casein	14.5g
Agar	14.0g
Papaic digest of soybean meal	5.0g
NaCl	5.0g
Growth factors	1.5g
Sheep blood, defibrinated	50.0mL
Vancomycin solution	10.0mL

pH 7.3 ± 0.2 at 25°C

Source: This medium is available as a premixed powder from BD Diagnostic Systems.

Vancomycin Solution:

Composition per 10.0mL:

Vacnomycin	4.0mg

Preparation of Vancomycin Solution: Add vancomycin to distilled/deionized water and bring volume to 10.0mL. Mix thoroughly. Filter sterilize.

Preparation of Medium: Add components, except sheep blood and vancomycin solution, to distilled/deionized water and bring volume to 940.0mL. Mix thoroughly. Gently heat and bring to boiling. Autoclave for 15 min at 15 psi pressure–121°C. Cool to 50°–55°C. Aseptically add sterile sheep blood and sterile gentamicin solution. Mix thoroughly. Pour into sterile Petri dishes or distribute into sterile tubes.

Use: For the isolation of *Streptococcus and Enterococcus* spp. from a variety of clinical specimens.

Trypticase Soy Agar Yeast Extract (TSAYE)

Composition per liter:

Pancreatic digest of casein	17.0g
Agar	15.0g
Yeast extract	6.0g
NaCl	5.0g
Papaic digest of soybean meal	3.0g
K_2HPO_4	2.5g
Glucose	2.5g

pH 7.3 ± 0.2 at 25°C

Preparation of Medium: Add components to distilled/deionized water and bring volume to 1.0L. Mix thoroughly. Gently heat and bring to boiling. Distribute into tubes or flasks. Autoclave for 15 min at 15 psi pressure–121°C. Pour into sterile Petri dishes or leave in tubes.

Use: For the cultivation and maintenance of a wide variety of heterotrophic microorganisms. For the isolation and cultivation of *Listeria monocytogenes*.

Trypticase Soy Broth with 10m*M* Glucose

Composition per liter:

Pancreatic digest of casein	17.0g
NaCl	5.0g
Papaic digest of soybean meal	3.0g
K_2HPO_4	2.5g
Glucose	1.8g

pH 7.3 ± 0.2 at 25°C

Preparation of Medium: Add components to distilled/deionized water and bring volume to 1.0L. Mix thoroughly. Distribute into tubes or flasks. Autoclave for 15 min at 15 psi pressure–121°C.

Use: For the cultivation of a variety of fastidious and nonfastidious microorganisms from clinical and nonclinical specimens.

Trypticase Soy Broth with 10 m*M* Glucose (ATCC Medium 1189)

Composition per liter:

Pancreatic digest of casein	18.0g

Papaic digest of soybean meal 6.0g
NaCl 6.0g
Glucose 1.0g

pH 7.3 ± 0.2 at 25°C

Source: This medium without glucose is available as a premixed powder from BD Diagnostic Systems.

Preparation of Medium: Add components to distilled/deionized water and bring volume to 1.0L. Mix thoroughly. Gently heat and bring to boiling. Distribute into tubes or flasks. Autoclave for 15 min at 15 psi pressure–121°C. Do not overheat. Pour into sterile Petri dishes or leave in tubes.

Use: For the isolation and cultivation of *Staphylococcus aureus* subsp. *aureus*

Trypticase Soy Broth without Glucose

Composition per liter:

Pancreatic digest of casein 17.0g
NaCl 5.0g
Papaic digest of soybean meal 3.0g
K_2HPO_4 .... 2.5g

pH 7.3 ± 0.2 at 25°C

Source: This medium is available as a premixed powder from BD Diagnostic Systems.

Preparation of Medium: Add components to distilled/deionized water and bring volume to 1.0L. Mix thoroughly. Distribute into tubes or flasks. Autoclave for 15 min at 15 psi pressure–121°C.

Use: For the cultivation of a wide variety of microorganisms when the presence of carbohydrate is undesirable.

Trypticase Soy Broth with 1.5% NaCl

Composition per liter:

Pancreatic digest of casein 17.0g
NaCl 15.0g
Papaic digest of soybean meal 3.0g
K_2HPO_4 .... 2.5g
Glucose 2.5g

pH 7.3 ± 0.2 at 25°C

Preparation of Medium: Add components to distilled/deionized water and bring volume to 1.0L. Mix thoroughly. Distribute into tubes or flasks. Autoclave for 15 min at 15 psi pressure–121°C.

Use: For the cultivation and maintenance of *Pasteurella* species.

Trypticase Soy Broth with Sheep Blood (LMG Medium 189)

Composition per liter:

Pancreatic digest of casein 17.0g
NaCl 5.0g
Papaic digest of soybean meal 3.0g
K_2HPO_4 .... 2.5g
Glucose 2.5g
Sheep blood, defibrinated 50.0mL

Preparation of Medium: Add components, except sheep blood, to distilled/deionized water and bring volume to 950.0mL. Mix thoroughly. Autoclave for 15 min at 15 psi pressure–121°C. Cool to 45°–50°C. Aseptically add 50.0mL of sterile, defibrinated sheep blood. Mix thoroughly. Aseptically distribute into sterile tubes or flasks.

Use: For cultivation and maintenance of *Bartonella bacilliformis.*

Trypticase Soy Broth with Tobramycin

Composition per liter:

Pancreatic digest of casein 17.0g
NaCl 5.0g
Papaic digest of soybean meal 3.0g
K_2HPO_4 .... 2.5g
Glucose 2.5g
Agar 1.0g
Tobramycin solution 10.0mL

pH 7.3 ± 0.2 at 25°C

Tobromycin Solution:
Composition per 10.0mL:

Tobramycin 8.0mg

Preparation of Tobromycin Solution: Add tobramycin to distilled/deionized water and bring volume to 10.0mL. Mix thoroughly. Filter sterilize.

Preparation of Medium: Add components, except tobramycin solution, to distilled/deionized water and bring volume to 990.0mL. Mix thoroughly. Autoclave for 15 min at 15 psi pressure–121°C. Cool to 45°–50°C. Aseptically add sterile tobramycin solution. Mix thoroughly. Aseptically distribute into sterile tubes.

Use: For the cultivation and maintenance of *Serratia marcescens.*

Trypticase Soy Broth with Tween 80

Composition per liter:

Pancreatic digest of casein 17.0g
NaCl 5.0g
Papaic digest of soybean meal 3.0g
K_2HPO_4 .... 2.5g
Glucose 2.5g
Agar 1.0g
Tween 80 .... 1.0mL

pH 7.3 ± 0.2 at 25°C

Preparation of Medium: Add components to distilled/deionized water and bring volume to 1.0L. Mix thoroughly. Distribute into tubes or flasks. Autoclave for 15 min at 15 psi pressure–121°C.

Use: For the cultivation and maintenance of *Corynebacterium genitalium*.

Trypticase Soy Broth with Yeast Extract

Composition per liter:

Pancreatic digest of casein	17.0g
Yeast extract	6.0g
NaCl	5.0g
Papaic digest of soybean meal	3.0g
K_2HPO_4	2.5g
Glucose	2.5g

pH 7.3 ± 0.2 at 25°C

Preparation of Medium: Add components to distilled/deionized water and bring volume to 1.0L. Mix thoroughly. Gently heat and bring to boiling. Distribute into tubes or flasks. Autoclave for 15 min at 15 psi pressure–121°C.

Use: For the cultivation and maintenance of a wide variety of heterotrophic microorganisms.

Trypticase Tellurite Agar Base

Composition per liter:

Agar	20.0g
Pancreatic digest of casein	10.0g
Peptic digest of animal tissue	10.0g
NaCl	5.0g
Glucose	2.0g
Serum	50.0mL
Chapman tellurite solution	10.0mL

pH 7.5 ± 0.2 at 25°C

Source: This medium is available as a premixed powder from BD Diagnostic Systems.

Chapman Tellurite Solution:

Composition per 100.0mL:

K_2TeO_3	1.0g

Preparation of Chapman Tellurite Solution: Add K_2TeO_3 to distilled/deionized water and bring volume to 100.0mL. Mix thoroughly. Filter sterilize.

Caution: Potassium tellurite is toxic.

Preparation of Medium: Add components, except serum and Chapman tellurite solution, to distilled/deionized water and bring volume to 940.0mL. Mix thoroughly. Gently heat and bring to boiling. Autoclave for 15 min at 15 psi pressure–121°C. Cool to 45°–50°C. Aseptically add sterile serum and sterile Chapman tellurite solution. Sheep serum, rabbit serum, or human serum may be used. Mix thoroughly. Pour into sterile Petri dishes.

Use: For the selective isolation of microorganisms from clinical specimens, especially from the nose, throat, and vagina.

Tryptone Bile Glucuronide Agar, Harlequim (Harlequin TBGA)

Composition per liter:

Tryptone	20.0g
Agar	15.0g
Bilie salts no. 3	1.5g
X-glucuronide	0.075g

pH 7.2 ± 0.2 at 25°C

Source: This medium is available from lab m.

Preparation of Medium: Add components to distilled/deionized water and bring volume to 1.0L. Mix thoroughly. Allow to soak for 10 min. Autoclave for 15 min at 15 psi pressure–121°C. Cool to 45°–50°C. Pour into sterile Petri dishes.

Use: For the simple enumeration of *E. coli* without the need for membranes, or preincubation. The medium has been modified by the addition of a chromogenic substrate to detect the β-glucuronidase, which is highly specific for *E. coli*. The advantage of the chromogenic substrate is that it requires no UV lamp to visualise the reaction, and it is concentrated within the colony, facilitating easier enumeration in the presence of other organisms, or when large numbers are present on the plate.

Tryptone Bile X-glucuronide Agar, Chromocult (Chromocult TBX) (ChromocultTryptone Bile X-glucuronide Agar)

Composition per liter:

Peptone	20.0g
Agar	15.0g
Bile salts no. 3	1.5g
X-β-D-glucuronide	0.075g

pH 7.2 ± 0.2 at 25°C

Source: This medium is available from Merck.

Preparation of Medium: Add components to distilled/deionized water and bring volume to 1.0L. Mix thoroughly. Autoclave for 15 min at 15 psi pressure–121°C. Cool to 45°–50°C. Pour into sterile Petri dishes.

Use: For the differentiation of *E. coli* from other coliforms. The presence of the enzyme β-D-glucuronidase differentiates most *E. coli* spp. from other coliforms. *E. coli* absorbs the chromogenic substrate 5-bromo-4-chloro-3-indolyl-β-D-glucuronide (X-β-D-glucuronide). The enzyme β-glucuronidase splits the bond between the chromophore 5-bromo-4-chloro-3-indolyle- and the β-D-glucuronide. *E. coli* colonies are colored blue-green. Growth of accompanying Gram-positive flora is largely inhibited by the use of bile salts The prepared medium is clear and yellowish.

Tryptone Phosphate Broth

Composition per liter:

Pancreatic digest of casein	20.0g
NaCl	5.0g
K_2HPO_4	2.0g
KH_2PO_4	2.0g
Tween 80	15.0mL

pH 7.0 ± 0.2 at 25°C

Preparation of Medium: Add components to distilled/deionized water and bring volume to 1.0L. Mix thoroughly. Distribute into tubes or flasks. Autoclave for 15 min at 15 psi pressure–121°C.

Use: For the cultivation of enteropathogenic *Escherichia coli.*

Tryptone Soya Agar

Composition per liter:

Agar	15.0g
Pancreatic digest of casein	15.0g
NaCl	5.0g
Pancreatic digest of soybean meal	5.0g

pH 7.3 ± 0.2 at 25°C

Source: This medium is available as a premixed powder from Oxoid Unipath.

Preparation of Medium: Add components to distilled/deionized water and bring volume to 1.0L. Mix thoroughly. Gently heat and bring to boiling. Distribute into tubes or flasks. Autoclave for 15 min at 15 psi pressure–121°C. Pour into sterile Petri dishes or leave in tubes.

Use: For the cultivation and maintenance of a wide variety of microorganisms.

Tryptone Soya Broth

Composition per liter:

Pancreatic digest of casein	17.0g
NaCl	5.0g
Pancreatic digest of soybean meal	3.0g
K_2HPO_4	2.5g
Glucose	2.5g

pH 7.3 ± 0.2 at 25°C

Source: This medium is available as a premixed powder from Oxoid Unipath.

Preparation of Medium: Add components to distilled/deionized water and bring volume to 1.0L. Mix thoroughly. Distribute into tubes or flasks. Autoclave for 15 min at 15 psi pressure–121°C.

Use: For the cultivation of a wide variety of microorganisms.

Tryptone Soya Broth with 0.1% Tween 80

Composition per liter:

Pancreatic digest of casein	15.0g
Papaic digest of soybean meal	5.0g
NaCl	5.0g
Tween 80	1.0g

Preparation of Medium: Add components to distilled/deionized water and bring volume to 1.0L. Mix thoroughly. Distribute into tubes or flasks. Autoclave for 15 min at 15 psi pressure–121°C.

Use: For the cultivation of *Corynebacterium genitalium* and *Corynebacterium pseudogenitalium.*

Tryptone Water Broth (Tryptone Broth)

Composition per liter:

Pancreatic digest of casein	10.0g
NaCl	5.0g

pH 7.5 ± 0.2 at 25°C

Source: This medium is available as a premixed powder from Oxoid Unipath.

Preparation of Medium: Add components to distilled/deionized water and bring volume to 1.0L. Mix thoroughly. Distribute into tubes or flasks. Autoclave for 15 min at 15 psi pressure–121°C.

Use: For the cultivation of production of indole by microorganisms.

Tryptone Yeast Extract Agar

Composition per liter:

Pancreatic digest of casein	10.0g
Agar	2.0g
Yeast extract	1.0g
Bromcresol Purple	0.04g
Carbohydrate solution	100.0mL

pH 7.0 ± 0.2 at 25°C

Carbohydrate Solution:

Composition per 100.0mL:

Carbohydrate	10.0g

Preparation of Carbohydrate Solution: Add carbohydrate to distilled/deionized water and bring volume to 100.0mL. Glucose or mannitol may be used. Mix thoroughly. Filter sterilize.

Preparation of Medium: Add components, except carbohydrate solution, to distilled/deionized water and bring volume to 900.0mL. Mix thoroughly. Adjust pH to 7.0. Gently heat and bring to boiling. Distribute into tubes in 13.5mL volumes. Autoclave for 20 min at 10 psi pressure–115°C. Cool to 45°–50°C. Aseptically add 1.5mL of carbohydrate solution to each tube. Mix thoroughly. Solidify agar quickly by placing tubes in ice water.

Use: For the cultivation and differentiation of *Staphylococcus aureus* based on glucose and mannitol fermentation. Bacteria that ferment the added carbohydrate turn the medium yellow.

Tryptone Yeast Extract Agar 1

Composition per liter:

Agar	20.0g
Pancreatic digest of casein	10.0g
Yeast extract	5.0g
K_2HPO_4	4.4g
Glucose	2.0g
NaCl	2.0g

pH 7.2 ± 0.2 at 25°C

Preparation of Medium: Add components to distilled/deionized water and bring volume to 1.0L. Mix thoroughly. Adjust pH to 7.2. Gently heat and bring to boiling. Distribute into tubes or flasks. Autoclave for 15 min at 15 psi pressure–121°C. Pour into sterile Petri dishes or leave in tubes.

Use: For the cultivation and maintenance of a wide variety of heterotrophic bacteria.

Tryptophan 1% Solution (Trypticase 1% Solution) (Tryptone 1% Solution)

Composition per liter:

Pancreatic digest of casein	10.0g

pH 7.0 ± 0.2 at 25°C

Source: This medium is available as a premixed powder from BD Diagnostic Systems.

Preparation of Medium: Add components to distilled/deionized water and bring volume to 1.0L. Mix thoroughly. Distribute into tubes or flasks. Autoclave for 15 min at 15 psi pressure–121°C.

Use: For the differentiation of bacteria, especially members of the Enterobacteaceae, based on their production of indole.

Tryptose Agar

Composition per liter:

Agar	15.0g
Pancreatic digest of casein	10.0g
Peptic digest of animal tissue	10.0g
NaCl	5.0g
Glucose	1.0g

pH 7.2 ± 0.2 at 25°C

Source: This medium is available as a premixed powder from BD Diagnostic Systems.

Preparation of Medium: Add components to distilled/deionized water and bring volume to 1.0L. Mix thoroughly. Gently heat and bring to boiling. Distribute into tubes or flasks. Autoclave for 15 min at 15 psi pressure–121°C. Pour into sterile Petri dishes or leave in tubes.

Use: For the cultivation and maintenance of fastidious aerobic and facultative microorganisms.

Tryptose Agar with Citrate

Composition per liter:

Agar	15.0g
Pancreatic digest of casein	10.0g
Peptic digest of animal tissue	10.0g
Sodium citrate	10.0g
NaCl	5.0g
Glucose	1.0g

pH 7.2 ± 0.2 at 25°C

Preparation of Medium: Add components to distilled/deionized water and bring volume to 1.0L. Mix thoroughly. Gently heat and bring to boiling. Distribute into tubes or flasks. Autoclave for 15 min at 15 psi pressure–121°C. Pour into sterile Petri dishes or leave in tubes.

Use: For the cultivation and maintenance of fastidious aerobic and facultative microorganisms, including *Brucella* species and streptococci.

Tryptose Agar with Sheep Blood (ATCC Medium 546)

Composition per liter:

Agar	15.0g
Tryptose	10.0g
NaCl	5.0g
Beef extract	3.0g
Sheep blood, defibrinated	50.0-100.0mL

pH 7.3 ± 0.2 at 25°C

Preparation of Medium: Add components, except sheep blood, to distilled/deionized water and bring volume to 900–950mL. Mix thoroughly. Gently heat and bring to boiling. Distribute into tubes. Autoclave for 15 min at 15 psi pressure–121°C. Cool to 45°–50°C. Aseptically add sterile sheep blood. Mix thoroughly. Pour into sterile Petri dishes in 17.0mL volumes or distribute into sterile tubes.

Use: For the cultivation and maintenance of a wide variety of fastidious microorganisms.

Tryptose Agar with Thiamine

Composition per liter:

Agar	15.0g
Pancreatic digest of casein	10.0g
Peptic digest of animal tissue	10.0g
NaCl	5.0g
Glucose	1.0g
Thiamine·HCl	5.0mg

pH 7.2 ± 0.2 at 25°C

Preparation of Medium: Add components to distilled/deionized water and bring volume to 1.0L. Mix thoroughly. Gently heat and bring to boiling. Distribute into tubes or flasks. Autoclave for 15 min at 15 psi pressure–121°C. Pour into sterile Petri dishes or leave in tubes.

Use: For the cultivation and maintenance of fastidious aerobic and facultative microorganisms, including *Brucella* species and streptococci.

Tryptose Blood Agar

Composition per liter:

Agar	12.0g
Tryptose	10.0g
NaCl	5.0g
Beef extract	3.0g
Sheep blood, defibrinated	70.0mL

pH 7.2 ± 0.2 at 25°C

Source: This medium is available as a premixed powder from Oxoid Unipath.

Preparation of Medium: Add components, except sheep blood, to distilled/deionized water and bring volume to 930.0mL. Mix thoroughly. Gently heat and bring to boiling. Autoclave for 15 min at 15 psi pressure–121°C. Cool to 45°–50°C. Aseptically add sterile sheep blood. Mix thoroughly. Pour into sterile Petri dishes in 17.0mL volumes or distribute into sterile tubes.

Use: For the cultivation and maintenance of a wide variety of fastidious microorganisms.

Tryptose Blood Agar Base with Yeast Extract

Composition per liter:

Agar	15.0g
Tryptose	10.0g
NaCl	5.0g
Beef extract	3.0g
Yeast extract	1.0g
Sheep blood, defibrinated	50.0mL

pH 7.3 ± 0.2 at 25°C

Source: This medium is available as a premixed powder from BD Diagnostic Systems.

Preparation of Medium: Add components, except sheep blood, to distilled/deionized water and bring volume to 950.0mL. Mix thoroughly. Gently heat and bring to boiling. Autoclave for 15 min at 15 psi pressure–121°C. Cool to 45°–50°C. Aseptically add sterile sheep blood. Mix thoroughly. Pour into sterile Petri dishes in 17.0mL volumes or distribute into sterile tubes.

Use: For the cultivation and maintenance of a wide variety of fastidious microorganisms.

Tryptose Broth

Composition per liter:

Pancreatic digest of casein	10.0g
Peptic digest of animal tissue	10.0g
NaCl	5.0g
Glucose	1.0g

pH 7.2 ± 0.2 at 25°C

Source: This medium is available as a premixed powder from BD Diagnostic Systems.

Preparation of Medium: Add components to distilled/deionized water and bring volume to 1.0L. Mix thoroughly. Distribute into tubes or flasks. Autoclave for 15 min at 15 psi pressure–121°C.

Use: For the cultivation of fastidious aerobic and facultative microorganisms, including streptococci.

Tryptose Broth

Composition per liter:

Pancreatic digest of casein	10.0g
Peptic digest of animal tissue	10.0g
NaCl	5.0g
Glucose	1.0g
Thiamine·HCl	5.0mg

pH 7.2 ± 0.2 at 25°C

Source: This medium is available as a premixed powder from BD Diagnostic Systems.

Preparation of Medium: Add components to distilled/deionized water and bring volume to 1.0L. Mix thoroughly. Distribute into tubes or flasks. Autoclave for 15 min at 15 psi pressure–121°C.

Use: For the cultivation of fastidious aerobic and facultative microorganisms.

Tryptose Broth with Citrate

Composition per liter:

Pancreatic digest of casein	10.0g
Peptic digest of animal tissue	10.0g
Sodium citrate	10.0g
NaCl	5.0g
Glucose	1.0g
Thiamine·HCl	5.0mg

pH 7.2 ± 0.2 at 25°C

Preparation of Medium: Add components to distilled/deionized water and bring volume to 1.0L. Mix thoroughly. Distribute into tubes or flasks. Autoclave for 15 min at 15 psi pressure–121°C.

Use: For the isolation and cultivation of a variety of fastidious aerobic microorganisms, especially *Brucella* species, from clinical sources.

Tryptose Cycloserine Dextrose Agar

Composition per liter:

Agar	20.0g
Tryptose	15.0g
Pancreatic digest of soybean meal	5.0g
Yeast extract	5.0g
Ferric ammonium citrate	1.0g
Cycloserine solution	10.0mL

pH 7.6 ± 0.2 at 25°C

Cycloserine Solution:

Composition per 10.0mL:

D-Cycloserine	0.4g

Preparation of Cycloserine Solution: Add D-cycloserine to distilled/deionized water and bring volume to 10.0mL. Mix thoroughly. Filter sterilize.

Preparation of Medium: Add components, except cycloserine solution, to distilled/deionized water and bring volume to 990.0mL. Mix thoroughly. Gently heat and bring to boiling. Autoclave for 15 min at 15 psi pressure–121°C. Cool to 45°–50°C. Aseptically add sterile cycloserine solution. Mix thoroughly. Pour into sterile Petri dishes or distribute into sterile tubes.

Use: For the isolation and cultivation of *Clostridium* species, especially *Clostridium botulinum*.

Tryptose Phosphate Agar

Composition per liter:

Tryptose	20.0g
Agar	15.0g
NaCl	5.0g
Na_2HPO_4	2.5g
Glucose	2.0g

pH 7.3 ± 0.2 at 25°C

Preparation of Medium: Add components to distilled/deionized water and bring volume to 1.0L. Mix thoroughly. Gently heat and bring to boiling. Distribute into tubes or flasks. Autoclave for 15 min at 15 psi pressure–121°C. Pour into sterile Petri dishes or leave in tubes.

Use: For the cultivation and maintenance of *Erysipelothris tonsillarum.*

Tryptose Phosphate Broth

Composition per liter:

Tryptose	20.0g
NaCl	5.0g
Na_2HPO_4	2.5g
Glucose	2.0g

pH 7.3 ± 0.2 at 25°C

Source: This medium is available as a premixed powder from BD Diagnostic Systems and Oxoid Unipath.

Preparation of Medium: Add components to distilled/deionized water and bring volume to 1.0L. Mix thoroughly. Distribute into tubes or flasks. Autoclave for 15 min at 15 psi pressure–121°C. Prior to the inoculation of anaerobic microorganisms, place tubes of sterile medium in a 100°C bath for 15 min and cool undisturbed.

Use: For the cultivation of a variety of fastidious bacteria.

Tryptose Phosphate Broth, Modified

Composition per liter:

Enzymatic digest of casein	20.0g
NaCl	5.0g
Na_2HPO_4	2.5g
Glucose	2.0g

pH 7.3 ± 0.2 at 25°C

Source: This medium is available as a premixed powder from BD Diagnostic Systems.

Preparation of Medium: Add components to distilled/deionized water and bring volume to 1.0L. Mix thoroughly. Distribute into tubes or flasks. Autoclave for 15 min at 15 psi pressure–121°C. Prior to the inoculation of anaerobic microorganisms, place tubes of sterile medium in a 100°C bath for 15 min and cool undisturbed.

Use: For the cultivation of a variety of fastidious microorganisms, including pneumococci, streptococci, and meningococci.

Tryptose Sulfite Cycloserine Agar (TSC Agar)

Composition per liter:

Tryptose	15.0g
Agar	14.0g
Pancreatic digest of soybean meal	5.0g
Yeast extract	5.0g
Ferric ammonium citrate	1.0g
$Na_2S_2O_5$	1.0g
Cycloserine solution	10.0mL

pH 7.6 ± 0.2 at 25°C

Cycloserine Solution:
Composition per 10.0mL:

D-Cycloserine	0.4g

Preparation of Cycloserine Solution: Add cycloserine to distilled/deionized water and bring volume to 10.0mL. Mix thoroughly. Filter sterilize.

Preparation of Medium: Add components, except cycloserine solution, to distilled/deionized water and bring volume to 990.0mL. Mix thoroughly. Gently heat and bring to boiling. Autoclave for 15 min at 15 psi pressure–121°C. Cool to 45°–50°C. Aseptically add sterile cycloserine solution. Mix thoroughly. Pour into sterile Petri dishes.

Use: For the presumptive identification and enumeration of *Clostridium perfringens.*

Tryptose Sulfite Cycloserine Agar (TSC Agar)

Composition per liter:

Tryptose	15.0g
Agar	14.0g
Beef extract	5.0g
Pancreatic digest of soybean meal	5.0g
Yeast extract	5.0g
Ferric ammonium citrate	1.0g
$Na_2S_2O_5$	1.0g
Egg yolk emulsion	50.0mL
Cycloserine solution	10.0mL

pH 7.6 ± 0.2 at 25°C

Source: This medium is available as a premixed powder from Oxoid Unipath.

Egg Yolk Emulsion:
Composition:

Chicken egg yolks	11
Whole chicken egg	1

Preparation of Egg Yolk Emulsion: Soak eggs with 1:100 dilution of saturated mercuric chloride solution for 1 min. Crack eggs and separate yolks from whites. Mix egg yolks with 1 chicken egg.

Cycloserine Solution:
Composition per 10.0mL:

D-Cycloserine	0.4g

Preparation of Cycloserine Solution: Add cycloserine to distilled/deionized water and bring volume to 10.0mL. Mix thoroughly. Filter sterilize.

Preparation of Medium: Add components, except cycloserine solution and egg yolk emulsion, to distilled/deionized water and bring volume to 940.0mL. Mix thoroughly. Gently heat and bring to boiling. Autoclave for 15 min at 15 psi pressure–121°C. Cool to 45°–50°C. Aseptically add sterile cycloserine solution and egg yolk emulsion. Mix thoroughly. Pour into sterile Petri dishes.

Use: For the presumptive identification and enumeration of *Clostridium perfringens*.

Tryptose Sulfite Cycloserine Agar with Polymyxin and Kanamycin

Composition per liter:

Tryptose	15.0g
Agar	14.0g
Beef extract	5.0g
Pancreatic digest of soybean meal	5.0g
Yeast extract	5.0g
Ferric ammonium citrate	1.0g
$Na_2S_2O_5$	1.0g
Antibiotic solution	10.0mL

pH 7.6 ± 0.2 at 25°C

Antibiotic Solution:
Composition per 10.0mL:

D-Cycloserine	0.4g
Polymyxin B sulfate	0.03g
Kanamycin sulfate	0.012g

Preparation of Antibiotic Solution: Add components to distilled/deionized water and bring volume to 10.0mL. Mix thoroughly. Filter sterilize.

Preparation of Medium: Add components, except antibiotic solution, to distilled/deionized water and bring volume to 990.0mL. Mix thoroughly. Gently heat and bring to boiling. Autoclave for 15 min at 15 psi pressure–121°C. Cool to 45°–50°C. Aseptically add sterile antibiotic solution. Mix thoroughly. Pour into sterile Petri dishes.

Use: For the isolation and enumeration of *Clostridium perfringens* from clinical specimens.

Tryptose Sulfite Cycloserine Agar without Egg Yolk (TSC Agar without Egg Yolk)

Composition per liter:

Tryptose	15.0g

Agar 14.0g
Beef extract 5.0g
Pancreatic digest of soybean meal 5.0g
Yeast extract 5.0g
Ferric ammonium citrate 1.0g
$Na_2S_2O_5$ 1.0g
Cycloserine solution 10.0mL
pH 7.6 ± 0.2 at 25°C

Cycloserine Solution:
Composition per 10.0mL:
D-Cycloserine 0.4g

Preparation of Cycloserine Solution: Add cycloserine to distilled/deionized water and bring volume to 10.0mL. Mix thoroughly. Filter sterilize.

Preparation of Medium: Add components, except cycloserine solution, to distilled/deionized water and bring volume to 990.0mL. Mix thoroughly. Gently heat and bring to boiling. Autoclave for 15 min at 15 psi pressure–121°C. Cool to 45°–50°C. Aseptically add sterile cycloserine solution. Mix thoroughly. Pour into sterile Petri dishes.

Use: For the presumptive identification and enumeration of *Clostridium perfringens*.

TSC Agar, Fluorocult (Fluorocult TSC Agar) (Fluorocult Tryptose Sulfite Cycloserine Agar) (Tryptose Sulfite Cycloserine Agar, Fluorocult)

Composition per liter:
Agar 15.0g
Tryptose 15.0g
Peptone from soymeal 5.0g
Yeast extract 5.0g
$Na_2S_2O_5$ 1.0g
Ammonium ferric citrate 1.0g
D-Cycloserine 0.2g
4-Methylumbelliferyl-phosphate disodium salt 50.0mg

Source: This medium is available from Merck.

Preparation of Medium: Add components to distilled/deionized water and bring volume to 1.0L. Mix thoroughly. Autoclave for 15 min at 15 psi pressure–121°C. Cool to 45°–50°C. Pour into sterile Petri dishes.

Use: For the isolation and enumeration of the vegetative and spore forms of *Clostridium perfringens*. D-Cycloserine inhibits the accompanying bacterial flora and causes the colonies which develop to remain smaller. 4-Methylumbelliferyl-phosphate (MUP) is a fluorogenic substrate for the alkaline and acid phosphatase. The acid phosphatase is a highly specific indicator for *C. perfringens*. The acid phosphatase splits the fluorogenic substrate MUP forming 4-methylumbelliferone which can be identified as it fluorescence in long wave UV light. Thus a strong suggestion for the presence of *C. perfringens* can be obtained. The acid phosphatase splits the fluorogenic substrate MUP forming 4-methylumbelliferone which can be identified as it fluorescence in long wave UV light, providing a strong suggestion for the presence of *Clostridium perfringens*.

TSN Agar (Trypticase Sulfite Neomycin Agar)

Composition per liter:
Pancreatic digest of casein 15.0g
Agar 13.5g
Yeast extract 10.0g
Na_2SO_3 .... 1.0g
Ferric citrate 0.5g
Neomycin sulfate 0.05g
Polymyxin sulfate 0.02g
Buffered thioglycolate solution 50.0mL
pH 7.2 ± 0.2 at 25°C

Source: This medium is available as a premixed powder from BD Diagnostic Systems.

Buffered Thioglycolate Solution:
Composition per 50.0mL:
Buffer solution 35.0mL
Sodium thioglycolate solution 15.0mL

Preparation of Buffered Thioglycolate Solution: Combine components. Mix thoroughly. Autoclave for 15 min at 15 psi pressure–121°C. Cool to 45°–50°C.

Buffer Solution:
Composition per 100.0mL:
Na_2CO_3 .... 28.0g
K_2HPO_4 .... 5.7g

Preparation of Buffer Solution: Add components to distilled/deionized water and bring volume to 100.0mL. Mix thoroughly.

Thioglycolate Solution:
Composition per 100.0mL:
Sodium thioglycolate 13.3g

Preparation of Thioglycolate Solution: Add sodium thioglycolate to distilled/deionized water and bring volume to 100.0mL. Mix thoroughly.

Preparation of Medium: Add components, except buffered thioglycolate solution, to distilled/deionized water and bring volume to 950.0mL. Mix thoroughly. Gently heat and bring to boiling. Autoclave for 12 min at 13 psi pressure–118°C. Do not overheat. Cool to 45°–50°C. Aseptically add buffered thioglycolate solution. Mix thoroughly. Pour into sterile Petri dishes or distribute into sterile tubes.

Use: For the selective isolation of *Clostridium perfringens.*

Tween 80 Hydrolysis Broth

Composition per liter:

Na_2HPO_4	5.79g
NaH_2PO_4	3.53g
Neutral Red	0.02g
Tween 80 solution	5.0mL

pH 7.0 ± 0.2 at 25°C

Tween 80 Solution:

Composition per 50.0mL:

Tween 80	10.0g

Preparation of Tween 80 Solution: Add Tween 80 to distilled/deionized water and bring volume to 50.0mL. Mix thoroughly. Autoclave for 15 min at 15 psi pressure–121°C.

Preparation of Medium: Add components, except Tween 80 solution, to distilled/deionized water and bring volume to 995.0mL. Mix thoroughly. Adjust pH to 7.2. Autoclave for 15 min at 15 psi pressure–121°C. Aseptically add 5.0mL of sterile Tween 80 solution. Mix thoroughly. Aseptically distribute into sterile tubes or flasks.

Use: For the differentiation of *Mycobacterium* species. Strains that hydrolyze Tween 80 within 5 days turn the medium pink to red.

Tween 80 Hydrolysis Broth

Composition per 125.0mL:

Neutral Red	0.1g
Solution 1	38.9mL
Solution 2	61.1mL
Tween 80 solution	25.0mL

pH 7.0 ± 0.2 at 25°C

Solution 1:

Composition per 400.0mL:

KH_2PO_4	22.7g

Preparation of Solution 1: Add KH_2PO_4 to distilled/deionized water and bring volume to 400.0mL. Mix thoroughly.

Solution 2:

Composition per 400.0mL:

Na_2HPO_4	23.8g

Preparation of Solution 2: Add Na_2HPO_4 to distilled/deionized water and bring volume to 400.0mL. Mix thoroughly.

Tween 80 Solution:

Composition per 50.0mL:

Tween 80	10.0g

Preparation of Tween 80 Solution: Add Tween 80 to distilled/deionized water and bring volume to 50.0mL. Mix thoroughly. Autoclave for 15 min at 15 psi pressure–121°C.

Preparation of Medium: Add components, except Tween 80 solution, to distilled/deionized water and bring volume to 975.0mL. Mix thoroughly. Adjust pH to 7.2. Autoclave for 15 min at 15 psi pressure–121°C. Aseptically add 25.0mL of sterile Tween 80 solution. Mix thoroughly. Aseptically distribute into sterile tubes or flasks.

Use: For the differentiation of *Mycobacterium* species. Strains that hydrolyze Tween 80 within 5 days turn the medium pink to red.

Tween 80 Hydrolysis Broth

Composition per 102.5mL:

$NaHPO_4$ (0.066*M* solution)	61.1mL
KH_2PO_4 (0.066*M* solution)	38.9mL
Neutral Red (0.1% solution)	2.0mL
Tween 80	0.5mL

pH 7.0 ± 0.2 at 25°C

Preparation of Medium: Combine components. Mix thoroughly. Distribute into tubes or flasks. Autoclave for 15 min at 15 psi pressure–121°C.

Use: For the differentiation of *Mycobacterium* species. Strains that hydrolyze Tween 80 within 5 days turn the medium pink to red.

Tween 80 Hydrolysis Medium

Composition per liter:

Agar	12.0g
Peptone	10.0g
Tween 80	10.0g
NaCl	5.0g
$CaCl_2$	0.1g

pH 7.2–7.4 at 25°C

Preparation of Medium: Add components to distilled/deionized water and bring volume to 1.0L. Mix thoroughly. Gently heat and bring to boiling. Distribute into tubes or flasks. Autoclave for 15 min at 15 psi pressure–121°C. Pour into sterile Petri dishes.

Use: For the cultivation and differentiation of *Pseudomonas* species based on their ability to hydrolyze Tween 80. Bacteria that hydrolyze Tween 80 appear as colonies surrounded by an opaque zone.

TY Medium

Composition per liter:

Agar	15.0g
Pancreatic digest of casein	5.0g
Yeast extract	3.0g
$CaCl_2 \cdot 6H_2O$	1.3g

Preparation of Medium: Add components to distilled/deionized water and bring volume to 1.0L. Mix thoroughly. Gently heat and bring to boiling. Distribute into tubes or flasks. Autoclave for 15 min at 15 psi pressure–121°C. Pour into sterile Petri dishes or leave in tubes.

Use: For the cultivation of a wide variety of bacteria.

TY Salt Medium

Composition per liter:

NaCl	10.0g
Pancreatic digest of casein	10.0g
Yeast extract	5.0g

pH 7.0 ± 0.2 at 25°C

Preparation of Medium: Add components to distilled/deionized water and bring volume to 1.0L. Mix thoroughly. Distribute into tubes or flasks. Autoclave for 15 min at 15 psi pressure–121°C.

Use: For the cultivation of a wide variety of bacteria.

TYGS Medium (Tryptone Yeast Extract Glucose Salt Medium)

Composition per liter:

Agar	15.0g
Pancreatic digest of casein	10.0g
NaCl	8.0g
Yeast extract	1.0g
$CaCl_2 \cdot 2H_2O$ solution	100.0mL
Glucose solution	100.0mL

$CaCl_2 \cdot 2H_2O$ Solution:

Composition per 100.0mL:

$CaCl_2 \cdot 2H_2O$	0.3g

Preparation of $CaCl_2 \cdot 2H_2O$ Solution: Add the $CaCl_2 \cdot 2H_2O$ to distilled/deionized water and bring volume to 100.0mL. Mix thoroughly. Filter sterilize.

Glucose Solution:

Composition per 100.0mL:

D-Glucose	1.0g

Preparation of Glucose Solution: Add glucose to distilled/deionized water and bring volume to 100.0mL. Mix thoroughly. Filter sterilize.

Preparation of Medium: Add components, except $CaCl_2 \cdot 2H_2O$ solution and glucose solution, to distilled/deionized water and bring volume to 800.0mL. Mix thoroughly. Gently heat and bring to boiling. Autoclave for 15 min at 15 psi pressure–121°C. Cool to 45°–50°C. Aseptically add the sterile $CaCl_2 \cdot 2H_2O$ solution and sterile glucose solution. Mix thoroughly. Pour into sterile Petri dishes or distribute into sterile tubes.

Use: For the cultivation and maintenance of a variety of bacteria.

TYI-S-33 Medium

Composition per liter:

Pancreatic digest of casein	20.0g
Yeast extract	10.0g
NaCl	2.0g
L-Cysteine·HCl	1.0g
Ascorbic acid	0.2g
Ferric ammonium citrate	22.8mg
Bovine serum, heat inactivated	100.0mL
Special 107 vitamin mix	100.0mL
Buffer solution	50.0mL
Glucose solution	50.0mL

pH 6.8 ± 0.2 at 25°C

Glucose Solution:

Composition per 50.0mL:

Glucose	10.0g

Preparation of Glucose Solution: Add glucose to distilled/deionized water and bring volume to 50.0mL. Mix thoroughly. Filter sterilize.

Buffer Solution:

Composition per 50.0mL:

K_2HPO_4	1.0g
KH_2PO_4	0.6g

Preparation of Buffer Solution: Add components to distilled/deionized water and bring volume to 50.0mL. Mix thoroughly. Adjust pH to 6.8. Autoclave for 15 min at 15 psi pressure–121°C. Cool to 25°C.

Special 107 Vitamin Mix:

Composition per 120.0.0mL:

Solution 1	1.2mL
Solution 2	0.4mL
Solution 3	0.4mL
Solution 4 vitamins	100.0mL

Preparation of Special 107 Vitamin Mix: Aseptically combine 1.2mL of sterile solution 1, 0.4mL of sterile solution 2, 0.4mL of sterile solution 3, and 100.0mL of sterile solution 4 vitamins. Bring volume to 120.0mL with sterile distilled/deionized water.

Solution 1:

Vitamin B_{12}	40.0mg

Preparation of Solution 1: Add vitamin B_{12} to distilled/deionized water and bring volume to 100.0mL. Mix thoroughly. Filter sterilize.

Solution 2:

Composition per 100.0.0mL:

DL-6,8-Thioctic acid, oxidized	100.0mg
Absolute ethanol	100.0mL

Preparation of Solution 2: Add DL-6,8-thioctic acid to absolute ethanol and bring volume to 100.0mL. Mix thoroughly. Filter sterilize.

Solution 3:

Tween 80	50.0g
Absolute ethanol	100.0mL

Preparation of Solution 3: Add Tween 80 to absolute ethanol and bring volume to 100.0mL. Mix thoroughly. Filter sterilize.

Solution 4 Vitamins:

Composition per liter:

Vitamin B_{12}	10.0mg
Calcium D-(+)-pantothenate	2.3mg
Choline chloride	1.25mg
Riboflavin	0.7mg
Calciferol (vitamin D_2)	0.25mg
Vitamin A, crystallized alcohol	0.25mg
p-Aminobenzoic acid	0.125mg
i-Inositol	0.125mg
Thiamine·HCl	0.025mg
Niacin	0.0625mg
Niacinamide	0.0625mg
Pyridoxal·HCl	0.0625mg
Pyridoxine·HCl	0.0625mg
Biotin	0.025mg
α-Tocopherol phosphate, disodium salt	0.025mg
Folic acid	0.025mg
Menadione (vitamin K_3)	0.025mg

Preparation of Solution 4 Vitamins: Add components to distilled/deionized water and bring volume to 1.0L. Mix thoroughly. Filter sterilize.

Preparation of Medium: Add components, except heat-inactivated bovine serum, special 107 vitamin solution, glucose solution, and buffer solution, to distilled/deionized water and bring volume to 700.0mL. Mix thoroughly. Autoclave for 15 min at 15 psi pressure–121°C. Aseptically add 100.0mL of sterile, heat-inactivated bovine serum, 100.0mL of sterile special 107 vitamin solution, 50.0mL of sterile glucose solution, and 50.0mL of sterile buffer solution. Mix thoroughly. Aseptically distribute into sterile tubes or flasks.

Use: For the cultivation of *Entamoeba* species.

TYM Basal Medium, Modified 1

Composition per liter:

Pancreatic digest of peptone	20.0g
Yeast extract	10.0g
Maltose	5.0g
L-Cysteine·HCl	1.0g
K_2HPO_4	0.8g
KH_2PO_4	0.8g
Agar, noble	0.5g
L-Ascorbic acid	0.2g

pH 7.8 ± 0.2 at 25°C

Preparation of Medium: Add components, except agar, to distilled/deionized water and bring volume to 1.0L. Mix thoroughly. Adjust pH to 7.8 with NaOH. Add the agar. Mix thoroughly. Gently heat and bring to boiling. Distribute into screw-capped tubes. Autoclave for 15 min at 15 psi pressure–121°C.

Use: For the cultivation of *Trichomonas* species.

TYM Basal Medium, Modified 2

Composition per liter:

Pancreatic digest of casein	20.0g
Yeast extract	10.0g
Maltose	5.0g
L-Cysteine·HCl	1.0g
K_2HPO_4	0.8g
KH_2PO_4	0.8g
Agar, noble	0.5g
L-Ascorbic acid	0.2g
Lamb serum, heat inactivated	300.0mL

pH 7.0 ± 0.2 at 25°C

Preparation of Medium: Add components, except agar and lamb serum, to distilled/deionized water and bring volume to 700.0mL. Mix thoroughly. Adjust to pH 7.0 with NaOH. Add the agar. Mix thoroughly. Gently heat and bring to boiling. Autoclave for 15 min at 15 psi pressure–121°C. Cool to 25°C. Aseptically add 300.0mL of sterile, heat-inactivated lamb serum. Mix thoroughly. Aseptically distribute into sterile screw-capped tubes. Use soon after preparation.

Use: For the cultivation of *Trichomonas vaginalis* and *Hypotrichomonas* species.

TYM Basal Medium, Modified 4

Composition per liter:

Pancreatic digest of casein	20.0g
Yeast extract	10.0g
Maltose	5.0g
L-Cysteine·HCl	1.0g
K_2HPO_4	0.8g
KH_2PO_4	0.8g
Agar, noble	0.5g

L-Ascorbic acid 0.2g
Horse serum, heat inactivated 300.0mL
pH 6.0 ± 0.2 at 25°C

Preparation of Medium: Add components, except agar and horse serum, to distilled/deionized water and bring volume to 700.0mL. Mix thoroughly. Adjust pH to 6.0. Add the agar. Mix thoroughly. Gently heat and bring to boiling. Autoclave for 15 min at 15 psi pressure–121°C. Cool to 50°–55°C. Aseptically add 300.0mL of heat-inactivated horse serum. Mix thoroughly. Aseptically distribute into screw-capped tubes.

Use: For the cultivation of *Tritrichomonas foetus* and *Trichomonas vaginalis*.

U Agar Plates (*Ureaplasma* Agar Plates) (MES Agar)

Composition per 100.2mL:
Base agar 65.0mL
Horse serum 20.0mL
Yeast dialysate 10.0mL
MES (2-*N*-Morpholinoethane sulfonic acid) buffer solution 3.0mL
Penicillin solution 2.0mL
Urea solution 0.2mL
pH 5.5 ± 0.2 at 25°C

Base Agar:
Composition per liter:
Papaic digest of soybean meal 20.0g
Agarose 10.0g
NaCl 5.0g
Phenol Red (2% solution) 1.0mL

Preparation of Base Agar: Add components to distilled/deionized water and bring volume to 1.0L. Mix thoroughly. Gently heat and bring to boiling. Adjust pH to 7.3. Autoclave for 15 min at 15 psi pressure–121°C. Cool to 45°–50°C.

Yeast Dialysate:
Composition per 10.0mL:
Yeast, active dried 450.0g

Preparation of Yeast Dialysate: Add active, dried yeast to distilled/deionized water and bring volume to 1250.0mL. Gently heat and bring to 40°C. Autoclave for 15 min at 15 psi pressure–121°C. Put into dialysis tubing. Dialyze against 1.0L of distilled/deionized water for 2 days at 4°C. Discard tubing and its contents. Autoclave dialysate for 15 min at 15 psi pressure–121°C. Store at –20°C.

MES Buffer Solution:
Composition per 100.0mL:
MES (2-*N*-Morpholinoethane sulfonic acid) buffer 19.52g

Preparation of MES Buffer Solution: Add MES buffer to distilled/deionized water and bring volume to 100.0mL. Mix thoroughly. Adjust pH to 5.5. Filter sterilize.

Penicillin Solution:
Composition per 10.0mL:
Penicillin 100,000U

Preparation of Penicillin Solution: Add penicillin to distilled/deionized water and bring volume to 10.0mL. Mix thoroughly. Filter sterilize.

Urea Solution:
Composition per 100.0mL:
Urea 6.0g

Preparation of Urea Solution: Add urea to distilled/deionized water and bring volume to 100.0mL. Mix thoroughly. Filter sterilize.

Preparation of Medium: To 65.0mL of cooled, sterile base agar, aseptically add 10.0mL of sterile yeast dialysate, 20.0mL of horse serum, 2.0mL of sterile penicillin solution, 3.0mL of sterile MES buffer solution, and 0.2mL of sterile urea solution. Mix thoroughly. Pour into 10mm × 35mm Petri dishes in 5.0mL volumes. Allow plates to stand overnight at 25°C to remove excess surface moisture.

Use: For the isolation and cultivation of *Ureaplasma* species.

U Broth (*Ureaplasma* Broth)

Composition per 99.5mL:
Base agar 65.0mL
Horse serum 20.0mL
Yeast dialysate 10.0mL
Penicillin solution 2.0mL
MES (2-*N*-Morpholinoethane sulfonic acid) buffer solution 1.0mL
Na_2SO_3 solution 1.0mL
Urea solution 0.5mL
pH 5.5 ± 0.2 at 25°C

Base Agar:
Composition per liter:
Papaic digest of soybean meal 20.0g
Agarose 10.0g
NaCl 5.0g
Phenol Red (2% solution) 1.0mL

Preparation of Base Agar: Add components to distilled/deionized water and bring volume to 1.0L. Mix thoroughly. Gently heat and bring to boiling. Adjust pH to 7.3. Autoclave for 15 min at 15 psi pressure–121°C. Cool to 45°–50°C.

Yeast Dialysate:

Composition per 10.0mL:

Yeast, active dried 450.0g

Preparation of Yeast Dialysate: Add active, dried yeast to distilled/deionized water and bring volume to 1250.0mL. Gently heat and bring to 40°C. Autoclave for 15 min at 15 psi pressure–121°C. Put into dialysis tubing. Dialyze against 1.0L of distilled/deionized water for 2 days at 4°C. Discard tubing and its contents. Autoclave dialysate for 15 min at 15 psi pressure–121°C. Store at –20°C.

Penicillin Solution:

Composition per 10.0mL:

Penicillin 100,000U

Preparation of Penicillin Solution: Add penicillin to distilled/deionized water and bring volume to 10.0mL. Mix thoroughly. Filter sterilize.

MES Buffer Solution:

Composition per 100.0mL:

MES (2-*N*-Morpholinoethane sulfonic acid) buffer 19.52g

Preparation of MES Buffer Solution: Add MES buffer to distilled/deionized water and bring volume to 100.0mL. Mix thoroughly. Adjust pH to 5.5. Filter sterilize.

Na_2SO_3 Solution:

Composition per 10.0mL:

Na_2SO_3 .... 0.126g

Preparation of Na_2SO_3 Solution: Add Na_2SO_3 to distilled/deionized water and bring volume to 10.0mL. Mix thoroughly. Filter sterilize.

Urea Solution:

Composition per 100.0mL:

Urea 6.0g

Preparation of Urea Solution: Add urea to distilled/deionized water and bring volume to 100.0mL. Mix thoroughly. Filter sterilize.

Preparation of Medium: To 65.0mL of cooled, sterile base agar, aseptically add 10.0mL of sterile yeast dialysate, 20.0mL of horse serum, 2.0mL of sterile penicillin solution, 1.0mL of sterile MES buffer solution, 1.0mL of sterile Na_2SO_3 solution, and 0.5mL of sterile urea solution. Mix thoroughly. Pour into 10mm × 35mm Petri dishes in 5.0mL volumes. Allow plates to stand overnight at 25°C to remove excess surface moisture. Use within 48h.

Use: For the isolation and cultivation of *Ureaplasma urealyticum.*

U9 Broth (Urease Color Test Medium)

Composition per 101.6mL:

U9 base 95.0mL

Horse serum, unheated.................................. 5.0mL

Penicillin G solution 1.0mL

Urea solution.. 0.5mL

Phenol Red solution....................................... 0.1mL

pH 6.0 ± 0.2 at 25°C

U9 Base:

Composition per 100.0mL:

NaCl 0.63g

Pancreatic digest of casein............................ 0.425g

Papaic digest of soybean meal....................... 0.075g

K_2HPO_4 .... 0.063g

Glucose 0.063g

KH_2PO_4 .... 0.02g

Preparation of U9 Base: Add components to distilled/deionized water and bring volume to 100.0mL. Mix thoroughly. Adjust pH to 5.5 with 1*N* HCl. Autoclave for 15 min at 15 psi pressure–121°C. Cool to 45°–50°C.

Penicillin G Solution:

Composition per 10.0mL:

Penicillin G .... 0.63 g

Preparation of Penicillin G Solution: Add penicillin G to distilled/deionized water and bring volume to 10.0mL. Mix thoroughly. Filter sterilize.

Urea Solution:

Composition per 30.0mL:

Urea... 3.0g

Preparation of Urea Solution: Add urea to distilled/deionized water and bring volume to 30.0mL. Mix thoroughly. Filter sterilize.

Phenol Red Solution:

Composition per 10.0mL:

Phenol Red.. 0.1g

Preparation of Phenol Red Solution: Add Phenol Red to distilled/deionized water and bring volume to 10.0mL. Mix thoroughly. Filter sterilize.

Preparation of Medium: To 95.0mL of cooled, sterile U9 base, aseptically add 5.0mL of sterile horse serum, 1.0mL of sterile penicillin G solution, 0.5mL of sterile urea solution, and 0.1mL of sterile Phenol Red solution. Mix thoroughly. Aseptically distribute into sterile tubes or flasks.

Use: For the isolation and identification of T-strain mycoplasmas from clinical specimens, especially *Ureaplasma urealyticum.* T-mycoplasmas are the only members of the *Mycoplasma* group known to contain urease. Bacteria with urease activity turn the medium dark pink.

U9 Broth with Amphotericin B

Composition per 101.6mL:

U9 base	95.0mL
Horse serum, unheated	5.0mL
Antibiotic solution	1.0mL
Urea solution	0.5mL
Phenol Red solution	0.1mL

pH 6.0 ± 0.2 at 25°C

U9 Base:

Composition per 100.0mL:

NaCl	0.63g
Pancreatic digest of casein	0.425g
Papaic digest of soybean meal	0.075g
K_2HPO_4	0.063g
Glucose	0.063g
KH_2PO_4	0.02g

Preparation of U9 Base: Add components to distilled/deionized water and bring volume to 100.0mL. Mix thoroughly. Adjust pH to 5.5 with 1*N* HCl. Autoclave for 15 min at 15 psi pressure–121°C. Cool to 45°–50°C.

Antibiotic Solution:

Composition per 10.0mL:

Penicillin G	0.63g
Amphotericin B	2.5mg

Preparation of Antibiotic Solution: Add penicillin G and amphotericin B to distilled/deionized water and bring volume to 10.0mL. Mix thoroughly. Filter sterilize.

Urea Solution:

Composition per 30.0mL:

Urea	3.0g

Preparation of Urea Solution: Add urea to distilled/deionized water and bring volume to 30.0mL. Mix thoroughly. Filter sterilize.

Phenol Red Solution:

Composition per 10.0mL:

Phenol Red	0.1g

Preparation of Phenol Red Solution: Add Phenol Red to distilled/deionized water and bring volume to 10.0mL. Mix thoroughly. Filter sterilize.

Preparation of Medium: To 95.0mL of cooled, sterile U9 base, aseptically add 5.0mL of sterile horse serum, 1.0mL of sterile antibiotic solution, 0.5mL of sterile urea solution, and 0.1mL of sterile Phenol Red solution. Mix thoroughly. Aseptically distribute into sterile tubes or flasks.

Use: For the isolation and identification of T-strain mycoplasmas from clinical specimens, especially *Ureaplasma urealyticum*. T-mycoplasmas are the only members of the *Mycoplasma* group known to contain urease. Bacteria with urease activity turn the medium dark pink.

U9B Broth

Composition per 102.1mL:

U9 base	95.0mL
Horse serum, unheated	5.0mL
Penicillin G solution	1.0mL
Urea solution	0.5mL
L-Cysteine·HCl·H_2O solution	0.5mL
Phenol Red solution	0.1mL

pH 6.0 ± 0.2 at 25°C

U9 Base:

Composition per 100.0mL:

NaCl	0.63g
Pancreatic digest of casein	0.425g
Papaic digest of soybean meal	0.075g
K_2HPO_4	0.063g
Glucose	0.063g
KH_2PO_4	0.02g

Preparation of U9 Base: Add components to distilled/deionized water and bring volume to 100.0mL. Mix thoroughly. Adjust pH to 5.5 with 1*N* HCl. Autoclave for 15 min at 15 psi pressure–121°C. Cool to 45°–50°C.

Penicillin G Solution:

Composition per 10.0mL:

Penicillin G	0.63 g

Preparation of Penicillin G Solution: Add penicillin G to distilled/deionized water and bring volume to 10.0mL. Mix thoroughly. Filter sterilize.

Urea Solution:

Composition per 30.0mL:

Urea	3.0g

Preparation of Urea Solution: Add urea to distilled/deionized water and bring volume to 30.0mL. Mix thoroughly. Filter sterilize.

L-Cysteine·HCl·H_2O Solution:

Composition per 50.0mL:

L-Cysteine·HCl·H_2O	1.0g

Preparation of L-Cysteine·HCl·H_2O Solution: Add urea to distilled/deionized water and bring volume to 50.0mL. Mix thoroughly. Filter sterilize.

Phenol Red Solution:

Composition per 10.0mL:

Phenol Red	0.1g

Preparation of Phenol Red Solution: Add Phenol Red to distilled/deionized water and bring volume to 10.0mL. Mix thoroughly. Filter sterilize.

Preparation of Medium: To 95.0mL of cooled, sterile U9 base, aseptically add 5.0mL of sterile horse serum, 1.0mL of sterile penicillin G solution, 0.5mL of sterile urea solution, 0.5mL of sterile L-cysteine·HCl·H_2O solution, and 0.1mL of sterile Phenol Red solution. Mix thoroughly. Aseptically distribute into sterile tubes or flasks.

Use: For the isolation and identification of T-strain mycoplasmas from clinical specimens, especially *Ureaplasma urealyticum*. T-mycoplasmas are the only members of the *Mycoplasma* group known to contain urease. Bacteria with urease activity turn the medium dark pink.

U9C Broth

Composition per 102.0mL:

U9C base	90.0mL
Horse serum, unheated	10.0mL
Penicillin G solution	1.0mL
L-Cysteine·HCl·H_2O solution	0.5mL
Urea solution	0.3mL
GHL tripeptide solution	0.1mL
Phenol Red solution	0.1mL

pH 6.0 ± 0.2 at 25°C

U9C Base:

Composition per 100.0mL:

NaCl	0.85g
Pancreatic digest of casein	0.25g
Papaic digest of soybean meal	0.15g
K_2HPO_4	0.12g
Glucose	0.12g
$MgCl_2·6H_2O$	0.2g
Yeast extract	0.1g
KH_2PO_4	0.02g

Preparation of U9C Base: Add components to distilled/deionized water and bring volume to 100.0mL. Mix thoroughly. Adjust pH to 5.5 with 2*N* HCl. Autoclave for 15 min at 15 psi pressure–121°C. Cool to 45°–50°C.

Penicillin G Solution:

Composition per 10.0mL:

Penicillin G	0.63 g

Preparation of Penicillin G Solution: Add penicillin G to distilled/deionized water and bring volume to 10.0mL. Mix thoroughly. Filter sterilize.

Urea Solution:

Composition per 30.0mL:

Urea	3.0g

Preparation of Urea Solution: Add urea to distilled/deionized water and bring volume to 30.0mL. Mix thoroughly. Filter sterilize.

L-Cysteine·HCl·H_2O Solution:

Composition per 50.0mL:

L-Cysteine·HCl·H_2O	1.0g

Preparation of L-Cysteine·HCl·H_2O Solution: Add urea to distilled/deionized water and bring volume to 50.0mL. Mix thoroughly. Filter sterilize.

GHL Tripeptide Solution:

Composition per 10.0mL:

GHL tripeptide	0.2mg

Preparation of GHL Tripeptide Solution: Add GHL tripeptide (glycyl-L-histidyl-L-lysine acetate) to distilled/deionized water and bring volume to 10.0mL. Mix thoroughly. Filter sterilize.

Phenol Red Solution:

Composition per 10.0mL:

Phenol Red	0.1g

Preparation of Phenol Red Solution: Add Phenol Red to distilled/deionized water and bring volume to 10.0mL. Mix thoroughly. Filter sterilize.

Preparation of Medium: To 90.0mL of cooled, sterile U9C base, aseptically add 10.0mL of sterile horse serum, 1.0mL of sterile penicillin G solution, 0.3mL of sterile urea solution, 0.5mL of sterile L-cysteine·HCl·H_2O solution, 0.1mL of sterile GHL tripeptide solution, and 0.1mL of sterile Phenol Red solution. Mix thoroughly. Aseptically distribute into sterile tubes or flasks.

Use: For the isolation and identification of T-strain mycoplasmas from clinical specimens, especially *Ureaplasma urealyticum*. T-mycoplasmas are the only members of the *Mycoplasma* group known to contain urease. Bacteria with urease activity turn the medium dark pink.

Urea Agar (Urease Test Agar) (Urea Agar Base, Christensen)

Composition per liter:

Urea	20.0g
Agar	15.0g
NaCl	5.0g
KH_2PO_4	2.0g
Peptone	1.0g
Glucose	1.0g
Phenol Red	0.012g

pH 6.8 ± 0.2 at 25°C

Source: This medium is available as a premixed powder from BD Diagnostic Systems.

Preparation of Medium: Add components, except agar, to distilled/deionized water and bring volume to 100.0mL. Mix thoroughly. Filter sterilize. Add agar to distilled/deionized water and bring volume to 900.0mL. Mix thoroughly. Gently heat and bring to boiling. Autoclave for 15 min at 15 psi pressure–121°C. Cool to 50°C. Aseptically add the 100.0mL of sterile basal medium. Mix thoroughly. Distribute into sterile tubes. Allow tubes to solidify in a slanted position.

Use: For the differentiation of a variety of microorganisms, especially members of the Enterobacteriaceae, aerobic actinomycetes, streptococci, and nonfermenting Gram-negative bacteria, on the basis of urease production.

Urea Agar Base

Composition per liter:

Agar	15.0g
NaCl	5.0g
Na_2HPO_4	1.2g
Peptone	1.0g
Glucose	1.0g
KH_2PO_4	0.8g
Phenol Red	0.012g
Urea solution	50.0mL

pH 6.8 ± 0.2 at 25°C

Source: This medium is available as a premixed powder from Oxoid Unipath.

Urea Solution:

Composition per 100.0mL:

Urea	40.0g

Preparation of Urea Solution: Add urea to distilled/deionized water and bring volume to 100.0mL. Mix thoroughly. Filter sterilize.

Preparation of Medium: Add components, except urea solution, to distilled/deionized water and bring volume to 950.0mL. Mix thoroughly. Gently heat and bring to boiling. Autoclave for 20 min at 10 psi pressure–115°C. Cool to 50°C. Aseptically add 50.0mL of sterile urea solution. Mix thoroughly. Pour into sterile Petri dishes or distribute into sterile tubes. Allow tubes to solidify in a slanted position.

Use: For the detection of *Proteus* species based on rapid urease activity and the identification of other members of the Enterobacteriaceae based on urease activity. Urease-positive bacteria turn the medium pink.

Urea Broth 10B for *Ureaplasma urealyticum*

Composition per 100.5mL:

PPLO broth without Crystal Violet	70.0mL
Horse serum, unheated	20.0mL
Fresh yeast extract solution	10.0mL
L-Cysteine·HCl·H_2O	0.5mL
CVA enrichment	0.5mL
Urea solution	0.4mL
Phenol Red	0.1mL

PPLO Broth without Crystal Violet:

Composition per 900.0mL:

Beef heart, solids from infusion	16.1g
Peptone	3.25g
NaCl	1.61g

Preparation of PPLO Broth without Crystal Violet: Add components to distilled/deionized water and bring volume to 900.0mL. Adjust pH to 5.5 with 2*N* HCl. Autoclave for 15 min at 15 psi pressure–121°C. Cool to 37°C.

Fresh Yeast Extract Solution:

Composition per 100.0mL:

Baker's yeast live, pressed, starch-free	25.0g

Preparation of Fresh Yeast Extract Solution: Add the live Baker's yeast to 100.0mL of distilled/deionized water. Autoclave for 90 min at 15 psi pressure–121°C. Allow to stand. Remove supernatant solution. Adjust pH to 6.6–6.8.

L-Cysteine·HCl·H_2O Solution:

Composition per 50.0mL:

L-Cysteine·HCl·H_2O	1.0g

Preparation of L-Cysteine·HCl·H_2O Solution: Add L-cysteine·HCl·H_2O to distilled/deionized water and bring volume to 50.0mL. Mix thoroughly. Filter sterilize.

CVA Enrichment:

Composition per liter:

Glucose	100.0g
L-Cysteine·HCl·H_2O	25.9g
L-Glutamine	10.0g
Adenine	1.0g
L-Cystine·2HCl	1.0g
Nicotinamide adenine dinucleotide	0.25g
Cocarboxylase	0.1g
Guanine·HCl	0.03g
$Fe(NO_3)_3$	0.02g
p-Aminobenzoic acid	0.013g
Vitamin B_{12}	0.01g
Thiamine·HCl	3.0mg

Preparation of CVA Enrichment: Add components to distilled/deionized water and bring volume to 1.0L. Mix thoroughly. Filter sterilize.

Urea Solution:

Composition per 30.0mL:

Urea	3.0g

Preparation of Urea Solution: Add urea to distilled/deionized water and bring volume to 30.0mL. Mix thoroughly. Filter sterilize.

Preparation of Medium: Aseptically combine the components, except the PPLO broth without Crystal Violet. Aseptically add this mixture to the cooled, sterile PPLO broth without Crystal Violet. Mix thoroughly. Aseptically distribute into sterile tubes or flasks.

Use: For the cultivation and maintenance of *Ureaplasma urealyticum* and other *Ureaplasma* species. Urease-positive bacteria turn the medium to peach orange.

Urea Broth Base

Composition per liter:

NaCl	5.0g
Na_2HPO_4	1.2g
Peptone	1.0g
Glucose	1.0g
KH_2PO_4	0.8g
Phenol Red	0.012g
Urea solution	50.0mL

pH 6.8 ± 0.2 at 25°C

Source: This medium is available as a premixed powder from Oxoid Unipath.

Urea Solution:

Composition per 100.0mL:

Urea	40.0g

Preparation of Urea Solution: Add urea to distilled/deionized water and bring volume to 100.0mL. Mix thoroughly. Filter sterilize.

Preparation of Medium: Add components, except urea solution, to distilled/deionized water and bring volume to 950.0mL. Mix thoroughly. Autoclave for 20 min at 10 psi pressure–115°C. Cool to 50°C. Aseptically add 50.0mL of sterile urea solution. Mix thoroughly. Aseptically distribute into sterile tubes or flasks.

Use: For the differentiation of members of the Enterobacteriaceae based on urease production. Urease-positive bacteria turn the medium pink.

Urea R Broth (Urea Rapid Broth)

Composition per liter:

Urea	20.0g
Yeast extract	0.1g
Na_2HPO_4	0.095g
KH_2PO_4	0.091g
Phenol Red	0.01g

pH 6.9 ± 0.2 at 25°C

Source: This medium is available as a prepared medium from BD Diagnostic Systems.

Preparation of Medium: Add components to distilled/deionized water and bring volume to 1.0L. Mix thoroughly. Filter sterilize. Aseptically distribute into sterile tubes or flasks.

Use: For the differentiation of members of the Enterobacteriaceae based on the rapid detection of urease activity. Urease-positive bacteria turn the medium cerise.

Urea Semisolid Medium

Composition per liter:

Solution A	400.0mL
Solution B	50.0mL

Solution A:

Composition per 400.0mL:

Pancreatic digest of casein	6.0g
Yeast extract	2.8g
NaCl	1.0g
Agar	0.3g
L-Cystine	0.1g
Thioglycollic acid	0.12mL

pH 7.2 ± 0.2 at 25°C

Preparation of Solution A: Add components to distilled/deionized water and bring volume to 400.0mL. Mix thoroughly. Gently heat and bring to boiling. Autoclave for 15 min at 15 psi pressure–121°C. Cool to 60°C.

Solution B:

Composition per 50.0mL:

Urea	8.0g
Na_2HPO_4	3.8g
KH_2PO_4	3.64g
Yeast extract	0.04g
Phenol Red	4.0mg

Preparation of Solution B: Add components to distilled/deionized water and bring volume to 50.0mL. Mix thoroughly. Filter sterilize.

Preparation of Medium: Aseptically combine 400.0mL of sterile solution A and 50.0mL of sterile solution B. Mix thoroughly. Aseptically distribute into sterile screw-capped tubes in 7.0mL volumes. Pass the tubes into an anaerobic chamber containing 85% N_2 + 10% H_2 + 5% CO_2 for 60 min. Close screw caps tightly.

Use: For the cultivation and differentiation of anaerobic bacteria based on their production of urease. Bacteria that produce urease turn the medium bright red.

Urea Test Broth

Composition per liter:

Urea	20.0g
Na_2HPO_4	9.5g
KH_2PO_4	9.1g
Yeast extract	0.1g
Phenol Red	0.01g
Urea solution	100.0mL

Urea Solution:

Composition per 100.0mL:

Urea	20.0g

Preparation of Urea Solution: Add urea to distilled/deionized water and bring volume to 100.0mL. Mix thoroughly. Filter sterilize.

Preparation of Medium: Add components, except urea solution, to distilled/deionized water and bring volume to 900.0mL. Mix thoroughly. Autoclave for 15 min at 15 psi pressure–121°C. Cool to 45°–50°C. Aseptically add sterile urea solution. Mix thoroughly. Aseptically distribute into sterile tubes in 3.0mL volumes.

Use: For the cultivation and differentiation of members of the Enterobacteriaceae and aerobic actinomycetes based on their production of urease. Bacteria that produce urease turn the medium bright red.

Urea Test Broth

Composition per 99.6mL:

H broth base	85.0mL
Horse serum	10.0mL
Penicillin solution	2.0mL
MES (2-*N*-Morpholinoethane sulfonic acid) buffer solution	1.0mL
Na_2SO_3 solution	1.0mL
Urea solution	0.5mL
Phenol Red solution	0.1mL

pH 7.2 ± 0.2 at 25°C

H Broth Base:

Composition per liter:

NaCl	5.0g
Pancreatic digest of casein	5.0g
Peptone	5.0g
Beef extract	3.0g
K_2HPO_4	2.5g
Glucose	1.0g

Preparation of H Broth Base: Add components to distilled/deionized water and bring volume to 1.0L. Mix thoroughly. Gently heat and bring to boiling. Distribute into tubes in 4.0mL volumes. Autoclave for 15 min at 10 psi pressure–115°C. Cool to 45°–50°C.

Penicillin Solution:

Composition per 10.0mL:

Penicillin	100,000U

Preparation of Penicillin Solution: Add penicillin to distilled/deionized water and bring volume to 10.0mL. Mix thoroughly. Filter sterilize.

MES Buffer Solution:

Composition per 100.0mL:

MES (2-*N*-Morpholinoethane sulfonic acid) buffer	19.52g

Preparation of MES Buffer Solution: Add MES buffer to distilled/deionized water and bring volume to 100.0mL. Mix thoroughly. Adjust pH to 5.5. Filter sterilize.

Na_2SO_3 Solution:

Composition per 10.0mL:

Na_2SO_3	0.126g

Preparation of Na_2SO_3 Solution: Add Na_2SO_3 to distilled/deionized water and bring volume to 10.0mL. Mix thoroughly. Filter sterilize.

Urea Solution:

Composition per 100.0mL:

Urea	6.0g

Preparation of Urea Solution: Add urea to distilled/deionized water and bring volume to 100.0mL. Mix thoroughly. Filter sterilize.

Phenol Red Solution:

Composition per 10.0mL:

Phenol Red	0.1g

Preparation of Phenol Red Solution: Add Phenol Red to distilled/deionized water and bring volume to 10.0mL. Mix thoroughly. Filter sterilize.

Preparation of Medium: To 85.0mL of cooled sterile H broth base, aseptically add 10.0mL of sterile horse serum, 2.0mL of sterile penicillin solution, 1.0mL of MES buffer solution, 1.0mL of Na_2SO_3 solution, 0.5mL of urea solution, and 0.1 mL of sterile Phenol Red solution. Mix thoroughly. Aseptically distribute into test tubes in 3.0mL volumes.

Use: For the cultivation and differentiation of *Ureaplasma* species based on their production of urease.

Urease Test Broth (Urea Broth)

Composition per liter:

Urea	20.0g
Na_2HPO_4	9.5g
KH_2PO_4	9.1g

Yeast extract 0.1g
Phenol Red 0.01g

pH 6.8 ± 0.2 at 25°C

Source: This medium is available as a premixed powder from BD Diagnostic Systems.

Preparation of Medium: Add components to distilled/deionized water and bring volume to 1.0L. Mix thoroughly. Filter sterilize. Aseptically distribute into sterile tubes or flasks.

Use: For the differentiation of organisms, especially the Enterobacteriaceae, on the basis of urease production. Urease-positive bacteria turn the medium pink.

UVM *Listeria* Enrichment Broth (University of Vermont *Listeria* Enrichment Broth)

Composition per liter:

NaCl 20.0g
Na_2HPO_4 .... 9.6g
Pancreatic digest of casein 5.0g
Peptic digest of animal tissue 5.0g
Beef extract 5.0g
Yeast extract 5.0g
KH_2PO_4 .... 1.35g
Esculin 1.0g
Nalidixic acid 40.0mg
Acriflavine·HCl 12.0mg

pH 7.2 ± 0.2 at 25°C

Preparation of Medium: Add components to distilled/deionized water and bring volume to 1.0L. Mix thoroughly. Distribute into tubes or flasks. Autoclave for 15 min at 15 psi pressure–121°C.

Use: For the selective isolation of *Listeria monocytogenes.*

UVM Modified *Listeria* Enrichment Broth (University of Vermont Modified *Listeria* Enrichment Broth)

Composition per liter:

NaCl 20.0g
Na_2HPO_4 .... 9.6g
Pancreatic digest of casein 5.0g
Peptic digest of animal tissue 5.0g
Beef extract 5.0g
Yeast extract 5.0g
KH_2PO_4 .... 1.35g
Esculin 1.0g
Nalidixic acid 20.0mg
Acriflavine·HCl 12.0mg

pH 7.2 ± 0.2 at 25°C

Source: This medium is available as a premixed powder from BD Diagnostic Systems.

Preparation of Medium: Add components to distilled/deionized water and bring volume to 1.0L. Mix thoroughly. Distribute into tubes or flasks. Autoclave for 15 min at 15 psi pressure–121°C.

Use: For the selective isolation of *Listeria monocytogenes.*

V Agar

Composition per liter:

Agar 13.5g
Pancreatic digest of casein 12.0g
Peptone 10.0g
Peptic digest of animal tissue 5.0g
NaCl 5.0g
Beef extract 3.0g
Yeast extract 3.0g
Cornstarch 1.0g
Human blood, anticoagulated 50.0mL

pH 7.4 ± 0.2 at 25°C

Source: This medium is available as a prepared medium from BD Diagnostic Systems.

Preparation of Medium: Add components, except human blood, to distilled/deionized water and bring volume to 950.0mL. Mix thoroughly. Gently heat and bring to boiling. Distribute into tubes or flasks. Autoclave for 15 min at 15 psi pressure–121°C. Cool to 50°C. Aseptically add 50.0mL of human blood. Mix thoroughly. Pour into sterile Petri dishes or leave in tubes.

Use: For the isolation and differentiation of *Gardnerella vaginalis* from clinical specimens. Plates are incubated under an atmosphere with 3-10% CO_2. *Gardnerella vaginalis* appears as small white colonies with diffuse β-hemolysis.

V-8™ Agar

Composition per liter:

Agar 20.0g
$CaCO_3$ 4.0g
V-8 canned vegetable juice 200.0mL

pH 7.3 ± 0.2 at 25°C

Preparation of Medium: Add components to distilled/deionized water and bring volume to 1.0L. Mix thoroughly. Gently heat and bring to boiling. Distribute into tubes or flasks. Autoclave for 15 min at 15 psi pressure–121°C. Pour into sterile Petri dishes or leave in tubes.

Use: For the isolation and cultivation of *Actinomadura* species.

V-8 Agar

Composition per liter:

Agar 15.0g
$CaCO_3$ 2.0g
V-8 canned vegetable juice 200.0mL

Preparation of Medium: Add components to distilled/deionized water and bring volume to 1.0L. Mix thoroughly. Gently heat and bring to boiling. Distribute into tubes or flasks. Autoclave for 15 min at 15 psi pressure–121°C. Pour into sterile Petri dishes or leave in tubes.

Use: For the cultivation of numerous yeasts and filamentous fungi.

Veal Infusion Agar

Composition per liter:

Agar 15.0g
Veal, infusion from 10.0g
Pancreatic digest of casein 5.0g
Peptic digest of animal tissue 5.0g
NaCl 5.0g

pH 7.4 ± 0.2 at 25°C

Source: This medium is available as a premixed powder from BD Diagnostic Systems.

Preparation of Medium: Add components to distilled/deionized water and bring volume to 1.0L. Mix thoroughly. Gently heat and bring to boiling. Distribute into tubes or flasks. Autoclave for 15 min at 15 psi pressure–121°C. Pour into sterile Petri dishes or leave in tubes.

Use: For the cultivation and maintenance of a variety of microorganisms. Can be used for the cultivation of fastidious microorganisms when enriched with blood or serum.

Veal Infusion Agar (ATCC Medium 521)

Composition per liter:

Veal, infusion from 500.0g
Agar 15.0g
Pancreatic digest of casein 5.0g
Peptic digest of animal tissue 5.0g
NaCl 5.0g

pH 7.4 ± 0.2 at 25°C

Source: This medium is available as a premixed powder from BD Diagnostic Systems.

Preparation of Medium: Add components to distilled/deionized water and bring volume to 1.0L. Mix thoroughly. Gently heat and bring to boiling. Distribute into tubes or flasks. Autoclave for 15 min at 15 psi pressure–121°C. Pour into sterile Petri dishes or leave in tubes.

Use: For the cultivation and maintenance of a variety of microorganisms. Can be used for the cultivation of fastidious microorganisms when enriched with blood or serum.

Veal Infusion Broth

Composition per liter:

Veal, infusion from 10.0g
Pancreatic digest of casein 5.0g
Peptic digest of animal tissue 5.0g
NaCl 5.0g

pH 7.4 ± 0.2 at 25°C

Source: This medium is available as a premixed powder from BD Diagnostic Systems.

Preparation of Medium: Add components to distilled/deionized water and bring volume to 1.0L. Mix thoroughly. Gently heat and bring to boiling. Distribute into tubes or flasks. Autoclave for 15 min at 15 psi pressure–121°C. Use freshly prepared solution.

Use: For the cultivation of streptococci and other microorganisms.

Veal Infusion Broth with Horse Serum

Composition per liter:

Veal, infusion from 500.0g
Pancreatic digest of casein 5.0g
Peptic digest of animal tissue 5.0g
NaCl 5.0g
Horse serum, heat inactivated 100.0mL

pH 7.4 ± 0.2 at 25°C

Preparation of Medium: Add components, except horse serum, to distilled/deionized water and bring volume to 900.0mL. Mix thoroughly. Gently heat and bring to boiling. Autoclave for 15 min at 15 psi pressure–121°C. Cool to 50°C. Aseptically add 100.0mL of horse serum. Mix thoroughly. Aseptically distribute into tubes or flasks. Use freshly prepared solution or boil without mixing prior to use.

Use: For the cultivation and maintenance of *Streptococcus pyogenes.*

Veillonella Agar

Composition per liter:

Agar 15.0g
Pancreatic digest of casein 5.0g
Yeast extract 3.0g
Sodium thioglycolate 0.75g
Vancomycin 7.5mg

Basic Fuchsin ... 2.0mg
Sodium lactate (60% solution) ... 21.0mL
pH 7.5± 0.2 at 25°C

Source: This medium is available as a premixed powder from BD Diagnostic Systems.

Caution: Basic Fuchsin is a potential carcinogen and care must be taken to avoid inhalation of the powdered dye and contact with the skin.

Preparation of Medium: Add components, except vancomycin, to distilled/deionized water and bring volume to 1.0L. Mix thoroughly. Gently heat and bring to boiling. Distribute into tubes or flasks. Autoclave for 15 min at 15 psi pressure–121°C. Cool to 50°C. Aseptically add vancomycin. Mix thoroughly. Pour into sterile Petri dishes or leave in tubes.

Use: For the selective isolation and cultivation of *Veillonella* species.

Veillonella Medium

Composition per liter:

Pancreatic digest of casein ... 5.0g
Yeast extract ... 3.0g
Tween 80 ... 1.0g
Glucose ... 1.0g
Sodium thioglycolate ... 0.75g
Sodium lactate (60% solution) ... 21.0mL
pH 7.5 ± 0.2 at 25°C

Preparation of Medium: Add components to distilled/deionized water and bring volume to 1.0L. Mix thoroughly. Adjust pH to 7.5 with K_2CO_3. Distribute into tubes or flasks. Autoclave for 15 min at 15 psi pressure–121°C.

Use: For the cultivation and maintenance of *Veillonella* species.

Veillonella Medium, DSM

Composition per liter:

Sodium lactate (60% solution) ... 7.5g
Pancreatic digest of casein ... 5.0g
Yeast extract ... 3.0g
Tween 80 ... 1.0g
Glucose ... 1.0g
Sodium thioglycolate ... 0.75g
Putrescine ... 3.0mg
Resazurin ... 1.0mg
pH 7.5 ± 0.2 at 25°C

Preparation of Medium: Prepare and dispense medium anaerobically under 100% N_2. Add components to distilled/deionized water and bring volume to 1.0L. Mix thoroughly. Adjust pH to 7.5 with K_2CO_3. Anaerobically distribute into tubes or flasks. Autoclave for 15 min at 15 psi pressure–121°C.

Use: For the cultivation and maintenance of *Veillonella parvula* and other *Veillonella* species.

Veillonella Selective Medium

Composition per liter:

Pancreatic digest of casein ... 5.0g
Yeast extract ... 3.0g
Tween 80 ... 1.0g
Sodium thioglycolate ... 0.75g
Sodium lactate (50% solution) ... 25.0mL
Streptomycin solution ... 10.0mL
pH 6.6 ± 0.2 at 25°C

Streptomycin Solution:

Composition per 10.0mL:

Streptomycin ... 5.0mg

Preparation of Streptomycin Solution: Add streptomycin to distilled/deionized water and bring volume to 10.0mL. Mix thoroughly. Filter sterilize.

Preparation of Medium: Add components, except streptomycin solution, to distilled/deionized water and bring volume to 990.0mL. Mix thoroughly. Gently heat and bring to boiling. Adjust pH to 6.6 with K_2CO_3. Autoclave for 15 min at 15 psi pressure–121°C. Cool to 45°–50°C. Aseptically add sterile streptomycin solution. Mix thoroughly. Aseptically distribute into sterile tubes or flasks.

Use: For the cultivation of *Veillonella* species.

Vibrio Agar

Composition per liter:

Sucrose ... 20.0g
Agar ... 15.0g
NaCl ... 10.0g
Sodium citrate·$2H_2O$ ... 10.0g
$Na_2S_2O_3 \cdot 5H_2O$ ... 6.5g
Oxgall ... 5.0g
Yeast extract ... 5.0g
Pancreatic digest of casein ... 4.0g
Proteose peptone ... 3.0g
Sodium deoxycholate ... 1.0g
Sodium lauryl sulfate ... 0.2g
Water Blue ... 0.2g
Cresol Red ... 0.02g
pH 8.5 ± 0.2 at 25°C

Preparation of Medium: Add components to distilled/deionized water and bring volume to 1.0L. Mix thoroughly. Adjust pH to 8.5. Gently heat and bring to boiling. Do not autoclave. Pour into sterile Petri dishes or distribute into sterile tubes.

Use: For the isolation and cultivation of the *Vibrio cholerae.*

Vibrio parahaemolyticus Sucrose Agar (VPSA)

Composition per liter:

NaCl	30.0g
Agar	15.0g
Sucrose	10.0g
Yeast extract	7.0g
Tryptose	5.0g
Pancreatic digest of casein	5.0g
Bile salts no. 3	1.5g
Bromthymol Blue	0.025g

pH 8.6 ± 0.2 at 25°C

Preparation of Medium: Add components to distilled/deionized water and bring volume to 1.0L. Mix thoroughly. Gently heat and bring to boiling. Do not autoclave. Cool to 50°C. Pour into sterile Petri dishes in 20.0mL volumes. Allow plates to dry before using.

Use: For the isolation, cultivation, and differentiation of *Vibrio parahaemolyticus. Vibrio parahaemolyticus* and *Vibrio vulnificus* appear as blue to green colonies. Other *Vibrio* species appear as yellow colonies.

Violet Peptone Bile Lactose Broth

Composition per liter:

Lactose	10.0g
Peptone	10.0g
Bile salts	5.0g
Gentian Violet	0.04g

pH 7.6 ± 0.2 at 25°C

Preparation of Medium: Add components to distilled/deionized water and bring volume to 1.0L. Mix thoroughly. Gently heat and bring to boiling. Distribute into tubes or flasks. Autoclave for 15 min at 15 psi pressure–121°C. Pour into sterile Petri dishes or leave in tubes.

Use: For the selective cultivation of members of the Enterobacteriaceae.

Viral Transport Medium (VTM)

Composition per 104.1mL:

Bovine serum albumin	0.5g
Veal infusion broth	100.0mL
Amphotericin B solution	2.0mL
Gentamicin	1.0mL
Phenol Red	0.4mL
Vancomycin	0.2mL

pH 7.4 ± 0.2 at 25°C

Veal Infusion Broth:

Composition per liter:

Veal, infusion from	500.0g
NaCl	5.0g
Pancreatic digest of casein	5.0g
Peptic digest of animal tissue	5.0g

Preparation of Veal Infusion Broth: Add components to distilled/deionized water and bring volume to 1.0L. Mix thoroughly. Distribute into tubes or flasks. Autoclave for 15 min at 15 psi pressure–121°C. Use freshly prepared solution.

Amphotericin B Solution:

Composition per 10.0mL:

Amphotericin B	2.5g

Preparation of Amphotericin B Solution: Add amphotericin B to distilled/deionized water and bring volume to 10.0mL. Mix thoroughly. Filter sterilize.

Gentamicin Solution:

Composition per 10.0mL:

Gentamicin	0.5g

Preparation of Gentamicin Solution: Add gentamicin to distilled/deionized water and bring volume to 10.0mL. Mix thoroughly. Filter sterilize.

Vancomycin Solution:

Composition per 10.0mL:

Vancomycin	0.5g

Preparation of Vancomycin Solution: Add vancomycin to distilled/deionized water and bring volume to 10.0mL. Mix thoroughly. Filter sterilize.

Preparation of Medium: To 100.0mL of sterile veal infusion broth, aseptically add bovine serum albumin, Phenol Red, amphotericin B solution, gentamicin solution, and vancomycin solution. Mix thoroughly. Dispense 2.0mL of medium into serum vials. Store at 4°C and use for up to 2 months.

Use: For the maintenance and transport of specimens suspected of being virally infected.

Vitamin B_6 Blood Agar (ATCC Medium 860)

Composition per liter:

Agar	15.0g
Pancreatic digest of casein	15.0g
Papaic digest of soybean meal	5.0g
NaCl	5.0g
Sheep blood, defibrinated	50.0mL

pH 7.3 ± 0.2 at 25°C

Preparation of Medium: Add components, except sheep blood, to distilled/deionized water and bring volume to 1.0L. Mix thoroughly. Gently heat and bring to boiling. Autoclave for 15 min at 15 psi pressure–121°C. Cool to 45°–50°C. Aseptically add sterile, defibrinated sheep blood. Pour into sterile Petri dishes or distribute into sterile tubes.

Use: For the cultivation and maintenance of fastidious microorganisms, especially *Streptococcus* species.

Vitamin B_6 Blood Agar with Pyridoxal-HCl (ATCC Medium 1511)

Composition per liter:

Agar	15.0g
Pancreatic digest of casein	15.0g
Papaic digest of soybean meal	5.0g
NaCl	5.0g
Pyridoxal·HCl	0.01g
Sheep blood, defibrinated	50.0mL

pH 7.3 ± 0.2 at 25°C

Preparation of Medium: Add components, except sheep blood, to distilled/deionized water and bring volume to 1.0L. Mix thoroughly. Gently heat and bring to boiling. Autoclave for 15 min at 15 psi pressure–121°C. Cool to 45°–50°C. Aseptically add sterile, defibrinated sheep blood. Pour into sterile Petri dishes or distribute into sterile tubes.

Use: For the cultivation and maintenance of fastidious microorganisms, especially *Streptococcus* species.

VL Agar with Blood

Composition per liter:

Agar	20.0g
Tryptone	10.0g
NaCl	5.0g
Yeast extract	5.0g
Beef extract	2.0g
Glucose	2.0g
L-Cysteine·HCl	0.3g
Sheep blood or horse blood	100.0mL

pH 7.4 ± 0.2 at 25°C

Preparation of Medium: Add components, except sheep blood, to distilled/deionized water and bring volume to 900.0mL. Mix thoroughly. Adjust pH to 7.4. Gently heat and bring to boiling. Autoclave for 15 min at 15 psi pressure–121°C. Cool to 50°–55°C. Warm sheep blood to 50°C. Aseptically add 100.0mL of sterile sheep blood or 100.0mL of sterile horse blood. Mix thoroughly. Pour into sterile Petri dishes or distribute into sterile tubes.

Use: For the cultivation and maintenance of *Bacterionema helcogenes, Bacteroides nodosus, Bacteroides pyogenes, Bacteroides salivosus, Bacteroides suis, Bifidobacterium adolescentis, Bifidobacterium bifidum, Bifidobacterium breve, Bifidobacterium longum, Campylobacter coli, Campylobacter concisus, Campylobacter fetus, Campylobacter hyointestinalis, Campylobacter jejuni, Campylobacter lari, Campylobacter mucosalis, Campylobacter* species, *Campylobacter sputorum, Capnocytophaga gingivalis, Capnocytophaga ochracea, Capnocytophaga sputigena, Clostridium colinum, Clostridium difficile, Clostridium* species, *Clostridium spiroforme, Falcivibrio grandis, Falcivibrio vaginalis, Fusobacterium simiae, Gardnerella vaginalis, Leptotrichia buccalis, Pectinatus frisingensis, Peptostreptococcus anaerobius, Peptostreptococcus asaccharolyticus, Peptostreptococcus indolicus, Peptostreptococcus magnus, Peptostreptococcus micros, Peptostreptococcus prevotii, Peptostreptococcus tetradius, Propionibacterium acnes, Propionibacterium avidum, Propionibacterium granulosum, Propionibacterium lymphophilum,* and *Tonsillophilus suis.*

VL Medium

Composition per liter:

Pancreatic digest of casein	10.0g
Agar	6.0g
NaCl	5.0g
Yeast extract	5.0g
Meat extract	2.0g
Glucose	2.0g
L-Cysteine·HCl·H_2O	0.3g
Antibiotic solution	10.0mL

pH 7.4 ± 0.2 at 25°C

Antibiotic Solution:

Composition per 10.0mL:

Kanamycin	0.1g
Vancomycin	7.5mg

Preparation of Antibiotic Solution: Add components to distilled/deionized water and bring volume to 10.0mL. Mix thoroughly. Filter sterilize.

Preparation of Medium: Add components, except antibiotic solution, to distilled/deionized water and bring volume to 990.0mL. Mix thoroughly. Gently heat and bring to boiling. Autoclave for 15 min at 15 psi pressure–121°C. Cool to 45°–50°C. Aseptically add sterile antibiotic solution. Mix thoroughly. Aseptically distribute into sterile tubes or flasks.

Use: For the isolation and cultivation of *Bacteroides* species.

Vogel and Johnson Agar

Composition per liter:

Agar	16.0g
Pancreatic digest of casein	10.0g
D-Mannitol	10.0g
Glycine	10.0g

Yeast extract	5.0g
K_2HPO_4	5.0g
LiCl	5.0g
Phenol Red	0.025g
K_2TeO_3 solution	20.0mL

pH 7.2 ± 0.2 at 25°C

Source: This medium is available as a premixed powder from BD Diagnostic Systems and Oxoid Unipath.

K_2TeO_3 Solution:
Composition per 100.0mL:

K_2TeO_3	1.0g

Preparation of K_2TeO_3 Solution: Add K_2TeO_3 to distilled/deionized water and bring volume to 100.0mL. Mix thoroughly. Filter sterilize.

Caution: Potassium tellurite is toxic.

Preparation of Medium: Add components, except K_2TeO_3 solution, to distilled/deionized water and bring volume to 980.0mL. Mix thoroughly. Gently heat and bring to boiling. Autoclave for 15 min at 15 psi pressure–121°C. Cool to 45°–50°C. Aseptically add 20.0mL of sterile K_2TeO_3 solution. Mix thoroughly. Pour into sterile Petri dishes or distribute into sterile tubes.

Use: For the detection of coagulase-positive *Staphylococcus aureus.*

VP Medium (Voges-Proskauer Medium)

Composition per liter:

Peptone	7.0g
K_2HPO_4	5.0g
Glucose	5.0g

pH 6.9 ± 0.2 at 25°C

Preparation of Medium: Add components to distilled/deionized water and bring volume to 1.0L. Mix thoroughly. Adjust pH to 6.9. Distribute into tubes in 3.0mL volumes. Autoclave for 15 min at 15 psi pressure–121°C.

Use: For the cultivation and differentiation of bacteria based on their ability to produce acetoin.

VRB Agar, Fluorocult (Fluorocult VRB Agar)

Composition per liter:

Agar	13.0g
Lactose	10.0g
Peptone from meat	7.0g
NaCl	5.0g
Yeast extract	3.0g
Bile salt mixture	1.5g
4-Methylumbelliferyl-β-D-glucuronide	0.1g
Neutral Red	0.03g
Crystal Violet	0.002g

pH 7.4 ± 0.2 at 25°C

Source: This medium is available from Merck.

Preparation of Medium: Add components to distilled/deionized water and bring volume to 1.0L. Mix thoroughly. Heat in a boiling water bath or in free flowing steam with frequent stirring until completely dissolved. Do not boil for more than 2 min. Do not autoclave. Do not overheat. Pour into sterile Petri dishes. The plates are clear and dark red.

Use: For the detection and enumeration of coliform bacteria, in particular *E. coli.* Crystal violet and bile salts largely inhibit the growth of Gram-positive accompanying bacterial flora. Lactose-postitive colonies show a color change to red of the pH indicator. *E. coli* colonies schow a fluorescence under UV light. Lactose-negative Enterobacteriaceae are colorless. Lactose-positive colonies are red and often surrounded by a turbid zone due to the precipitation of bile acids. Light blue fluorescing colonies denote *E. coli.*

VRB MUG Agar (Violet Red Bile Lactose MUG Agar)

Composition per liter:

Agar	13.0g
Lactose	10.0g
Meat peptone	7.0g
NaCl	5.0g
Yeast extract	3.0g
Bile salt mixture	1.5g
4-Methylumbelliferyl-β-D-glucuronide	0.1g
Neutral Red	0.03g
Crystal Violet	0.002g

pH 7.4 ± 0.2 at 37°C

Source: This medium is available from Fluka, Sigma-Aldrich.

Preparation of Medium: Add components to distilled/deionized water and bring volume to 1.0L. Mix thoroughly. Gently heat while stirring and bring to boiling. Autoclave for 15 min at 15 psi pressure–121°C. Cool to 50°C. Pour into sterile Petri dishes.

Use: For the detection and enumeration of coliform bacteria, in particular *E. coli.* Gram-positive accompanying flora are extensively inhibited by Crystal Violet and bile salts. A color-change to red indicates lactose-positive colonies, within which *E. coli* can be demonstrated by fluorescence in the UV.

VRE Agar

Composition per 1004mL:

Tryptone	20.0g

Agar	10.0g
Yeast extract	5.0g
NaCl	5.0g
Sodium citrate	1.0g
Aesculin	1.0g
Ferric ammonium citrate	0.5g
NaN_3	0.15g
Selective supplement solution	4.0mL

pH 7.0 ± 0.2 at 25°C

Source: This medium is available as a premixed powder from Oxoid Unipath.

Selective Supplement Solution:
Composition per 4.0mL:

Meropenum	1.0mg
Vancomycin	6.0mg

Preparation of Selective Supplement Solution: Add components to distilled/deionized water and bring volume to 4.0mL. Mix thoroughly. Filter sterilize.

Preparation of Medium: Add components, except selective supplement solution, to distilled/deionized water and bring volume to 1.0L. Mix thoroughly. Gently heat while stirring and bring to boiling. Autoclave for 15 min at 15 psi pressure–121°C. Cool to 50°C. Aseptially add 4.0mL selective supplement solution. Mix thoroughly. Pour into sterile Petri dishes.

Use: For the isolation of vancomycin resistant enterococci (VRE) from clinical samples. Nonresitant enterococci containing the *Van* C genes will not grow on this medium.The selective supplement suppresses growth of Gram negative bacteria and *E. gallinarum*. The medium contains an indicator system to detect the growth of aesculin-hydrolysing organisms. Enterococci produce black zones around the colonies from the formation of black iron phenolic compounds derived from aesculin-hydrolyis products and ferrous iron.

VRE Agar

Composition per 1004mL:

Tryptone	20.0g
Agar	10.0g
Yeast extract	5.0g
NaCl	5.0g
Sodium citrate	1.0g
Aesculin	1.0g
Ferric ammonium citrate	0.5g
NaN_3	0.15g
Selective supplement solution	4.0mL

pH 7.0 ± 0.2 at 25°C

Source: This medium is available as a premixed powder from Oxoid Unipath.

Selective Supplement Solution:
Composition per 4.0mL:

Gentamicin	512.0mg

Preparation of Selective Supplement Solution: Add gentamicin to distilled/deionized water and bring volume to 4.0mL. Mix thoroughly. Filter sterilize.

Preparation of Medium: Add components, except selective supplement solution, to distilled/deionized water and bring volume to 1.0L. Mix thoroughly. Gently heat while stirring and bring to boiling. Autoclave for 15 min at 15 psi pressure–121°C. Cool to 50°C. Aseptially add 4.0mL selective supplement solution. Mix thoroughly. Pour into sterile Petri dishes.

Use: For the isolation of high level aminoglycoside resistant enterococci (HLARE) from clinical samples. Nonresitant enterococci containing the *Van* C genes will not grow on this medium.The selective supplement suppresses growth of Gram negative bacteria and *E. gallinarum*. The medium contains an indicator system to detect the growth of aesculin-hydrolysing organisms. Enterococci produce black zones around the colonies from the formation of black iron phenolic compounds derived from aesculin-hydrolyis products and ferrous iron.

VRE Broth

Composition per 1004mL:

Calf brain infusion solids	12.5g
Proteose peptone	10.0g
Beef heart infusion solids	5.0g
NaCl	5.0g
Na_2HPO_4	2.5g
Glucose	2.0g
Selective supplement solution	4.0mL

pH 7.4 ± 0.2 at 25°C

Source: This medium is available as a premixed powder from Oxoid Unipath.

Selective Supplement Solution:
Composition per 4.0mL:

Meropenum	2.0mg

Preparation of Selective Supplement Solution: Add meropenum to distilled/deionized water and bring volume to 4.0mL. Mix thoroughly. Filter sterilize.

Preparation of Medium: Add components, except selective supplement solution, to distilled/deionized water and bring volume to 1.0L. Mix thoroughly. Gently heat while stirring and bring to boiling. Autoclave for 15 min at 15 psi pressure–121°C. Cool to 50°C. Aseptially add 4.0mL selective supplement solution. Mix thoroughly. Aseptically distribute into sterile tubes.

Use: For the isolation of high level aminoglycoside resistant enterococci (HLARE) from clinical samples. Nonresitant enterococci will not grow on this medium. The selective supplement suppresses growth of Gram negative bacteria and *E. gallinarum*.

Wagatsuma Agar

Composition per 1050.0mL:

NaCl	70.0g
Agar	15.0g
Mannitol	10.0g
Peptone	10.0g
K_2HPO_4	5.0g
Yeast extract	3.0g
Crystal Violet	1.0mg
Red blood cells	50.0mL

pH 8.0 ± 0.2 at 25°C

Red Blood Cells:
Composition per 100.0mL:

Blood, human or rabbit	100.0mL

Preparation of Red Blood Cells: Mix freshly drawn human or rabbit blood with anticoagulant and an equal volume of sterile 0.85% saline solution. Centrifuge cells at 4000 × g at 4°C for 15 min. Pour off saline and wash two more times with sterile saline. After last wash, pour off saline and resuspend cells to their original volume.

Preparation of Medium: Add components, except blood, to distilled/deionized water and bring volume to 1.0L. Mix thoroughly. Adjust pH to 8.0. Place in a steam bath for 30 min. Do not autoclave. Cool to 45°–50°C. Add 50.0mL of washed red blood cells. Mix thoroughly. Pour into sterile Petri dishes. Dry plates before using.

Use: For the cultivation and detection of thermostable hemolysin of *Vibrio parahaemolyticus* by the Kanagawa reaction.

Wallerstein Medium

Composition per 4.225L:

Malachite Green	0.75g
Egg yolk emulsion	3.125L
Glycerol	100.0mL

pH 6.75 ± 0.2 at 25°C

Egg Yolk Emulsion:
Composition:

Chicken egg yolks	66
Whole chicken egg	6

Preparation of Egg Yolk Emulsion: Soak eggs with 1:100 dilution of saturated mercuric chloride solution for 1 min. Crack eggs and separate yolks from whites. Mix egg yolks with 6 chicken eggs.

Preparation of Medium: Add components to distilled/deionized water and bring volume to 1.0L. Mix thoroughly. Distribute into tubes. Autoclave for 15 min at 15 psi pressure–121°C.

Use: For the isolation of *Mycobacterium* species other than *Mycobacterium leprae.*

Wickerham Broth

Composition per 100.0mL:

Carbohydrate	10.0g
Yeast nitrogen base	100.0mL

Yeast Nitrogen Base, 10X:
Composition per liter:

Glucose	10.0g
KH_2PO_4	1.0g
$MgSO_4 \cdot 7H_2O$	0.5g
NaCl	0.1g
$CaCl_2 \cdot 2H_2O$	0.1g
DL-Methionine	0.02g
DL-Tryptophan	0.02g
L-Histidine·HCl	0.01g
Inositol	2.0mg
H_3BO_3	0.5mg
$ZnSO_4 \cdot 7H_2O$	0.4mg
$MnSO_4 \cdot 4H_2O$	0.4mg
Thiamine·HCl	0.4mg
Pyridoxine	0.4mg
Niacin	0.4mg
Calcium pantothenate	0.4mg
p-Aminobenzoic acid	0.2mg
Riboflavin	0.2mg
$FeCl_3$	0.2mg
$Na_2MoO_4 \cdot 4H_2O$	0.2mg
KI	0.1mg
$CuSO_4 \cdot 5H_2O$	0.04mg
Folic acid	2.0μg
Biotin	2.0μg

pH 4.5 ± 0.2 at 25°C

Preparation of Yeast Nitrogen Base: Add components to distilled/deionized water and bring volume to 1.0L. Mix thoroughly.

Preparation of Medium: To 100.0mL of yeast nitrogen base, add 10.0g of carbohydrate. Mix thoroughly. Filter sterilize. Aseptically distribute 0.5mL into tubes containing 4.5mL of sterile, distilled/deionized water.

Use: For the cultivation and differentiation of bacteria based on carbohydrate assimilation.

Wickerham Broth

Composition per 100.0mL:

KNO_3	0.78g
Yeast carbon base	100.0mL

pH 4.5 ± 0.2 at 25°C

Yeast Carbon Base:

Composition per liter:

Glucose	10.0g
KH_2PO_4	1.0g
$MgSO_4 \cdot 7H_2O$	0.5g
NaCl	0.1g
$CaCl_2 \cdot 2H_2O$	0.1g
DL-Methionine	0.02g
DL-Tryptophan	0.02g
L-Histidine·HCl	0.01g
Inositol	2.0mg
H_3BO_3	0.5mg
$ZnSO_4 \cdot 7H_2O$	0.4mg
$MnSO_4 \cdot 4H_2O$	0.4mg
Thiamine·HCl	0.4mg
Pyridoxine	0.4mg
Niacin	0.4mg
Calcium pantothenate	0.4mg
p-Aminobenzoic acid	0.2mg
Riboflavin	0.2mg
$FeCl_3$	0.2mg
$Na_2MoO_4 \cdot 4H_2O$	0.2mg
KI	0.1mg
$CuSO_4 \cdot 5H_2O$	0.04mg
Folic acid	2.0μg
Biotin	2.0μg

Preparation of Yeast Carbon Base: Add components to distilled/deionized water and bring volume to 1.0L. Mix thoroughly.

Preparation of Medium: To 100.0mL of yeast carbon base, add 0.78g of KNO_3 (or peptone). Mix thoroughly. Filter sterilize. Aseptically distribute 0.5mL into tubes containing 4.5mL of sterile distilled/deionized water.

Use: For the cultivation and differentiation of bacteria based on nitrate assimilation.

Wickerham Broth with Raffinose

Composition per 100.0mL:

Raffinose	20.0g
Yeast nitrogen base	100.0mL

Yeast Nitrogen Base, 10X:

Composition per liter:

Glucose	10.0g
KH_2PO_4	1.0g
$MgSO_4 \cdot 7H_2O$	0.5g
NaCl	0.1g
$CaCl_2 \cdot 2H_2O$	0.1g
DL-Methionine	0.02g
DL-Tryptophan	0.02g
L-Histidine·HCl	0.01g
Inositol	2.0mg
H_3BO_3	0.5mg
$ZnSO_4 \cdot 7H_2O$	0.4mg
$MnSO_4 \cdot 4H_2O$	0.4mg
Thiamine·HCl	0.4mg
Pyridoxine	0.4mg
Niacin	0.4mg
Calcium pantothenate	0.4mg
p-Aminobenzoic acid	0.2mg
Riboflavin	0.2mg
$FeCl_3$	0.2mg
$Na_2MoO_4 \cdot 4H_2O$	0.2mg
KI	0.1mg
$CuSO_4 \cdot 5H_2O$	0.04mg
Folic acid	2.0μg
Biotin	2.0μg

pH 4.5 ± 0.2 at 25°C

Preparation of Yeast Nitrogen Base: Add components to distilled/deionized water and bring volume to 1.0L. Mix thoroughly.

Preparation of Medium: To 100.0mL of yeast nitrogen base, add 20.0g of raffinose. Mix thoroughly. Filter sterilize. Aseptically distribute 0.5mL into tubes containing 4.5mL of sterile distilled/deionized water.

Use: For the cultivation and differentiation of bacteria based on carbohydrate assimilation.

Wilkins-Chalgren Agar

Composition per liter:

Agar	15.0g
Gelatin peptone	10.0g
Pancreatic digest of casein	10.0g
NaCl	5.0g
Yeast extract	5.0g
Glucose	1.0g
L-Arginine	1.0g
Sodium pyruvate	1.0g
Hemin	5.0mg
Vitamin K_1 (Menadione)	0.5mg

pH 7.1 ± 0.2 at 25°C

Source: This medium is available as a premixed powder from BD Diagnostic Systems.

Preparation of Medium: Add components to distilled/deionized water and bring volume to 1.0L. Mix thoroughly. Gently heat and bring to boiling. Distribute into tubes or flasks. Autoclave for 15 min at 15 psi pressure–121°C. Cool to 50°–55°C. Add antibiotic to be assayed; varying concentrations of antibiotics are used. Mix thoroughly. Pour into sterile Petri dishes or leave in tubes.

Use: For the cultivation and maintenance of anaerobic bacteria. For standardized antimicrobic susceptibility testing to determine the minimum inhibitory concentrations of antimicrobics for anaerobic bacteria.

Wilkins-Chalgren Anaerobe Agar

Composition per liter:

Agar	10.0g
Pancreatic digest of casein	10.0g
Gelatin peptone	10.0g
NaCl	5.0g
Yeast extract	5.0g
Glucose	1.0g
L-Arginine	1.0g
Sodium pyruvate	1.0g
Hemin	5.0mg
Menadione	0.5mg
Defibrinated blood	50.0mL
Tween 80	1.0mL

pH 7.1 ± 0.2 at 25°C

Source: This medium is available as a premixed powder from Oxoid Unipath.

Preparation of Medium: Add components, except defibrinated blood, to distilled/deionized water and bring volume to 950.0mL. Mix thoroughly. Gently heat and bring to boiling. Distribute into tubes or flasks. Autoclave for 15 min at 15 psi pressure–121°C. Cool to 50°–55°C. Aseptically add 50.0mL of defibrinated blood. Mix thoroughly. Pour into sterile Petri dishes or leave in tubes.

Use: For the cultivation of nonsporulating anaerobes.

Wilkins-Chalgren Anaerobe Agar with GN Supplement

Composition per liter:

Agar	10.0g
Pancreatic digest of casein	10.0g
Gelatin peptone	10.0g
NaCl	5.0g
Yeast extract	5.0g
Glucose	1.0g
L-Arginine	1.0g
Sodium pyruvate	1.0g
Hemin	5.0mg
Menadione	0.5mg
Defibrinated blood	50.0mL
GN anaerobe selective supplement	20.0mL

pH 7.1 ± 0.2 at 25°C

Source: This medium is available as a premixed powder from Oxoid Unipath.

GN Anaerobe Selective Supplement

Composition per 20.0mL:

Nalidixic acid	10.0mg
Hemin	5.0mg
Sodium succinate	2.5mg
Vancomycin	2.5mg
Menadione	0.5mg

Preparation of GN Anaerobe Selective Supplement: Add components to distilled/deionized water and bring volume to 20.0mL. Mix thoroughly. Filter sterilize.

Preparation of Medium: Add components, except defibrinated blood and GN anaerobe selective supplement, to distilled/deionized water and bring volume to 900.0mL. Mix thoroughly. Gently heat and bring to boiling. Distribute into tubes or flasks. Autoclave for 15 min at 15 psi pressure–121°C. Cool to 50°–55°C. Aseptically add 20.0mL of GN anaerobe selective supplement and 50.0mL of defibrinated blood. Bring volume to 1.0L with distilled/deionized water. Mix thoroughly. Pour into sterile Petri dishes or leave in tubes.

Use: For the selective isolation of Gram-negative anaerobes.

Wilkins-Chalgren Anaerobe Agar with NS Supplement

Composition per liter:

Agar	10.0g
Pancreatic digest of casein	10.0g
Gelatin peptone	10.0g
NaCl	5.0g
Yeast extract	5.0g
Glucose	1.0g
L-Arginine	1.0g
Sodium pyruvate	1.0g
Hemin	5.0mg
Menadione	0.5mg
Defibrinated blood	50.0mL
NS anaerobe selective supplement	20.0 mL
Tween 80	1.0mL

pH 7.1 ± 0.2 at 25°C

Source: This medium is available as a premixed powder from Oxoid Unipath.

NS Anaerobe Selective Supplement:

Composition per 20.0mL:

Sodium pyruvate	1.0g
Nalidixic acid	0.01g
Hemin	5.0mg
Menadione	0.5mg

Preparation of NS Anaerobe Selective Supplement: Add components to distilled/deionized water and bring volume to 20.0mL. Mix thoroughly. Filter sterilize.

Preparation of Medium: Add components, except defibrinated blood and NS anaerobe selective supplement, to distilled/deionized water and bring volume to 900.0mL. Mix thoroughly. Gently heat and bring to boiling. Distribute into tubes or flasks. Autoclave for 15 min at 15 psi pressure–121°C. Cool to 50°–55°C. Aseptically add 20.0mL of NS anaerobe selective supplement and 50.0mL of defibrinated blood. Bring volume to 1.0L with distilled/deionized water. Mix thoroughly. Pour into sterile Petri dishes or leave in tubes.

Use: For the selective isolation of nonsporulating anaerobes.

Wilkins-Chalgren Anaerobe Broth (Anaerobe Broth, MIC)

Composition per liter:

Pancreatic digest of casein	10.0g
Gelatin peptone	10.0g
NaCl	5.0g
Yeast extract	5.0g
Glucose	1.0g
L-Arginine	1.0g
Sodium pyruvate	1.0g
Hemin	5.0mg
Menadione	0.5mg

pH 7.1 ± 0.2 at 25°C

Source: This medium is available as a premixed powder from BD Diagnostic Systems and Oxoid Unipath.

Preparation of Medium: Add components to distilled/deionized water and bring volume to 1.0L. Mix thoroughly. Distribute into tubes or flasks. Autoclave for 15 min at 15 psi pressure–121°C.

Use: For the cultivation and antimicrobial susceptibility (MIC) testing of anaerobic bacteria.

Wilson Blair Base

Composition per liter:

Agar	30.0g
Proteose peptone no. 3	10.0g
Glucose	10.0g
Beef extract	5.0g
NaCl	5.0g
Selective reagent	70.0mL
Brilliant Green (1% solution)	4.0mL

pH 7.3 ± 0.2 at 25°C

Selective Reagent:

Composition per 100.0mL:

Solution 1	100.0mL
Solution 2	100.0mL
Solution 3	100.0mL
Solution 4	20.2mL

Preparation of Selective Reagent: Combine 100.0mL of solution 1, 100.0mL of solution 2, 100.0mL of solution 3, and 20.2mL of solution 4. Mix thoroughly. Gently heat to boiling until a slate-grey color develops. Cool to 50°C.

Solution 1:

Composition per 100.0mL:

$NaHSO_3$	40.0g

Preparation of Solution 1: Add $NaHSO_3$ to 100.0mL of distilled/deionized water. Mix thoroughly.

Solution 2:

Composition per 100.0mL:

NaH_2PO_4	21.0g

Preparation of Solution 2: Add NaH_2PO_4 to 100.0mL of distilled/deionized water. Mix thoroughly.

Solution 3:

Composition per 100.0mL:

Bismuth ammonium citrate	12.5g

Preparation of Solution 3: Add bismuth ammonium citrate to 100.0mL of distilled/deionized water. Mix thoroughly.

Solution 4:

Composition per 20.2mL:

$FeSO_4$	0.96g

Preparation of Solution 4: Add $FeSO_4$ to 20.0mL of distilled/deionized water. Add 0.2mL of concentrated HCl. Mix thoroughly.

Preparation of Medium: Add components, except selective reagent and Brilliant Green solution, to distilled/deionized water and bring volume to 976.0mL. Mix thoroughly. Gently heat and bring to boiling. Distribute into tubes or flasks. Autoclave for 15 min at 15 psi pressure–121°C. Cool to 50°C. Aseptically add selective reagent and Brilliant Green solution. Mix thoroughly. Pour into sterile Petri dishes or leave in tubes.

Use: For the isolation and cultivation of *Salmonella*, especially *Salmonella typhi*.

Winge Agar

Composition per liter:

Glucose	20.0g
Agar	15.0g
Yeast extract	3.0g

pH 7.2 ± 0.2 at 25°C

Preparation of Medium: Add components to distilled/deionized water and bring volume to 1.0L. Mix thoroughly. Adjust pH to 7.2. Gently heat and bring to boiling. Distribute into tubes or flasks. Autoclave for 15 min at 15 psi pressure–121°C. Pour into sterile Petri dishes or leave in tubes.

Use: For the cultivation and maintenance of *Candida albicans* and *Candida tropicalis*.

Wolin-Bevis Medium

Composition per liter:

Agar	20.0g
$(NH_4)_2SO_4$	1.0g
KH_2HPO_4	1.0g
Glucose	0.25g
L-Histidine·HCl	0.25g
Tween 80 (polysorbate 80)	3.0mL

pH 5.4 ± 0.2 at 25°C

Preparation of Medium: Add components to distilled/deionized water and bring volume to 1.0L. Mix thoroughly. Gently heat and bring to boiling. Distribute into tubes or flasks. Autoclave for 15 min at 15 psi pressure–121°C. Pour into sterile Petri dishes or leave in tubes.

Use: For the enhanced production of chlamydospores by *Candida albicans*.

Worfel-Ferguson Agar

Composition per liter:

Sucrose	20.0g
Agar	15.0g
NaCl	2.0g
Yeast extract	2.0g
K_2SO_4	1.0g
$MgSO_4·7H_2O$	0.25g

pH 6.5 ± 0.2 at 25°C

Preparation of Medium: Add components to distilled/deionized water and bring volume to 1.0L. Mix thoroughly. Gently heat and bring to boiling. Distribute into tubes or flasks. Autoclave for 15 min at 15 psi pressure–121°C. Pour into sterile Petri dishes or leave in tubes.

Use: For the detection of capsule production by *Klebsiella*. For serological detection of the Neufeld (Quellung) reaction.

Xanthine Agar

Composition per liter:

Solution 1	900.0mL
Solution 2	100.0mL

pH 7.0 ± 0.2 at 25°C

Solution 1:

Composition per 900.0mL:

Agar	15.0g
Pancreatic digest of gelatin	5.0g
Beef extract	3.0g

Preparation of Solution 1: Add components to distilled/deionized water and bring volume to 900.0mL. Mix thoroughly. Gently heat and bring to boiling.

Solution 2:

Composition per 100.0mL:

Xanthine	4.0g

Preparation of Solution 2: Add xanthine to distilled/deionized water and bring volume to 100.0mL. Mix thoroughly. Gently heat and bring to boiling.

Preparation of Medium: Combine solutions 1 and 2. Mix thoroughly. Distribute into tubes or flasks. Autoclave for 15 min at 15 psi pressure–121°C. Pour into sterile Petri dishes or leave in tubes.

Use: For the differentiation of aerobic *Actinomycete* species. Clearing around a colony indicates utilization of xanthine. *Streptomyces* species utilize xanthine; most *Nocardia* and *Actinomadura* species do not utilize xanthine.

XL Agar Base (Xylose Lysine Agar Base)

Composition per liter:

Agar	13.5g
Lactose	7.5g
Sucrose	7.5g
L-Lysine	5.0g
NaCl	5.0g
Xylose	3.5g
Yeast extract	3.0g
Phenol Red	0.08g
Thiosulfate-citrate solution	20.0mL

pH 7.5 ± 0.2 at 25°C

Source: This medium is available as a premixed powder from BD Diagnostic Systems.

Thiosulfate-Citrate Solution:

Composition per 100.0mL:

$Na_2S_2O_3$	34.0g
Ferric ammonium citrate	4.0g

Preparation of Thiosulfate-Citrate Solution: Add components to distilled/deionized water and bring volume to 100.0mL. Mix thoroughly.

Preparation of Medium: Add components, except thiosulfate-citrate solution, to distilled/deionized water and bring volume to 980.0mL. Mix thoroughly. Gently heat while stirring and bring to boiling. Distribute into tubes or flasks. Autoclave for 10 min at 14 psi pressure–118°C. Cool to 55°C. Aseptically add 20.0 mL of the sterile thiosulfate-citrate solution. Mix thoroughly. Pour into sterile Petri dishes or leave in tubes.

Use: For the isolation, cultivation, and differentiation of enteric pathogens. Nonfermenting xylose/lactose/sucrose bacteria appear as red colonies. Xylose-fermenting, lysine-decarboxylating bacteria appear as red colonies. Xylose-fermenting, lysine-nondecarboxylating bacteria appear as opaque yellow colonies. Lactose or sucrose-fermenting bacteria appear as yellow colonies.

XL Agar Base

Composition per liter:

Agar	15 g
Lactose	7.5g
Sucrose	7.5g
L-Lysine	5.0g
NaCl	5.0g
Xylose	3.75g
Yeast extract	3.0g
Phenol Red	0.08g
Thiosulfate-citrate solution	20.0mL

pH 7.4 ± 0.2 at 25°C

Source: This medium is available as a premixed powder from BD Diagnostic Systems.

Thiosulfate-Citrate Solution:
Composition per 100.0mL:

$Na_2S_2O_3$	34.0g
Ferric ammonium citrate	4.0g

Preparation of Thiosulfate-Citrate Solution: Add components to distilled/deionized water and bring volume to 100.0mL. Mix thoroughly.

Preparation of Medium: Add components, except thiosulfate-citrate solution, to distilled/deionized water and bring volume to 980.0mL. Mix thoroughly. Gently heat while stirring and bring to boiling. Distribute into tubes or flasks. Autoclave for 10 min at 14 psi pressure–118°C. Cool to 55°C. Aseptically add 20.0 mL of the sterile thiosulfate-citrate solution. Mix thoroughly. Pour into sterile Petri dishes or leave in tubes.

Use: For the isolation, cultivation, and differentiation of enteric pathogens. Nonfermenting xylose/lactose/sucrose bacteria appear as red colonies. Xylose-fermenting, lysine-decarboxylating bacteria appear as red colonies. Xylose-fermenting, lysine-nondecarboxylating bacteria appear as opaque yellow colonies. Lactose or sucrose-fermenting bacteria appear as yellow colonies.

XLD Agar (Xylose Lysine Deoxycholate Agar)

Composition per liter:

Agar	13.5g
Lactose	7.5g
Sucrose	7.5g
$Na_2S_2O_3$	6.8g
L-Lysine	5.0g
NaCl	5.0g
Xylose	3.5g
Yeast extract	3.0g
Sodium desoxycholate	2.5g
Ferric ammonium citrate	0.8g
Phenol Red	0.08g

pH 7.5 ± 0.2 at 25°C

Source: This medium is available as a premixed powder from BD Diagnostic Systems and Oxoid Unipath.

Preparation of Medium: Add components to distilled/deionized water and bring volume to 1.0L. Mix thoroughly. Gently heat and bring to boiling. Do not overheat. Distribute into tubes or flasks. Autoclave for 15 min at 15 psi pressure–121°C. Pour into sterile Petri dishes or leave in tubes. Plates should be poured as soon as possible to avoid precipitation.

Use: For the isolation and differentiation of enteric pathogens, especially *Shigella* and *Providencia* species. Nonfermenting xylose/lactose/sucrose bacteria appear as red colonies. Xylose-fermenting, lysine-decarboxylating bacteria appear as red colonies. Xylose-fermenting, lysine-nondecarboxylating bacteria appear as opaque yellow colonies. Lactose or sucrose-fermenting bacteria appear as yellow colonies.

Xylose Lactose Tergitol 4 (XLT-4)

Composition per 1004.6mL:

Agar	18.0g
Lactose	7.5g
Sucrose	7.5g
$Na_2S_2O_3$	6.8g
Lysine	5.0g
NaCl	5.0g
Xylose	3.75g
Yeast extract	3.0g
Proteose peptone	1.6g
Ferric ammonium citrate	0.8g
Phenol Red	0.08g
Selective supplement solution	4.6mL

pH 7.4 ± 0.2 at 25°C

Source: This medium is available as a premixed powder from Oxoid Unipath.

Selective Supplement Solution:
Composition per 100.0mL:

Tergitol 4	Proprietary

Preparation of Selective Supplement Solution: Available as premixed solution.

Preparation of Medium: Add components, except selective supplement solution, to distilled/deionized water and bring volume to 1.0L. Mix thoroughly. Add 4.6mL of selective supplement solution. Mix thoroughly. Gently heat while stirring and bring to boiling. Do not autoclave. Cool to 50°C. Mix thoroughly. Pour into sterile Petri dishes.

Use: For the isolation and identification of Salmonellae from clinical specimens.The presence of the selective agent, Tergitol 4, in this medium inhibits many organisms that can be problematic on other plating media. In addition, biochemical and pH changes within the medium allow *Salmonella* spp. (black colonies) to be differentiated from organisms, such as *E. coli* (yellow colonies) and *Shigella* spp. (red colonies). The enhanced selectivity of XLT-4 Agar reduces the need for further identification procedures, saving time and money, and results in fewer false presumptive positive colonies when compared to other *Salmonella* plating media.

Xylose Sodium Deoxycholate Citrate Agar

Composition per liter:

Agar	12.0g
Xylose	10.0g
Sodium citrate	5.0g
$Na_2S_2O_3·5H_2O$	5.0g
Beef extract	5.0g
Peptone	5.0g
NaCl	2.5g
Sodium deoxycholate	2.5g
Ferric ammonium citrate	1.0g
Neutral Red (1% solution)	2.5mL

pH 7.5 ± 0.2 at 25°C

Preparation of Medium: Add components to distilled/deionized water and bring volume to 1.0L. Mix thoroughly. Gently heat and bring to boiling for 20 sec. Do not autoclave. Cool to 45°–50°C. Pour into sterile Petri dishes.

Use: For the cultivation of *Salmonella* species and some *Shigella* species.

Y 1 Adrenal Cell Growth Medium

Composition per 101.0mL:

Ham's F-10 medium	90.0mL
Fetal bovine serum	10.0mL
Penicillin-streptomycin solution	1.0mL

pH 7.0 ± 0.2 at 25°C

Ham's F-10 Medium:

Composition per liter:

NaCl	7.4g
$NaHCO_3$	1.2g
Glucose	1.1g
$NaH_2PO_4·H_2O$	0.29g
KCl	0.28g
L-Arginine·HCl	0.21g
L-Glutamine	0.15g
$MgSO_4·7H_2O$	0.15g
Sodium pyruvate	0.11g
KH_2PO_4	0.08g
$CaCl_2·2H_2O$	0.04g
L-Cystine·2HCl	0.04g
L-Histidine·HCl·H_2O	0.02g
L-Lysine·HCl	0.02g
L-Asparagine-H_2O	0.01g
L-Aspartic acid	0.01g
L-Glutamic acid	0.01g
L-Leucine	0.01g
L-Proline	0.01g
L-Serine	0.01g
L-Alanine	8.9mg
Glycine	7.5mg
D-Phenylalanine	5.0mg
L-Methionine	4.5mg
Hypoxanthine	4.1mg
L-Threonine	3.6mg
L-Valine	3.5mg
L-Isoleucine	2.6mg
L-Tyrosine	1.8mg
Vitamin B_{12}	1.4mg
Folic acid	1.3mg
Phenol Red	1.2mg
Thiamine·HCl	1.0mg
$FeSO_4·7H_2O$	0.8mg
Choline chloride	0.7mg
D-Calcium pantothenate	0.7mg
Thymidine	0.7mg
Niacinamide	0.6mg
L-Tryptophan	0.6mg
Isoinositol	0.5mg
Riboflavin	0.4mg
Lipoic acid	0.2mg
Pyridoxine·HCl	0.2mg
$ZnSO_4·7H_2O$	0.03mg
Biotin	0.02mg
$CuSO_4·5H_2O$	3.0μg

Preparation of Ham's F-10 Medium: Add components to distilled/deionized water and bring volume to 1.0L. Mix thoroughly.

Penicillin-Streptomycin Solution:
Composition per 100.0mL:

Streptomycin	0.5g
Penicillin G	500,000U

Preparation of Penicillin-Streptomycin Solution: Add components to distilled/deionized water and bring volume to 100.0mL. Mix thoroughly. Filter sterilize.

Preparation of Medium: Aseptically combine components. Filter sterilize. Store at 4-5°C.

Use: For the cultivation of Y-1 mouse adrenal tissue culture cells used for the detection of heat-labile toxin (LT) produced by enterotoxigenic strains of *Escherichia coli.* LT causes the conversion of elongated fibroblast-like cells into round, refractile cells.

Y 1 Adrenal Cell Growth Medium

Composition per 580.0mL:

Ham's F-10 medium	500.0mL
Fetal bovine serum	75.0mL
Penicillin-streptomycin solution	5.0mL

pH 7.0 ± 0.2 at 25°C

Ham's F-10 Medium:
Composition per liter:

NaCl	7.4g
$NaHCO_3$	1.2g
Glucose	1.1g
$NaH_2PO_4 \cdot H_2O$	0.29g
KCl	0.28g
L-Arginine·HCl	0.21g
L-Glutamine	0.15g
$MgSO_4 \cdot 7H_2O$	0.15g
Sodium pyruvate	0.11g
KH_2PO_4	0.08g
$CaCl_2 \cdot 2H_2O$	0.04g
L-Cystine·2HCl	0.04g
L-Histidine·HCl·H_2O	0.02g
L-Lysine·HCl	0.02g
L-Asparagine-H_2O	0.01g
L-Aspartic acid	0.01g
L-Glutamic acid	0.01g
L-Leucine	0.01g
L-Proline	0.01g
L-Serine	0.01g
L-Alanine	8.9mg
Glycine	7.5mg
D-Phenylalanine	5.0mg
L-Methionine	4.5mg
Hypoxanthine	4.1mg
L-Threonine	3.6mg
L-Valine	3.5mg
L-Isoleucine	2.6mg
L-Tyrosine	1.8mg
Vitamin B_{12}	1.4mg
Folic acid	1.3mg
Phenol Red	1.2mg
Thiamine·HCl	1.0mg
$FeSO_4 \cdot 7H_2O$	0.8mg
Choline chloride	0.7mg
D-Calcium pantothenate	0.7mg
Thymidine	0.7mg
Niacinamide	0.6mg
L-Tryptophan	0.6mg
Isoinositol	0.5mg
Riboflavin	0.4mg
Lipoic acid	0.2mg
Pyridoxine·HCl	0.2mg
$ZnSO_4 \cdot 7H_2O$	0.03mg
Biotin	0.02mg
$CuSO_4 \cdot 5H_2O$	3.0μg

Preparation of Ham's F-10 Medium: Add components to distilled/deionized water and bring volume to 1.0L. Mix thoroughly.

Penicillin-Streptomycin Solution:
Composition per 100.0mL:

Streptomycin	0.5g
Penicillin G	500,000U

Preparation of Penicillin-Streptomycin Solution: Add components to distilled/deionized water and bring volume to 100.0mL. Mix thoroughly. Filter sterilize.

Preparation of Medium: Aseptically combine components. Filter sterilize. Store at 4°–5°C.

Use: For the cultivation of Y-1 mouse adrenal tissue culture cells used for the detection of cholera enterotoxin (CT) produced by enterotoxigenic strains of *Vibrio cholerae* or *Vibrio mimicus.* CT causes the conversion of elongated fibroblast-like cells into round, refractile cells.

Yeast Extract Agar

Composition per liter:

Agar	15.0g
Peptone	9.5g
Yeast extract	7.0g
Beef extract	5.0g
NaCl	5.0g

pH 7.0 ± 0.2 at 25°C

Preparation of Medium: Add components to distilled/deionized water and bring volume to 1.0L. Mix thoroughly. Gently heat and bring to boiling. Distribute into tubes or flasks. Autoclave for 15 min at 15 psi pressure–121°C. Pour into sterile Petri dishes or leave in tubes.

Use: For the cultivation of *Aeromonas salmonicida.*

Yeast Extract Agar

Composition per liter:

Agar	20.0g
Yeast extract	1.0g
Buffer solution	2.0mL

pH 6.0 ± 0.2 at 25°C

Buffer Solution:

Composition per 400.0mL:

KH_2PO_4	60.0g
Na_2HPO_4	40.0g

Preparation of Buffer Solution: Add 40.0g of Na_2HPO_4 to 300.0mL of distilled/deionized water. Mix thoroughly. Add 60.0g of KH_2PO_4. Mix thoroughly. Adjust pH to 6.0.

Preparation of Medium: Add components to distilled/deionized water and bring volume to 1.0L. Mix thoroughly. Autoclave for 15 min at 15 psi pressure–121°C. Pour into sterile Petri dishes.

Use: For the identification of *Histoplasma capsulatum*, *Blastomyces dermatitidis*, and *Coccidioides immitis*.

Yeast Extract Agar

Composition per liter:

Agar	15.0g
Proteose peptone	10.0g
NaCl	5.0g
Yeast extract	3.0g

Preparation of Medium: Add components to distilled/deionized water and bring volume to 1.0L. Mix thoroughly. Gently heat and bring to boiling. Distribute into tubes or flasks. Autoclave for 15 min at 15 psi pressure–121°C. Pour into sterile Petri dishes or leave in tubes.

Use: For the cultivation of a variety of heterotrophic microorganisms.

Yeast Extract Sodium Lactate Medium

Composition per liter:

Agar	15.0g
Pancreatic digest of casein	10.0g
Yeast extract	10.0g
Sodium lactate	10.0g
KH_2PO_4	2.5g
$MnSO_4$	5.0mg

pH 7.0 ± 0.2 at 25°C

Preparation of Medium: Add components to distilled/deionized water and bring volume to 1.0L. Mix thoroughly. Gently heat and bring to boiling. Distribute into tubes or flasks. Autoclave for 15 min at 15 psi pressure–121°C. Pour into sterile Petri dishes or leave in tubes.

Use: For the isolation, cultivation, and maintenance of *Propionibacterium* species.

Yeast Extract Sucrose Agar (YESA) (ATCC Medium 2125)

Composition per liter:

Sucrose	20.0g
Agar	15.0g
Yeast extract	4.0g
KH_2PO_4	1.0g
$MgSO_4 \cdot 7H_2O$	0.5g

pH 6.2 ± 0.2 at 25°C

Preparation of Medium: Add components to distilled/deionized water and bring volume to 1.0L. Mix thoroughly. Adjust pH to 6.2. Gently heat and bring to boiling. Distribute into tubes or flasks. Autoclave for 15 min at 15 psi pressure–121°C. Pour into sterile Petri dishes or leave in tubes.

Use: For the cultivation and maintenance of various fungi.

Yeast Extract Sucrose Agar

Composition per liter:

Sucrose	150.0g
Agar	20.0g
Yeast extract	20.0g
$MgSO_4 \cdot 7H_2O$	0.5g
Trace metal solution	1.0mL

pH 6.5 ± 0.2 at 25°C

Trace Metal Solution:

Composition per 100.0mL:

$ZnSO_4 \cdot 7H_2O$	1.0g
$CuSO_4 \cdot 5H_2O$	0.5g

Preparation of Trace Metal Solution: Add components to 100.0mL distilled/deionized water. Mix thoroughly.

Preparation of Medium: Add components to distilled/deionized water and bring volume to 1.0L. Mix thoroughly. Adjust pH to 6.5. Gently heat and bring to boiling. Distribute into tubes or flasks. Autoclave for 15 min at 15 psi pressure–121°C. Pour into sterile Petri dishes or leave in tubes.

Use: For the cultivation and maintenance of various fungi.

Yeast Fermentation Medium

Composition per liter:

Peptone	7.5g
Yeast extract	4.5g

Bromthymol Blue (1.6% solution)................. 1.0mL
Carbohydrate solution.................................... 1.0mL

Carbohydrate Solution:
Composition per 10.0mL:
Carbohydrate..0.6g

Preparation of Carbohydrate Solution: Add carbohydrate to distilled/deionized water and bring volume to 10.0mL. Glucose, maltose, lactose, galactose, or trehalose may be used. If raffinose is used, prepare a 12% solution. Mix thoroughly. Filter sterilize.

Preparation of Medium: Add components, except carbohydrate solution, to distilled/deionized water and bring volume to 1.0L. Mix thoroughly. Gently heat and bring to boiling. Distribute in 2.0mL volumes into test tubes that contain an inverted Durham tube. Autoclave for 15 min at 15 psi pressure–121°C. Cool to 45°–50°C. Aseptically add 1.0mL of sterile carbohydrate solution. Mix thoroughly.

Use: For the cultivation and differentiation of yeast based on carbohydrate fermentation patterns. Yeasts that can ferment a specific carbohydrate turn the medium yellow.

Yeast Glucose Agar

Composition per liter:
Agar 15.0g
Pancreatic digest of gelatin..............................7.75g
Beef extract...4.75g
Yeast extract...2.5g
K_2HPO_4..2.5g
Glucose 1.0g
pH 7.0 ± 0.2 at 25°C

Preparation of Medium: Add components to distilled/deionized water and bring volume to 1.0L. Mix thoroughly. Gently heat and bring to boiling. Distribute into tubes or flasks. Autoclave for 15 min at 15 psi pressure–121°C. Pour into sterile Petri dishes or leave in tubes.

Use: For the cultivation and maintenance of a wide variety of bacteria.

Yeastrel Agar

Composition per liter:
Agar 15.0g
Peptone...9.5g
Yeastrel...7.0g
Lab-lemco (meat extract)...................................5.0g
NaCl 5.0g
pH 7.0 ± 0.2 at 25°C

Source: Lab-lemco is available from Oxoid Unipath. Yeastrel is produced by Mapleton's Foods Ltd., Moss Street, Liverpool and is available from health food shops worldwide.

Preparation of Medium: Add components to distilled/deionized water and bring volume to 1.0L. Mix thoroughly. Gently heat and bring to boiling. Distribute into tubes or flasks. Autoclave for 15 min at 15 psi pressure–121°C. Pour into sterile Petri dishes or leave in tubes.

Use: For the cultivation of *Aeromonas salmonicida.*

YEPD Medium

Composition per liter:
Agar 20.0g
Glucose 20.0g
Peptone 20.0g
Yeast extract..10.0g

Preparation of Medium: Add components to distilled/deionized water and bring volume to 1.0L. Mix thoroughly. Gently heat and bring to boiling. Distribute into tubes or flasks. Autoclave for 15 min at 15 psi pressure–121°C. Pour into sterile Petri dishes or leave in tubes.

Use: For the cultivation of a variety of heterotrophic microorganisms.

Yersinia Selective Agar Base

Composition per liter:
Mannitol..20.0g
Peptone 17.0g
Agar 12.5g
Proteose peptone..3.0g
Yeast extract...2.0g
Sodium pyruvate...2.0g
NaCl 1.0g
Sodium desoxycholate..0.5g
$MgSO_4 \cdot 7H_2O$ 0.01g
Neutral Red..0.03g
Crystal Violet...1.0mg
Selective supplement 6.0mL
pH 7.4 ± 0.2 at 25°C

Source: This medium is available as a premixed powder from BD Diagnostic Systems and Oxoid Unipath.

Selective Supplement:
Composition per 6.0mL:
Cefsulodin...15.0mg
Irgasan...4.0mg
Novobiocin 2.5mg
Ethanol 2.0mL

Preparation of Selective Supplement: Aseptically add components to 4.0mL of distilled/deionized water and 2.0mL of ethanol. Mix thoroughly.

Preparation of Medium: Add components to distilled/deionized water and bring volume to 1.0L. Mix thoroughly. Gently heat and bring to boiling. Distribute into tubes or flasks. Autoclave for 15 min at 15 psi pressure–121°C. Cool to 50°C. Aseptically add selective supplement. Mix thoroughly. Pour into sterile Petri dishes or leave in tubes.

Use: For the isolation and enumeration of *Yersinia enterocolitica* from clinical specimens.

YI-S Medium

Composition per liter:

YI broth	880.0mL
Bovine serum, heat inactivated	100.0mL
Vitamin mixture 18	20.0mL

Source: Vitamin mixture 18 is available from Biofluids, Inc., Rockville, MD.

YI Broth:

Composition per liter:

YI base stock	780.0mL
10× Glucose buffer stock	100.0mL

YI Base Stock:

Composition per 780.0mL:

Yeast extract	30.0g
L-Cysteine·HCl	1.0g
NaCl	1.0g
Ascorbic acid	0.2g
Ferric ammonium citrate	228.0mg

10× Glucose Buffer Stock:

Composition per 100.0.0mL:

Glucose	10.0g
K_2HPO_4	1.0g
KH_2PO_4	0.6g

Preparation of 10× Glucose Buffer Stock: Add components to distilled/deionized water and bring volume to 100.0mL. Mix thoroughly. Filter sterilize.

Preparation of YI Base Stock: Add components to 600.0mL of distilled/deionized water. Mix thoroughly. Bring volume to 780.0mL with distilled/deionized water. Adjust pH to 6.8 with 1*N* NaOH. Distribute in 78.0mL aliquots to 100.0mL screw-capped bottles. Autoclave for 15 min at 15 psi pressure–121°C. Cool to room temperature.

Preparation of YI Broth: Aseptically add 10.0mL of 10× glucose buffer stock to 78.0mL of cooled YI base stock. Adjust osmolarity with NaCl to 380.0milliosmols/kg.

Preparation of Medium: Aseptically add 2.0mL of vitamin mixture 18 and 10.0mL of heat-inactivated bovine serum to 88.0mL of YI broth. Distribute in 13.0mL aliquots to 16 x 125mm screw-capped test tubes. Store at 4°C in the dark with the caps screwed on tightly. Use within 96h.

Use: For the cultivation of *Entamoeba* species.

YPC Medium

Composition per liter:

Agar	15.0g
Proteose peptone	15.0g
Yeast extract	5.0g
KH_2PO_4	4.0g
Sucrose	2.5g
Glucose	2.0g
L-Cystine	0.5g
Na_2SO_3	0.2g

pH 7.2 ± 0.2 at 25°C

Preparation of Medium: Add components to distilled/deionized water and bring volume to 1.0L. Mix thoroughly. Gently heat and bring to boiling. Distribute into tubes or flasks. Autoclave for 15 min at 15 psi pressure–121°C. Pour into sterile Petri dishes or leave in tubes.

Use: For the cultivation of *Pasteurella multocida.*

YPNC Medium

Composition per liter:

Agar	18.0g
NaCl	2.92g
KH_2PO_4	0.596g
Yeast extract	0.5g
Sodium hydrogen glutamate (pH 6.0)	0.37g
K_2HPO_4	0.107g
NH_4Cl	0.107g
$MgSO_4 \cdot 7H_2O$	0.049g
Glucose solution	10.0mL
Trace metals solution	1.0mL

pH 6.0 ± 0.2 at 25°C

Glucose Solution:

Composition per 10.0mL:

Glucose	4.0g

Preparation of Glucose Solution: Add glucose to distilled/deionized water and bring volume to 10.0mL. Mix thoroughly. Filter sterilize.

Trace Metals Solution:

Composition per liter:

EDTA	50.0g

$ZnSO_4 \cdot 7H_2O$	22.0g
$CaCl_2$	5.54g
$MnCl_2 \cdot 4H_2O$	5.06g
$FeSO_4 \cdot 7H_2O$	4.99g
$(NH_4)_6 Mo_7O_{14} \cdot H_2O$	1.10g
$CoSO_4 \cdot 5H_2O$	1.57g
$CoCl_2 \cdot 6H_2O$	1.61g

Preparation of Trace Metals Solution: Add components, one at a time, to distilled/deionized water and bring volume to 1.0L. Mix thoroughly. Filter sterilize.

Preparation of Medium: Add components, except glucose solution and trace metals solution, to distilled/deionized water and bring volume to 989.0mL. Mix thoroughly. Gently heat and bring to boiling. Adjust pH to 6.0 with KOH. Autoclave for 15 min at 15 psi pressure–121°C. Cool to 50°C. Aseptically add 10.0mL of sterile glucose solution and 1.0mL of sterile trace metals solution. Mix thoroughly. Pour into sterile Petri dishes or aseptically distribute into sterile tubes.

Use: For the cultivation and maintenance of a variety of *Cryptococcus* species.

Agars

Below are some agars used as solidifying agents in various media.

Agar Bacteriological (Agar No. 1)
: An agar with low calcium and magnesium. Available from Oxoid Unipath.

Agar, Bacto
: A purified agar with reduced pigmented compounds, salts, and extraneous matter. Available from Difco, BD Diagnostic Systems.

Agar, BiTek™
: Agar prepared as a special technical grade. Available from Difco, BD Diagnostic Systems.

Agar, Flake
: A technical grade agar. Available from Difco, BD Diagnostic Systems.

Agar, Grade A
: A select grade agar containing minerals. Available from BD Diagnostic Systems.

Agar, Granulated
: A high grade granulated agar that has been filtered, decolorized, and purified. Available from BD Diagnostic Systems.

Agarose
: A low sulfate neutral gelling fraction of agar that is a complex galactose polysaccharide of near neutral charge.

Agar, Purified
: A very high grade agar that has been filtered, decolorized, and purified by washing and extraction of refined agars. It has reduced mineral content. Available from BD Diagnostic Systems.

Agar Technical (Agar No. 3)
: A technical grade agar. Available from Difco, BD Diagnostic Systems, and Oxoid Unipath.

Ionagar
: A purified agar. Available from Oxoid Unipath.

Noble Agar
: An agar that has been extensively washed and is essentially free of impurities. Available from Difco, BD Diagnostic Systems.

Purified Agar
: An agar that has been extensively washed and extracted with water and organic solvent. Available from Difco, BD Diagnostic Systems, and Oxoid Unipath.

Peptones

Below is a list of some of the peptones that are used as ingredients in various media.

Acidase™ Peptone
: A hydrochloric acid hydrolysate of casein. It has a nitrogen content of 8% and is deficient in cystine and tryptophan. Available from BD Diagnostic Systems.

Bacto Casitone
: A pancreatic digest of casein. Available from Difco, BD Diagnostic Systems.

Bacto Peptamin
: A peptic digest of animal tissues. Available from Difco, BD Diagnostic Systems.

Bacto Peptone
: An enzymatic digest of animal tissues. It has a high concentration of low molecular weight peptones and amino acids. Available from Difco, BD Diagnostic Systems.

Bacto Proteose Peptone
: An enzymatic digest of animal tissues. It has a high concentration of high molecular weight peptones. Available from Difco, BD Diagnostic Systems.

Bacto Soytone
: A enzymatic hydrolysate of soybean meal. Available from Difco, BD Diagnostic Systems.

Bacto Tryptone
: A pancreatic digest of casein. Available from Difco, BD Diagnostic Systems.

Bacto Tryptose
: An enzymatic hydrolysate containing numerous peptides including those of higher molecular weights. Available from Difco, BD Diagnostic Systems.

Biosate™ Peptone
: A hydrolysate of plant and animal proteins. Available from BD Diagnostic Systems.

Casein Hydrolysate
: A hydrolysate of casein prepared with hydrochloric acid digestion under pressure and neutralized with sodium hydroxide. It contains total nitrogen of 7.6% and NaCl of 28.3%. Available from Oxoid Unipath.

Gelatone
: A pancreatic digest of gelatin. Available from Difco, BD Diagnostic Systems.

Gelysate™ Peptone
: A pancreatic digest of gelatin deficient in cystine and tryptophan and which has a low carbohydrate content. Available from Oxoid Unipath.

Lactoalbumin Hydrolysate
A pancreatic digest of lactoalbumin, a milk whey protein. It has high levels of amino acids. It contains total nitrogen of 11.9% and NaCl of 1.4%. Available from Difco, BD Diagnostic Systems, and Oxoid Unipath.

Liver Digest Neutralized
A papaic digest of liver that contains total nitrogen of 11.0% and NaCl of 1.6%. Available from Oxoid Unipath.

Mycological Peptone
A peptone that contains total nitrogen of 9.5% and NaCl of 1.1%. Available from Oxoid Unipath.

Myosate™ Peptone
A pancreatic digest of heart muscle. Available from BD Diagnostic Systems.

Neopeptone
An enzymatic digest of protein. Available from Difco, BD Diagnostic Systems.

Peptone Bacteriological Neutralized
A mixed pancreatic and papaic digest of animal tissues. It contains total nitrogen of 14.0% and NaCl of 1.6%. Available from Difco, BD Diagnostic Systems, and Oxoid Unipath.

Peptone P
A peptic digest of fresh meat that has a high sulfur content and contains total nitrogen of 11.12% and NaCl of 9.3%. Available from Difco, BD Diagnostic Systems, and Oxoid Unipath.

Peptonized Milk
A pancreatic digest of high grade skim milk powder. It has a high carbohydrate and calcium concentration. It contains total nitrogen of 5.3% and NaCl of 1.6%. Available from Oxoid Unipath.

Phytone™ Peptone
A papaic digest of soybean meal. It has a high vitamin and a high carbohydrate content. Available from BD Diagnostic Systems.

Polypeptone™ Peptone
A mixture of peptones composed of equal parts of pancreatic digest of casein and peptic digest of animal tissue. Available from BD Diagnostic Systems.

Proteose Peptone
A specialized peptone prepared from a mixture of peptones that contains a wide variety of high molecular weight peptides. It contains total nitrogen of 12.7% and NaCl of 8.0%. Available from Difco, BD Diagnostic Systems, and Oxoid Unipath.

Proteose Peptone No. 2
An enzymatic digest of animal tissues with a high concentration of high molecular weight peptones. Available from Difco, BD Diagnostic Systems.

Proteose Peptone No. 3
An enzymatic digest of animal tissues. It has a high concentration of high molecular weight peptones. Available from Difco, BD Diagnostic Systems.

Soya Peptone
A papaic digest of soybean meal with a high carbohydrate concentration. It contains total nitrogen of 8.7% and NaCl of 0.4%. Available from Oxoid Unipath.

Soytone
A papaic digest of soybean meal. Available from Difco, BD Diagnostic Systems, and Oxoid Unipath.

Special Peptone
A mixture of peptones, including meat, plant and yeast digests. It contains a wide variety of peptides, nucleotides, and minerals. It contains total nitrogen of 11.7% and NaCl of 3.5%. Available from Oxoid Unipath.

Thiotone™ E Peptone
An enzymatic digest of animal tissue. Available from BD Diagnostic Systems.

Trypticase™ Peptone
A pancreatic digest of casein. It has a very low carbohydrate content and a relatively high tryptophan content. Available from BD Diagnostic Systems.

Tryptone
A pancreatic digest of casein. It contains total nitrogen of 12.7% and NaCl of 0.4%. Available from Oxoid Unipath.

Tryptone T
A pancreatic digest of casein with lower levels of calcium, magnesium, and iron than tryptone. It contains total nitrogen of 11.7% and NaCl of 4.9%. Available from Difco, BD Diagnostic Systems, and Oxoid Unipath.

Tryptose
An enzymatic hydrolysate containing high molecular weight peptides. It contains total nitrogen of 12.2% and NaCl of 5.7%. Available from Difco, BD Diagnostic Systems, and Oxoid Unipath.

Meat and Plant Extracts

Below is a list of some of the meat and plant extracts that are used as ingredients in various media.

Bacto Beef

A desiccated powder of lean beef. Available from Difco, BD Diagnostic Systems.

Bacto Beef Extract

An extract of beef (paste). Available from Difco, BD Diagnostic Systems.

Bacto Beef Extract Desiccated

An extract of desiccated beef. Available from Difco, BD Diagnostic Systems.

Bacto Beef Heart for Infusion

A desiccated powder of beef heart. Available from Difco, BD Diagnostic Systems.

Bacto Liver

A desiccated powder of beef liver. Available from Difco, BD Diagnostic Systems.

Lab-Lemco

A meat extract powder. Available from Oxoid Unipath.

Liver Desiccated

Dehydrated ox livers. Available from Oxoid Unipath.

Malt Extract

A water soluble extract from germinated grain dried by low temperature evaporation. It has a high carbohydrate content. It contains total nitrogen of 1.1% and NaCl of 0.1%.

Growth Factors

Many microorganisms have specific growth factor requirements that must be included in media for their successful cultivation. Vitamins, amino acids, fatty acids, trace metals, and blood components often must be added to media. Most often mixtures of growth factors are used in microbiological media. Acid hydrolysates of casein commonly are used as sources of amino acids. Extracts of yeast cells also are employed as sources of amino acids and vitamins for the cultivation of microorganisms. Many media, particularly those employed in the clinical laboratory, contain blood or blood components that serve as essential nutrients for fastidious microorganisms. X factor (heme) and V factor (nicotinamide adenine dinucleotide) often are supplied by adding hemoglobin (BBL and Difco, BD Diagnostic Systems), IsoVitaleX (BD Diagnostic Systems), and Supplement VX (Difco, BD Diagnostic Systems).

Below is a list of some of the growth factors that are used as ingredients in microbiological media.

Bacto Casamino Acids

A mixture of amino acids formed by acid hydrolysis of casein. Available from Difco, BD Diagnostic Systems.

Bacto Vitamin Free Casamino Acids

A mixture of amino acids formed by acid hydrolysis of casein that is free of vitamins. Available from Difco, BD Diagnostic Systems.

Bovine Albumin

Bovine albumin fraction V 0.2% in 0.85% saline solution. Available from BD Diagnostic Systems.

Bovine Blood, Citrated

Calf blood washed and treated with sodium citrate as an anticoagulant. Available from BD Diagnostic Systems.

Bovine Blood, Defibrinated

Calf blood treated to denature fibrinogen without causing cell lysis. Available from BD Diagnostic Systems.

Campylobacter Growth Supplement

Sodium pyruvate, sodium metabisulfite, and $FeSO_4$.

Castenholtz Salts

Agar, $NaNO_3$, Na_2HPO_4, KNO_3, nitrilotriacetic acid, $MgSO_4 \cdot 7H_2O$, $CaSO_4 \cdot 2H_2O$, NaCl, $FeCl_3$, $MnSO_4$, H_3BO_3, $ZnSO_4$, $CoCl_2 \cdot 6H_2O$, Na_2MoO_4, $CuSO_4$, and H_2SO_4.

CVA Enrichment

Glucose, L-cysteine·HCl·H_2O, vitamin B_{12}, L-glutamine, L-cystine·2HCl, adenine, nicotinamide adenine dinucleotide, cocarboxylase, guanine·HCl, $Fe(NO_3)_3$, *p*-aminobenzoic acid, and thiamine·HCl.

Cysteine Sulfide Reducing Agent

L-Cysteine·HCl·H_2O and $Na_2S \cdot 9H_2O$.

Dubos Medium Albumin

Albumin fraction V, glucose, and saline solution. Available from Difco, BD Diagnostic Systems.

Dubos Oleic Albumin Complex

Alkalinized oleic acid, albumin fraction V, and saline solution. Available from Difco, BD Diagnostic Systems.

Egg Yolk Emulsion

Chicken egg yolks and whole chicken egg. Available from Difco, BD Diagnostic Systems, and Oxoid Unipath.

Egg Yolk Emulsion, 50%

Chicken egg yolks, whole chicken egg, and saline solution. Available from Difco, BD Diagnostic Systems.

EY Tellurite Enrichment

Egg yolk suspension with potassium tellurite. Available from Difco, BD Diagnostic Systems and Oxoid Unipath.

Fresh Yeast Extract Solution
: Live, pressed, starch-free, hydrolyzed Baker's yeast.

Fildes Enrichment
: A peptic digest of sheep or horse blood that is a rich source of growth factors including hemin and nicotinamide adenine dinucleotide. Available from BD Diagnostic Systems and Oxoid Unipath.

Hemin Solution
: Hemin and NaOH.

Hemoglobin
: Dried bovine hemoglobin. Used to provide hemin required by many fastidious microorganisms. Available from BD Diagnostic Systems.

Hemoglobin Solution 2%
: Provides hemin required by many fastidious microorganisms. Available from BD Diagnostic Systems.

Hoagland Trace Element Solution, Modified
: H_3BO_3, $MnCl_2{\cdot}4H_2O$, $AlCl_3$, $CoCl_2$, $CuCl_2$, KI, $NiCl_2$, $ZnCl_2$, $BaCl_2$, Na_2MoO_4, $SeCl_4$, $SnCl_2{\cdot}2H_2O$, $NaVO_3{\cdot}H_2O$, KBr, and LiCl.

Horse Blood, Citrated
: Horse blood washed and treated with sodium citrate used as an anticoagulant. Available from BD Diagnostic Systems.

Horse Blood, Defibrinated
: Horse blood treated to denature fibrinogen without causing cell lysis. Available from BD Diagnostic Systems and Oxoid Unipath.

Horse Blood, Hemolysed
: Horse blood treated to lyse cells. Available from Oxoid Unipath.

Horse Blood, Oxalated
: Horse blood treated with potassium oxalate as an anticoagulant. Available from Oxoid Unipath.

Horse Serum
: Horse blood is allowed to clot at 2°–8°C so that the serum separates; the serum is filter sterilized. Serum usually is inactivated by heating to 56°C for 30 min to eliminate lipases that would cause degradation of lipids and inactivation of complement. Available from Difco, BD Diagnostic Systems, and Oxoid Unipath.

IsoVitaleX® Enrichment
: Glucose, L-cysteine·HCl, L-glutamine, L-cystine, adenine, nicotinamide adenine dinucleotide, vitamin B_{12}, thiamine pyrophosphate, guanine·HCl, $Fe(NO_3)_3{\cdot}6H_2O$, *p*-aminobenzoic acid, and thiamine·HCl. Available from BD Diagnostic Systems.

Legionella Agar Enrichment
: L-Cysteine and ferric pyrophosphate. Available from Difco, BD Diagnostic Systems.

Legionella BCYE Growth Supplement
: ACES buffer/KOH, ferric pyrophosphate, L-cysteine-HCl, and α-ketoglutarate. For the enrichment of *Legionella* species. Available from Oxoid Unipath.

Leptospira Enrichment
: Lyophilized pooled rabbit serum containing hemoglobin that provides long chain fatty acids and B vitamins for growth of *Leptospira* species. Available from Diagnostic Systems.

Middlebrook ADC Enrichment
: NaCl, bovine albumin fraction V, glucose, and catalase. The albumin binds free fatty acids that may be toxic to mycobacteria. Available from BD Diagnostic Systems.

Middlebrook OADC Enrichment
: NaCl, bovine albumin, glucose, oleic acid, and catalase. The albumin binds free fatty acids that may be toxic to mycobacteria; the enrichment provides oleic acid used by *Mycobacterium tuberculosis* for growth. Available from BD Diagnostic Systems.

Mycoplasma Enrichment without Penicillin
: Horse serum, fresh autolysate of yeast—yeast extract, and thallium acetate. Provides cholesterol and nucleic acids for growth of *Mycoplasma* species. The thallium selectively inhibits other microorganisms. Available from BD Diagnostic Systems.

Mycoplasma Supplement
: Yeast extract and horse serum. Available from Difco, BD Diagnostic Systems.

Nitsch's Trace Elements
: $MnSO_4$, H_3BO_3, $ZnSO_4$, Na_2MoO_4, $CuSO_4$, $CoCl_2{\cdot}6H_2O$, and H_2SO_4.

Oleic Albumin Complex
: NaCl, bovine albumin fraction V, and oleic acid. The albumin binds free fatty acids that may be toxic to mycobacteria and the enrichment provides oleic acid that is used by *Mycobacterium tuberculosis* for growth. Available from BD Diagnostic Systems.

PPLO Serum Fraction
Serum fraction A. Available from Difco, BD Diagnostic Systems.

Rabbit Blood, Citrated
Rabbit blood washed and treated with sodium citrate as an anticoagulant. Available from BD Diagnostic Systems.

Rabbit Blood, Defibrinated
Rabbit blood treated to denature fibrinogen without causing cell lysis. Available from BD Diagnostic Systems.

RPF Supplement
Fibrinogen, rabbit plasma, trypsin inhibitor, and potassium tellurite. For the selection and nutrient supplementation of *Staphylococcus aureus*. Available from Oxoid Unipath.

Sheep Blood, Citrated
Sheep blood washed and treated with sodium citrate as an anticoagulant. Available from BD Diagnostic Systems.

Sheep Blood, Defibrinated
Sheep blood treated to denature fibrinogen without causing cell lysis. Available from Oxoid Unipath.

SLA Trace Elements
$FeCl_2 \cdot 4H_2O$, H_3BO_3, $CoCl_2 \cdot 6H_2O$, $ZnCl_2$, $MnCl_2 \cdot 4H_2O$, $NiCl_2 \cdot 6H_2O$, $CuCl_2 \cdot 2H_2O$, $Na_2MoO_4 \cdot 2H_2O$, and $Na_2SeO_3 \cdot 5H_2O$.

Soil Extract
African Violet soil and Na_2CO_3.

Supplement A
Yeast concentrate with Crystal Violet. Available from Difco, BD Diagnostic Systems.

Supplement B
Yeast concentrate, glutamine, coenzyme, cocarboxylase, hematin, and growth factors. Available from Difco, BD Diagnostic Systems.

Supplement C
Yeast concentrate. Available from Difco, BD Diagnostic Systems.

Supplement VX
Essential growth factors V and X. Available from Difco, BD Diagnostic Systems.

VA Vitamin Solution
Nicotinamide, thiamine·HCl, *p*-aminobenzoic acid, biotin, calcium pantothenate, pyridoxine·2HCl, and cyanocobalamin.

Vitamin K_1 Solution
Vitamin K_1 and ethanol.

Vitox Supplement
Glucose, L-cysteine·HCl, L-glutamine, L-cystine, adenine sulfate, nicotinamide adenine dinucleotide, cocarboxylase, guanine·HCl, $Fe(NO_3)_3 \cdot 6H_2O$, *p*-aminobenzoic acid, vitamin B_{12}, and thiamine·HCl. Available from Oxoid Unipath.

Yeast Autolysate Growth Supplement
Yeast autolysate fractions, glucose, and $NaHCO_3$. Available from Oxoid Unipath.

Yeast Extract
A water soluble extract of autolyzed yeast cells. Available from BD Diagnostic Systems, and Oxoid Unipath.

Yeastolate
A water soluble fraction of autolyzed yeast cells rich in vitamin B complex. Available from Difco, BD Diagnostic Systems.

Selective Components

Below is a list of selective components that are used to inhibit the growth of nontarget microorganisms and favor the growth of specific organisms. Selective media are especially useful in the isolation of specific microorganisms from mixed populations.

Ampicillin Selective Supplement
Ampicillin. Used in media for the selection of *Aeromonas hydrophila*. Available from Oxoid Unipath.

Anaerobe Selective Supplement GN
Hemin, menadione, sodium succinate, nalidixic acid, and vancomycin. For the selection of Gram-negative anaerobes. Available from Oxoid Unipath.

Anaerobe Selective Supplement NS
Hemin, menadione, sodium pyruvate, and nalidixic acid. For the selection of nonsporulating anaerobes. Available from Oxoid Unipath.

Bacillus cereus Selective Supplement
Polymyxin B. For the selection of *Bacillus cereus*. Available from Oxoid Unipath.

Bordetella Selective Supplement
Cephalexin. For the selection of *Bordetella* species. Available from Oxoid Unipath.

Brucella Selective Supplement
Polymyxin B, bacitracin, cycloheximide, nalidixic acid, nystatin, and vancomycin. For the selection of *Brucella* species. Available from Oxoid Unipath.

Campylobacter Selective Supplement Blaser-Wang
Vancomycin, polymyxin B, trimethoprim, amphotericin B, cephalothin. For the selec-

tion of *Campylobacter* species. Available from Oxoid Unipath.

Campylobacter Selective Supplement Butzler
Bacitracin, cycloheximide, colistin sulfate, sodium cephazolin, and novobiocin. For the selection of *Campylobacter* species. Available from Oxoid Unipath.

Campylobacter Selective Supplement Preston
Polymyxin B, rifampicin, trimethoprim, and cycloheximide. For the selection of *Campylobacter* species. Available from Oxoid Unipath.

Campylobacter Selective Supplement Skirrow
Vancomycin, trimethoprim, and polymyxin B. For the selection of *Campylobacter* species. Available from Oxoid Unipath.

CCDA Selective Supplement
Cefoperazone and amphotericin B. For the selection of *Campylobacter* species. Available from Oxoid Unipath.

Cefoperazone Selective Supplement
Cefoperazone. For the selection of *Campylobacter* species. Available from Oxoid Unipath.

CFC Selective Supplement
Cetrimide, fucidin, and cephaloridine. For the selection of pseudomonads. Available from Oxoid Unipath.

Chapman Tellurite Solution
Potassium tellurite 1% solution. Available from Difco, BD Diagnostic Systems.

Chloramphenicol Selective Supplement
Chloramphenicol. For the selection of yeasts and filamentous fungi. Available from Oxoid Unipath.

Clostridium difficile Selective Supplement
D-Cycloserine and cefoxitin. For the selection of *Clostridium difficile*. Available from Difco, BD Diagnostic Systems, and Oxoid Unipath.

CN Inhibitor
Cesulodin and novobiocin. It inhibits enteric Gram-negative microorganisms. Available from BD Diagnostic Systems.

CNV Antimicrobic
Colistin sulfate, nystatin, and vancomycin. Available from Difco, BD Diagnostic Systems.

CNVT Antimicrobic
Colistin sulfate, nystatin, vancomycin, and trimethoprim lactate. Available from Difco, BD Diagnostic Systems.

Colbeck's Egg Broth
Egg emulsion and saline solution. Formerly available from Difco, BD Diagnostic Systems—replaced with egg emulsion.

Fraser Supplement
Ferric ammonium sulfate, nalidixic acid, and Acriflavin hydrochloride. For the selection of *Listeria* species. Available from Oxoid Unipath.

Gardnerella vaginalis Selective Supplement
Gentamicin sulfate, nalidixic acid, and amphotericin B. For the selection of *Gardnerella vaginalis*. Available from Oxoid Unipath.

GC Selective Supplement
Yeast autolysate, glucose, Na_2HCO_3, vancomycin, colistin methane sulfonate, nystatin, and trimethoprim. For the selection of *Neisseria* species. Available from Oxoid Unipath.

Helicobacter pylori Selective Supplement Dent
Vancomycin, trimethoprim, cefulodin, and amphotericin B. For the selection of *Helicobacter pylori*.

Kanamycin Sulfate Selective Supplement
Kanamycin sulfate. For the selection of enterococci. Available from Oxoid Unipath.

LCAT Selective Supplement
Lincomycin, colistin sulfate, amphotericin B, and trimethoprim. For the selection of *Neisseria* species. Available from Oxoid Unipath.

Legionella BMPA Selective Supplement
Cefamandole, polymyxin B, and anisomycin. For the selection of *Legionella* species. Available from Oxoid Unipath.

Legionella GVPC Selective Supplement
Glycine, vancomycin hydrochloride, polymixin B sulfate, and cycloheximide. For the selection of *Legionella* species. Available from Oxoid Unipath.

Legionella MWY Selective Supplement
Glycine, polymyxin B, anisomycin, vancomycin, Bromthymol B, and Bromcresol Purple. For the selection of *Legionella* species. Available from Oxoid Unipath.

Listeria Primary Selective Enrichment Supplement
Nalidixic acid and acriflavin. For the selection of *Listeria* species. Available from Oxoid Unipath.

Listeria Selective Enrichment Supplement
Nalidixic acid, cycloheximide, and acriflavin. For the selection of *Listeria* species. Available from Oxoid Unipath.

Listeria Selective Supplement MOX
Colistin and moxalactam. For the selection

of *Listeria monocytogenes*. Available from Oxoid Unipath.

Listeria Selective Supplement Oxford
Cycloheximide, colistin sulfate, acriflavin, cefotetan, and fosfomycin. For the selection of *Listeria* species. Available from Oxoid Unipath.

Modified Oxford Antimicrobic Supplement
Moxalactam and colistin sulfate. Available from Difco, BD Diagnostic Systems.

MSRV Selective Supplement
Novobiocin. For the selection of *Salmonella*. Available from Oxoid Unipath.

Mycoplasma Supplement G
Horse serum, yeast extract, thallous acetate, and penicillin. For the selection of *Mycoplasma* species. Available from Oxoid Unipath.

Mycoplasma Supplement P
Horse serum, yeast extract, thallous acetate, glucose, Phenol Red, Methylene Blue, penicillin, and *Mycoplasma* broth base. For the selection of *Mycoplasma* species. Available from Oxoid Unipath.

Mycoplasma Supplement S
Yeast extract, horse serum, thallium acetate, and penicillin. Available from Difco, BD Diagnostic Systems.

Oxford Antimicrobic Supplement
Cycloheximide, colistin sulfate, acriflavin, cefotetan, and fosfomycin. Available from Difco, BD Diagnostic Systems.

Oxgall
Dehydrated fresh bile. For the selection of bile tolerant bacteria. Available from Difco, BD Diagnostic Systems.

Oxytetracycline GYE Supplement
Oxytetracycline in a buffer. For the selection of yeasts and filamentous fungi. Available from Oxoid Unipath.

PALCAM Selective Supplement
Polymyxin B, acriflavin hydrochloride, and ceftazidime. For the selection of *Listeria monocytogenes*. Available from Oxoid Unipath.

Perfringens OPSP Selective Supplement A
Sodium sulfadiazine. For the selection of *Clostridium perfringens*. Available from Oxoid Unipath.

Perfringens SFP Selective Supplement A
Kanamycin sulfate and polymyxin B. For the selection of *Clostridium perfringens*. Available from Oxoid Unipath.

Perfringens TSC Selective Supplement A
D-Cycloserine. For the selection of *Clostridium perfringens*. Available from Oxoid Unipath.

Sodium Desoxycholate
Sodium salt of desoxycholic acid. Available from Difco, BD Diagnostic Systems.

Sodium Taurocholate
Sodium salt of conjugated bile acid—75% sodium taurocholate and 25% bile salts. For the selection of bile tolerant bacteria. Available from Difco, BD Diagnostic Systems.

STAA Selective Supplement
Streptomycin sulfate, cycloheximide, and thallous acetate. For the selection of *Brochothrix thermosphacta*. Available from Oxoid Unipath.

Staph/Strep Selective Supplement
Nalidixic acid and colistin sulfate. For the selection of *Staphylococcus* species and *Streptococcus* species. Available from Oxoid Unipath.

Streptococcus Selective Supplement COA
Colistin sulfate and oxolinic acid. For the selection of *Streptococcus* species. Available from Oxoid Unipath.

Sulfamandelate Supplement
Sodium sulfacetamide and sodium mandelate. For the selection of *Salmonella* species. Available from Oxoid Unipath.

Tellurite Solution
A solution containing potassium tellurite. Inhibits Gram-negative and most Gram-positive microorganisms. It is used for the isolation of *Corynebacterium* species, *Streptococcus* species, *Listeria* species, and *Candida albicans*. Available from BD Diagnostic Systems.

Tinsdale Supplement
Serum, potassium tellurite, and sodium thiosulfate. For the selection of *Corynebacterium diphtheriae*. Available from Oxoid Unipath.

V C A Inhibitor
Vancomycin, colistin, anisomycin, and trimethoprim. Inhibits most Gram-negative and Gram-positive bacteria and yeasts. It is used for the isolation of *Neisseria* species. Available from BD Diagnostic Systems.

V C A T Inhibitor
Vancomycin, colistin, anisomycin, and trimethoprim lactate. Inhibits most Gram-negative and Gram-positive bacteria and yeasts.

It is used for the isolation of *Neisseria* species. Available from BD Diagnostic Systems and Oxoid Unipath.

V C N Inhibitor
: Colistin, vancomycin, and nystatin. Inhibits most Gram-negative and Gram-positive bacteria and yeasts. It is used for the isolation of *Neisseria* species. Available from BD Diagnostic Systems and Oxoid Unipath.

V C N T Inhibitor
: Colistin, vancomycin, nystatin, and trimethoprim lactate. Inhibits most Gram-negative and Gram-positive bacteria and yeasts. It is used for the isolation of *Neisseria* species. Available from BD Diagnostic Systems and Oxoid Unipath.

Yersinia Selective Supplement
: Cefsulodin, irgasan, and novobiocin. For the selection of *Yersinia enterocolitica*. Available from Oxoid Unipath.

pH Buffers

. The pH using phosphate buffers is established by using varying volumes of equimolar concentrations of Na_2HPO_4 and NaH_2PO_4.

pH	Na_2HPO_4 (mL)	NaH_2PO_4 (mL)
5.4	3.0	97.0
5.6	5.0	95.0
5.8	7.8	92.2
6.0	12.0	88.0
6.2	18.5	81.5
6.4	26.5	73.5
6.6	37.5	62.5
6.8	50.0	50.0
7.0	61.1	38.9
7.2	71.5	28.5
7.4	80.4	19.6
7.6	86.8	13.2
7.8	91.4	8.6
8.0	94.5	5.5

Differential Components

Below is a list of some commonly used pH indicators and their color reactions.

pH Indicator	pH Range	Acid Color	Alkaline Color
m-Cresol Purple	0.5–2.5	Red	Yellow
Thymol Blue	1.2–2.8	Red	Yellow
Bromphenol Blue	3.0–4.6	Yellow	Blue
Bromcresol Green	3.8–5.4	Yellow	Blue
Chlorcresol Green	4.0–5.6	Yellow	Blue
Methyl Red	4.2–6.3	Red	Yellow

pH Indicator	pH Range	Acid Color	Alkaline Color
Chlorphenol Red	5.0–6.6	Yellow	Red
Bromcresol Purple	5.2–6.8	Yellow	Purple
Bromthymol Blue	6.0–7.6	Yellow	Blue
Phenol Red	6.8–8.4	Yellow	Red
Cresol Red	7.2–8.8	Yellow	Red
m-Cresol Purple	7.4–9.0	Yellow	Purple
Thymol Blue	8.0–9.6	Yellow	Blue
Cresolphthalein	8.2–9.8	Colorless	Red
Phenolphthalein	8.3–10.0	Colorless	Red

Preparation of Media

The ingredients in a medium are usually dissolved and the medium is then sterilized. When agar is used as a solidifying agent the medium must be heated gently, usually to boiling, to dissolve the agar. In some cases where interactions of components, such as metals, would cause precipitates, solutions must be prepared and occasionally sterilized separately before mixing the various solutions to prepare the complete medium. The pH often is adjusted prior to sterilization, but in some cases sterile acid or base is used to adjust the pH of the medium following sterilization. Many media are sterilized by exposure to elevated temperatures. The most common method is to autoclave the medium. Different sterilization procedures are employed when heat-labile compounds are included in the formulation of the medium.

Autoclaving

Autoclaving uses exposure to steam, generally under pressure, to kill microorganisms. Exposure for 15 min to steam at 15 psi—121°C is most commonly used. Such exposure kills vegetative bacterial cells and bacterial endospores. However, some substances do not tolerate such exposures and lower temperatures and different exposure times are sometimes employed. Media containing carbohydrates often are sterilized at 116°–118°C in order to prevent the decomposition of the carbohydrate and the formation of toxic compounds that would inhibit microbial growth. Below is a list of pressure–temperature relationships.

Pressure—psi	Temperature—°C
0	100
1	101.9
2	103.6
3	105.3
4	106.9
5	108.4
6	109.8

Pressure—psi	Temperature—°C
7	111.3
8	112.6
9	113.9
10	115.2
11	116.4
12	117.6
13	118.8
14	119.9
15	121.0
16	122.0
17	123.0
18	124.0
19	125.0
20	126.0
21	126.9
22	127.8
23	128.7
24	129.6
25	130.4

Tyndallization

Exposure to steam at 100°C for 30 min will kill vegetative bacterial cells but not endospores. Such exposure can be achieved using flowing steam in an Arnold sterilizer. By allowing the medium to cool and incubate under conditions where endospore germination will occur and by repeating the 100°C–30 mi e exposure on three successive days the medium can be sterilized because all the endospores will have germinated and the heat exposure will have killed all the vegetative cells. This process of repetitive exposure to 100°C is called tyndallization, after its discoverer John Tyndall.

Inspissation

Inspissation is a heat exposure method that is employed with high protein materials, such as egg-containing media, that cannot withstand the high temperatures used in autoclaving. This process causes coagulation of the protein without greatly altering its chemical properties. Several different protocols can be followed for inspissation. Using an Arnold sterilizer or a specialized inspissator, the medium is exposed to 75°–80°C for 2 h on each of three successive days. Inspissation using an autoclave employs exposure to 85°–90°C for 10 min achieved by having a mixture of air and steam in the chamber, followed by 15 min exposure during which the temperature is raised to 121°C using only steam under pressure in the chamber; the temperature then is slowly lowered to less than 60°C.

Filtration

Filtration is commonly used to sterilize media containing heat-labile compounds. Liquid media are passed through sintered glass or membranes, typically made of cellulose acetate or nitrocellulose, with small pore sizes. A membrane with a pore size of 0.2mm will trap bacterial cells and, therefore, sometimes is called a bacteriological filter. By preventing the passage of microorganisms, filtration renders fluids free of bacteria and eukaryotic microorganisms, that is, free of living organisms, and hence sterile. Many carbohydrate solutions, antibiotic solutions, and vitamin solutions are filter sterilized and added to media that have been cooled to temperatures below 50°C.

Caution about Hazardous Components

Some media contain components that are toxic or carcinogenic. Appropriate safety precautions must be taken when using media with such components. Basic fuchsin and acid fuchsin are carcinogens and caution must be used in handling media with these compounds to avoid dangerous exposure that could lead to the development of malignancies. Thallium salts, sodium azide, sodium biselenite, and cyanide are among the toxic components found in some media. These compounds are poisonous and steps must be taken to avoid ingestion, inhalation, and skin contact. Azides also react with many metals, especially copper, to form explosive metal azides. The disposal of azides must avoid contact with copper or achieve sufficient dilution to avoid the formation of such hazardous explosive compounds. Media with sulfur-containing compounds may result in the formation of hydrogen sulfide which is a toxic gas. Care must be used to ensure proper ventilation. Media with human blood or human blood components must be handled with great caution to avoid exposure to human immunodeficiency virus and other pathogens that contaminate some blood supplies. Proper handling and disposal procedures must be followed with blood-containing as well as other media that are used to cultivate microorganisms.

Printed and bound by CPI Group (UK) Ltd, Croydon, CR0 4YY
23/10/2024
01778250-0016